Cave Conservation and Restoration

2006 Edition

© Val Hildreth-Werker

Helictites in Left Hand Tunnel, Carlsbad Cavern

Cave Conservation and Restoration

2006 Edition

Edited by
Val Hildreth-Werker
and Jim C. Werker

National Speleological Society, Inc. Huntsville, Alabama U.S.A.

In cooperation with the National Cave and Karst Research Institute, Inc.
Carlsbad, New Mexico U.S.A.

Cave Conservation and Restoration

2006 Edition

Published by the National Speleological Society, Inc.
2813 Cave Avenue
Huntsville, Alabama 35810-4431 U.S.A.
Phone: 256-852-1300
Fax: 256-851-9241
E-mail: <nss@caves.org> Web site: <http://www.caves.org/>

Published in cooperation with the National Cave and Karst Research Institute, Inc., Carlsbad, New Mexico U.S.A.

Library of Congress Catalog Card Number: 2002106017

ISBN: 1-879961-15-6

Conservation Alert: Many resources found in caves are fragile and extremely easy to damage—including speleothems (cave formations), cave minerals, cave flora and fauna, paleontological remains, archaeological evidence, and historical artifacts. Delicate soda straws and eccentric helictites, for example, that possibly took untold thousands of years to create can be destroyed in an instant by a careless or thoughtless human act. The conservation ethic of the National Speleological Society—and other speleological associations worldwide—is to study spelean resources, to perpetuate natural speleological processes, and to enjoy the wonder of speleothems and other cave values in their natural setting, without disturbing them in any way. Cave conservation is more a state of mind than a set of rules. Use common sense and take the action that will cause the least damage to the cave and its unique resources.

Front Cover: Cavers work on pearl restoration, flowstone cleaning, and silt pumping in Pellucidar, Lechuguilla Cave, New Mexico. Clockwise from top: Jim Werker, Justin Shaw, Phyllis Hamer, and Val Hildreth-Werker. © Allan Cobb and Val Hildreth-Werker

Frontispiece: Helictites in Left Hand Tunnel, Carlsbad Cavern, New Mexico. © Val Hildreth-Werker

Back Cover: (Left) Jim Werker gathers pieces to repair a broken stalagmite in Slaughter Canyon Cave, Carlsbad Caverns National Park, New Mexico. On a sloping flowstone surface, nylon climbing rope is used to secure both Jim and the pieces during the repair. (Right) The repaired stalagmite stands again in Slaughter Canyon Cave. © Val Hildreth-Werker

Dedication

This book is dedicated …
To cavers of the Earth who give
labor-intensive days of their lives in volunteer service
to protect, conserve, explore, and restore caves and karst.

To our moms who taught us
to be good stewards of Earth's resources
and to respect God's signature
in natural processes.

In memory of Ivy C. Werker, 1902–1976
In honor of Gene M. Hildreth, 1928–

To a man who mentored many cavers
in their commitments to conservation
and devoted his enthusiasm
to the protection of caves and karst.
He served in leadership roles for the
National Speleological Society,
American Cave Conservation Association,
National Cave and Karst Management Symposia,
Journal of Cave and Karst Studies,
and the International Speleological Union.
We value his personal support and encouragement
for this publication.

In memory of Dr. George N. Huppert, NSS#7717, 1945–2001

To these cavers who created legacies
that will continue to benefit cave conservation.

In memory of Russell H. Gurnee, NSS#1907, 1922–1995
In memory of Fred L. Wefer, NSS#10845, 1945–1999
In memory of Pat Copeland, NSS#22361, 1942–1999
In memory of Noble Stidham, NSS#26240, 1938–2002
In memory of Ransom W. Turner, NSS#49208, 1959–2005

Contributors

Contributing Authors

James F. Baichtal	Thorne Bay, Alaska
Hazel A. Barton	Highland Heights, Kentucky
Jonathan B. Beard	Springfield, Missouri
Thomas B. Bemis	Carlsbad, New Mexico
Barbara Bilbo	Socorro, New Mexico
Michael Bilbo	Socorro, New Mexico
Hans G. Bodenhamer	Kalispell, Montana
Penelope J. Boston	Socorro, New Mexico
Debbie C. Buecher	Tucson, Arizona
Paul Burger	Carlsbad, New Mexico
Alvin D. Collier	Olathe, Colorado
George M. Crothers	Lexington, Kentucky
Robert R. Currie	Hendersonville, North Carolina
Joseph C. Douglas	Nashville, Tennessee
Harvey R. DuChene	Centennial, Colorado
William R. Elliott	Jefferson City, Missouri
Joseph H. Fagan	Blacksburg, Virginia
John A. French	Huntsville, Alabama
James R. Goodbar	Carlsbad, New Mexico
Val Hildreth-Werker	Hillsboro, New Mexico
Rodney D. Horrocks	Hot Springs, South Dakota
George N. Huppert	La Crosse, Wisconsin
Jay R. Jorden	Celina, Texas
Ronal C. Kerbo	Denver, Colorado
Jim "Crash" Kennedy	Austin, Texas
Kathleen H. Lavoie	Plattsburgh, New York
Thomas Lera	Falls Church, Virginia
Douglas M. Medville	Reston, Virginia
Paul J. Meyer	Huntsville, Alabama
Stephen R. Mosberg, MD	Vienna, West Virginia
Diana E. Northup	Albuquerque, New Mexico
Marc Ohms	Hot Springs, South Dakota
Rick Olson	Cave City, Kentucky
William D. Orndorff	Blacksburg, Virginia
Dale L. Pate	Carlsbad, New Mexico
Victor J. Polyak	Albuquerque, New Mexico
Paula P. Provencio	Albuquerque, New Mexico
John E. Roth	Cave Junction, Oregon
Patricia E. Seiser	Vienna, West Virginia
Bernard W. Szukalski	Redlands, California
Michael Ray Taylor	Arkadelphia, Arkansas
Rickard S. Toomey III	Smiths Grove, Kentucky
Jerry L. Trout	Tucson, Arizona
George Veni	San Antonio, Texas
Jim C. Werker	Hillsboro, New Mexico
John M. Wilson	Richmond, Virginia

Photographers and Illustrators

Jonathan B. Beard	Springfield, Missouri
Gary C. Berdeaux	Park City, Kentucky
Hans G. Bodenhamer	Kalispell, Montana
Penelope J. Boston	Socorro, New Mexico
Dave Bunnell	Angels Camp, California
Sherry Cady	Portland, Oregon
Allan B. Cobb	Austin, Texas
Keith Christenson	Falls Church, Virginia
Bonnie Curnock	Shell Knob, Missouri
Tom Dotter	Golden, Colorado
Sheryl Ducummon	Austin, Texas
Joseph H. Fagan	Blacksburg, Virginia
Gus Frederick	Silverton, Oregon
Val Hildreth-Werker	Hillsboro, New Mexico
Kevin Glover	Talihina, Oklahoma
Rodney D. Horrocks	Hot Springs, South Dakota
Louise D. Hose	Carlsbad, New Mexico
Kenneth Ingham	Albuquerque, New Mexico
David H. Jagnow	Albuquerque, New Mexico
Jon Jasper	Orem, Utah
Keith Johnson	Oakland, California
Jay R. Jorden	Celina, Texas
Jim "Crash" Kennedy	Austin, Texas
Kathleen H. Lavoie	Plattsburgh, New York
Ben Maxwell	Baker, Nevada
David R. McClurg	Vallejo, California
Janet L. McClurg	Vallejo, California
Larry McLaughlin	Boulder, Colorado
Diana E. Northup	Albuquerque, New Mexico
Robert Nymeyer	Carlsbad, New Mexico
Rene Ohms	Custer, South Dakota
Rick Olson	Cave City, Kentucky
Arthur N. Palmer	Oneonta, New York
Dale L. Pate	Carlsbad, New Mexico
Victor J. Polyak	Albuquerque, New Mexico
Matt Reese	Southport, Indiana
Douglas S. Soroka	Perkasie, Pennsylvania
Michael N. Spilde	Albuquerque, New Mexico
Rickard S. Toomey III	Smiths Grove, Kentucky
Jerry L. Trout	Tucson, Arizona
Merlin D. Tuttle	Austin, Texas
Jim C. Werker	Hillsboro, New Mexico
George Veni	San Antonio, Texas

Design and Production

Gus Frederick	Silverton, Oregon
David R. McClurg	Vallejo, California
G. Thomas Rea	Cloverdale, Indiana

Reviewers

Stanley D. Allsion	Carlsbad, New Mexico
Paula J. Bauer	Carlsbad, New Mexico
Jonathan B. Beard	Springfield, Missouri
Thomas B. Bemis	Carlsbad, New Mexico
Penelope J. Boston	Socorro, New Mexico
Evelyn Bradshaw	Fredericksburg, Virginia
Paul Burger	Carlsbad, New Mexico
Rane L. Curl	Ann Arbor, Michigan
Donald D. Davis	Denver, Colorado
Dick DesJardins	Sandia Park, New Mexico
Jill DesJardins	Sandia Park, New Mexico
Joseph C. Douglas	Nashville, Tennessee
Keith Dunlap	Greenwood, Indiana
Joel Despain	Three Rivers, California
Harvey R. DuChene	Centennial, Colorado
William R. Elliott	Jefferson City, Missouri
Scott Fee	Birmingham, Alabama
William S. Frantz	Los Gatos, California
James R. Goodbar	Carlsbad, New Mexico
Phyllis Hamer	Tucson, Arizona
Merideth A. Hildreth	Hillsboro, New Mexico
Val Hildreth-Werker	Hillsboro, New Mexico
Karmen Hopkins	Little Rock, Arkansas
Dwain Horrocks	Woodland Hills, Utah
Rodney D. Horrocks	Hot Springs, South Dakota
George N. Huppert	La Crosse, Wisconsin
Kenneth Ingham	Albuquerque, New Mexico
Patricia L. Jablonsky	Devil's Tower, Wyoming
David H. Jagnow	Albuquerque, New Mexico
Cheryl S. Jones	McLean, Virginia
Laura LaVonne Jorden	Albuquerque, New Mexico
Jay R. Jorden	Celine, Texas
Ray Keeler	Glendale, Arizona
Jim "Crash" Kennedy	Austin, Texas
Ronal C. Kerbo	Denver, Colorado
Lynn G. Kleina	Perkasie, Pennsylvania
Thomas Lera	Falls Church, Virginia
Mary Laycock	Hot Springs, South Dakota
David R. McClurg	Vallejo, California
Janet L. McClurg	Vallejo, California
Gail McCoy	Saratoga, California
Douglas M. Medville	Reston, Virginia
Paul J. Meyer	Huntsville, Alabama
William Mixon	Austin, Texas
Steven R. Mosberg	Vienna, West Virginia
Diana E. Northup	Albuquerque, New Mexico
Marc Ohms	Hot Springs, South Dakota
Rick Olson	Cave City, Kentucky
Arthur N. Palmer	Oneonta, New York
Margaret V. Palmer	Oneonta, New York
Don A. Paquette	Martinsville, Indiana
Dale L. Pate	Carlsbad, New Mexico
Jim Pisarowicz	Custer, South Dakota
G. Thomas Rea	Cloverdale, Indiana
Jason Richards	Carlsbad, New Mexico
John E. Roth	Cave Junction, Oregon
Ira D. Sasowsky	Akron, Ohio

Patricia E. Seiser	Vienna, West Virginia
Harley G. Shaw	Hillsboro, New Mexico
Gordon L. Smith	Floyds Knobs, Indiana
Douglas S. Soroka	Perkasie, Pennsylvania
Michael N. Spilde	Albuquerque, New Mexico
Bernard W. Szukalski	Redlands, California
Steve Taylor	Champaign, Illinois
Jerry L. Trout	Tucson, Arizona
Ransom W. Turner	La Luz, New Mexico
George Veni	San Antonio, Texas
Lydia Webster	Hillsboro, New Mexico
Jim C. Werker	Hillsboro, New Mexico
Betty J. Wheeler	Saint Paul, Minnesota
Mike Wiles	Newcastle, Wyoming
Patricia N. Woodruff	Hillsboro, New Mexico
And one anonymous reviewer	

Contents

Part One
Introduction

Foreword

Preface

Part Two
Cave Conservation, Management, and Ethics

Section A Identifying and Protecting Cave Resources

Section B Developing Cave Management Programs

Section C Improving Caver Ethics

Part Three

Cave Restoration

Section A Introducing Cave Restoration

Section B Organizing Cave Projects

Section C Restoring Cave Passages

Section D Restoring Speleothems

Part Four

Speleothem Repair

Section A Introducing Speleothem Repair

Section B Repairing Speleothems

Part 1
Introduction

Foreword

Endless Caves and Lost Stalagmites

Ronal C. Kerbo

Where Will Our Spirits Lead?

When did people stop seeking caves for shelter?
When did we stop thinking of caves as sacred places?

When did we start using claustrophobia as an excuse
not to go as far as our spirits will lead?

When did we reject the comforting velvet dark of caves
beckoning us forward toward the unknown?

When did we replace our awe with a fear of the great
underground wildernesses—and how have we allowed
caves to become arenas for extreme-sport escapades?

How can we reconnect with the natural underground world
and build ethics that evolve from understanding caves as fragile
systems created and sustained by natural processes?

Understanding Caves as Systems

Before we can develop a conservation ethic for caves we must understand that caves are the result of natural processes that are neither magical, mysterious, nor supernatural. Caves form through processes as natural as rainwater running into cracks in the rock, lava flowing down slope, or ocean waves slamming into coastal rock outcrops. These processes can be observed, measured, and understood by anyone willing to take a little time to read and observe.

As you make your way through this book, written by members of the National Speleological Society and edited as a labor of love by Val Hildreth-Werker and Jim C. Werker, you will come to understand that both the conservation of caves and the ethics of cavers have changed over the years. You will find out that leaving things as you find them may not be as easy as it first seems.

You will read about how we have come to understand that our excursions into caves, for whatever reasons—exploration, survey, photography, science, or recreation—can make profound and lasting detrimental impacts on these very fragile systems in ways that weren't even recognized 30 or even 20 years ago. You will begin to see why some feel that caves can be protected only by keeping discoveries secret and by not revealing the existence of new caves to anyone.

Our conservation ethics must be born out of understanding caves as natural processes—as systems that are greater than the sums of their parts.

I hope this book will help deepen our understanding of caves and pave the way for new modes of thinking.

Human Impact, an Endless Challenge

As a caver who started out in southeastern New Mexico and far west Texas, I can think of no better example of how humans impact caves than to recall the story of Endless Cave.

In 1968, Joe Spencer from Oklahoma and I accompanied Jerry Trout on a cave photography trip. Jerry took us on a whirlwind trip through the major passages and chambers of Endless Cave before we finally stopped in the Totem Pole Room to begin our photography. We were impressed with the complexity of what was really a rather small cave. In fact we made four more trips over the next two months, just to relocate and explore the places Jerry had revealed to us on that first trip.

During those years, the cave entrance was not gated as it is now. Some distance into the cave a narrow pinch of piled rocks disguised the entryway into more fragile areas.

As we explored the cave, we became more impressed with its beauty and more protective of its contents. As we learned more about its history, we realized that we were only seeing a shameful remnant of the cave's former glory.

Speleothem is a general term for cave formation. The word embraces all stalactites, stalagmites, helictites, flowstone, moonmilk, flowers, and other secondary mineral deposits in caves.

Early Exploration of Endless

Endless Cave was first extensively explored in the 1930s by a group from Carlsbad, New Mexico. One member of this group was Robert Nymeyer, author of the book *Carlsbad, Caves, and a Camera.* Nymeyer writes of his initial trips into the cave and the great beauty discovered there, saying that Endless soon became his favorite cave.

He tells of going away to school and returning to find that his favorite cave had been vandalized by a rockhound who polished and sold its speleothems.

In his book, Nymeyer (1978, page 87) had this to say about the day that one of his caving friends showed him the wanton destruction of Endless Cave:

> *He reached up on a shelf and handed me an enlarged photograph. It took me a while to recognize the scene. Then I realized it was the Soda Straw Room [in Endless]. But nothing I remembered was there—just bare walls and ceiling and a few remnants of the coral fountains and broken edges of the lily pads. Gone were the soda straws and the gleaming white stalactites and the rows of chocolate brown draperies that had festooned the walls. I realized my hands were shaking. I felt like a little boy holding his favorite toy that had been smashed by the bully next door. The only difference was that the little boy could cry. I never went back to Endless Cave.*

Endless Cave is an example of telling one-person-too-many about a cave location.

For the 1978 National Cave Management Symposium held in Carlsbad, New Mexico, two photographs were posted prominently by the doors leading to the Symposium meeting rooms. The photos were simply captioned: "Endless Cave: circa 1930s" and "Endless Cave: 1960s."

The picture depicting the cave in the 1930s was one of Bob Nymeyer's early photographs and the other was a Jerry Trout photograph made in the same place a quarter century later. The difference between the two pictures was startling and no one could pass them without commenting on the damage shown in the later photo.

As cavers and resource managers studied the contrasting pictures, they too did not cry—but I don't think they would have wanted to return to Endless Cave either, had they been there before the looting.

Figure 1. This photo, taken in about 1934 by Bob Nymeyer, shows the original beauty of the Soda Straw Room in Endless Cave. Looting by vandals destroyed the pristine room. According to Nymeyer, the walls were covered with rippling brown flowstone curtains and hundreds of pencil-thin stalactites hung over the water. The white soda straw on the left was estimated to be more than 3 meters (10 feet) long.

Figure 2. This photo was taken in about 1961 by Jerry Trout. The room is now called the Green Lake Room, not the Soda Straw Room. All the soda straws are gone—only broken stumps remain of the hundreds of soda straws, stalactites, draperies, and other speleothems—the result of vandalism before the cave was gated and access was controlled.

Chain of Vandalism

During my caving career I have seen and heard references to Endless Cave as an example of sharing a cave location with one too many people. Had one of the original explorers not taken the rockhound vandal to the cave, it might have survived in much better condition.

But most people who go into caves do not, unfortunately, keep secrets. It may start small—this knowing about a cave. At first only one or two friends know—then three or four have been told. After a time, the informal information network takes over and soon too many people know where the new cave is, or they have at least heard rumors.

Away we go with a hearty “High-ho, Silver”—ropes, hard hats, and lights in hand—for *we* are the real cavers and we know the way. We must go and see, for we are not like the three guys I once met deep underground in another New Mexico cave in about 1968.

They Tried to Steal the Candle Table

I was again caving with my friend Joe Spencer. This time we were in a cave high in the Guadalupe Mountains of southeastern New Mexico.

Surrounded by crystalline beauty and that deep silence common to many desert caves, I rustled around in my pack searching for my extra charge of carbide and decided I was long past due for something to eat.

I shed my pack and extracted a small white candle from its depths. I pulled off my hard hat and touched the flame of my carbide lamp to the candle's wick, then melted the other end of the candle just enough to stick

Jerry L. Trout

Figure 3. In 1961, the Candle Table was still intact with the unusual candle-like speleothem in place on the coke table shelfstone. Bob and Jerry Trout enjoyed making photographs in this beautiful room hidden inside a New Mexico cave.

it to the top of my helmet. I placed my hard hat on the cave floor in front of me and extinguished the flame of my headlamp so that I could eat by candlelight and save my precious carbide for the trip out.

Sitting quietly on the cave floor in the warm glow of my little candle, I was now ready to eat and relax into the darkness that is always just beyond the reach of my light. I carefully made sure no trash was left in the cave. I was very conservation-minded. I always tried to *take nothing but pictures, leave nothing but footprints, and kill nothing but time.*

As I sat quietly watching the wax from my candle drip and run down the sides of my helmet I had no idea that three guys were trying to remove a well-known speleothem called the Candle Table from a chamber deeper in the cave.

As we ate, the three flashlight cavers were sweating their way through another passage seeking the "best route" to the entrance. Finally, they came to where Joe and I were having our lunch. They asked us where to find the best route out of the cave. We told them the best way back was the same way they had come.

After they left, Joe and I finished eating, packed up, and went down a nearby slope to check out a crawl under the breakdown. We were soon almost directly beneath the three spelunkers who were having a heated debate over something they had left behind.

It didn't take a lot of imagination to guess that they were trying to steal speleothems.

> *"We'd better get out of here now and leave that thing for later,"* said the first.
>
> *"Those two guys might not like the idea of us taking that thing out of the cave,"* replied another.
>
> Then, we heard one dissenting voice, *"Damn it, I want that thing in my backyard. It'll look great."*

But the others insisted, *"Leave it for later, we'll come back when those guys are gone."*

We decided to find out what they were up to. It didn't take a lot of imagination to guess that they were trying to steal speleothems. We started back toward the main passage. The three must have heard us rustling about below them and headed for the surface. Joe and I tried to get back to the main passage quickly, but instead popped out further into the cave, just at the right spot to see the "thing" they had been talking about.

Lying on the passage floor were the broken remains of the Candle Table, a beautiful stalagmite with rimstone encircling its top. The table-shaped formation had been broken off at the base and the three had managed to carry it only some 50 feet from its former spot in a chamber called the Table Room.

We headed for the cave entrance, but the three had already reached their vehicle and sped away before we could catch up.

Joe and I then trudged back into the cave and returned the table to its proper place, leaving a note about the incident. A few days later we helped Jerry Trout (who is now the National Cave Coordinator for the USDA Forest Service) repair and restore the stalagmite.

D & J McClurg Collection

Figure 4. Look carefully at this photo of the still beautiful peach-colored column rising from the coke table shelfstone in a New Mexico cave. You will discover countless broken speleothems descending from the ceiling. This breakage dates from before this delicate section of the cave was gated to help stop further vandalism.

Ever Evolving Ethics

There you have it. Back in the 1960s I was not only a dedicated disciple of the National Speleological Society's motto (all of that *take nothing, leave nothing, and kill nothing* philosophy), but I was also a defender of the flame, a keeper of the truth, and certainly a most true crusader.

Had we caught those three guys, I would have been happy to take them on half a cave dive or on a one-way trip down a deep pit. I was properly outraged, morally offended, righteously indignant, and knew deep in my heart that they would all three come to no good in the future.

What did it matter that right after our encounter with the Candle Table vandals, I did several things that no careful caver would consider doing now.

When we finished lunch, I dusted food crumbs off my jeans onto the cave floor. I jerked the candle off my helmet and flicked the excess wax onto the cave floor. I turned toward a pile of bat guano and urinated into the darkness. I was secure in the knowledge that I was Mr. Conservation.

After all, bats pee in caves. Those little food crumbs I brushed off might come in handy to a starving cave cricket. You couldn't even see the candle wax—besides, it's inert stuff anyway, isn't it? Now where is a good place to bury my spent carbide…?

Today we know more about how humans can impact cave systems. We've changed our ways, and we now discourage activities that were common practice until only a few years ago.

Changing Times

As the poet, Robert Zimmerman (aka Bob Dylan) says, "The Times They Are A-Changin'." In fact things have changed quite a lot. In recent years, the most fascinating and exciting discoveries about caves involve microorganisms that dwell in the deep. Research conducted on cave microbes is

> Research conducted on cave microbes is revealing that compounds from these organisms may have significant pharmaceutical and industrial applications.

> Today we know more about how humans can impact cave systems. We've changed our ways, and we now discourage activities that were common practice until only a few years ago.

revealing that compounds from these organisms may have significant pharmaceutical and industrial applications.

Microscopic organisms also may play important roles in the development of speleothems. We now understand that neglecting to clean mud from our cave gear (or transferring mud between various caves) can contaminate and severely affect cave microbial communities. Human impact on previously uninvestigated microorganisms can initiate a ripple effect and eventually disrupt the food chain of a cave ecosystem.

It is no longer good enough just to remind each other to *take only pictures, leave only footprints, and kill nothing but time*. We must become sensitive enough to these environments that we automatically use minimum-impact behaviors.

For example, when we take a photograph, we shouldn't become so engrossed in the endeavor that we neglect to look at our surroundings and damage other speleothems just to take a picture.

We should make sure that our footsteps do not mar pristine flowstone, disrupt the ecology of cavern pools, or crush fragile floor features. Killing time has been expanded to *not killing* whole ecosystems out of ignorance.

In this book, the traditional motto has been modified to reflect changing ethics:

Take nothing but conservation-wise photos.
Leave nothing but careful footprints on established trails.
Kill nothing but time.

Embracing Cave Systems

As I think about the responsibility we shoulder when we enter caves, I'm of two minds—hesitating to ever enter a cave again, or realizing that by being sensitive as I traverse a cave, I can greatly enhance my experience.

Inside a cave, we become intimately aware of the unique environment that surrounds us. We are immersed in wilderness so profound that new discoveries may await us at every turn of a passage. Novel ideas may stalk us at the portal to a new chamber. Exciting discoveries of previously unidentified organisms may be in store.

Our conservation ethics must be born out of understanding caves as natural processes—as systems that are greater than the sums of their parts. For far too many years, those of us who are the most intimate with caves, from recreational cavers to scientists, have focused almost exclusively on the nonliving, geologic components of cave systems.

Our personal ethics, as well as laws, regulations, and policies, reflect our focus on the physical contents of caves rather than on how entire cave systems are made up of interdependent components. We can no longer afford to ignore karst processes.

We need to develop aggressive educational programs, strategies, guidelines, and policies that can help us understand, protect, conserve, and interpret the overall processes of cave and karst systems.

I don't want to diminish the importance of protecting the physical contents of caves (both natural and cultural). But we must emphasize the importance of looking beyond single components and learning to view caves as ecological systems.

I hope that this book will help deepen our understanding of caves and pave the way for new modes of thinking. We can no longer ignore these interdependent processes as we develop better methods in cave and karst conservation management.

Outreach and Education Do More than Laws

We have laws that help protect caves and their contents—the Federal Cave Resources Protection Act of 1988 and various state laws. None of the current laws place priority on the need to understand caves and the processes by which they are created. No current legisla-tion ensures that caves will be considered in terms of environmental ethics. Promoting ethics and clarifying the reasons for conservation are not within the scope of laws.

That important work of establishing an acceptable conservation ethic must still be accomplished one step at a time by patient cavers—collectively the members of the National Speleological Society, American Cave Conservation Association, Cave Research Foundation, Karst Waters Institute, National Cave and Karst Research Institute, and others who have devoted their energies to speleological research, education, and public outreach. There have been many positive changes over the last 35 years—changes accomplished by the persistent spirits of individual and collective cavers.

We must be ever vigilant or we will lose the best of ourselves as well as the great and small caves of the world.

D & J McClurg Collection

Figure 5. While still impressive, the Breast of Venus Room in Endless Cave is only a shadow of its former glory. Stones on the floor mark a trail through the room so cavers can avoid walking on areas now restored to a near-original condition by volunteer cleanup crews. These volunteers washed off the many layers of mud tracked in by countless visitors. The caver sitting within the trail provides scale for the stalagmite and column.

Cited Reference

Nymeyer R. 1978. *Carlsbad, Caves and a Camera.* Teaneck, New Jersey: Zephyrus Press. 318 p.

Author's Note

Forcing people to develop an environmental ethic by dint of law rather than by educational processes is a skewed attitude that will ultimately alienate us from the very people who could make a real difference, the public at large.

The Federal Cave Resources Protection Act of 1988 may have forced federal land managers to cover their collective bureaucratic backsides and make lists of caves. However, at this time no one has invoked our existing laws to address the issue of so-called extreme-sport enthusiasts using caves and lava tubes as underground climbing gyms, stress-challenge venues, and sites for media-promoted eco-challenge events.

These extreme activities certainly disrupt the biotic communities of caves, destroy speleogens, damage mineral deposits, and have no true justification other than for the participants' ego satisfaction. With land managers' reluctance to use the Federal Cave Resources Protection Act for fear of lawsuits, how long before sport enthusiasts start climbing directly on America's speleothems? (Climbing on speleothems is the current practice in the caves of Thailand, a land beyond the reach of our laws.)

At the mere suggestion of not using caves as sports gyms, the climbers whine to their advocacy organization and use U.S. discrimination laws to combat conservation ethics in the world of speleology. There is a lesson here for cavers—it is clearly time to promote low-impact caving ethics and reach out in advancing awareness of conserving cave and karst resources.

Preface

Vision for This Book

The vision to create this book was conceived a decade ago. In Carlsbad, New Mexico, at the 1993 National Cave Management Symposium, Jim Werker and Jerry Trout presented papers on speleothem repair. A caver from the audience stood up and stated:

> *You two guys should put together a book on formation repair.*

Trout and Werker glanced at each other and shrugged. But the idea kept nudging them and the two guys kept talking. They took the bait and started to think about organizing a manual on formation repair. While composing a list of repair subjects, it grew apparent that cave restoration techniques should also be addressed. The topic list doubled.

Both Trout and Werker spend a good deal of their lives outdoors—they both prefer being in caves rather than sitting at a computer wrestling with a book. So they decided to cajole other cavers into the joy of writing many sections for the book. Their hope was to speed up the project.

While the concept of a simple book on speleothem repair was growing into a labyrinth with a life of its own, real life events were moving along for both Trout and Werker. Jerry moved from New Mexico to Arizona and was named the USDA Forest Service National Cave Coordinator.

Back in New Mexico, Jim and Val found each other in a cave, started working together on restoration, repair, and photomonitoring, eventually fell in love, and tied the knot. After they married, Jim and Val launched the daunting task of editing a multi-author technical book.

> All the authors of this book are cavers—each is intently aware that few absolutes exist when it comes to cave management practices. The only absolute fact is that no two caves are alike.

Content Doubled Again

Meanwhile, fresh, new ideas were churning for the book. Naturally, a how-to book on restoration and repair should include strategies for intervention and ways to prevent damage. The content doubled again, and cave scientists from various disciplines came aboard.

While chapters were growing in the wombs of cavers' computers, more surprises were in store. The authors obviously wanted to present state-of-the-art philosophy and methods, but new research was exploding with new speleological data. The new information led to improvements in restoration techniques and current best practices had to be redefined for many of the topics.

During the editing process, we realized that many chapters were skirting around the edges of important conservation concerns. In the speleological community it is often counterproductive to profess definitive recommendations about cave practices, even when the stated objective is to define current best practice.

All the authors for this book are cavers—each is intently aware that few absolutes exist when it comes to cave management practices. The only absolute fact is that no two caves are alike. Cave characteristics and contents vary tremendously. Likewise, conservation concerns and techniques vary widely.

Caves differ in their genesis, fragility, and impact management needs. Conservation and restoration tasks must be planned and executed in accordance with the unique values of each cave.

Caves, karst, and psuedokarst form through an intriguing variety of geologic processes. Caves differ in their genesis, fragility, and impact management needs.

In some cave and karst systems, water frequently flows and rejuvenates passages. In other caves, water occasionally flows and promises only slight potential of erasing human impact. And in caves located beneath arid surface terrain, water rarely flows, leaving passages and speleothems dry, fragile, and easily damaged. Adjacent caves can house extremely different resources. Each cave should be managed according to its own unique characteristics.

It follows that restoration tasks must be planned and executed in accordance with the unique values of any given cave. Footprints, trails, and conservation decisions depend on the nature of each cave. The inherent diversity makes the tasks of cave conservation, restoration, and repair ever changing and always challenging.

The team of NSS authors, reviewers, and editors present the speleological community with the best of what we know at this time. Be assured, recommendations will change. New scientific information will lead to new conclusions. Best practices will evolve. In this book we hope cavers will find useful state-of-the-art information, but not the last word in cave restoration.

We present what we know today as the current best practices. We encourage all to keep exploring new passages and improving techniques through new layers of research and experience. We urge implementation of better methods as new information comes to light.

About This Book

In Walt Kelly's 1971 Earth Day cartoon Pogo proclaimed:
We have met the enemy and he is us.

The enemy is us. Humans are the nemeses of cave systems. Pogo's insight aptly serves our subsurface world of caves and cavers and introduces the ubiquitous message of this book.

Cave conservation is a lot like treating a dreaded chronic disease. With human affliction, once you recognize that you have a constant or recurring condition, you have the obvious opportunity to make necessary changes and take really good care of yourself.

Cave conservation is similar. In the rich archival darkness of cave and karst systems, human presence creates the malady of negative impact. Once we recognize the symptoms, we have the manifest opportunity to make changes, influence ethics, and take really good care of our Under Earth. This book is about realizing those opportunities—it is about taking care of caves and the resources within.

Although mainly a how-to manual on cave restoration and repair, this volume is also about becoming more conscious of the impacts that create needs for conservation actions. Caves do fine when left on their own to follow nature's intended path. But interference from people can produce irreparable alterations in cave systems.

Harm is certainly caused by detrimental behaviors from individuals who visit the underground, but damage is also incited by municipal, industrial, and agricultural enterprises on the surface. The intent of this manual is to help all become more mindful of conditions that call for human conservation intervention and resource management.

For every decision related to a cave, the foremost concern should be the perpetuation of speleological processes, values, and resources. As cavers, cave managers, landowners, project leaders, and speleologists, what factors should we consider when making decisions about cave systems? Crafted to explore questions and formatted to use as a field manual, this book describes the current best practices in cave conservation and management. Information is presented in easy-to-identify sections.

Representatives from various scientific disciplines within speleology explain their conservation and preservation concerns. Tools and proven methods are presented for cave restoration and speleothem repair. Encouraging low-impact ethics, this book covers techniques that any caver can use in protecting, understanding, restoring, and conserving cave environments.

The further intent of this book is to encourage all to focus beyond individual areas of interest and expertise—to become more mindful of the often overlooked, intricately interrelated aspects of ecological balance and resource values in cave systems. Increased awareness of subterranean values may bring us to deeper stewardship of Earth's balances. Understanding the mysteries of the subsurface may also help us reach beyond in responsibly exploring the secrets of extraterrestrial realms.

Ultimately, if we can avoid unnecessary impacts to caves—if we can move safely and softly—we will minimize the need for restoration.

Caves do fine when left on their own to follow nature's intended path. But interference from people can produce irreparable alterations in cave systems.

Mainly a how-to manual on cave restoration and repair, this volume is also about becoming more conscious of the impacts that create needs for conservation actions.

How to Use This Book

Sections in this volume focus on techniques for accomplishing cave conservation, restoration, speleothem repair, and minimum-impact ethics. We expect this publication to be used as a manual or handbook—it will be hauled in packs, tossed in backseats, and found on library shelves.

Sections in this volume focus on techniques for accomplishing cave conservation, restoration, speleothem repair, and minimum-impact ethics.

Cave Conservation and Restoration is divided into four parts with contributions from more than 40 authors. The introductory material is followed by three major parts divided into chapters. Easily located headings and subheadings, cross-refernces, and the index are designed to guide readers to desired information.

Cross-references throughout the text refer readers to other chapters where concepts and techniques are more thoroughly explained. In addition to black-and-white photos and illustrations within the chapters, selected photos are in a full-color insert. These photos are also cross-referenced.

Units of measure are written in both metric and English throughout this volume. Because construction materials are sold in English units in the United States, dimensions for construction materials deviate from the standard metric/English format used elsewhere in this book.

Contributions were critiqued and revised through peer review processes. A team of style and copy editors performed final reviews. Coordinating the effort, Jim and Val reviewed and edited all submissions.

In composing a manual of this nature, redundancy factors into the equation. Because we expect most readers will browse to find subjects for the moment, some repetitious material intentionally remains. To those who read straight through, we ask your tolerance for the redundancy inherent to our subject matter.

The editors, contributing authors, and reviewers offer this handbook to the caves, cavers, and speleologists of the world. We hope all of you will find helpful information in this volume.

- Cavers, cave managers, and cave owners
- Commercial cave personnel
- People involved in cave restoration projects
- Cavers interested in formation repair
- Government land managers
- Natural resource personnel
- International environmental conservation community
- Libraries in areas with cave interests
- University libraries
- NSS internal organizations, such as regions, grottos, conservancies, and task forces
- Speleological organizations in other countries

Acknowledgements

We gratefully acknowledge the support and funding of the National Speleological Society, Inc. in publishing this volume, and the National Institute of Cave and Karst Research for providing additional funding to support the full-color inserts.

We extend deep graditude to the many authors, photographers, illustrators, reviewers, and graphic artists who participated in this publication. Each task and each stage leading to publication was accomplished through their volunteer efforts.

We extend our thanks to long-time cave conservation leaders in the National Speleological Society (NSS)—Albert A. Krause, George N. Huppert, Robert R. Stitt, and John M. Wilson—from the beginning, each individually took the time to encourage the continuation of our conservation efforts.

David H. Jagnow served as the NSS Conservation Division Chief and NSS Conservation Chairman from 1996 to 2001, and we extend gratitude to him for initiating our first proposals to publish this book through the NSS. Without Dave's support of this work and commitment to the NSS, this publication might have never been.

We gratefully acknowledge the commitment of NSS Executive Committee Members who served during the time we were composing and editing this manual. With their continuing support, this publication came to fruition. The NSS Conservation Division works most closely with the NSS Administrative Vice President (AVP). We thank Steve Ormeroid, Tom Lera, Thom Engel, Cheryl Jones, and Doug Medville who all served in the NSS AVP position during our work on this book—we appreciate the energy, honesty, and forthright character of each in supporting this cave conservation publication. The NSS Executive Vice President (EVP) is tasked with overseeing NSS publications—Gordon Birkhimer, Don Paquette, and Ray Keeler served in the EVP position while we were building this book—their continuing enthusiasm, support, and encouragement helped bring this publication to reality.

Also, we thank all the NSS Executive Officers and Board Members who supported this long-awaited publication with encouragement and budget commitment. Much thanks to NSS Presidents Bill Tozer, Scott Fee, Mike Hood, Fred Wefer, and President Pro Tem Hazel Medville for their leadership. Our thanks go to Tom Rea, Scott Fee, and Colin Gatlin, who served as NSS Operations Vice Presidents (OVP), for supporting and promoting NSS book publications and sales. We extend thanks to Paul and Lee Stevens for their decades of commitment to NSS resources and to David Irving and Ted Kayes for their continuing support.

Much gratitude goes to the NSS Special Publications Committee. We sincerely thank David and Janet McClurg for design, layout, and the tremendous work they contributed in the early stages of preparing this publication (and many NSS publications before this one). Through several decades of sincere commitment, David McClurg's continuing role in NSS publications has brought many important speleological books to press.

We extend deep gratitude to Tom Rea, current Chairman of the NSS Special

Publications Committee. We thank Tom for his timely responses to a myriad of pop-up questions along the way, for his forthright common sense, and for his persistence in seeing it through and making it right. Many thanks go to both Tom Rea and Gail McCoy for their consistent copy edits.

We extend sincere kudos and appreciation to Gus Frederick for his focused commitment of time, graphic energy, and design expertise. Gus completed the design, layout, and corrections, prepared the photos and the color insert section, reworked illustrations for digital layout and formatting, and coordinated the digital prepress for this publication.

We thank Bat Conservation International (BCI) for permitting the use of Merlin D. Tuttle's photographs of bat species and other BCI photographs in this publication. We thank Jim Kennedy for facilitating our work and cooperation with BCI.

We extend much gratitude to Emily Davis and Michael Warner of Speleobooks for providing reference materials and making certain we had all the latest cave conservation publications at our fingertips.

To several friends, we extend especially deep appreciation. This publication reached completion because of their always-ready support, encouragement, time, expertise, and consultation. The following people deserve enormous thanks for their instant readiness at the other end of the phone or email.

We express especially big thanks to four writer friends who live in our charming, tiny town of Hillsboro, New Mexico. Through this book, they have become special contributors to speleological literature. We thank Harley G. Shaw for his thorough review of every chapter, his ruthless slashing of redundancy, and his rearrangements for improved organization. We thank Lydia Webster for her copyediting expertise and her amazing knack of coming up with quick fixes for awkward sentences. We thank Merideth A. Hildreth for her time, motivation, and perseverance in tediously comparing manuscripts, marking corrections, and persistently proofing chapters. We thank Patricia N. Woodruff for her eagle eye and care in proofreading the final formatted manuscript.

We appreciate Gus Frederick more-than-we-can-express for his successful rescue and marathon effort. We especially thank Gus for being a superb team player, a great guest in our home, and a persistent proponent of a get-it-done-right spirit. The caves thank you and we thank you for your time and devotion to making it all happen.

To Tom Rea we extend grateful appreciation for his unwavering support in this project—for quick replies, definitive action, and get-it-done commitment. Thank you for reviewing the manuscript and seeing this publication through the press process. Thank you for believing the NSS needs more books.

To our immediate family, relatives, and extended family of friends who shared their long-suffering support and enduring tolerance, we extend love and thank yous. We especially thank Merideth for becoming our chief copy editor and bottle washer. Thank you for your Cuna Cueva Collaborative spirit in giving quiet, nourishing support and many weeks of work on this project.

To Billy and June Hoffman we give heartfelt gratitude for generously providing quiet days, warm hospitality, and supportive friendship during many important weeks in the stretches of completing this book.

To Jim "Crash" Kennedy, Ron Kerbo, and George Veni—who were always ready to answer questions, find references, and provide timely forthright reviews—we express deep gratitude.

We appreciate the support of friends at Carlsbad Caverns National Park. We thank Dale Pate, Stan Allison, Paula Bauer, Tom Bemis, Paul Burger, and Jason Richards for data and review.

To Penelope J. Boston we extend copious gratitude for graciously giving several weekends of tedious collaboration along with much phone time along the way to tweak sentences and concepts.

We extend big, warm thank yous to Diana E. Northup and Kenneth Ingham who shared their home, computers, and purring cats for many overnights, plus a grueling marathon weekend of laptops, keyboards, and edits at a critical stage during the development of this book. Many thanks to the marathon reviewers—Diana Northup, Kenneth Ingham, Dick Desjardens, Jill Desjardens, Mike Spilde, Penny Boston, and Ariel Boston.

We also extend sincere appreciation to Diana E. Northup for her untiring commitment in assuring the accuracy and completeness of references in this volume. She used the following resources to check most of the references in the chapters and sidebars:

- Books: OCLC Online Computer Library Center, Inc. produced the Worldcat database, which includes more than 52 million records for books, maps, proceedings, and other resources from more than 37,000 member libraries in the world. Information about conference proceedings was reformatted using data from the OCLC Worldcat in order to make it easier to find these references through interlibrary loan.
- Journal articles: References to articles were checked in the most appropriate subject database, including SciSearch, ArtIndex, BIOSIS, Georef.

In formatting the references and citations, we used a modified form of the reference style guide recommended by the Council of Biology Editors (CBE) in *Scientific Style and Format: The CBE Manual for Authors, Editors, and Publishers*, Sixth Edition, 1994, prepared by the Council of Biology Editors Style Manual Committee. Their sixth edition recommends both general and scientific publication style and formats for journals, books, and other types of publications. Minor reference style modifications deemed appropriate for this publication include italicized titles, complete journal titles instead of abbreviations, and Web addresses for some entries.

Rather than listing trademarks here, they are included throughout the text. While we have tried to include all registered (or applied for) trademarks, any omission of such marks from any product is regretted and is not intended as an infringement on such trademarks. Please contact us so we are sure to correct any omissions in future editions.

Not the Last Word— But a Call to the Future

The alternative to negative impact is no entry—no recreation, no science, no exploration, no survey, no maps, and no photographs.

As cavers, our goal should be balance—we want to limit detrimental consequences while we continue to visit caves. Underground environments can be protected and preserved if we learn to cave softly—and if we learn to minimize human impacts, both aboveground and below in the subterranean realm.

The authors, reviewers, and editors of this manual invite all cavers to explore ways to reduce or eliminate the need for cave and karst restoration. Albert Einstein summed it up with this admonition:

Problems cannot be solved at the same level of awareness that created them.

That's the environmentally correct way of saying we can't fix a mess at the same level of ignorance that allowed it to happen.

People do create impact in caves—but we can change habits, reduce impacts, and learn ways to mitigate cave damage.

For decades cavers promoted a follow-the-footsteps-of-others ethic. In the past, we would place our feet inside the original footprints of others to avoid causing new harm in pristine areas. More people are caving. Footprint pathways are expanding into multiple scars across cave floors. Out of necessity, ethics are changing.

Today, rather than retracing former impact, cavers are developing gentle ways to erase human damage. Minimum-impact etiquette is encouraging cavers to avoid creating new scars.

Even the traditional caver motto is changing—*Take nothing but pictures. Leave nothing but footprints. Kill nothing but time.* Many are now augmenting the axiom to reflect low-impact ethics.

Take nothing but conservation-wise photos.
Leave nothing but careful footprints on established trails.
Kill nothing but time.

Cave safely ... cave softly,
Val Hildreth-Werker and Jim C. Werker, 2006

Part 2
Conservation, Management, Ethics

Section A—Identifying and Protecting Cave Resources

Current Best Practices

Val Hildreth-Werker

This book presents cave conservation methods and philosophy from an assortment of speleologists and cave managers. In composing this work, the authors, reviewers, and editors have collaborated in refining the methods that are presented as current best conservation practices. The editors have sorted through techniques, evaluated concepts, and coordinated statements to present consensus and describe the current best practices in cave management, cave conservation, and cave restoration.

The first objective in cave conservation is to avoid creating new problems. First, do no harm—*primum non nocere*. Evaluate the situation from all angles and consult with trained speleologists before launching into any cave conservation effort. Before deciding how to change it, fix it, clean it, or remediate the conservation problem, be sure to explore approaches, research current management practices, and gather information about materials that are reasonably safe for long-term use in caves.

No two cave systems are alike and the demands for protection vary tremendously. Evaluate each cave individually and avoid becoming stymied in a cycle of standards. For any action in a cave, do no harm—and if that sometimes means doing nothing, or sometimes means returning to an old standard because the current best practice doesn't fit a situation, go with the common sense solution that best protects the cave. For cave dilemmas, the demand for immediate protection of resources is often vitally important to long-term conservation.

Many cave systems are extremely sensitive to human impacts. As scientists and cavers learn more about underground environments, it is immensely clear how little is known. The body of knowledge in speleology is rapidly expanding with new information gleaned through advanced technologies.

Scientific disciplines are replete with topics for new spelean studies. As new scientific facts are integrated with cave management, improved best practices continually evolve.

There is one constant factor in cave conservation—mistakes lead to new methods. Many practices of the past are now rejected because of detrimental effects (Hamilton-Smith and others 1998). There is much to learn about the scientific mysteries of Earth's subterranean systems—conservation practices are refined as new answers emerge.

There is one constant factor in cave conservation—mistakes lead to new methods. Many practices of the past are now rejected because of detrimental effects.

The term *best practice* (or world's best practice) is a standard buzzword for management. (See best management practices, page 34.) People are sometimes suspicious or contemptuous of the term (Spate and others 1998). However, for cave conservation, discussing standards based on the current best information is valid for two reasons.

First, adding the word *current* in front of best practice reminds us that improved methods are always on the horizon.

Second, the term inherently spawns the all-important process of questioning, "What are the current best conservation practices?" Thus, the phrase encourages research, evolution of ideas, and advances in methodology.

Current best practice in cave conservation and management is not an end product, but rather a conscious process of defining and enhancing standards. The current best practices presented in this volume are principles intended as springboards for discussion. Best practices for protecting cave and karst resources are not prescriptions for all situations.

Current best practice in cave conservation and management is not an end product, but rather a conscious process of defining and enhancing standards.

Cited References

Hamilton-Smith E, McBeath R, Vavryn D. 1998. Best practice in visitor management. In: Smith DW, editor. *Cave and Karst Management in Australasia XII*. Carlton South, Victoria, Australia: Australasian Cave and Karst Management Association. p 85-96. [Also in: Australasian Cave and Karst Management Association Journal. 1997. 27:33-43.]

Spate A, Hamilton-Smith E, Little L, Holland E. 1998. Best practice and tourist cave engineering. In: Smith DW, editor. *Cave and Karst Management in Australasia XII*. Carlton South, Victoria, Australia: Australasian Cave and Karst Management Association. p 97-109.

Additional Reading on Cave and Karst Management

A comprehensive list of references concerning the development of cave and karst management is included in this volume. (See Appendix 3, Cave and Karst Management References. page 531.)

Section A—Identifying and Protecting Cave Resources

Resource Inventory: A Tool for Cave Science, Management, and Restoration

Harvey R. DuChene

The exploration and study of caves has steadily increased in popularity over the last 50 years. This interest is shown by growth in the membership of the National Speleological Society and similar organizations around the world. The growing number of cavers results in increased pressure on known caves. Damage to fragile resources is an almost inevitable consequence of visitation. It is now recognized that we must take care of caves and find new ways to conserve, protect, study, and enjoy them.

To manage caves effectively, their extent, contents, and significance must be assessed.

In 1988, Congress enacted the Federal Cave Resources Protection Act. (See Appendix 1, page 507, for the complete text of the Act. Also see cave laws, page 220.)

This Act mandates the identification of significant caves on federal lands and prohibits:

> *...the willful disturbance, altering, removal, sale, receipt, and possession of ... cave resources obtained illegally on Federal lands.*

The Federal Cave Resources Protection Act of 1988 instructs that significant caves are to be considered in the preparation or implementation of any land-management plan. Through this Act, federal agencies, including the Bureau of Land Management, US Fish and Wildlife Service, USDA Forest Service, National Park Service, and US Geological Survey, are mandated to manage and protect caves.

To manage caves effectively, their extent, contents, and significance must be assessed. This evaluation requires the methodical gathering, sorting, and interpreting of cave data. The first cave inventory projects led to development of methods for acquiring this information.

Inventory projects are active on both private and federal lands. Many caves managed by the Bureau of Land Management, the National Park Service, or the USDA Forest Service, as well as independent cave projects or conservancies, have active inventory programs. The information presented in this chapter is based on the Lechuguilla Cave Mineral Inventory, a project that has been collecting and analyzing geological data in Carlsbad Caverns National Park since 1990.

Cave inventory is the systematic observation and recording of significant features found within a cave.

What Is Cave Inventory?

Cave inventory is the systematic observation and recording of significant features found within a cave. An inventory may include many types of data on the archaeology, biology, chemistry, hydrology, geology, history, mineralogy, paleontology, speleogenesis, and impacts of modern human use. The amount and type of information collected depends on several factors: the purpose of the project; the nature and complexity of the cave; and technical, financial, personnel, and temporal limitations.

Cave inventory is not a new concept. Scientists have been cataloging biota, archaeological sites, and fossil deposits in caves since at least the 1700s. These systematic studies are essential for understanding processes, geological history, and biological communities in caves. Since the adoption of the Federal Cave Resources Protection Act of 1988, inventories have expanded to include the cataloging of all significant attributes in a cave.

Cave inventory is not a new concept. Scientists have been cataloging biota, archaeological sites, and fossil deposits in caves since at least the 1700s.

Cave mapping is a popular, systematic method for collecting data from caves. Early cave maps were simple, commonly showing passage walls but little or no internal detail. Over time, cave maps have evolved to include other information such as streams, breakdown, formations, bedrock geology, and other characteristics deemed significant by the survey crew and cartographer. The amount of detail included on a map often depends on the observational skills, drawing ability, and patience of the survey sketcher.

The term *cave survey* usually refers to linear and angular measurements made inside of cave passages. This mathematical data is used to describe the location, form, and extent of a cave. Typically, a cave survey is used to make a map of a cave. *A cave resource inventory* lists the contents of a cave. Cave cartography—making maps of caves—has become an art form with cartographers striving to include as much inventory information as possible while making maps that are faithful depictions of the cave.

Why Do Inventory?

Information is the key to appreciating and understanding caves and their contents—understanding cave systems is the key to managing and protecting resources. The amount of detail and specific type of information required depends on the reasons for conducting an inventory. For some projects, a reconnaissance inventory is adequate. Inventory for scientific studies, restoration projects, and management of large systems requires more detailed data collection.

The depth and breadth of knowledge about various aspects of the cave environment varies widely among map sketchers. Thus, the data included on maps, especially of large and complex systems, is inherently inconsistent. This is particularly true where many sketchers are responsible for mapping a large cave.

The cave inventory process utilizes checklists that help survey teams and sketchers gather consistent information. These lists itemize the features found in the cave and provide a mechanism for documenting attributes that are rare or significant. The objectives for a cave inventory project should be well defined to determine the type of information and the amount of detail to be included.

Most cave inventories do not require specialists in geology, biology, or other sciences, but they do require a basic understanding of the features likely to be encountered in a cave. Some inventories of a highly technical or scientific nature may require active involvement of specialists.

Inventories as Tools

The following examples describe how cavers, managers, and scientists use inventories as tools for project design.

Cave Science. Basic cave inventories are used by scientists and other researchers to locate areas for study. Once inventory data is stored in a database, information can be accessed quickly. The use of inventory data can lessen human impact on caves. By providing a reference to recorded information, researchers may not need to enter the cave to learn about the resources. For example, a researcher interested in water sampling can use inventory data to locate cave pools. Without a good cave inventory, the researcher would probably have to visit multiple sites to determine if any were suitable for study.

Management. Cave inventories provide information for making educated management decisions about cave resources. When making choices about access through cave passages, cave managers can use inventory databases to locate sensitive archaeological or biological resources (for example, historical artifacts, bat roosts, aquatic troglobites, or habitat for threatened species). Managers can use inventory data to protect important resources by designating travel routes or surface activities that mitigate negative impact.

The objectives for a cave inventory project should be well defined to determine the type of information and the amount of detail to be included.

Restoration. Cave inventories can pinpoint information on areas or passages that have been negatively impacted and need restoration. Using inventory data, managers can locate these areas quickly and can also compare the potential impacts of proposed restoration tasks to the potential benefits of restoration.

For example, restoration is proposed for an area that was muddied by traffic and then closed to visitation. Cave managers can use inventory data to help evaluate the sensitivity or fragility of the area. The data can provide information for determining the size of restoration teams—perhaps only two or three skilled individuals should take on the task or the job may require caver groups working in multi-day shifts. Or it might be decided the damaged area should be left as it is and no restoration effort should be initiated. Databases can also be used to identify multiple restoration objectives that can be accomplished by one team to maximize productivity and minimize additional human impact.

Resources for Inventory

Many information sources exist that can help project managers design inventory programs. Primary sources include published literature specific to the topic or area covered by the inventory. (See Appendix 2, listings of resources and agencies for archaeological, biological, geological, and speleological information, page 515.) Information may also be gathered from local experts, libraries, colleges, universities, and from local, state, and federal government agencies.

Information on scientific aspects of a specific cave or cave region may be available within the caving community. Members of the National Speleological Society (NSS) have formed special interest groups within the Society called Sections. Members of the various NSS Sections (geology/geography, history, biology, paleontology, conservation/management and others) can provide helpful information for the design of inventory projects.

Other information sources include the National Speleological Society Library, Cave Research Foundation, American Cave Conservation Association, Karst Waters Institute, and the National Cave and Karst Research Institute. (See Appendix 2.)

Geological maps and reports are available for many locations and may be obtained from state geological surveys or the US Geological Survey. Maps provide information on the geology of the area surrounding the cave as well as information on topography, surface drainage, rock types, geologic formations, minerals, and fossils.

Information about archaeology may be obtained from the state offices of archaeology. At the national level, the Society for American Archaeology has contact information for local experts.

Biological information is available through professional wildlife biologists at state, federal, and private agencies (for example, Karst Waters

Institute, Missouri Department of Conservation, Illinois Natural History Survey, and Bat Conservation International, Inc).

Hands-on training familiarizes nonspecialists with the features of cave inventory. This may be accomplished "on the job" by mixing new personnel with experienced leaders on inventory teams. It is also helpful for the person or group managing the inventory to prepare a photographic guide to specific features found in a cave, especially those features that are uncommon or unique to the region.

Types of Inventory

Qualitative or Quantitative

Inventory can be either qualitative or quantitative. Qualitative data are easier to obtain, and provide information on the geographic distribution of features without specific details on rarity, abundance, density, or other measurable attributes within a given cave. Qualitative inventory information from several survey stations can be compiled onto one notebook page as shown in Figure 1.

When these data are combined with abundance estimates indicating whether a feature is rare, common, or abundant, as shown in Figure 2, the project is a quantitative inventory.

Levels of Detail

The amount of detail incorporated into an inventory depends on the objectives of the project. In order of increasing detail, inventory projects may be classified in the following categories:

Hands-on training familiarizes nonspecialists with the features of cave inventory.

- Reconnaissance
- General-purpose
- Project-specific

Reconnaissance Inventory. The purpose of a reconnaissance inventory is to gather general information about a cave to determine if more detailed study is warranted. Reconnaissance may be a simple compilation of features without specific information on distribution or abundance.

Quantitative data on features may be included in a reconnaissance inventory, but usually only for a small portion of the cave. A reconnaissance project should be conducted with care because it is the foundation for more comprehensive projects and it may be the only summary of scientific information for a cave.

General-Purpose Inventory. During a general-purpose inventory, the team collects information on the identity, condition, and distribution of all the significant attributes of a cave. This is the most common type of inventory and is applicable to most caves. A general-purpose project may follow a reconnaissance project if the cave merits additional study.

Project-Specific Inventory. A project-specific inventory involves collecting detailed information on a cave or a specific part of a cave. It may be qualitative, but more often is quantitative because the investigator is interested not only in the identity and distribution of features, but also in the abundance, intensity, or frequency.

Project-specific inventories may focus on certain features of a cave, such as archaeological artifacts or cave biota. This type of data collection usually supports a larger project, such as restoration, biological or geological studies, archaeological excavation, or the design of a management or development plan.

Performing an Inventory

Gathering good inventory information requires careful preparation and use of the correct tools. Essential items include prepared inventory forms for recording data, a notebook to hold the forms, a map of the cave, pencils, and other supplies. (See tool list at end of chapter.)

Preparing the Notebook

The basic inventory tool is a notebook with forms designed for recording data from a specific cave or project. To prepare the inventory forms, it may be necessary to perform a reconnaissance trip to create a list of the features. If the investigators are familiar with the cave, this list may be constructed from memory. (See inventory list at end of chapter.)

Once the list of features is made it should be sorted by category, with items in each category listed alphabetically for quick reference (Figure 1). The form is printed on pages designed to fit in a standard surveyors loose-

To prepare the inventory forms, it may be necessary to perform a reconnaissance trip to create a list of the features. If the investigators are familiar with the cave, this list may be constructed from memory.

Imaginary Cave Date: Speleothems: Calcite Columns Draperies Flowstone Helictites Soda straws Stalactites Stalagmites Cave Fill: Breakdown Gravel Guano Mud Sand Silt Fossils: Bones Skull Archaeology: Torch fragments Biota: Bats Beetles Isopods Fish Crayfish Spiders	Team:

Figure 1. This is an example of a simple inventory form.

leaf notebook. For some caves, the list may be long enough to require several notebook pages. In wet caves, special water-resistant paper may be appropriate.

Each category includes several related features. For example, the category for *Calcite Speleothems* includes flowstone, helictites, soda straws, stalactites, and stalagmites. Several blank lines are left at the end of each category so that additional features can be added to the list. Blank pages should also be included for verbal descriptions, drawings of unusual features, documentation of photographs, site-specific directions, and other information that does not fit the structure of the inventory form.

Once the list of features is made it should be sorted by category, with items in each category listed alphabetically for quick reference.

Collecting the Data

In most cases, the cave to be inventoried has been surveyed. If no survey and map of the cave exists, inventory data may be collected in conjunction with the cave mapping data (the cave survey). In caves that have been previously surveyed, the inventory team should have a copy of the map with the names and locations of the survey stations clearly marked.

Referring to the map will speed the process of locating survey stations

Imaginary Cave
Date: 8/26/99
Survey Station: A1
Team: Joe Hodag, Sally Caver, Jack Karstman

	Rare	Common	Abundant	Comments
Speleothems:				
Calcite				
Columns	☐	☐	☐	
Draperies	☐	☐	☐	
Flowstone	☐	☐	☑	brownish-red
Helictites	☑	☐	☐	transparent. 1-6 cm long
Soda straws	☐	☐	☐	
Stalactites	☐	☑	☐	cream to yellow, .5-1 m long
Stalagmites	☐	☑	☐	cream to yellow, 15 cm tall
Cave Fill:				
Breakdown	☐	☑	☐	blocks 30cm to 1m
Gravel	☐	☐	☐	
Guano	☐	☐	☐	
Mud	☐	☐	☐	
Sand	☐	☐	☐	
Silt	☐	☐	☐	
Fossils:				
Bones	☐	☐	☐	
Skull	☐	☐	☐	
Archaeology:				
Torch fragments	☐	☐	☐	
Biota:				
Bats	☐	☐	☐	
Beetles	☐	☐	☐	
Isopods	☐	☐	☐	
Fish	☐	☐	☐	
Crayfish	☐	☐	☐	
Spiders	☐	☐	☐	

Figure 2. This example of an inventory form is designed for collecting data on the type and abundance of cave features. Note that one copy of the form is needed for each survey station.

and collecting data. It is helpful if the cave contains either permanent or temporary survey station markers. (See station markers, page 189.) If stations are not marked, surveying tools may be used to relocate and temporarily mark the stations.

Normally, the area to be examined at a station extends from the midpoint between stations A and B to the midpoint between stations B and C. This midpoint rule works well where the distance between stations is 12 meters (about 40 feet) or less. Where the distance is greater than 12 meters, it may be necessary to add supplemental survey stations, especially if the cave has an abundance of features that should be included in the inventory.

It may be appropriate in quantitative inventories to divide cave passage into equal-length segments so that the collected data is statistically valid. However, equally spaced inventory sites must be added to the survey data if the inventory information is to be displayed on computer-generated maps.

Data Collecting Scenario

To systematically collect data, (assuming the survey stations in the cave are labeled) the inventory team goes to station A-1 and carefully examines walls, ceiling, and floor while discussing what each person is seeing. Once all of the significant features have been identified, the bookkeeper records the data on the inventory form.

If no survey and map of the cave exists, inventory data may be collected at the same time the cave is surveyed.

At station A-1, the passage is decorated with calcite speleothems, including stalactites, stalagmites, flowstone, and helictites. The floor of the passage is covered with breakdown blocks and mud, and the remains of a torch are found next to one of the walls.

The bookkeeper finds the name of each feature on the inventory form and writes the name of the station next to the feature as shown in Figure 3. When all of the features near Station A-1 have been identified and recorded in the notebook, the team moves to station A-2 and repeats the process.

At station A-2, there are columns, draperies, flowstone, stalactites, and stalagmites, and the floor is covered with mud and breakdown blocks. A small red beetle is seen on the floor of the passage.

The bookkeeper makes the appropriate entries as shown in Figure 4, and adds notes regarding color and size of the insect. Short notes can be included directly on the inventory page. Longer notes are written on blank pages with a reference to the appropriate survey station. When the team is satisfied that everything of significance at this site has been recorded, it moves to the next survey station.

At station A-3, the walls are covered with flowstone, helictites, and *popcorn*. The floor consists of breakdown and mud. Notice that popcorn does not appear on the prepared form so it is necessary to add it to the speleothem category as shown in Figure 5. The bookkeeper writes in *popcorn* on one of the blank lines under the heading for speleothems and makes an entry showing that popcorn is present at station A-3.

At station A-4, features include flowstone, helictites, soda straws, and some unusual crystals next to a small pool. The pool is in a flowstone basin, is partly filled with water, and has a thin layer of silt on the bottom. A photograph is made of the crystals and the pool using a metric ruler for scale.

The bookkeeper adds a category for crystals and records the name of the survey station next to the appropriate categories on the inventory form as shown in Figure 6. A note next to the line for crystals indicates that a photo-graph of the pool and crystals was taken. At the bottom of the page, a short note describes the pool, gives the dimensions and depth, and notes that the bottom of the pool is covered with silt.

Any time extraordinary, rare, or unusual features are found, written descriptions and sketches should be made or photographs should be taken.

(See photodocumentation, page 204.) A blank page in the notebook is used to record these descriptions. Written descriptions should include the direction and distance of the feature relative to the nearest survey station. Descriptions and directions should be clear and concise so that the site can be easily relocated. If the feature is particularly significant, it is advisable to add a survey station for the site.

Any time extraordinary, rare, or unusual features are found, written descriptions and sketches should be made or photographs should be taken.

An inventory team quickly becomes familiar with these techniques and with the attributes of the cave. Typically, cave attributes are similar from one station to the next, with only one or two items dropped off as others are added. Once the basic attributes are identified, which usually occurs at the first two or three stations, the inventory process becomes smooth and efficient and the team is able to move rapidly through the cave to complete the project.

When inventory data are not proprietary, the field forms should be copied for all members of the inventory team and for the cave owner or manager. This helps ensure that data are not lost and provides team members with a sense of accomplishment.

Imaginary Cave Date: 8/26/99	Team: Joe Hodag, Sally Caver, Jack Karstman
Speleothems:	
Calcite	
Columns	
Draperies	
Flowstone	A1
Helictites	A1
Soda straws	
Stalactites	A1
Stalagmites	A1
Cave Fill:	
Breakdown	A1
Gravel	
Guano	
Mud	A1
Sand	
Silt	
Fossils:	
Bones	
Skull	
Archaeology:	
Torch fragments	A1
Biota:	
Bats	
Beetles	
Isopods	
Fish	
Crayfish	
Spiders	

Figure 3. A sample inventory form showing data entered for station A-1.

Transcribing Inventory Data

Accurately transcribing inventory results into a computer database is as important as gathering the basic information. Entering survey information and inventory data into a computer database simplifies duplication and distribution, eases sorting and analysis, and enables automatic plotting of inventoried features on multiple versions of the cave map. The database will become the working repository for the inventory data and can be linked to computer mapping software to create displays.

Computer formats tend to be ephemeral and storage media often fail. It is wise to print a paper copy of the entire database and to save the database as a delimited raw text (ASCII) file. Hard copies of original inventory sheets need to be archived to ensure future access. Because an original database can eventually become unreadable, computer records should be copied into newer programs every few years.

Many database programs for home computers are adequate for storing and manipulating inventory data. It is advisable to check with the owner or

Computer formats tend to be ephemeral and storage media often fail, so it is wise to print a paper copy of the entire database and to save the database as a delimited raw text (ASCII) file.

Imaginary Cave Date: 8/26/99	Team: Joe Hodag, Sally Caver, Jack Karstman
Speleothems:	
Calcite	
Columns	A2
Draperies	A2
Flowstone	A1, A2
Helictites	A1
Soda straws	
Stalactites	A1, A2
Stalagmites	A1, A2
Cave Fill:	
Breakdown	A1, A2
Gravel	
Guano	
Mud	A1, A2
Sand	
Silt	
Fossils:	
Bones	
Skull	
Archaeology:	
Torch fragments	A1
Biota:	
Bats	
Beetles	A2 (red-brown 4mm long)
Isopods	
Fish	
Crayfish	
Spiders	

Figure 4. This sample inventory form contains data for station A-2.

manager of the cave to see if specific software and formats are required.

Database Formatting

In the simplest case, a table with a record for each survey station is created. The record is labeled with the name of the station (A-1 is used in this chapter). Each record includes the X, Y, and Z coordinates of the station as computed from the raw survey data, as well as a list of all the features at that location.

A data-entry screen that allows clicking on feature names can help to avoid typing identical entries for every station where they occur. To make data entry easier and less error-prone, the data-entry screen should be similar in arrangement to the inventory forms. While notes from the inventory forms will not usually be useful in computer processing, provision should be made to copy them into the record in order to archive all data in one place.

A data-entry screen that allows clicking on feature names can help to avoid typing identical entries for every station where they occur.

If the database is *relational*, it is possible to separate the data into two or more tables. Well-developed inventory databases contain many tables that specify types of data, such as biological inventory, mineralogical

Imaginary Cave Date: 8/26/99	Team: Joe Hodag, Sally Caver, Jack Karstman
Speleothems:	
Calcite	
Columns	A2
Draperies	A2
Flowstone	A1, A2, A3
Helictites	A1, A3
Soda straws	
Stalactites	A1, A2
Stalagmites	A1, A2
Popcorn	A3
Cave Fill:	
Breakdown	A1, A2, A3
Gravel	
Guano	
Mud	A1, A2, A3
Sand	
Silt	
Fossils:	
Bones	
Skull	
Archaeology:	
Torch fragments	A1
Biota:	
Bats	
Beetles	A2 (red-brown 4mm long)
Isopods	
Fish	
Crayfish	
Spiders	

Figure 5. This sample inventory form contains data entered for station A-3. Note that a new category for *popcorn* has been added to the list of speleothems.

inventory, and so on.

The first table contains a record for each station name and contains only the geographical coordinates. Other tables contain records for each inventory feature and include a list of the stations at which that feature occurs. The station names in the inventory data table must be identical to those in the survey data table.

The program can then find the coordinates of feature locations by relating the two tables using the station names recorded in each. Properly designed, either the single table or related tables can be efficient and easy to use. The map drawing or geographical information systems (GIS) software used to display the data may dictate whether more tables are appropriate.

If the database is *relational*, it is possible to separate the data into two or more tables. Well-developed inventory databases contain many tables that specify types of data, such as biological inventory, mineralogical inventory, and so on.

Displaying the Data

Creating multiple versions of a cave map by hand is tedious and time-consuming. Personal computers have simplified this process, facilitating

Imaginary Cave	**Team:** Joe Hodag, Sally Caver, Jack Karstman
Date: 8/26/99	
Speleothems:	
Calcite	
Columns	A2
Draperies	A2
Flowstone	A1, A2, A3, A4
Helictites	A1, A3
Soda straws	
Stalactites	A1, A2
Stalagmites	A1, A2
Popcorn	A3
Crystals	A4 (photo #1)
Cave Fill:	
Breakdown	A1, A2, A3
Gravel	
Guano	
Mud	A1, A2, A3
Sand	
Silt	
Fossils:	
Bones	
Skull	
Archaeology:	
Torch fragments	A1
Biota:	
Bats	
Beetles	A2 (red-brown 4mm long)
Isopods	
Fish	
Crayfish	
Spiders	
	Note: At A4 there is pool next to left wall (going in) dimensions are 15 cm X 48 cm X 5 cm deep. Bottom covered with thin layer of brown silt. (See photo #1)

Figure 6. This sample inventory form has data entered for station A-4. A new category for *crystals* has been added, with a note indicating that a photograph of the crystals was taken.

the production of a series of maps with different features. Layers of information can be created and superimposed with digital technology.

Putting all the available data on a single map usually results in a cluttered presentation that is difficult to understand and interpret. Thus, it is useful to create several versions of a cave map with each version featuring different attributes. For example, one map may display bedrock geology, another might show speleothems, and yet another could show the distribution of biota.

Cave mapping programs have evolved to include specialized GIS tools that can store and display both survey and inventory data. Geographic data systems are designed to work with databases and computer aided design (CAD) programs. GIS software relates inventory data, photographs, and other documentation tools to specific locations in the cave.

A wide variety of software is available for cave mapping and for displaying and managing inventory data.

Importantly, the software also relates inventory data to other layers in the GIS database, potentially identifying relationships and patterns previously unrecognized. There are a number of CAD and cave mapping software packages available. Selection of mapping software depends on the needs and financial resources of the user, and on the requirements of the cave owner or manager.

A wide variety of software is available for cave mapping and for displaying and managing inventory data. These include specialized cave survey programs as well as commercial CAD and GIS software. Many include capabilities to incorporate various database and geographic data formats. Increasingly, GIS is being used to query and manage cave inventory data, to understand the spatial relationships within the cave, and also to identify relationships with the external environment.

Additional Reading

General

Northup DE, Mobley ED, Ingham, KL III, Mixon WW. 1998. *A Guide to the Speleological Literature of the English Language 1794-1996*. St. Louis (MO): Cave Books. 539 p.

Mineralogy

Hill CA, Forti P. 1997. *Cave Minerals of the World*. 2nd ed. Huntsville (AL): National Speleological Society. 463 p.

Speleology

Klimchouk AB, Ford DC, Palmer AN, Dreybrodt W, editors. 2000. *Speleogenesis: Evolution of Karst Aquifers*. Huntsville (AL): National Speleological Society. 527 p.

Surveying

Dasher GR. 1994. *On Station: A Complete Handbook for Surveying and Mapping Caves*. Huntsville (AL): National Speleological Society. 242 p.

Potential Items for Cave Inventory Lists

Speleothems:

Aragonite
- Anthodites
- Bushes
- Frostwork

Calcite
- Bell canopies
- Coatings
- Crusts
- Crystals
  - Nailhead spar
  - Dogtooth spar
- Draperies
- Drip pit linings
- Flowstone
- Helictites
- Mammillaries
- Rafts
- Rims
- Rimstone dams
- Shelfstone
- Shields
- Soda straws
- Splash rings
- Stalactites
- Stalagmites
- Trays

Sulfate
Barite
- Crystals
- Massive
- Stalactites

Celestite
- Coating
- Crystals

Gypsum
- Crust
- Crystals
- Needles

Flowers

Moonmilk:
- Balloons
- Coatings

Minerals:
- Aragonite
- Barite
- Calcite
- Gypsum
- Nitrates
- Quartz

Fill Material:
- Clay
- Gravel
- Guano
- Ice
- Mud
- Sand
- Loess
- Topsoil

Bedrock Features:
- Bedding planes
- Breccia
- Limestone beds
- Sandstone beds
- Shale beds
- Strike and dip of beds
- Faults
- Joints

Paleontological Features:

Bedrock Fossils
- Algal deposits
- Brachiopods
- Bryozoans
- Cephalopods
- Coral
- Crinoids
- Echinoids
- Gastropods
- Pelecypods
- Scaphopods
- Sponges

Secondary Fossil Deposits
- Pleistocene bones
- Trace fossils
- Footprints
- Scratch marks
- Scat
- Nests
- Dens

Speleogenetic Features:
- Boxwork
- Drip pits
- Karren
  - Pothole karren
  - Rillenkarren
  - Spitzkarren
- Scallops
- Stream slots
- Water lines

Biota:

Vertebrates
- Amphibians
- Bats
- Birds
- Fish
- Mammals
- Reptiles

Invertebrates
- Amphipods
- Isopods
- Beetles
- Centipedes
- Crayfish
- Earthworms
- Flies
- Leeches
- Millipedes
- Moths
- Psuedoscorpions
- Scorpions
- Snails
- Spiders

Microbiota
- Coatings
- Filaments

Water:
- Dripping water
- Pools
- Seeps
- Streams
  - Pools
  - Raceways
  - Riffles
  - Springs
  - Swallets
- Wet surfaces

Archaeological Features:

Cultural Artifacts
- Carbon blackened
  - Walls
  - Ceilings
- Ceremonial items
- Chips and flakes
- Hearths
- Pottery
- Pottery sherds
- Projectile points
- Petroglyphs
- Pictographs
- Sandals
- Tools
- Torch fragments

Human Remains
- Bones
- Burials
- Coprolites
- Scat

Historical Features:
- Habitation
- Historical signatures
- Mining equipment
- Nitrate leaching tools
- Stills

Damage:

(Restoration Targets)
- Batteries
- Broken formations
- Carbide dumps
- Chalk marks
- Contemporary graffiti
- Flagging material
- Human waste
- Mold blooms
- Mud tracks
- Off-trail footprints
- Trail markers needed
- Soiled formations
- Trails
- Contemporary trash

Tools for Cave Inventory

- Map of the cave
- Inventory notebook (surveyors book)
- Custom-designed waterproof pages listing features to be inventoried
- Pencils
- Eraser
- Magnifying glass (not a loupe or hand lens)
- Scale (metric)
- Camera and film
- Compass
- Clinometer
- Surveyors tape
- Indelible marking pencil
- Surveyors flagging tape (for temporary station markers)

Section A—Identifying and Protecting Cave Resources

Biological Dos and Don'ts for Cave Conservation and Restoration

William R. Elliott

Coexisting harmoniously with a cave's ecology is extremely important for planning cave restoration projects, protecting cave life, and creating low-impact guidelines for cavers. This chapter targets the conservation of cave macrobiology—cave life that you can see with the unaided eye. Conservation of microscopic cave-dwelling organisms is discussed in the chapter by Boston, Northup, and Lavoie. (See microbial habitats, page 61.)

> This chapter targets the conservation of cave macrobiology—cave life that you can see with the unaided eye.

Before restoring a cave to a more natural state, consider how altered the cave is and define realistic goals for the restoration project. Is it a show cave with many years of accumulated change and little hope of complete restoration, or is it a wild cave that is not so ecologically disturbed? In a show cave we might ask if *nuisance* species are present—perhaps cyanobacteria thrive near electric lights; perhaps *exotic* (nonnative) species live in cave lint along trails; or perhaps surface wildlife is attracted to artificial food sources in the cave. No cave is ever completely restored to its former aesthetic or ecological state. Proper biological inventory and project planning will increase the success of restoration efforts.

For both show caves and wild caves, ask questions before planning projects:

- Do we know what native species should be in the cave?
- How can we restore the cave to a more normal aesthetic and ecological state without harming the native species?
- To what time period or prior natural condition should we restore this cave—10 years ago, 100 years, pre-European?

Restoration Dos

Follow a few simple principles to keep restoration goals in sight:

- Plan the project's scope and ask basic questions about the desired outcome.
- Enlist the help of a cave biologist to compile a species list for the cave. (If no species list is available, use lists for similar caves in the area as a guide.)
- Ask the cave biologist to lead a bioinventory trip into the cave, update the species list, and make notes on cave microhabitats and sensitive areas before the restoration commences.
- Ask the biologist to examine cave lint and manmade woodpiles that you may want to remove. These areas can be *biological magnets* for native cave species because of the shelter, food, and higher moisture content in the lint or woodpile (Elliott 1992, 1997).
- Schedule the cleaning of special habitat areas over an extended time—cleanup in stages will allow fauna to migrate to other areas

Best Management Practices
William R. Elliott

For cavers and cave managers, best management practices (BMPs) are guidelines for protecting cave or karst resources. BMPs originated as a method to help land managers comply with environmental regulations.

Future BMP revisions will be based on scientific research, monitoring results, and technological improvements. The following are examples of current BMPs associated with caves and karst.

Sometimes we walk a thin line between diverting water away from sinkholes and maintaining recharge.

It may be good to divert polluted runoff from streets and parking lots away from sinkholes and caves, but it may be detrimental to divert all runoff and diminish sources of water. Elliott (2004) provides a karst BMP

(Continued on next page)

and biofilms to recover. (A *biofilm* is a thin layer of sediment, organic matter, and bacteria or fungi, on which small arthropods, like millipedes, feed. See biofilms, page 68.)

- Remove glass, metal, wiring, and other trash if it has been checked for small cave dwellers. Gently brush or blow small creatures from the trash onto the floor, or have a cave biologist examine them. Some old trash is historical—contact a historical archaeologist to look things over. Historical items from early tourism have been lost to the good intentions of cleanup crews. (See historical preservation, pages 110, 114, and 334.)
- In show caves, various methods have been used to control photosynthetic organisms that grow near electric lights. Sodium hypochlorite (household bleach), formaldehyde, hydrogen peroxide, and steam generators have been used to temporarily reduce algae and other lamp flora. However, these methods have no lasting effect and tend to indiscriminately kill other cave-dwelling biota. (See algae control, page 349.) Growth can be better controlled through selection of light wavelength, reduction in wattage, or light-shielding and redirection. Current best practice is moving toward solutions provided by new lighting technologies. (See lamp flora control, page 343.)

Restoration Don'ts

- Don't disturb bats or other wildlife while restoring the cave.
- Don't disturb or remove anything from a biological study site.
- Don't indiscriminately use chemicals. In rare cases, if manufactured chemicals are deemed appropriate, then completely remove the residue and byproducts from the cave. (See anthropogenic chemicals, page 57.)
- Don't contaminate adjacent areas with water containing sediments, chemicals, wastes, or nutrients.
- Don't remove cave lint or manmade woodpiles in one step, unless a qualified cave biologist has determined that the area has few or no inhabitants. Drastic removal may *trap out* the small creatures that have migrated into the area and reproduced there over many decades. Consider leaving a small residue of lint or wood in hidden areas at the site if small numbers of creatures are using it (Elliott 1992, 1997).
- Don't remove natural food sources for cave life, such as sticks and leaves, or corpses and droppings of wood rats, mice, raccoons, ringtails, bats, crickets, frogs, and others. Such food sources may present opportunities for cave biologists to study species rarely seen or not previously known in the cave. Removal of naturally occurring organics may be justified for scientific sampling or for restoration of the natural nutrient balance in the cave.
- Don't install cave gates, modify the entrance, or move anything within the cave if it could significantly alter air or water flow, or hinder the movements of bats and other wildlife (Elliott 1997). Digging into an entranceless cave would be an exception, but conscientious gating decisions should be made in favor of conservation. (See virgin digs, page 261.) Consult the latest gating designs of the American Cave Conservation Association, Bat Conservation International, and the National Speleological Society (Vories and others 2004), and use experienced cave gate designers and welders. (See cave gating, page 147.)

Conservation Dos

Cavers need a few guidelines to coexist harmoniously with cave life during caving trips:

- Learn more about cave life, especially sensitive species and *species of concern* (rare, threatened, or endangered species). Entering some caves or areas during the wrong season or ignoring posted warnings may be a violation of state or federal law, even on private land. The NSS Web site <http://www.caves.org> or members of the NSS Biology Section and the Biospeleology Web site <http://www.utexas.edu/tmm/sponsored_sites/biospeleology/> are good sources of information.
- Report sightings or provide photos of cave wildlife to the cave owner or manager and, if possible, to scientists who may be documenting species records in the area. Scientists at state or federal agencies, universities, or museums often handle the reporting.
- Be careful in cave entrances in the spring, when eastern phoebes, vultures, swallows, and other birds may be nesting and rearing young. Inform others so they can avoid accidentally disturbing wildlife anywhere in the cave.
- Mark narrow paths through cave passages to avoid trampling habitats for cave-dwelling species, speleothems, or other features. Consider placing small, well-positioned signs in the cave to warn others of bat roosts, crayfish, or cavefish habitat, and other sensitive areas. (See trails and signs in caves, page 175.)
- Go equipped to carry out all personal waste products—urine, solid waste, vomit, spit, and trash. Cave communities are often adapted to low levels of nutrients, and we should not add to the mix. (See human waste, pages 71, 125, and 276; also see packaging human waste, page 269.)
- Work with others to determine the maximum number of people per cave entry and the maximum number of trips per year (or other time interval). This limitation would be the carrying capacity for human visitation to a cave. The term *carrying capacity*, borrowed from ecological science, means the population size of a species that an area can continuously support. In some caves, carrying capacities for humans should be established and monitored. It is not easy to calculate the number of cavers or trips that can be tolerated by a cave. Some cave stewards set reasonable limits and carefully observe the results over long periods of time. Cave registers and permit systems are effective monitoring tools when coupled with good bioinventories (Elliott 1997). (See cave inventories, page 19.)
- Become more aware of current best management practices (BMPs) for caves and karst. (See BMP sidebar, page 34; also see current best practices, page 17 and strategies, page 127.)

Conservation Don’ts

- Don’t allow people to take pets, horses, or vehicles into cave entrances.
- Don’t make loud noises. Shrill or loud voices and sounds will disturb nesting birds, roosting bats, and other wildlife.
- Don’t build fires near cave entrances or inside caves. Fires and smoke can kill bats, birds, insects, and other animals.
- Don’t smoke in caves.

(Continued)

that applies to many areas.

Another BMP advocates the use of retention/filtration basins to clean up routine runoff and to retain spills before they go underground (Olson and Schaefer 2002).

Forested buffer zones are necessary for the health of some cave systems. For example, in the midwestern U.S., forested riparian corridors should be maintained and human visitation should be monitored to help protect gray bat populations in caves.

Karst management guidelines have been published in various government and consulting reports (Elliott 2000). Cave management guidelines are usually based on inventory information and are specifically designed for the protection needs of each cave. (See cave inventories, page 19.)

Cavers need a few guidelines to coexist harmoniously with cave life during caving trips.

- Don't use toxic chemicals in caves. If manufactured chemicals are deemed appropriate, then completely remove residue and byproducts from the cave. (See anthropogenic chemicals, page 57.)
- Don't disturb bats while caving. Various bat species use caves for nurseries, hibernation, and bachelor colonies. Two especially sensitive times in the United States are May through July, when many bats give birth and rear young, and November through April, during hibernation. These times vary depending on the latitude and the species. (See hibernating bats and nursery colonies, page 43.)
- Don't heavily disturb streams, especially those with *stygobites* (cave-adapted fish, crayfish, salamanders, or other aquatic species). Cavers should go single-file or stay out of the water if travel excessively disturbs the bottom.
- Don't drastically change water flow patterns, drain pools, or make alterations that could affect stream scouring or sedimentation, especially in areas inhabited by *stygobites* (aquatic troglobites). For instance, blowing out or draining a rimstone dam could strand or kill cave life. (See hydrologic modifications, page 128.)
- Don't collect cave life unless you possess the proper permit for scientific studies, and have the owner's or managing agency's written permission. For example, in Missouri, regardless of other federal, state, local, or private permitting requirements, you must obtain a Wildlife Collector's Permit from the Missouri Department of Conservation to collect any cave life. Other states have similar laws.

Examples of Ecological Disturbance in Show Caves

A show cave sometimes accumulates unwanted nutrients and materials that gradually disturb the cave's ecological balance. Structures and development modifications influence changes in temperature, airflow, and humidity.

A show cave sometimes accumulates unwanted nutrients and materials that gradually disturb the cave's ecological balance. Except for bat caves with large amounts of guano, caves are typically nutrient-poor environments and the cave-adapted species living there are adapted to a slow way of life.

Some show caves eventually lose their native wildlife as a result of human disturbance. Structures and development modifications influence changes in temperature, airflow, and humidity. Cyanobacteria and mosses often grow around show cave lights. Accumulations of cave lint, trash, wood, and even human waste can influence habitat changes. In a few extreme cases, show caves have been contaminated by leaking septic systems and sewer lines, causing overgrowths of bacteria and fungi, and resulting in invasions of exotic species that replaced the native cave species.

Electric Lights

Algal growth in show caves is stimulated by electric lights and restoration is often difficult and time-consuming (Aley 1972). In Carlsbad Caverns National Park, Aley and others (1985) conducted studies on exotic flora and microflora and identified 26 species of algae, plus moss protonema and two ferns.

Hundreds of species of algae were found growing in the cave. Approximately 70% of the organisms identified were cyanobacteria, 20% were green algae, and 10% were moss protonema. Many of the algal genera and three of the algal species in Carlsbad Cavern have also been found growing in total darkness in other caves. (See algae, page 344.) Algal growth, once abundantly established due to artificial lighting, will not quickly disappear when it is deprived of light. Light intensities were measured at a number of the sites in Carlsbad Caverns.

In alcoves, the minimum light intensity threshold for algal growth had a mean value of 17.2 lux (1.6 footcandles) with a standard deviation of 7.5 lux. In nonalcove sites (most situations) the mean threshold was 47.4 lux (SD = 20.5 lux). At Frozen Niagara in Mammoth Cave, algae and moss protonema were steam-cleaned from ceiling formations (Aley 1997) but the growth returned.

Show caves have used either sodium hypochlorite (household bleach such as Clorox®) diluted to a 10% solution, or a 3% to10% solution of hydrogen peroxide, or steam generators for controlling photosynthetic flora. (See algae control, page 349.)

However, bleaching, cleaning, or steaming the algae are inadequate solutions for several important reasons:

- The growth of algae and other lamp flora is only temporarily mitigated.
- Other cave organisms are compromised in the process.
- Even depriving the algae of light is inadequate for controlling growth.

Bleaching and cleaning treatments may make the cave look better in the short term, but they are almost certainly not ecologically sustainable—other cave organisms are easily harmed during these procedures. (See bleach, page 349.)

Several new lighting technologies are providing improved methods for avoiding inappropriate over-stimulation of photosynthetic species in show caves. (See lamp flora control, page 343.)

Exotic (introduced) invertebrate communities may also be encouraged by the presence of electric lights. Signature Pool, located near the trail in Carlsbad Cavern, is a microcosm of moss, algae, eyed copepods, and eyed flatworms (*Phagocata* sp.), that were probably introduced. The small community is driven by a light bulb hanging over the pool (Elliott 1997, 2000).

Algae bleaching and cleaning treatments may make the cave look better in the short term, but they are almost certainly not ecologically sustainable.

Woodpiles

In Carlsbad Cavern Elliott (1997) studied old woodpiles that probably were remnants of structures installed in the 1920s. Most of the wood was depleted of nutrients and no longer supported visible fungal or bacterial growth. However, some wood was up to 80% water by weight, and served as a moist substrate. Some piles contained almost no invertebrates, while others were a haven for several species. Elliott recommended removing most of the piles over an extended time and leaving a small residue for fauna to utilize.

W.R. Elliott

Figure 1. William R. Elliott examined an old woodpile off the trail in Carlsbad Cavern, 1996. The wood was depleted of nutrients, contained no inhabitants, and was scheduled for removal. Woodpiles in other areas contained arthropods and were retained. Some remnant wood contained up to 80% water and served as moist substrate material for certain invertebrate species.

Cave Lint

Lint studies have yielded interesting results in Carlsbad Caverns. In connection with annual lint camp cleanups, Jablonsky and others (1995) placed tagged lint (treated to be clearly visible under ultraviolet) on the visitor trail, and found that it moved up to 100 meters down the trail. Testing also showed that much lint moved to the edge or completely off the trail. Under a microscope, samples of cave lint contained synthetic and natural fibers, dirt, wood, insect parts, human hair, animal fur, fungi, processed tobacco, paper, and other debris. Unidentified mites were also seen in some of the samples (PL Jablonsky, personal communication). (See lint and dust, page 351.)

Food Service

The National Park Service instituted tighter housekeeping controls on the food service operators inside Carlsbad Cavern. They now remove the contents of all trash barrels by the end of the day to combat the large number of raccoons that invaded the cave at night (DL Pate, personal communication). Crickets are unnaturally distributed in the cave because they are attracted to food in the area called the Lunch Room, 750 feet below the surface (Elliott 2000). The National Park Service management plan advocates removing the in-cave food facilities (NPS 1996).

Sewage Leaks

Sewage leaking from pipes or septic tanks may cause ecological imbalance in caves. In Carlsbad Cavern, a sewage leak infiltrated to the cave passages,

Figure 2. This fungal overgrowth was caused by a sewage leak into Carlsbad Cavern. Broken tile sewer lines caused the problem, even though this location was hundreds of feet below the surface.

causing localized fungal growths fed upon by swarms of fungus gnats—the gnats, in turn, supported a local community of spiders (Elliott 1997). (See infrastructure and infiltration, page 237.)

In a bizarre 1993 case in the midwestern U.S., earthworms came out of the cave walls and rocks fell out of the ceiling of a show cave over an extended time period. The cave management thought that the earthworms were causing the problems and should be exterminated.

Systematic observation and sampling revealed that the earthworms were following infiltrating sewage into the cave from broken sewer lines and forgotten, leaking, septic tanks built on top of the cave (Elliott 2000).

The epikarst in that area, though mantled with good soil, was highly transmissive. Dye traces proved that the cave was cross-connected to many input points on the surface, including septic systems. Bacteriological sampling showed the presence of sewage-associated bacteria (including *E. coli, Salmonella,* and *Shigella*) in many air and water samples from the cave. Two European exotic species of earthworm, *Eiseniella tetraedra* and *Dendrobaena rubida*, were found crawling on wet surfaces in the cave, and were actually eating softer, marly beds of rock, which probably were laden with bacteria.

W.R. Elliott

Figure 3. The manure earthworm, *Eisenia foetida*, was identified in a polluted, midwestern show cave. Sewage infiltration brought large numbers of two earthworm species into the cave. (See page 1 of color section.)

There was little or no odor in the cave, but airborne bacteria levels were high enough to cause concern about prolonged exposure. Clay banks in

W.R. Elliott

Figure 4. Green and gray mold colonies on shelf, probably *Penicillium*, were found on a shelf in a sewage-polluted, midwestern show cave. (See page 1 of color section.)

W.R. Elliott

Figure 5. Hothouse millipedes *Oxidus gracilis*, on a ledge in a polluted midwestern show cave. Large numbers of this species in a cave indicate nutrient enrichment.

some areas were festooned with rich and colorful fungal gardens, which were inhabited by invading hothouse millipedes and sow bugs. No troglobites could be found in the cave.

After these initial findings, the owners initiated intensive maintenance and repair of the septic system. They also took steps to prevent infiltration of nutrients from the theme park above the cave. The earthworms were not killed, and were considered friends because they consumed much of the contamination. After about two years of work, the management reported that the cave was returning to normal biologically, and that rockfall (which probably was related to both high water infiltration rates and earthworm activity) had decreased to a low level.

Human overuse of cave passages may cause disturbance even of tolerant species and compaction of substrates with loss of microhabitats for small, cryptic species.

Examples of Ecological Disturbance in Wild Caves

Trampling Species and Compacting Habitats

Overuse by visitors may cause disturbance even of tolerant species and compaction of substrates with loss of microhabitats for small, cryptic species. Many troglobites are *thigmotactic*, that is they like to hide under rocks. Though the effect of trampling on such species has not been studied extensively, cavers should constantly be aware of what's under their feet. It stands to reason that confining impacts to marked routes will compact less area and will result in healthier habitat for invertebrates (Elliott 2000).

Bats Are Sensitive

Some bats are sensitive to human disturbance as slight as traveling past their hibernation or nursery roosts. Disturbance can cause bats to abandon their young and the cave. Summer or winter disturbance is critical since only a few caves have suitable microclimates for colonial bats (Mohr 1972; Poulson 1976). (See bat disturbance, page 43.)

Disturbance and arousal of hibernating bats can cause 10 to 30 days of fat to be used (Brady 1982). Several such disturbances can cause bats to starve before their springtime exit. The gray bat, *Myotis grisescens*, and the Indiana bat, *Myotis sodalis*, are especially sensitive to disturbance; both are endangered species. (See endangered bat species, pages 47–50.)

A critical question concerns the appropriate historical period to target in ecological restoration for bats. Define the most appropriate natural condition for restoration of a cave that serves (or served) as a bat roosting site. Many bat colonies have declined drastically within the last few decades.

Some bats are sensitive to human disturbance as slight as traveling past their hibernation or nursery roosts.

With proper management, some roosts may return to the maximum past population as measured by ceiling stains, but most colonies do not fully recover (Toomey and others 2002). Some species of bats can be protected by bat-friendly structures. For example, fly-over gates or fences are necessary for some bat species, while properly designed bat-compatible gates are appropriate for species that will fly through (Vories and others 2004).

In one study, the population of aquatic invertebrates and eastern pipistrelle bats, *Pipistrellus subflavus*, increased greatly in Mushroom Cave, Missouri, after a bat-friendly gate was installed to limit human traffic. (D Ashley and D Drees, unpublished results). Plan carefully because some bat species (for example, Mexican free-tails) will not return if a gate is installed. (See cave gating, page 147.)

Toxic Fumes

Fumes can be harmful to cave life and humans. For example, welding fumes can be very dangerous, especially if the metal being welded contains zinc or other elements. Welding in caves should be preceded by a study of air movements to see if contaminants will naturally blow out of the cave entrance, or if a temporary exhaust system should be used (Elliott 1995).

Elliott (1997) improvised an exhaust system for welding fumes during the construction of a bat gate in Gorman Cave, Texas. However, the system was overwhelmed by cigarette-smoking workers who created a cloud of particulates and harmful vapors (nicotine is not only harmful to humans, it also works as a contact insecticide as well as a vapor insecticide). Always consider the potential effects of fumes and chemical vapors.

Summary

To avoid damaging the cave's ecology, a few dos and don'ts can be followed during restoration projects and other trips to caves. Some of the rules come from common sense and fit easily into the general caving ethic, but some concepts require increased knowledge about cave life. Experienced cave biologists are available and willing to help. Most of them started out as cavers, and they enjoy working with cavers. For the address of the NSS Biology Section, see the current National Speleological Society Members Manual or visit the NSS Web site at <http://www.caves.org>.

For information and photos, visit the Biospeleology Web site (Elliott 2002) at <http://www.utexas.edu/tmm/sponsored_sites/biospeleology/>. For further information on the many threats to cave life as well as cave and karst management issues, see Elliott (2000) posted on that Web site.

Remember that cave life is part of the cave itself, not something that came along after cavers discovered the passages. Some cave-adapted species have lived underground for millions of years. Some species have probably lived underground longer than the age of the caves that they currently inhabit.

Many bat species have been using caves for hundreds of thousands, if not millions, of years. Many cave-dwellers have long life spans, slow reproductive rates, and adaptations to low-food situations. They are very vulnerable to disturbance and sensitive to alterations in their environment.

Another significant fact is that caves serve as important refuges for common surface species during drought and winter. A few examples are pickerel frogs, salamanders, overwintering moths, raccoons, and sometimes bears. Without undisturbed cave habitat, even these common species can decline.

Most cavers would never knowingly harm a cave animal. To be responsible cavers, we must consider the possible future consequences of our actions and remember that we are only temporary visitors to someone else's home.

Remember that cave life is part of the cave itself, not something that came along after cavers discovered the passages.

Cited References

Aley T. 1972. Control of unwanted plant growth in electrically lighted caves. *Caves and Karst* 14:33-35.

Aley T, Aley C, Rhodes D. 1985. Control of exotic plant growth in Carlsbad Caverns, New Mexico. In: Vandyke JE, editor. *Proceedings of the 1984 National Cave Management Symposium. Missouri Speleology* 25(1-4):159-171.

Aley T. 1997. Caves in crisis. In: *1997 Yearbook of Science and the Future, Encyclopaedia.* Chicago (IL): Britannica, Inc. p 116-133.

Brady JT. 1982. The status of the Indiana bat (*Myotis sodalis*). In: Wilson RC, Lewis JJ, editors. *National Cave Management Symposia Proceedings, Carlsbad, New Mexico 1978 and Mammoth Cave, Kentucky 1980.* Oregon City (OR): Pygmy Dwarf Press. p 127-132.

Elliott WR. 1992. An introduction to biospeleology. In: Rea GT, editor. *Caving Basics: A Comprehensive Guide for Beginning Cavers.* 3rd revised edition. Huntsville (AL): National Speleological Society. p 141-149.

Elliott WR. 1995. Air monitoring during construction of a cave gate. In: Pate DL, editor. *Proceedings of the 1993 National Cave Management Symposium: Carlsbad, New Mexico, October 27-30, 1993.* [Huntsville (AL)]: National Cave Management Symposium Steering Committee. p 45-51.

Elliott WR. 1997. *A Survey of Ecologically Disturbed Areas in Carlsbad Cavern, New Mexico.* Report to Carlsbad Caverns National Park. 10 p.

Elliott WR. 2000. Conservation of the North American cave and karst biota. In: Wilkens H, Culver DC, Humphreys WF, editors. *Subterranean Ecosystems*. Ecosystems of the World 30. Amsterdam: Elsevier. Chapter 34, p 665-689. Internet. <http://www.utexas.edu/tmm/sponsored_sites/biospeleology/>. Accessed 2006 May 6.

Elliott WR. 2002 . Biospeleology: The Biology of Caves, Karst, and Groundwater. Internet <http://www.utexas.edu/tmm/sponsored_sites/biospeleology/>. Accessed 2006 May 6.

Elliott WR. 2004. Protecting caves and cave life. In: Culver DC, White WB, editors. *The Encylopedia of Caves*. Elsevier Science. p 458-467.

Jablonsky P, Kraemer S, Yett B. 1995. Lint in caves. In: Pate DL, editor. *Proceedings of the 1993 National Cave Management Symposium: Carlsbad, New Mexico, October 27-30, 1993.* [Huntsville (AL)]: National Cave Management Symposium Steering Committee. p 73-81.

Mohr CE. 1972. The status of threatened species of cave-dwelling bats. *Bulletin National Speleological Society* 34(2):33-47.

National Park Service. 1996. *Final General Management Plan/Environmental Impact Statement.* Carlsbad Caverns National Park, New Mexico. 287 p.

National Speleological Society. 2005. *2005 Members Manual. NSS News*. 184 p. [Editors' note: The *NSS Members Manual* is updated annually.]

Olson RA, Schaefer JL. 2002. Planned spill retention and runoff filtration structures on Interstate 65. Abstract. In: Rea GT, editor. *Proceedings of the 15th National Cave Management Symposium: Tucson, Arizona, October 16-19, 2001.* Coronado National Forest: USDA Forest Service. p 190.

Poulson TL. 1976. Management of biological resources in caves. In: *Proceedings of the National Cave Management Symposium, Albuquerque, New Mexico, 1975.* Albuquerque (NM): Speleobooks. p 46-52.

Toomey, RS III, Coburn ML, Olson RA. 2002. Paleontological evaluation of past use of caves by the Indiana bat: a significant tool for restoration of hibernacula. In: Kurta A, Kennedy J, editors. *The Indiana Bat: Biology and Management of an Endangered Species*. Austin (TX): Bat Conservation International. p 79-85.

Vories KC, Throgmorton D, Harrington A, editors. 2004. *Proceedings of Bat Gate Design: A Technical Interactive Forum held in Austin, Texas, March 4-6, 2002*. Carbondale (IL): USDOI Office of Surface Mining, Alton, Illinois, and the Coal Research Center, Southern Illinois University at Carbondale. 434 p. Proceedings available on the Internet <http://www.mcrcc.osmre.gov/bats>. Accessed 2004 March 30.

To be responsible cavers, we must consider the possible future consequences of our actions and remember that we are only temporary visitors to someone else's home.

Section A—Identifying and Protecting Cave Resources

Do Not Disturb Hibernating Bats or Nursery Colonies

Debbie C. Buecher

In the United States, caving has increased in popularity during the last three decades. Unfortunately, human visitation in caves, even by the most conscientious cavers, gradually leaves negative impacts. Because we love our caves there is increasing awareness among the caving community toward performing ongoing restoration or building secure gates to maintain the integrity of caves we visit. However, we must consider the negative implications that our restoration efforts may have on the cave organisms adapted to this unique environment.

Roosting Sites and Nursery Colonies

Bats are extremely intolerant of human intrusion into their roosting sites (Mann and others 2002, Tuttle 1979) and roost disturbance over time can negatively impact population size (Mohr 1972). (See bat sensitivity, page 40.) Some people mistakenly believe that bats can use any cave or portion of a cave as their daily roosting area. Unfortunately this is not the case (Kunz 1982). Bats choose sites because of a constrained range of tolerant temperature and humidity requirements—conditions that help insure their survival (McNab 1982).

There are two extremely critical times in a bat's life when the roosting site is particularly important, during reproduction and during hibernation (Twente 1955). The location where female bats give birth and rear their young is called a maternity roost. For most temperate bat species that use caves, mating occurs in the fall, the bats then hibernate until late spring, whereupon the females begin gestation. Most bats have only one young per year, born in early-to-mid summer (depending on the species). Pups are totally dependent on "mom's" rich milk for about 4 to 6 weeks (Hill and Smith 1984). Adult female bats require a warm area of the cave with very high humidity to insure the rapid growth of their nursing (altricial) young (Betts 1997; Williams and Brittingham 1997).

When females with young are disturbed, there are two possible scenarios. In one case, the females may attempt to flee with their young pups attached to them. In the panic that ensues, the young that do not already fly (nonvolant) can be dropped onto the floor. Many species of bats have difficulty retrieving a fallen youngster and the pup will perish (McCracken 1989).

If the females are successful in

Figure 1. Mexican free-tailed bat *(Tadarida brasiliensis)* pups clustered on a cave ceiling.

flying with their young, the alternative roosting site they choose may be less optimal for the survival of the young bats. Female bats choose the best site they can find and, barring disturbance from "predatory" humans, they will be loyal to that maternity site their entire lives, as will their daughters, granddaughters, and so on (Hill and Smith 1984; Sidner 1997).

Because a maternity colony is particularly vulnerable to disturbance, we must never perform restoration within a maternity site when the bats are in residence (American Society of Mammalogists 1992). Fortunately cave restoration is not usually a seasonally critical activity and can be performed when the bats have departed for the season.

There are two extremely critical times in a bat's life when the roosting site is particularly important, during reproduction and during hibernation.

Winter Hibernation

Another critical period in a bat's life history, when disturbance can reduce survival, is during winter hibernation. When winter arrives and the insects disappear, bats have two options available to them. They can migrate southward to an area where the insects are still somewhat abundant. Or they can sleep (hibernate) through the period of reduced food resources (McNab 1982).

Bats that hibernate put on additional fat reserves in the fall in preparation for the winter months of reduced metabolic activity (McNab 1982). Once they are ready to hibernate, bats seek out a roost with an optimally cold but humid microclimate. Different species of bats require different conditions, which is why bats can be found dispersed throughout a cave that is used as a hibernaculum (Hill and Smith 1984; Tuttle and Stevenson 1978).

Hibernation in bats is characterized by reduced oxygen consumption, greatly reduced heart rate (approximately 1/40th normal), and a body temperature close to the ambient cave temperature. Under these reduced metabolic rates, the bats slowly draw from their fat reserves and survive the long period of dormancy. Unfortunately, the fat reserves are finite and have evolved to last the necessary period of time the bat is asleep (Thomas 1996).

Research shows that even nontactile disturbance (lights, sound, and so on) can trigger a slow arousal in a hibernating bat. It can take upwards of one hour for a bat to awaken enough to fly. In addition, it appears that the arousal of a single bat may disturb other bats asleep in the area, producing a cascading effect from one human disturbance (Thomas 1995). It is in this way that bats "burn up" critical fat reserves each time they awaken because it takes metabolic energy to warm the bat, reserves that were intended to last throughout the winter.

Figure 2. Indiana bats *(Myotis sodalis)* clustered on a cave wall.
© Merlin D. Tuttle, BCI

Studies show that the metabolic energy required for a single arousal from deep sleep is equivalent to 10 to 30 days of uninterrupted hibernation (Thomas and others 1990). After a number of episodes of human disturbance, a bat can literally starve to death before insects are available in the spring. Therefore, arousals can be the greatest factor in depleting hibernation energy budgets (Johnson and others 1998). This is particularly critical for the young of that year. Once pups learn to fly and employ echolocation effectively enough to capture insect prey in flight, it is late enough in the season that it is often difficult to store adequate fat reserves to survive the rigors of hibernation (Thomas and others 1990). In fact, the first winter is such an arduous period for yearlings that there is about 45% mortality for these animals (Sidner 1997). However, for all ages, multiple disturbance episodes can result in overwintering bat mortality due to energy shortfalls.

Although conservation and restoration is becoming an important aspect of caving activity in the U.S., we as cavers must always consider the cave life that may be impacted by our efforts. When caving, never jeopardize the bats, invertebrates, or other organisms that dwell in caves—it is truly their habitat that we are invading. It is our obligation to cave responsibly and to be active stewards for the resource.

Bats are extremely intolerant of human intrusion into their roosting sites. When caving, never jeopardize the bats, invertebrates, or other organisms that dwell in caves—it is truly their habitat that we are invading.

Cited References

American Society of Mammalogists. 1992. Guidelines for the protection of bat roosts. *Journal of Mammalogy* 73(3):707-710.

Betts BJ. 1997. Microclimate in Hell's Canyon mines used by maternity colonies of *Myotis yumanensis*. *Journal of Mammalogy* 78(4):1240-1250.

Brigham RM. 1993. The implications of roost sites for the conservation of bats. *Provincial Museum of Alberta Natural History, Occasional Paper Number 19*. [Brandon, Manitoba] 19:361-365.

Hill JE, Smith JD. 1984. *Bats: A Natural History*. Austin (TX): University of Texas Press. 243 p.

Johnson SA, Brack V, Rolley RE. 1998. Overwinter weight loss of Indiana bats (*Myotis sodalis*) from hibernacula subject to human visitation. *American Midland Naturalist* 139:255-261.

Kunz TH. 1982. Roosting ecology of bats. In: Kunz TH, editor. *Ecology of Bats*. New York: Plenum Press. p 1-55.

Mann SL, Steidl RJ, DaltonVM. 2002. Effects of cave tours on breeding *Myotis velifer*. *Journal of Wildlife Management* 66(3):618-624.

McCracken GF. 1989. Cave conservation: special problems of bats. *National Speleological Society Bulletin* 51:47-51.

McNab BK. 1982. Evolutionary alternatives in the physiological ecology of bats. In: Kunz TH, editor. *Ecology of Bats*. New York: Plenum Press. p 151-200.

Mohr CE. 1972. The status of threatened species of cave-dwelling bats. *National Speleological Society Bulletin* 34(2):33-47.

Sidner RM. 1997. Studies of bats in southeastern Arizona with emphasis on aspects of life history of *Antrozous pallidus* and *Eptesicus fuscus*. PhD Thesis. Tucson (AZ): University of Arizona.

Thomas DW, Dorais M, Bergeron JM. 1990. Winter energy budgets and cost of arousals for hibernating little brown bats, *Myotis lucifugus*. *Journal of Mammalogy* 71(3):475-479.

Thomas D. 1995. Hibernating bats are sensitive to nontactile human disturbance. *Journal of Mammalogy* 76(3):940-946.

Tuttle MD, Stevenson DE. 1978. Variation in the cave environment and its biological implications. In: Zuber R, Chester J, Rhodes D, editors. *National Cave Management Symposium Proceedings: Big Sky, Montana, October 3-7, 1977*. Albuquerque (NM): Adobe Press. p 108-121.

Tuttle MD. 1979. Status, causes of decline and management of endangered gray bats. *Journal of Wildlife Management* 43:1-17.

Twente JW. 1955. Some aspects of habitat selection and other behavior of cavern-dwelling bats. *Ecology* 36:706-732.

Williams LM, Brittingham MC. 1997. Selection of maternity roosts by big brown bats. *Journal of Wildlife Management* 61(2):359-368.

Additional Reading

American Society of Mammalogists. 1992. Guidelines for the protection of bat roosts. *Journal of Mammalogy* 73(3):707-710.

Bat Conservation International (BCI). BCI home page. <http://www.batcon.org>. Direct link to full articles from *Bats Online*. <http://www.batcon.org/batsmag/batindex.html>.

Brown C. 1996. Bat habitat in caves: What are the caver's responsibilities? *NSS News* 54(23):52-53.

Mohr CE. 1976. The protection of threatened cave bats. In: Aley T, Rhodes D, editors. *1976 National Cave Management Symposium Proceedings Albuquerque, New Mexico, October 6-10, 1975*. Albuquerque (NM): Speleobooks. p 57-62.

Tuttle MD. 2000. Where the bats are, Part III: caves, cliffs, and rock crevices. *Bats* 18(1):6-11.

Section A—Identifying and Protecting Cave Resources

Federally Listed Threatened and Endangered Bat Species of Importance to Caves and Mines

Robert R. Currie

Six North American bats are listed as endangered under the Endangered Species Act of 1973, as amended. All of these federally listed species are dependent upon caves or abandoned mines during all or part of the year:

- The Indiana bat *(Myotis sodalis)*, a species that is currently undergoing a serious population decline, uses caves or mines for hibernation.
- The gray bat *(Myotis grisescens)* is dependent upon cold caves or mines during hibernation and warm caves or mines during the summer maternity season.
- The Virginia big-eared bat *(Corynorhinus townsendii virginianus)* is restricted to small populations in four eastern states and uses caves or mines year-round.
- The Ozark big-eared bat *(Corynorhinus townsendii ingens)* is the rarest of the endangered bats and is dependent on caves year-round. Historically, it was found in three states, Arkansas, Oklahoma, and Missouri. It has apparently been extirpated from Missouri, and only about 2,000 Ozark big-eared bats remain in Arkansas and Oklahoma. Although only one mine roost for this species is currently known, it could potentially be found in some of the abandoned mines located just south and west of its currently known distribution.
- The Mexican and lesser long-nosed bats *(Leptonycteris nivalis* and *Leptonycteris curasoae yerbabuenae)* are migratory nonhibernating species found in the southwestern U.S. and Mexico. Both species are integral components of southwestern desert ecosystems; caves and mines provide essential roosting habitat for them.

This chapter targets the conservation of federally listed bats that depend on cave and mine habitats.

Threats to all these species include several important factors:

Figure 1. Special photographic techniques record the Mexican long-nosed bat *(Leptonycteris nivalis)* in flight.

- Roosting and foraging habitat destruction and alteration
- Chemical contamination of their food supply
- Human disturbance at their summer and winter roosts

Intensive disturbance of the bats at their maternity and/or hibernation caves has increased the importance of protecting and maintaining bat access to mines. Without this protection it will be difficult to meet the long-term protection and recovery goals for these endangered species.

Endangered Species Act of 1973

The Endangered Species Act was enacted in 1973, by the 100th Congress of the United States. Section 2 of the Act states that the purposes of the Act are "to provide a means whereby the ecosystems upon which endangered species and threatened species depend may be conserved, to provide a program for the conservation of such endangered and threatened species..."

This is a noble objective that continues to be a valid, although sometimes problematic goal for all involved in implementation of the Act. The Act defines an endangered species as "any species which is in danger of extinction throughout all or a significant portion of its range." A threatened species is "any species which is likely to become an endangered species within the foreseeable future throughout all or a significant portion of its range."

Figure 2. Three Indiana bats *(Myotis sodalis)* roost on a cave wall.

© Merlin D. Tuttle, BCI

Critical habitat has been formally designated for some listed bats that occur in areas impacted by cave visitation and active and abandoned mine programs. Critical habitat is defined as " (i) the specific areas within the geographical area occupied by the species, at the time it is listed in accordance with the provisions of Section 4 of this Act, on which are found those physical or biological features, (I) essential to the conservation of the species and (II) which may require special management considerations or protection; and (ii) specific areas outside the geographical area occupied by the species at the time it is listed in accordance with the provisions of Section 4 of the Act, upon a determination by the Secretary that such areas are essential for the conservation of the species."

Section 4 of the Act establishes the process the Departments of Interior and Commerce must use in identifying endangered and threatened species, designating critical habitat, and developing recovery plans.

Section 7 of the Act prohibits federal agencies from undertaking, permitting, authorizing or funding any activity that will jeopardize the continued existence of federally listed species. This Section also requires federal agencies to be proactive and use their programs to enhance the status of federally listed species.

Section 9 of the Act prohibits taking a listed species without a permit issued under Section 10 of the Act. *Take* is defined by regulations promulgated to implement the Act to mean "...to harass, harm, pursue, hunt, shoot, wound, kill, trap, capture, or collect (a listed species), or to attempt to engage in any such conduct."

Critical habitat has been formally designated for some listed bats that occur in areas impacted by cave vistitation and by active and abandoned mine programs.

Six Federally Listed Endangered Bats

There are six federally listed endangered bats that occur within the continental U.S. In implementing abandoned mined land reclamation activities and other mine and cave programs, federal and state agencies must insure that all of their activities are in compliance with Section 7 of the Act and that these activities do not violate Section 9.

Indiana bat *(Myotis sodalis)*

The Indiana bat was listed on March 11, 1967, as an endangered species throughout its range. Critical habitat which includes most of its important hibernation sites was formally designated on September 24, 1976. A recovery plan for the species was issued on October 14, 1983. This plan is currently under revision and an Agency Draft Indiana Bat Revised Recovery Plan was published in March 1999 (US Fish and Wildlife Service 1999).

The Indiana bat is a medium-sized bat with a wingspan of about 280 millimeters (11 inches) and a weight of 5–ll grams (0.2–0.4 ounce). It is differentiated from other species in the genus by its smaller foot, short toe hairs, keeled calcar, and fur texture and coloration. It occurs in the eastern U.S. from North Carolina west to Oklahoma and north to Iowa, Michigan, and Vermont.

There are six federally listed endangered bats that occur within the continental U.S.

During the winter the Indiana bat hibernates in cold caves and mines, 4–8°C (40–46°F), in the central portion of its range. In summer the species disperses from its hibernation sites to form small (30–300 females with young) maternity colonies.

These colonies roost under the sloughing bark of dead and dying trees and under the exfoliating bark of live trees, such as shagbark hickory. Roosts are found in riparian, bottomland hardwood and upland forests

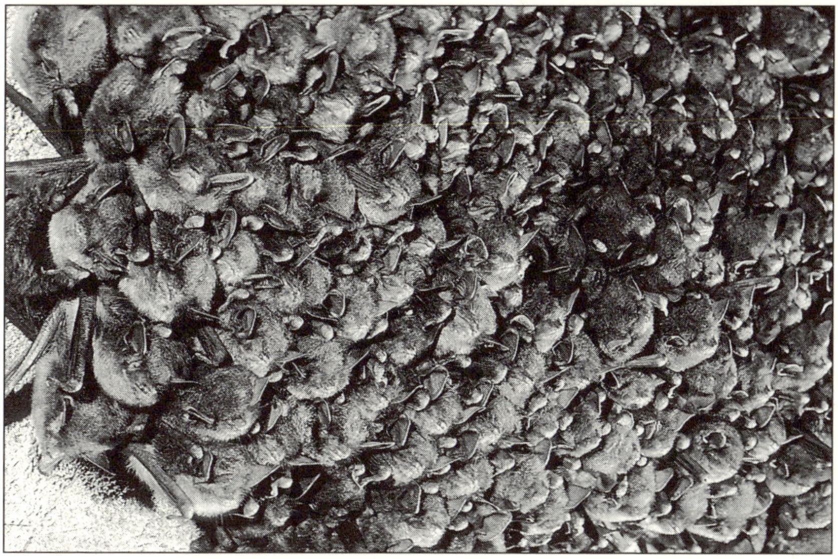

Figure 3. Indiana bats *(Myotis sodalis)* cluster together in a cave. (See page 4 of color section.)

(Barbour and Davis 1969; US Fish and Wildlife Service 1999). Excellent photographs and generalized range maps for the Indiana bat and all of the other bats that occur in the U.S. can be found in the booklet on bats titled *Bats of the United States* (Harvey and others 1999). This booklet is available, free of charge, from the US Fish and Wildlife Service's Field Office in Asheville, North Carolina.

The current draft of the Indiana bat recovery plan identifies a series of tasks that should determine what is causing the current decline and permit more effective recovery of the species.

Historically, the primary threat to the species was believed to be disturbance at its hibernation sites. Early emphasis of recovery efforts was to protect these sites with suitable gates or fences to control human access and thereby eliminate disturbance. Despite these efforts the species continues to decline. At the present time the cause of this decline is unknown. Potential explanations include currently unidentified changes with the species' summer habitat, inappropriate protection efforts at hibernation sites, and/or pesticides. The current draft of the Indiana bat recovery plan identifies a series of tasks that should determine what is causing the current decline and permit more effective recovery of the species. The Indiana bat has experienced a serious decline over the past 40 years. We estimate that in 1960 there were approximately 808,505 Indiana bats, by 1980 the population had declined to about 589,120 individuals, and during the 1995–1997 survey period only 353,185 were found (US Fish and Wildlife Service 1999).

Abandoned mines are extremely important to the continued existence of the Indiana bat. Two abandoned mines were designated as Critical Habitat for the species in 1976, and the species has since been found in numerous abandoned mines throughout its range. Most of the mines used by the species are hard rock mines or quarries. However, in 1981, John MacGregor (USDA Forest Service, personal communication 1981) observed the Indiana bat in an abandoned coal mine in Kentucky and the potential thus exists for this species to depend upon abandoned coal mines.

Gray bat (*Myotis grisescens*)

The gray bat was listed on April 28, 1976, as endangered throughout its range. No critical habitat has been designated for the species. The Gray Bat Recovery Plan was issued on July 1, 1982 (US Fish and Wildlife Service 1982).

The gray bat is slightly above average size for the genus, and the gray bat is easily distinguished from other members of the genus by its uniformly gray fur and the attachment point of the wing membrane to the foot. Its wingspan is about 305 millimeters (12 inches) and it weighs 5–10 grams (0.2–0.4 ounce). The gray bat is primarily found in the cave regions of Alabama, Kentucky, Tennessee, Arkansas, and Missouri; however, small populations also occur in Kansas, Indiana, Illinois, Oklahoma, and Florida.

The gray bat is dependent upon caves or mines all year. During the winter it primarily hibernates in cold caves in the heart of its range. During the summer the females disperse out to suitable warm caves and other cave-like structures. Foraging habitat is primarily along large- to medium-sized streams, rivers, and reservoirs. Although most foraging takes place over open water, the species occasionally feeds in wooded areas adjacent to their primary foraging areas (Barbour and Davis 1969; US Fish and Wildlife Service 1982).

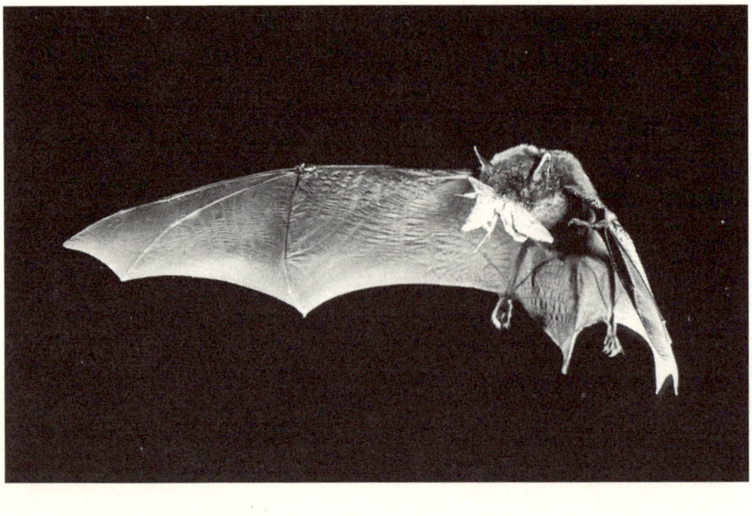

Figure 4. A gray bat *(Myotis grisescens)* is catching an insect in flight.

© Merlin D. Tuttle, BCI

The primary threat to the gray bat at the time it was listed was human disturbance at its summer and winter roost sites. Other factors that caused the decline that led to its addition to the federal list included loss of roost sites to commercialization and reservoir construction. Persistent pesticides such as DDT probably also played a role in the decline of the species (US Fish and Wildlife Service 1982).

Since 1982 the severe declines that resulted in the federal listing of the species have been reversed by positive conservation actions undertaken by states, federal agencies, and conservation groups such as Bat Conservation International, The Nature Conservancy, American Cave Conservation Association, National Speleological Society, and their dedicated members. All appropriate agencies have taken part in this effort but some, such as the Missouri Department of Conservation and the Tennessee Valley Authority deserve special mention. Because of these conservation activities we may be at the point where the species may qualify for downlisting to threatened status.

The gray bat primarily uses caves for its roost sites. It does however, readily use manmade structures whenever these provide the right microclimate and are protected from disturbance. Gray bats have been found roosting in abandoned coal mines, bridges, culverts, and dams. Any abandoned mine within the range of the species that has the appropriate temperature and humidity could support the species.

Virginia big-eared and Ozark big-eared bats *(Corynorhinus)*

The genus *Corynorhinus* is the most distinctive group of species found in the eastern U.S. They are similar in size to the gray bat but all have distinctive, large ears that are not found on any other bats in the eastern U.S. Two subspecies of Townsend's big-eared bat (Ozark and Virginia big-eared bats) are listed as endangered. The closely related Rafinesque's big-eared bat *(Corynorhinus rafinesquii)* is easily distinguished by gray colored dorsal fur. Both subspecies of Townsend's big-eared bat have brownish colored dorsal fur.

The genus *Corynorhinus* is the most distinctive group of species found in the eastern U.S. They are similar in size to the gray bat but all have distinctive, large ears that are not found on any other bats in the eastern U.S.

Virginia big-eared bat *(Corynorhinus townsendii virginianus)*

The Virginia big-eared bat was listed as endangered throughout its range on November 30, 1979. Critical habitat was designated at the time the species was listed and included many of its most important roost sites. A recovery plan was prepared for the species on May 8, 1984 (US Fish and Wildlife Service 1984).

The Virginia big-eared bat is a medium-sized bat, weighing 7–12 grams (0.2–0.4 ounce), with forearms measuring 39–48 millimeters (1.5–1.9

© Merlin D. Tuttle, BCI

Figure 5. In 2005, the Virginia big-earred bat *(Corynorhinus townsendii virginianus)* was designated the official state bat of Virginia. (See page 4 of color section.)

inches) long. Total body length is 98 millimeters (3.8 inches), the tail is 46 millimeters (1.8 inches), and the hind foot is 11 millimeters (0.4 inch) long. This bat's long ears, over 2.5 centimeters (1.0 inch), and facial glands on either side of the snout are quite distinctive. Fur is light to dark brown in color. The only other eastern bats that resemble the Virginia big-eared bat are Rafinesque's big-eared bat *(C. rafinesquii)* and the Ozark big-eared bat. Rafinesque's big-eared bat has toe hairs that extend beyond the end of the toes and the dorsal fur is gray rather than brown. The belly fur of Rafinesque's big-eared bat is white or whitish rather than light brown or buff (Schmidly 1991; Barbour and Davis 1969).

This subspecies is found in Kentucky, North Carolina, Virginia, and West Virginia. The Virginia big-eared bat roosts in caves and mines year-round. During the winter it hibernates in cold caves and mines and during the summer the females establish maternity colonies in warm caves or mines.

The primary threat to this subspecies is disturbance at its roost sites. It seems to be more susceptible to disturbance than other endangered bats. There are several instances of colonies abandoning favored roostsites after only one intensive disturbance (John MacGregor, personal communication 2000; Barbour and Davis 1969). When disturbance is eliminated, the species will usually return to its favored roost after a few years. The current population of the Virginia big-eared bat is estimated to be 18,442 individuals; the estimated total population in 1996 was 15,360 individuals. At the time the species was listed the population was thought to contain only a few thousand individuals (Traci Wethington, Kentucky Department of Fish and Wildlife Resources, personal communication 2000; Craig Stihler, West Virginia Department of Natural Resources, personal communication 2000; Rick Reynolds, Virginia Department of Game and Inland Fisheries, personal communication 2000; Chris McGrath, North Carolina Wildlife Resources Commission, personal communication 2000).

The Virginia big-eared bat roosts in caves and mines year-round. During the winter it hibernates in cold caves and mines and during the summer the females establish maternity colonies in warm caves or mines.

This subspecies has a limited distribution. Its microhabitat requirements for roost sites are specific and any site that meets these requirements, whether it is natural or manmade, can support the species. An abandoned mine in North Carolina supports a small population of the Virginia big-eared bat. This mine is one of the best hibernation sites in the state and if the mine can be protected from the regular human disturbance that it now receives, the population should dramatically increase. The largest known population (about 1,700 bats) of the closely related Rafinesque's big-eared bat uses an abandoned series of mines in the North Carolina portion of Great Smoky Mountains National Park during both the summer and the winter.

Ozark big-eared bat *(Corynorhinus townsendii ingens)*

The Ozark big-eared bat was listed as endangered throughout its range on November 30, 1979. No critical habitat has been designated for the species. The most recent recovery plan for the Ozark was released on March 28, 1995 (US Fish and Wildlife Service 1995).

This subspecies is very similar to the Virginia big-eared bat in appearance and habitat requirements. Historically it was found in Arkansas, Missouri, and Oklahoma. It is believed to have been extirpated from Missouri.

The current threats to the Ozark big-eared bat are believed to be low population numbers, human disturbance, and loss of habitat. When this subspecies was listed, only a few hundred individuals were known to exist. The current estimated population of the Ozark big-eared bat is about 1,800 bats in Arkansas and Oklahoma (Steve Hensley, US Fish and Wildlife Service, personal communication 2000).

All other members of the genus *Corynorhinus* readily use abandoned mines when these are available and are suitable. Any mines found within the range of the species could, it they provide suitable conditions, support the species. Michael J. Harvey (personal communication 2000) reports that a few individuals have been observed in an abandoned lead mine in Arkansas.

Lesser long-nosed bat *(Leptonycteris curasoae yerbabuenae)*
The lesser long-nosed bat was listed as endangered throughout its range on September 30, 1988. No critical habitat has been designated for the species. A recovery plan for the species was released on May 4, 1994 (US Fish and Wildlife Service 1994a).

The lesser long-nosed bat is a migratory, nonhibernating species that feeds almost exclusively on nectar, pollen, and fruit of columnar desert cacti and agave plants. It is a medium-sized bat that weighs 20–25 grams (0.7–0.9 ounce) and has a wingspan of about 406 millimeters (16 inches). Fur color is gray to reddish brown dorsally and brownish ventrally. Seasonally the bats move very long distances. Their distribution appears to be directly related to food supply and the availability of suitable roost sites (US Fish and Wildlife Service 1994a).

The lesser long-nosed bat is a migratory, nonhibernating species that feeds almost exclusively on nectar, pollen, and fruit of columnar desert cacti and agave plants.

In the U.S., the species is found in Arizona and New Mexico. It also occurs in Mexico and Central America.

The lesser long-nosed bat inhabits warm caves and mines year-round. The species is an important component of southwestern desert ecosystems. They pollinate agaves and several of the columnar cacti such as the saguaro. Later they return and feed on the fruits of the cacti and then play a

© Merlin D. Tuttle, BCI

Figure 6. Lesser long-nosed bat *(Leptonycteris curasoae yerbabuenae)* pollinating.

role in the dispersal of seeds.

This species is vulnerable to disturbance at its cave and mine roost sites and to loss and change in the composition of the desert flora that provides its food supply. The current population level of this species is much larger now than at the time it was listed; however, it is still considered to be vulnerable (US Fish and Wildlife Service 1994).

Lesser long-nosed bats pollinate agaves and several of the columnar cacti such as the saguaro. Later they return and feed on the fruits of the cacti and then play a role in the dispersal of seeds.

The lesser long-nosed bat is very dependent upon abandoned mines as roost sites and loss of these roosts would seriously impact the species. Six of the eight roost sites in Arizona and New Mexico are mines. Several of the known Mexican winter roost sites are also mines. If we are to protect the lesser long-nosed bat, the known roost sites must be protected and the mines scheduled for closure must be evaluated for potential use by this species.

Mexican long-nosed bat *(Leptonycteris nivalis)*
The Mexican long-nosed bat was listed as endangered throughout its range on September 30, 1988. No critical habitat has been designated for the species. A recovery plan for the Mexican long-nosed bat was released in September 1994 (US Fish and Wildlife Service 1994b).

The Mexican long-nosed bat is slightly larger than the lesser long-nosed bat with a wingspan of about 432 millimeters (17 inches). It also has more brownish-colored fur. In the U.S. it occurs in New Mexico and Texas. It is primarily a Mexican and Central American species with its range barely extending into the Big Bend area of Texas and the southwest corner of New Mexico.

The habitat and threats to the continued existence of the Mexican long-nosed bat are similar to those listed for the lesser long-nosed bat. It is however, a much rarer species.

The largest known U.S. site for the species is a cave in Big Bend National Park, Texas. Because the Mexican long-nosed bat's habitat requirements are similar to those for the lesser long-nosed bat, mines may play a similar role in their survival and recovery.

Summary

Caves and abandoned mines are extremely important to the conservation and recovery of most bats that are currently listed as endangered species under the Endangered Species Act. Caves supporting significant populations of these endangered bats must be managed for species protection. Closure of abandoned mines, reclamation of abandoned mined land, renewed mining, and new mines can all adversely affect these endangered species.

Federal agencies, state agencies implementing federal programs, as well as state agencies, private organizations, and individuals that need some form of federal authorization or permit for their activities must comply with the provisions of Section 7 of the Act. Everyone must ensure that their activities do not violate Section 9 of the Act.

Bats are a unique, vulnerable, and valuable part of naturally functioning ecosystems. Past human activities have pushed many cave and mine dependent bats to the brink of extinction. To reverse these declines and to provide for their long-term protection and recovery, we must incorporate impact analysis and proactive bat conservation measures into all cave and mine activities. If we do not, the recovery and eventual delisting of these bats will be difficult, if not impossible.

Cited References

Barbour RW, Davis WH. 1969. *Bats of America.* Lexington (KY): The

University Press of Kentucky. 286 p.
Harvey MJ, Altenbach JS, Best TL. 1999. *Bats of the United States.* Little Rock (AR): Arkansas Game and Fish Commission. 64 p.
US Fish and Wildlife Service. 1982. *Gray Bat Recovery Plan.* Twin Cities (MN): US Fish and Wildlife Service. 94 p.
US Fish and Wildlife Service. 1984. *A Recovery Plan for the Ozark Big-eared Bat and the Virginia Big-eared Bat.* Twin Cities (MN): US Fish and Wildlife Service. 113 p.
US Fish and Wildlife Service. 1994a. *The Lesser Long-nosed Bat Recovery Plan.* Albuquerque (NM): US Fish and Wildlife Service. 45 p.
US Fish and Wildlife Service. 1994b. *Mexican Long-nosed Bat* (Leptonycteris nivalis) *Recovery Plan.* Albuquerque (NM): US Fish and Wildlife Service. 91 p.
US Fish and Wildlife Service. 1995. *Ozark Big-eared Bat* [Plecotus (Corynorhinus) townsendii ingens] *Revised Recovery Plan.* Tulsa (OK): US Fish and Wildlife Service. 51 p.
US Fish and Wildlife Service. 1999. *Agency Draft Indiana Bat* (Myotis sodalis) *Revised Recovery Plan.* Ft. Snelling (MN): US Fish and Wildlife Service. 53 p.
Schmidly, DJ. 1991. *The Bats of Texas.* College Station (TX): Texas A&M University Press. 188 p.

Section A—Identifying and Protecting Cave Resources

Anthropogenic and Foreign Chemicals in Caves

Penelope J. Boston

What's the difference between your kitchen floor and a cave? On the kitchen floor, you want to get rid of visible dirt and lower the populations of microorganisms and unwanted invertebrates. In a cave, after careful evaluation, you want to get rid of visible negative impacts while keeping the environment safe for the native microbes and larger cave organisms.

Throughout this book, we recommend using *no* commercial chemicals for restoration and repair—unless the product is *demonstrably* harmless to the biota of a specific cave.

Types of Chemical Agents

When planning to address contemporary graffiti, get rid of unwelcome microbes like algae, or repair physical damage to speleothems, many potential chemical products come to mind. These chemicals fit into several categories:

- Anthropogenic (human-manufactured) chemicals not found in nature
- Products derived from a "natural" source but not native to the cave environment
- Chemicals that are native to the cave environment

Of the bewildering array of products available, which are safe and effective for use in caves?

Throughout this book, we recommend using *no* commercial chemical agents for restoration and repair unless the product is *demonstrably* harmless to the biota of a specific cave.

Any agent should be quickly and thoroughly removed from the cave. Over time, degradation of materials left behind in caves and outgassing from residual deposits of cleaning chemicals or glues can cause tremendous harm to cave biota, habitat, delicate speleothems, and minerals.

Damage from Chemical Compounds

What factors influence the potential damage that a particular compound can do?

- **First**. Various caves differ in their sensitivity to materials. Caves with many or large openings to the surface or a high volume of flowing water may be able to purge themselves of harmful traces of products in solution or outgassing of deleterious compounds.

- **Second**. Caves with no flowing water and caves with small openings are effectively *closed systems* and do not have the capacity to flush out harmful chemicals or their breakdown products. Obviously, a small amount of material getting into a vigorously flowing cave stream is less harmful than the same amount of material flowing into a tiny, closed pool with no outlet. The biological uniqueness is likely to be higher in that tiny pool because of long isolation from other influences. Thus, the potential damage to native microbial populations is much greater and more significant in isolated cave passages.

A product labeled "biodegradable" may be a great choice for aboveground household tasks ... but biodegradable simply means the product is "edible" to some microorganisms.

- **Third**. A compound that may be suitable for use on the surface may not be the best choice for cave use. For example, a product labeled "biodegradable" may be a great choice for aboveground household tasks. But biodegradable simply means the product is "edible" to some microorganisms. Introduction of such a compound into a cave means that you are upsetting the natural nutrient balance of the cave and feeding some organisms at the likely expense of others.

- **Fourth**. Some compounds (for example, certain glues, epoxies, paints, and plastics) continue to emit gasses over long periods of time. Some of these gasses are toxic and some could be inappropriate nutrient sources for organisms. Additionally, nonwater-based cleaners use organic solvents (for example, toluene, trichloroethane, or kerosene) that are toxic to many cells and a potential food source for others.

- **Fifth**. Consider the introduction of surface microorganisms into caves. Human cavers carry invading microbial populations into isolated cave passages on soiled boots, clothing, and gear. Introduced microbes are much more likely to use inadvertently introduced chemical compounds as nutrients. Surface organisms are adapted to highly fluctuating environments and may easily supplant the native species. Cave natives are typically adapted to ultra-low nutrient and relatively unchanging environments.

Soaps, Detergents, and Cleaning Agents

Soaps and detergents can be disastrous in caves. They make surfaces more wettable, help dissolve oils, and make other organic molecules more soluble. All of these properties adversely affect the protecting biofilms that organisms make to trap and store nutrients, protect themselves against desiccation, and make local microenvironments most compatible with their needs (Costerton and others 1994; Ben-Ari 1999). Soaps and detergents also have organic components, and may provide nutrients for some nonresident microbes.

Soaps and detergents can be disastrous in caves. They make surfaces more wet-table, help dissolve oils, and make other organic molecules more soluble.

Organisms have preferred levels of acidity or alkalinity (expressed as pH values) in which they like to live. A pH of 7.0 is considered neutral. Numbers less than 7 indicate acidity, those larger than 7 indicate alkalinity.

Most cleaning chemicals are highly alkaline:

- Household ammonia (typically pH 11.5)
- Common chlorine bleach (typically pH 12.5)
- Common oven cleaners (typically pH 13.5)

Others are at the opposite or acidic end of the pH spectrum:

- Muramic acid (typically pH 3.5)
- Muriatic acid (trade name for product often used in spas and swimming pools, typically 31.5% hydrochloric acid in water with pH less than 1.0; acidity varies with concentration; chemical name is hydrochloric acid, pH 0.0)

Compounds with pH values far from the natural pH value of a particular microbial habitat can impair growth or even kill the organisms. Luckily, most caves formed in carbonates have a good buffering capacity that minimizes the effects of acidic pH extremes.

Nevertheless, application of acidic compounds can have deleterious effects on small spatial scales. Alkaline compounds are typically very aggressive oxidants that destroy organic matter, both living and dead.

Cave Restoration Chemicals

For speleothem repair, always use museum-grade epoxies and adhesives. The recommended products are formulated to achieve minimal degradation and outgassing. Used in caves since the early 1980s, these compounds have proven to be relatively safe for use in subterranean environments. (See cave-safe materials, page 172 and page 445.)

Unfortunately, there are few chemicals native to caves that are appropriate for cave restoration. *Cave water* is a potentially safe cleaning agent if there is enough water for restoration purposes. (See sources for cave restoration water, page 393.)

Ideally, cleaning water is removed from the cave. If removal is impractical, then an acceptable alternative may be to dispose of restoration water in the trail only if the dirty water cannot find its way into pools or onto speleothems or other features. Water contaminated with any anthropogenic agent must be removed from the cave.

Any agent should be quickly, thoroughly removed from the cave. Over time, degradation of materials left behind can cause tremendous harm to cave biota, habitat, delicate speleothems, and minerals.

Cited References

Ben-Ari ET. 1999. Not just slime: Beneath the slippery exterior of a microbial biofilm lies a remarkably organized community of organisms. *Bioscience* 49(9):689.

Costerton JW, Lewandoski Z, DeBeer D. 1994. Biofilms, the customized microniche. *Journal of Bacteriology* 176:2137-42.

Section A—Identifying and Protecting Cave Resources

Protecting Microbial Habitats: Preserving the Unseen

Penelope J. Boston, Diana E. Northup, and Kathleen H. Lavoie

A vast, unseen world of life covers all the surfaces that we clean and restore in caves (Lavoie and others 2000). This world encompasses the microbial realm—those organisms that you usually can't see because they are *microscopic*—that is, too small to be seen with the naked eye. *Microorganisms* (microbes) include bacteria (both archaeal and bacterial domains), fungi, some algae and protozoa, and viruses. Fungi, algae, and some bacteria become visible to the unaided human eye when they occur in large numbers. They often cover cave walls or other surfaces with a thin coating (green in the case of algae, or other colors in the case of fungi and bacteria, Figures 1a, 1b, and 1d).

This chapter discusses the effects of restoration on *native* microbes that inhabit caves as we work to remove *alien* microbes, contemporary graffiti, tracked-in mud, or to remediate other human-introduced changes. Telling the difference between resident and nonresident microorganisms can be difficult, and varies by situation. Algae and mosses growing near an entrance or under a skylight are native (Figure 1a), while the same plants growing around a light in a show cave, are usually *exotic*. Fungi are usually not native in areas with extremely low organic content, but may be native in areas that are richer in organics.

This chapter discusses the effects of restoration on *native* microbes that inhabit caves as we work to remove *alien* microbes, contem-porary graffiti, tracked-in mud, or to remediate other human-introduced changes.

You usually will *not* see most of the microbes that live in caves. When you do, they will typically be visible as splashes of color (Figures 1b, 1c, 3a, and 4b). The visual display can be either the microbial colonies themselves or their byproducts (for example, characteristic mineral deposits—some forms of manganese or iron oxides, Figure 7b).

Macroscopically visible microorganisms can display a wide range of physical characteristics. Following are a few examples of microbes at work in caves:

- Bubble-gum pink fungi
- Patches of fluffy white mold
- Millimeter- to centimeter-sized white colonies with pink centers floating on pools
- Whole walls in lava tube caves glowing with silver or gold actinomycete bacteria (Figure 3a)
- Slimy bacterial strings (Figure 2a)
- Mounds of mold covering natural organic remains like dead bats or animal scat (Figures 4a and 4b)
- Patches of colorful fungi and slime on spilled food, old candle wax, or vomit (Figure 4c)

Terms for Organisms Native to Caves

- Resident
- Indigenous
- Autochthonous

Terms for Organisms Nonnative to Caves

- Nonresident
- Exotic
- Alien
- Introduced
- Allochthonous

Bacteria and fungi that cavers may see in caves include forms that are common aboveground (Northup and others 1994), such as the bread molds (*Rhizopus, Mucor, Penicillium,* and *Aspergillus*); *Histoplasma capsulatum* (the fungus that causes histoplasmosis, a lung disease often contracted in caves); and bacteria such as the filamentous actinomycetes.

Figure 1. Cave Microorganisms Visible to the Naked Eye

Penelope J. Boston

Figure 1a. A moss garden grows in Four Windows Cave (lava tube), El Malpais National Monument, New Mexico. (See page 2 of color section.)

Kevin Glover

Figure 1b. Penny Boston collects actinomycete bacteria from a wall in Ft. Stanton Cave, New Mexico.

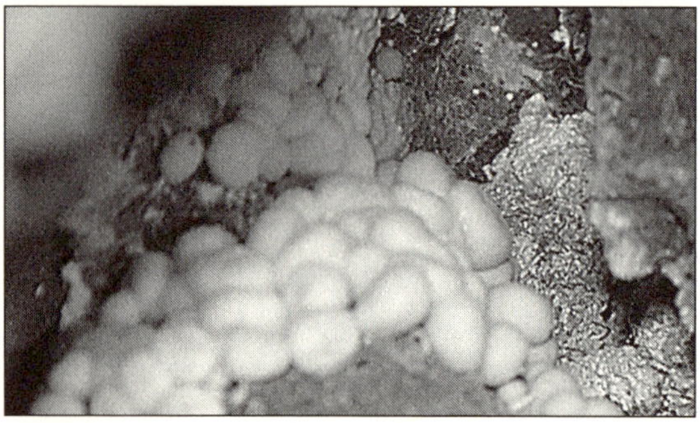

Penelope J. Boston

Figure 1c. Unusual slime molds, fungi, yeasts, and bacterial assemblages grow in Four Windows Cave, El Malpais, New Mexico. (See page 2 of color section.)

Diana E. Northup

Figure 1d. Ladybugs massing on an algae-covered boulder below Carlsbad Cavern entrance, New Mexico. (See page 4 of color section.)

Figure 2. Cave Organisms in a Rich Cave Food Web

© Kenneth Ingham

Figure 2a. Snottites (stalactite-shaped bacterial slime colonies) form on selenite crystals in Cueva de Villa Luz, Mexico.

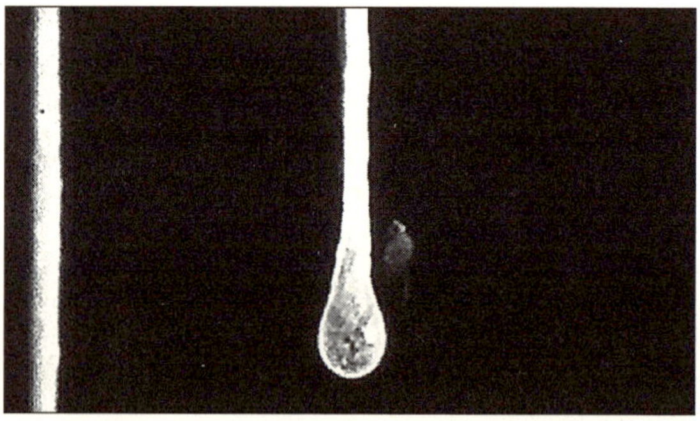

Louise D. Hose

Figure 2b. Midges are visible on a snottite hanging in Cueva de Villa Luz, Mexico.

© Kenneth Ingham

Figure 2c. Amblyopygid observed in Cueva de Villa Luz, Mexico.

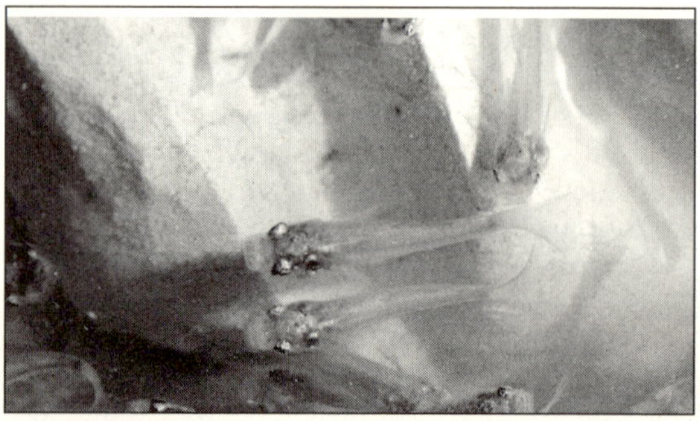

© Kenneth Ingham

Figure 2d. Fish live in a stream within Cueva de Villa Luz, Mexico. (See page 4 of color section.)

Figure 3. Who's at Home in Cave Habitats?

© Kenneth Ingham

Figure 3a. Actinomycete bacteria shine on a lava wall in Michael Wester's caving light. Four Windows Cave, El Malpais National Monument, New Mexico.

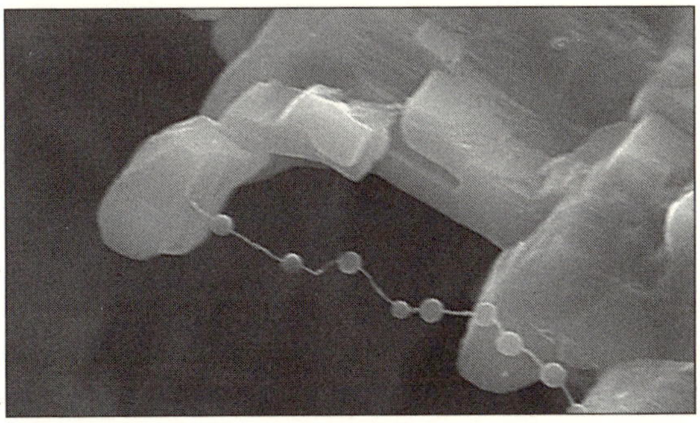

Spilde & Northup

Figure 3b. Scanning electron micrograph of microorganisms on a calcite speleothem. Spider Cave, New Mexico.

Spilde & Northup

Figure 3c. Scanning electron micrograph of microorganisms on lava mud. Cape Verde Islands, eastern Atlantic ocean.

Figure 4. Degradation of Organic Remains

Kathleen H. Lavoie

Figure 4a. "Marshmallow cricket." A cricket carcass is consumed by white fungi, giving it the appearance of a puffy marshmallow. Mammoth Cave, Kentucky.

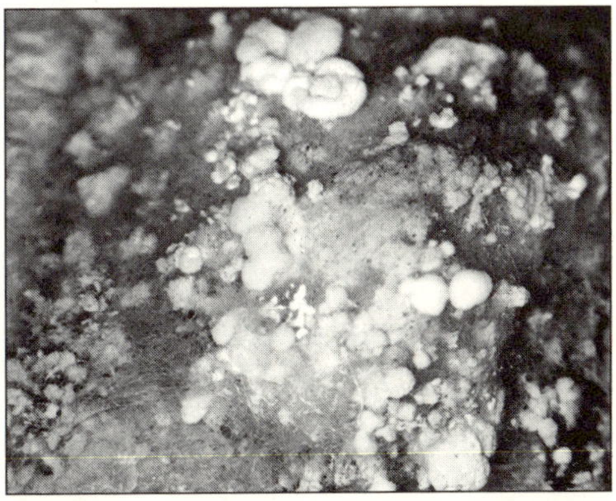

Diana E. Northup

Figure 4b. Bat carcass is consumed by yellow mold on a wall near Bat Cave in Carlsbad Cavern. (See page 1 of color section.)

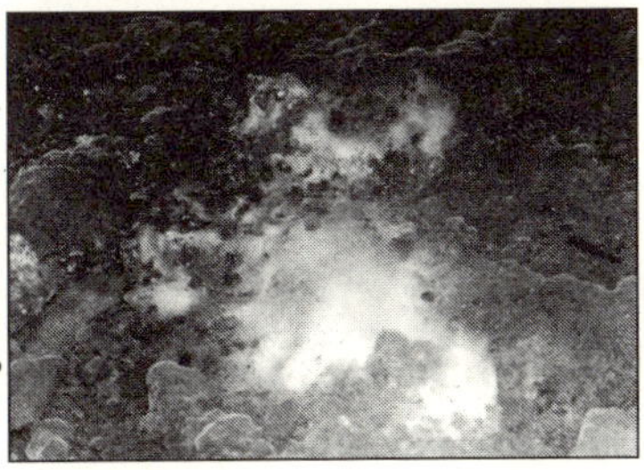

© K. Ingham & D.E. Northup

Figure 4c. Human vomit covered with fungi was found along visitor trail. Carlsbad Cavern, New Mexico. (See page 1 of color section.)

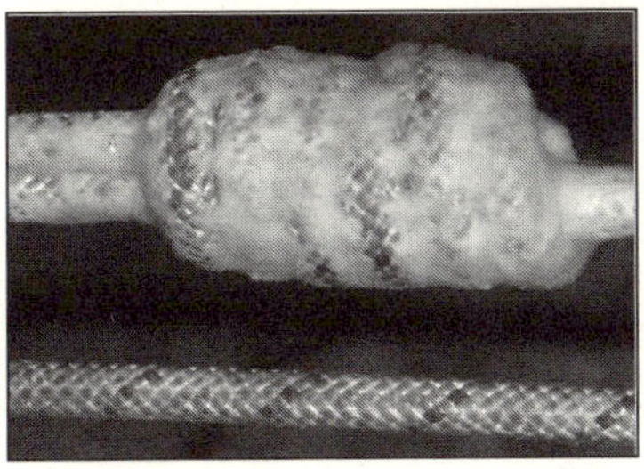

© Val Hildreth-Werker

Figure 4d. Fungi consumes a caving rope left in La Cueva de las Barrancas, New Mexico. (See page 1 of color section.)

Michael N. Spilde

Figure 4e. Scanning electron micrograph of fungi entwined in cave rope fibers. Lechuguilla Cave, New Mexico.

Figure 5. On a Speck of Mud

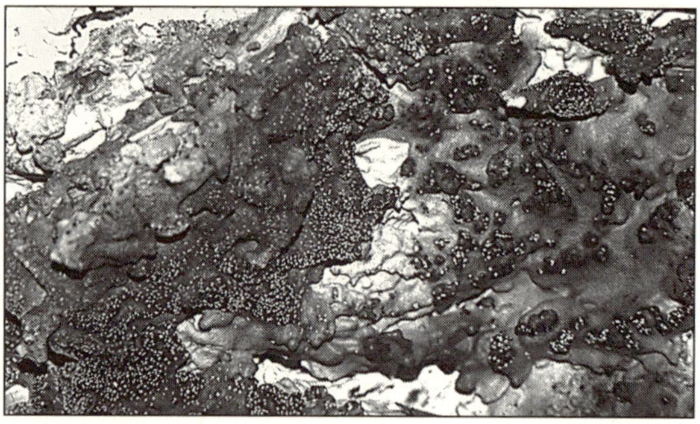
Louise D. Hose

Figure 5a. Actinomycetes grow on gypsum paste (pH 1.0) in Cueva de Villa Luz, Mexico. (See page 2 of color section.)

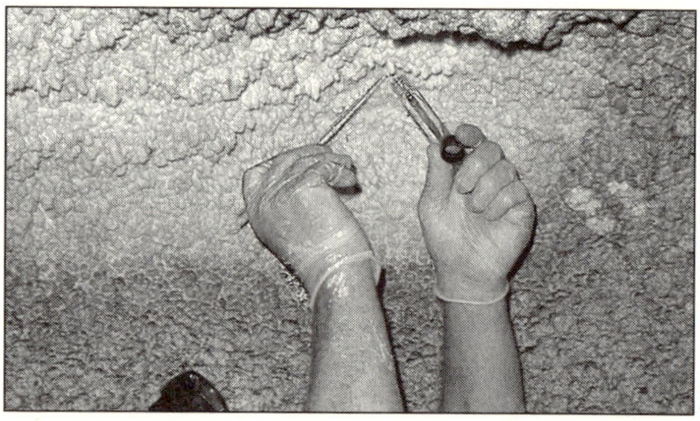
© Kenneth Ingham

Figure 5b. Collecting moonmilk paste that has the consistency of Crisco® Spider Cave, New Mexico. (See page 2 of color section.)

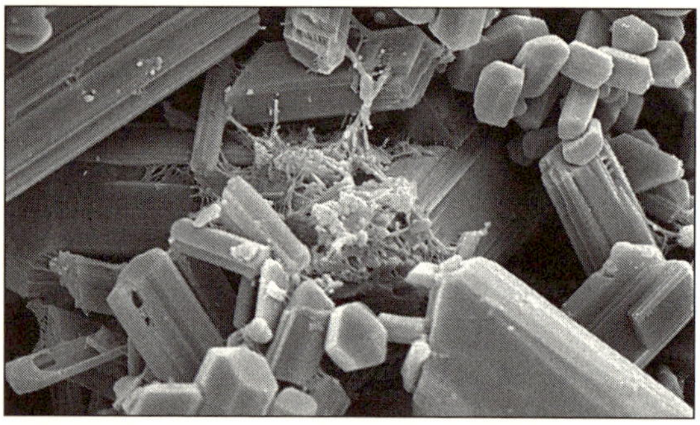
Douglas S. Soroka

Figure 5c. Scanning electron micrograph of gypsum paste. Cueva de Villa Luz, Mexico.

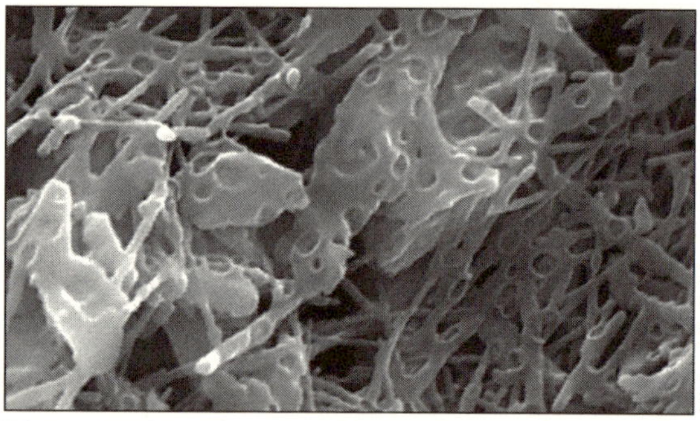
Michael N. Spilde

Figure 5d. Scanning electron micrograph of Crisco-like moonmilk. Spider Cave, New Mexico.

Figure 6. Examples of Biofilms in Caves

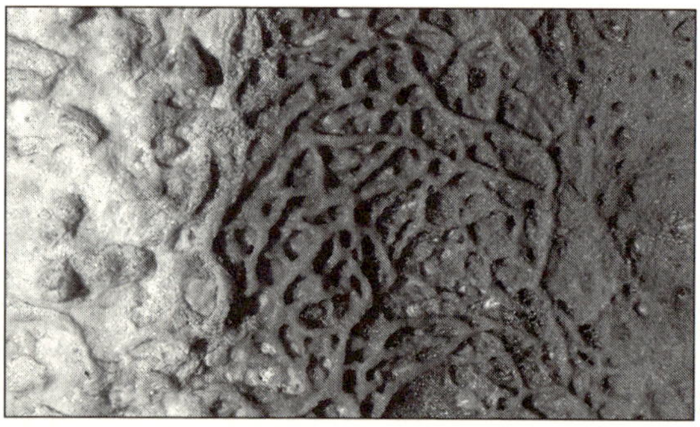
© Kenneth Ingham

Figure 6a. Biovermiculations (microbial mud mats) on walls in Cueva de Villa Luz, Mexico. (See page 2 of color section.)

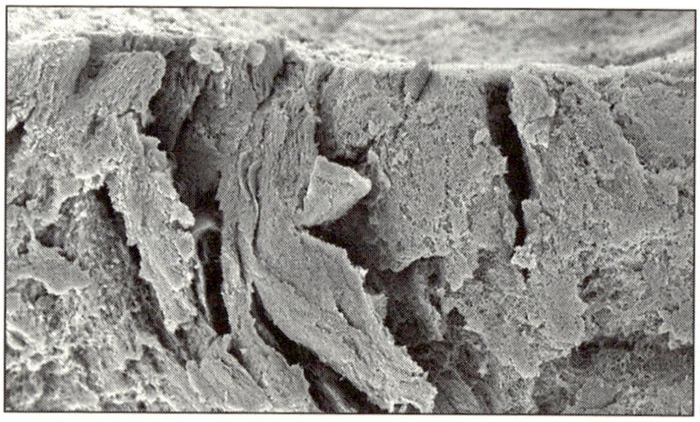
Douglas S. Soroka

Figure 6b. Microbial mats (known as "phlegm-ball mats") in hydrogen sulfide springs in Cueva de Villa Luz, Mexico.

Larry McLaughlin

Figure 6c. U-loops discovered in Lechuguilla Cave, New Mexico.

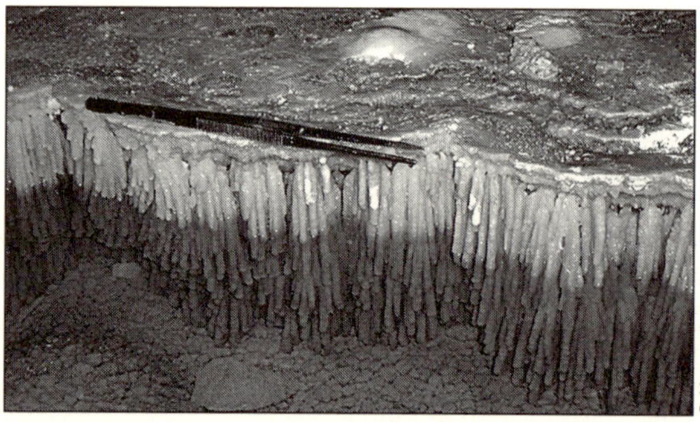
© Kenneth Ingham

Figure 6d. Pool fingers in Lower Cave, Carlsbad Cavern, New Mexico. (See page 2 of color section.)

Figure 7. More Examples of Biofilms in Caves

Sherry Cady

Figure 7a. Colorful surface microbial mats found in hydrothermal springs at Yellowstone National Park, Wyoming. (See page 3 of color section.)

David H. Jagnow

Figure 7b. Conspicuous biofilms like this black iron mat may contain significant microbial populations, extracellular materials like polysaccharide slimes and enzymes, and precipitated minerals like iron oxides. Coldwater Cave, Iowa. (See page 3 of color section.)

© Kenneth Ingham

Figure 7c. Wearing protective gear, Penny Boston collects a sample from conspicuous dangling biofilms in Cueva de Villa Luz, Mexico. (See page 3 of color section.)

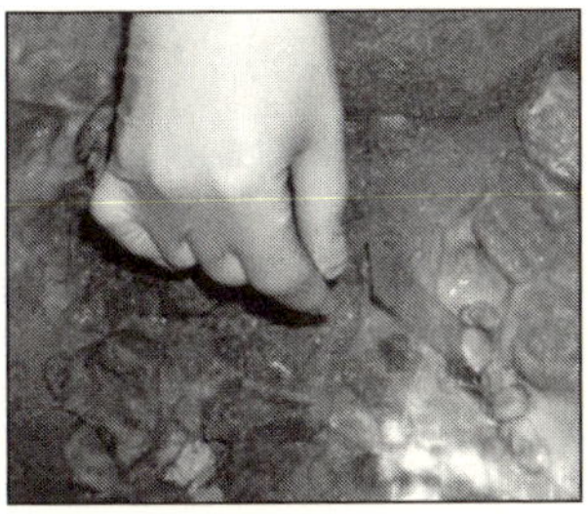

© Val Hildreth-Werker

Figure 7d. Invisible bio-films on flowstone and other moist cave surfaces are enough to support cave-adapted invertebrates. La Cueva de las Barrancas, New Mexico. (See page 3 of color section.)

© Val Hildreth-Werker

Figure 7e. Mike Spilde collects a sample from colorful encrustations of manganese and iron oxides that can contain rich microbial diversity. Lechuguilla Cave, New Mexico. (See page 3 of color section.)

Actinomycetes give caves their musty odor and can provide spectacular displays in lava tubes and some other caves (Figures 1b, 3a, and 5a). Walls covered with actinomycetes gleam as if they have been coated in silver when a caver's headlamp reflects off thousands of beads of water on the colonies (Figure 3a). Some of the fungi and bacteria are contaminants from humans or human artifacts and should be removed (Figures 4c and 4d), while others are native to caves and deserve the protection afforded to other wildlife. Knowing the difference can be difficult, even for experts.

Why Should Cavers Care About Microbes?

It's difficult to care about something you can't see, but microbes in caves play many important roles in nature and may produce compounds of untold value to humans. Study of microorganisms is critical to our understanding of fundamental biology, the way biogeochemical cycles work on Earth, and the way rocks and minerals on our small rocky planet are precipitated and weathered.

It's difficult to care about something you can't see, but microbes in caves play many important roles in nature and may produce compounds of untold value to humans.

Organisms That Decompose and Recycle

Microbes, especially bacteria and fungi, play crucial roles as decomposers in all ecosystems on Earth. The familiar food chain production system (sunlight energy trapped by plants, plants eaten by herbivores, herbivores eaten by carnivores) often neglects to mention the important decomposition subsystem of food chains. As dead organic materials (withered vegetation, animal wastes, and carcasses) accumulate in the environment, decomposer organisms recycle it back into CO_2 (carbon dioxide), H_2O (water), and nutrients that enter biogeochemical cycles and return to the production system. All life on Earth would cease without the action of decomposer microbes. Many ecosystems on Earth have food chains based on the decomposition of organic material (Schaechter 1997; Needham and others 2000; Wackett and others 2001). Examples include soil, large rivers, the deep ocean, some lakes, and most caves.

Food Chains Based on Sulfur Compounds

Microbes that use organic matter provide the base of the simplified food chain found in most caves. If these microbes are harmed or removed, higher organisms further up the food chain will be impacted or might die. Sometimes there are unusual microorganisms that form the base of food chains in caves where inorganic chemicals provide major sources of energy, such as the large input of sulfur compounds into Cueva de Villa Luz in Mexico (Lavoie and others 1998; Hose and others 2000).

In Villa Luz, microbes in springs and the stream produce large amounts of biomass. This biomass supports an aquatic food chain with abundant cave fish and aquatic invertebrates, including midge larvae (Boston and others 2000). A rich terrestrial ecosystem relies on the flying adult midges hatched from the stream that feed spiders and other predatory invertebrates (Figure 2). Cave systems with food chains partially based on hydrogen sulfide include the following:

- Cueva de Villa Luz in Mexico (Lavoie and others 1998; Hose and others 2000)
- Movile in Romania (Sarbu and others 1996)
- Frasassi Cave in Italy (Sarbu and others 2000)
- Submarine caves off the coast of Italy (Southward and others 1996; Mattison and others 1998)
- Sulphur River in Parker Cave, Kentucky (Angert and others 1998)
- Lower Kane Cave in Wyoming (Engel and others 2003)
- Cesspool Cave in Virginia (Engel and others 2001)

Microbes use and transform rock in their growth processes, helping to form some cave deposits and eroding others away.

Microbial Involvement in Speleothems

Microbes use and transform rock in their growth processes, helping to form some cave deposits and eroding others away. There is an increasing awareness that microbes are involved in speleothem formation (Northup and others 1997b; Boston and others 2001; Northup and Lavoie 2001).

The discovery of U-loops and webulites in Lechuguilla Cave (Cunningham and others 1995) (Figure 6c) and demonstration of microbial effects on pool finger formation in Hidden Cave, New Mexico (Melim and others 2001) (Figure 6d) show us dramatic examples of how bacteria may produce some of the spectacular formations we see in caves.

Other evidence demonstrates the involvement of bacteria and fungi in the formation of some types of stalactites, iron and manganese oxides, moonmilks (Figures 5b and 5d), saltpeter, and sulfur compounds (Cunningham 1991; Cunningham and others 1995; Northup and others 2000; Northup and Lavoie 2001; Spilde and others 2001). Study of microbial involvement in speleothem formation is expanding rapidly (Sasowsky and Palmer 1994; Northup and others 1997b; Northup and Lavoie 2001), drawing from the discoveries in the geomicrobiology of noncave subsurface environments (Amy and Haldeman 1997) and other surface environments (Banfield and Nealson 1997).

Potential Pharmaceutical and Industrial Benefits

Perhaps the strongest inducement (often cited to the general public) to preserve microbial communities in caves is their potential medical and other commercial benefits. Poulson and coworkers (1986) found a relationship between production of antibiotics and environmental stress in a passage within Great Onyx Cave, Mammoth Cave National Park, Kentucky. Their findings support the idea that an extreme environment increases microbial competition, leading to greater production of novel organic compounds.

In studies of microbial byproducts produced by bacteria from Lechuguilla Cave in New Mexico, Mammoth Cave in Kentucky, and lava tubes in Hawaii, Larry Mallory (unpublished data) has shown that compounds produced by cave bacteria may be lethal to cancer cells.

Materials derived from such microorganisms may also have industrial applications. Bacteria are especially good at breaking chemical bonds that are difficult and expensive for human chemists to break. *Bioremediation* refers to biologically based cleanup of hazardous materials in the environment. Bacteria are often chosen for bioremediation of oil spills, toxic waste dumps, and other harmful chemical spills. They act by digesting or otherwise chemically changing the hazardous materials into less hazardous byproducts (Shannon and Unterman 1993). Bacteria are the "superheroes" of hazardous substance cleanup.

Prevent Potential Harm

Careful consideration is essential in the choice of cleaning methods and techniques to protect important native microbial communities in caves. Great harm can be done by the wrong choices. For example, scrubbing moist flowstone and cave walls that are covered with an invisible (but possibly essential) coating of bacteria may retard or destroy speleothem development. If common cleaning chemicals that kill bacteria and fungi are allowed to enter a cave pool that contains unique bacteria, we could destroy the potential to discover new organisms with new chemistries or an important future medical compound. Throughout this book, the liberal use of caution and "educated" common sense is recommended when considering the impacts of human activities on microbes in caves.

Because of the vital roles that microbes play in cave ecosystems, we remind the reader that manufactured chemical agents can be indiscriminate biocides in cave environments. (See anthropogenic chemicals, page 57.) Physical methods such as steam or germicidal ultraviolet (UV) light have essentially the same destructive impact on native microbial communities. A far better approach is to prevent disturbances to cave ecosystems that result in the need for aggressive restoration efforts.

Critical Microbial Habitats

Microbes live everywhere in caves, but some habitats are more likely to contain native bacteria including those that are important in the formation of some speleothems. Unusual mineral environments such as iron and manganese oxides and native sulfur are perfect environments for bacteria that use those compounds as energy sources. Cave pearls, actively forming speleothems, and pools are other critical environments. Muds, moonmilk, vermiculations, saltpeter, and gypsum paste also harbor interesting microbes (Figures 3, 5, 6, and 7). In most circumstances, such mineral environments should receive little or no restoration and cleanup.

Careful consideration is essential in the choice of cleaning methods and techniques to protect important native microbial communities in caves. Great harm can be done by the wrong choices.

Perhaps the strongest inducement (often cited to the general public) to preserve microbial communities in caves is their potential medical and other commercial benefits.

Biofilms

Individual microbes are very small and invisible to the unaided human eye, but most of them don't live alone in nature (Costerton and others 1994; Ben-Ari 1999). They live in complex communities composed of many different types of microorganisms and are sometimes visible in layers or patches known as *biofilms* (Figure 6). Often the organisms are embedded in sticky or slimy materials that they use for protection and to trap nutrients. The sticky materials also frequently trap sediment particles and can precipitate mineral grains.

Communities composed of many different types of microorganisms and are sometimes visible in layers or patches known as *biofilms.*

Complex biofilm structures can be large and obvious. For example, the colorful microbial mats that line the spillways from geysers in Yellowstone National Park are easily seen and catch our attention with their beautiful hues (Figure 7a).

A very rich biofilm, 2 to 4 centimeters (0.75 to 1.5 inches) thick, described as "phlegm" by discoverer Doug Soroka, lines the throats of sulfur springs entering Cueva de Villa Luz (Boston and others 2001) (Figure 6b). There are many examples of more subtle biofilms. The plaque that you brush off your teeth every day is a biofilm. The slick and slimy feel of rocks in a streambed is due to biofilm.

We are now recognizing that many vermiculations (worm-like mud deposits on smooth cave walls) may actually be *biovermiculations*, microbial mats composed of many types of microorganisms, slime, and mud particles (Figure 6a). The characteristic light, fluffy soil associated with saltpeter deposits in caves in the southeastern U.S. and in Europe has been attributed to microbial action. The slimy nature of muds may also be attributable to microbial biofilm. These less obvious (and more difficult to identify) biofilms are the types of films we expect to find in the moist, but low nutrient environments of many caves.

Macroscopic Aesthetics Versus Microscopic Health

Restoration is no longer restoration when it degenerates merely to cave housecleaning. The goal is not simply to "make the cave look nice" in narrow terms. The object of restoration is to maintain the cave's natural state or to return the cave to a former natural state that existed before impact by human activities.

Restoration is no longer restoration when it degenerates merely to cave housecleaning.

The critical question is, "How can we help this cave environment be as close to its natural conditions as possible?" Sometimes, this will mean doing nothing at all rather than inflicting undesirable side effects that might result from cleaning or restoration. Some situations may require restoration—for example, features that have been heavily impacted by vandalism or contamination. This is a judgement call—however, if in doubt, it is often a better policy to err on the side of doing less.

Cave pearls are a good example of difficult decision making in restoration. When cleaning pearls, it is imperative to avoid using any kind of scrubbing device that could remove an invisible but essential biofilm. Also, it is best not to remove water surrounding pearls because of the presence of bacteria that may contribute to pearl growth. The degree to which microbes contribute to the formation of cave pearls is a new area of research, but preliminary evidence suggests that microorganisms may play a role. (See cave pearls, page 76; cave pearl restoration techniques, page 425.)

Unintentional Negative Effects of Restoration on Microbial Populations

Effects of Cleaning Chemicals

The very notion of cleaning, as we commonly understand it, involves the idea of getting rid of "germs." Of course, disease-causing microbes in our kitchens are one thing, but in caves we are trying to *preserve* the interesting native microbial life forms that we find there.

One very important caveat with the use of any cleaning program in any setting is to consider the possible effects on humans. Protective clothing, especially gloves, should always be worn. Wearing masks or providing ventilation must be considered. In cave restoration, we have to ask, "What are the effects of common cleaning and restoration chemicals on native cave microbiota and animal life?" (Makela and others 1991; Williams and others 1994; Wingender and others 1998).

Acidity and Alkalinity. Besides the effects on the humans applying the chemicals, we must also assess the impact on microbial communities. Microbes all have preferred levels of acidity or alkalinity (expressed as pH values) in which they like to live. Most cleaning chemicals are highly alkaline, such as household ammonia (pH 11.5), bleach (pH 12.5), and oven cleaners (pH 13.5). Others, such as muramic acid (pH 3.5) and muriatic acid (pH varies), are at the opposite, acidic end of the pH spectrum.

Anything with an extreme pH will disturb the natural pH value of the microbes' surroundings and reduce their optimum growth or kill them. Luckily, carbonate caves have a good buffering capacity that minimizes the effects of acidic pH extremes. Nevertheless, application of acidic compounds can have deleterious effects on small spatial scales. Alkaline compounds are not made less alkaline by carbonates except by dilution and are typically very aggressive oxidants that destroy organic matter, both living and dead.

What about chlorine bleach? In cave habitats, Enemy Number One is chlorine bleach. This includes Clorox® or any brand of bleach that contains sodium hypochlorite or calcium hypochlorite.

No Chlorine Bleach. For cave habitats, Enemy Number One is chlorine bleach. This includes Clorox® or any brand of bleach that contains sodium hypochlorite or calcium hypochlorite. Bleach is a very strong oxidant chemical that literally chews up all organic material. Save it for the bathroom bowl if you must, but don't use it anywhere near possible natural microbial populations in caves. Chlorine bleach leaves an oxidizing residue even when it is dry and outgasses chlorine gas.

Algae, the scourge of show caves, is a special problem. The use of bleach has been the accepted method for controlling algal and cyanobacterial growth around lights in show caves, but we recommend prevention via nonphotosynthetically active wavelength selection wherever possible (Olson 2006, page 343 in this volume).

Hydrogen Peroxide. For cleaning unsightly algal mats, a better option than chlorine bleach may be hydrogen peroxide, H_2O_2 (Grobbelarr 2000). This chemical is the peroxide or "extra oxide" of water—that is H_2O with an extra O. When H_2O_2 comes in contact with surfaces, it oxidizes organic materials and (in sufficiently high concentrations) is a contact poison for microbes. Remember that hydrogen peroxide is not as strong an oxidant as chlorine bleach—therefore, longer contact times and higher concentrations are necessary to kill the nonnative microbes that you may wish to remove (Makela and others 1991; Wingender and others 1998; Grobbelarr 2000).

However, unlike chlorine bleach, the breakdown products of hydrogen peroxide (water and free oxygen gas) are not likely to harm cave microorganisms. Most microbes produce H_2O_2 as a byproduct of cellular metabolism so they already have mechanisms to deal with it. Some who have tried hydrogen peroxide have found it ineffective—however, this may be the result of inadequate contact time and/or concentration (Boston 2006, page 349 in this volume). Use of hydrogen peroxide must be appropriate to the situation, and some care should be taken to prevent it from coming into contact with any potential microbial habitats or allowing it to slowly leach off into pools.

Common household hydrogen peroxide is sold as a 3% solution in water. However, up to a 30% concentration can be purchased from chemical supply companies. Concentrated solutions are a potential danger to the user and must be treated with caution (as with any commercial chemical) per the label and the Material Safety Data Sheet (MSDS) that is available from the National Institute of Occupational Health and Safety (NIOSH) online at <http://www.cdc.gov/niosh/npg/npg.html>.

Soaps and detergents can be disastrous in caves, stripping away protective biofilms and serving as a rich growth medium for non-resident microbes.

No Soap. Soaps and detergents can be disastrous in caves. We use them in daily life for making surfaces more wettable and causing oils and other materials to solubilize and be carried away by water. Their ability to do this also works on microbes.

Soap and detergent can remove the natural protective layers of various biofilm materials that microbes typically secrete around themselves. These layers protect the microbes from adversities in their environments, often trap and store nutrients, and help maintain their preferred pH (Costerton and others 1994; Ben-Ari 1999). If we strip away the biofilm with soaps and detergents, the microbes will die. Soaps and detergents also have organic components, and may serve as a rich growth medium for nonresident microbes in caves.

No Solvents. Cleaners that are not based on water use organic solvents such as toluene, trichloroethane, kerosene, and others. Many organic solvents are toxic to living cells. In some cases, they can also be a food source to exotic microbes from the outside. While you might conclude that giving food to hungry microbes could be a good thing, remember … we are trying to *preserve the balance* between the different populations of native microbes and not preferentially feed some while poisoning others. The signs asking people to avoid feeding the animals in the zoo can be equally applied to the tiny, microscopic zoo in a cave.

Physical Effects

We have already mentioned the undesirable effects of scrubbing walls or flowstone that may be covered with biofilms or microbial communities. We don't want to become the Godzillas of the microbial world, wiping out whole communities with a single swipe of our restoration toothbrushes!

Pressure sprays. Use of high-pressure water sprays is also potentially destructive—pressurized sprays can take out both microbes and larger (but still scarcely visible) invertebrate life. Use of high-pressure devices should be limited to extreme cases or heavily impacted areas in show caves. Remember the guidelines mentioned in this chapter about prime habitats for native cave microbes and give careful consideration before cleaning or restoring these types of areas. (See pressurized water, page 397.)

Compaction. A more subtle deleterious physical effect is compaction caused by travel on cave soils. Microbes and invertebrates live in soils and most need air and space (Stotzky and Norman 1964). When we walk on soils we compact them, killing organisms outright and making the soil an undesirable place to live. Keep to trails or previously compacted areas.

Other Contamination Dangers

Surface Contamination to Karst Hydrology

Contamination of water in karst terrain is a sad reality in most places in the world. Contamination ranges from direct input of toxic chemicals as in gasoline spills, to fecal contamination from agricultural runoff and leaking septic systems. In some cases, the immediate die-off of macroscopic species is evident. In other situations, the contaminating material may serve as a food source for nonnative species. Nonnative microbes may use oxygen or other compounds, thus changing the physical conditions in the cave and killing or inhibiting native organisms.

Appropriate wastewater treatment and control of nonpoint pollution sources are essential to the long-term well-being of karst systems. Areas that may be affected by toxic spills from trucks or trains should develop contingency plans to protect sensitive karst areas.

Contamination of water in karst terrain is a sad reality in most places in the world.

Human Health

Another aspect of contamination is the risk to cavers from harmful microbes. Many examples can be observed. The Red Lake case described in the next section shows possible harmful health effects from drinking contaminated water. A survey of flowing water in Mammoth Cave showed high levels of fecal coliforms (Lavoie, unpublished data), and the presence of a specific bacterial pathogen, *Salmonella* (Mallory and others, unpublished data). A cave under a visitors center in a mid-western state was so heavily impacted by polluted water from broken sewage lines that normal cuts and scrapes on skin quickly became infected (Lavoie, unpublished data). These are not rare incidences.

Trash and Human Waste

The introduction of food, hair, skin cells, urine, feces, household waste, and other trash in caves are amongst the most harmful events in the lives of native bacteria adapted to low nutrient cave conditions. Trash increases the amount of organic matter and changes the neighborhood so that the natives don't want to live there. Often the native microbes can't compete against introduced organisms that are adapted to a higher organic concentration and sometimes they are even poisoned by high levels of organic material (Poindexter 1981). Trash itself can be a source of invading microbes from the surface that preferentially grow on the materials introduced.

The introduction of food, hair, skin cells, urine, feces, household waste, and other trash in caves are amongst the most harmful events in the lives of native bacteria adapted to low nutrient cave conditions.

In all cases, pack out the organic material you bring on your back or in your body. (See packaging human waste, page 269.) A study of the impacts of urine dumping in caves showed significant negative effects including growth of nonresident microbes, odor production, possible corrosion of speleothems, changes in texture of soils and muds, and growth of a layer of microbes on the soil surface (Lavoie 1995).

Even lint can be a significant source of undesirable organics and there are special challenges associated with lint removal (Horrocks and Ohms 2006, page 351 in this volume).

In general, remove contemporary trash that you find, but carefully consider the possible historic significance of any old trash. Old tickets in the back of Great Onyx Cave from guided tours in the early 1900s have been left in place, while old gum wrappers and gum have been removed (K Lavoie, personal observation). Consult with experienced archaeologists and historians as part of your restoration plan.

Wood

Wood is particularly problematic in cleanups. Some wood may be in caves as the natural result of flooding or the opening and closing of entrances.

Other wood is brought in by humans, and may have significant historical or archaeological value. A single pole in a passage in Mammoth Cave might have been viewed by a caver as something unnatural that should be brought out. Patty Jo Watson recognized it as a scaling pole brought in by native peoples during historic explorations of the cave (Watson 1974).

Even if there is wood that was clearly brought in by humans and it has no historical value, exercise caution in removal. That wood has now become part of the food base in the cave and may be supporting large populations of microbes and invertebrates. If removal is necessary, it should be accomplished over a period of years to allow time for the community to shift to other resources (Elliott 2006, page 37 in this volume).

Freshly spent carbide is highly alkaline with a pH of over 11. This high pH can seriously impact the microbial population in underlying soils and water.

Carbide

Carbide is a human-introduced contaminant and has deleterious pH effects in caves. Freshly spent carbide is highly alkaline with a pH of over 11. This high pH can seriously impact the microbial population in underlying soils and water. The pH will gradually drop to neutral as the carbide waste reacts with carbon dioxide in the atmosphere, and the carbide will gradually lose its overt toxicity (Lavoie 1980). But spent carbide does look ugly forever, and should be removed whenever it is encountered unless there is particular historical significance that justifies its presence. (See carbide removal, page 411; also see lampblack, page 337.)

Reaeration of Compacted Soils

There is currently no perfect solution for remediation of crushed and compacted cave soils and muds. It is probably acceptable to lightly brush or gently comb the surface of soils to remove evidence of off-trail footprints. However, vigorous fluffing is probably bad because it can resuspend bacteria, mold, and fungal spores (the reproductive cells of these organisms) into the cave air, thus providing the opportunity to spread. Resuspended dust will also resettle, affecting aesthetics. Carefully comb through compacted materials to remove visible trash and organic debris. This action will also help rob undesirable microbe invaders of nutrient sources. (See trail maintenance, page 183.)

Mud banks along streams in many caves in the midwestern, southeastern, and eastern parts of the United States support populations of annelid worms, commonly known as earthworms. These worms are important in aerating the soil. However, sometimes populations may increase locally when there is a sudden input of organic contamination, such as leakage of sewage (Elliott 2006, page 38 in this volume). Cave crickets, using their ovipositiors to lay eggs in soils and mud, are frequently important sources of aeration.

Critical habitats for native cave microbes should be protected from physical scrubbing, contamination by lint or other particles, and from the introduction of dirty water or chemicals such as bleach.

Triage: Which Areas Should Be Left Unrestored to Save Microbes?

By now you're probably asking yourself, "How can I possibly know what to clean and what not to clean?" Don't be discouraged. There are no hard and fast rules but there are some guidelines that will help you make reasoned decisions. Critical habitats for native cave microbes should be protected from physical scrubbing, contamination by lint or other particles, and from the introduction of dirty water or chemicals such as bleach.

- **Isolated pools of water.** *Isolated* cave pools may have little organic carbon and may contain microbes adapted to lower nutrient levels. Microbes dwelling in the more extreme environments are more likely to produce interesting chemicals. Microbes in isolated or extreme environments are also more sensitive to human impacts than microbes in organic-rich environments. Isolated cave pools are fairly self-contained with limited ability to flush out potentially harmful inputs. (See cave pool restoration, page 416.) Pools in areas that routinely flood are often able to recover on their own with the flushing action of water.

- **Biofilms.** Flowstone or speleothems are frequently covered with a moist film or with actively flowing water. Microbial biofilms often form at interfaces between air, rock, and water.
- **Cave Streams.** Flowing water frequently cleans itself with flushing action over time. It is most important to stop or reduce input of contaminants to the stream.
- **Visible microbial communities.** Figures 1a, 1b, 1d, 2a, 2b, 3a, 5a, 5b, 6a, and 7b show examples of visible microbial communities. Figures 6c and 6d show examples of fossilized microbial communities and their associated textures.
- **Muds and pastes.** Muds and other unconsolidated materials like vermiculations, moonmilk, and gypsum paste are shown in Figures 5 and 6a.
- **Iron and manganese oxides.** Figures 7b and 7e show iron oxides. Iron oxides often form bright red-orange or red-brown coatings on limestone walls, while manganese oxides form black coatings.
- **Cave pearls.** Possibly resulting from bacterial processes (Gradzinski 1999), cave pearls may be harmed by cleaning. (See cave pearls, page 76 and page 425.)
- **Guano.** We hope you aren't cleaning up undisturbed bat poop—but if you are, don't. It's bad for your health as well as that of the microbes. Perhaps most importantly, bat guano is a natural feature of the cave and should be left alone. As well, cave crickets leave visible but subtle black deposits. Please don't remove it—cricket guano is the base for another food chain of cave-adapted animals (Poulson and others 1995; Poulson and others 1996).

Microbes that dwell in extreme environments are more likely to produce interesting chemicals. Microbes in isolated or extreme environments are also more sensitive to human impacts than microbes in organic-rich environments.

Case Studies: What Should You Do When...?

Here is a series of case studies to help readers develop a sense of how to assess sensitive situations and determine whether, how much, and when to intervene.

Mold Gardens in Lechuguilla Cave Camps

Occasionally cavers enter a camp in Lechuguilla Cave and find a garden growing—a garden of mold growing luxuriantly on food, skin particles, and hair left behind by the human campers. Mold patches are also sometimes visible along trails and in spots where people take breaks to eat and drink.

Cleaning up such patches presents two problems. Mold spores aren't healthy for the humans cleaning them up, and the chemicals used to clean up the molds are not healthy for native bacteria and fungi.

If you are allergic to mold, either don't participate in mold cleanup, or wear a HEPA filter respirator. Also, everyone should avoid patches of starkly white fungi associated with bird or bat guano; this may be *Histoplasma capsulatum*, the fungus that causes histoplasmosis. Even if you've been exposed to this fungus before, you can still get a new case of histoplasmosis if you are exposed to a large number of spores. (See histoplasmosis, page 277.)

We recommend using dampened paper towels or antimicrobial, unscented wipes to collect as much of the visible mold as possible. Cover offending blob with a wipe and carefully enclose and scoop the mold. Don't open up the wipe again. Deposit the wipe and all material in a plastic bag and remove it from the cave. Remember, mold spores scatter very easily so move slowly as you place the moist wipe or towel over the top of the fungal spot (Hildreth-Werker 2006, page 413 in this volume).

Chlorine bleach will kill off the good and the bad organisms and will leave harmful chemical residues. Simple removal of the contaminating organic material will usually suppress the growth of more mold. (This is not the same as the toilet bowl or kitchen sink where strong cleansers are necessary to reduce the spore load because the input of organic material is constant.)

Chlorine bleach will kill off the good and the bad organisms and will leave harmful chemical residues. Simple removal of the contaminating organic material will usually suppress the growth of more mold.

The best strategies are to encourage everyone to eat over plastic bags that catch the crumbs, to prevent candle wax spills, and to avoid leaving any human effluvium (vomit, saliva, feces, urine, and hair). Even a tiny crumb is a multi-day feast for a microbial community (Northup and Welbourn 1995).

Uncontaminated Pools: How to Keep Them That Way

Pristine cave pools often harbor microorganisms that are unknown to science and that may produce chemical substances useful to humans. The pools that produce the best results are those most isolated from the influence of humans (Mallory, unpublished data). We can limit human impact on approved cave pool drinking sources by using clean pitchers to collect water. In the process keep fingers out of the water. Siphons, which eliminate the need to approach the pool edge, may be an even better solution (Hildreth-Werker and Werker 1997). Carefully choose tubing and siphon materials to prevent the leakage of plasticizers into pools (Hunter and others 2004).

Keep boots away from pool edges to help prevent the introduction of mud, surface bacteria, and other foreign substances.

It is especially critical to prevent any water used for cleaning cave formations from draining back into pools.

Cave pools can be so beautiful and inviting. We just want to wade in and wash off all of that dirt, sweat, and grease—*please don't.* You harbor an amazing Noah's Ark of microbes that are native to human skin and hair. As you swim, literally countless bacteria will swirl off your body and into the pristine waters of the pool to be loosed on the unsuspecting native cave microbes. In addition, the organic compounds that wash off you (remember the sweat) will provide the invaders with a head start toward out-competing the natural microbial populations. Humans are giant microbial incubators, teeming with millions of bacteria and fungal spores living on giant hunks of organic matter (Mosberg 2006, page 276 in this volume). Please keep your microbial flora to yourself until you get into your shower at home.

Keep boots away from pool edges to help prevent the introduction of mud, surface bacteria, and other foreign substances.

Red Lake: No Good Solutions

What do you do when you may have a seriously contaminated area in a cave? This is the problem at Red Lake in Lechuguilla Cave, New Mexico. It has been used as one of the water sources for cavers on multi-day underground expeditions. An incident in which a number of people became ill led to our assaying the pool and surroundings for the presence of organisms that indicate contamination from feces and urine (Thompson and Boston 1997; Boston 1998). Previous comparisons of trails and camps to less impacted sites had shown that human-associated bacteria could be successfully confined to heavy use areas (Northup and others 1997a). Was something different about the Red Lake situation?

We performed a rapid series of screening experiments to look for *coliforms*. These bacteria inhabit the human intestine and are used as standard indicators of fecal contamination of water. The initial screening was positive. We then did a more complete study laying out a transect of sample sites through puddles and surfaces on route from the water source to the collection spigot at the end of a siphon tube. Again, our results showed coliform contamination of all surfaces and puddles except the main pool. We were relieved that the primary water body was uncontaminated. However, mud from boots was found along the top of a flowstone slope that leads down to the main pool. Streaks of mud were already visibly making their way to the pool. We immediately feared that contamination of the main pool might be imminent from the transport of mud with attached organisms draining down the wet flowstone and into the water.

What to do? We viewed this as an emergency operation to stop further contamination, thus justifying greater intervention than we would normally use. Using sponges to soak up all cleaning waters and hauling the dirty water away from the site, we cleaned the contaminated areas as best we could. The site was then closed to all human use, hoping that the cleaning and lack of new organic input would allow the coliforms to die off. We tested the site six months later and it was still positive for coliforms. We tested again another eight months later. Alas! The Red Lake area was still positive—as well, the main pool also tested positive for coliforms.

Why have the coliforms not died off in spite of lacking the rich organic food they were used to in the human digestive tract? At this writing, we do not have the answer and are still cautiously wrestling with this problem (Hunter 2001; Hunter and others 2004). Our working hypothesis is that the sticky, ultra fine-grained soils of the nearby Huapache Camp area and urine dumping spot have two properties that promote the persistence of coliform bacteria in the vicinity of Red Lake.

The only guides that can be used are knowledge of the situation, the minimum possible intervention, and (above all) caution.

First, the stickiness and texture of the soil make it adhere very well to boots and all other surfaces; hence it is very easy to track this material around. Second, it is known that clays protect both microorganisms and organic compounds from destruction by other factors in the environment. One or both of these properties (or something else that we haven't thought of yet) is preventing the loss of coliforms that we expected based on their behavior in surface systems. Study results by A. Hunter (2001; Hunter and others 2004) show that the particular type of tubing once used to siphon drinking water from the main pool was providing a home for filamentous *Hyphomicrobium*-like microorganisms. In turn, these organisms may have been involved in providing nutrients or other conditions suitable to the protection and growth of the contaminating coliforms.

Our current approach is to continue testing the Red Lake area until we find the area coliform free. At that point, we hope to recommend reopening the area for at least limited use. We will continue to monitor the coliform situation to see if recontamination occurs.

So, what is the moral of this story? Deciding what to clean and restore is context dependent. The only guides that can be used are knowledge of the situation, the minimum possible intervention, and (above all) caution.

Mammoth Cave Entrance Biomonitoring Study

Mammoth Cave, the longest cave in the world, has many entrances. Most of the entrances have been modified or are artificial, and have existed for decades. These entrances have caused the influx of cold, dry air in the winter, resulting in significant drying effects on formations, walls, and biota.

In a major restoration effort conducted by the National Park Service, entrances to Mammoth Cave have been retrofitted to restore more natural climatic conditions that are favorable to native invertebrates and microorganisms. Some entrances were fitted with airlock doors, others with baffles. Poulson and others (1996) determined baseline numbers, types, and distribution of organisms near entrances before and after retrofitting. While the main focus of the study was on the keystone species,

Hadenoecus subterraneus (a species of cave cricket), results can be extended to most organisms found near entrances. In most cases, the retrofitting had a positive or at least neutral effect on populations of cave crickets. In only one case, the effect was negative. Entrances became much warmer and wetter, which would also contribute to greater success for actinomycete bacteria and other microbes.

Restoration efforts like entrance retrofitting can have profound and complex effects on cave biota. It must be thoroughly thought through and studied before and after to monitor possible changes, either beneficial or deleterious.

Cave crickets have active populations of symbiotic microbes in their digestive tracts. The microbes produce enzymes that break down food for the cricket. In turn, the carcasses of deceased crickets provide organic matter for many microorganisms, especially fungi (Figure 4a). We know that cricket guano communities are very sensitive to external weather conditions. Influx of cold, dry air causes seasonal changes in organisms that colonize guano deposits (Poulson and others 1995). The moral of the tale is that restoration efforts like entrance retrofitting can have profound and complex effects on cave biota. It must be thoroughly thought through and studied before and after to monitor possible changes, either beneficial or deleterious.

Cave Pearls: When You Just Don't Know Enough

The last of our case studies describes attempts to decide what to do about cleaning cave pearls. These gloriously beautiful cave formations often become soiled by mud and debris that falls from boots and gear. Because the pearls occur in their own little depressions (nests), they sometimes become surrounded by a halo of dirt particles. The ring of dirt spoils their looks and can become incorporated into the pearls if they are still actively forming. Because cave pearls are such a beautiful and precious type of speleothem, there is temptation to go at it immediately with enthusiasm and get rid of that gunk.

However, tantalizing new evidence is pointing to the possible role of microbial biofilm in the very creation of these treasures (Gradzinski 1999). As scientists, we know it is much too premature to conclude that this is true. The evidence is not yet at hand. However as restorationists, if it is true, we certainly don't want to interfere with cave pearl formation while trying to make them look cleaner.

We have been wrestling with what to do in pristine situations and what to advise others. We are still in the experimental stages. So far, we think that completely sterile equipment should be used to remove the natural waters surrounding the pearls, and the pearl water should be placed in a sterile container. The dirt should be removed with sterile equipment (for example, sterile syringes and tweezers).

The *pearls should not be scrubbed* with anything for fear of disrupting the microbial biofilm that may be surrounding them. Importantly, pearls should not be touched with bare hands, but only with sterile surgical gloves. After the pearl nests are cleaned, the natural waters should be reintroduced (Hildreth-Werker 2006, page 425 in this volume). Does this all seem like a huge amount of work? It is. And we don't even know if it is necessary until we know more about the science.

What to do about this problem in your own restoration efforts? Choices in cave pearl restoration depend on the degree of site impact. We think it may be better to leave pristine cave pearls alone rather than use nonsterile equipment or harsh physical cleaning techniques.

On the other hand, some impacted pearl nests may require gentle restoration intervention to prevent further damage. As the scientific understanding of cave pearls increases, we can incorporate new knowledge into restoration work.

Special Techniques

The most important step in any restoration project is planning. Consider possible impacts to microbes as an important component of planning. What are the possible connections between contaminated areas and nearby passages? What is the safety risk to cavers when using any cleaning technique? Will restoration efforts increase contamination at the site? Will the area recover from cleaning efforts? Will any item of historical significance be lost? Plan carefully and consult experts as appropriate.

What are the possible connections between contaminated areas and nearby passages?

Sterilizing Tools

Use sterile tools and materials and make every effort to avoid introducing nonresident microbes into sensitive cave areas. Tools can be sterilized at home without a medical or laboratory *autoclave* (steam sterilizer) if you have an ordinary household pressure cooker:

- Prewrap the items in aluminum foil.
- Treat items to 15 minutes of steam under a pressure of 103.4 kPa (15 psi) to kill all microbes and resistant bacterial spores.
- The items will remain sterile in their aluminum foil wrappings as long as they are not opened or punctured.
- Instruments must be carefully unwrapped and touched only on the handle.
- Sterilization will allow you to work on the cave feature of interest without introducing contaminant bacteria.

Many plastic items will melt and deform if steam sterilized. Test for tolerance to this autoclaving treatment by wrapping an item securely in aluminum foil and subjecting it to pressure cooking. If the item melts, at least the mess will be contained in disposable foil.

Materials like Nalgene® are autoclavable. Nalgene containers are commonly sold at outdoor equipment stores and through laboratory suppliers. Pyrex® cookware can also be sterilized by this method.

Obtain presterilized syringes through medical supply houses or pharmacies. The paper wrappings on syringes must be kept dry to maintain sterility. Note also that syringes will melt and cannot be resterilized as described above.

The most important step in any restoration project is planning. Consider possible impacts. Plan carefully and consult experts as appropriate.

For sensitive applications, sterile surgical gloves are available through medical and pharmaceutical sources. Always use powder-free, nonlatex (for example, nitrile) surgical gloves for cave applications. (See gloves, page 433.)

Tools must be cleaned between uses. To clean organic material from restoration tools, wash with water and then soak in chlorine bleach (contact time is vital for disinfection). *But* remember to rinse many times before ever introducing those tools back into a cave again. Realize that the disinfected tools are *not* sterile.

Tools can also be cleaned in the cave by soaking in alcohol for at least 10 minutes. However, alcohol exposure does *not* sterilize things.

To sterilize metal tools, fully immerse the end that will come in contact with the cave surface into ethanol, grain alcohol, or Ever-Clear®. Ignite with a lighter or match. The surface will be sterilized by the process of burning the ethanol. Be careful that the alcohol does not drip onto skin or clothes and ignite, and hold tools so that the alcohol will not run backwards onto the hand.

Water Filtration and Replacement

Occasionally, water is removed from puddles or pools to remediate visible mudflow, sediment from boots, or debris from other unnatural sources

(Werker and Hildreth-Werker 2006, page 415 in this volume). After careful consideration, if water must be removed from a cave pool in an otherwise pristine area, sterile procedures and equipment are best. Contaminated water may need to be removed from the cave. If removal is not an option, then dispose of the contaminated water in a well-inspected area away from the restoration site. If possible allow cave pools to refill with their own natural drips.

When restoring cave passages, please use patience and watch out for our unseen friends. Although in daily life we give little heed to the myriad microbes that share our planet, in caves their role is crucial and irreplaceable.

Protective Garments

Wear clean clothes, clean boots or flowstone shoes, and powder-free, nonlatex surgical gloves to reduce the possibility of contaminating the site. Depending on the nature of the work, a surgical mask may be recommended. Consider wearing a HEPA filter face mask for your protection, especially if working around mold or other fungi, guano, bird dropping deposits, or in areas with known histoplasmosis. (See protocol, page 427.)

Protect the Unseen

When restoring cave passages, please use patience and watch out for our unseen friends. Although in daily life we give little heed to the myriad microbes that share our planet, in caves their role is crucial and irreplaceable. Remember, the microbes you care for today may be making the speleothems of tomorrow!

Acknowledgements

Thanks go to Dr. Lynn Kleina and anonymous reviewers for reviewing the manuscript. We are grateful to Michael Spilde for scanning electron micrographs of microorganisms. Thanks go to Sherry Cady, Kevin Glover, Val Hildreth-Werker, Louise Hose, Kenneth Ingham, David H. Jagnow, Larry McLaughlin, Doug Soroka, and the family of Kiym Cunningham for permission to use photographic images. We are grateful to the National Park Service at Mammoth Cave, Carlsbad Caverns, and Lechuguilla Cave for their cooperation and enthusiastic interest in our findings. We extend gratitude to the National Speleological Foundation for supporting water contamination research in Lechuguilla Cave. Diana E. Northup extends thanks to the Lindbergh Foundation for support of the trails and camps study. We thank Andi Hunter for her follow-up study of the coliform contamination in Lechuguilla as part of her graduate work. Lastly, we are grateful for help from Jim C. Werker, Val Hildreth-Werker, and other cavers on our own expeditions who have facilitated our study of the scientific issues involved in cave restoration.

Cited References

Amy PS, Haldeman DL. 1997. *Microbiology of the Terrestrial Subsurface*. Boca Raton (FL): CRC Lewis Publishers. 356 p.

Angert ER, Northup DE, Reysenbach A-L, Peek AS, Goebel GM, Pace NR. 1998. Molecular phylogenetic analysis of a bacterial community in Sulphur River, Parker Cave, Kentucky. *American Mineralogist* 83(11 pt 2):1583-1592.

Banfield JF, Nealson KH. 1997. Geomicrobiology: Interactions between microbes and minerals. *Reviews in Mineralogy Vol 35*. Washington DC: Mineralogical Society of America. 448 p.

Ben-Ari ET. 1999. Not just slime: Beneath the slippery exterior of a microbial biofilm lies a remarkably organized community of organisms. *Bioscience* 49(9):689.

Boston PJ. 1998. *Annual Report to the Cave Resources Office, Carlsbad Caverns National Park, Carlsbad, NM.* Unpublished technical report. 10 p.

Boston PJ. 2006. To bleach or not to bleach: Algae control in show caves. In: Hildreth-Werker V, Werker JC, editors. *Cave Conservation and Restoration*. Huntsville (AL): National Speleological Society. p 349-350. [This volume.]

Boston PJ, Soroka DS, Kleina LG, Lavoie KH, Spilde MN, Northup DE, Hose LD. 2000. A garden inside out: Microbial mats in springs, wall muds, and ceiling formations of a sulfur-dominated cave, Cueva de Villa Luz, Tabasco, Mexico. *Journal of Cave and Karst Studies* 62(3):198-9.

Boston PJ, Spilde MN, Northup DE, Melim LA, Soroka DS, Kleina LG, Lavoie KH, Hose LD, Mallory LM, Dahm CN, Crossey LJ, Schelble RT. 2001. Cave biosignature suites: Microbes, minerals and Mars. *Astrobiology Journal* 1(1):25-55.

Costerton JW, Lewandoski Z, DeBeer D. 1994. Biofilms, the customized microniche. *Journal of Bacteriology* 176:2137-42.

Cunningham KI. 1991. Organic and inorganic composition of colored corrosion residues: Lechuguilla Cave: Preliminary report. *NSS News* 49(8):252-254.

Cunningham KI, Northup DE, Pollastro RM, Wright W, LaRock EJ. 1995. Bacteria, fungi, and biokarst in Lechuguilla Cave, Carlsbad Caverns National Park, NM. *Environmental Geology* 25:2-8.

Elliott WR. 2006. Biological dos and don'ts for cave restoration and conservation. In: Hildreth-Werker V, Werker JC, editors. *Cave Conservation and Restoration*. Huntsville (AL): National Speleological Society. p 33-42. [This volume.]

Engel AS, Lee N, Porter ML, Stern LA, Bennett PC, Wagner M. 2003. Filamentous *Epsilonproteobacteria* dominate microbial mats from sulfidic cave springs. *Applied and Environmental Microbiology* 69(9):5503-5511.

Engel AS, Porter ML, Kinkle BK, Kane TC. 2001. Ecological assessment and geological significance of microbial communities from Cesspool Cave, Virginia. *Geomicrobiology Journal* 18(3):259-274.

Gradzinski M. 1999. Role of microorganisms in cave pearl formation. *Journal of Conference Abstracts, 11th Bathurst Meeting July 13th-15th, 1999.* Cambridge, UK 4. p 924.

Grobbelaar JU. 2000. Lithophytic algae: A major threat to the karst formation of show caves. *Journal of Applied Phycology* 12:309-315.

Hildreth-Werker V, Werker JC. 1997. Lechuguilla restoration: Techniques learned in the Southwest. *NSS News* 55(5):107-113.

Hildreth-Werker V. 2006. Restoration of delicate speleothems and cave pearls. In: Hildreth-Werker V, Werker JC, editors. *Cave Conservation and Restoration*. Huntsville (AL): National Speleological Society. p 423-426. [This volume.]

Hildreth-Werker V. 2006. Mold kits. In: Hildreth-Werker V, Werker JC, editors. *Cave Conservation and Restoration*. Huntsville (AL): National Speleological Society. p 413-414. [This volume.]

Horrocks RD, Ohms M. 2006. Cave lint and dust removal projects. In: Hildreth-Werker V, Werker JC, editors. *Cave Conservation and Restoration*. Huntsville (AL): National Speleological Society. p 351-362. [This volume.]

Hose LD, Palmer AN, Palmer MV, Northup DE, Boston PJ, Duchene HR. 2000. Microbiology and geochemistry in a hydrogen sulfide-rich karst environment. *Chemical Geology* 169:399-423.

Hunter AJ. 2001. *Environmental Disturbance of Oligotrophic Bacteria and Effects on Water Quality in Deep Karstic Pools.* Unpublished professional project report, Water Resources Program, University of New Mexico.

Hunter AJ, Northup DE, Dahm CN, Boston PJ. 2004. Persistent coliform contamination in Lechuguilla Cave pools. *Journal of Cave and Karst Studies* 66(3):102-110.

Lavoie KH. 1980. Toxicity of carbide waste to heterotrophic microorganisms in caves. *Microbial Ecology* 6:173-9.

Lavoie KH. 1995. The effects of urine deposition on microbes in cave soils. In: Pate DL, editor. *Proceedings of the 1993 National Cave Management Symposium: Carlsbad, New Mexico, October 27-30, 1993.* [Huntsville (AL)]: National Cave Management Symposium Steering Committee. p 302-11.

Lavoie K, Northup D, Boston P, Blanco-Montero C. 1998. Preliminary report on the biology of Cueva de Villa Luz, Tabasco, Mexico. *Journal of Cave and Karst Studies* 60(3):180.

Lavoie K, Northup D, Boston P. 2000. Sight unseen: Microbes in caves. *NSS News* 58(3):68-9.

Makela PM, Korkeala HJ, Sand EK. 1991. Effectiveness of commercial germicide products against the ropy slime-producing lactic acid bacteria. *Journal of Food Protection* 54(8):632-636.

Mattison RG, Abbiati M, Dando PR, Fitzsimons MF, Pratt SM, Southward AJ, Southward EC. 1998. Chemoautotrophic microbial mats in submarine caves with hydrothermal sulphidic springs at Cape Palinuro, Italy. *Microbial Ecology* 35:58-71.

Melim LA, Shinglman KM, Boston PJ, Northup DE, Spilde MN, Queen JM. 2001. Evidence for microbial involvement in pool finger precipitation, Hidden Cave, New Mexico. *Geomicrobiology Journal* 18(3):311-329.

Mosberg SR. 2006. Health and hygiene related to cave conservation. In: Hildreth-Werker V, Werker JC, editors. *Cave Conservation and Restoration.* Huntsville (AL): National Speleological Society. p 271-280. [This volume.]

Needham C, Hoagland M, McPherson K, Dodson B. 2000. *Intimate Strangers: Unseen Life on Earth.* Washington DC: American Society for Microbiology. 225 p.

Northup DE, Carr DL, Crocker MT, Cunningham KI, Hawkins LK, Leonard P, Welbourn WC. 1994. Biological investigations in Lechuguilla Cave. *National Speleological Society Bulletin* 56(2):54-63.

Northup DE, Welbourn WC. 1995. Conservation of invertebrates and microorganisms in the cave environment. In: Pate DL, editor. *Proceedings of the 1993 National Cave Management Symposium: Carlsbad, New Mexico, October 27-30, 1993.* [Huntsville (AL)]: National Cave Management Symposium Steering Committee. p 292-301.

Northup DE, Beck KM, Mallory LM. 1997a. Human impact on the microbial communities of Lechuguilla Cave: Is protection possible during active exploration? *Journal of Cave and Karst Studies* 59(3):166.

Northup DE, Reysenbach A-L, Pace NR. 1997b. Microorganisms and speleothems. In: Hill CA, Forti P, editors. *Cave Minerals of the World.* 2nd edition. Huntsville (AL): National Speoleological Society. p 261-266.

Northup DE, Dahm CA, Melim LA, Spilde MN, Crossey LJ, Lavoie KH, Mallory LM, Boston PJ, Cunningham KI, Barns SM. 2000. Evidence for geomicrobiological interactions in Guadalupe caves. *Journal of Cave and Karst Studies* 62(2):80-90.

Northup DE, Lavoie KH. 2001. Geomicrobiology of caves: A review. *Geomicrobiology Journal* 18(3):199-220.

Olson R. 2006. Control of lamp flora in developed caves. In: Hildreth-Werker V, Werker JC, editors. *Cave Conservation and Restoration*. Huntsville (AL): National Speleological Society. p 343-348. [This volume.]

Poindexter JS. 1981. Oligotrophy: Feast and famine existence. *Advances in Microbial Ecology* 5:63-89.

Poulson TL, Keith JH, Lavoie KH. 1986. Biological deserts under the caprock. *1985 Annual Report Cave Research Foundation.* p 24-25.

Poulson TL, Lavoie KH, Helf K. 1995. Long-term effects of weather on the cricket (*Hadenoecus subterraneus,* Orthoptera, Rhaphidophoridae) guano community in Mammoth Cave National Park. *American Midland Naturalist* 134(2): 226-236.

Poulson T, Lavoie K, Helf K. 1996. Effects of natural changes and entrance retrofitting on numbers and sizes of cave crickets. *Proceedings of the 5th Mammoth Cave National Park Science Conference: Science in the National Parks*. Mammoth Cave National Park (KY): Mammoth Cave National Park.

Sarbu SM, Kane TC, Kinkle BK. 1996. Chemoautotrophically based cave ecosystem. *Science* 272:1953-1955.

Sarbu SM, Galdenzi S, Menichetti M, Gentile G. 2000. Geology and biology of Grotte di Frasassi (Frasassi Caves) in Central Italy, an ecological multi-disciplinary study of a hypogenic underground karst system. In: Wilkens H, Culver DC, Humphreys WF, editors. *Subterranean Ecosystems*. Ecosystems of the World Vol 30. Amsterdam: Elsevier. p 359-378.

Sasowsky ID, Palmer MV. 1994. *Breakthroughs in Karst Geomicrobiology and Redox Geochemistry: Abstracts and Field-Trip Guide for the Symposium held February 16 through 19, 1994, Colorado Springs, Colorado.* Special Publication 1. Charles Town (WV): Karst Waters Institute. 111 p.

Schaechter E. 1997. *In the Company of Mushrooms: A Biologist's Tale.* Washington DC: American Society for Microbiology. 320 p.

Shannon MJR, Unterman R. 1993. Evaluating bioremediation. *Annual Review of Microbiology* 47:715-730.

Southward AJ, Kennicutt MC II, Alcalà-Herrera J, Abbiati M, Airoldi L, Cinelli F, Bianchi CN, Morri C, Southward EC. 1996. On the biology of submarine caves with sulphur springs: Appraisal of $^{13}C/^{12}C$ ratios as a guide to trophic relations. *Journal of the Marine Biological Association of the UK* 76:265-285.

Spilde MN, Brearley AJ, Papike JJ. 2001. Mn-oxide minerals from a terrestrial cave environment: Biomarkers for the search for life on Mars? *Lunar and Planetary Science, XXXII; Papers Presented to the Thirty-second Lunar and Planetary Science Conference; Abstracts of Papers Submitted to the Lunar and Planetary Science Conference.* 32:1454.

Stotzky G, Norman AG. 1964. Factors limiting microbial activities in soil. *Canadian Journal of Microbiology* 10:143-149.

Thompson SL, Boston PJ. 1997. *Urine Mitigation with Concurrent Power Generation for Lechuguilla Cave.* Unpublished technical report to Carlsbad Caverns National Park, Carlsbad, NM.

Wackett LP, Hershberger CD. 2001. *Biocatalysis and Biodegradation: Microbial Transformation of Organic Compounds*. Washington DC: American Society for Microbiology. 250 p.

Watson PJ. 1974. *Archaeology of the Mammoth Cave Area*. New York: Academic Press.

Werker JC and Hildreth-Werker V. 2006. Water filtration and cave pool restoration. In: Hildreth-Werker V, Werker JC, editors. *Cave Conservation and Restoration*. Huntsville (AL): National Speleological Society. p 415-418. [This volume.]

Williams HN, Quinby H, Romberg E. 1994. Evaluation and use of a low nutrient medium and reduced incubation temperature to study bacterial contamination in the water supply of dental units. *Canadian Journal of Microbiology* 40(2):127-131.

Wingender J, Griebe T, Flemming H-C. 1998. The role of biofilm and slime formation and the effect of chlorine, monochloramine, hydrogen peroxide and silver on *Pseudomonas aeruginosa*. *Water Supply* 16 (3/4):207-12.

Section A—Identifying and Protecting Cave Resources

Paleontology in Cave Conservation and Restoration

Rickard S. Toomey, III

Caves often serve as natural archival vaults and can protect valuable scientific information through the ages. Because a cave may contain potentially significant paleontological remains, it is important to consider these resources when planning and executing restoration or conservation projects.

> Caves often serve as natural archival vaults and can protect valuable scientific information through the ages.

What Are Paleontological Resources?

Remains or traces of once-living organisms are referred to as fossils or paleontological remains (paleo-remains). In this chapter, the terms fossils and paleontological remains are used interchangeably. Any kind of biota—plants, protists, or animals—may be fossilized.

When people think of fossils, they often think of bones, but bones are only one category among many varieties of fossils. Any animal or plant parts can be fossilized—teeth, shells, carapaces, leaves, pollen, or stems. Traces of animal activities, such as footprints, scratch marks, nests, middens, feces, or bite marks, can also remain behind as fossils.

Since a cave system may contain remains from several eras, paleontological resources may range in age from hundreds to hundreds of millions of years. For example, shells on a cave wall from an ancient seabed may be found a few feet from mammal skeletons that are thousands of years old or beside bat bones dating from a few hundred years.

Paleo-remains do not include materials that are associated with human activity. Artifacts and traces left by humans are classified either as archaeological remains or as contemporary trash. Archaeological materials sometimes provide significant information for anthropological studies of prehistoric and historic human activities. Thus, both archaeological and paleontological materials should be treated with care and respect.

Cavers often are the first people to encounter paleontological and archaeological materials in caves. Thus, cavers frequently initiate the first steps in protecting these sometimes irreplaceable, often fragile resources. For more information on the conservation and restoration implications of archaeological remains in caves, consult the archaeology chapter of this volume (page 93).

> Fossils are the primary sources of information on biodiversity and ecosystems of the past.

Why Are Fossils Important?

Fossils are important because they are the primary sources of information on biodiversity and ecosystems of the past. Scientists study fossils to learn how our planet has changed through time. Through fossils, we discover information about past plants, animals, climates, vegetation, and environments. In addition, paleo-remains can sometimes *tell time* geologically—for example, some fossils can be analyzed to determine the age of the rocks and sediment in which they occur. In a speleological context, experienced researchers may study paleo-remains to determine details about the developmental history of a cave, such as the presence of past entrances or patterns of past cave temperatures.

Paleo-remains are important considerations in cave conservation and restoration projects. The first reason may be obvious—the remains are nonrenewable cave resources and may be impacted by caving activities. Any remains should be considered, assessed, and managed in plans for cave restoration and conservation. The second reason is more subtle—recent paleo-remains may be used to set important targets for cave restoration, especially environmental restoration. Also, the presence and significance of paleo-remains may be a factor in determining whether a cave has significant resource value and in developing access policies and procedures. Cavers should become aware of the types of remains harbored in caves. Through education, cavers can avoid causing damage to remains, can help identify paleontological and archaeological resources, and can provide important information to experts who evaluate and study the remains.

Archived Antiquity
James F. Baichtal

A thought has come to me about the antiquity of the surfaces on which we cave. Excavating the bones of mammals that date from 8,000 to beyond 50,000 years has driven this home.

Those surfaces that we decide to track across or crawl through may be thousands if not hundreds of thousands of years old. We, as the first ones to go there, have an immense responsibility to carefully choose how we proceed (if at all) lest important data be lost.

I have found the places dug out by denning bears and the tracks of bears deep within caves.

I have seen soda straws broken from tight passages where the 12,000-year-old remains of a brown bear were later discovered; those formations he most likely broke on his last quest for a denning site.

I have seen the footprints of deer trapped in vertical pits where they eventually perished. What about hair from these animals? Is ancient data archived in the DNA of a cell or a microbe?

What can be learned from the charcoal torch of a past explorer?

What about the bones of small mammals?

These all are easily overlooked, broken, displaced, and sometimes erased by our exploration efforts.

We really need to thoughtfully consider what awaits us in these passages. We need to find ways to explore in a manner that ensures these resources are not lost.

Types of Remains Found in Caves

Some fossilized remains are found in the rocks in which caves form, while other paleo-remains consist of the fossils of organisms that entered after the cave formed. Fossils from the rocks are often found in the walls and ceilings of caves. The fossils of organisms that entered after the cave formed are found along the passages or in the sediments of the cave. Cavers should learn to recognize the paleo-remains typically found in their caving regions.

Fossils found in rocks are generally the same age as the rocks in which the cave formed—thus, fossils in cave rocks are usually much older than the cave itself.

Most caves form in limestone, and since limestone originally formed on the floors of ancient seas, fossils of sea creatures are often harbored in caves. Cavers often find the fossilized remains of marine animals such as coral, brachiopods, crinoids, clams, bryzoans, and other common sea-dwelling groups.

Less commonly, cavers may encounter rarer remains such as those of sharks or even dugongs in the walls of limestone caves. Typically these fossils are exposed in cave walls where the limestone was dissolved away during cave genesis.

Cavers also find fossil trees and other remains in lava flows where caves have formed. An entire fossilized rhinoceros was found in a lava cave.

As well, cavers occasionally find plant and animal fossils in beds of gypsum.

The other type of paleo-remains consists of organisms that entered sometime after the cave formed. These remains can be incorporated in the sediments of the cave or can be found on the surfaces of cave walls, floors, and ceilings, and in rocks that have spalled off the ceiling. Examples include bones, mummified remains, bat guano, ceiling stains from bat use, feces of various animals, middens, nests, footprints, animal scratches, plant parts, snail shells, parts from insects and other invertebrates (including cave-dwelling invertebrates like mites). Remains in this category are almost always much, much younger than the cave in which they are found.

Although both classes of remains should be treated as potentially significant, the first type is not unique to caves. Similar fossils can be found in surface exposures of the same limestone outside of the caves. However, the fossils protected by caves may be better preserved than those found on the surface since cave-forming processes can expose the fossils to less damage than surface processes. The second class of remains is unique to caves since their presence depends on the existence of the cave.

How Do Fossils End Up in Caves?

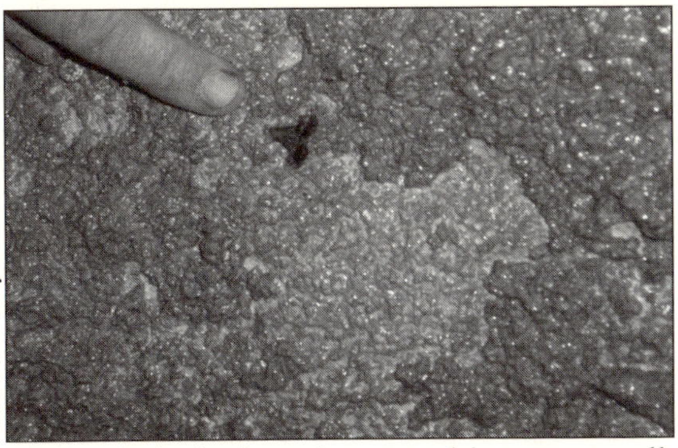

Rickard S. Toomey III

Figure 1. A shark tooth exposed in a cave wall. Fossils like this one can be impacted when cleaning cave walls.

© Val Hildreth-Werker

Figure 2. Tiny limestone caps top miniature mud hoodoos found on the floor of La Cueva de las Barrancas.

© Val Hildreth-Werker

Figure 3. In the lab, scanning electron microscopy of this hoodoo capstone from Barrancas revealed the miniscule marine fossils in Figure 4.

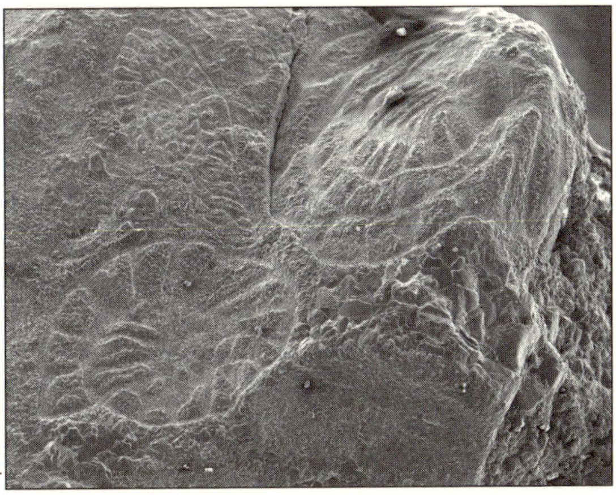

Spilde & Boston

Figure 4. Three Permian-age *Formaminifera* fossils decorate the top of a small piece of limestone that caps a miniature hoodoo-like structure from the mud floor in La Cueva de las Barrancas. The width of the electron micrograph is 1.8 millimeters.

Rick Olson

Figure 5. Gray fox skeleton on the floor of a cave alcove. Skeletal remains such as these can be found in many caves.

Rick Olson

Figure 6. Woodrat skull in a woodrat latrine. The nests and middens of woodrats often contain important paleontological remains near cave entrances. (See page 5 of color section.)

Rickard S. Toomey III

Figure 7. Raccoon skeleton on the floor of a cave passage. Unfortunately, this skeleton was stepped on by a careless caver.

Bones and other materials enter caves in a variety of ways. Some animals live in caves—bats, cave rats, a few carnivores, and various cave-dwelling invertebrates. When they die, their remains can become fossilized. While they are alive, animals also do things that produce traces of their activities. They may leave behind nests and beds, guano and feces, or staining and scratches on walls and ceilings. Animals may accidentally wander into a cave and die. They often leave remains that are more commonly found near pits or other hazards. Carnivores might drag their prey into caves and leave the remains of their dinners behind. Some cave-dwelling rodents—for example, woodrats, packrats, and porcupines—drag bones from outside into the cave. Or, animal and plant remains may be washed into a cave from the surface. These remains would be most common near an entrance, sinkhole, or other water inlet.

Understanding some of the factors that contribute to the scientific importance of fossils should help cavers determine what kind of infor-mation needs to be recorded or protected.

Where are Fossils Found in Caves?

As noted above, fossils of animals from the cave rock are most commonly found exposed on the cave walls and ceilings. In addition, some may weather out of the rock and may be incorporated into the cave sediment. The fossils of animals that enter the cave after it formed can be found in a somewhat wider variety of places. Bats and cave invertebrates are often found more widely ranging in caves than animals that are not cave adapted. However, following are some generalizations about where to find various animal remains.

Remains of animals such as woodrats, raccoons, bears, surface snails, insects, and other cave accidentals, are often found in caves, as are their nests and activity areas. Groups of bones or entire skeletons may remain, and sometimes mummified bodies are found in drier caves and passages. Rarely, a cave area will contain bones of a large number of animals, either one species or several. Larger animal bones and nests are usually found near modern entrances in the twilight area, or into the dark zone, or in the vicinity of ancient entrances. Occasionally, isolated skeletons are found in passages far from any known entrance, and these may represent individual animals that got lost and died. Individual bones can be present anywhere in a cave, especially if moved by water. Footprints and other trace evidence—scratchings, claw marks, bear beds, feces—are fairly rare, but can be very important.

Bats are among the most common of cave-dwelling animals; bats are also among the animals that range most widely through caves. For these reasons, it is not unusual to find bat bones and other bat-related materials in many areas of caves. Bones from multiple species are often found in roost sites, which may be modern or ancient.

Fossils of animals from the cave rock are most commonly found exposed on the cave walls and ceilings. In addition, some may weather out of the rock and may be incorporated into the cave sediment.

However, individual bat skeletons can occur anywhere in a cave. Skulls from individual bats are frequently seen in caves, but the small wing and body bones are often overlooked or have deteriorated.

Guano and roost staining are common in many caves, especially in defined areas of large caverns. (Figure 14.) (Also see Figure 1, page 150.)

Bat scratches on ceilings may be encountered deep into caves. These scratches, which are caused by bats trying to roost on soft ceiling rock, are very fragile and can easily be rubbed away.

A variety of cave invertebrates are found throughout many caves. These animals die in caves, and their bodies can be incorporated into cave sediments. These animals are so small and delicate they usually are not preserved as fossils, or if preserved, the remains are hard to find. However, several researchers have recently identified fossils of cave dwelling mites preserved in flowstone (Davis 1999; Polyak and others 2001).

Overall, paleo-remains are found in a variety of locations within caves. Some of the most common settings include the following list:

- Talus cones or sediments associated with current or former entrances

© Val Hildreth-Werker

Figure 8. This Pleistocene bat, discovered during a bat skull study, is well preserved in Land of Awes, Lechuguilla Cave. (See page 5 of color section.)

© Val Hildreth-Werker

Figure 9. Wearing flowstone shoes with covers, Tom Madison carefully makes only two points of contact off trail to return a bat skull that was collected during a scientific study in Lechuguilla Cave. The research proposal called for returning the skulls to original locations after the study was complete.

© Val Hildreth-Werker

Figure 10. The caver in the previous photo returned the skull of Bat #155 to its original position in Lip Service, Lechuguilla Cave, New Mexico.

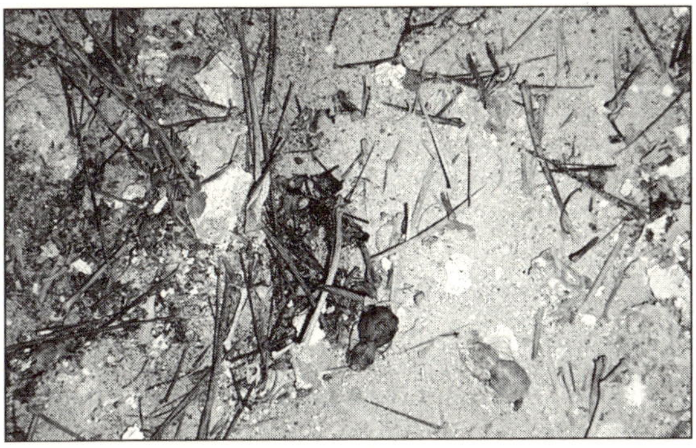

© Val Hildreth-Werker

Figure 11. These bat bones are clearly visible against the white calcite surface where they were found in Ogle Cave, Carlsbad Caverns National Park, New Mexico. (See page 5 of color section.)

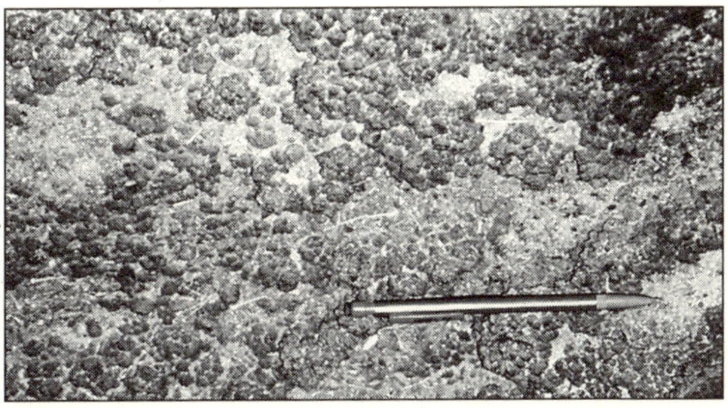

Rickard S. Toomey III

Figure 12. Bat bones in small, dry, rimstone pools. Delicate bones such as these can be easily destroyed by careless traffic. (See page 5 of color section.)

Rick Olson

Figure 13. Deposit of bat bones exposed in an old dig site in a Kentucky cave. The dig destroyed most of the deposit, but small portions remained for study.

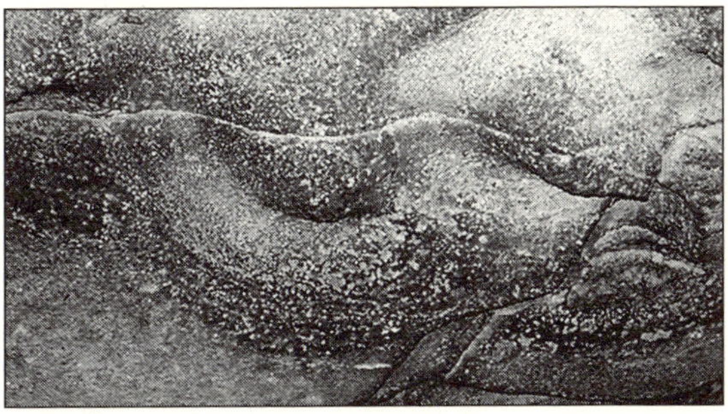

Rick Olson

Figure 14. Bat stain is visible on a cave ceiling. (See page 5 of color section.)

- The bottom of shafts and pits
- Beneath vertical cave entrances—at the bottom of sinks, pit entrances, and vertical drops
- Stream beds where sediments are deposited and in streams large enough to carry gravel
- Current or former bat roosts or animal dens
- Undisturbed passages with floors of fine-grained sand or firm mud where footprints are most commonly found
- Flowstone and dripstone deposits

How Can Cavers Protect Paleo-Resources?

All paleontological remains are not equally significant. From scientific, educational, and cave management viewpoints, some remains are highly significant while others are not particularly vital. In general, it will take a paleontologist, or caver experienced in paleontology, to assess the significance of any particular set of paleo-remains. It is prudent to treat any remains as potentially significant unless an expert informs you otherwise.

It is prudent to treat any remains as potentially significant unless an expert informs you otherwise. Paleontologists evaluate a variety of factors to determine whether a particular set of paleo-remains is significant.

Paleontologists evaluate a variety of factors to determine whether a particular set of paleo-remains is significant. Although the following list does not qualify the average person to determine the significance of fossils, it is included so the reader can understand how significance is assessed. Recognizing the factors that contribute to the scientific importance of fossils should help cavers determine what kind of information needs to be recorded or protected. Listed here are evaluation points that contribute to the significance of paleontological remains:

- The type of animal(s) or plant(s) represented
- The abundance of material
- The state of preservation
- The type of preservation
- The context (such as, in place or transported)
- Other deposits in the area
- Age of deposits
- Previous disturbance of the remains
- Whether the fossils address specific research questions

If paleo-remains are encountered in a cave, it is important to consider their protection in continuing cave exploration and in planning for cave conservation or restoration projects. The preservation of paleo-remains in caves starts with proper handling of the situation by the cavers who recognize the remains.

What should cavers do if they encounter potentially significant paleontological remains in a cave? The appropriate course of action is described here:

- Leave the remains in place. Do not handle or touch. Handling may damage the resource or change the orientation of the material.
- Record as much information as possible about the remains and their context.
- Photograph the material in context, if possible. Remember to place an object in the photos for scale.
- Mark the area so that other cavers can avoid damaging the remains.
- Report the material to the individual or group responsible for management of the cave.
- Bring the material to the attention of a paleontologist. You can contact the Paleontology Section of the NSS to find information about paleontologists who may be able to help.
- If bones or other remains are encountered while digging in a cave, discontinue the dig until the material can be evaluated for significance

and a plan for recovery of the material can be prepared, if necessary.

The rationale for leaving remains *in situ* is clear.

- **First**. In most states, it is illegal to remove paleo-remains without permission from the cave owner. If the cave is publicly owned in the United States (that is, managed by a federal, state, county, or local government agency), it is almost certainly illegal to collect the remains. Collecting paleontological materials from any cave is illegal in some countries.

- **Second**. The context in which remains occur is vital to assessing their significance. If the material is removed from its context, it can be difficult or impossible to establish that context later.

- **Third**. Most caving parties are not adequately equipped to safely remove remains and transport them out of the cave. An additional consideration is that removing bones and replacing them later may result in contamination of the bones (and the cave) by foreign microbes.

The more you are able to record about the remains, the better chance a paleontologist will have to properly assess the potential significance.

The following information should be documented in as much detail as is reasonably possible. The more you are able to record about the remains, the better chance a paleontologist will have to properly assess the potential significance:

- Note the location of the remains. Place them on your sketch if you are surveying—otherwise, record their position with respect to nearby survey stations or landmarks.
- Describe the material—bone, guano, footprints, bat staining—in as much detail as you reasonably can.
  - Note size and shape.
  - Sketch material (if possible or useful).
  - Photograph material (if possible or useful). Include an object for scale in the photos.
- Record quantity or extent of material.
- Document context of the find (in sediments, on surface, in niche on wall, in stream).
- State factors that might affect the survival of the material. (Is it in a stable location?)
- Note difficulties or special skills needed to obtain the remains (vertical work, extraordinarily tight passages, sumps, and other situations).

Planning for Paleontological Protection

When paleontological remains are encountered in a cave, it is important to include details for their protection when developing a conservation plan for the cave. An expert should examine the remains to determine how they can best be protected. Remains vary in their sensitivity to damage, and different actions pose varying risks to remains.

Remains vary in their sensitivity to damage, and different actions pose varying risks to remains.

For example, bones contained within a sediment bank may be adequately protected from being damaged by normal caving activity, but they could be impacted by digging. The same bones lying on a surface could be much more easily damaged, possibly even by simple travel in the passage. Bat staining on a ceiling may be fairly stable, but animal scratches on the wall might be easily obscured by any activity that affects the wall.

Other types of materials are often more easily damaged. For example, a single person walking carelessly can obliterate animal footprints on a smooth mud floor. Areas with remains that can be easily damaged should be marked so that people do not travel over those sections. Staying on a designated path minimizes potential damage to most cave resources, including paleo-remains. Protecting caves or cave passages with particularly significant paleo-remains may require restrictions on groups visiting the areas.

Significance of Paleo-Remains When Planning for Restoration

Paleo-remains are rarely a problem in planning cave restoration projects. Before restoration is initiated, the area should be investigated for remains. The floor should be examined to make sure bones, guano, or other remains are not trampled during restoration. With caution, damage can be avoided.

A more difficult situation involves subtle types of paleo-remains like staining and scratches that might occur on cave walls. Before removing graffiti, it is important to determine whether potentially significant animal scratchings are on the walls. This type of evidence can be easily obscured or eliminated by overzealous cleaning.

In caves that have been heavily visited, paleontological remains in the main travel areas may have been eliminated. However, important remains can often still be found undisturbed on ledges, under overhangs, in small notches and anastomoses, and under large pieces of breakdown.

Important remains can often still be found undisturbed on ledges, under overhangs, in small notches and anastomoses, and under large pieces of breakdown.

In a recent paleontological inventory of Mammoth Cave (Mammoth Cave National Park, Kentucky), we found that, although many areas had been visited for almost 200 years by millions of people, significant paleo-remains were still present along the edges of the passages and in areas that had been sheltered from easy direct contact (Toomey and others 2002).

Paleontological remains may also play another role in the restoration of cave habitats. Fossils help establish the historical conditions to which a cave might be restored. If the paleontological materials are comparatively young, less than about 1,000 years old, they can provide significant data on historical conditions that existed in the cave and, thus, information about changes in the cave conditions.

For example, recent remains might indicate that a large bat colony formerly inhabited the cave, and the remains can be studied to identify the type of bats and their season of use. Restoration of the colony might be defined as a target of the cave restoration project and human intrusion might be limited during the most likely season of occupation.

In addition, paleontological remains may provide information on previous meteorological conditions in the cave, such as temperature and airflow. Because many cave animals, particularly bats, choose very specific habitat conditions within a cave, the presence of these animal remains provides strong evidence of former conditions.

If, for example, a bat hibernaculum was found in an area that is no longer used, it may indicate that a significant change in the airflow has occurred. Perhaps an entrance or small opening has been closed that previously allowed cold air into the area, or, perhaps an enlarged opening has cooled or dried the area. Recent analysis of bat roosts near the historic entrance of Mammoth Cave, Kentucky, has provided this type of restoration information (Toomey and others 2002). A variety of information may be extremely useful in planning environmental restorations.

Summary

When caving or preparing for restoration, be aware that paleo-materials may be present in any cave. Take precautions not to disturb potential paleontological resources. Report significant finds to the cave management or owner, to the Paleontological Section of the National Speleological Society, and to other appropriate agencies or organizations. Any paleo-remains may be valuable in scientific research and should always be considered in the development of cave conservation plans.

Any paleo-remains may be valuable in scientific research and should always be considered in the development of cave conservation plans.

Cited References

Davis OK. 1999. Pollen and other microfossils in Pleistocene speleothems, Kartchner Caverns, Arizona. *Journal of Cave and Karst Studies* 61(2): 89-92.

Polyak VJ, Cokendolpher JC, Norton RA, Asmerom Y. 2001. Wetter and cooler late Holocene climate in the Southwestern United States from mites preserved in stalagmites. *Geology* 29(7): 643-646.

Toomey, RS III, Coburn ML, Olson RA. 2002. Paleontological evaluation of past use of caves by the Indiana bat: A significant tool for restoration of hibernacula. In: Kurta A, Kennedy J, editors. *The Indiana Bat: Biology and Management of an Endangered Species*. Austin (Texas): Bat Conservation International. p 79-85.

Additional Reading

Andrews P. 1990. *Owls, Caves, and Fossils: Predation, preservation, and accumulation of small mammal bones in caves, with an analysis of the Pleistocene cave faunas from Westbury-sub-Mendip, Somerset, UK.* Chicago (IL): University of Chicago Press. 231 p. [This book, although somewhat technical, provides one of the best introductions available to cave fossils and the process of fossilization in caves.]

Brain, CK. 1980. *The Hunters or the Hunted?: An Introduction to African Cave Taphonomy.* Chicago (IL): University of Chicago Press. 365 p. [Another somewhat technical book that is a classic introduction to how bones get into caves.]

Kurten B. 1976. *The Cave Bear Story: Life and Death of a Vanished Animal*. New York: Columbia University Press. 163 p. [An easy to read account of the study of cave bears that shows both the importance of cave fossils and what can be learned from them.]

Section A—Identifying and Protecting Cave Resources

Archaeology in Cave Conservation and Restoration

George M. Crothers

Archaeology in cave conservation and restoration shares many of the same problems and analytical procedures found in other chapters of this book. Rock art and historic writing (page 99) are types of archaeological remains. The processes that deposit and alter paleontological remains (page 83) are the same as those affecting archaeological remains. Chapters on those subjects should be considered prerequisite to this chapter, which is limited to comments regarding conservation and restoration as it applies to archaeology, and includes a discussion of the variety of remains found in caves.

When Contemplating a Restoration Project

When considering a conservation project there is no substitute for researching the history of the cave in question before work begins. That history includes official records—deeds, titles, written histories, news-papers, and past research projects—as well as oral accounts of the cave gleaned from landowners, local residents, cavers, and others. Of course, one has to evaluate the merits of both the official and unofficial records of any site, but careful historical research will enhance understanding of what might be found.

Experts should be consulted whenever possible regarding known or expected remains found in a cave. I have experienced a few cave conservation situations where an attitude prevailed that protecting archaeological remains might stand in the way of some "greater" good for the cave. Perhaps this is due to rivalry among disciplines (which in turn is probably due to the competition for limited funding dollars in the sciences) or simple mutual misunderstanding of goals. Involving archaeologists (and all specialists for that matter) early in the planning process can help avoid conflicts among conservators and archaeologists.

All cave conservation and restoration projects should institute a system to inventory and monitor the environmental conditions and the state of archaeological remains in the cave prior to and following any project. One must know the extent of archaeological remains or resources before initiating any action, and then follow up with monitoring of those remains or resources over extended periods. The type of inventory or monitoring, of course, will have to be tailored to the types of remains potentially impacted and the alterations that are planned. (See inventory, page 19; also see photomonitoring, page 207.)

When considering a conservation project there is no substitute for researching the history of the cave in question before work begins.

Restoration: An Event in the Archaeological Timeline

Humans are, of course, one of many animals that enter and use caves. Each trip into a cave has an impact, and leaves a record of that trip, no matter how subtle. It is that record that archaeologists study, especially as it recedes further into time. By piecing together the surviving evidence of that record, archaeologists construct the history of cave use to understand human fascination with caves, how caves affect human cultures, and how humans affect caves. Restoration, whether for aesthetic or ecological purposes, has an impact on the cave environment.

For an archaeologist, the act of restoration is but one in a long line of human events within a particular cave. Each new act has the potential to alter or obliterate evidence of past activities. Combine human activity with other processes in caves—structural breakdown, water movement, other animal activity, and organic decay—and the archaeological task of constructing a timeline of past events can be daunting.

To the archaeologist, context of a find is as important as the particular artifact itself.

Preserve Remains in Context

Archaeologists' particular analytical specialty is to study and document the remains or evidence of past human activity in its context. In other words, to the archaeologist, context of a find is as important as the particular artifact. That is why archaeologists go to elaborate lengths to document the context of the artifacts they dig up or collect before that context is destroyed.

By *context*, archaeologists mean the place where an artifact is found, and what that place tells us about when, where, how, and why an artifact was used. Recording the context of artifacts has many scales. An archaeologist may say that this artifact was found in a cave, or that this artifact was found in a particular cave, or that this artifact was found in this particular cave at this particular spot, or that this artifact was found in this particular cave at this particular spot associated with five other particular artifacts, and so on.

The types of questions asked about archaeological remains determine the level and detail of contextual information that is recorded. In the final product, archaeologists may not use all the data that they collect in their analysis. But once the context is destroyed (that is, the artifact is removed and its matrix destroyed), one cannot go back and collect additional data regarding its context. Hence, archaeologists are taught to err on the side of recording more data rather than less.

The foremost goal of archaeology is to preserve the remains of past human activity in their original context until such time comes that a thorough and rigorous research design can be created that will incorporate the study of those remains into larger research questions. Insofar as possible, archaeological excavation is designed to have minimal impact, and to leave a portion of the site for future study.

Once the artifact is removed and its matrix is destroyed, one cannot go back and collect additional data regarding the context.

However, current human activity continues to degrade previous archaeological evidence while creating new remains that may someday also be the subject of archaeological study. Consequently, archaeologists must continuously make decisions about what is significant to preserve and study. Such decisions apply to remains found aboveground or within caves.

Special Considerations in Caves

Caves require special considerations from archaeologists due to the subtle and fragile remains that are preserved underground.

Always leave artifacts in place and resist handling them. Do not collect artifacts (unless under the supervision of a permitted archaeologist). Photograph the remains if possible and accurately record their locations. There are exceptions to this rule—for example, if the remains are in imminent threat of being destroyed by natural processes or human activity. If possible, involve an archaeologist in such removal. Where this is not possible, take detailed notes of the location and context. Remove the items to a safe location only if there is imminent threat of destruction.

In many states, the removal of artifacts or paleontological remains from caves is illegal without a permit from the state and permission of the landowner. Make every effort to treat archaeological remains like all objects in caves—geological or biological—leave them as you find them.

Implications of Preservation, Conservation, and Restoration

Studies of the archaeological context of remains depends upon the geological and hydrological context in which those human artifacts and activities were introduced and how that context may have changed or may have been altered over time.

Preservation

Preservation implies that the archaeological remains are stable in their current environment—but stability depends on the geological context of the cave. Dry, upper-level passages offer more stability than wet sinks or stream passages. An artifact exposed in the cut bank of a stream passage could be destroyed or moved by the next flood episode.

The talus cone at the bottom of an entrance pit is a dynamic environment in which both archaeological and paleontological remains are often deposited together. Investigating the archaeological context of remains in such an environment incorporates the study of geological as well as archaeological processes.

Studies of the archaeological context of remains depends upon the geological and hydrological context.

Conservation

Conservation implies that policies should be implemented and procedures followed to minimize future impacts and to protect resources where they are found. This means reducing traffic, restricting access, confining pathways to designated trails, and generally caving softly. It is impossible to eliminate all impacts from human presence in a cave, but by being cognizant of the resources that are potentially present we can significantly reduce damage.

It became necessary, for example, to preserve the famous Paleolithic cave paintings in Lascaux, France, by closing the cave to the public and building a full size replica for tourists. Only a few individuals are granted access to the real Lascaux.

Restoration

Restoration implies that something can be returned to a former state. In reality, however, we are creating a new state that, based on our best information, approximates some state in the past. Envisioning that former state is the difficult part. It is also impossible to always anticipate all the consequences of any restoration act. Preserving an archaeological site in a cave and restoring the natural ecological processes to that cave may not always be compatible. The resource manager then must decide which is more important.

For example, reconfiguring the natural entrance to Mammoth Cave, Kentucky, with a bat-and-animal-friendly gate allowed a greater amount of airflow into the historic section of the cave. This changed the condensation zone, altered the amount of water dripping from the ceiling, and accelerated the deterioration of significant 19th century saltpeter remains. In an attempt to restore "natural" ecological conditions, the cave was subject to greater airflow than probably ever existed in the recent geological past. (Placing Plexiglas® over part of the gate to reduce the volume has since reduced airflow.)

Archaeological restoration usually refers to the study of structural remains that are excavated and analyzed in order to construct a replica of a structure. Until fairly recently, archaeological restoration was quite popular but it has largely fallen out of favor. It is much more common in public archaeological sites to uncover archaeological remains and interpret them in place, rather than attempt full-scale reconstructions.

Types of Archaeological Remains Found in Caves

Where caves are significant features of the landscape, humans were likely to have entered, explored, and exploited them in prehistoric as well as historic times. The types of archaeological remains potentially found in caves varies by region and period of use. Consult specialists for a particular region to obtain the best information for any given cave site (for example, Crothers and Watson 1993).

Mining Niter Dirt

In the eastern U.S., one of the most common historic uses of caves was mining of niter dirt (which was converted to potassium nitrate or saltpeter), especially during the Revolutionary War, the War of 1812, and the U.S. Civil War. These remains are quite conspicuous in Mammoth Cave, Kentucky; Big Bone Cave, Tennessee; and Organ Cave, West Virginia.

Thousands of smaller caves, however, were mined for nitrates, and evidence may be limited to mounds of niter dirt where the remains of wooden equipment have rotted away and a few names may be scratched on the cave walls.

> Where caves are significant features of the landscape, humans were likely to have entered, explored, and exploited them in prehistoric as well as historic times.

Prehistoric Uses of Caves

Evidence for prehistoric use of caves is generally much more subtle than evidence for historical use. For example, the earliest known record of cave exploration in eastern North America is a remote passage in a Tennessee cave with 274 complete foot impressions in the soft mud floor (Willey and others 2004). Associated with these footprints is a sparse scatter of torch charcoal that has been radiocarbon dated to between 5,600 and 4,870 BP (before present, calibrated to calendar years).

These footprints are still preserved only because the cavers who first entered this passage were keenly observant and steps were taken to limit caving activity in the passage. Even with these precautions, a few of the prints were obliterated by later cavers who ignored the warning signs and flagging tape, and blundered through the footprint passage.

Torch charcoal and charcoal marks from striking or stoking a burning bundle of torch material against the walls and ceilings of a cave (dry river cane, weed stalks, and small wooden sticks were common torch fuels) are telltale signs of prehistoric exploration or activity These remains are not always obvious to the untrained eye, and are easily obliterated by modern traffic, graffiti, or attempts to remove overlying graffiti.

Clay and Stone Mining

A common prehistoric use of caves was mining clay and stone for tool making. For example, chert (sometimes referred to as flint) is commonly found in caves and was a highly desirable tool stone. A number of caves with evidence of chert mining have been found, but one of the more spectacular is in Tennessee (Franklin 1999).

Wyandotte Cave, Indiana, contains well-documented evidence for the mining of aragonite from a large column, which was made into a number of fancy artifact types (Munson and Munson 1990; Tankersley and others 1990).

Gypsum, mirabilite, and probably epsomite were all mined from Mammoth and Salts Caves, Kentucky (Watson 1969, 1997), and gypsum was mined from Big Bone Cave, Tennessee (Crothers 1987). Evidence for mineral mining and stone quarrying—dug sediment and broken, scraped, or crushed mineral faces found on the walls and ceilings—may not be obvious to the untrained eye.

Prehistoric pictographs, petroglyphs, and "mud" glyphs have also been documented in a number of caves. (See rock art and historic writing for descriptions of these remains, page 100.) Rock art may be found in association with other activities (for example, chert and mineral mining), but it also appears to have been the primary activity in some caves. Mud Glyph Cave, Tennessee, is one of the more celebrated caves in the eastern U.S. where drawing was the primary prehistoric activity (Faulkner 1986).

> If a cave is known to have a rich history or prehistory of use, an archaeologist (preferably one with cave experience) should be consulted very early in any project.

Burial Caves

Another prehistoric use of caves was inhumation of the dead. Apparently, this practice was widespread in North America, but has been studied only in a few locations (Crothers and others 2002). Talus cones at the bottom of vertical shafts with openings to the surface have been documented as human burial repositories. In these cases, human and nonhuman bone may be commingled in the deposits, because pit caves often trap animals as well. (See paleontology, pages 83 and 86.)

All states have laws concerning the discovery of human remains that usually prohibit the removal of the remains until they are assessed by local law enforcement officers and the coroner's office. Burials may not be removed without proper permits in most states. The laws and procedures vary by state.

Summary

Archaeological remains are irreplaceable resources that inform us on the historic and prehistoric uses of caves. The remains are often subtle and not obvious to many cavers. If a cave is known to have a rich history or prehistory of use, then an archaeologist (preferably one with cave experience) should be consulted very early in any project to evaluate potential impacts and suggest ways to minimize those impacts.

Even if a cave is not known to have archaeological remains, but is suspected of having cultural value, it is advisable to have an archaeologist look for possible evidence in conjunction with restoration efforts. Once archaeological remains are identified and deemed significant (that is, worthy of preserving in their context), a plan to monitor long-term effects to those remains is advisable.

Rarely can we anticipate all the consequences of an action even under the best of intentions. Caves can be one of the most amazing receptacles for preserving archaeological remains, but they are also one of the most susceptible environments to unintended impacts.

> Rarely can we anticipate all the consequences of an action even under the best of intentions.

Cited References

Crothers GM. 1987. *An Archaeological Survey of Big Bone Cave, Tennessee, and Diachronic Patterns of Cave Utilization in the Eastern Woodlands* [Master's thesis]. Knoxville, University of Tennessee.

Crothers GM, Faulkner CH, Simek JF, Watson PJ, and Willey P. 2002. Woodland cave archaeology in eastern North America. In: Anderson DG, Mainfort RC Jr, editors. *The Woodland Southeast*. Tuscaloosa (AL): The University of Alabama Press. p 502-524.

Crothers GM, Watson PJ. 1993. Archaeological Contexts in Deep Cave Sites: Examples from the Eastern Woodlands of North America. In: Goldberg P, Nash DT, Petraglia MD editors. *Formation Processes in Archaeological Context*. Madison (WI): Prehistory Press. p 53-60.

Faulkner CH. 1986. *The Prehistoric Native American Art of Mud Glyph*

Cave. Knoxville (TN): The University of Tennessee Press.
Franklin JD. 1999. *The Rime of the Ancient Miners*. [Master's thesis]. Knoxville, University of Tennessee.
Munson PJ, Munson CA. 1990. *The Prehistoric and Early Historic Archaeology of Wyandotte Cave and Other Caves in Southern Indiana*. Indianapolis (IN): Historical Society.
Tankersley KB, Munson CA, Munson PJ, Shaffer NR, Leininger RK. 1990. The Mineralogy of Wyandotte Cave Aragonite, Indiana, and Its Archaeological Significance. In: Lasca NP, Donahue J, editors. *Archaeological Geology of North America*. Centennial Special Volume 4. Boulder (CO): Geological Society of America. p 219-230.
Watson PJ. 1969. *The Prehistory of Salts Cave, Kentucky*. Springfield (IL): Illinois State Museum.
Watson PJ. 1997. *Archaeology of the Mammoth Cave Area*. St. Louis (MO): Cave Books.
Willey P, Stolen J, Crothers G, Watson PJ. 2005. Preservation of prehistoric footprints in Jaguar Cave, Tennessee. *Journal of Cave and Karst Studies* 67(1):61-68.

Section A—Identifying and Protecting Cave Resources

Rock Art and Historic Writing in Caves: Restoration Implications

Barbara Bilbo and Michael Bilbo

Restoration projects have sometimes resulted in the unintentional removal of prehistoric or historic rock art and historic writing. Projects initiated to remove contemporary graffiti from dark zone caves, rock shelters, and other rock surfaces call for thorough site assessment. Restoration methods must be objectively planned to assure protection of cultural resources. This chapter cautions cavers to study vandalized sites for the presence of rock art and historic writings and to consult with historians and rock art specialists during planning processes.

Cultural sites are especially important when considering a cave for significance inventory and nomination. (See culural resource laws, page 110.) When a dark zone cave meets significance criteria for cultural resources, it may be eligible for nomination to the National Register of Historic Places (Advisory Council on Historic Preservation 1973).

Nonrenewable Cultural Resources Versus Contemporary Graffiti

Throughout the United States and the world there are rock art and historic writing sites that have been vandalized by graffiti. Examples included in this chapter are from the southwestern U.S. (Figure 1), but the patterns are similar elsewhere. Some writers state that all rock writings are graffiti. We insist that rock art and historic writing are cultural resources from past eras, whereas modern spray painted graffiti is usually considered vandalism.

The need to distinguish between cultural resources and contemporary graffiti became clear when Mike Bilbo observed that historic names in Fort Stanton Cave, New Mexico, may have been overlooked or interpreted as graffiti and were removed during a restoration project. In addition, statements in several journal articles, including those by Welsch (1993) and White (1993), seemed to disparage Native American rock art and equate it with graffiti.

Rock art and historic writing are significant nonrenewable cultural resources and must be recognized and inventoried along with other cave

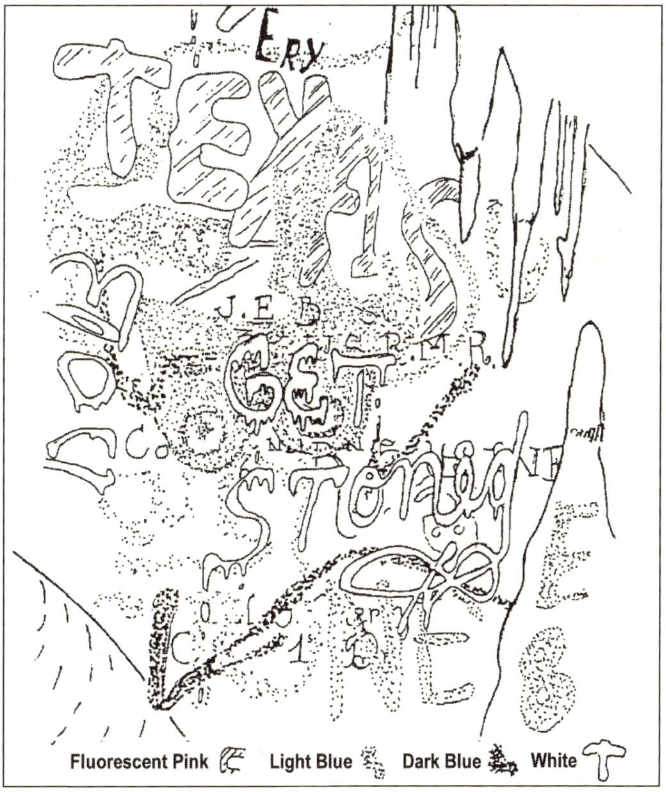

Bilbo & Bilbo

Figure 1. Vandalized cave on public land in southeastern New Mexico, now restored. Layers of spray paint covered the historic writings.

resources. Furthermore, contemporary writing of names and words or pictures in caves and on other rock surfaces has become unacceptable and unethical behavior for cavers. Since the late 1950s, the National Speleological Society has discouraged the writing of new graffiti in and around caves (Weaver 1992).

Rock art researcher James G. Bain (1978) stated that rock art in caves in the United States might be rare but believed that there are more sites. He asked that cavers learn to identify rock art and watch for it as they explore and survey caves. It is important to note that in the caving community, rock art, mud glyphs, and historic writings are being recognized more frequently, and their significance and need for preservation acknowledged. Ideally, in documenting rock art and historic writing sites, cavers should note that important information is conveyed in three significant ways—through content, style, and medium.

We usually associate petroglyphs and pictographs with the art of aboriginal cultures. Petroglyphs are rock engravings that are made by percussion or incision. Pictographs are rock paintings. Technically, historic writing can also be pictographic or petroglyphic. Historic writings in caves and other rock surfaces were usually created with axle grease, carbon black, charcoal, or pencil. They were painted by hand, incised with stone or metal, or made by percussion with stone or another object.

Recognizing Prehistoric and Historic Rock Art and Historic Writing

In the effort to remove graffiti from cave walls and other surfaces in natural settings, rock art and historic writing are sometimes overlooked because they are not easy to "see." Recognizing rock art, mud glyphs, or historic writing depends on learning what to look for. Seeing these resources depends on conditions such as incised writing or petroglyph depth, pictograph color, time of day, weather condition, light source, surface reflection or light source glare, mineral occlusion, drawing technique, and other factors.

Visiting known sites is a good way to see rock art and historic writing in natural settings. In North America there are a number of preserved and interpreted sites worth visiting.

- Writing on Stone Provincial Park, Alberta
- Pipestone National Monument, Minnesota
- Hueco Tanks State Historical Park, Texas
- El Morro National Monument, New Mexico
- Petroglyph National Monument, New Mexico
- Grimes Point Archaelogical Site, Nevada
- Scottsbluff National Monument, Nebraska

To find sites in your region, use libraries, museums, and the World Wide Web to research state parks and other preserves (Bilbo 1982; Faulkner 1986, 1992; Kirkland and Newcomb 1967; Schaafsma 1980, 1992).

When looking for historic writing and pictographs at a cave restoration site, scan the location for a few minutes using halogen or LED lighting. Study the patterns, shadows, textures, glare, forms, and lines which appear to be different from the natural surface. At graffitied surfaces, attempt to look beyond or under the graffiti. Portions of names, dates, phrases, or figures may be adjacent or sticking out from the edge of areas of spray painted, incised, or scratched graffiti. Look for layered markings and superimposed graffiti—for example, a 1763 date was found on top of an

Graffiti Versus Graffito

Editors' Note: The word graffiti is listed in most contemporary dictionaries as the plural form of graffito.

Graffito is defined as an inscription or drawing made on a public surface such as a wall, or on a natural or cultural surface.

Thus, the plural form, *graffiti*, indicates several inscriptions or drawings and should be used with plural verbs.

In this volume, if we were following the dictates of perfect English usage, sentences would read: "Graffiti are painted along the walls of Good Grammar Cave. One brightly colored graffito is inscribed at the remote end of the short passage and reads 'Have No Fear, Graffiti Rule.'"

However, after much debate among reviewers, we decided to go with common current usage in the pages of this book and we wrote graffiti sentences to reflect the way cavers actually talk about graffiti.

Cavers tend to say, "New graffiti is covering authentic historic signatures in Current Lingo Cave."

early 1800s date in Hubbards Cave, Tennessee. The 1763 date was obviously bogus (JC Douglas, personal communication 2004).

Pictograph colors tend to be flat earth tones of red, white, black, charcoal, yellow or brown. When painted, historic writing is often made in flat black media from sources such as axle grease, carbon black, soot, or charcoal. Recognition of incised writing and petroglyphs depends on rock surface weathering, or patina, through which designs are pounded, leaving the lighter rock interior exposed. Mud glyphs, incised by fingers or tools, are also important resources. Side lighting the cave or rock shelter walls, ceilings, and floors can aid in revealing the presence of cultural remains.

Rock surfaces, along with rock art and historic writing in caves and elsewhere, are gradually destroyed by natural processes. Moisture seeping down rock surfaces forms a thin mineral coating which occludes figures and words (Ralph and Sutherland 1979). Visibility of occluded figures varies under different lighting and angles of incidence. The rate of occlusion, including the development of flowstone in caves, varies widely from site to site and may develop more rapidly in caves than at surface sites. At the Natural Entrance Pictograph Site, Carlsbad Caverns National Park, New Mexico, most of the hunter-gatherer style art is occluded by calcium carbonate wash (Bilbo 1997). Few, if any, studies on occlusion rates on rock art or historic writing have been done. Occlusions may protect writing and rock art from removal during spray paint removal operations if special cleaning techniques are used (Bilbo and Ralph 1984; Ralph and Sutherland 1979).

See Weaver (1992) for a discussion of graffiti and rock art in caves, as well as biological marks such as animal footprints, bear claws, and human footprints. (Also see paleontology, page 83.)

Native American Rock Art

Rock art consists of pictographic (painted) and petroglyphic (pecked, carved, or incised) forms. Thousands of years of early Native American visual depictions have been recorded on boulders, rock faces, shelter walls, around cave entrances, and inside caves.

Rock art consists of pictographic (painted) and petroglyphic (pecked, carved, or incised) forms.

Native American petroglyphs and pictographs occur at numerous sites, many of which have been preserved and interpreted. However, the number of sites that are known, but undocumented or unprotected, are far more numerous.

The primary cultural time periods in the Southwest are: Paleo-Indian (± 12,000–8,000 BC); Archaic (± 8,000 BC–AD 300); Pueblo (± AD 300–1450); and Historic (beginning about AD 1520).

Representational and Abstract Styles

There are two basic rock art styles, *representational* and *abstract*. Both styles are subdivided into regional and local styles where distinct stylistic differences are noted. Figures usually consist of recognizable forms, such as human, animal, or plant forms, and include masks, human hands, human and animal footprints, and supernatural concepts (Figure 2). Abstract figures, possibly carryovers from early cultures, are also seen in representational rock art sites, but less frequently.

The abstract style in North America (Figure 3) is associated with nomadic hunting-and-gathering societies beginning more than 6,000 years ago (and may also date from the North American Paleo–Indian period, although Paleo–Indian abstract or representational figures are currently unidentified in North America). Figures include spirals, wavy lines, hourglass-shaped forms, rows of lines or dots, circles, crosses, stylized animal or human forms, and other geometrical shapes representing cultural concepts, and may include representations of the supernatural, cosmology, and other ideas (Schaafsma 1992), but which are unknown to modern observers. Abstract figures are sometimes recognized by older members of

Figure 2. Examples of Representational Style

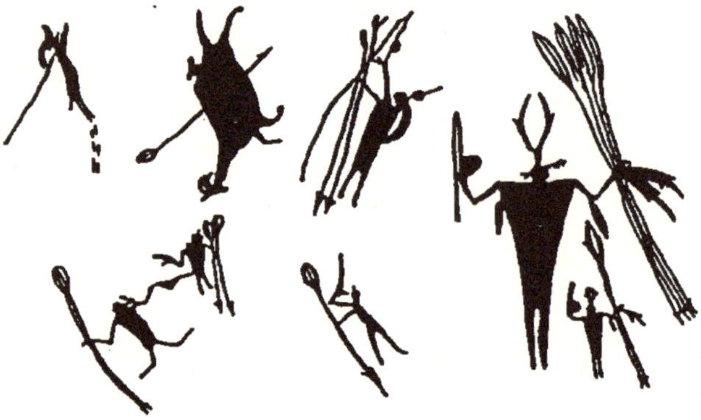

Schaafsma

Figure 2a. Probable Middle Archaic hunters and animals (from Schaafsma 1980).

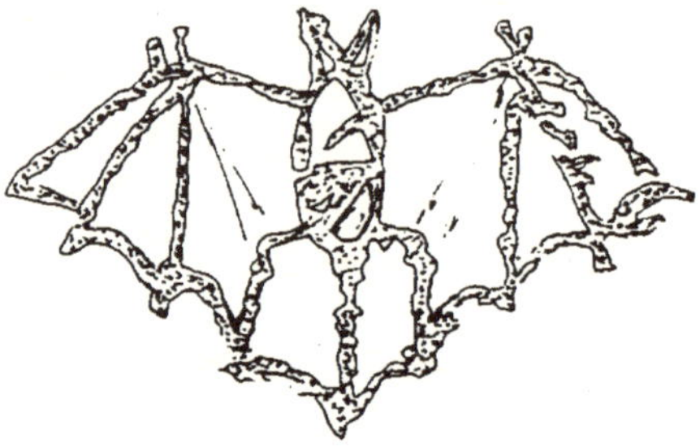

Bilbo & Bilbo

Figure 2b. Navaho bat (18th century) from northern New Mexico.

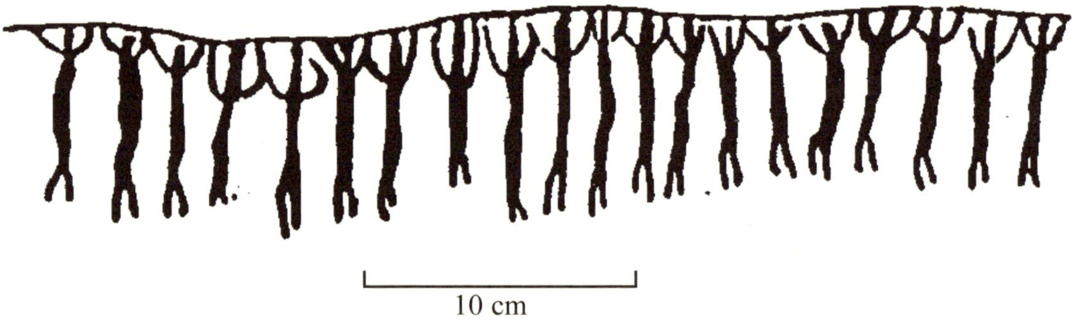

Bilbo & Bilbo

Figure 2c. Prehistoric line of human forms.

Faulkner & O'Blair

Figure 2d. Prehistoric mudglyphs, petroglyphs, and pictographs from caves in the Tennessee, Alabama, and Georgia regions (from Faulkner 1984, 1988; O'Blair 1996). The circular figures in the lower right are considered representational shields.

Kirkland & Newcomb

Figure 2e. Possible Comanche pictographs, ca.1580, from near the Pecos and Rio Grande rivers confluence in Texas (from Kirkland and Newcomb 1967).

Figure 3. Examples of Abstract Style

Kirkland & Newcomb

Figure 3a. Prehistoric abstract style from Lower Pecos River Region in Texas (from Kirkland and Newcomb 1967).

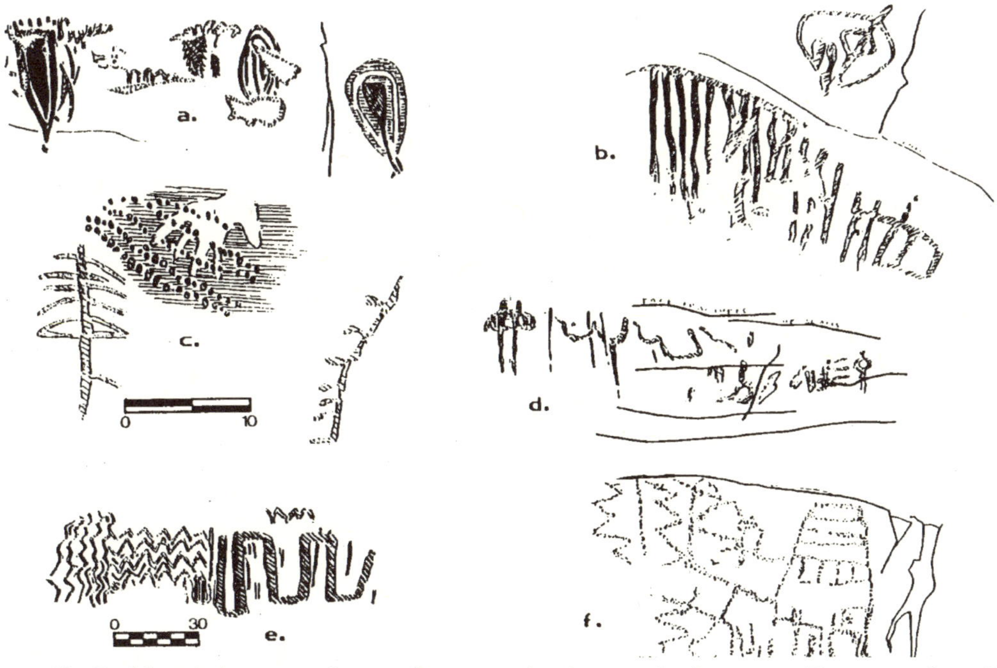

Bilbo & Bilbo

Figure 3b. Prehistoric hunter-gatherer, abstract style pictographs from west Texas, southern New Mexico, and northern Mexico.

modern Native American tribes, although the meanings are seldom revealed to nontribal people. Some contemporary cultures, such as the Hopi and Zuni tribes, Australian Aborigines, and African Bushmen still create both representational and abstract style rock art. Comparative studies of these cultures and their art gives some insight into early North American cultures.

Abstract is the most common rock art style in North America, but most people do not recognize it. In our years of experience we have encountered some archaeologists who do not acknowledge abstract figures as rock art. A few others do not view rock art as important archaeological artifacts. Unless a rock art site contains abstract figures of the same density and coloration as those in Figure 4 or includes the well-known representational style of Pueblo masks, flute players, and other popularized figures, it can be overlooked, especially when the abstract figures are faint or occluded.

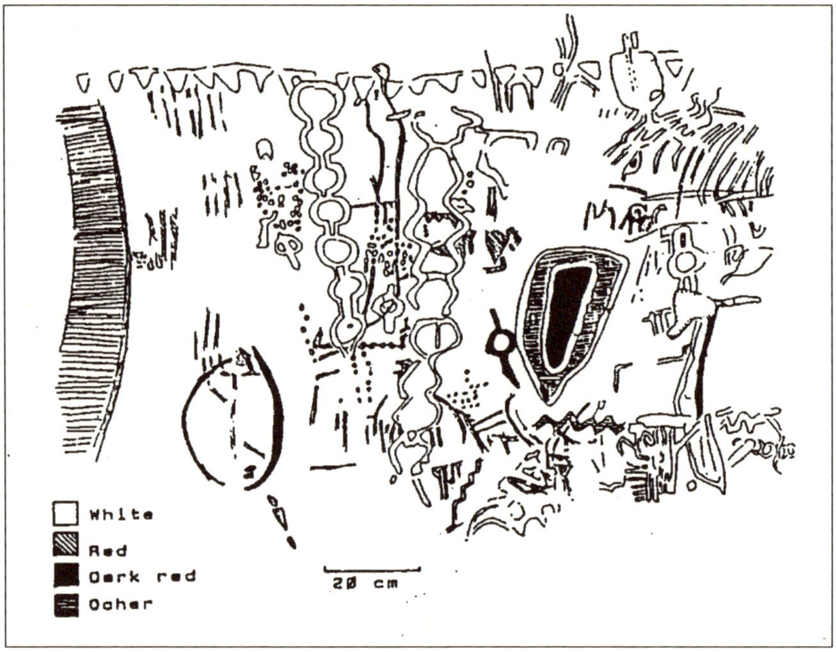

Bilbo & Bilbo

Figure 4. Detail of abstract style art from southern New Mexico that may date from prehistoric to historic periods.

Contemporary Native American Rock Art

Contemporary Native Americans continue the tradition of rock art in some regions. In New Mexico, the Zuni create both pictographs and petroglyphs—some are considered ritualistic while others are being done for artistic reasons. Some younger Zuni create figures depicting contemporary subjects, including cars, women, and "George loves Bettina." Techniques used are the traditional incising and painting, but the paint often comes from aerosol cans and has occasionally been superimposed over traditional figures. Other Zuni who had lost contact with tribal identity may have been defacing traditional sites (Young 1990). Should these activities be considered vandalism?

Other Native Americans also create contemporary rock art such as the Hopi who draw traditional rock art in the Chaco Canyon area of New Mexico and in Mesa Verde National Park, Colorado. These ritual sites or *traditional cultural properties* have been used for centuries. Similarly, the Ojibwa, in Canada, draw figures and inscribe their names at traditional sites to insure good health. It is conceivable that caves may be sites for some of these activities. Should a traditional figure created in recent decades by a Native American for ritual or religious purposes be considered graffiti?

Dark Zone Sites

More than 50 dark zone sites are known in North America (Faulkner 1986, 1992, personal communication 1993; Gurnee and Gurnee 1985; Greer and Greer 1996; Veni, personal communication 1993; Simek and others 2001; Simek and Cressler 2001).

Figures that archaeologists have dated as being thousands of years old have been found in caves throughout the world. Significant examples include the Mayan cave art in Central America (Greer and Greer, personal communication 1992) and the famous depictions of Pleistocene fauna in caves such as Lascaux in France, Altimira in Spain, and other sites.

Several black figures are visible in the dark zone of a cave in southern New Mexico. We examined the black figures and the surrounding wall area and saw that the dominant black forms were superimposed over faded black figures, and that beneath one was another figure in faded yellow, all occluded by a thin coating of calcium carbonate. As we continued to study the site under the "white" light of halogen bulbs, more occluded yellowish and reddish figures were observed on the cave wall. These faded figures had remained unobserved by most previous visitors due to the occlusion and to the "yellow" light of carbide lamps, flashlights, and Coleman or kerosene lanterns (Figure 5). Careful examination during our recording project revealed over 40 abstract figures or traces, many of them forms common not only in the Guadalupe Mountains but throughout southwestern North America (Bilbo and Bilbo 1993).

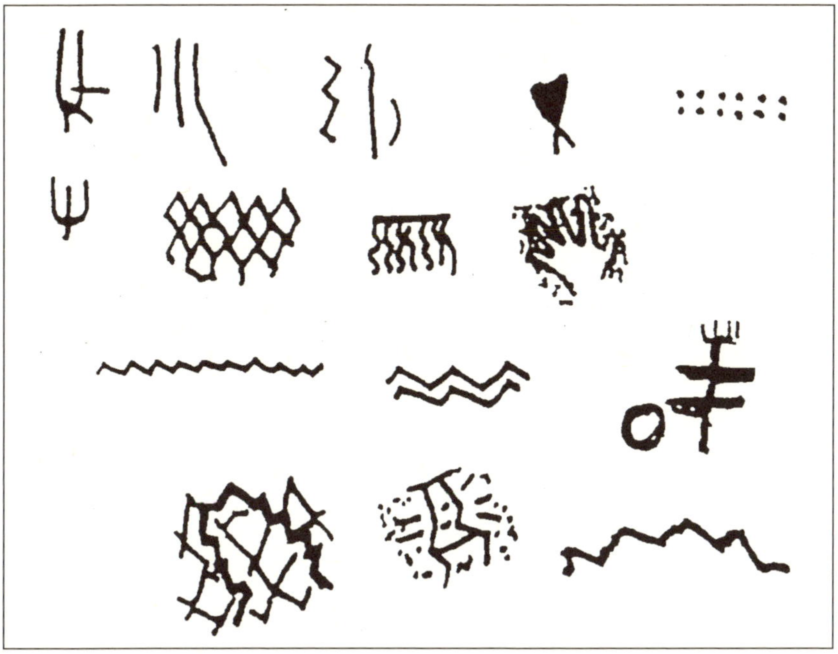

Bilbo & Bilbo

Figure 5. Hunter-gatherer abstract style from southern New Mexico.

Spanish and Euro-American Historic Writing

As a cultural artifact, writing found on natural rock surfaces within the United States is considered historic if it is 50 years old or more under guidelines of the Advisory Council of Historic Preservation (1973). (See 50-year rule, page 341.)

In North America writing was done by explorers, soldiers, and settlers—for example, Spaniards during and after the Spanish entrada in southwestern North America in the early 1500s (the beginning of the historic period), and Euro-Americans following settlement along the eastern coast of North

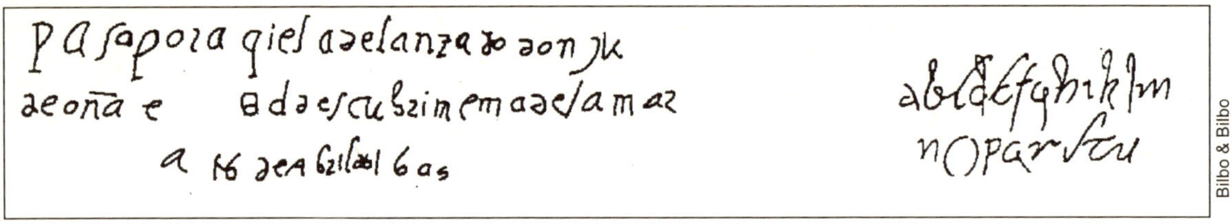

Figure 6a. Humanistic cursive, a substyle of Roman Cursive. Text on the left is New Mexico colonizer Juan de Oñate's 1605 inscription from El Morro National Monument, New Mexico (from SPMA 1989). Text on the right is an encyclopedic example.

America in the early 1600s. Historic writing was done in charcoal, carbon paint (lamp black), axle grease, pencil, and other media, or by carving or incising. The content may consist of a name or names, places of origin, dates, military affiliations, or other information. Historic writing may be situated in locations similar to rock art.

Writing Styles

Historic writing styles consist of four basic types. Variations exist due to rock textures and availability of implements or media.

- Roman Cursive
- Modern Roman
- Roundhand Script
- Block or Monumental

Roman Cursive. This is the manuscript printing characteristic of Spaniards and other Europeans and was used from about AD 1500 to as late as AD 1800. It is identified by curved appendages and slight hooks or loops on the ends of the appendages. Three cursive substyles may also be seen in Spanish inscriptions: humanistic cursive, black letter cursive, and black letter Gothic.

Humanistic cursive (Figure 6a) dates from AD 1450–1600 and consists of sloping and well-rounded letters with distinctive terminal blots, hooks, or small loops at the ends of ascending letters. Black letter cursive and Gothic (Figure 6b) are patterns of angular manuscript letters and were common from the Medieval period through the early Renaissance. Individual writers sometimes mixed these elements.

Modern Roman. Dating from about AD 1600 into recent times, this style (Figure 7) is identified by serifs, short lines stemming from and at an angle to the upper and lower end strokes of a printed letter.

Roundhand Script. This is an italic longhand style characterized by strongly slanted thick and thin strokes. Capital letters are characterized by distinctive sweeping flourishes and were contemporaneous with Modern Roman printing.

Block or Monumental. Most commonly seen after about AD 1900 is

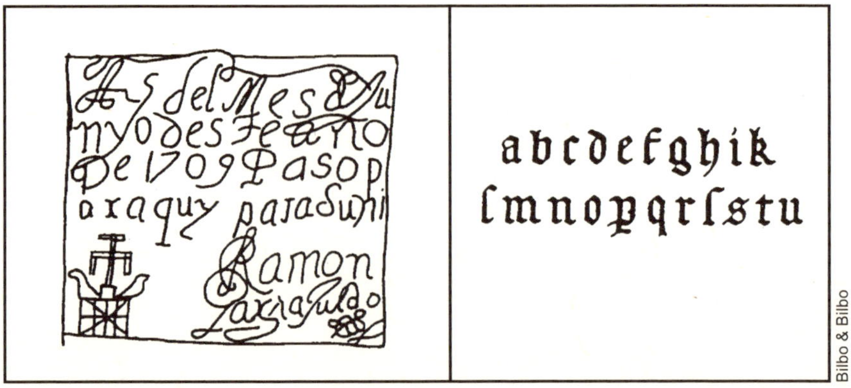

Figure 6b. Black letter Gothic, a substyle of Roman Cursive. Text and design on the left is inscription of Ramon Jurado Garcia, 1709, from El Morro National Monument, New Mexico (from SPMA 1989). Note the significant evolution of text during the 100 years between Figure 6a and 6b. Text on the right is an encyclopedic example.

Block or Monumental style (Figures 8a and 8b). This printing style lacks serifs and often has a heavy line weight. Writing styles in caves in the early 20th century are most frequently inscribed or drawn in these block letters, but Roman printing and Roundhand Script styles are also common. Jim

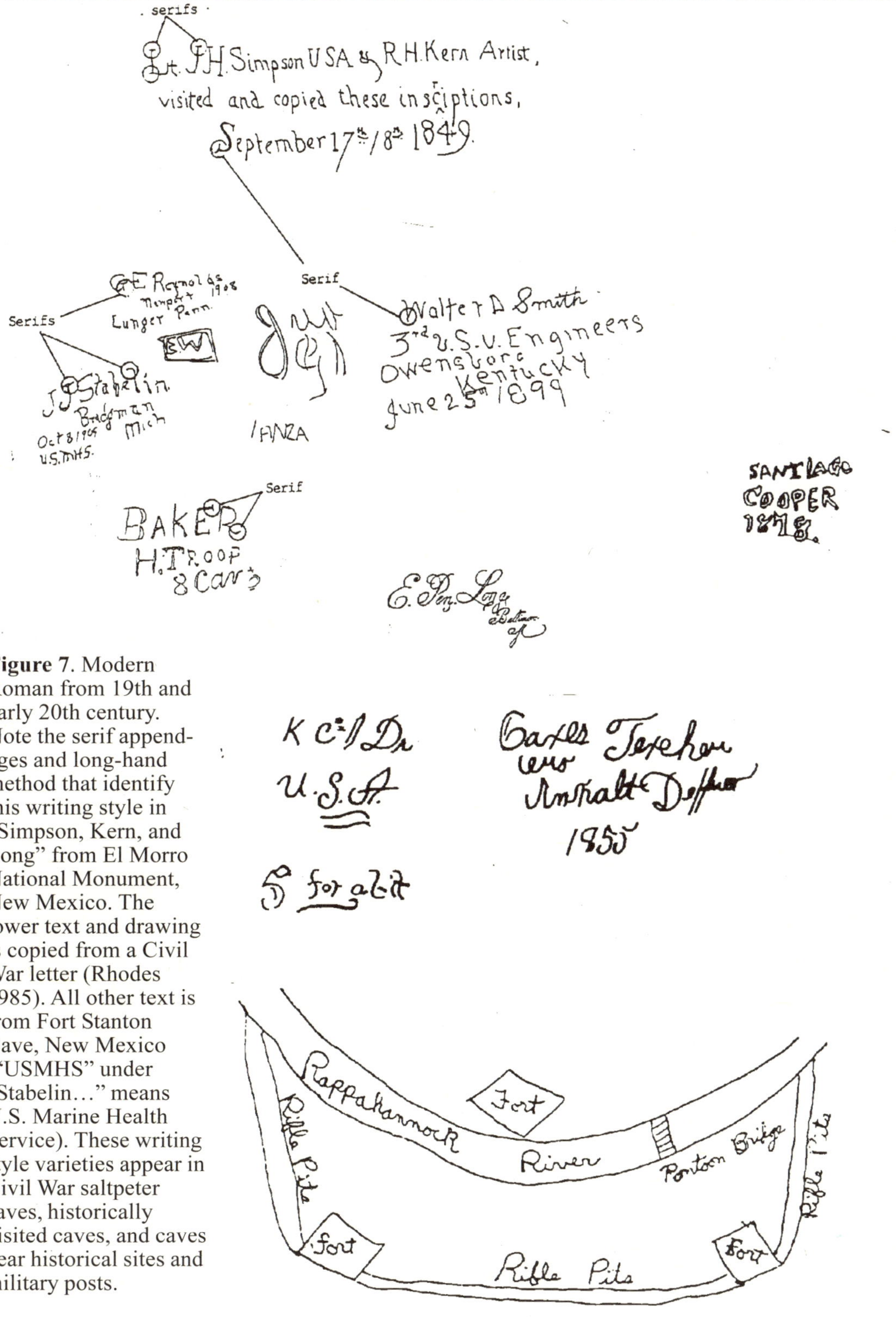

Figure 7. Modern Roman from 19th and early 20th century. Note the serif appendages and long-hand method that identify this writing style in "Simpson, Kern, and Long" from El Morro National Monument, New Mexico. The lower text and drawing is copied from a Civil War letter (Rhodes 1985). All other text is from Fort Stanton Cave, New Mexico ("USMHS" under "Stabelin…" means U.S. Marine Health Service). These writing style varieties appear in Civil War saltpeter caves, historically visited caves, and caves near historical sites and military posts.

Bilbo & Bilbo

White and other explorers of Carlsbad Caverns and nearby caves left their names and dates on cave walls in the 1920s in these styles. For more information on writing styles refer to encyclopedias and books on epigraphy.

Examples of Historic Writing

Some names recorded on rock walls are well known in history, such as the early Spanish colonizer, Don Juan de Oñate, who inscribed his name and the date, 1605, on the rock face now included in El Morro National Monument, New Mexico. Others left their names there as well, including several 19th century Euro-Americans. At most sites, however, are the names of common everyday people who indicated their passing. While their writing on rocks may seem insignificant, the opposite is often true. Research of an inscription can reveal unique and important histories.

For example, names and dates were made on a shelter wall by mid-19th century stagecoach travelers in what is now Hueco Tanks State Historical Park, located on the outskirts of El Paso, Texas (Figure 9). Santiago Cooper's

Bilbo & Bilbo

Figure 8a. Block or Monumental style from southern New Mexico.

Bilbo & Bilbo

Figure 8b. Block or Monumental style from southern New Mexico. Notice the holdover Roman signature detail in "Jim White" of Carlsbad Caverns fame between the "1916" and "EM" of "Clements."

1878 signature was probably written with axle grease. Local historical society records show that he was associated with stagecoach driving, but include no indication of when. We later learned that Cooper was in the Texas Rangers, Company C, stationed in El Paso, Texas, in 1877 (Gillett 1976, page 79). The location of Cooper's and other names written at the site also suggests a possible original stage stop location that had not been accurately identified in other records. More questions and associated research can be suggested by a few names, which except for their occurrence as historic writing on a rock surface, may not be part of known, written history.

Huntsville Grotto members were in the process of a cleanup project (Varnedoe and Lundquist 1993), when Varnedoe noticed "J C 'A' T Co A 6th RGT NY 1898" along with recent spray paint graffiti done by soldiers from nearby Redstone Arsenal. Having knowledge of area history, the authors knew that Redstone was not established until World War II. Research revealed that U.S. Army troops were in the area in 1898 to secure crops, probably for troop use during the Spanish–American War. Following their research, the restoration team removed the recent spray paint, but left the historic inscriptions intact.

Bilbo & Bilbo

Figure 9. Historical writing (possibly axle grease) at Hueco Tanks State Historical Park, Texas.

In a similar situation, Bilbo (1982) recorded names and dates from 1855 in Fort Stanton Cave, left by the U.S. Army 1st Dragoons (now the 1st Cavalry) and the 3rd U.S. Infantry. The names and dates had been incised on flowstone in Roundhand Script (much like the script in Figure 9). Research revealed that the names were soldiers in the regiments that built Fort Stanton, a 19th century U.S. Army post, and were probably inscribed within three months of their arrival in 1855.

The entrance to Crystal Crawl in Fort Stanton Cave was once covered by modern spray painted graffiti, which was successfully removed during restoration projects. In 1993, however, Mike Bilbo noticed faint remnants of an indecipherable name and the date of 1862 at the project location. The year 1862 is a significant date in the history of nearby Fort Stanton, as 1862 is the year that the post was reoccupied by Colonel Kit Carson and five companies of the 1st New Mexico Volunteers. Additional survey and inventory revealed that Fort Stanton Cave has at least eight interior areas

NSS Cave Vandalism Deterrence Reward Commission
Jay Jorden

The Cave Vandalism Deterrence Reward Commission of the National Speleological Society reviews all successful prosecutions of cave vandals with an eye toward publicizing the Society's conservation goals and rewarding diligence in support of those goals.

The commission promotes and administers an NSS reward given to the person(s) who provides information resulting in convictions for cave vandalism in the United States.

That includes convictions for damage or harm to caves, signs and gates, cave-dwelling organisms, cave-related vandalism, or malicious damage.

Commission members may recommend payment of a reward in the event of a conviction under any cave protection law of any state in the United States. (See cave laws, page 217.)

By rewarding such information, commissioners want to encourage the public to preserve and safeguard nonrenewable cave resources. (See commission page 224.)

For additional information, visit the NSS Web site <www.caves.org> and click on Conservation.

with historic writing, dating from 1855 to 1943. Names from the 1877 survey and mapping Wheeler Expedition, and the 1891 Great Divide expedition, correspond directly to period articles and reports. The Wheeler Expedition was one of four 19th century great western surveys, which resulted in the formation of the U.S. Geological Survey. Also, 1870s and 1880s names and dates from the 5th U.S. Infantry and the 6th and 8th U.S. Cavalry and turn-of-the-20th century U.S. Marine Health Service may be seen.

Phrases and slogans from the 19th and early 20th centuries may have little or no meaning in modern culture and should alert observers to the presence of historic writing. For instance, what is the meaning of *slavocrat* and how was it used? (This is a derogatory term used by Unionists toward secessionists just before the Civil War.)

Graffiti and Vandalism—Rock Art and Historic Writing Included?

The term *graffiti*, an Italian word for scribblings, was applied by archaeologists to early Roman words and phrases scratched or chalked on walls or buildings. (See sidebar note on graffiti grammar, page 100.) Contemporary graffiti includes names, political and social statements, hate and love messages, obscenities, sexual propositions, racial hate statements, gang identifiers and messages, pictures of cars, faces, and sexual organs.

Contemporary graffiti is a graphic expression of vandalism by using paint and other media, or by incising the graphic expression on natural or cultural surfaces, such as rock faces or building walls with no regard for adverse socioeconomic or cultural impacts. In what appears to be a macho trend, late 20th century and later vandals distribute trash on public land and others superimpose their names and statements with spray paint over earlier historic names and on prominent rock or structural features. Police department gang units can provide insight on gang graffiti. There are numerous books and papers on the sociocultural aspects of vandalism and graffiti, gang-related as well as that resulting from indiscriminate vandalism and casual ego trips.

Weaver (1992) discusses graffiti in caves and defines terms related to rock art in caves.

There are numerous sites in the western U.S. where cowboys and others, during the late 19th century and into the 20th century, incised cattle brands, railroad trains, names and dates, and other items into rock faces. These and other inscriptions, although considered historic and worth preserving under various laws, are sometimes referred to as *historic graffiti*. At the time these writings and drawings were made they could have been considered graffiti. However, these markings were done when very little in the way of current events was being recorded in newspapers, journals, or in other period media, and so should be considered a part of history that is not recorded elsewhere.

Laws Protecting Cultural Resources

Historic graffiti falls under the guidelines of various laws protecting cultural resources and might be considered for preservation. The Antiquities Act of 1906 provides protection of archaeological resources. (See Antiquities Act, page 221.) The National Historic Preservation Act of 1966 and the Archaeological Resources Protection Act of 1979 (ARPA) expand the language of the Antiquities Act by specifying objects significant to American history, architecture, archaeology, and culture that may have national, state, or local significance. It also specifies that protection is given to cultural resource sites on federal public lands (USDA Forest Service,

Bureau of Land Management, National Park Service, Corps of Engineers, federal wildlife refuges, tribal, and other such lands); to projects on federal, state, or private land for which there is *federal funding*; and to sites that are eligible for the National Register of Historic Places. Summaries of applicable programs are available on the Web at <http://laws.fws.gov/lawsdigest/historic.html>. Summaries are also available at <http://www.cr.nps.gov/museum/laws/lawregax.html>.

Hutt and others (1992) review preservation issues in depth. Artifacts, including historic writing, which are 50 years old or more are to be preserved under this Act. (See 50-year rule, page 341.) The ARPA defines archaeological resources on federal lands and contains guidelines for excavation or other investigation of possible Native American archaeological sites, with refer-ences to the American Indian Religious Freedom Act of 1978 and the Native American Graves Protection and Repatriation Act (NAGPRA) of 1990.

The Federal Cave Resources Protection Act of 1988 contains guidelines for the identification and protection of significant caves. (See cave protection laws, page 217.) The Act also allows for the withholding of information concerning the locations of nominated caves. Cultural resources are one of six criteria that can form the basis for significant cave nominations under this Act. (See Appendix 1, Federal Cave Resources Protection Act of 1988, page 507.)

States and other jurisdictions also have laws regarding cultural resource protection, which in some cases may apply to sites on private and state land. A gray area will emerge in the coming decades as the definition of historic artifacts begins to merge with the content of modern recording media. There may be a case for amending the definition and its context.

Vandalism

Vandalism is the willful or ignorant destruction of cultural or natural objects. There are two forms of vandalism, unintentional and intentional.

Unintentional vandalism is caused by ignorance about what constitutes resources protected by law, or by one's own definition of vandalism. For example, someone not "seeing" rock art figures or not receiving education through interpretive means, may spray paint names on, or build a fire next to, a panel of rock art figures and cause either obliteration by paint or soot, or rock spalling due to heat expansion.

Intentional vandalism consists of power or "turf" messages and personal identifiers, created as a reaction against authority or the "establishment," for group or individual recognition, as a vindictive act, or simply because there is nothing better to do (Bilbo 1987). (See Figures 10a and 10b.)

Vandalism and Cultural Resources

Despite public education efforts, spray paint graffiti is all too common in urban areas and in natural settings, including caves. In a small shelter in southeastern New Mexico the words "ALIEN WILL" are incised just above five red rock art figures. According to a forensic psychologist, (one who deals with the social and legal aspects of criminals, gangs, and other fringe groups) the words are believed to be the work of some followers of beliefs regarding space aliens, crystal power, and so forth. Possibly the same ideas went into the incising of flying saucers and the alteration of some figures on

Figure 10a. Example of gang graffiti in the U.S.

Bilbo & Bilbo

the extensive prehistoric to historic rock art panels at Comanche Gap, near Santa Fe, New Mexico (Figure 10b). Is this intentional vandalism?

Mid-to-late 20th century graffiti, much of it consisting of roadside name painting, or urban secretive or illegible statements, or personal identifiers and gang territorial messages, will persist into the future. But does modern graffiti need to be preserved? Do the Antiquities Act and other federal, state, and area laws need to be amended to provide for exclusions of gang or roadside graffiti and contemporary rock art done by Native Americans?

Far more is known and documented about gang graffiti than may ever be known about the creators of historic writing or historic rock art or prehistoric rock art. Usually the best we can describe about rock art is that a

Figure 10b. Intentional vandalism in northern New Mexico.

Bilbo & Bilbo

culture from a stated time frame defined by archaeological evidence, may be the creators of a given rock art site or style. In some instances, it can be assigned to a known group, but the creators of most rock art are essentially unknown and in a problematic time period. Contemporary Native American rock art may be studied more easily because the authors are still present.

We became concerned about rock art preservation when we read a statement by Welsch (1993) stating that "petroglyph is a fancy word for graffiti." He went on to say that "some were probably the product of too much time and leftover paint," confusing petroglyphs with pictographs.

White (1993) stated that "… rock art is interesting and sometimes historically important." He then used the term "rock art" for modern graffiti, stating that "human beings have been obsessed with marking our passage in time and space … and that early cultures in North America created rock art as … perhaps an early illiterate signature."

Prehistoric Native American cultures, while lacking the written languages of modern Western cultures, were not illiterate for their time. Their written communication consisted of visual statements like Mayan and Egyptian hieroglyphics, which have been translated into modern meanings.

The fact that most Native American rock art figures have uncertain or unknown meaning to us does not make them illiterate marks. Consider that many Native American cultures may have originated in Asia, where pictographic writing is in use even now. For example, Chinese writing, which dates back over 5,000 years and has been modified over time, still consists of pictograms and pictorial statements depicting concepts, objects, or other meanings. Early rock art researchers called rock art by the perhaps better term *picture-writing*. The letter-by-letter word construction based on Arabic script used for writing by western cultures is not the only measure of literacy. Inferring that rock art is an "illiterate mark" is an ethnocentric bias that should be avoided.

The letter-by-letter word construction based on Arabic script used for writing by western cultures is not the only measure of literacy. Inferring that rock art is an "illiterate mark" is an ethnocentric bias that should be avoided.

Rock art and historic writing, along with other archaeological artifacts, are cultural resources and must be identified, inventoried, and preserved under current laws and regulations. Special-interest groups have sometimes acted in ways that lack consideration of the original nature of a site. Prehistoric rock art figures in several New Mexico shelters were scraped away or had crosses incised over the sites, or both, during the 1700s by Spanish priests who were exorcising what they believed to be pagan messages (CM Carrillo, personal communication 1989). To Native Americans this was sacrilegious vandalism. Due to the date, however, the priest's activity could be considered evidence of a significant historic event. At the time, was it vandalism or graffiti?

In the 1980s a fundamentalist Christian sect used enamel paint to state their religious opinions over the figures on several of rock art panels in New Mexico (Figure 10b). Such actions should be considered graffiti and just as unsightly as other roadside graffiti. It is also insulting to present-day Native American people. But some view such religious actions to be sincere and legitimate.

Protecting Cultural Landscapes

There are different attitudes toward preservation of cultural and natural landscapes among regions, cultural groups, and even individuals. Those who value the cultural artifacts need to communicate the reasons for their approach and actively support conservation of sites. Concerned individuals can help by supporting and participating in the management and protection of sites, and by participating in educational activities about the significance and value of preserving cultural history (ARARA 1977, page 17).

Where can the line be drawn between graffiti and religious statements applied to rock surfaces? In the United States, the First Amendment and the American Indian Freedom Act protect one's right to practice religion. Other nations have similar laws. A discussion on the difficulty in addressing this aspect of graffiti is presented in the ARARA conference proceedings of the Symposium on Rock Art Conservation and Protection (ARARA 1977, page 84–86).

Others do not seem to value rock art and historic writings as part of cultural history. Welsch (1993) and White (1993) both appear to disparage rock art. Welsch (1993) considers that some rock art may be "symbolic, mystic, or artistic" but may be more likely to be power statements or personal identifiers.

Perhaps for prehistoric cultures, for which there is no written history as defined by western cultures, the visual statements along with other artifacts (potsherds, lithics, middens, tepee rings, pueblos, and so forth) help us know something about early cultures. Historic period rock art and historic writing are interesting and valuable additions to written history because they may document a name or an idea at a particular place and time.

Cultural resources should be protected and preserved, not only because there are laws saying so, but also because they are the basis of history. Cultural artifacts in caves have gained recognition due to the publication in popular books and journals of the significant prehistoric paintings in Lascaux and other French caves, at Altimira in Spain, and in the Mayan region of Central America. However, the minor occurrences of cultural artifacts, including rock art and historic writing, should also be recognized as part of mankind's history.

Examples of Well-Intentioned Restoration Projects

Restoration projects intended to remove modern graffiti have had varying results. Some have included looking for and recording historic or prehistoric writings or art before graffiti cleaning began, but others have not.

In 1969 at the direction of Texas Parks and Wildlife headquarters, sandblasting was used to remove "graffiti" from a rock shelter wall in Hueco Tanks State Historical Park, Texas. Mike Bilbo and Kay Sutherland, both rock art specialists familiar with most of some 5,000 figures at Hueco Tanks, were present when the cleanup project began. They asked that the work be stopped and were referred to the Austin, Texas, headquarters.

A well-intentioned restoration project early in 1992 focused only on the physical environment of Cave du Mayriers, France. However, significant portions of 15,000 year-old bison paintings were removed.

The sandblasting was stopped, but a significant portion of a large rock art panel had been destroyed (Bilbo 1983) (Figure 11). Several baseline studies of rock art and historic writing were then initiated due to the new recognition of the park's resources. But in 1984, while working at the park for the summer, Mike Bilbo observed a graffiti cleaning project that was conducted with minimal instructions or supervision, resulting in the loss of seven historic period figures at another site on the park. Forrest Kirkland (1967) had recorded many of the known figures at Hueco Tanks—thus, most of what was lost during the 1984 cleaning effort had been recorded.

A well-intentioned restoration project early in 1992 focused only on the physical environment of Cave du Mayriers, France (Art News 1992; Weaver 1992). However, significant portions of 15,000 year-old bison paintings were removed. No photographs or drawings were done. The group's leaders stated that the cave had not been designated as an historic site and so they were not aware it was protected, even though the presence of documented Paleolithic pictographs are widely known in the same part of France, Lascaux among them.

Debris or "junk" removal as part of a cave cleanup project also should be evaluated and included in the resource inventory. Some of it may be historic, with little or no documentation in existing historic records. In Carlsbad Cavern and in Ogle Cave, New Mexico, relics of guano mining activity during the 1930s and 1940s remain in place and have been inventoried and preserved.

In Matthews Cave, Alabama, Varnedoe and Lundquist (1993) removed military debris, including a barrel of diesel fuel, a 90-millimeter shell and other "junk" related to U.S. Army activity in the area during World War II. The artillery shell casings can be dated and indicate military activity though no other historical evidence or records have been found.

The preceding case-studies illustrate several basic principles for avoiding restoration problems:

- Pre-project planning
- Effective communication within an organization, agency, or group
- Recognition and acceptance of cultural resources and the importance of their preservation

Restoration Management Implications

The landowner or manager must determine whether or not graffiti is to be removed and must recognize that a vandalized, trashed out, and spray-painted cave may actually contain significant resources. An understanding of what constitutes graffiti, as opposed to what is historically and culturally significant, is a critical part of both inventory and restoration planning

Kirkland & Newcomb

Bilbo & Bilbo

Figure 11. Historic representational art from Hueco Tanks State Historical Park, Texas. Top drawing, before sandblasting in 1969 (from Kirkland and Newcomb 1967). Bottom drawing, the result of mechanical sandblasting in 1969.

processes. Removing graffiti may cause excessive impact on the site and restoration or resource management plans may include decisions not to restore sensitive vandalized sites. Each site must be judged separately based on what is in the best interest for the resources of the specific site. Various resource management tools can provide aid when writing prescriptions for each site. (See cave and karst management, Appendix 3, page 531.)

Resource Inventories and Cave Surveys

Restoring caves to as natural a condition as possible is not always the overall goal in cave projects. Identification and protection of cultural resources, including historic writings, rock art, and mud glyphs should also be considered—but these important resources can easily be overlooked or ignored.

Restoration plans should involve tying cultural resource sites into cave surveys, photographing, and recording the content, style, and media of all rock art and writing before graffiti is removed. Also note layered markings and superimposed graffiti during inventory activities.

Photographs can document the relationship of graffiti to rock art or historic writing. Photos are essential for study and evaluation before restoration work begins. Photographs, drawings, and field notes describing the site and media should also document rock art and writing in the unfortunate event anything is removed during restoration. (See photodocumentation, page 204.) Because entrance areas are frequent sites for rock art, historic writing, and contemporary graffiti, cave entrances should be carefully assessed and included in documentation. Cultural sites should always be tied into the cave survey to pinpoint locations.

Resource inventories should thoroughly describe cave attributes. Site locations should be coordinated with cave survey data. (See cave inventories, page 19.) Inventories, surveys, photographs, and other documentation are essential tools for planning restoration projects, coordinating cave rescue efforts, and recording the presence of natural hazards or hazardous materials.

Environmental Assessments

During restoration planning, the potential impacts of project methods should always be considered. On federal lands, an environmental assessment (EA) may be required to document descriptions of methods, evaluations of projected impacts, and recommendations for preservation, mitigation, or restoration actions.

For example, if the use of chemical agents is proposed, the effects on physical cave attributes, cave species, and habitats should be evaluated. An EA may also make recommendations for avoiding overzealous restoration efforts—for example, perhaps every speck of paint should *not* be removed from every pore of rock so that a more natural appearance results rather than totally scrubbed and discolored surfaces. For some sites, recommendations may include disguising paint remnants with mud from the cave floor—while for other cave sites, leaving paint specks exposed may be prescribed.

> On federal lands, an Environmental Assessment (EA) may be required to document descriptions of methods, evaluations of projected impacts, and recommendations for preservation, mitigation, or restoration actions.

Testing the effects of a particular method may also be part of an EA. For example, during field experiments between 1991 and 1995 Bilbo observed the wetting, drying, and integrity of cave surfaces when experimenting with several different mixtures of gypsum and water to make a naturally appearing cover for modern incised lines at one historic writing site in Fort Stanton Cave, New Mexico.

Similarly, in 1995 mud was prepared by Mike Bilbo and Val Hildreth-Werker from weathered gypsum and limestone dirt particles from the floor of Crocketts Cave and used to cover enamel paint graffiti to test the effect over a long term. At this writing the layer remains intact and appears very natural. (See camouflaging poultice, page 341.)

Effects of various restoration methods are included in ARARA (1977), Bilbo (1987, 1983), Bilbo and Ralph (1984), Morion (1989), Rhodes (1976), Ralph and Sutherland (1979), and White (1993). In addition, annual *NSS News* conservation issues and the NSS Conservation Committee Web site <http://www.caves.org/committee/conservation/> should be consulted for the latest techniques and cautions.

Many chemical techniques may adversely affect cave ecology, and the appearance and integrity of rock surfaces. (See anthropogenic chemicals, page 57.) Water and careful handwork may be the best choice as recommended by Goodbar and Hildreth-Werker in this volume (page 333.)

Consult with Experts

Every state has a State Historic Preservation Office (SHPO) or other agency responsible for archaeology and historic sites. Federal preservation and protection laws apply only to federal land in any state. The same laws will apply to state or local land if the state has incorporated them into state statutes (most states have) or when federal funds are being used for a project. Standardized recording methods and recording forms are available through SHPO agencies, local archaeological societies, and university archaeology departments.

These sources can supply or identify archaeology and history consultants for restoration projects. Information provided to these organizations as a result of restoration and recording projects should be considered proprietary. If individuals or groups are identified on the Internet, they should be considered in light of their affiliations with acceptable institutions and organizations.

Cave managers and restoration project leaders should become aware of agency functions and the assistance they can offer. Professional archaeologists and historians may participate in a restoration project as cultural resource consultants or project leaders. Often, good assistance comes from local or regional archaeological societies, whose membership may include one or more avocational rock art specialists. Likewise, local historical societies may have members who can identify historic writing and artifacts. Seek experts with experience in cultural material found in caves.

Every state has a State Historic Preservation Office (SHPO) or other agency responsible for archaeology and historic sites. Assistance may also come from local or regional archaeological societies, whose membership may include one or more avocational rock art specialists.

The importance of consultants is illustrated by the work of rock art specialists from Rupestrian CyberServices in Flagstaff, Arizona. While cataloging 41 known rock art sites in Hueco Tanks State Historical Park, three times that number were identified using digital enhancing techniques. This consultant information has brought emphasis to the importance of Hueco Tanks in the Native American cultural history of southwestern North America.

Equally important is that about three-quarters of the known 5,000 prehistoric and historic rock art figures and panels of historic writing at Hueco Tanks have been impacted by vandalism, campfires, rock climbers, weather, and wasp nests. The resulting inventory and mapping will provide the basis for determining the progress of continuing impacts at this important prehistoric to historic site on the outskirts of El Paso, Texas (Texas Parks and Wildlife Department 1999).

Public Outreach and Education

Educational outreach programs, such as Leave No Trace® and Tread Lightly®, teach people to be aware of the importance of preserving cultural artifacts and the natural environment. Efforts include interpretive signs and brochures and presentations by caving volunteers, specialists, and land managers to caving groups, schools, organizations, and governmental bodies at on-site and off-site locations (Bilbo and Ralph 1984).

The effectiveness of this approach depends on the presence and consistent availability of volunteers or agency interpretive staffs to monitor sites and plan and conduct public outreach activities. An effective cave management plan should include not only provisions for restoration and maintenance but also public education. It is through public outreach that the occurrence of vandalism and graffiti can be reduced or eliminated (Anderson 1977).

An effective cave management plan must include not only provisions for restoration and maintenance but public education as well.

Cited References

Advisory Council on Historic Preservation. 1973. *Procedures for the Protection of Historic and Cultural Properties, Title 36, Chapter viii, Part 800.* U.S. Government Printing Office. <http://archnet.asu.edu/archnet/topical/crm/usdocs/36cfr60.html>.

American Rock Art Research Association (ARARA). 1977. Preserving our rock art heritage. *Proceedings from the Symposium on Rock Art Conservation and Protection. American Rock Art Research Association, Occasional Paper 1*, ARARA. San Miguel (CA): American Rock Art Research Association.

Anderson DG. 1977. Archeology and speleology; the case for conservation. *NSS Bulletin* 39(2):55-63.

Anon. 1992. Blemished bisons. *Art News* 91(2):50.

Bain, JG. 1978. Paleolithic rock art in the United States (Abstract). In: Selected abstracts of papers, 1977 NSS Convention, Alpena, Michigan. *NSS News* 40(3):97.

Bilbo M. 1982. Historic names in Fort Stanton Cave. *Texas Caver* 27(6):113-117.
Bilbo M. 1983. Texas Parks and Wildlife Department Memo, Subject: *Graffiti Cleaning Techniques*. From Mike Bilbo, Interpretive Specialist to Ronald Ralph, Texas Parks and Wildlife Department Cultural Resource Manager, Austin, Texas.
Bilbo M. 1987. Hueco Tanks and the cost of conservation. *Southwestern Caver* 25(1):9-12.
Bilbo M. 1997. Carlsbad Caverns Natural Entrance Pictograph Sites 1 and 2. In: Freers SM, editor. *American Indian Rock Art Volume 23*. American Rock Art Research Association. p 41-47.
Bilbo M, Bilbo B. 1993. Pictographs of Slaughter Canyon Cave, Carlsbad Caverns National Park, New Mexico. *The Artifact* 31(1):59-70.
Bilbo M, Ralph R. 1984. *Rock Art Preservation in a Recreational Environment*. Manuscript prepared for annual American Rock Art Research Association symposium.
Bilbo M, Sutherland K. 1986. Stylistic evolution of rock art in the Jornada region. *The Artifact* 24(3):11-29.
Faulkner CH. 1986. *The Prehistoric Native American Art of Mud Glyph Cave*. Knoxville (TN): University of Tennessee Press. 124 p.
Faulkner CH. 1992. Prehistoric pictographs in southeastern caves. *NSS News* 50(6):144-145.
Gillett JB. 1976. *Six Years with the Texas Rangers 1875-1881*. Reprint edition. Lincoln (NE): University of Nebraska Press. 259 p.
Greer J, Greer M. 1996. *Dark Zone Rock Art in Surratt Cave, a Ceremonial Site in Central New Mexico*. Manuscript on file, Archaeological Records Management Section, Museum of New Mexico, Santa Fe. [Editors' note, see also: Greer J, Greer M. 1997. Dark zone rock art in Surratt Cave, a deep cave in central New Mexico. In: Freers SM, editor. *American Indian Rock Art Volume 23*. American Rock Art Research Association.]
Gurnee R, Gurnee J. 1985. *Cave Rock Art in the United States and Canada*. Manuscript of a Paper Given at International Rock Art Congress, Havana, Cuba.
Hutt S, Jones EW, McCallister ME. 1992. *Archaeological Resource Protection*. Washington, DC: Preservation Press, National Trust for Historic Preservation. 179 p. [Editors' note, see also: Hutt S and others. 1995. *Archaeological Resource Protection*. John Wiley and Sons, Inc.]
Kirkland F, Newcomb WW Jr. 1967. *The Rock Art of Texas Indians*. Austin (TX): University of Texas Press. 239 p.
Morion TL and others. 1989. Special Cave Restoration Issue. *NSS News* 47(2):32-47.
National Speleological Society. 1993. News Note: Cave Mountain Cave, West Virginia graffiti cleanup project. *NSS News* 51(4):114.
Patterson A. 1992. *A Field Guide to Rock Art Symbols of the Greater Southwest*. Boulder (CO): Johnson Books. 256 p.
Ralph R, Sutherland K. 1979. Pictograph conservation at Hueco Tanks State Historical Park: An initial experiment. In: Beckett PH, Wiseman RN, editors. *Jornada Mogollon Archaeology. Proceedings of the First Jornada Conference*. Las Cruces (NM): New Mexico State University. p 273-291.
Rhodes D. 1976. Cave restoration: a(n) (im)possible task. *NSS News* 34(11):200-202.
Schaafsma P. 1980. *Indian Rock Art of the Southwest*. Santa Fe (NM): School of American Research. 379 p. [Editors' note: see also Schaafsma P. 1980. *Indian Rock Art of the Southwest*. Albuquerque (NM): University of New Mexico Press.]

Schaafsma P. 1992. *Rock Art in New Mexico*. Fully revised and expanded edition. Santa Fe (NM): Museum of New Mexico Press. 175 p.
Simek JF, Cressler A, Faulkner CH, Ahlman TM, Creswell B, Franklin JD. 2001. The context of late prehistoric cave art: The art and archaeology of 11th Unnamed Cave, Tennessee. *Southeastern Archaeology* 20(2)[winter 2001]:142-153.
Simek JF, Cressler A. 2001. Issues in the study of prehistoric cave art. *Midcontinental Journal of Archaeology* 26(2)[fall 2001]:233-250. [Editors' note: This entire issue of MJA is about cave archaeology.]
Southwest Parks and Monuments Association (SPMA) and National Park Service. 1989. *El Morro National Monument Trail Guide*. Tucson (AZ): Southwest Parks and Monuments Association. 4 p.
Sutherland K. 1992. *Rock Paintings at Hueco Tanks State Historical Park*. Austin (TX): Texas Parks and Wildlife Department. 21 p. [Editors' note, see also: Sutherland K. 1991. Same title and publisher.]
Texas Parks and Wildlife Department. 1999. Computer-enhanced photos uncover Hueco Tanks rock art. *TPW News* Sept 13, 1999.
Varnedoe WW Jr, Lundquist CA. 1993. Matthews Cave, a profitable project for the Huntsville Grotto. *NSS News* 51(9):251-252.
Weaver HD. 1992. Glyph talk and graffiti. *Ozark Speleograph*. Sept 1992. p 36.
Welsch RL. 1993. Off the wall: Kilroy strikes again. *Natural History* 102(5):30,32.
White CM. 1993. The big "E" challenge: Large scale graffiti removal. *Recnote*. R-93-1:1-4. Vicksburg (MS): U.S. Corps of Engineers, Waterways Experiment Station.
Young MJ. 1990. *Signs from the Ancestors: Zuni Cultural Symbolism and Perception of Rock Art*. Albuquerque (NM): University of New Mexico Press. 308 p. [Editors' note, see also: Young MJ. 1988 same title and publisher.]

Additional Reading

Bilbo B, Bilbo M. 1995. Management implications of historic writing and rock art in caves. In: Pate DL, editor. *Proceedings of the 1993 National Cave Management Symposium: Carlsbad, New Mexico, October 27-30, 1993*. [Huntsville (AL)]: National Cave Management Symposium Steering Committee. p 233-264.
Douglas JC. 1998. Discovering history in American caves. *NSS News* 56(9):261-63.
Douglas JC, Smith MO, Simek JF. 2000. Identification and analysis of a Civil War soldier's name in South Carthage Cave, Tennessee. *Tennessee Caver* Spring:7-13.
Douglas JC. 2001. Historic preservation at Hubbard's Cave: Inventory and management of cultural resources. In: Rea GT, editor. *Proceedings of the 14th National Cave and Karst Management Symposium: Chatanooga, Tennessee, Ocotober 19-22, 1999*. Chattanooga (TN): Southeastern Cave Conservancy. p 46-50.
Gillieson D. 1996. *Caves: Processes, Development, and Management*. Cambridge (MA): Blackwell Publishers. p 237-267.
Goodbar JR, Hildreth-Werker V. 2006. Cave graffiti: The writing is on the wall. In: Hildreth-Werker V, Werker JC, editors. *Cave Conservation and Restoration*. Huntsville (AL): National Speleological Society. p 333-342. [This volume.]
Hagan S, Sutton M. 1991. Cave restoration can cause damage. *NSS News* 49(6):170-171.
Hubbard DA Jr. 1995. Cave conservation and cave clean-ups: Not always one in the same. *NSS News* 53(2):31.
Smith MO. 2003. *Civil War Cave Writings*. Unpublished ms presented at 2003 NSS Convention in Porterville, California. See Convention abstracts.

Section A—Identifying and Protecting Cave Resources

Karst Hydrology: Protecting and Restoring Caves and Their Hydrologic Systems

George Veni

Cavers tend to be conscientious. We try to tread softly through passages to limit our impact. We clean up and restore caves that have been impacted by others. We fight to preserve and protect caves and their contents from outside impacts like urbanization. We work to improve our restoration and protection methods, and, through vehicles like this book, share that information as much as possible.

Many of the adverse impacts a cave may suffer and the means to prevent or alleviate them are determined by the cave's hydrology. This chapter provides hydrologic information and guidelines to assist cavers in protecting and restoring caves. It teaches the basics of how caves form and how water moves through caves and their surrounding landscapes. The chapter also examines common hydrologic problems and impacts on caves, and what problems can be solved by individual and group actions.

Many of the adverse impacts a cave may suffer and the means to prevent or alleviate them are determined by the cave's hydrology.

The following sections are meant to reach cavers of all experience levels. References are cited for those wanting details. Specific recommendations are included, but the focus is on general principles to help guide cavers through situations that cannot be covered within this chapter.

The Basics of Karst Hydrology

How Water Enters, Moves Through, and Exits Caves

The movement of water through caves is closely tied to the question of how caves form. Moore and Sullivan (1997) provide a good basic overview, while White (1988), Ford and Williams (1989), and Klimchouk and others (2000) offer highly detailed information. Gillieson (1996) offers less detail but more emphasis on cave management.

Despite their wide variety of origins, caves or cave areas can be classified in one of three groups: carbonate, evaporite, and pseudokarst (Figure 1). Caves in carbonate rocks form primarily in limestone, but some also occur in dolomite and marble. Caves in evaporite rocks usually form in gypsum but also in halite in exceptionally arid climates. Caves in both carbonate and evaporite rocks form primarily by water dissolving away the bedrock. The landscapes where such solutional processes are dominant are called karst.

It is beyond the scope of this brief chapter to discuss all cave types in detail. Limestone caves will be emphasized since they are the most common.

The typical limestone cave begins to form where water enters the rock along a fracture or bedding plane and slowly flows downward and laterally until discharged from a spring at a lower elevation. While pure water has little ability to dissolve limestone, water entering the ground is charged with carbon dioxide from the atmosphere and soil to form carbonic acid. Over millennia, the weak acid enlarges fractures and bedding planes. As the openings become larger, water drains more efficiently. These increasing

Pseudokarst
George Veni

Caves and karst-like features that do not form primarily by the dissolution of the rock are called pseudokarst.

Several types of pseudokarst features occur. Major types include the following:

- **Sea Caves**. Formed where waves pound on cliffs and preferentially enlarge fractures, zones of softer rock, or areas where wave action is concentrated.

(Continued on next page)

volumes of water then enlarge the openings at faster rates. This process self-accelerates. Eventually, one flow path toward the spring dominates the local drainage pattern and captures flows from smaller channels. When it becomes large enough for human exploration, we call that conduit a cave.

In the more common situation, a cave map looks like a branching surface stream. The tips of the hydrologic network typically include fractures, sinkholes, and swallets that capture surface water and route it underground. In the subsurface, each branch flows downstream to join other branches, eventually forming limbs and then the trunk of the underground drainage network which discharges from a spring (*angular* and *curvilinear* passages, Figure 1).

Geologic and hydrologic factors often prevent the development of such ideal flow systems, which usually occur in relatively flat-lying, highly-fractured rocks that sit atop relatively impermeable rocks.

Local geologic factors frequently affect cave development. Low fracture frequency and/or continuation of the limestone deep below spring levels result in cave networks that extend deep below the water table. Caves in dipping rocks may branch asymmetrically, capturing most water from the updip direction.

Figure 1 shows some other examples of how hydrogeologic factors that create a cave relate to the cave's shape.

- Braided or anastomotic passage patterns indicate slow or ponded conditions.
- Rectilinear mazes may suggest development by flooding or seepage through a caprock.
- Ramiform and spongework patterns, such as in Carlsbad Caverns, did not form epigenically by water flowing down from the surface, but hypogenically by rising hydrogen sulfide gas mixing with groundwater to form sulfuric acid, which in turn dissolved the limestone to create caves.

Figure 1. A summary chart of how cave morphology reflects hydrogeologic origin (from Palmer 1991).

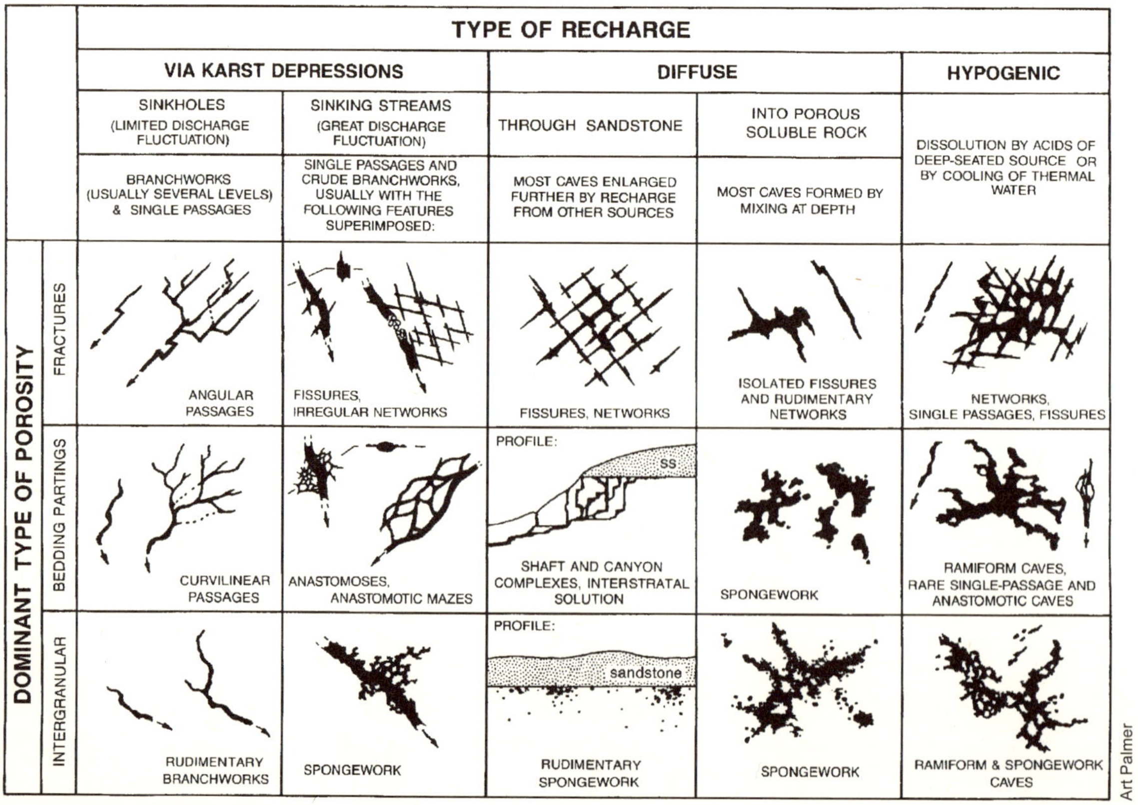

Beyond the Cave: Water in the Drainage Basin

For most of the 20th century, scientists argued whether caves form above, below, or at the water table, or even whether water tables actually exist in karst. We now know that caves form in all of those situations, and that karst water tables do exist. Caves are integral parts of karst aquifers and are a part of an interconnected series of voids that transmit water down through the vadose zone (the area above the water table) and into the phreatic zone (the water-saturated area below the water table), and eventually out of the spring. Aquifers are reservoirs of water held underground in whatever voids exist in the rocks and soil. In karst, these voids make up complicated networks of fractures, bedding planes, some pore spaces, and of course, caves.

The area that feeds all water into a stream, spring, or cave is called the drainage basin. Valley ridgelines define the drainage basin boundaries for surface streams. All water that enters the basins must eventually flow through these streams. Groundwater drainage basins do not always conform to the boundaries of surface basins. Karst groundwater basins are notorious for crossing below surface basin boundaries, but they still function by similar principles. Rather than surface water flowing down a topographic surface, groundwater flows down from "ridges" to "valleys" or troughs in the water table where flow is concentrated. Significant cave streams often flow along these troughs.

Subsurface flow converges on large conduits because of their greater ability to transmit water. Even hypogenic caves, which form independently of water entering the ground from the surface, capture local flow paths from the surface to form drips, domes, pools, and streams. Figure 2 illustrates how water that seeps or floods into sinkholes, fractures, and sinking streams flows down troughs in the water table to merge into single large cave streams, and discharges from the same spring. It should therefore be clear that caves, plus the rate, volume, and quality of water that flows through them, and the materials carried in the water, directly reflect the conditions and activities on the surface in the cave's drainage basin.

Cave Chemistry and Speleothems

Some of the water which enters caves is responsible for creating speleothems. Hill and Forti (1997) provide the authoritative review of speleothems and cave minerals. While there are over 250 minerals and dozens of types and subtypes of speleothem forms, calcite and gypsum speleothems are by far the most common and are those discussed here.

As water moves through a rock, it dissolves minerals along the way and carries the ions in solution. At some point, the water may become supersaturated with respect to a particular mineral, meaning it is carrying more of the dissolved material than it can hold, and so it deposits the minerals.

The amount of carbon dioxide (CO_2) in the water primarily determines how much dissolved calcite the water will hold, and where and at what rate the calcite will be released. The more CO_2 in the water, the more calcite it can keep in solution. Calcite is most commonly deposited where sufficient CO_2 is released from the water to result in calcite supersaturation. Typical locations include where water emerges from the limestone wall, releasing carbon dioxide into the cave atmosphere which has far less CO_2, or where turbulence releases CO_2 by splashing the water onto a stalagmite or running it over a flowstone slope or rimstone dam.

Evaporation also plays a part in the deposition of calcite but is far more important in the development of gypsum speleothems and is their usual cause of supersaturation. While calcite does not dissolve in pure water, gypsum is soluble in water and thus more chemically vulnerable to changes in cave environments. Consequently, gypsum speleothems form primarily in dry passages that no longer contain active streams.

(Continued)

- **Tectonic Caves**. Humanly enterable fractures that occur where large sections of bedrock separate, such as by one side slumping down a hillside or valley.

- **Talus Caves**. Humanlyenterable spaces that occur beneath and within piles of large fallen rocks.

- **Suffosion Features**. These include caves and sinkhole-like depressions that form by the localized downward movement of fine, unconsolidated material through or below locally denser or better-cemented material.

- **Volcanic Pseudo-karst**. Lava tubes are the most extensive type of volcanic pseudokarst. Lava tubes form where molten lava drains from beneath a cooling, solidifying lava flow. (See lava tube caves, page 133.)

While these features form by radically different processes than karst caves, they often bear hydrologic similarities such as turbulent flow through large conduits and little to no filtration of contaminants. Much of the information presented in this chapter will also apply to pseudokarst areas.

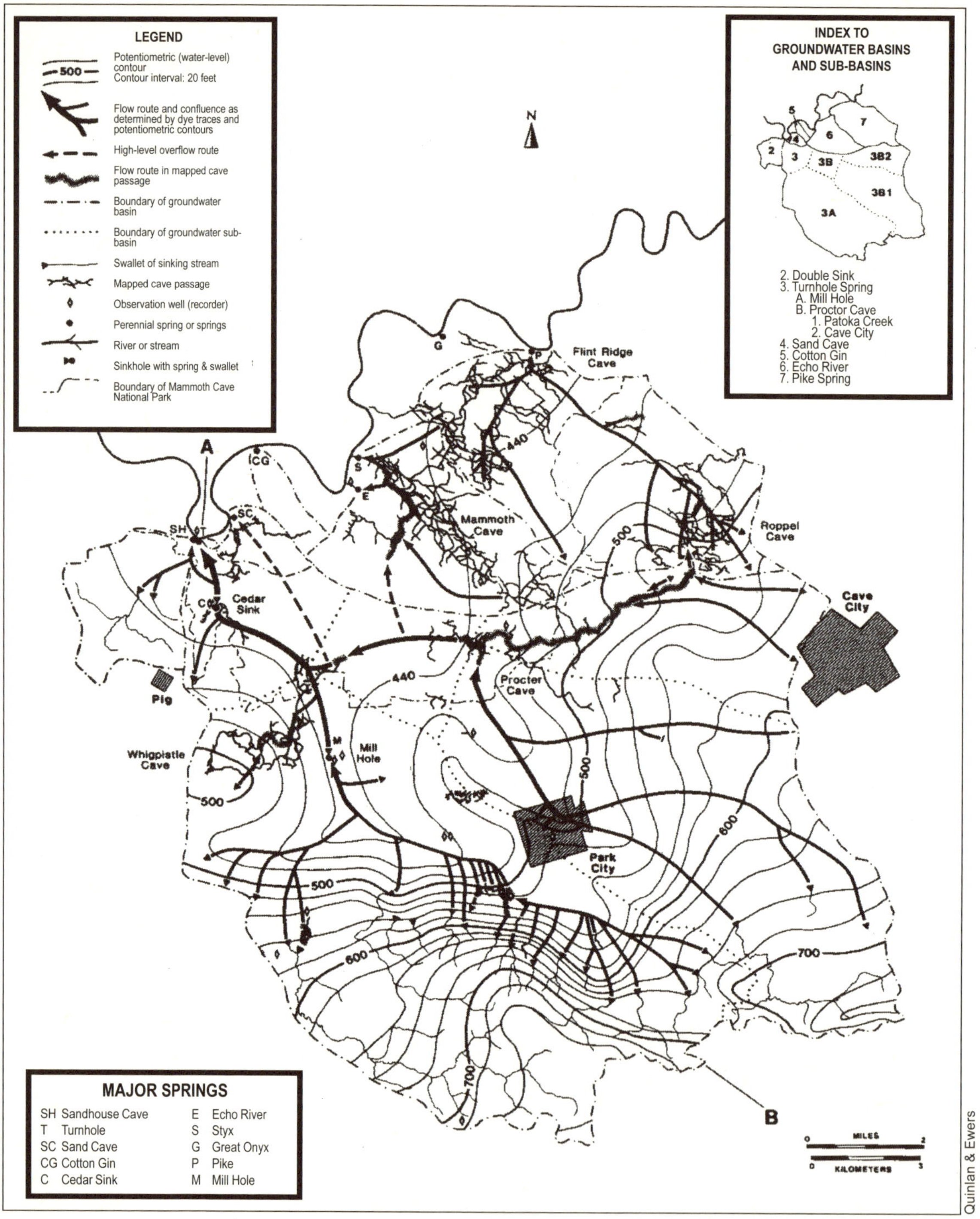

Figure 2. Map of the Turnhole Spring groundwater drainage basin and some smaller adjacent basins (from Quinlan and Ewers 1989).

The ages of speleothems are frequently misunderstood. Ages are often estimated based on an assumed constant growth rate, which is rarely the case. Each speleothem is different, reflecting the unique water chemistry and flow rate along each flow path, as well as soil and climatic factors on the surface. All of these factors change with time, and thus change the conditions of speleothem development.

Generally, the larger speleothems are the oldest, but not always; they may just reflect rapid mineral deposition, while nearby small speleothems, which form by slower deposition rates, may in fact be much older. Under certain unusual conditions, speleothems can grow up to several centimeters within a year. Typically, several centimeters of growth requires hundreds to thousands of years. Thus, any impacts on speleothems are effectively permanent.

The Sensitivity of Karst Systems to Contamination and Modification

Among the various types of aquifers around the world, karst aquifers are the most sensitive to groundwater contamination and the most easily impacted by modification of the surface landscape. Drew and Hötzl (1999) describe these impacts in detail. Caves, solutionally enlarged fractures, and other voids create highly permeable features so that the volume of water discharged from karst wells and springs are the greatest volumes on record. However, this same permeability also allows easy access for pollutants. Karst aquifers can respond so quickly to surface activities, that in many ways, they function as direct extensions of the surface landscapes. The conduit systems in karst allow for the rapid transmission of contaminants with effectively no filtration. Studies repeatedly show that when contaminants are present in karst areas, they invariably reach and adversely impact their aquifers.

Among the various types of aquifers around the world, karst aquifers are the most sensitive to groundwater contamination and the most easily impacted by modification of the surface landscape.

The close hydrologic connection with the surface also makes karst aquifers sensitive to physical changes in the landscape. Increased flooding, decreased runoff, sedimentation, and erosion on the surface are often mirrored by changes in caves and the general behavior of the karst aquifer. Well pumping and artificial recharge of water can also dramatically change aquifer conditions on a scale and rate that is much greater than in other groundwater systems. Few effective engineering solutions have been developed to prevent many of the problems which are unique to karst. It is therefore vital that people living and working in karst areas avoid creating problems in the first place. Watson and others (1997) and Veni and DuChene (2001) offer broad, yet concise guidelines for the protection and management of karst areas.

Common Hydrologic Problems: How Cavers Can Avoid, Find, and Resolve Them

The many types of activities that threaten caves and karst areas fall into two groups: those resulting from pollution and those caused by physical modification of the karst system. This section addresses the most common problems and provides guidance on what cavers can do to resolve them, beginning with problems most directly related to cavers and ending with those over which cavers have no direct control. Hydrologic issues related to cave restoration activities are discussed at the end of this section.

Groundwater Pollution

Exploration Pollution

"Pack it in. Pack it out." This motto has frequently been used by cavers to state that we should carry our trash out of caves. Spent batteries or carbide are especially hazardous to groundwater. They should be removed in leak-proof containers and disposed of properly. Too many stories are still told of how bags or bottles of spent carbide ruptured or exploded. Let's do the job right and use appropriate containers.

Human Waste. Unfortunately, urine and feces are sometimes deposited in caves. In hydrologically inactive caves, these wastes should instead be packaged and removed whether or not the caves are decorated with speleothems. Certain areas in such caves tend to become restrooms which accumulate wastes. Even if buried in sediments, the wastes may migrate into the local groundwater. In a concentrated amount they could prove harmful. (See human waste, pages 35, 71, and 276; also see packaging waste, page 269.)

In hydrologically active caves, while unappealing, relatively small amounts of human wastes left in caves will get flushed out of the system or mixed with animal wastes washed in from the surface and will cause no significant impact.

If camping in or near a cave entrance and packing out wastes is not an option, cavers should dig latrines as far from the entrance as practical in deep, vegetated soil where biological activity will more rapidly decompose the wastes. Latrines should not be placed where they may be rapidly eroded by surface runoff. If possible, they should be placed down slope from springs, sinkholes, or cave entrances.

Contaminated water that soaks into cave sediments should not be discounted as harmless; many species burrow and lay eggs in the sediments and can be injured or killed by acids or bleach.

Bathing and Washing Gear. Bathing and washing camping and caving equipment in hydrologically active caves is not a problem from a general groundwater contamination standpoint. However, in hydrologically inactive caves with only isolated pools, contact with the pools should be avoided. Cavers may use them as untreated water supplies and could become ill. Additionally, the pools could be biologically significant and harmed by caver-introduced bacteria. (See pools page 74.)

Tracer Tests. Dye tracing has been used by cavers to delineate drainage basins and hydrologic connections between caves, sinkholes, and springs. However, as increasing numbers of tracer tests are conducted, state and municipal authorities are increasingly regulating their use. Cavers could be fined for tracer testing without an appropriate permit.

One important reason for regulation is to reduce cross-contamination, where one tracer test unknowingly picks up dye from a second test and produces an inaccurate interpretation. Since tracer tests are important in decisions on land, water, and waste management, undiscovered interference in the results could have major repercussions. While most dyes have low toxicities, some are more hazardous and should not be used if they may reach a spring or well used for drinking water.

Cavers who wish to dye trace should first check their state regulations, check carefully for other possible traces in the area, inform potentially affected people of their trace, and learn the proper methods, amounts, and types of dyes prior to injecting them into the karst. Eckenfelder (1996) provides guidance on designing and implementing sound tracer studies.

Restoration Pollution

Like some pollutants resulting from exploration, pollutants produced during the restoration of caves are not usually produced in sufficient volume to harm an aquifer's groundwater quality. However, these pollutants can severely impact water quality in localized areas within caves to the detriment of cave fauna and cavers who may use that water for drinking.

Contaminated water that soaks into cave sediments should not be discounted as harmless; many species burrow and lay eggs in the sediments and can be injured or killed by acids or bleach running off walls and speleothems after removing graffiti and algae. Too much water flowing into normally dry passages can be equally destructive to cave ecosystems.

The use of chemicals in caves should be avoided or minimized. If applied, all chemicals should be thoroughly rinsed and diluted with water. Clean towels and sponges should be placed immediately downslope of restored surfaces to capture as much runoff as possible. Runoff containing chemicals should be placed into water-tight sealable bottles, then removed from the cave and deposited into a municipal sewage system for proper treatment. (See anthropogenic chemicals, page 57.)

Rural Pollution

Caves located in rural areas have the best chance to avoid human-produced contaminants. While such contaminants do occur in rural areas, they do not usually occur in the volume and severity found in urban and industrial areas.

The drainage basins of many rural caves include areas of light to intensive agriculture. Extensive activities, such as livestock grazing, result in higher than natural bacteriological contamination of caves (Figure 3). Such operations cannot be easily relocated, but they pose far less of an impact than runoff from feedlots or barnyards.

Caves located in rural areas have the best chance to avoid human-produced contaminants.

Landowner Assistance. Many rural landowners recognize the hazard to their own drinking water supplies and do not place septic systems and barnyards near caves and sinkholes. Cavers can assist landowners by guiding such activities away from surface drainages that feed into caves or from areas and sinkholes directly above caves that are identified during cave mapping. Cavers can especially help by discouraging owners from dumping trash into caves and sinkholes and by cleaning up existing dumps. In some areas, municipal or regional agencies may offer funding or other support to volunteers and owners who clean up such sites. (See sinkhole cleanout, page 381.)

Residents of rural areas should be made aware of the special sensitivity of karst to pollution. Cavers can offer publications and photographs showing how pesticides and fertilizers used in some agriculture and pollutants in trash dumps may contaminate karst aquifers. Rural residents' standard water purification method of sedimentation in storage tanks, natural die-off of some pathogens, and chlorine treatment of any remaining organisms may not adequately cleanse the water.

Urban and Industrial Pollution

Urban and industrial areas produce pollutants in the greatest volume, variety, and toxicity, and they survive the longest in the environment. While it is tempting to suggest plugging caves and sinkholes, or at least diverting drainage away from them to prevent groundwater contamination, such actions will often prove ineffective. Polluted runoff can enter karst aquifers in greatest volume via large entrances or sinkholes, but significant pollution can also flow underground through small sinkholes, solutionally enlarged fractures, and other less obvious pathways. Protection of individual karst features alone is not a sound method for aquifer protection.

While it is tempting to suggest plugging caves and sinkholes, or at least diverting drainage away from them to prevent groundwater contamination, such actions will often prove ineffective.

Pollution Prevention Strategies. Protection of aquifer water quality requires excluding contaminants from the karst area and preventing those present from entering the surface water or groundwater. Prevention is not

George Veni

Figure 3. Livestock in a sinkhole pond.

easy. Sedimentation and filtration basins are among a group of measures called *best management practices* used to capture and remove contaminants. (See best management practices, page 34; also see current best practices, page 17.)

However, it is important to remember that *best* does not mean *effective*. Their efficiency varies widely according to the specific method used and the pollutant treated, with common contaminant removal rates less than 40% (for example, Tenney and others 1995).

Veni (1999) proposed a strategy for assessing potential impacts in karst and found where there are insufficient data for definitive assessments, limiting impervious cover to 15% of the area of the drainage basin should preserve groundwater quality. This percentage correlates to the level of urbanization where a broad suite of studies found that significant ecological and water quality degradation begins to occur (Schueler 1994).

Pollution Prevention Coalitions. The water quality problems which exist in urban karst areas are well beyond the ability of most cavers to solve individually. As groups, cavers should learn to work cooperatively with landowners and regulatory agencies, yet also be prepared for adversarial relationships with regulators who may lack the authority or impetus for certain protective actions without sufficient public outcry.

In many cases, the best solution is to purchase critical caves and drainage areas for protection from urbanization. Examples exist throughout the country of volunteer groups pushing major land purchases made by local, state, and federal agencies, by cave conservancies and organizations like The Nature Conservancy, and even by businesses looking to develop good public images and tax write-offs for conservation easements.

In many cases, the best solution is to purchase critical caves and drainage areas for protection from urbanization.

In working with these groups, cavers should be careful to not oversell their technical knowledge and expertise if they don't have an actual specialist in their group. However, cavers should be equally careful not to undersell their perspective. Most regulators and many nonkarst geologists lack cavers' important information and knowledge from inside the aquifer. Maps, photos, and videos demonstrating how water and contaminants move through caves can be worth far more to regulators and judges than the opinions of well-paid consultants who never venture underground into a karst aquifer. (See photodocumentation, page 204.)

In working with these groups, cavers should be careful to not oversell their technical knowledge and expertise if they don't have an actual specialist in their group. However, cavers should be equally careful not to undersell their perspective.

Land purchases will not be practical in many situations, in which case cavers should work with agencies and organizations to educate the public and promote sound voluntary stewardship of the karst by public and private landowners. Conservation easements, tax relief incentives, and support for ecologically sound yet profitable land uses can also be promoted to reduce impacts in karst areas and drainage basins.

Hydrologic Modifications

Exploration Modifications

To further exploration, cavers have drained sumps, diverted streams, notched rimstone dams, and built dams, among other actions that hydrologically changed caves. The possible consequences of each action are too numerous to individually discuss and rely too much on specific local conditions beyond the scope of this chapter.

Most hydrologic changes from exploration tend to be short-term and thus probably of no significant impact. The impacts of long-term or permanent changes should be closely considered before implementation, and tested, if possible, with temporary modifications that produce the same effects. Distinguishing between short-term and long-term modifications will need to be done case-by-case. The impacts of hydrologic modifications from exploration can be addressed as two general groups.

Hydrologic Impacts. One type of impact is the hydrologic change in the cave and aquifer. Water in cave streams flows to springs and/or wells and may be increased or decreased by dams or diversion. Removal of natural dams (breakdown, rimstone, or otherwise) may result in turbid (muddy) flow for some time as sediment once trapped behind the dam is washed out and affects the stream's use as a water supply. Speleothem growth and development may change in affected parts of the cave. Dam construction would have opposite effects, increasing flooding and sedimentation of upstream areas, and causing some sediment erosion in downstream areas.

Biological Impacts. Hydrologic modifications can also adversely affect a cave's ecosystem. Permanent draining of sumps or flooding of passages may isolate populations and disrupt feeding and reproductive behavior. Changes in water depth, temperature, turbidity, velocity, flood levels, and flood frequency may also harm ecosystems adapted to specific conditions. (See conservervation don'ts, page 35.)

Water Chemistry Modifications

Discounting the introduction of contaminants into a cave and the hydrologic modifications discussed above, a significant hydrochemical change occurs in some show caves which divert water from one part of the cave to another to "reactivate" inactive speleothems or provide constant flow to waterfalls. The idea is that since water from the cave is used, speleothems will grow and the cave will be kept "natural." The problem is that the chemistry of waters throughout a cave may vary dramatically. The water diverted to reactivate a speleothem, especially if diverted from a flowing stream, may be chemically undersaturated and hence would begin to dissolve the speleothem. Certainly, wet speleothems are more sparkly and attractive than dry ones, but dry speleothems are also natural features of caves.

Surface Modifications

The effects of modifications on the land surface provide additional examples of how closely caves and karst aquifers are tied to activities on the surface.

The effects of modifications on the land surface provide additional examples of how closely caves and karst aquifers are tied to activities on the surface.

Vegetation and Soil Loss. Vegetation and soil are important to preserving the reservoirs of shallow groundwater that sustain speleothem growth and pools in the caves below. While vegetation tends to use water stored in the soil, it also holds the soil in place, protecting it from erosion and high evaporation of its water. Significant loss of vegetation invariably leads to the loss of soil, soil moisture, and moisture in the caves. The chemical changes in the water are less predictable. Less soil and vegetation means less CO_2. The resulting changes in water flow rates into the cave and reaction rates of the water with the limestone may prompt, halt, or have no measurable effect on speleothem growth. Eroded soil also often ends up deposited in and sometimes plugs caves and sinkholes.

Increases in Surface Water Runoff. Decreases in soil and vegetation on the surface result in greater and faster runoff of rainwater. Impervious covers like concrete and asphalt allow no infiltration or storage, and all rainwater runs off into streams. Increased runoff results in higher and more frequent flooding of streams. If the streams lead underground, caves may be flooded beyond their normal levels, destroying habitat for cave species adapted to dry conditions in those areas, dissolving speleothems, depositing sediments in some areas, and eroding old sediment deposits in other areas that might hold paleontologically or archaeologically significant materials. Gypsum speleothems, even above the new flood levels, could potentially be dissolved by flood-induced increases in humid air flow.

While the quarrying or mining away of a cave are obvious impacts on cave hydrology, major adverse impacts are possible in caves not directly touched by the operations.

Quarries and Mines. Quarrying is the complete removal of the rock for use, such as the removal of limestone for building stone, concrete, and gravel. Mining is the selective removal of material, such as minerals and fuels, from within the larger bedrock mass. Quarrying occurs by digging huge open pits, while mining can be through open pits, tunneling, or well drilling.

While the quarrying or mining away of a cave are obvious impacts on cave hydrology, major adverse impacts are possible in caves not directly touched by the operations. The most common problem is dewatering or lowering of the natural water table, which can drain cave streams, stop springflows, and cause sinkholes to collapse (Figure 4). Another severe problem is groundwater contamination, especially from acids and metal-laced waters discharging into streams or the ground from mining activities.

Dams. Dams are built to hold water, which is extremely difficult to accomplish in karst areas. Caves and sinkholes below and near dam sites are plugged with concrete to keep the water from leaking out of the reservoir. Nonetheless, water levels and flooding in caves upstream and downstream of the dam will increase because most of the caves and karst features will not be found and plugged. Those above the dam will be inundated if they are located beneath the reservoir; caves further away and at higher elevations may have their lower levels flooded by the elevated water table produced by the dammed lake. Caves and springs downstream of the dam will see a marked increase in flow from the leakage that will invariably occur despite the best efforts to block it.

Regulations. All of the surface modifications described above are likely beyond the ability of individual cavers to affect. Fortunately, tougher regulations and more in-depth research prior to the permitting of such operations are decreasing the frequency and intensity of these problems. Cavers working with the permitting process may help regulators with the important measures listed here.

- Steer operations away from the most sensitive areas.
- Enact tough protective measures if sensitive areas cannot be avoided.
- Perhaps limit already permitted activities if new information shows the potential for severe impacts that were previously unknown.

As discussed in the above section on pollution, conservation actions that preserve, purchase, or consider the entire drainage basin will be most

George Veni

Figure 4. Sinkholes may collapse from several factors that modify their hydrology. Groundwater contamination can result if a collapse breached sewage and other pollutant-bearing pipelines.

effective in protecting caves and aquifers from activities that modify large sections of the karst surface.

Hydrology and Cave Restoration

Restoration activities have mostly beneficial hydrologic impacts on caves; adverse impacts are few and minor. The removal of unnatural rock and sediment piles is important to restoring natural water and sediment flow through caves. Trash removal improves water quality and ecological integrity.

Concentrated sprays of water are effective in cleaning many cave surfaces. (See pressurized water for restoration, page 397.) Unfortunately, bleach is commonly used in water sprays to remove algae in show caves. Fungi have been cleaned from gypsum crusts with a bleach mist solution, followed by rinsing with water the next day (Rogers 1995). Until new methods are available, bleach should be minimized in concentration, volume, and area because it indiscriminately kills cave organisms. (See bleach, page 349; also see lamp flora control, page 343.) Research is also needed for better ways of cleaning gypsum speleothems. Spray cleaning dissolves gypsum and frequent cleaning may prove harmful. (See gypsum cleaning, page 419.)

Restoration activities must be planned to avoid adverse impacts from the use of inappropriate chemicals and materials, or appropriate materials used in inappropriate ways. Other chapters in this book describe cave-safe supplies and methods in detail. Luckily, the quantity of materials and the areas of restoration are generally small; improper techniques and supplies will usually not have adverse cave-wide hydrologic impacts, and there will be time for corrective measures.

Restoration activities have mostly benefi-cial hydrologic impacts on caves. Adverse impacts are few and minor.

In *hydrologically active caves*, there is probably no substantial difference in impact by using cave water instead of noncave water to clean speleothems, so long as the water source is unpolluted and unchlorinated. (See sources for restoration water, page 393.) Water brought into a cave that is significantly hotter or colder than the cave's temperature should be allowed to equalize with the cave if rare species may be present. The short-term and small-range application of the water should make any differences in chemistry inconsequential, especially if the restoration water is captured after use and removed from the cave. Water from within a cave may be the most convenient, but use it with caution if few pools exist and are slow to replenish. Filtering the used water for placement back into the pools is an option assuming chemicals have not been added to the water and the pool has no rare organisms and was not completely drained. (See filtration and pool restoration, page 415.) Filtered restoration water could be used for additional washing if water is scarce. If the cave is dry and has a sensitive microbiological community, water from surface streams could introduce new microorganisms and should not be used. (See isolated pools, page 73; see uncontaminated pools, page 74; also see water choices, page 338.)

Summary

The hydrologic protection and management of caves requires consideration of a broad range of factors. They necessitate understanding of small features within caves to karst aquifer drainage basins that extend far beyond a cave's known limits. Small-scale considerations typically involve aesthetics and biological management. Large scale factors include small-scale concerns, but mainly involve issues of groundwater quality and quantity, plus land use management.

As urbanization, mining, and other intensive uses of karst areas increase, cavers will need to extend their efforts in cave restoration and protection to include the landscapes in which they occur. They will need to coordinate with and lobby regulatory agencies for general land resource protection. It is vital for cavers to work with other groups to purchase and preserve karst drainage areas in perpetuity as the only assured means of conserving our fragile and unique underground resources.

Cited References

Drew D, Hötzl H, editors. 1999. *Karst hydrogeology and human activities: Impacts, consequences and implications. International Contributions to Hydrogeology* 20. Rotterdam: A.A. Balkema Publishers. 322 p.

Eckenfelder, Inc. 1996. *Guidelines for wellhead and springhead protection area delineation in carbonate rocks.* Region 4, US Environmental Protection Agency, EPA 904-B-97-003.

Ford DC, Williams PW. 1989. *Karst geomorphology and hydrology.* London: Unwin Hyman. 601 p.

Gillieson DS. 1996. *Caves: Processes, Development, Management.* Oxford: Blackwell Publishers. 324 p.

Hill CA, Forti P, editors. 1997. *Cave Minerals of the World.* 2nd ed Huntsville (AL): National Speleological Society. 463 p.

Klimchouk (Klimchuk) AB, Ford DC, Palmer AN, Dreybrodt W, editors. 2000. *Speleogenesis: Evolution of Karst Aquifers.* Huntsville (AL): National Speleological Society. 527 p.

Moore GW, Sullivan N. 1997. *Speleology: Caves and the Cave Environment.* Rev 3rd ed St. Louis (MO): Cave Books. 176 p.

Palmer AN. 1991. Origin and morphology of limestone caves. *Geological Society of America Bulletin* 103(1):1-21.

Quinlan JF, Ewers RO. 1989. Subsurface drainage in the Mammoth Cave area. In: White WB, White EL, editors. *Karst Hydrology: Concepts from the Mammoth Cave Area.* New York: Van Nostrand Reinhold. p 65-103.

Rogers N. 1995. Mammoth Cave restoration field camp. *NSS News* 53(2):44-45.

Schueler TR. 1994. The importance of imperviousness. *Watershed Protection Techniques* 1(3):100-111.

Tenney S, Barrett ME, Malina JF Jr, Charbeneau RJ, Ward GH. 1995. *An Evaluation of Highway Runoff Filtration Systems.* Technical Report CRWR-265. Austin (TX): Center for Research in Water Resources, Bureau of Engineering Research, University of Texas at Austin. 133 p.

Veni G. 1999. A geomorphological strategy for conducting environmental impact assessments in karst areas. *Geomorphology* 31:151-180. Reprinted in: Giardino JR, Marston RA, Morisawa M, editors. 1999. *Changing the Face of the Earth—Engineering Geomorphology: Proceedings of the 28th Binghamton Symposium in Geomorphology, held 28 August to 3 September 1997 in Bologna.* Amsterdam: Elsevier Publishers. p 151-180.

Veni G, DuChene H, editors. 2001. *Living with Karst: A Fragile Foundation.* AGI Environmental Awareness Series, no 4. Alexandria (VA): American Geological Institute. 64 p. <http://www.agiweb.org/environment/publications/karst.html>, <http://www.agiweb.org/environment/publications/karst.pdf >. Accessed 2005 June 21.

Watson J, Hamilton-Smith E, Gillieson D, Kiernan K, editors. 1997. *Guidelines of Cave and Karst Protection.* Cambridge: International Union for Conservation of Nature and Natural Resources. 53 p.

White WB. 1988. *Geomorphology and Hydrology of Karst Terrains.* New York: Oxford University Press. 464 p.

Section A—Identifying and Protecting Cave Resources

Protecting Lava Tube Caves

Victor J. Polyak and Paula P. Provencio

Basaltic lava flows worldwide contain beautiful and impressive lava tube caves. Although just as intriguing and delicate as limestone caves, there is a perception that lava tube caves have not been as thoroughly studied and protected as limestone and dolostone caves.

Outdoor recreational activities such as caving have become increasingly popular over the last few decades. As a result, there are equally accelerating impacts to natural and cultural resources in caves. Prior to the Federal Cave Resources Protection Act of 1988, there was undoubtedly less incentive to manage and protect most caves, and lava tube caves were no exception. (See cave laws, page 217; for complete text of the Act, see Appendix 1, page 507.) There was also no measure for determining the impact to these caves, mainly because little was known about the lava tube caves and their resources. Since the Act was implemented, management emphasis of publicly owned lava tube caves has shifted from recreation to conservation (Nieland 1992). There is now a need to more thoroughly define lava tube caves and their resources so the most appropriate management plans can be administered.

What should be preserved in a lava tube cave? This is the fundamental question that ultimately leads to the protection of lava tube caves and their features. *Caves of Washington* by William R. Halliday (1963) and *An Illustrated Glossary of Lava Tube Features* by Charles Larson (1993) are monographs that point out what exists in these caves in terms of uniqueness and worthiness of preservation. While the lava features in lava tube caves are often small and difficult to see, they are truly unique and usually very fragile. Once damaged, these features do not recover and are not easily restored.

Lava tube caves can be quite diverse and complex, as indicated in the *Proceedings of the 6th International Symposium on Vulcanospeleology* (Hong 1992), published by the National Speleological Society. Braids, mazes, multi-levels, vertical entrances, crawlways, and immense boreholes are descriptors of lava tube cave passages. Mineralogical deposition in these caves can be equally impressive. This chapter offers a brief synopsis of the uniqueness of lava tube caves with emphasis on preservation, conservation, and protection.

This chapter offers a brief synopsis of the uniqueness of lava tube caves with emphasis on preservation, conservation, and protection.

Resources in Lava Tube Caves

Lava tube caves are formed in basalt flows. Unlike limestone caves, they form rapidly (over weeks, months, or years). Those interested in the origin and character of lava tubes can refer to Greeley (1987, 1971, 1972), Peterson and Swanson (1974), Allred and Allred (1997), Rogers and Mosch (1997), and others. The timing of the speleogenesis of lava tube caves can simply be determined by dating the basalt bedrock of the cave.

During the active lifetime of the lava tube, remelting of the surfaces produces lava features that mimic speleothems commonly found in limestone caves, such as stalagmites and stalactites. These are primary speleogenetic features (formed when the cave formed). Secondary speleothems like those found in limestone caves can also form in lava tubes. Because the speleogenetic features and secondary speleothems found in lava tube caves are generally small and often not easily seen, they are vulnerable to damage by visitation.

Lava Features

Visitors often overlook lava features because of their dark color, or because the features are not well defined. Common lava features in caves are lava stalactites, stalagmites, helictites, columns, flowstone, coralloids, grooves (flow lines), shelves, ceiling cusps, linings, lava falls, dams, levees, gutters, and benches (Larson 1993). Cavers tend to be most familiar with lava features that resemble typical speleothems such as stalactites and stalagmites.

Although most of these speleothem-looking lava features are small and delicate, some can be quite large and impressive. For example, a lava column 7.6 meters (29.4 feet) high is located in a Korean lava tube cave (Hong 1992). Lava stalagmites can exceed 3 meters (10 feet) in height. Small stalagmites and other lava features (for example, peanut-sized lava bubbles) on cave floors are very vulnerable to damage by visitation (Figure 3). Even shelves and benches that appear to be physically robust can be damaged easily (Figure 2).

Lava features should receive as much recognition and protection as speleothems in limestone caves. Lava helictites seem to be as unique and rare as carbonate helictites. Lava stalagmites, in most cases, are extremely delicate and are much rarer than carbonate stalagmites. Continued lack of protection for these features will make them exceedingly rare.

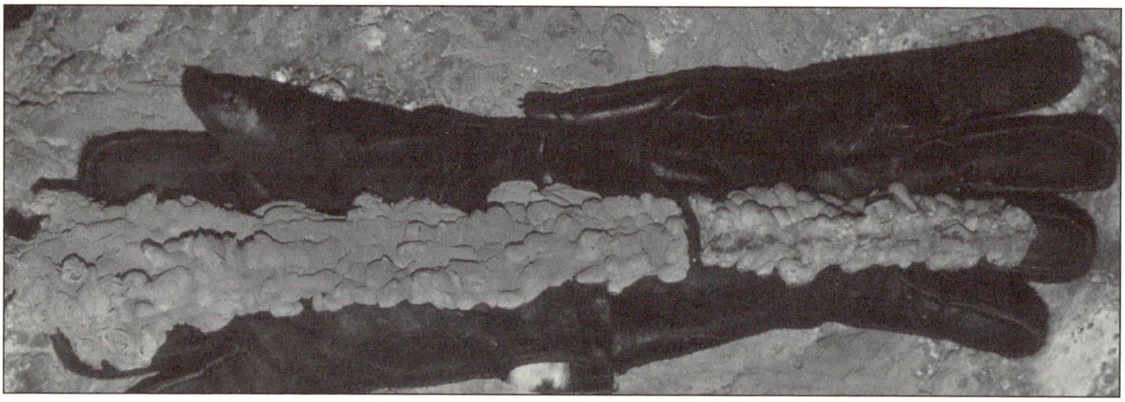

Victor Polyak

Figure 1. This lava stalagmite was probably broken by early cave visitors (shown here on top of a glove). These features are not easily noticed with carbide or weak electric lights.

Mineral and Depositional Resources

Lava tube caves usually contain secondary salts, carbonates, and silicates. The salts usually consist of gypsum, epsomite, thernardite, and mirabilite. These form delicate speleothem types such as crust and moonmilk. Calcite is common in lava tube caves, also as coralloids, crust, and moonmilk. Silicates such as amorphous silica and poorly formed magnesium clays (trioctahedral smectite) are also found in lava tube caves in crust, coralloids, and moonmilk. Oxides are observed in lava tube caves as iron-oxides (probably hydrous) which commonly coat or stain the walls.

Ice decorates many lava tube caves, creating spectacular speleothems. Found in caves located in cooler climates, some ice speleothems can be very large and long-lived if the cave temperature perennially stays below 0ºC (32ºF). Other ice speleothems are seasonal, melting during the summer and fall, and reforming during the winter and spring. Some ice deposits have potential to reveal important historic or climatic information (Dickfoss and others 1997).

Figure 2. It is not known whether this broken lava shelf was damaged by people climbing on it, or whether it broke naturally. In any case, these types of features are prone to damage by careless visitors.

Dark spots or lines on lava surfaces (isolated and discrete features) may be formed by dripping water or ice melt and airflow, and appear to be spots where the original lava surface is left clean. These secondary features are not speleothems or lava features, and may seem unimportant to most visitors in lava tube caves. As insignificant as these floor spots might seem, they may provide information regarding past cave climate. Since they seem to result from dripping water, they may also be microhabitats for unusual or important life.

Another interesting secondary feature found in many lava tube caves is *rock flour*. This powdery material often can be associated with primary features such as linings, and is a weathering product.

Biologic Resources

Scientific research of lava tubes has accelerated over the past 40 years with NASA becoming interested in habitats for extraterrestrial life and potential human shelter. (See cited references for microbiology, pages 78–82.) Caves have interesting biology and lava tube caves are no exception.

Many traditional cave studies in biology relate to bats. Bats frequently use lava tubes, and bat habitation of lava tube caves can be significant. For instance, one of the larger bat colonies in the southwestern United States resides in an impressive lava tube cave. Since the twilight zones of these caves are commonly extensive, other animals also frequently use lava tube caves.

The ecological balance of lava tube caves is often quite fragile.

The ecological balance of lava tube caves is often quite fragile. For example, beautiful carpets of moss develop near entrances and below skylights in the lava tube caves of El Malpais National Monument. These are important habitats for invertebrates and microorganisms. Silvery or golden-colored slimes are often observed on the walls and ceilings of lava tube caves—these reflective slimes result from colonies of bacteria such as actinomycetes (Northup and Welborn 1997). Microbes may also be partly responsible for the origin of secondary deposits such as moonmilk and vermiculations. (See biofilms page 68.)

Cultural Resources

Caves made good shelters for people of early cultures. Lava tube caves were used as burial or religious sites, or as places of refuge (La Plante 1992; Sinoto 1992). Important cultural materials still exist in many lava tube caves (Cresswell 1999). Religious resources are precious to native cultures and are receiving greater respect and protection. Cultural materials in lava tube caves are difficult to protect because the materials, if still present, are typically exposed and visible on the bedrock floors. To ensure preservation of these resources, cavers should report potential cultural sites to the appropriate archaeologists and keep the cave locations confidential.

Protecting Lava Resources: Restriciting Visitation and Establishing Trails

Lava Tube Cave Entrances

Establishing restrictions (permits, signs, gates, trails) in lava tube caves is not straightforward because the entrances are usually too large to gate, and entrance areas often harbor a number of different resources.

Establishing restrictions (permits, signage, gates, trails) in lava tube caves is not straightforward because the entrances are usually too large to gate, and entrance areas often harbor a number of different resources. Many lava tube cave entrances form niches for plants and animals not common elsewhere in the region surrounding the caves. This is well discussed by Northup and Welbourn (1997) and citations within. For instance, lava tube caves in the southwestern United States provide havens for moss, grass, or fern gardens, which are often located at entrances or below skylights. These *gardens* are habitats for mites, flies, microbes, and other creatures. Visitors need guidance to avoid trampling or disrupting these fragile ecosystems. Who knows what future studies of these sites might reveal?

Archaeological materials are sometimes located on the surface near cave entrances. Rock walls that were constructed as shelters are often observed. Long tree logs are found at the mouth of collapses near entrances to lava tube caves and were sometimes used during historic or prehistoric times as bridges or ladders to enter the caves. Disturbing any artifacts destroys the cultural history of these caves.

Lava Tube Cave Passages

The cave floor is the area most prone to damage from visitation, especially if visitors are unaware of fragile cave resources. Permanent damage can occur even by lightly walking across a delicate area. For instance, sometimes there are abundant spots and lines on a lava tube floor. These floor features are often hard to see and might seem insignificant, but they may harbor important climate and microbial information. Floor deposits can be easily obliterated by just a few visits to a lava tube cave if trails are not established and properly used.

Small lava features on cave floors, walls, or ceilings are easily damaged by unmanaged visitation (Figure 3). Large features can be tempting structures for curious and energetic visitors. For instance, some shelves are very well formed and large enough to walk or crawl on, and may be easily broken (Figure 2) if guidance and restrictions are not in place. Speleothem deposits such as moonmilk, coralloids, and crusts are also prone to inadvertent or intentional damage by visitors. Many deposits seemingly having little scientific or historic value—but even mounds of rock flour should be protected for aesthetic reasons and for future scientific research.

Impacts from Emergency Situations

Search-and-rescue groups often have no experience in caves and no special knowledge about lava tube cave resources—consequently, personnel may inadvertently damage a cave during a search. Much damage can be prevented if an organized caving group has inventories and maps of area caves. Familiar with caves, and equipped with the proper gear and maps,

Victor Polyak

Figure 3. This small lava stalagmite with a lava straw welded to it is in the center of a crawl passage. Increased visitation will undoubtedly damage this special little candlestick stalagmite, which is less than 10 centimeters (4 inches) in height. Also note lava "peanuts" (bubbles) on the cave floor surrounding the stalagmite.

cavers can assist search-and-rescue efforts with increased efficiency, thoroughness, and minimum-impact stewardship.

Until we are content with "virtual cave visits," trails will have to suffice as a means for protecting important areas of caves from frequent visitation. Many areas in lava tubes contain fragile and unique features. Probably the best way to protect these features in caves that receive frequent visitation is to establish trails.

Developing trails that will not contribute to contamination is a challenge. Trail development, approached as a science, will hopefully be a significant topic of speleology in the near future; appropriate materials for cave trail construction and placement strategies need to be addressed. (See trails, page 175.) Conservation management through restricting visitation, establishing trails, and posting signs to limit impact can help protect the natural and cultural resources.

Conservation management through restricting visitation, establishing trails, and posting signs to limit impact can help protect the natural and cultural resources.

Potential Hazards of Lava Tube Caves

Skylights

Skylights are intriguing openings in the ceilings of lava tube caves that almost always have an associated vertical component. An important conservation message—*Be Prepared*—is especially important in avoiding the inherent hazards of lava tube caves. On the surface, small skylights are not easily seen. Hiking at night, or running in areas with lava tube skylights can be dangerous. Also, the thickness of cave roofs around skylights can be very thin. Approach lava tube cave skylights with great caution, or better yet, simply avoid walking near them.

Approach lava tube cave skylights with great caution, or better yet, simply avoid walking near them.

Ceiling Collapse

Many lava tube caves have ceilings that are partially collapsed. While these caves seem solid and stable, blocks are obviously falling from the ceiling over time. Frost wedging is one of the culprits. Caves that fit this description and receive moderate visitation should be monitored periodically for evidence of freshly fallen blocks. In this type of cave, avoid touching the ceiling. Managing agencies might want to carefully consider allowing open public access to caves with partially collapsed ceilings.

Hypothermia and Histoplasmosis

Many lava tube caves are cold. Basalt is a good insulator and lava tube caves often form effective cold air sinks. Hypothermia is the chilling of the body's core temperature and commonly results from caving. Halliday (1974) and Mosberg (2006, page 271 in this volume) describe in detail how hypothermia can fatally affect cavers. Prepare properly for cold cave environments and avoid hypothermia.

- Dress warmly. As well, anticipate changes in surface weather.
- Take along a small thermos of hot fluids to drink.
- Carry snacks with high fat content (for example, summer sausage and nuts).
- Ask for the assistance of fellow cavers if you feel chilled.
- Never underestimate your situation. Take precautions at the first sign of shivering. (See hypothermia, page 271.)

Other health hazards are common for cavers visiting lava tubes. Deposits of guano and rodent middens may create a cave environment conducive to diseases such as histoplasmosis. Travel carefully to avoid stirring up dust or spores, and wear respirators for protection in dry, dusty caves. (See histoplasmosis, page 277.)

Conservation Management

Rapid growth of the world population is contributing to increased human impacts on caves in at least two ways.

- Encroachment by housing and business development.
- Accelerated visitation by people.

Lava tube caves are no exception. Since the Federal Cave Resources Protection Act of 1988, under-funded federal agencies in the United States are increasingly relying on volunteers for assistance. In lava tubes, cavers are helping by surveying the caves, inventorying the cave resources, and organizing cave cleanups.

In the United States, volunteers are providing an important service, but the responsibility of saving and preserving the significant publicly owned lava tube caves ultimately belongs to the appropriate federally employed managers (Nieland 1992). (See cave management tools, page 229.) Managing human curiosity and the will to explore on one end and malicious intentions on the other is a great challenge. (See managing caves or cavers, page 237.) Controlling urban and industrial encroachment in some areas will be an even greater challenge in the quest to preserve the world's beautiful lava tube caves.

In lava tubes, cavers are helping by surveying the caves, inventorying the cave resources, and organizing cave cleanups.

Acknowledgements

We are grateful to members of the Sandia Grotto who are participating in a survey and inventory project that directly supports protection of the lava tube caves in El Malpais National Monument. We also thank John Lujan, Herschel Schulz, and the staff of El Malpais National Monument for supporting the ongoing research of lava tube caves in the Monument. This volunteer effort has enabled us to offer the information herein regarding the protection of lava tube caves.

Cited References

Allred K, Allred C. 1997. Development and morphology of Kazumura Cave, Hawaii. *Journal of Cave and Karst Studies* 59(2):67-80.

Cresswell LT. 1999. Archaeology of the Snake River Plain, Idaho. In: Kline T, editor. *Idaho, the Underground Gem*. The 1999 National Speleological Society Convention Guidebook. Filer Idaho. p 23-25.

Dickfoss PV, Betancourt JL, Thompson LG, Turner RM, Thronstrom S. 1997. History of ice at Candelaria Ice Cave. In: Mabery K, editor. *Natural History of El Malpais National Monument.* New Mexico Bureau of Mines & Mineral Resources, Bulletin 156. p 91-112.

Greeley R. 1971. Observations of actively forming lava tubes and associated structures. *Modern Geology* 2(3):207-223.

Greeley R. 1972. Additional observations of actively forming lava tubes and associated structures. *Modern Geology* 3(3):157-160.

Greeley R. 1987. The role of lava tubes in Hawaiian volcanoes. In: Decker RW, Wright TL, Stauffer PH, editors. *Volcanism in Hawaii.* United States Geological Survey Professional Paper 1350 (Vol 2). Reston (VA): U.S. Geological Survey. p 1589-1602.

Halliday WR. 1963. *Caves of Washington.* State of Washington Division of Mines and Geology, Information Circular 40. Olympia (WA): State of Washington, Department of Conservation, Division of Mines and Geology. 132 p.

Halliday WR. 1974. *American Caves and Caving: Techniques, Pleasures, and Safeguards of Modern Cave Exploration.* New York: Harper & Row. 348 p.

Hong SH. 1992. Caves in Cheju Island, Korea. In: Rea GT, editors. *6th International Symposium on Vulcanospeleology: Hilo, Hawaii, August 1991*. Huntsville (AL): National Speleological Society. p 168-169.

Larson CV. 1993. *An Illustrated Glossary of Lava Tube Features*. Western Speleological Survey, Bulletin 87. Vancouver (WA): Carlie & Jo Larson. 56 p.

La Plante M. 1992. Recently discovered Hawaiian religious and burial caves. In: Halliday WR, Rea GT, editors. *6th International Symposium on Vulcanospeleology: Hilo, Hawaii, August 1991*. Huntsville (AL): National Speleological Society. p 7-9.

Mosberg SR. 2006. Health and hygiene related to cave conservation. In: Hildreth-Werker V and Werker JC, editors. *Cave Conservation and Restoration*. Huntsville (AL): National Speleological Society. p 271-280. [This volume.]

Nieland J. 1992. Inventory, evaluation, and management of publicly owned caves in the western United States and the impact of the Federal Cave Resources Protection Act. In: Halliday WR, Rea GT, editors. *6th International Symposium on Vulcanospeleology: Hilo, Hawaii, August 1991*. Huntsville (AL): National Speleological Society. p 273-279.

Northup DE, Welbourn WC. 1997. Life in the twilight zone–Lava tube ecology. In: Mabery K, editor. *Natural History of El Malpais National Monument*. New Mexico Bureau of Mines & Mineral Resources, Bulletin 156. Socorro (NM): New Mexico Bureau of Mines & Mineral Resources. p 69-82.

Peterson DW, Swanson DA. 1974. Observed formation of lava tubes. *Studies in Speleology* 2(6):209-222.

Rogers BW, Mosch CJ. 1997. In the basement–Lava tube origins and morphology. In: Mabery K, editor. *Natural History of El Malpais National Monument*. New Mexico Bureau of Mines and Mineral Resources, Bulletin 156. Socorro (NM): New Mexico Bureau of Mines and Mineral Resources. p 61-68.

Sinoto YH. 1992. Hawaiian use of lava tube caves and shelters. In: Halliday WR, Rea GT, editors. *6th International Symposium on Vulcanospeleology: Hilo, Hawaii, August 1991*. Huntsville (AL): National Speleological Society. p 3-6.

Section B—Developing Cave Management Programs

Defining Limits and Protocol at Kartchner Caverns

Rickard S. Toomey, III

It may be politically correct for cavers to think we can somehow *manage* natural voids under Earth's surface. But is it accurate to make this audacious claim? Skeptically, we ask how humans can possibly manage cave environments.

Any kind of management involves defining limits and protocols for human actions.

Usually, managing caves goes hand in hand with protecting the resources from damage by people. Resource managers develop philosophies and administrative procedures to support the conservation and protection of cave values.

The goals involving protection are addressed through an expanding array of creative management objectives. Valid administrative options range from extreme secrecy, to gate installations, permitting systems, and sometimes, show cave development. Through three decades of development history, all of these management tactics were applied in the protection of Kartchner Caverns State Park, Arizona.

© Arizona State Parks

Figure 1. The beautiful speleothems in the Throne Room at Kartchner Caverns State Park® spawned both the need to protect the cave and the opportunity to make it an educational destination for tourists.

Discovery of Kartchner

In 1974 Randy Tufts and Gary Tenan discovered a cave on the eastern slope of the Whetstone Mountains about ten miles southwest of Benson, Arizona. The cave they found was wet, well decorated, and in pristine condition. The find caused them to worry about how to protect such a cave. They were familiar with numerous examples of caves that were badly damaged after their existence became public knowledge.

The pair of explorers examined methods to protect the cave from damage. They first opted for extreme secrecy. Their quest to keep it quiet extended to having

© Val Hildreth-Werker

Figure 2. Gleaming flowstone and exquisite soda straws remain pristine even in the new areas under development at Kartchner Caverns State Park®.

anyone they took to the cave sign a secrecy agreement. By 1977 they had determined that secrecy alone was not sufficient. The cave's location near a major highway, and the fact that unauthorized people were learning about it, clearly indicated it was time for another plan.

Conservation by Development

Inspired by a paper written by Russell Gurnee (1966) they considered another possibility—conservation by development. Under this concept, to assure the long-term protection of the cave, it would be developed as a show cave. Although development of the cave would result in significant impact (from trails, lights, and tourists), it would also provide for controlled access that might prevent other types of damage and impacts. In 1978 Tufts and Tenan approached the landowners, the James A. Kartchner family, about the cave, and began exploring the possibility of a private enterprise show cave.

In 1985, Tufts, Tenan, and the Kartchners decided that development through a public entity was a more appropriate solution. In 1988 after numerous secret negotiations through members of the state legislature and with the help of the Nature Conservancy, the state of Arizona purchased the cave and about 560 acres of land around it. The cave became the James and Lois Kartchner Caverns State Park under the stewardship of Arizona State Parks and its director Ken Travous. For a more comprehensive history of the cave see Tufts and Tenan (1999).

From the beginning, the goal of Arizona State Parks was—to the greatest extent possible—to develop the cave in an environmentally and speleologically sensitive manner (Travous and Ream 2001). They took many complicated steps to succeed in that objective. One of the first and most important steps was to document a series of baseline studies before initiating any development of the cave.

> Conservation by development—to assure the long-term protection of Kartchner Caverns, it was developed as a show cave.

These investigations provided the necessary information for the responsible planning of development. Studies provided the raw data that were needed to limit development and tours and define development protocols. Arizona Conservation Projects, Inc. (ACPI), under the management of Robert Buecher, was awarded the contract to perform two years of predevelopment studies. Starting in the spring of 1989, research included both the surface ecosystem (weather, geology, hydrology, vegetation, mammals, amphibians, and reptiles) and the cave environment (mapping, geology, hydrology, climate, mineralogy, invertebrate biology, algae, fungi, and bats).

The results included an extensive final report (ACPI 1992) and numerous subsidiary reports and maps (approximately 328-page report with an additional 590 pages in appendices). The studies also led to a symposium held during the annual NSS national convention, and the papers from that symposium were published together as the *Journal of Cave and Karst Studies*, Volume 61 Number 2. Overall, Kartchner is among the most thoroughly studied caves in the world. Baseline investigations yielded more predevelopment data than had been achieved for any previous show cave.

Following predevelopment studies, the planning and development process was as thorough as state of the art allowed. The primary objectives were to provide for cave protection and quality visitor experience. The planning process used the predevelopment studies, as well as consultation with speleologists and other experts, to design an environmentally friendly facility and tour experience highlighting safety, education, and entertainment.

Potential impacts from surface facilities and cave facilities were examined in the planning process. The surface facilities were located to cause minimum impact on the cave. Cave facilities were designed to protect the cave and its resources and to provide adequate visitor access (including barrier-free access to the cave tour).

Designs for access and trails at Kartchner Caverns employ cave protection techniques pioneered at other show caves. These include trail placement to minimize impact, use of airlocks, construction of lint curbs, and the separation of trail lighting from feature lighting. One innovation used in trail design and construction in Kartchner features the combination of an integrated trail base and lint curbs with a sump system. This design allows trails to be washed down and the water and trail debris to be pumped out of the cave.

Designs for access and trails at Kartchner Caverns employ cave protection techniques pioneered at other show caves.

In November 1999, the first section of the cave (the Rotunda and Throne Room) opened for tours. An important part of protecting the cave was to keep the tour group size small. Each tour is limited to 20 guests. Two guides are assigned to each tour, a lead guide at the front of the visitors and a trailing guide to bring up the rear. The two guide system allows for cave protection, improved interpretation, and improved visitor safety. There are 25 tours each day in the Rotunda–Throne section. This limits visitation to 500 guests entering the cave each day.

Development of another section of the cave (the Big Room area) is complete and tours began in November of 2003. Because the tours going through this section walk much closer to delicate cave features, stricter limits on tour size are indicated for this area. Tours in this area are limited to about 20 tours per day with 15 people on each tour.

The Big Room poses additional challenges that were not encountered with the Rotunda–Throne Room area. The Big Room is the summer home to a southwestern cave myotis (*Myotis velifer*) colony of between 1,000 and 2,000 bats. In order to protect the bats, cave development work in the Big Room ceased between May and mid-September. Studies were performed to predict potential bat reaction to various tour scenarios (Mann and others 2002). As a result of the studies, Arizona State Parks decided that the Big Room section will be open for tours only between mid-October and mid-April. Bat populations are monitored for potential impacts from development. So far, none have been observed.

Rickard S. Toomey III

Figure 3. Trail features, including integration of the impervious trail surface, lint curbs, and a sump system, allow the trails to be washed at Kartchner Caverns State Park®. The water and trail debris is pumped out of the cave. Kartchner Caverns also employs a specific Cave Unit staff to monitor cave conditions, maintain the cave tour infrastructure, and take the lead in cave protection.

Well-conceived development and limits on size and numbers of tours only go so far in protecting the cave. Conservation by development represents a compromise—some impact on the cave is allowed in order to prevent other potentially more damaging uncontrolled impact.

Arizona State Parks made two crucial commitments to insure continued protection of Kartchner Caverns:

- Continue study and monitoring to lessen or prevent impacts.
- Adjust limits, protocols, and operations as conditions change.

To accomplish these commitments, Arizona State Parks created a specific staff (the Cave Unit) to monitor cave conditions, to open and close the cave each day, and to do maintenance for the cave tour infrastructure. This arrangement keeps these tasks from falling between the duties of tour guides and the park maintenance staff. In addition, it creates a core group with commitment to those tasks. Arizona State Parks also created a Cave Resources Manager within its Resources Management Section. That person is in charge of cave research and monitoring, and recommends operational changes to better protect the cave.

Kartchner Caverns has been developed according to the best current practices. Solutions from the most advanced show caves in the world have been implemented.

Cigna (2002) noted the following about Kartchner Caverns upon assessing the condition of the cave one year after it opened for tours:

> *It must be emphasized that Kartchner Caverns has been developed according to the best standard, because each particular solution adopted in the most advanced show caves in the world have been implemented. This is one of the greatest successes ever obtained in this field and should be taken as an example for any further development of a tourist cave.*

Whether the careful, thorough approach is sufficient to protect the cave from long-term impacts remains to be seen. Over time, other caves will develop additional solutions and improvements on those implemented at Kartchner Caverns. In addition, limits and protocols at Kartchner will be altered as conditions merit due to improved knowledge and technology. Ultimately, Kartchner Caverns will serve as an important test of "conservation by development" as applied to caves.

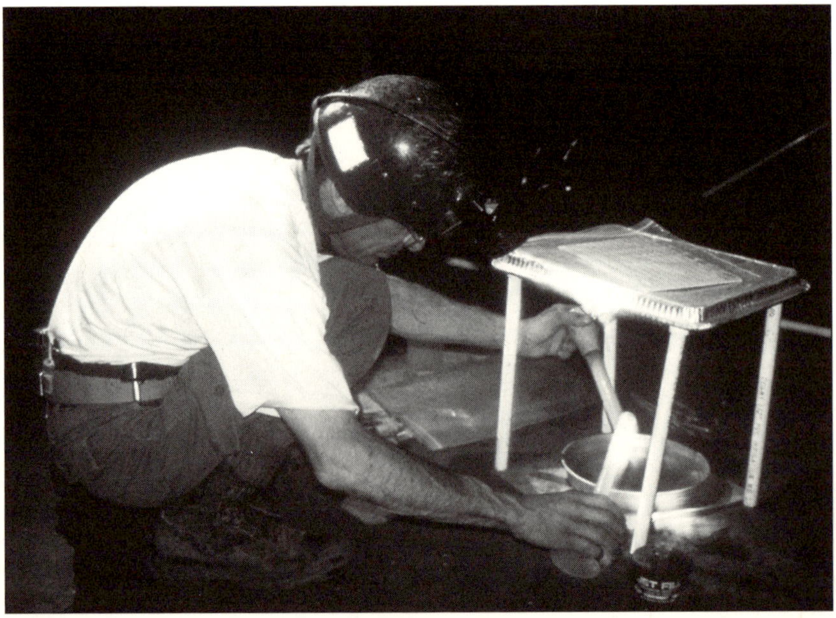

Figure 4. A member of the Cave Unit at Kartchner Caverns State Park® checks the evaporation at one of the cave's environmental monitoring stations. Ongoing monitoring of the cave's environmental condition is vital to understanding and mitigating impacts from development.

Cited References

Arizona Conservation Projects, Inc. 1992. *Final Report: Environmental and Geological Studies for Kartchner Caverns State Park*. Report to Arizona State Parks Board. 328 p.

Cigna AA. 2002. Modern trends in cave monitoring. *Acta Carsologica* 31(1):35-54.

Gurnee RH. 1966. Conservation through commercialization. In: *Actes du IVe Congrs International de Splologie en Yougoslavie—Proceedings of the 4th International Congress of Speleology in Yogoslavia: Postojna, Ljubljana, Dubrovnik, 12-26 IX 1965*. Ljubljana: The Congress [1962]. p 109-114.

Mann SL, Steidl RJ, Dalton VM. 2002. Effects of cave tours on breeding *Myotis velifer. Journal of Wildlife Management* 66(3):618-624.

Travous KE, Ream JP. 2001. Developing and managing an environmentally responsible tourist cave. *Proceedings of the 13th International Congress of Speleology, Brasilia*. Paper 086/S6, Abstract 269.

Tufts R, Tenen G. 1999. Discovery and history of Kartchner Caverns, Arizona. *Journal of Cave and Karst Studies* 61(2):44-48.

Ultimately, Kartchner Caverns will serve as an important test of "conservation by development" as applied to caves.

Additional Reading

Buecher RH, Hill CA, Hose LD, Pisarowicz, JA. 1999. *Kartchner Caverns: Research Symposium*. Huntsville (AL): National Speleologial Society. 87 p. Published in: *Journal of Cave and Karst Studies* 61(2):40-123.

Graf CG. 1990. Kartchner Caverns State Park: A geologic showpiece. *Arizona Geology* 20(1):2-15.

Hill CA. 1997. Kartchner Caverns, Arizona. In: Hill CA, Forti P, editors. *Cave Minerals of the World,* 2nd edition. Huntsville (AL): National Speleological Society. p 350-354.

Negri S. 1998. *Nature's Underground Wonderland: Kartchner Caverns*. Phoenix (AZ): Arizona Highways Book Division. 32 p.

Travous K. 1995. Kartchner Caverns: A gift of nature. *Arizona Highways* 71:4-13.

Section B—Developing Cave Management Programs

On Cave Gates

Jim Kennedy

Cave gates. There are hardly two words that polarize cavers as much as these. Even the most vocal anti-gate cavers admit that gates serve an important function in protecting irreplaceable cave assets, and in reducing the liability of cave owners. Yet all too often land managers turn to gates as quick and easy solutions to complex cave management problems.

Cave gates can be an important part of a comprehensive cave management plan, but there is much more to gating a cave that just welding steel. This chapter will *not* tell you everything you need to know about gates and gating, but it will give you an overview of the planning, design, building, and monitoring process and will direct you to additional expert resources.

Is a Gate Needed?

If a gate is needed, it should have minimum negative impact on the cave.

First, determine if a gate is truly necessary. Since a gate is a somewhat permanent structure that requires great expenditures of resources and may negatively impact the cave environment, it should be installed only after careful planning and design. Other protective methods may be more efficient or effective and should be explored first.

Other protective measures for cave habitats include but are not limited to the items in the following list:

- Administrative closures
- Signage
- Fencing
- Redirecting trails
- Public education
- Protective stewardship
- Electronic surveillance

While carefully designed and constructed gates have minimum effect on the cave environment, poorly placed gates can be very detrimental to the cave and its resources. If a gate is needed, it should have minimum impact on the cave.

Units of Measurement

Editors' Note: In this chapter, dimensions for materials deviate from the standard metric/ English format used elsewhere in this volume because construction materials are usually sold in English units in the United States.

Types of Protective Closures

Next decide on an appropriate gate design. In this section, the term *cave gate* is used for any type of lockable barricade that prevents human access to the cave, including fences, doors, and bars. Some types of closures, such as a simple chain across a passage restriction, are less secure than others.

The majority of this chapter focuses on various types of *bat-friendly* horizontal bar gates, which are suitable for most situations and are very secure. In rare instances that require an environmental seal, such as a newly opened cave or section of cave with no natural entrance, bat-friendly gates would be inappropriate. In those cases, *air-lock gates* may be necessary to prevent drying air currents and contamination by outside organisms or materials such as mud.

Bat Friendly Gates

Most cave gating scenarios call for a bat-friendly gate. Fortunately, there are many types of gates that incorporate bat-friendly features. Standard bat-friendly gates are designed with widely spaced uprights and 5 3/4-inch (146-millimeter) spacing between horizontal bars. The actual design depends on the amount of human vandalism pressure, the bat species present, and the way the bats use the cave. For instance, we must be aware that some species of bats do not tolerate cave gates at all, and others only at certain times in their life-cycle. The size and angle of the cave entrance may also dictate innovative adaptations of the *standard bat gate* designs. (See drawings for the horizontal bar gate, Figure 3.)

After carefully choosing a location and initiating the actual construction, observe the effectiveness and impact of the gate over time. If the gate is creating negative impacts, quickly modify or remove it. *Routine maintenance tasks* should be planned before commencing the actual construction. Maintenance schedules may be required to repaint the gate if necessary, remove sticks and leaves or flood debris, change locks before they stop working, and remove rocky debris that accumulates around the gate. Signs, fences, and gates are also susceptible to vandalism, and repairs or replacement may be necessary.

> Editor's Note: If a cave gate will change air currents that originally flowed through breakdown or small openings, then measure the natural airflow before the gate installation begins. Design and construct the gate to mimic the original airflow. (See virgin digs, page 261.)

Selection of Protection Method

Before installing a gate at a cave entrance, many factors must be considered. Issues to examine can be divided into two broad categories.

- Evaluate the cave resources themselves.
- Assess the level of threat to the cave resources.

Obviously, an easily accessible cave is more in need of protection than a rarely visited cave in a remote wilderness area. Likewise, a cave with a wealth of speleothems, important biota, or archaeological and paleontological remains, is more in need of protection than a small, featureless, relatively sterile cave. We believe that all caves have value. But how do we determine what is significant and threatened?

Ideally, a complete resource inventory is done for the cave in question, with periodic monitoring up to the time of the actual gating. In reality, this rarely happens. Even caves that have been known and visited for decades hardly ever have simple baseline data, like temperature and invertebrate studies.

Often a gate is planned because the cave owner or manager is reacting to a crisis—the discovery of a rare and threatened resource, advanced loss of cave resources, sharply increased visitation, or liability concerns. No matter what the impetus for protection, we should consider all users and resources when designing a gate or other type of protective closure.

Five Possible Scenarios

This process can be illustrated by a hypothetical example. Assume that we have five caves on a 1,000-acre (405-hectare) parcel of land.

Cave 1. This cave is located on a remote back corner of the property, accessible only by fording a shallow river. It is backed by several hundred contiguous acres of forest under other ownership. It has a few thousand feet of passage, some fun climbs, and ancient bear den sites.

Cave 2. This is a shallow, 25-foot (8-meter) pit leading to 300 feet (90 meters) of easy canyon and crawlway. This cave is very near a road, and an obvious trail leads to its entrance. No bats or other obvious wildlife have been noted, but the temperatures are very cold, even in the summer.

Cave 3. This cave is on a distant hillside and has a small obscure opening that leads through breakdown and crawls to a fairly large room. Endangered bats hibernate in this cave during the winter.

Cave 4. This is a large, well-known system with several horizontal entrances. Several entrances have obvious trails leading to them, and one entryway is small, torturous, and rarely used. There are many delicate and unusual speleothems in this cave, and damage has been steadily increasing for many years.

Cave 5. This is a small crawl cave with records of endangered invertebrates. Because it is near the fourth cave, it is often mistaken as an entrance to Cave 4 and receives unnecessary traffic.

What to do with these? Gating all the entrances would be time-consuming and expensive, would likely aggravate those people currently visiting them, and might cause overflow problems in neighboring caves. We already have some resource information on the five caves, so we can prioritize their significance. We also have information on the level of disturbance and threats to these caves, so we can determine the level of urgency for protecting each one. Now we have to determine exactly how we will protect each cave.

> Gating all entrances would be time-consuming and expensive, would likely aggravate those people currently visiting them, and might cause overflow problems in neighboring caves.

Cave	Significance	Threats
1	paleontology, recreation, pristine	few due to difficult access
2	possible bats in winter, recreation	highly visible, liability (pit)
3	bats in winter	small, hard to find, rarely visited
4	recreation, speleothems, possible invertebrates	heavy traffic, increasing damage
5	invertebrates	unintentional traffic from Cave 4

Set Up a Table to Prioritize Actions

Cave 5. This cave appears to have an urgent need for protection because of its endangered fauna and the unintentional traffic. This reality would need to be weighed against the population size of the invertebrates, and the numbers of those species in other caves. Since this is a relatively small cave with a well-known entrance and no bats, a gate could be appropriate.

Cave 4. This cave needs a more thorough resource inventory. Its proximity to Cave 5 indicates a likely connection. Although it is viewed primarily as a recreational cave, the possibility of finding endangered invertebrates there is high. There are too many unknowns at this time to make a good decision. Perhaps the entrances can be gated. A small internal gate might allow access to only part of the cave. Signs and a permit system might reduce the number of visitors to a sustainable level. We need to know more.

Cave 3. This cave might be categorized toward the opposite end of the spectrum. Rarely visited and obscure, it faces no immediate threats. The only critical time of year is winter when the bats are hibernating. Winter visits could be curtailed simply by doing public education through the local grottos. Because the entrance is obscure, a gate or signs might draw unnecessary, detrimental attention to the cave.

Cave 2. This cave presents a different challenge. It is easily accessible and

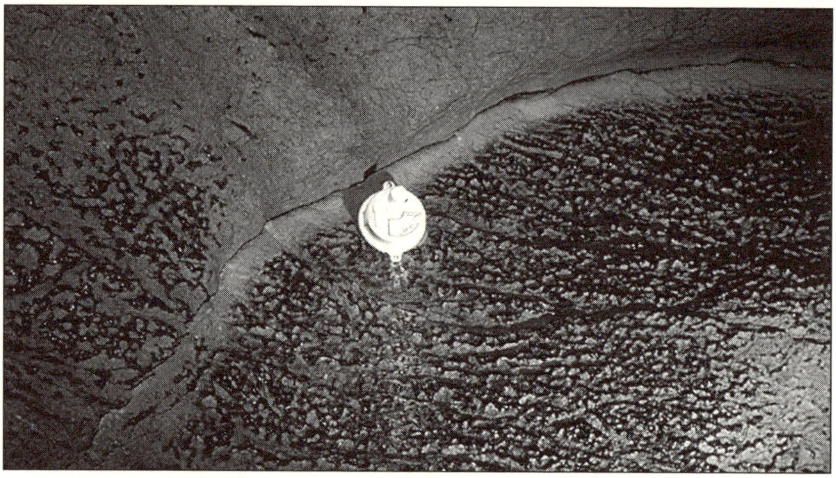

Figure 1. Bat roost stains on cave walls provide evidence of bat population even when bats are not present. In the image, a 3-inch (80-millimeter) HOBO® Pro data logger is used for scale. (See page 5 of color section.)

well known, so rerouting the trail would make little difference. A combination of educational signage and a bat-friendly fence could prove beneficial, and would not detract from the aesthetics of the pit. If the fence is repeatedly damaged, and if the cave is suitable for bats, a cupola-style bat gate could be installed over the entrance (Figure 4). Since temperatures are suitable for hibernating bats, we might conclude that bats are no longer in that cave due to disturbance, so fencing or gating should allow for their eventual recolonization. A thorough in-cave survey for old guano or roost stains would help with this decision (Figure 1). As with any site where there is a strong history of visitation, the reputation for open access must be broken, even if it means patrolling the site and arresting violators.

Cave 1. This is a relatively pristine wild cave. However, traffic may increase if other nearby caves are gated. The paleontological resources are very vulnerable. A permit system, combined with increased caver education, might work here if the location is protected by the terrain and the remoteness of the site. If natural site protection is not adequate, the cave might need a gate. Since the threats are not immediate, protective efforts for this cave are not as urgent.

Careful assessment of a cave's resources and threats is necessary before installing any protective device on a cave.

Summary of Assessment

Careful assessment of a cave's resources and threats is necessary before installing any protective device on a cave—particularly more permanent structures like gates. Public input from concerned user groups should be solicited, especially if those groups oppose closure and may damage or destroy protection efforts.

It is essential that gates and other protective structures be continually monitored, not only for structural damage, but also for their impact on the cave ecosystem. Gates, culverts, or fences that cause a negative effect should be modified or removed. Cave gating is not a quick Band-Aid approach to cave management. Gating is merely one tool a cave manager can use. Maintenance schedules should be established because gates need attention and review after installation.

Certain types of protective efforts may have an opposite effect than that intended. For instance, several species of North American cave-dwelling bats do not tolerate any type of gate at all. Some species only tolerate gates during one part of their life cycle and not at other times of the year.

Always consult experts early in the planning stages of any gating project and be sure to get the most current gate design recommendations through Bat Conservation International, the National Speleological Society, and the American Cave Conservation Association.

Location and Design

Placement of a fence or gate is as critical as the actual design of the structure. Poorly located gates may increase flood damage to the cave, accumulate debris, restrict airflow, and restrict movement of bats or other wildlife. Poorly placed gates may also be more susceptible to natural damage or vandalism, and may increase predation at the cave. Much depends on the size, shape, and orientation of the opening, but in general, bat gates should not be situated in natural passage constrictions, and fences should not interfere with the flight path at the entrance.

It must be stressed that cave gating is not a cookie-cutter management technique. Simply because a cave has bats does not mean that one can dust off a gate design and build it in the cave mouth. But even if a cave does not have bats, the cave may need a bat-friendly gate. The approach to protecting each cave should be based on the configuration of the cave itself, the species using it, the season bats occupy it, the proximity to civilization, and so on.

It must be stressed that cave gating is not a cookie-cutter management technique.

There is not a one-size-fits-all solution to cave protection. Poor gate design or placement can render the cave unsuitable for bats. Consult the experts listed in the resources section at the end of this chapter.

Gate Location

As mentioned above, cave gates should not interfere with the natural flow of air, water, nutrients, or wildlife to and from the cave. Gates should never be in a constricted part of the passage.

The bottom of an entrance slope should also be avoided since it will catch debris that will pile up against the gate. In cave entrances that have inflowing streams this can be a very serious problem. The gate on the North Entrance of Bat Cave (Carter County, Kentucky) failed in the spring of 1996 as flood debris lodged against the gate, backing up water until the increased pressure finally collapsed the gate. The resultant flood pulse destroyed many low-roosting Indiana bats, a federally listed endangered species. (See Indiana bats, page 49.)

Predation Dangers

Predation can also increase dramatically because of badly located gates. Most bat predators rely on vision when hunting, so gates in the daylight or twilight zone may enhance the predators' foraging success. When bats slow down to negotiate the gate bars, or back up behind the gate waiting their turn to pass through, they are easily captured by enterprising raccoons, ringtails, and feral cats.

Gates installed beyond the twilight zone eliminate the predators' advantage. The old gate to the lower entrance of Sinnett–Thorn Mountain Cave (Pendleton County, West Virginia) had piles of Virginia big-eared bat wings around it from the nightly predations of local house cats. The gate was removed in October 1998 and a new gate was built in a tall area approximately 75 feet (23 meters) further in, despite having to maneuver the steel and equipment through a crawlway. The new gate, in the dark zone, has eliminated the predator problems.

Cupola or Cage Gates for Vertical Entrances

Vertical or near-vertical entrances pose their own set of problems. A horizontal gate at such an entrance accumulates debris, makes a perfect feeding platform for predators, and is very difficult for most bats to negotiate.

To solve these problems, a raised gate called a *cupola gate* or *cage gate* can be used. Generally, the longer and narrower the opening,

© 1989 Jim Kennedy, BCI

Figure 2. This is a poorly designed gate, constructed of 1-inch (25-millimeter) round bars. It is not very secure—the bars may be easily bent and the welds are small. The small rectangular openings in the narrow vertical entrance make the gate difficult for bats to fly through. On this type of platform gate constructed in a vertical entrance, branches and leaves can collect to restrict airflow and light.

Figure 3. Idealized Sequence of Horizonal Bar Gate Construction Front and Side Views

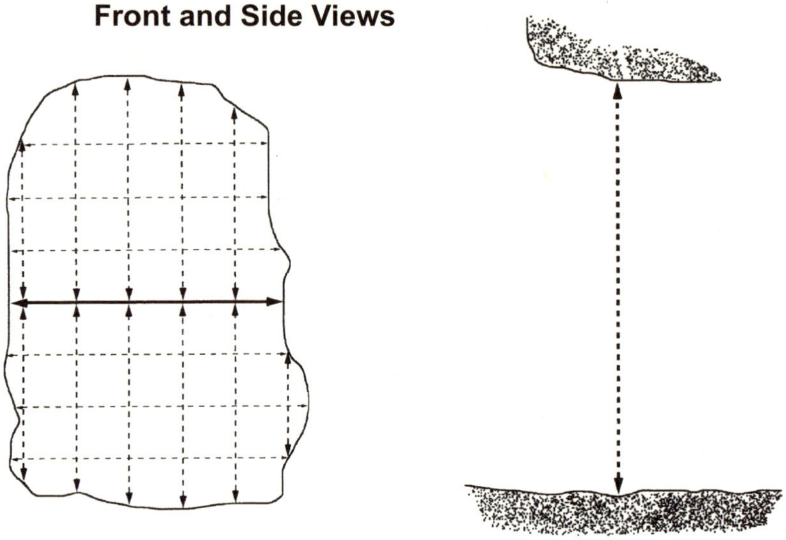

Figure 3a. Measurements of the gate location, taken at regular intervals from a horizontal (level) line.

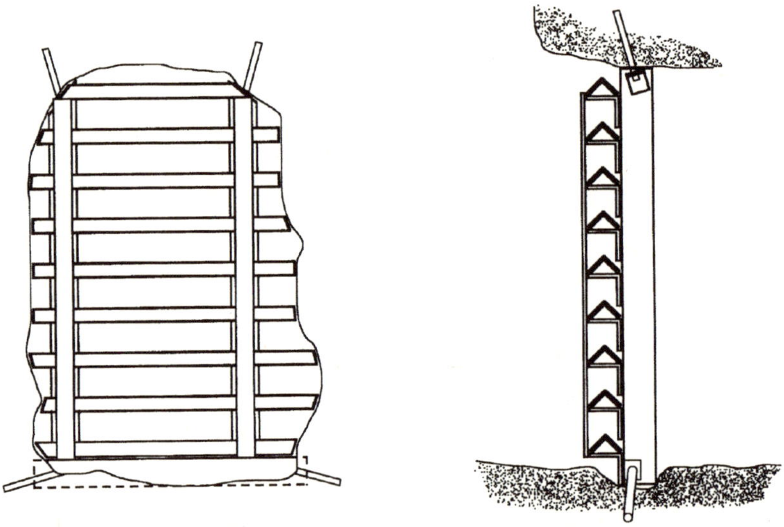

Figure 3b. Scale drawing of finished gate, used to estimate materials.

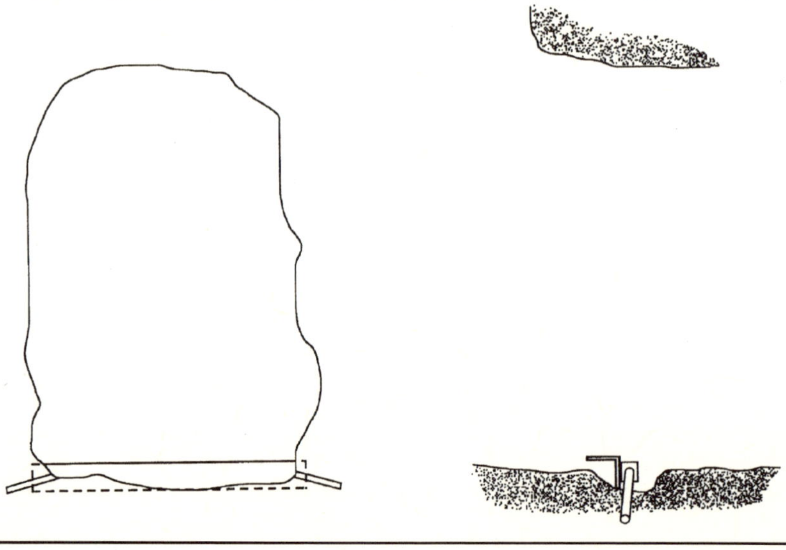

Figure 3c. Trenching and securing of sill.

Diagrams © Jim Kennedy, BCI

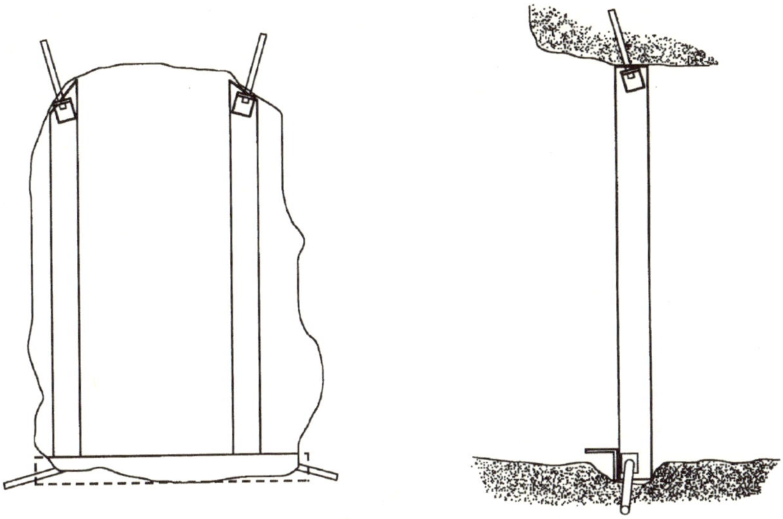

Figure 3d. Installation and securing of uprights.

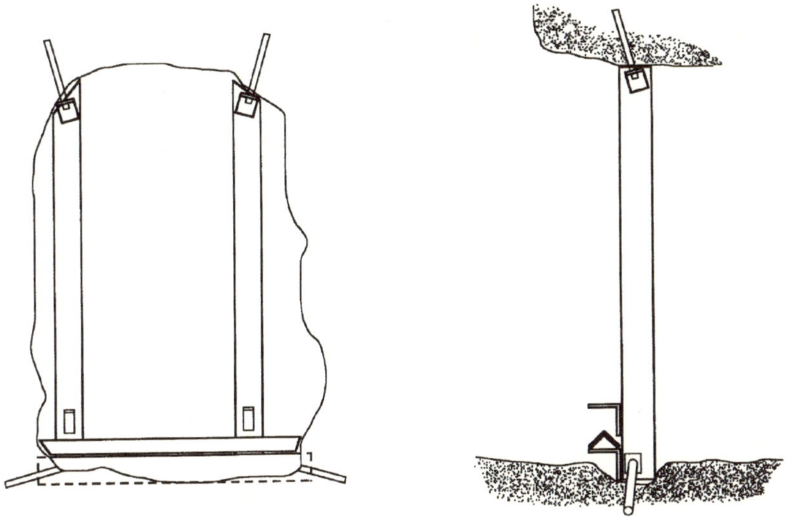

Figure 3e. Installation of bottom bar and hangers for second bar.

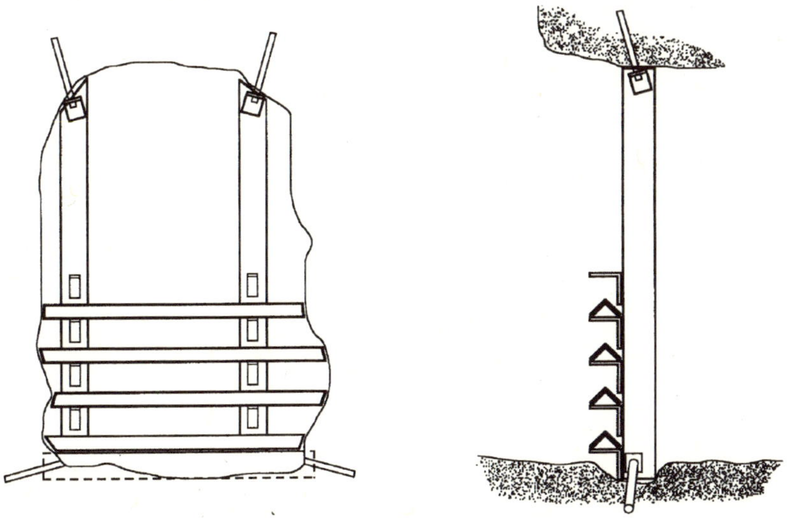

Figure 3f. Continuation of hangers and horizontal bars (for clarity, removable bar is not shown).

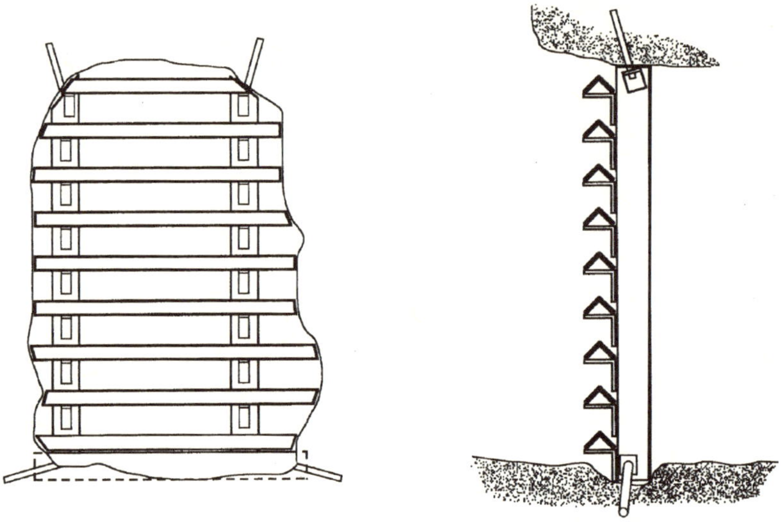

Figure 3g. Completion of horizontals.

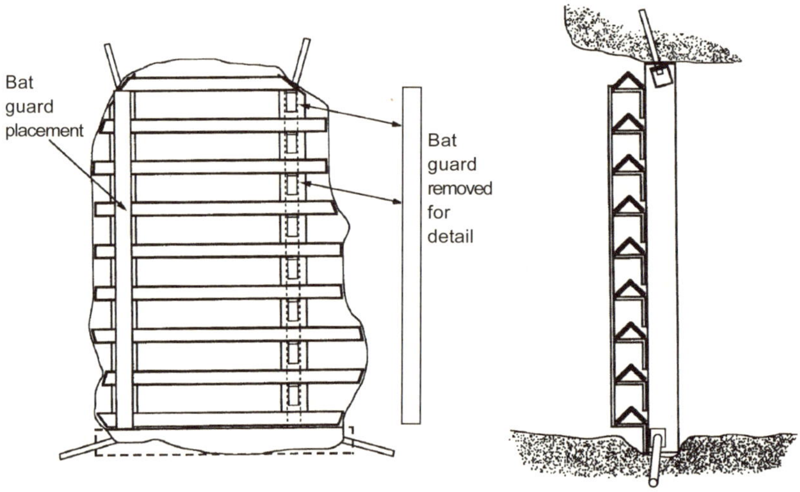

Figure 3h. Placement of bat guards.

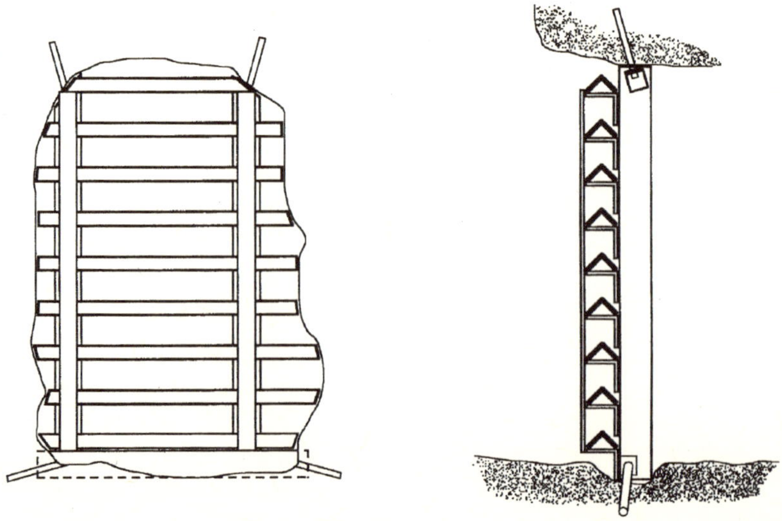

Figure 3i. Completed gate.

Diagrams © Jim Kennedy, BCI

the larger and taller the cupola gate should be in order to give the bats adequate space to gain altitude and avoid predators. Cupola gates are not practical for very large openings, and fencing may be the only option.

For vertical entrances with very short drops, a standard gate may be installed deeper within the cave where the passage begins to be more horizontal (when the vertical entrance itself is not a liability concern).

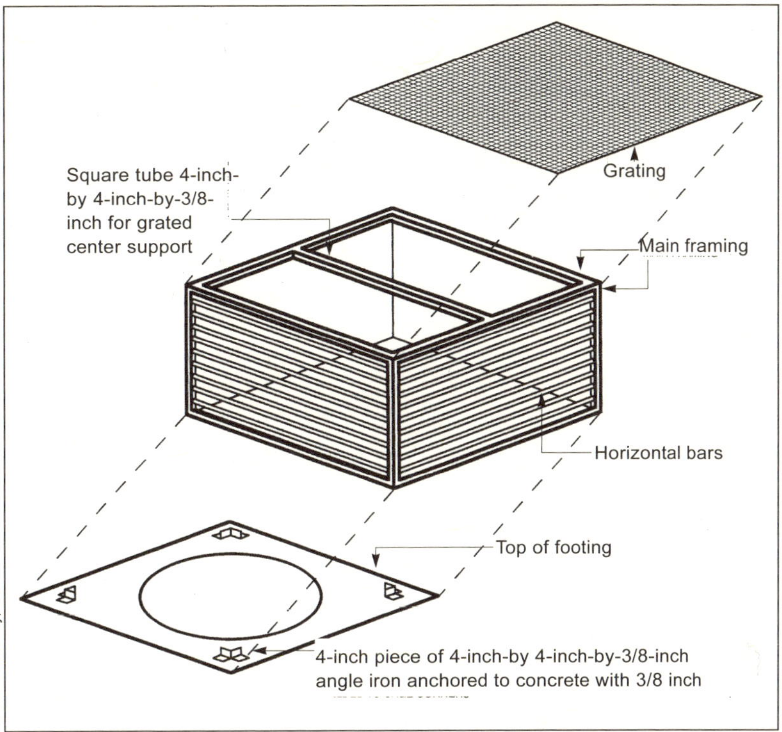

Figure 4. A cupola or cage gate is often used in vertical or near vertical-entrances.

Chute or "Window" Gates

A recent innovation, since the late 1990s, is the chute gate, sometimes called a window gate. An otherwise standard horizontal gate is modified with a rectangular opening boxed in with additional angle iron and expanded metal mesh. This design allows sufficient opening for emerging bats and makes it very difficult for trespassers to breach the opening. The chute is usually angled to make it more difficult for humans to enter. This particular gate design is especially useful in caves with large bat populations, such as gray bat maternity colonies, which have entrances that are too small for traditional half gates or flyover gates. Because of the weight extending out from the main (standard) part of the gate and the resulting mechanical stresses, extra attention is needed in the design and construction to prevent future cracked welds and gate failure. Chute gates have been used successfully on numerous Alabama, Kentucky, Missouri, and Tennessee caves, and are well accepted by bats.

Figure 5. Chute gate at McDowell Cave, Missouri.

Nonstandard Gates

Caves that are entirely unsuitable for bats (as opposed to sites where bats are not currently found) may be candidates for gates that are not bat-friendly. However, the bat-friendly design is the preferred solution for most caves, except those that have no natural entrance and require some sort of environmental seal. Sometimes the availability of materials and volunteer labor, or the lack of adequate funds will dictate construction of a nonstandard (not bat-friendly) gate. Nonstandard gates are almost always poor substitutions.

Educational Signage

All finished gates require signs stating the purpose of the gate and contact numbers for more information. The penalty for entering the cave or vandalizing the gate can be written in small print, but this should not be the focus of the sign because it is often taken as a dare by would-be vandals.

Figure 6. Typical bat gate (not to scale).

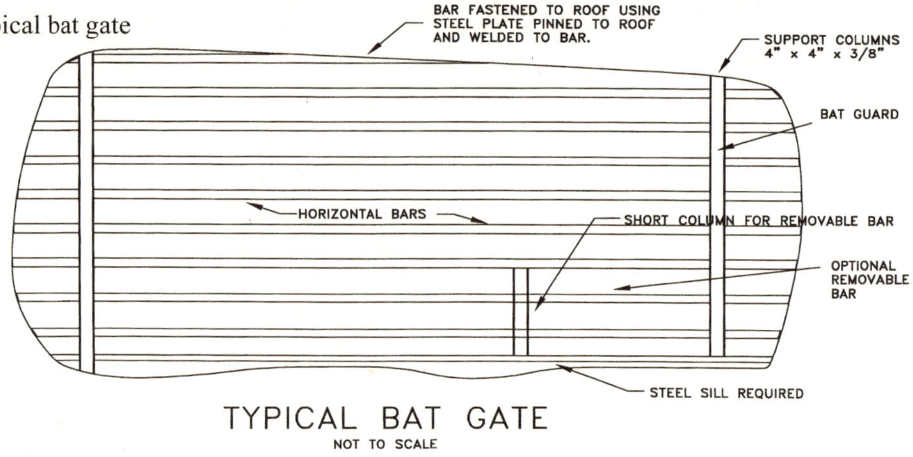

Figure 7. Horizontal bar spacing, typical detail (not to scale).

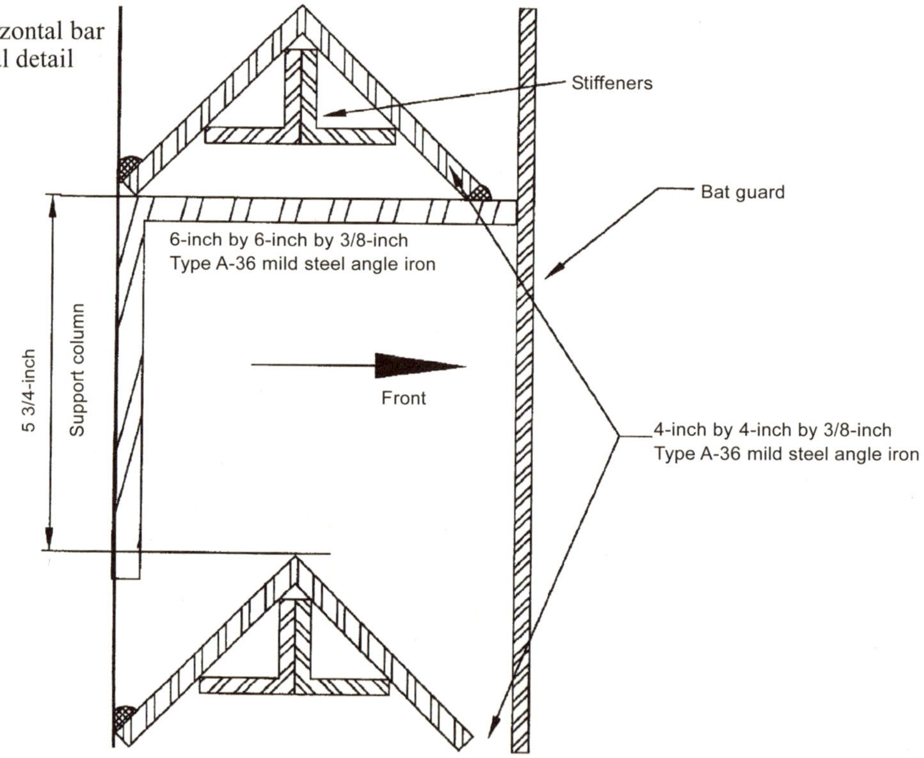

Figure 8. Stiffener detail horizontal bars (not to scale).

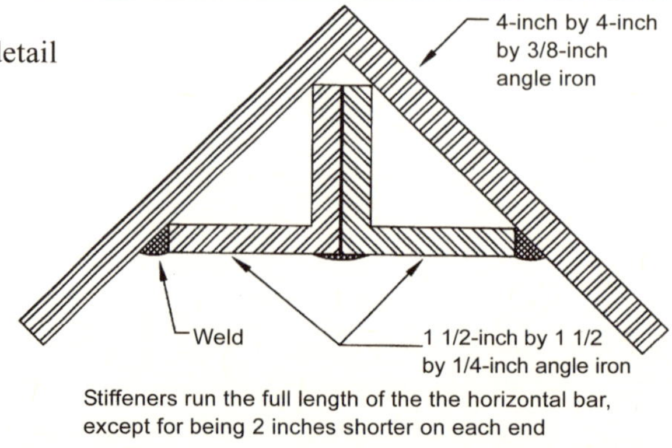

Diagrams © Jim Kennedy, BCI

Figure 9. Half or "flyover" gate. Note the removable bar at bottom right. Expanded metal mesh covers the overhanging top portion, making it extremely difficult to climb over. Coating the overhang with grease also helps repel trespassers.

Educational material is less antagonistic. Signs themselves sometimes become collectors' items, or are needlessly damaged by thwarted cave visitors. Permanent signs mounted inside the gate where they can be read, but are out of harms way, will last longer. Paper and wooden signs are highly susceptible to weather, decomposition, and the gnawing teeth of rodents. Metal or plastic signs are preferred. (See protective signs in caves, page 183.)

Permanent signs mounted inside the gate where they can be read but are out of harms way will last longer.

Construction Logistics

This cave gating chapter is no substitute for a more complete cave gating manual or training workshop. While it covers the rudiments of cave gating to assist resource managers in making better informed decisions, it is too brief to help with actual design and installation. Nevertheless, here we provide information for better planning of gating projects. Further assistance is readily available on request. (See cave gating resources, page 161.)

Timing

Construction should take place during seasons when human activity is least disturbing to the cave resources. For bat caves, this means the work must be done when the bats are absent. Some caves may be used as both summer and winter roosts, which leaves only short periods in the spring and fall for construction. Seasonal temperature variations may also cause reversals in the cave's airflow. If the cave is drawing in air, it may be necessary to install temporary plastic curtains inside the construction site to keep smoke and noxious welding fumes out of the cave. (See toxic fumes, page 40.)

Construction should take place during seasons when human activity is least disturbing to the cave resources.

Materials

Ordering adequate materials depends on accurate estimates of the area to be covered. Gate construction projects require accurate measurements and scale drawings of the finished gate. Materials should be ordered well in advance of the actual gating and may need to be stored off-site in a secured area before being transported to the cave. Always order a little extra for emergencies. (See materials, page 167; also see managing projects, page 305.)

Supplies

To help ensure completion of the project, carefully calculate welding gases, welding rods, grinding wheels, and other expendables. It is much easier to return unused supplies, or save them for the next project, than to run out before the new gate is finished.

Tools

The remoteness of the site will dictate the type of tools needed, but almost every gating project requires an electric generator or two to run the welders, grinders, and lights. Most projects need the following equipment:

- Electric generator(s)
- Extension cords
- Oxy-acetylene torches (with spare tips and regulators)
- Chipping hammers and wire brushes
- Tape measures, levels, and squares
- Ladders (for tall gates)
- C-clamps
- Portable work lights
- Hammer drills and hand-held grinders
- Digging and rock breaking tools to prepare the site

Also, provide the following safety equipment at cave gating sites:

- Rakes and water for fire control
- Buckets
- Welding vest, hood, and gloves
- Cutting goggles, and so on

Every cave gating work plan needs to address the protection of the cave and surrounding site as well as the safety of the workers involved.

Always plan for things to break, so have backups on site or readily available. Other equipment such as come-alongs, pulleys, chainsaws, all terrain vehicles, or in extreme cases, a helicopter may be needed to move materials to the site.

Short, 8-foot (2.5-meter) lengths of 1-inch (25-millimeter) tubular nylon webbing tied in loops make excellent carry handles for moving lengths of steel. Tools should be color-coded or labeled so they get back to their proper owners. Be careful to keep track of tools and equipment. Tools are especially easy to lose in or around the cave area.

Transportation of Materials

Many ingenious methods have been developed for moving materials to cave sites. Rarely can the delivery truck drive to the cave mouth. For long hauls, caver power may suffice, given a large enough workforce. Animal power (horses, mules, and burros) is sometimes used. All-terrain vehicles are sometimes used in nonwilderness areas with adequate trails. Boats or rafts may be necessary along rivers or lakes.

Materials may even need to be airlifted in extremely rugged terrain. Airlifts are sometimes accomplished with the cooperation of a local military reserve unit (the project may be used as a training mission). But during the course of most projects, all materials must be carried by hand. Keep in mind that a 20-foot (6-meter) length of 4-inch (10-centimeter) angle iron, 3/8-inch (9.5-millimeter) thick, weighs about 196 pounds (89 kilograms). Avoid pinched fingers and crushed toes by keeping safety in mind.

Personnel

The gate designer should oversee construction—this person is most important in any gating project. Currently, there are very few people in North America with the experience needed for all but the simplest jobs. (See contact list, page 165.)

Next comes the welder, who may be an agency employee, a volunteer, or a person hired specifically for the project. Depending on the size of the gate and the amount of work necessary, it is usually good to have several welders (people and machines) available to make the work go faster and to offer rest breaks.

Gating projects also need one or more welding assistants, anticipating the next piece to be cut, handing tools, taking measurements, and generally facilitating the workflow so that no one is standing around idle.

Finally, a project needs sufficient labor to prepare the site, carry items from the cutting area to the gate location, carry the steel from the drop point to the work area, and clean up afterwards. These workers can be hired with the welder, be provided by the responsible agency or organization, or be volunteers such as local cavers.

In several gating projects, prison labor was arranged for much of the heavy work. Using volunteers is beneficial because it involves the cave's user groups, educates them about purposes for the gate, and lessens potential for opposition and future vandalism to the gate.

Don't forget to take care of the safety and well-being of your workers. Provide plenty of food and drinks, and give adequate recognition after the project is finished.

Safety

Every cave gating work plan needs to address the protection of the cave and surrounding site as well as the safety of the workers involved. Prevent ground fires from starting at the work site. It may be necessary to temporarily remove dead leaves or grasses in the areas where cutting, welding, and grinding occur. As a precaution, have plenty of water and fire fighting tools (rakes and shovels) on hand. An Indian Pump or chemical fire extinguisher is also handy.

All workers should wear leather work boots, preferably steel-toed, as well as leather gloves, hardhats (caving helmets are fine), long pants, long-sleeved shirts or coveralls, protective eyewear, as well as hearing protection, especially when working around the welders, torches, and grinders. Caution all workers not to look directly at the torch flame or welding arc. Brief the crew on hot metal, heavy objects, potential dangers from the tanks of welding gases, and any other hazards specific to the site (loose rock, steep slopes, poison ivy, and the like). Keep a well-stocked first-aid kit on site. Also, be aware of the dangers of exhaustion, dehydration, hypothermia, and heat-related illnesses. Be sure the team takes breaks, eats during the day, and keeps hydrated.

Site Restoration

It may be difficult, but try to minimize disturbance of rocks, vegetation, and ground cover during steel hauling and other work. Natural contours should be restored after the gating is completed, unless the work on the cave entrance includes retuning it to a former historic configuration in an attempt to restore internal conditions. Sites may need revegetation, and trails may need to be blocked to divert casual hikers from the cave. All trash should be picked up and removed, including all scrap metal and as much welding waste as possible, including welding rod stubs. Cave gates, after painting (if necessary), should blend in rather than attract attention.

If an entrance was previously modified or enlarged, gating processes may provide a perfect time to restore the entrance to a former ecological state. Keep in mind that, relatively speaking, caves are short-lived geologic features that constantly change.

Entrances open and close naturally during the life of some caves, sometimes repeatedly. Choosing the historic baseline configuration is sometimes a judgement call based on the special resources for which the site is actively managed. For declining populations of endangered Indiana bats, for instance, we would aim for restoration to a time frame of pre-

European settlement, but post-Pleistocene.

Locks and Removable Locking Bars

Since the main purpose of a cave gate is to secure the site from intrusion, the choice of locking mechanism is critical. Many modern gates now dispense with hinged doors entirely and use removable locking bars. The removable bars can be secured with standard padlocks or with specially keyed bolts, similar to automotive locking lugnuts. Removable bars have several advantages:

Many modern gates now dispense with hinged doors entirely, and use removable locking bars.

- Removable bars are easy to construct.
- They disguise the obvious entry point.
- They eliminate the use of moving parts.
- They reduce maintenance tasks.

All padlock mechanisms must be designed to protect the lock from damage. Locks should be inspected regularly and replaced at the first sign of trouble or failure. No gate is completely vandal proof, but the idea of building a strong gate secured by a weak lock is ridiculous. If the cave is worth gating, make it as secure as possible.

Monitoring and Maintenance

So, you have finished building the gate and restored the entrance zone to a natural appearance. Job well done, right? More like job half done. There are no guarantees that the gate will accomplish your objectives despite your most careful planning.

Instead of helping maintain or restore the cave's ecosystem, a gate may cause further problems. Only long-term monitoring and assessment will tell. For bat caves this entails nightly and seasonal observations to monitor and ensure the bats' behavior is unchanged and uninterrupted. For other critters, monitoring might involve population estimates via specific sampling techniques. Monitoring requirements also point out the need to establish good baseline data before gate installation so comparisons can be made with postgating data. At the minimum, temperature and airflow should be recorded, but observations of moisture and humidity, animal distribution, and nutrient flow are also useful.

If the gate is not doing its job, then it should be modified or removed. Many bat caves gated in the 1970s and early 1980s were thought to be protected and were largely ignored thereafter. Continuing bat population declines puzzled researchers, who *believed* the caves were protected and looked for other reasons to explain decreases.

If the gate is not doing its job, then it should be modified or even removed.

Recent advances in gating knowledge show that the gates themselves were causing negative impacts on the caves because they were poorly designed or placed, or because the entrance was modified during the gating process. In an extreme case, the temperature of the cave was raised by as much as 5°F (2.8°C). Temperatures were restored and the population began to increase when the original gate was replaced with a better-positioned and better-designed closure. Monitoring programs are now initiated early in gating projects to identify and correct bad situations before human modification results in tragedy.

Gates must also be monitored for the inevitable breaching attempts. Certain segments of our society delight in trying to break into places where access is denied. Proper signage will go a long way toward educating most of the public about the reasons the cave was gated. Signs should point visitors to more information and contacts for access. Gaining the trust and cooperation of user groups and local cavers during the planning and construction processes will also alleviate potential animosity and break-in attempts.

Repairing Damage

Any damage to the gate should be repaired immediately—otherwise, you will be repairing more damage and dealing with illegal entries. When design flaws and weaknesses are discovered, you have the obvious opportunity to modify the gate and make it stronger. As noted gate expert Roy Powers says, "We have to keep one step ahead of the vandals." Be careful not to negatively impact the cave environment with security modifications.

Recurring vandalism may require increased security measures, such as surveillance. Sometimes trustworthy local cavers can be named as volunteer cave stewards who can provide much-needed manpower for patrolling the site. A well-publicized arrest of trespassers vandalizing a posted cave gate makes a wonderful deterrent to other would-be lawbreakers. Many other clever techniques have been utilized to deter vandalism, including fake monitors and signs announcing (usually nonexistent) alarm systems. Real alarms can also be used, triggering a dispatch to the agency office or local law-enforcement authorities.

To get hands-on training, participate in one of the Cave Gating Seminars co-sponsored by the American Cave Conservation Association, Bat Conservation International, the US Fish and Wildlife Service, and the USDA Forest Service.

Cave Gating Resources

If, after reading this, you feel overwhelmed and want to stay as far away from cave gating issues as possible, RELAX! There are several sources of excellent assistance available to help you. Modern, bat-friendly cave gates (also called zero-airflow-reduction bat gates) are the result of many years of experimentation and development, supplemented by field observation, strength testing, and wind tunnel testing.

The design presented in this chapter is the standard accepted by most federal and state agencies that manage caves, and by organizations such as The Nature Conservancy and Bat Conservation International. The leading force behind bat-friendly gate development has been the American Cave Conservation Association, particularly Roy Powers. Detailed drawings may be requested from them. Across the country, there are examples of many adaptations showing varying degrees of success. Successful gate designs provide entrance security and avoid the blockage of airflow, water, nutrients, and animals.

Current Books on Cave Gate Design

Bat-friendly gate designs are also widely used for closing abandoned mines. Mines and caves are similar, but not equivalent management concerns. Mines usually lack the complex ecosystems and recreational values that caves offer, and mines often pose bigger liability problems. Mines are extremely short-lived in comparison to caves. Stabilizing or closing mine entrances to achieve desired conditions does not have the ramifications that such actions cause in undisturbed caves. Bat Conservation International (BCI) has produced a free booklet, *Bats and Mines*, that discusses in detail the suitability of mines as habitat, addresses the dangers associated with them, and includes full plans for both standard and cupola gates. The booklet also offers excellent template forms for conducting external and internal summer and winter bat site assessments. (In the additional reading list for this chapter, see Tuttle and Taylor 1998.)

The authors and editors know of no modern, comprehensive, published gate plans for caves that have no bats or other vertebrates. *Trap-door gates* and *air-lock gates* are common in several parts of the United States but are usually built by local experts.

A detailed book on cave gating has been developed by the US Fish and

Wildlife Service, the USDOI Office of Surface Mining, Bat Conservation International, the American Cave Conservation Association, the National Speleological Society, and numerous other sponsors. It includes the entire proceedings from the groundbreaking conference on cave and mine protection options held in Austin, Texas, in March 2002. It is available through the National Speleological Society and covers the entire gating process in detail. (In the additional reading list for this chapter, see Vories and others 2004.)

Modern, bat-friendly cave gates (also called zero-airflow-reduction bat gates) are the result of many years of experimentation and development, supplemented by field observation, strength testing, and wind tunnel testing.

Cave Gating Seminars

To get hands-on training, participate in one of the Cave Gating Seminars cosponsored by the American Cave Conservation Association, Bat Conservation International, the US Fish and Wildlife Service, and the USDA Forest Service. These workshops combine evening slide lectures and discussions with hands-on gate building experience. The small group residential setting teaches design and placement philosophy, covers design options and case studies, and offers an opportunity to interact with some of the most knowledgeable cave gaters in the country. Contact the American Cave Conservation Association or Bat Conservation International for dates and locations of upcoming workshops.

Cave Gate Contractors

There are also several private individuals and firms that will contract gate-building projects. The best of these have many years experience or are graduates of the Cave Gating Seminar. Names of those known to be knowledgeable and reliable can also be obtained from Bat Conservation International or the American Cave Conservation Association. (See the contact list at the end of this chapter, page 165.)

Summary

Cave gating is only one form of cave protection. It should not be undertaken without sufficient study and planning. There are many types of gates and the manager should choose the type that best protects the resources within the cave and best fits the cave configuration.

Planning, construction, and follow-up activities are time and resource intensive. Gating projects may require a lot of manpower and other resources, including volunteers as well as specialized equipment. There are several sources of expertise and possible funding assistance for gating projects. Gating experts should always be contacted before any work begins.

There are several sources of expertise and possible funding assistance for gating projects.

Additional Reading

Anon. 1985. Volunteers protect Tennessee bats. *American Caves* 1(2):8-9.

Anon. 1993. James Cave receives new gates for bats. *Bats* 11(3):4.

Anon. 1993. BCI helps protect Pennsylvania mine for bats. *Bats* 11(4):16-17.

Anon. 1995. New ACCA gates protect Kentucky and Tennessee caves. *American Caves* 8(1):10-11.

Anon. 1997. Gates, gates, and more gates! *American Caves* 10(1):6-7.

Bat Conservation International (BCI). BCI home page. <http://www.batcon.org>. Direct link to full articles from *Bats Online*. <http://www.batcon.org/batsmag/batindex.html>. Accessed 2004 March 30.

Bobo K, Greene J. The gating of Wolf River Cave. *NSS News* 58(10):283, 286, 294.

Brady JT. 1985. Cave management for the endangered Indiana bat (Myotis sodalis) and gray bat (Myotis grisescens). In: Thornton H, Thornton J, editors. *1982 Cave Management Symposia Proceedings*. Richmond (VA): American Cave Conservation Association. p 86-95.

Brown CB. 1996. Bat habitat in caves: What are the caver's responsibilities? *NSS News* 54(2-3):52-53.

Burghardt JE. 1997. Bat-compatible closures of abandoned underground mines in National Park System units. In: Brandt JE, editor. *1997 Proceedings of the American Society for Surface Mining and Reclamation*. Lexington (KY): American Society for Surface Mining and Reclamation. p 184-195.

Currie RR. 2002. Response to gates at hibernacula. In: Kurta A, Kennedy J, editors. *The Indiana Bat: Biology and Management of an Endangered Species*. Austin (TX): Bat Conservation International. p 86-99.

Dutko R. 1994. Protected at last, the Hibernia Mine. *Bats* 12(3):3-5.

Elliott WR. 1996. The evolution of cave gating. *American Caves* 9(2):9-15.

Elliott WR. 2001. Project reports 2000: Missouri caves. *American Caves* 14(3):6-9.

Elliott WR. 2001. *Cave Gating Criteria*. Unpublished document for the Missouri Department of Conservation. 5 p.

Fletcher MR. 1985. Endangered bat protection at Buffalo National River, Arkansas. In: Vandyke JE, editor. *1984 National Cave Management Symposium Proceedings. Missouri Speleology* 25(1-4):147-150.

Forbis S, Haskins V, Foster DG. 1996. Protecting the ancient history of Crumps Cave. In: Rea GT, editor. *1995 National Cave Management Symposium Proceedings: Spring Mill State Park, Mitchell, Indiana, October 25-28, 1995*. Indianapolis (IN): Indiana Karst Conservancy. p 88-92.

Hathorn J, Thornton J. 1987. The common sense guide to cave gates. In: *ACCA Cave Management Series* 1(3):23-45. Horse Cave (KY): American Cave Conservation Association, Inc.

Hathorn J, Thornton J. 1993. Cave gates: Design and construction considerations. In: Foster DL, Foster DG, Snow MN, Snow RK, editors. *1991 National Cave Management Symposium Proceedings: Bowling Green, Kentucky, October 23-26, 1991*. Horse Cave (KY): American Cave Conservation Association, Inc. p 359-363.

Hopper HL. 1999. Cave gating: Three success stories. *American Caves* 12(3):6-9.

Hunt G, Stitt RR. 1975. *Cave Gating*. Huntsville (AL): National Speleological Society. 60 p. [*Cave Gating* (2nd edition). 1981. Huntsville (AL): National Speleological Society. Editors' note: Use both editions for historical reference only—these gate designs are outdated.]

Ludlow ME, Gore JA. 2000. Effects of a cave gate on emergence patterns of colonial bats. *Wildlife Society Bulletin* 28(1):191-196.

MacGregor J. 1993. Responses of winter populations of the federal endangered Indiana bat (Myotis sodalis) to cave gating in Kentucky. In: Foster DL, Foster DG, Snow MN, Snow RK, editors. *1991 National Cave Management Symposium Proceedings: Bowling Green, Kentucky, October 23-26, 1991*. Horse Cave (KY): American Cave Conservation Association. p 364-370.

Martin KW, Puckette WL, Hensley SL, Leslie DM Jr. 2000. *Proceedings of the Oklahoma Academy of Science* 80:133-137.

Murphy M. 1993. Restoring Coach Cave. *Bats* 11(3):3-5.

Nieland J. 1996. Washington's Christmas Tree Cave gated–Bats protected. *American Caves* 9(2):16-17.

Olson R. 1996. This old cave–The ecological restoration of the Historic Entrance ecotone of Mammoth Cave, and mitigation of visitor impact. In: *Proceedings of the Fifth Annual Mammoth Cave Science Conference*. Mammoth Cave (KY): National Park Service and Cave Research Foundation. p 87-95.

Powers RD. 1985. General cave gate considerations. In: Thornton H, Thornton J, editors. *1982 National Cave Management Symposium Proceedings*. Richmond (VA): American Cave Conservation Association. p 77-79.

Powers RD. 1993. Design improvements for gating bat caves. In: Foster DL, Foster DG, Snow MN, Snow RK, editors. *1991 National Cave Management Symposium Proceedings: Bowling Green, Kentucky, October 23-26, 1991*. Horse Cave (KY): American Cave Conservation Association. p 356-358.

Powers RD. 1996. A study of acoustical confusion. In: Rea GT, editor. *1995 National Cave Management Symposium Proceedings: Spring Mill State Park, Mitchell, Indiana, October 25-28, 1995*. Indianapolis (IN): Indiana Karst Conservancy. p 274-276.

Quinn M, Petterson J. 1997. A grand effort in the Grand Canyon. *Bats* 15(3):4-7.

Raesly RL, Gates JE. 1986. Winter habitat selection by north temperate cave bats. *American Midland Naturalist* 118(1):15-31.

Richter AR, Humphrey SR, Cope JB, Brack V Jr. 1993. Modified cave entrances: Thermal effect on body mass and resulting decline of endangered Indiana bats (*Myotis sodalis*). *Conservation Biology* 7(2):407-415.

Roebuck BE, Vakili A, Roebuck L. 2001. Cave gate airflow disturbance: a qualitative study. In: Rea GT, editor. *Proceedings of the 14th National Cave and Karst Management Symposium: Chattanooga, Tennessee, October 19-22, 1999*. Chattanooga (TN): Southeastern Cave Conservancy. p 169-175.

Roebuck L. 2000. Gating in Tennessee: A plethora of summer bat protection projects. *American Caves* 14(1):6-8.

Stihler CW. 2000. Gates constructed to protect bats in West Virginia caves. *American Caves* 13(2):15-16.

Toomey, RS III, Coburn ML, Olson RA. 2002. Paleontological evaluation of past use of caves by the Indiana bat: A significant tool for restoration of hibernacula. In: Kurta A, Kennedy J, editors. *The Indiana Bat: Biology and Management of an Endangered Species*. Austin (TX): Bat Conservation International. p 79-85.

Tuttle MD. 1977. Gating as a means of protecting cave-dwelling bats. In: Aley T, Rhodes D, editors. *National Cave Management Symposium Proceedings: Albuquerque, New Mexico, October 6-10, 1975*. Albuquerque (NM): Speleobooks. p 77-82

Tuttle MD, Stevenson DE. 1978. Variation in the cave environment and its biological implications. In: Zuber R, Chester J, Gilbert S, Rhodes D, editors. *1997 National Cave Management Symposium Proceedings: Big Sky, Montana, October 3-7, 1977*. Albuquerque (NM): Adobe Press. p 108-121.

Tuttle MD. 1985. Joint effort saves vital bat cave. *Bats* 2(4):3-4.

Tuttle MD. 1997. A Mammoth discovery. *Bats* 15(4):3-5.

Tuttle MD, Taylor DAR. 1998. *Bats and Mines*. Resource Publication (Bat Conservation International) No 3. Revised edition. Austin (TX): Bat Conservation International.

Tuttle MD. 2000. Where the bats are, Part III: Caves, cliffs, and rock crevices. *Bats* 18(1):6-11.

Tuttle MD, Kennedy J. 2002. Thermal requirements during hibernation. In: Kurta A, Kennedy J, editors. *The Indiana Bat: Biology and Management of an Endangered Species*. Austin (TX): Bat Conservation International. p 68-78.

Twente JW Jr. 1955. Some aspects of habitat selection and other behavior of cavern-dwelling bats. *Ecology* 36(4):706-732.

Vories KC, Throgmorton D, Harrington A, editors. 2004. *Proceedings of*

Bat Gate Design: A Technical Interactive Forum held in Austin, Texas, March 4-6, 2002. Carbondale (IL): USDOI Office of Surface Mining, Alton, Illinois and the Coal Research Center, Southern Illinois University at Carbondale. 434 p. Proceedings available on the internet <http://www.mcrcc.osmre.gov/bats/>. Accessed 2004 March 30.

Werker JC. 2005. Materials considerations for cave installations. In: Hildreth-Werker V, Werker JC, editors. *Cave Conservation and Restoration*. Huntsville (AL): National Speleological Society. p 171-178.

White DH, Seginak JT. 1987. Cave gate designs for use in protecting endangered bats. *Wildlife Society Bulletin* 15:445-449.

Contacts for Additional Information

American Cave Conservation Association (Roy Powers or Dave Foster)
PO Box 409
Horse Cave KY 42749-0409
502-786-1466
E-mail: <acca@caveland.net>
Web: <http://www.cavern.org/>

ACCA Northwest Chapter (Jim Nieland)
12178 Lewis River Rd
Ariel WA 98603-9745
360-247-3946
360-247-3901 fax
E-mail: <jnieland@fs.fed.us>

Bat Conservation International (Jim Kennedy)
PO Box 162603
Austin TX 78716-2603
512-327-9721, extension 17
512-327-9724 fax
E-mail: <jkennedy@batcon.org>
Web: <http://www.batcon.org/>

US Fish and Wildlife Service (Bob Currie)
160 Zillicoa Street
Asheville NC 28801-1038
828-258-3939, extension 224
828-258-5330 fax
E-mail: <robert_currie@fws.gov/>

Section B—Developing Cave Management Programs

Materials Considerations for Cave Installations

Jim C. Werker

What materials are safe and cost effective for long-term use in the harsh environments of caves? Characteristics of common construction materials are described in this chapter. Emphasis is placed on materials that are considered corrosion resistant, tough, and readily available. Other exotic materials, though their characteristics may be beneficial to the protection of underground habitats, tend to be expensive, less available, and sometimes require special fabrication procedures.

Before planning and designing cave construction projects, evaluate the material options and fabrication requirements. Several reference manuals are especially useful when considering options.

- *Machinery's Handbook* (Industrial Press)
- *Manual of Steel Construction* (American Institute of Steel Construction, Inc.)
- *Electrode Pocket Guide* (Airco)
- *Ryerson Steels Stock List and Data Book and Ryerson Special Metals Data Book* (Joseph T. Ryerson and Son, Inc.)

Most materials reference books are updated regularly and contain useful information for any construction project. For example, *Machinery's Handbook* contains sections on types and properties of materials, welding specifications, and finishes. The strength of materials section in *Machinery's Handbook* provides simplified mathematical formulas to use in calculating material strengths.

What materials are safe and cost effective for long-term use in the harsh environments of caves? Characteristics of common construction materials are described in this chapter.

Stainless Steels

Chromium–nickel austenitic steels are commonly known as stainless steels. Both names refer to the same family of steel materials. The corrosion resistance and toughness of stainless steels make them highly suitable for cave installations. The longevity of stainless is ten-fold that of mild steels. Even though stainless is more costly than other materials, it is increasingly used to replace wood and steel structures in caves.

Stainless steel life expectancy and corrosion resistance far exceed the characteristics of most other materials. Any of the chromium–nickel austenitic steels offer good characteristics for cave installations, however there are important fabrication requirements for these stainless steel materials.

There are a number of weldable chromium–nickel austenitic steels on the market. If the project is fabricated in a controlled, clean welding shop, any of the weldable austenitic stainless steels will work well. Find complete listings of the characteristics of austenitic steels in *Machinery's Handbook*. For austenitic stainless steels, welding requirements call for shielded gas metal arc welding processes to minimize the potential for carbide precipita-

The corrosion resistance and toughness of stainless steels make them highly suitable for cave installations.

tion; for example, tungsten inert gas (TIG), or metal inert gas (MIG) processes are often recommended.

Generally, stainless cannot be properly fabricated in the field unless special welding equipment is used to purge the welds with argon to minimize corrosion-inducing carbide precipitation. If welding must be done at the cave site, chromium–nickel austenitic steel types 304L, 316L, or 321 are recommended. These steels are less vulnerable to the harmful carbide precipitation that enhances weld corrosion (rust) tendencies.

For field applications, optional construction methods may apply. For example, predrill and countersink holes in a fabrication shop, transport the pieces to the site, then use bolts to assemble the pieces. Holes for the rivet-like bolting can be drilled in the shop. If bolts are used for stainless gate designs, after installing stainless bolts with countersunk heads, flatten the heads with a hammer, then grind the nuts to a cylindrical shape so they will not accept a vandal's wrench.

Manganal®

A high-manganese, austenitic, work-hardening steel that is currently used in some cave gates is available under the trade name Manganal. Typically, the chemical composition is manganese (12.00–14.00%) and carbon (1.00–1.25%). Manganal bars, plates, and castings are used for high-impact industrial applications. Cost of these high-manganese materials tends to run two to three times that of mild steel (Stulz–Sickles Steel 2002).

Characteristics of Manganal® include high-strength, ductility, toughness, and substantial longevity.

Manganal is used in extreme wear conditions and is hardened by impact, hammering, and abrasion. This surface characteristic is known as work hardening. As this alloy work hardens, strength and resistance increase (Amodt and Mesch 2004). In other steels (for example, carburized or casehardened), the depth of hardness is fixed. When Manganal is subjected to wear, the surface toughens and the material remains ductile underneath.

Characteristics include high strength, ductility, toughness, and substantial longevity. Corrosion resistance (resistance to rust and attack by acids) is about the same as ordinary steels. Manganal is extremely tough when work hardened and may tolerate harsh environments. Functional mine rails made from this material are over 100 years old (Louis Arnodt, personal communication). Cave gates constructed of Manganal can deter vandals using hacksaws. However, power tools or cutting torches can breach the closures. Manganal is used for wear resistance in jail bars, bulldozers, scrapers, stone chutes, and military equipment.

Continuous high temperature can embrittle the high-manganese, austenitic steels. In electric arc welding processes, no local area should remain at visible red heat for more than two or three minutes. If there is build up with multiple layers from weld passes, the welder should either skip weld, or weld intermittently to reduce localized heat.

Manganal is a good choice for field fabrication if the high cost is justified. Preferred applications tend to be in remote sites where minimal acidic conditions exist and where vandals cannot easily use power tools. Manganal is durable and has excellent longevity characteristics.

Structural Steels

Because mild steels (structural cold-rolled steels) corrode quickly, these materials should be carefully evaluated before installing within cave environments. However, structural steels often provide cost-effective solutions for cave gates. When constructing cave gates, stainless steel life expectancy and corrosion resistance far exceed the characteristics of other steel materials. However, since repetitious vandalism is a key issue at some cave sites, and since replacing stainless is expensive, cost may dictate the use of structural steel for gate construction. If the environment is too harsh for mild steels to survive, the site conditions may dictate that stainless be selected for its increased life expectancy.

The most common grade of structural steel (sometimes called mild steel) is ASTM A-36. The high-strength structural steels ASTM A-529 and A-440 have high carbon content for strength but they are no more durable than mild steels. Corrosion-resistant, high-strength steels have one advantage over the mild varieties in that they are more difficult to vandalize.

Mild steels are easy to fabricate, readily available, and cost less than most other options. Mild steel is available in a variety of structural shapes that are easily welded and fabricated in the field. For gate construction at cave entrances, the life expectancy is 50 to 100 years. However, within caves, the life expectancy of structural steels may be tremendously reduced by the environmental conditions, and habitat may be compromised by rapid degradation of the material.

Since repetitious vandalism is a key issue at some cave sites, and since replacing stainless is expensive, cost may dictate the use of structural steel for gate construction.

Aluminum

Aluminum can deteriorate rapidly and the degradation may introduce toxins into cave habitats. For example, an aluminum ladder left in a cave located in the arid southwestern U.S. literally deteriorated to a pile of scrap in less than 20 years (Werker 2003). Aluminum carabiners left in caves for varying time intervals rapidly show signs of pitting and corrosive deterioration (Storage 1994). Aluminum will probably work for gates placed in dry, nonalkaline environments. However, aluminum is easy to vandalize because, generally, it is not as strong as steel.

When aluminum structures are exposed to the atmosphere, a thin, invisible oxide skin forms immediately and protects the surface from additional oxidation. This self-protecting characteristic gives aluminum its high resistance to corrosion unless it is exposed to some substance or condition that destroys the oxide coating. Alkalies are among the few substances that will attack the oxide skin—thus, the alkaline conditions of most limestone caves will cause aluminum to corrode.

When aluminum is placed in direct contact with other metals, the presence of an electrolyte (moist conditions or high humidity) will cause galvanic corrosion of the aluminum at the contact points. Dissimilar metals in direct contact with each other within a moist environment will definitely induce corrosion.

If dissimilar metals are already in place, nylon or neoprene washers can be used to separate or isolate the materials from each other and delay the corrosion. (A moisture film may eventually form across barriers and corrosion will continue.)

Depending on the site conditions, protective coatings may increase the life expectancy of aluminum. Chromate coating can be brushed on in the field, but anodizing must be done at a coating lab. However, once any kind of coating is breached by a scratch or nick, the integrity of the surface is compromised and the electrochemical process begins to produce pitting and deterioration as the metal corrodes (Spate and others 1998).

Because aluminum is especially susceptible to both vandalism and corrosion, it is usually a less desirable material for cave applications.

Aluminum can deteriorate rapidly and the degradation may introduce toxins into cave habitats. For example, an aluminum ladder left in a cave located in the arid south-western U.S. literally deteriorated to a pile of scrap in less than 20 years.

Concrete

Concrete works well in most environments. It is resistant to chemical and corrosive attack and has extremely good longevity characteristics. Structures

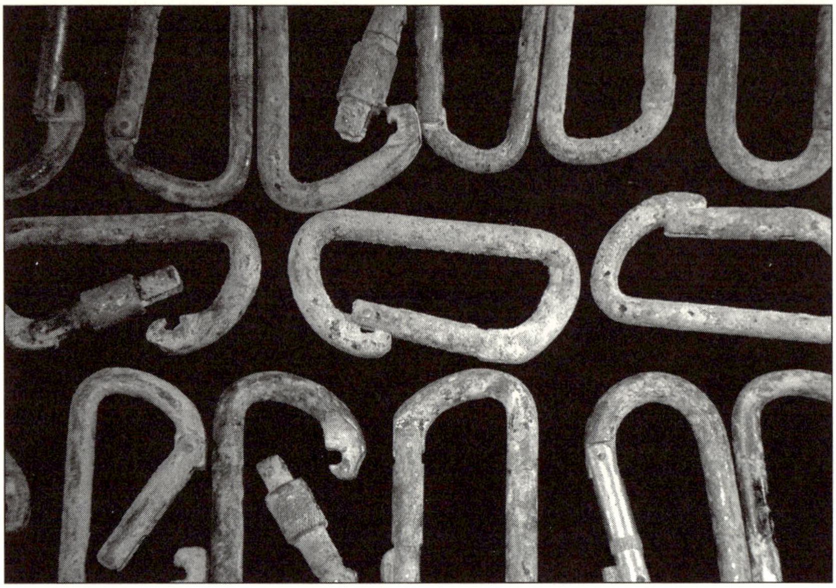

© Val Hildreth-Werker

Figure 1. These aluminum carabiners were left on a traverse in Virgin Cave, New Mexico. After only six months in the cave, pitting and corrosion was abundant. Material strength was greatly reduced, and some of the biner gates no longer closed. Stainless steel carabiners are recommended for long-term use in cave environments. (See page 6 of color section.)

Concrete works well in most environments. It is resistant to chemical and corrosive attack and has extremely good longevity characteristics.

built with 3,000 psi (20,680 kilopascal) concrete reinforced with rebar will deter vandals and will hold up for many decades in most circumstances. The chemical composition of concrete resembles the natural composition of some cave passages and it introduces few toxins to cave systems. Cement, which is basically burned limestone, is also known as Portland cement. Sand, gravel, and water are added to cement products to make concrete.

An Australian paper by Spate and others (1998) recommends applying a protective membrane between natural floors, rock, or flowstone before pouring concrete. Also, fine sand is recommended for covering microgours and sharp surfaces and can be vacuumed away if the concrete is removed in the future.

If concrete is chosen for cave projects, use high tensile strength cement with low-content calcium hydroxide, which will last longer in most cave environments. The calcium hydroxide in cement products is very soluble and it will quickly dissolve and redeposit to form new "soda straws" or "flowstone" below concrete structures (Horrocks and Roth 2006, page 369 in this volume). (For additional information on cement choices, see the speleothem repair section (page 476).

Culvert and Pipe

Culverts and pipes installed as cave access structures can be made from a variety of materials. Several material types are addressed below. Be aware that culverts or pipes may not be the best option for protecting bat colonies and habitats for other animals. Small diameter flyways and absence of natural barriers can set the stage for easy predation. Culverts and pipes can change the natural airflow of a cave, alter air temperatures and relative humidity, and impede or increase the presence of biota. However, where cave entrances have been enlarged for human access, a culvert may provide the best protection.

Galvanized Steel

For decades, galvanized steel culvert has been used in roadway construction where it seems to function well. However, in caves, galvanized culvert may deteriorate rapidly. For example, a galvanized culvert installed in Lechuguilla Cave in 1986 showed visible signs of degradation by 1994, and had severely deteriorated by the time it was replaced in 2000 (Werker

© Val Hildreth-Werker

Figure 2. In 1986, after digging open the entrance, a galvanized culvert with a solid steel gate was installed in Lechuguilla Cave. After only a few years, it was visibly rusting and deteriorating. (See page 6 of color section.)

2004).

As the zinc coating of galvanized steel degrades, it may generate toxins that are harmful to flora, fauna, and perhaps to speleothems (Jameson and Alexander 1996; Spate and others 1998). Welding galvanized material results in noxious fumes that can be hazardous to human health and cave-dwelling animals. The breakdown, outgassing, and deterioration of galvanized steel culvert result in byproducts that may be especially toxic to bats and other cave-dwelling biota.

Plastics

Little is known about the degradation processes of plastics that are used in cave environments. Polyvinyl chloride (PVC) used in water lines and air conduits tends to become brittle over time. PVC also outgasses potentially harmful substances. Until studies further define the longevity and degradation characteristics of PVC and various plastics placed in subterranean environments, other construction materials are preferred for cave use.

© Val Hildreth-Werker

Figure 3. In 2002, the rusted culvert was replaced. A stainless steel culvert with an air lock entry was installed in Lechuguilla Cave. (See page 6 of color section.)

Recycled Materials Developed for Environmental Use

Advances in fabrication technologies have provided recycled construction products that are intended to be environmentally friendly. These materials are generally designed for aboveground applications. Time will tell how well these products endure subterranean environments. Timpanogos Caves National Monument, Oregon Caves National Monument, and various other sites are using products made from recycled materials for in-cave installations. The durability, degradation, and outgassing characteristics of these innovative materials may prove to be appropriate for underground applications in cave environments.. (See anthropogenic chemicals, page 57.)

Archival Epoxies and Adhesives

All glues break down over time. Many epoxies and adhesives are known to be composed of detrimental materials and display degradation characteristics that will destroy or harm biota.

Any new product should be checked by research chemists and biologists who understand the cave environments where use is proposed. What are the components of the agent and how will the degradation and outgassing characteristics of those compounds affect cave-dwelling biota, ecosystems, chemistry, water quality, or the minerals of a cave system? Arrange for a chemist or a materials engineer to evaluate long-term degradation and potential chemical interactions. Get the federally regulated Material Safety Data Sheets (MSDS) and understand recommended safety precautions. Wise product choices for cave environments are based on thorough research. Updated information is often available on the Web. (See MSDS, page 70.)

Epoxies, bonding agents, and quick glues found in neighborhood hardware and variety stores are certainly easier to obtain than the archival-quality adhesives, but the mass market products can introduce toxins and nasty agents into caves. All products will break down over time, but archival products are formulated to do less harm than those on the general market.

All products will break down over time, but archival products are formulated to do less harm than those on the general market.

To augment safe travel in caves, stainless steel bolts are sometimes installed. For outdoor applications, adhesives have been used in the installation of climbing bolts. However, an archival-grade epoxy may be the best choice for installing bolts in cave systems because most consumer epoxies have not been studied in underground environments.

When adhesives are necessary, use archival products, museum-grade epoxies, or pure forms of cyanoacrylate adhesives. All products will degrade over time, but some compounds are formulated to be more archival than others. The following recommended adhesives, used in subterranean applications over the past several decades, appear to be relatively safe for long-term use in cave environments.

Epon 828® With Versamid® or Epi-cure® 3234 (TETA)

- Epon 828 epoxy has been successful in underground environments for decades.
- Epon 828 epoxy resin combined with Versamid 40 curing agent works for dry surfaces, even in humid cave environments.
- For wet applications, Versamid 28 hardener cures more efficiently.
- Epon 828 with Versamid 25 curing agent will bond underwater.
- The Epon family of archival adhesives will develop strong bonds with shear strengths up to 6,000 pounds per square inch—6,000 psi (41,370 kilopascal).
- Curing time can take 24–72 hours and sometimes longer. Shrinkage is minimal. The bond is resistant to a broad range of chemicals.
- Mixing ratio is typically 1:1–one part Epon to one part Versamid (50:50 mix). However, for a more rapid drying time, use just a little more hardener for a 40:60 mix.
- Epon 828 and the Versamid hardeners were lab tested for long-term underground use at the US Department of Energy Nevada Test Site and have proved successful in cave applications for several decades.
- These products are available from the Shell® Chemical Company and through local chemical or plastic product suppliers. However, the Versamid curing agents are increasingly difficult to locate and fabrica-

tors in the plastics industry are recommending a replacement Shell product, Epi-cure 3234, which is a highly concentrated curing agent with bonding properties and archival characteristics similar to Versamid.
- Epi-cure 3234 (TETA) is now the recommended curing agent for Epon 828 and is typically mixed in a 12:1 ratio, twelve parts Epon to one part TETA.

Hot Stuff® Super T and Special T

Fast-drying cyanoacrylate adhesives are useful for repairing soda straws, helictites, thin draperies, and other delicate speleothems or small applications in caves. Hot Stuff adhesives are industrial-strength products containing a very pure form of cyanoacrylate that remains clear when it cures. Oil is not added to lengthen shelf life (as it is in many other instant glues).

- Hot Stuff Super T is often used in paleontological applications and model-building and works well for small repairs in caves.
- Hot Stuff Special T is a more viscous product, formulated to fill in gaps and is extremely useful for cave applications.
- In most caves, Hot Stuff cures in 30–90 seconds, but bonding can be accelerated with NCF-Mild Accelerator. Shear strength is weakened by accelerated curing times, but cyanoacrylate does not bond quickly in large quantities. One spritz from a spray pump bottle of Hot Stuff NCF-Mild will cause chain reaction bonding.
- Hot Stuff Super T, Special T, and NCF-Mild Accelerator can be purchased in model-building stores, quality woodworking shops, museum supply catalogs, and through a few sources on the Web.

Finishes and Paints

Finishes applied to the surfaces of materials are intended to enhance longevity but may introduce contaminants to cave environments. All paints and finishes deteriorate over time. Flaking onto the habitat floor is obviously detrimental, but there are more subtle degradation characteristics.

Outgassing and chemical deterioration of finishing products may introduce potentially toxic materials to cave systems. Increased corrosion may be generated if a paint layer is compromised—severe corrosion may result when an electrolyte, perhaps a simple film of moisture on a scratched finish, reacts with the large metal anode area under the paint (Spate and others 1998). Research is needed to investigate the potential benefits and harms of various finishes applied to installations within caves and near cave entrances.

Finishes applied to the surfaces of materials are intended to enhance longevity but may introduce contaminants to cave environments.

Common Sense

Function and cost are important components of materials selection for cave environments. Characteristics of materials and finishes are vital factors in any design and planning process. Assessment of site environmental conditions and evaluation of longevity criteria are essential to project planning. For each material historically used in caves—stainless steels, Manganal, mild structural steels, concrete, aluminum, galvanized steel, plastic products, epoxies, adhesives, paints, and other finishes—there are beneficial as well as disadvantageous characteristics. Specific knowledge about species that use the cave, habitat requirements, chemical compositions within the cave, and general common sense must dictate design and material selection.

The arduous tasks of investigating the multitude of material choices, evaluating their varying characteristics, analyzing the costs, considering the potential for vandalism, and understanding the inherent longevity of the materials may appear overwhelming. First, evaluate the site, the habitat, and the purposes for any structure or installation. Simplify the goals, state the site objectives, and then allow common sense to dictate material choices and construction techniques. When planning projects to protect caves, the priority is to be realistic about the habitat, the site requirements, and the budget. Keep it simple and remember that opting for no construction is sometimes the best decision.

Cited References

Airco. No date. *Electrode Pocket Guide*. Airco Welding Products. Union (NJ). 161 p.

Amodt LA, Mesch MR. 2004. Round bar Manganal® steel "jail bar" bat gate. In: Vories KC, Throgmorton D, Harrington A, editors. *Proceedings of Bat Gate Design: A Technical Interactive Forum held in Austin, Texas, March 4-6, 2002*. Carbondale (IL): USDOI Office of Surface Mining, Alton, Illinois and the Coal Research Center, Southern Illinois University at Carbondale. 434 p. Proceedings available on the internet <http://www.mcrcc.osmre.gov/bats>. Accessed 2004 March 30.

American Institute of Steel Construction, Inc. 1973. *Manual of Steel Construction, 7th Edition*. New York. 999 p.

Horrocks RD, Roth JE. 2006. Removing artificial fill from developed caves. In: Hildreth-Werker V, Werker JC, editors. *Cave Conservation and Restoration*. Huntsville (AL): National Speleological Society. p 367–380. [This volume.]

Jameson RA, Alexander EC Jr. 1996. Zinc leaching from galvanized steel in Mystery Cave, Minnesota. In: Rea GT, editor. *Proceedings of the 1995 National Cave Management Symposium: Spring Mill State Park, Mitchell, Indiana, October 25-28, 1995*. Indianapolis (IN): Indiana Karst Conservancy. p 178-186.

Joseph T. Ryerson and Son, Inc. 1967. *Ryerson Special Metals Data Book*. Denver (CO). 288 p.

Joseph T. Ryerson and Son, Inc. 1987. *Ryerson Steels Stock List and Data Book*. Denver (CO). 432 p.

Oberg E, Jones FD, Horton HL. 1985. *Machinery's Handbook, 22nd Revised Edition*. Ryffel HH, Geronimo JH, editors. New York: Industrial Press, Inc. 2512 p. [Editors' note: *Machinery's Handbook* has been updated continuously since 1914. See Oberg, Jones, Horton, Ryffel, McCauley, Heald, Hussain. 2000. *Machinery's Handbook*, 26th Revised Edition. New York: Industrial Press, Inc. 2640 p.]

Spate A, Hamilton-Smith E, Little L, Holland E. 1998. Best practice and tourist cave engineering. In: Smith DW, editor. *Cave and Karst Management in Australasia XII*. Carlton South, Victoria (AU): Australasian Cave and Karst Management Association. p 97-109.

Storage B. 1994. Techniques and safety: Aging carabiners. *NSS News*, 52(9):277-279.

Stulz–Sickles Steel Company. 2002. *FAQ About Manganol*. <www.stulzsicklessteel.com/faq.html>. Accessed 2004 March 30.

Werker, JC. 2004. Characteristics of materials used in cave and mine gates. In: Vories KC, Throgmorton D, Harrington A, editors. 2004. *Proceedings of Bat Gate Design: A Technical Interactive Forum held in Austin, Texas, March 4-6, 2002*. Carbondale (IL): USDOI Office of Surface Mining, Alton, Illinois and the Coal Research Center, Southern Illinois University at Carbondale. 434 p. Proceedings available on the internet <http://www.mcrcc.osmre.gov/bats/>. Accessed 2004 March 30.

Section B—Developing Cave Management Programs

Trail Delineation and Signage in Caves: Reduce Visitor Impact

Val Hildreth-Werker, James R. Goodbar, and Jim C. Werker

Why do we need trails in caves? Trails help reduce cumulative impacts in show caves and in undeveloped caves. When visitors are confined to defined pathways, features outside the trails are better preserved. Trail designation is an excellent tool for managing sensitive cave resources.

Show cave routes should provide durable, safe walking surfaces and definite boundaries for visitors. Trails in wild caves should confine travel impact and help protect cave values. Designated routes and delineated trails can improve cave protection and enhance resource management in several important ways.

- Direct travel to durable surfaces.
- Mark safe routes through caves.
- Reduce expansion by marking both sides of trails.
- Prevent new, inappropriate routes.
- Minimize compaction of soils that provide habitat for cave life.
- Preserve cave values by routing visitors a safe distance from sensitive or fragile resources, cultural materials, speleothems, wildlife, and habitat.

Management objectives frequently focus on preventing harm to cave values. Clearly marked trails through cave passages help confine human impacts to areas that are already impacted. Strategically placed signs convey conservation information and encourage low-impact protocol. This chapter focuses on trails and signs in undeveloped caves.

This chapter focuses on establishing and maintaining trails and signs in undeveloped cave passages.

Trails in Undeveloped Caves

Do wild or undeveloped caves really need marked trails? Cavers are defining routes through many wild caves because allowing people to wander leads to damage and eventual destruction. Like walking on side-walks, or following paths in the wilderness, cave trails can make travel more efficient and help protect resources.

Population increases in urban areas result in more people looking for adventure outside of cities—thus, more untutored people are entering caves. Increased visitation typically leads to increased resource damage. Grottos and caver educators simply cannot stay ahead of the growing recreation demands. To help inform cave visitors and encourage good caving ethics, it is time to implement more on-site training tools like trails and educational signage.

It seems surprising, but in many undeveloped caves most people will comply and remain on the trails. When cave visitors understand the importance of staying on already compacted routes, they tend to act with respectful stewardship and remain within the trails.

First, explore whether the path through a cave needs to be marked. Trail designation is not appropriate for all caves.

Is there an obvious route that most people will follow? Is the pathway visible to untrained visitors moving through darkness with only a flashlight or a helmet light?

In the past, reflective paints, tapes, and adhesive stickers were applied to plastic spoons, popsicle sticks, rocks, aluminum strips, plastic stakes, PVC tubes, and other support devices.

What is the appropriate material for marking trails? Surveyors flagging tape works very well in some environments, but it is not the right choice for all caves. In caves that are prone to flooding, flagging tape may be useless, while rocks may work great along trail edges. Stream passage caves with few formations and delicate areas may not need marked trails. However, if there is some compelling reason for taped trails in a stream passage, one option might be to remove the flagging during flood seasons.

Caves with large chambers have different trail needs than those with long, tight crawls. Heavily decorated caves may need trails that help minimize impacts by limiting traffic to a narrow path. Protecting cave passages that contain cultural or historical materials may make defined pathways necessary.

Some cave passages need the extra protection afforded by marked trails, and some don't. Ask this question—will marked trails help reduce cumulative visitor impacts?

Materials for Marking Trails in Undeveloped Caves

What materials work best for marking trails? New answers for trail delineation materials will emerge as better materials become available. Only a few trail-marking materials fit the standards of current best practice.

- Cave rocks
- Surveyors *non*biodegradable flagging tape
- High visibility nylon or polypropylene cord

In the past, reflective paints, tapes, and adhesive stickers were applied to plastic spoons, popsicle sticks, aluminum strips, plastic stakes, PVC tubes, rocks, and other support devices. Though these markers were innovative for their time, some have caused impacts that we can avoid in the future.

For example, globs of clay holding markers in place have streaked and discolored flowstone surfaces. Over time, paint may turn into a pile of

Figures 1a (left) and **1b** (closeup, right). Reflectors mounted on PVC rings appear to be durable in an Idaho lava tube cave environment. But in some cave systems, these materials will rapidly degrade and the byproducts may cause harm to cave-dwelling life. Unfortunately, intermittent marking systems also tend to encourage visitors to wander off-trail between the markers.

flakes. Supports made of wood may mold and rot. In many cave environments, aluminum will degrade, leaving aluminum oxide and other contaminants. Most paints, adhesives, and many plastic products deteriorate quickly in cave environments, and the byproducts are likely to be detrimental to native biota in cave systems. (See chapters on biology, page 33; microbiology, page 61; and materials, page 167.)

© Val Hildreth-Werker

Figure 2. Aluminum strip and red adhesive marker. Though an innovative and simple trail marking device, this marker will deteriorate rapidly in many cave environments, leaving aluminum oxide and other contaminants in the cave system.

Techniques for Marking Trails in Undeveloped Caves

When trails are not marked through caves, responsible cavers choose previously impacted, durable surfaces for travel and avoid leaving new footprints.

Traveling on previously compacted routes helps preserve cave sediments, small floor speleothems, spelean habitat, and invertebrate populations beyond the pathway. Trail routing and delineation also helps protect overhead speleothems and wall decorations, cultural and paleontological resources, and other cave values.

Carefully plan routes for travel on durable surfaces. Protect sensitive resources. Facilitate visitor safety and rest. Allow cavers to enjoy, study, and photograph the magnificent features found in caves.

With intermittent trail marking techniques such as reflectors, rock cairns, or short strips of flagging, visitors tend to wander off-trail between the markers and cause broader impacts.

If footprints are visible beyond the designated paths, others will follow and trails will expand. One of the most significant aspects of evolving cave etiquette is the request that cavers stay on designated trails.

In the past, when we found a footstep going into pristine territory, it was okay to place a foot precisely inside each original imprint, and follow the footprints to retrace the path. But, as more people followed, too many feet made new, ever-widening trails. Today, rather than retrace a step, we erase it and redirect traffic to designated routes. Leaving footprints behind will only invite others to follow. (See footstep ethic, page 409.)

By removing visible traces of human travel outside of trails, the invitation to follow is removed, and future damage is mitigated. Observe caver patterns and mark user-friendly pathways that make it safe and logical for cave visitors to stay within the trails. Once established, trail systems must be managed—rock borders and taped boundaries require maintenance.

Rocks for Marking Cave Trails

Although rock-lined trails may be aesthetically more appealing than trails marked with surveyors flagging tape, rocks are not always available. In most cases, rocks should *not* be imported from outside a cave.

- Border both sides of the trail to best protect cave values.
- Rocks marking trails should be as continuous as the supply of rocks in a given cave will allow.
- Rocks may also be used to define and protect special floor features.
- Rock cairns may be appropriate for marking travel routes in some caves. However, cairns provide little benefit for protecting a cave from unnecessary impact.
- Avoid setting rocks along an edge where they may dislodge and fall on speleothems or cavers below.
- Avoid using broken speleothems to line cave trails. Broken speleothems should be gathered and stored in a relatively safe location

Trail flagging takes a bit of skill. There tends to be more to the process than might be expected.

Figure 3. Rocks found within the cave were placed along the border of this trail through the Breast of Venus Room in Endless Cave, New Mexico.

Figure 4. A rock border helps to protect this fossil, which is adjacent to a muddy cave trail.

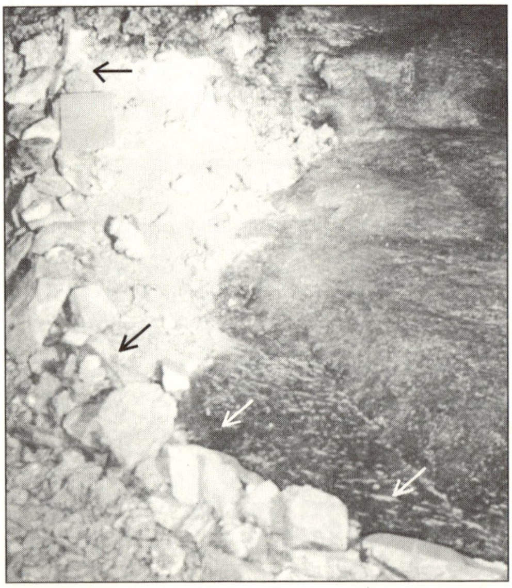

Figure 5. Flagging tape is laced between trail-defining rocks to aid visibility and increase visitor compliance. (See page 8 of color section.)

Photos © Val Hildreth-Werker

in the cave, away from traffic, and near the area where they were originally located. Tucking the pieces away in an off-trail location may help preserve them for potential repair and may remove the temptation for visitors to haul them out of the cave.

- Rocks sometimes don't command the attention of visitors. In heavily traveled caves, Joe Spelunker does not always notice the rocks edging trails—it seems that people have to be trained to see the rocks. The simple technique of weaving a continuous length of surveyors tape into a rock border accomplishes two things. First, the flagging demands attention and defines the rocks as trail delineation. Second, the practice potentially teaches visitors to begin noticing other trail borders lined only with rocks.

Flagging Tape to Delineate and Communicate

Trail flagging takes a bit of skill. There tends to be more to the process than expected. It is good to define and practice techniques before entering the cave and then work in small groups with an experienced trail-flagging leader directing the process.

Surveyors flagging tape is relatively inexpensive and readily available through hardware stores, home improvement outlets, and outdoor or forestry suppliers. Flagging tape is efficient to carry in a cave pack. Each roll of tape typically contains 61 meters (200 feet). Be sure to purchase only *non*biodegradable surveyors flagging tape made of polyvinyl, polypropylene, or nylon.

Some cavers may profess that flagged trails are obtrusive, detract from the wilderness experience, and disrupt the aesthetic scene. However, the benefits of flagging may outweigh these drawbacks. Cave resources outside the delineated pathways are less likely to be damaged—thus, overall wilderness quality, aesthetic value, and scientific potential are better preserved. (See trail delineation, page 296.)

Like lines painted on highways, continuous lines of neatly positioned flagging on each side of the trail may free cavers from the constant focus of trying to stay within the designated boundaries. If there are no highway lines painted on a stretch of road, then drivers are forced to focus most of their attention toward remaining on the road.

Test Flagging Materials and Avoid Problems

Based on several decades of observation and successful use in caves, the *non*biodegradable surveyors flagging tape is recommended for marking trails. Nylon or polypropylene cord is also successful in some cave environments. This section explores several potential problems that accompany flagging tape. More scientific research is needed to verify the future of trail marking materials for caves.

Before deploying thousands of feet of surveyors

tape, small strips should first be placed in the proposed cave passages and tested for several months. We cannot assume that plastics are inert to cave bacteria—some types of surface bacteria live on hydrocarbons. Pack rats, other rodents, and some invertebrates may gnaw on any flagging tape and then distribute it in small bits throughout cave passages.

Use *non*biodegradable products to mark trails. Biodegradable flagging tape should not be used in caves. The products labeled as biodegradable can be yummy and toxic to cave organisms. (See biodegradable, page 58.) All flagging products will degrade over time, and critters may chew on any of it, but the nonbiodegradable tapes typically last longer than the biodegradable varieties.

Under unusual chemical conditions, the polyvinyl or nylon tapes may degrade at an inconveniently rapid rate. Or, in some cases, erosive factors may deteriorate the tape—for example, flagging tape may be broken to tiny bits by the simple force of water droplets repeatedly striking the same spot (which is usually resolved by moving the flagging away from the drip).

If any of these problems persist in your caves, you'll obviously have to avoid flagging tape and come up with plan B. Even with these factors, nonbiodegradable polyvinyl or nylon flagging tape has surprisingly good longevity and durability when used for marking trails in many cave environments. Continuous lengths of high visibility nylon or polypropylene cord have also proved successful for delineating both sides of cave trails.

Tips for Flagging Cave Trails

- Use continuous strips of surveyors flagging tape to delineate both sides of pathways, trails, and routes in caves that have appropriate environments and surfaces.

- Only use *non*biodegradble flagging tape made of polyvinyl, polypropylene, or nylon. Biodegradable products may be harmful to cave-dwelling biota.

- Intermittent strips of trail flagging are not recommended. Cavers tend to inadvertently walk off the trail in search of the next piece of trail marking. Strips of flagging tape drop down holes and get strewn along cave passages.

- An 0.5-meter (18-inch) trail width usually allows enough space for efficient travel and for keeping bodies and heavy packs inside the trail. Trails should be wide enough for cavers to negotiate the path safely, but not so wide that two people can walk side by side.

- Give crawls 0.75–1.25 meters (2–4 feet) of trail width depending on how wide cavers need to spread hands and knees inside the trails to avoid scraping the ceilings.

- Anything within the flagged pathway is eventually discolored, trampled, or flattened. All features inside the trail corridor will be destroyed. (In sensitive areas, narrow trails may help cavers avoid speleothems located on the floor and may encourage slow and deliberate movement.)

- When a sturdy stalagmite has historically provided an obvious and necessary handhold, consider including part of it inside the flagged trail.

- Where there is a choice between a high route and a low route, mark the trail along the low route to help prevent the dislodging of upper soil

Flagging Tape Color Schemes

Tom Bemis

Various color schemes for flagging tape are used in the U.S.

Specific color/pattern combinations can reduce confusion by designating leads, scientific sites, or significant features.

A system of predetermined colors may indicate special mineralogical, paleontological, biological, or cultural sites. Specific color schemes may also mark contemporary human-use sites (for first aid, gear caches, rest areas, camps, and so on).

Confusion and damage can occur when flagging schemes are not communicated. Wind and Jewel Caves, for example, use blue/white striped flagging to designate delicate areas, while Lechuguilla Cave uses blue/white stripes to indicate unchecked leads. Imagine the damage that might occur if a Lechuguilla Caver went into Jewel and was not informed of the difference. At this point, it is impractical for either cave to change flagging schemes in order to standardize.

The two main purposes of flagging are to attract attention and convey information to the caver. The choice of color is not important as long as the meaning is clear. Even one single color of flagging throughout the cave is sufficient, if it is plainly labeled with a permanent marker.

(Continued on next page)

(Continued)

Lechuguilla Cave Accepted Colors:
- Orange—trail
- Red/White—sensitive or delicate area
- Blue—survey station
- Blue/White—lead
- White—note or sign
- Yellow/Black—science station
- Pink/Black—bones (bat, paleo, or others)

Possible Additional Colors:
- Yellow/White—biohazard (histoplasmosis, Rabies, or others)
- Blue/Black—drinking water source
- Pink—waste station
- Red—contaminated water
- Orange/White—camp

and rocks that may impact the area below.

- On steep slopes where travel tends to create erosion, zigzagging the path into switchbacks helps reduce erosion and sets the stage for safer travel as the trail becomes worn.

- On climbs and slopes, the flagged trail should be wide enough to accommodate various climbing styles. Where possible, include appropriate footholds and handholds within the trail delineation. The goal is to keep hands and feet inside the flagged routes and stay safe.

- Create rest areas along the way for cavers to stretch out (and eat if permitted).

- Where it is appropriate, designate staging areas for photographers to set up tripods and still remain on the trails. Obtain special permission and use low-impact techniques if there is specific need to step off the trail. (See photography, page 212.)

- When trails cannot be delineated through chimneys, mazes, and climbs, suspend an occasional strip of flagging tape from the ceiling or wall to make the route easier to follow. Tie flagging at eye level. When cavers don't have to search around for routes, impact is reduced.

- Use a different color surveyors tape (red/white stripe is recommended) to flag off sensitive areas. Use flagging to enhance protection and appreciation of delicate or unusual features. It is important to use a flagging color scheme that is communicated to and understood by all who use the cave. (See flagging colors in sidebar, page 179.)

- Solid-colored flagging tape should not come in contact with wet flowstone surfaces because the colors bleed. However, white-backed flagging does not bleed. For example, red/white candy-striped flagging can be used if the white side is placed directly on flowstone surfaces. (See flowstone restoration, page 401.)

Figure 6. Flagging tape looped around a rock with a twist or knot secured underneath makes a tidy flagging technique. The red/white candy-stripe warns cavers to beware of delicate features just outside the trail.

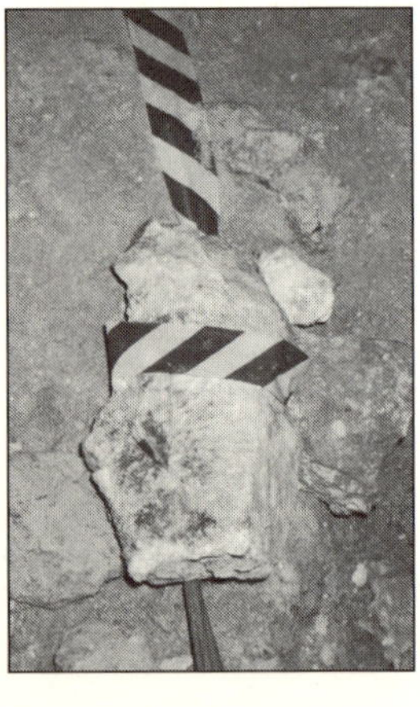

- Place all flagging tape low, at the trail level and simply tuck, tie, or neatly wrap and secure at frequent intervals along the trail edge. Without tie downs, misplaced footsteps will snag the flagging out of position.

- Natural protrusions make the least obtrusive flagging tie downs or anchor points. Use subtle wraps around rocks and bedrock features to secure tape positioning.

- Where anchor points are not available, a single wrap and twist around an occasional fist-to-football-sized rock secures the tape and prevents the unsightly appearance of stray and straggling flagging, caught and dragged by inattentive footsteps.

- When placing or maintaining flagging tape, make it tidy.
- Tie small uniform knots.
- Cut ends short.
- Don't leave dangles.

- Position trail flagging slightly inside the path, *not* directly on the fragile edges, *not* on crystals, *not* on pristine speleothems, and *not* on unusual features.

Figure 7. Double lengths of continuous flagging help preserve cave values beyond the trail boundaries. Without trail delineation, this room in Deep Seas Camp of Lechuguilla Cave would likely be trampled flat.

Figure 8. Double strips of continuous flagging confine impact to a limited pathway. This trail is flagged wider than 0.5 meter (18 inches) to accommodate safe travel and handholds along the slope. (See page 8 of color section.)

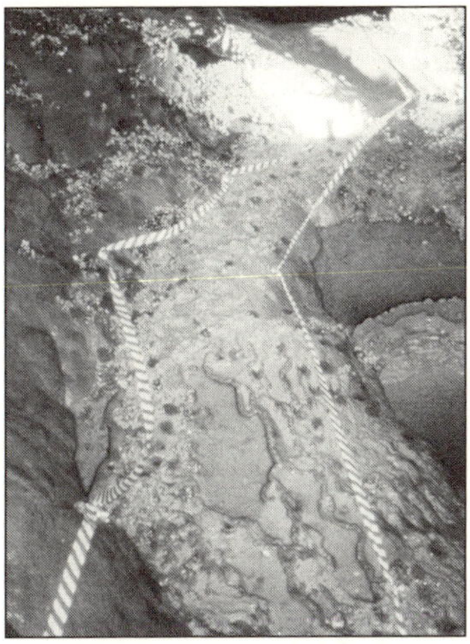

Figure 9. Using natural protrusions for tie-downs, double lines of continuous flagging help protect the larger flowstone area. (See page 9 of color section.)

Figure 10. Trail flagging along with a laminated sign—DO NOT GO OFF MAIN TRAIL. These management tools inform cavers and encourage protection of cave passages.

Figure 11. On climbs, wider flagged areas accommodate handholds and varying climbing styles. (See page 8 of color section.)

Figure 12. To reduce further impact, a caver stands in a flagged posing area for a portrait in Lechuguilla's Chandelier Ballroom. (See page 8 of color section.)

Figure 13. Hanging tape defines the route through this gypsum maze. In pristine passages, cavers are encouraged to wear clean gloves and flowstone shoes. (See page 9 of color section.)

Figure 14. To prevent color-bleeding, carefully place the white side of flagging tape down against wet flowstone surfaces. (See page 9 of color section.)

Figure 15. Wider areas are flagged for resting. All gear remains within the trails. (See page 8 of color section.)

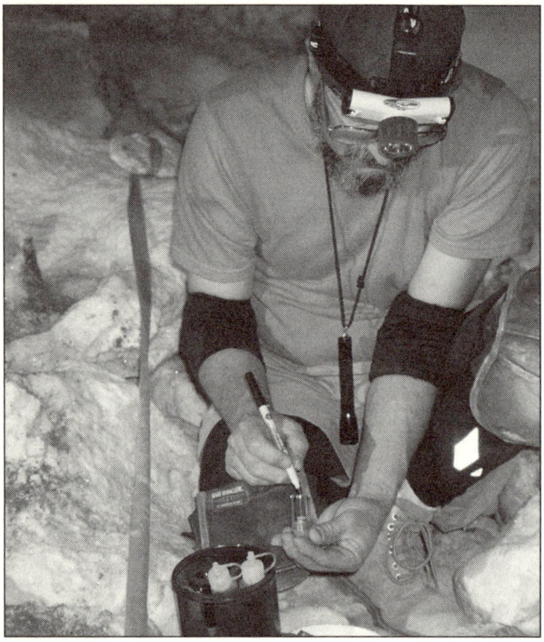

Figures 16 and 17. While cavers work or rest, they remain within the delineated trail and carefully keep all gear within the boundaries.

Figure 18. A whisk broom gently combs away footsteps outside the trail. Leaving off-trail footsteps encourages more to follow. (See page 9 of color section.)

Photos © Val Hildreth-Werker

Figure 19. A caver replaces blue/white flagging with red/white at the trail level through the Boogins in La Cueva de las Barrancas, New Mexico. (See page 9 of color section.)

- Cavers are required to stay within the flagged boundaries during all routine travel and resting in the cave. Whether eating, napping, or moving, everything is expected to stay in the trail—packs, gloves, bodies, and all gear stays inside the flagging. This practice limits impact to the already trampled zones within the boundaries defined by the flagging tape.

- If small strips of flagging or other materials are used for marking temporary routes in a cave, remove it on the way out. Never leave personal trail markers in a cave. Never leave personal cairns set up in a cave.

Signs are often placed at cave entrances to explain laws, permit requirements, safety precautions, behavior expectations, species, habitat, cave resource education, and contact information.

Trail Maintenance

In caves where trail definition is appropriate, the flagging, rocks, or other trail markers must be maintained. The characteristics of the cave environment and the nature of visitation will dictate maintenance schedules for realigning rock borders, repositioning flagging, restoring sections of flagging, and updating signs.

- When flagging becomes brittle, it should be replaced. In some caves, flagging will last up to a decade—in other caves, it should be replaced every year or two.
- When stray footprints are found outside of delineated trails, or when paths lead to nowhere—gently erase, comb, or camouflage to restore the natural appearance of clastic sediments and cave surfaces. Use gentle combing motions with lightweight, nylon brushes or whisk brooms to avoid stirring up dust and spores while erasing footprints. (See compacted soils, microbiology chapter, page 72).
- Deep footprints in some types of rock flour or sand can be camouflaged with natural sediments and soils taken from alongside the trail. Gentle air puffs from a clean turkey basting tool will erase footprints in some powdery substrates.
- If boot soles make unsightly scuff marks on flowstone, erase the marks before new layers of calcite make them permanent. Ugly scuffs can sometimes be removed with the very soft scrubbing side of a restaurant-grade sponge. (See soft scrubbers, page 406.)
- If more damage is likely to occur by trying to restore an off-trail footstep, leave it alone, perhaps place an informative sign, and hope people learn that one footstep should not invite more to follow.

Figure 20. Laminated signs warn cavers—DO NOT GET OFF MAIN TRAIL—the second sign tells cavers the area is closed for scientific study.

© Val Hildreth-Werker

Signs in Undeveloped Caves

Signs are often placed at cave entrances to explain laws, permit requirements, safety precautions, behavior expectations, resource education, and contact information for cave management and for the National Speleological Society. Educational material tends to do more good than signs emphasizing restrictions (penalties are sometimes included in small print). Permanent signs installed inside gates tend to last longer and draw less vandalism. (See educational signage, page 155.)

Signage is made of all sorts of materials, but painted metal or polyvinyl signs are the most durable for undeveloped cave entrances. (Stainless steel is a preferred mater-

Figure 21. A computer-generated, laminated sign warns cavers about pristine areas ahead and gives specific instructions about clothing and caving gear.

Figure 22. A warning is written with indelible ink on flagging tape stretched across the trail—STOP! BRUSH OFF BODY-HELMET-GEAR-REAR!

Figure 23. Two signs explain precise instructions for the water siphon tube and spigot that were once at Deep Secrets in Lechuguilla Cave—DO NOT LIFT SIPHON ABOVE FLAGGING TAPE—DO NOT TOUCH NOZZLE. (On multi-day Lechuguilla expeditions, cavers routinely drink tested cave water from designated sites.)

Photos © Val Hildreth-Werker

ial, but is rarely an affordable option for cave entrance signs. See materials, page 167; common sense, page 173.)

Small laminated paper signs have been used along the trails of many undeveloped caves. Create simple, easy-to-read signs with a computer and completely seal all four edges with lamination material (heat-press lamination is recommended). As long as there are no gaps or holes in the lamination, these signs survive surprisingly well for up to a decade, depending on the cave environment. Or, use an indelible Sharpie® to write instructions and information on flagging tape. Rigid sheets of PVC or Mylar® also work well for signs and survey station markers. (See survey stations, page 189.)

Flagging and signage can significantly reduce impact in a cave. Signs provide important information for the uninformed and serve as a reminder for the forgetful or tired caver.

Examples of Signage

- Stay inside flagged trail
- Brush off helmet and clothes before entering pristine area ahead
- Surgical gloves only
- Boots off here—flowstone shoes only
- Move slowly—extremely fragile speleothems ahead
- Packs off—formations overhead

Conclusion

Mark travel routes in wild caves? For some, it's a tough message. But more people are caving and many fragile, irreplaceable cave resources are being destroyed. To mitigate human impacts in some caves, it may be necessary to confine travel to designated pathways and keep cavers off the less impacted surfaces.

In the ideal subterranean world, no trails and no restoration would be necessary. We would know how to prevent negative impacts. But in our real world of caves and cavers, we must apply practical observation and foresight to avoid further damaging underground resources. Trails through some cave passages should be clearly marked to help us leave nothing but careful footprints on established trails.

Leave nothing but careful footprints on established trails.

Acknowledgements

The Editors gratefully acknowledge Tom Bemis, Jason Richards, Ransom Turner, and George Veni for contributing materials to the development of this chapter.

Additonal Reading

Buecher RH. 1995. Footprints, routes, and trails. In: Rea GT, editor. *Proceedings of the 1995 National Cave Management Symposium: Spring Mill State Park, Mitchell, Indiana, October 25-28, 1995.* Indianapolis (IN): Indiana Karst Conservancy. p 47-50.

Hildreth-Werker V, Werker JC. 1997. Lechuguilla restoration: Techniques learned in the Southwest. *NSS News* 55(5):107-113.

Hildreth-Werker V, Werker JC. 1999. Restoration, trail designation, and microbial preservation in Lechuguilla Cave. In: Stitt RR, editor. *Proceedings National Cave Management Symposium 1997: October 7-10 in Bellingham, Washington, and Chilliwack and Vancouver Island, B.C., Canada.* National Cave Management Symposium Steering Committee. p 181-189.

Veni G. 1997. Speleothems: Preservation, display, and restoration. In: Hill C, Forti P, editors. *Cave Minerals of the World.* 2nd edition. Hunstville (AL): National Speleological Society. p 301-309.

Section B—Developing Cave Management Programs

Conservation-Minded Cave Surveying

Patricia E. Seiser

While the accuracy of cave survey and quality of cave maps have improved over the years, more attention should be focused on the conservation ethics of cave surveying. Several fine technical books on cave mapping are available, but none address survey and exploration from a conservation perspective.

Currently, the practice of "survey as you go" is applied extensively in the exploration of virgin caves and newly discovered passageways. Preservation and conservation should begin with original exploration. Cave surveying itself is not inherently a preservation activity. The survey process can be destructive, or it can be performed in a responsible manner that limits impact. The very act of entering a cave has an impact on it, but survey can be the foundation for protecting a cave system. By exploring, describing, and documenting cave passages, we can identify ways to protect a cave and the surface above it.

While the accuracy of cave survey and quality of cave maps have improved over the years, more attention should be focused on the conservation ethics of cave surveying.

The goal is to think and act with minimal impact. By thinking ahead and discussing options, survey teams can often assess potential impacts and determine the appropriate actions. Cave explorers and surveyors should make this a point of honor—when the survey is finished, future visitors will experience the cave in as near-to-pristine condition as when the first explorers entered.

Preparing for Conservation-Style Surveying

Minimum-impact surveying requires preparation. Surveying is more than just being able to read instruments. A cave surveyor must understand the roles of all team members, use the instruments properly, comprehend the requirements for recording data and sketching, and know how to read maps and line plots. The explorer and surveyor must also be a competent caver, capable of carefully moving through the various obstacles that are encountered while traveling through a cave.

The National Speleological Society has published an excellent book on cave survey and mapping, *On Station* by George Dasher (1994). This is a valuable text for learning how to survey, but as is true of all activities, expertise only comes with practice. Learn to read instruments and sketch on the surface, then practice in a cave. However, not all caves are suitable for practice. Training exercises should not cause the kind of damage you are trying to prevent.

Each member of the survey team is responsible for the quality of the survey and notes. It is important that all members of the survey team understand the basics of sketching so that team members can review the notes with the sketcher. Periodic in-cave review of the sketch and survey notes will help avoid repeat trips to resolve inaccuracies.

Before the survey trip, the size of the team and duties of its members need to be determined. Basic survey functions are sketching, route finding, and instrument reading, but additional tasks—such as flagging trails, photodocumenting, or inventorying cave resources—may be planned. Keep in mind that individuals on a team can perform multiple tasks, and some tasks can be exchanged among team members during the course of the survey.

The goal should be to accomplish the survey with the least impact. In a simple, horizontal cave passage, it may be possible to use as few as two people. In more complicated situations as many as six may be used, three to run the actual survey and three to rig ropes and do additional documentation of the cave.

The quality of the team has a direct influence on cave impact and on the quality of the survey. Team members should have the necessary skills for the cave, and they should be able to focus on the goal of the trip and not be prone to borehole fever, scooping, or excessive solo lead-checking.

Aborted surveys result in additional trips and increased impact on the cave. On the other hand, being too hard core can result in a tired and careless team. Decide in advance how much time is going to be spent surveying.

In the exploration of virgin caves, "survey as you go" has become the accepted practice. For every foot of new passage, documentation should be generated. By concentrating on minimizing impact during the exploration and surveying process, a survey team can lay the groundwork for conservation and preservation of cave resources.

Exploration and Survey Illness

Patricia E. Seiser

Borehole Fever and Scooper Syndrome are exploration illnesses.

The discovery of huge, virgin passage often creates a burning desire to run off and see what there is to see. Borehole fever is highly contagious and typically spreads to the entire survey team. When the fever strikes, maintaining a disciplined survey attitude is usually difficult if not impossible. Often the best remedy is to spend a few minutes presurveying the area (with minimum-impact in mind) and then returning to the survey.

Scooper syndrome may infect one or more members of a survey team. The caver takes off exploring for an extended period. This essentially hijacks the survey process while the rest of the team awaits the return of the errant caver. The scooper often reports "nothing to survey," despite an extended absence. This is an increasingly rare illness and is usually cured by earnest group feedback. Aspiring explorers afflicted with a severe case should not be allowed to participate in the exploration and survey process, or at least not without vigilant supervision.

Survey illnesses cover a wide range of symptoms, from serious episodes of survey of the crawling dead, sketcher burnout, instrument reader dyslexia, and contortionist bends, to simple cases of whineritis.

Minimum-Impact Surveying

Here are some guidelines for minimum-impact surveying. The discussion should not be considered complete—there is always room for new ideas and methods. Since cave environments vary, not all considerations will apply to all caves.

Survey It or Not? Thorough surveying does not necessarily mean that every passage must be physically surveyed. (See survey, page 258.) In fragile passages, it may be appropriate for one person to examine a delicate area and determine if it should be surveyed and, if so, select the least-damaging route. This presurvey may include estimating readings and measurements.

Using estimates, fragile areas can be drawn on the sketch without the whole team entering the area. In many cases, the decision is readily apparent; in others, it may be necessary for the group to discuss the question. What is reasonable and prudent? Will future visitors go there anyway, or will they too turn back? In some cases, the final decision will be made by the stewards of the cave, even though this may require an additional trip to the area. Why perform survey at the cost of a delicate area that could have been sketched from a distance and left virgin? In some caves, stewardship goals may require only a survey of the major passages to provide information about the general layout of the cave. While this leaves unchecked leads on the map, it serves to preserve areas of the cave for future management objectives.

Reduce Gear. Exploration and survey can require a lot of gear. Carrying heavy packs and ropes through a cave has consequential impacts, especially in passages that are delicate or tight. It is rarely necessary to carry all the gear all the time. Leave large packs, climbing or camping gear, and ropes at logical and safe locations along the route and retrieve when necessary. Use pockets or small packs to carry surveying equipment and supplies (of course, always carry a spare light and batteries). Surveying is a slow process, and when something is needed, a gear cache is usually not far away. Even if heavy gear eventually has to be carried though the new passage, it can be hauled with less impact when it is not set down at every station to conduct the survey.

Working with Photographers. Cave photography serves a variety of purposes ranging from personal scrapbooks, to publication, to documentation. In order to avoid the impacts of dedicated photo trips, much photography can be done during survey trips without being overly disruptive. Indeed, many published cave photographers also serve as instrument readers or sketchers on highly productive survey teams. Before entering the cave, survey team members need to discuss the use of cameras, including the purposes and limits of photography.

Survey stations should never be located to meet a photographer's needs, but rather to best suit low-impact surveying and caving. Nor should existing survey markers be moved for the purpose of a photograph. In some caves, specific photomonitoring points have been established as part of an impact-monitoring program and their locations may be included in the survey. (See photomonitoring, page 207.)

(Continued on next page)

(Continued)

Essentially these illnesses are all derived from staying at the survey process too long and exceeding the capacity of one or more surveyors to participate effectively. Establish a stop time for exploration and then stick to it. Inordinately tired explorers and surveyors are a danger to themselves and a threat to cave resources, adversely impacting the cave during their stumbling exodus.

Knowing when to quit is important for conservation-style survey.

Survey Stations. It is important to recognize that the survey line, in terms of best defining the cave passage, may not be the same route as the trail through the passage. Additional impact on the cave should be evaluated when running the survey along a different route. When the survey line and the trail coincide, placement decisions for survey stations should take into account whether they will be destroyed by travelers on the trail (and whether there will ever be a need to use those stations again).

The type of station marker used should be appropriate for the type of cave passage. Flagging tape or reflective Mylar® tags are successful in many caves. (See cave-safe materials, page 185.) The visibility of the station marker is important because it does little good if surveyors must walk all over looking for a station marker.

When selecting stations, evaluate potential damage that may occur while reading instruments for both foresights and backsights. A convenient site for a station may not be the best choice in terms of minimizing impact. Whenever possible, avoid placing survey markers on delicate or active formations. When selecting station sites, also consider potential tie-ins to other survey lines and the possible need for a station near a point of scientific interest.

The Sketcher. It is important that the sketcher is capable of creating an accurate, readable sketch as well as clear notes. Poor sketching often results in repeat visits that add unnecessary impacts. Illegible notes create difficulties in data entry, reporting, and checking accuracy of sketches. Many survey projects require sketchers to meet specific guidelines and standards before being allowed to sketch. This helps ensure the quality of the survey.

It is common practice for the sketcher to be the team leader during the surveying process. The sketcher sets the pace of the survey. Instrument readers may have to wait for the sketcher to complete a thorough record of the site. In addition to the sketch and survey data, the sketcher may be required to take notes to document hazards, rope positions and lengths, flagging locations, or other pertinent information. Reviewing the survey book with team members during the survey process can help ensure the quality of sketches and notes.

It is common practice for the sketcher to be the team leader during the surveying process. The sketcher sets the pace of the survey.

Sketching can be a tedious process. Sketchers need to be aware of body position at all times, and take care to limit impact when sitting or stretching.

Flagging. Surveyors flagging tape is used for a variety of purposes: survey stations, protection of fragile resources, designation of scientific stations, lead marking, trail marking, and signage. Flagging tape comes in a variety of colors and patterns and survey teams should be aware of the color conventions for a particular cave. (See color schemes for flagging tape, page 179.)

In passages that are subject to damage, it is important to clearly mark the path. Trail delineation may be one of the duties of the surveying team. It is helpful to note in the survey book where trail flagging is placed. This information can be recorded on the sketch or in the notes section. (See trail marking, page 176–183.)

Instrument Reading. Accurate instrument reading is essential for quality surveys. Computer programs can quickly identify survey errors when surveys contain loops. However, the site of the error can usually only be roughly identified, and a resurvey of the whole area may be required, resulting in additional impacts. The use of both foresight and backsight readings provides an in-cave check for reducing errors in both reading and recording. To reduce the chance of mistakes in hearing or writing, the sketcher should repeat back readings after recording the data.

Instrument reading can create significant impact on cave resources. The sign of a good instrument reader is the ability to recognize that the least impacting position may also be the most uncomfortable one. It is important to constantly be aware of body position and the potential to damage small deposits and features. Take care to reduce contact with the cave as much as possible (Figures 1 and 2). Understanding and using correct instrument-placement offsets from a station can limit impact to delicate features, yet still produce accurate readings. To protect the cave or to get a better reading, helmets may need to be removed.

It is a good idea to have a bandana covering the head to reduce hair and sweat left-behind on the cave walls. In some situations, it may be necessary for all team members to carry a set of clean clothing (including knee pads, elbow pads, and gloves) to change into before surveying.

The Survey Tape. Reading a survey tape is usually a low-impact activity. However, the potential for unnecessary impact occurs in reeling out, reeling in, or stretching the tape. Surveyors need to be aware of where the tape is running and what it might catch on and destroy (Figure 3). Take appropriate actions to avoid such problems.

Taking a Break. Explorers need to be aware of their surroundings and actions when resting or waiting for team members to complete a task. If a trail or resting site has been established, surveyors need to be sure that their bodies and their gear are contained within the boundaries. Avoid leaning on formations and walls (even if the walls are speleothem free). It is easy to transfer sweat and dirt to areas that were previously pristine. Cavers should be as careful and alert to body and gear positioning when resting as they are when moving through the cave (Figure 4).

After the Survey

The responsibilities of the exploration and survey team do not end once the survey trip is completed. Surveying the cave, gathering data, and photodocumenting the resources do little for cave management and protection unless the information is properly reported.

A comprehensive trip report and the finalized survey notes should be submitted to cave owners or managers upon completion of a survey trip. In addition to describing the survey trip, the report should document problems or hazards posed by the particular cave passage mapped. The report should also include a list of equipment required for future exploration and survey, and a lead list to facilitate future exploration. The trip report should also contain recommendations for preservation, conservation, or restoration. Copies

The Zen of Cave Survey

Patricia E. Seiser

... the warrior is pledged to protect whatever is lovely, vulnerable, and truly precious.
–Rick Fields, *The Awakened Warrior*

What is the Zen of cave survey?

It is the understanding that a cave exists on its own terms. It might be described as becoming one with the cave. See the cave passage as it is, envision a survey line through it, and then leave the passage as little disturbed as possible.

Zen understanding of the cave involves the whole of the cave environment and its resources—the formations, biology, hydrology, historical and cultural features—then evaluating the possible consequences of survey actions.

The Zen of cave survey requires a state of consciousness in which cavers choose routes and perform survey, yet thoughtfully reduce their impacts.

Surveyors must implement what has already been learned through the experiences of others. The caving community develops the knowledge, ethics, and expertise for caving nondestructively.

Surveyors need to implement these low-impact techniques.

It is important to think of minimal impact from the start, to be proactive, not merely reactive, and to prevent damage that can result from exploration and survey practices.

(Continued on page 192)

© Val Hildreth-Werker

Figure 1. Protect speleothems and cave walls by removing the helmet to read instruments in tight locations. Contain hair and sweat with a bandana or other head gear.

© Val Hildreth-Werker

Figure 2. This caver is using an instrument offset to reduce contact with the fragile surface of the cave wall covered in colorful, fluffy corrosion residue.

© Val Hildreth-Werker

Figure 3. To prevent the survey tape from damaging formations, surveyors carefully lift the tape high over the speleothems.

Figure 4 (left). Impact is reduced when cavers are careful to stay inside the delineated trails. Gear should never be set outside the trail boundaries.

Figure 5 (right). During survey and exploration, choose routes through new passages to preserve pristine qualities. Trail designation is typically assigned to survey teams. (See page 9 of color section.)

Photos © Val Hildreth-Werker

(Continued)

Cave explorers and surveyors can view themselves as survey warriors—guardians protecting caves through their actions. By serving as guardians, there is *active* participation in the conservation ethic instead of *enforced* participation in the ethic. The latter usually delivers only lip service or is applied only for survey and not for other caving pursuits.

Conscientious surveyors not only learn and pass along new concepts and practices, but they also apply low-impact techniques to all caves they enter regardless of the condition.

of appropriate photographs should be submitted in a timely manner.

The aim of the survey is to produce a quality map that documents the cave. If the cave is large, an interim exploration-quality map should be produced as a survey tool. It may be necessary to provide surveyors with line plots for in-cave use to prevent unnecessary impact to the cave.

At the completion of the survey, or during the survey if it is an extended project, plans should be made for removing or maintaining flagging and any equipment left in the cave. Any restoration projects should be finished.

The ultimate goal of the survey is to produce maps, photographs, and reports that clearly document the cave, significant features, established trails, and passages set aside for preservation or scientific study. The cave steward can use this information in developing a management plan that will best protect the cave, and if possible, allow for continued access by other cavers and scientists in the years to come.

Conclusion

Those who choose to explore and map a cave have the responsibility of protecting it. Careful choices can minimize the impacts of survey and exploration. The resulting maps can help provide for the protection of caves and conservation of surface regions above those caves for future generations.

The goal is to act with minimum impact. By implementing conscientious surveying techniques, we increase the potential for others to experience the cave in as near-pristine state as possible. The quality of exploration and survey that occurs now will determine the quality of the legacy that cave explorers and surveyors leave for the future.

Additional Reading

Thomson C, Taylor R. 1991. The art of cave mapping. *Missouri Speleology* 31:1–181.

Dasher G. 1994. *On Station: A Complete Handbook for Surveying and Mapping Caves*. Huntsville (AL): National Speleological Society. 240 p.

Day A. 2002. *Cave Surveying. Cave Studies Series 11*. Buxton (UK): British Cave Research Association. 40 p.

Ellis B. 1976. *Surveying Caves*. Bridgwater, Somerset (UK): British Cave Research Association. 88 p.

Section B—Developing Cave Management Programs

Visitor Impact Mapping in Caves

Hans Bodenhamer

Many cave managers rely on impact mapping and photomonitoring as tools to monitor nonrenewable resources. Visitor impact mapping is used to quantify damage and augment photomonitoring. When combined, impact mapping and photomonitoring provide extremely detailed information about mineral formations, bone deposits, floor surfaces, human impacts, and so on. (See photomonitoring, page 207–210.)

Visitor impact mapping involves pinpointing damages on detailed maps. Impacts caused by human visitation are added to existing cave maps. Also any fragile, undamaged resources are detailed and quantified. Different types of visitor impact monitoring can be established to suit a variety of management needs (Bodenhamer 1995). In this chapter, two types of monitoring are discussed:

- Visitor impact point mapping
- Visitor impact area mapping

Visitor impact mapping involves documenting undamaged resources as well as damage caused by human visitation. Locations are pinpointed on a detailed map of the cave.

Visitor Impact Point Mapping

Visitor impact point mapping locates damaged and fragile resources using numbered points. Each point is drawn onto a map of the cave based on its

FIGURE 1

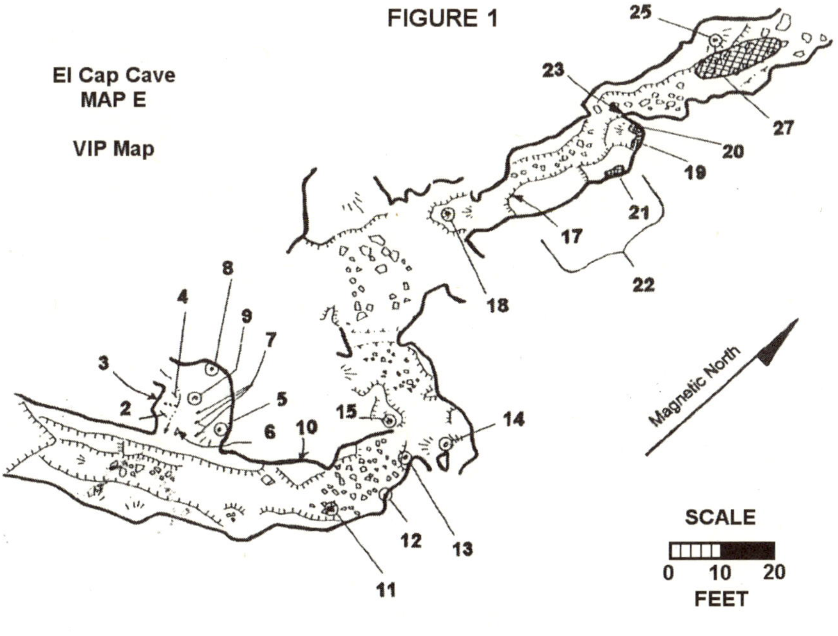

Hans Bodenhamer

Table 1
Impact Point Descriptions
El Cap Cave
Map E

Point Number	Description
2	Ten mud stalagmites, each about 2 inches in diameter. Five are about 10 inches high and five are about 2 inches high. No signs of any visitor damage.
3	Three mud stalagmites, each is about 1 inch in diameter and 4 inches high. No damage.
4	Five mud stalagmites, each is about 1 inch in diameter and 1 inch high.
5	Beautiful *pristine* mud surface with red-orange coating and white speckles.
6	One boot skid mark in silt surface.
7	Small depressions in silt surface, three sets of closely spaced, 1/2 inch diameter depressions. Probably finger (or hand) prints.
8	Strange feature -- thin, white, irregular coating which covers area less than 1 foot diameter.
9	Large depression in silt surface (about 2 feet in diameter). May be the result of visitors kneeling, falling, sitting, or placing a pack in this area. Depression may also be natural.
10	Hand traffic on edge of silt deposit has slightly altered the character of the surface. On top of ledge there are three areas with obvious finger depression marks. Other than these three areas top of ledge is *pristine*.
11	Blackcoated rock. Some wear and spalling probably from visitor traffic. Throughout this passage (Map E) these coated rocks are present in a density of about 5 to 15%. Some seem to have been worn a bit by visitor traffic.
12	Resolutioned flowstone on corner. No damage due to visitors.
13	Pristine silt and cobble deposit in alcove. Deposit has white speckles. No visitor damage to deposit.

relative position to mapped features, and a written description of the damaged (or fragile and at-risk) resource is recorded. Photos can augment the descriptions.

An example of a *visitor impact point map* (VIPM) is presented in Figure 1, and some of the corresponding point descriptions are included in Table 1. By studying a VIPM map, descriptions, and photographs the cave manager can evaluate and compare the conditions of resources (at the time of mapping and in subsequent updates) at specific points throughout the cave.

Visitor impact area mapping documents areas that have been impacted by visitation. Areas are drawn onto the cave map based on their relative position to mapped features.

Visitor Impact Area Mapping

Visitor impact area mapping locates areas that have been impacted by visitation. Areas are drawn onto the cave map based on their relative position to mapped features. Each area is classified and color-coded according to the severity of impacts within it. A sample *visitor impact area map* (VIAM) for floor surfaces is presented in Figure 2.

On VIAMs, visitor impacts may fall into five different classes:

- Pristine
- No observable impacts
- Light impacts
- Heavy impacts
- Severe impacts

Areas classified as *pristine* have floor surfaces that are extremely fragile and it is obvious that visitors have never walked there.

Areas that are classified as *no observable impacts* indicate that visitors may have walked there, but floor surfaces are of such a nature that impacts cannot be detected, even under close visual inspection.

Areas that are classified with *light, heavy,* or *severe impacts* are defined

FIGURE 2

El Cap Cave
VIA Map for
Floor Surface

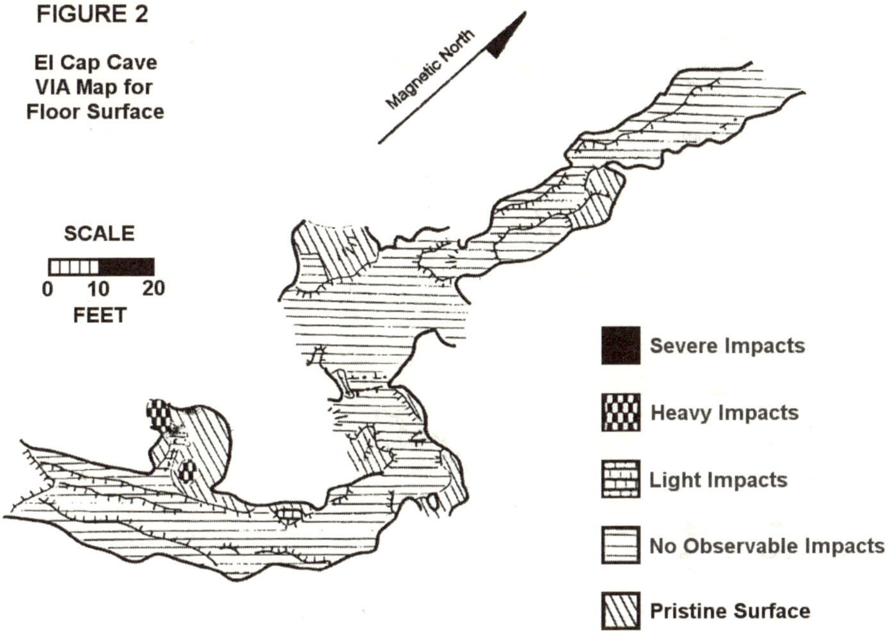

Hans Bodenhamer

Table 2
Explanation of Impact Classes for Different Types of Floor Surfaces

Impact Class	Types of Floor Surfaces: Silt, Mud, or Sand	Rounded Cobbles or Angular Rock	Bedrock or Flowstone	All Types of Floor Surfaces
Light	Light brushing of surface covering less than 25% Faint depressions covering less than 25%	Mud smears cover less than 50% of tops	Mud smears cover less than 25%	¼ in. or less debris layer rolled onto surface
Heavy	Trenching less than ½ in. deep Brushing of more than 25% Noticable depressions covering 25 to 50%	Mud smears cover 50 to 100% of tops	Mud smears cover 25 to 50%	¼ in. to 1 in. debris layer rolled onto surface
Severe	Trenching greater than½ in. deep Depressions ½ in. or greater cover 50 to 100%	Mud is deposited in layers ¼ in. or greater Cobbles (or Rocks) are rolled to side to form trench	Mud smears cover 50 to 100% Mud is deposited in thick layers ¼ in. or greater Surface has been chipped or broken	Greater than 1 in. debris layer rolled onto surface

according to the types of floor surfaces. Table 2 presents impact classifications for several different types of floors. A map indicating the types of floor surfaces for the same portion of the cave shown in Figure 2 is presented in Figure 3.

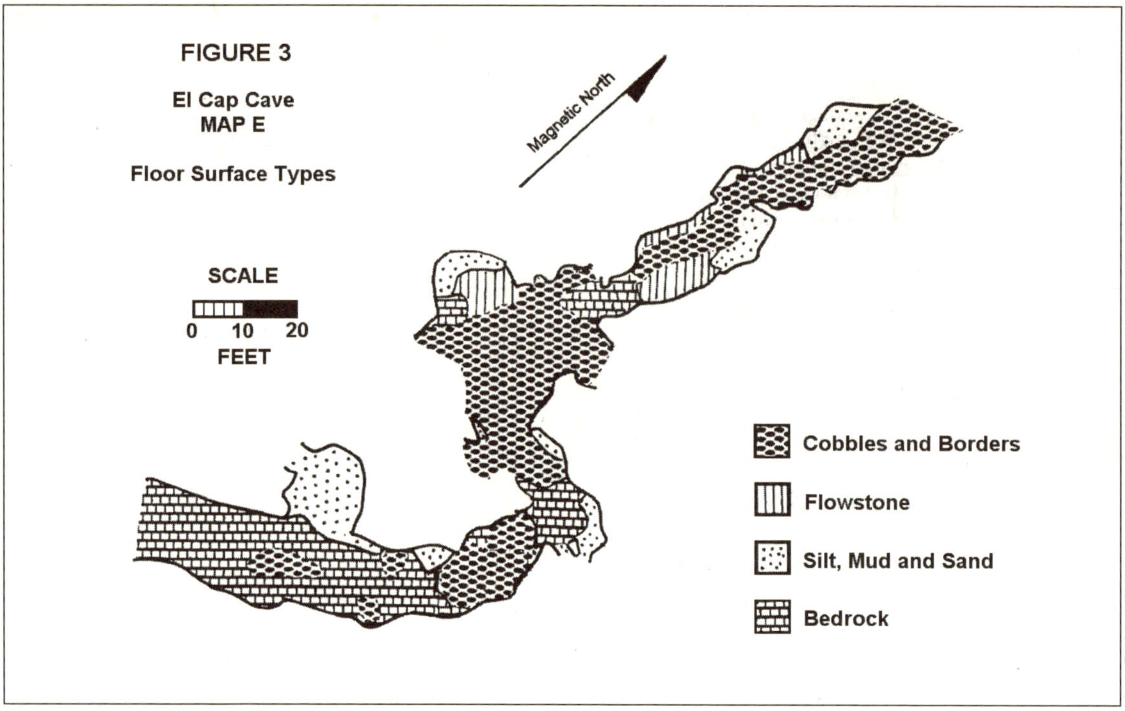

Visitor impact mapping is used to evaluate and compare resource conditions of different caves within a region.

All the mapping tools described here coordinate and give the cave manager an overall pattern for the condition of resource areas as well as specific impact information for points throughout the mapped portion of the cave:

- Visitor impact point map
- Table of visitor impact point descriptions
- Visitor impact area map
- Map of floor surface types
- Table of impact classes for types of floor surfaces

Visitor Impact Mapping for Evaluating and Comparing Caves within a Region

Visitor impact mapping can be used to evaluate and compare resource conditions of different caves within a region. Figures 4, 5, and 6 show VIAMs for floor surfaces in the entrances of three caves located in one region. The floor within each cave consists of thick deposits of pollen and powdered sulfate minerals. The presence of extinct animal bones on the floor suggests that the surfaces in each cave may be over 8,000 years old.

Rex Cave. Shown in Figure 4, Rex Cave has received a cumulative visitation of about 500 people. Most of the visitors were recreationists. *Severe* impacts are limited to a small area, which is near the north wall about 12 meters (40 feet) from the entrance. These impacts were the result of visitors digging a pattern into floor sediments by dragging their feet. *Heavy* impacts cover the largest portion of the floor. *Light* impacts are in low crawl areas. Finally, only a few small areas near walls and under low ceilings are classified as *no observable* impacts.

Kappen Cave. Shown in Figure 5, Kappen is similar to Rex Cave in regard to visitation. The entrance is highly visible from nearby hiking trails, yet access is relatively difficult. Cumulative visitation to Kappen Cave is about 300 visitors (very similar to cumulative visitation to Rex Cave). However,

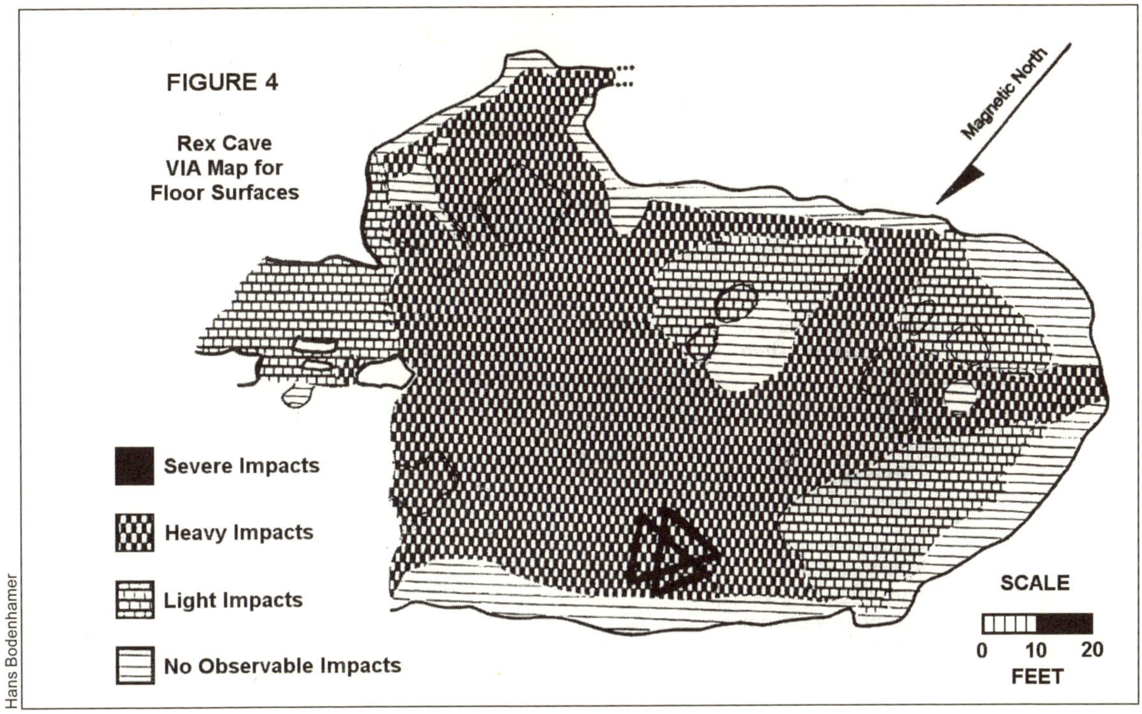

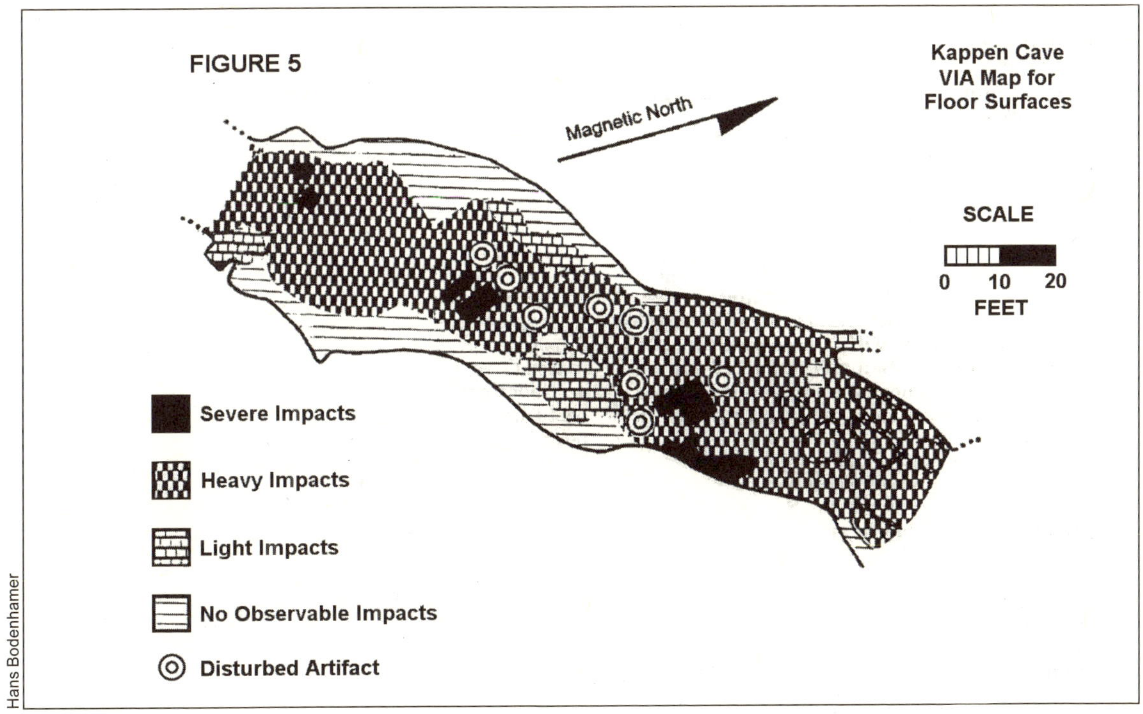

unlike Rex Cave, about 25% of the visitors to Kappen Cave were affiliated with research activities. The amount of floor surface that has received *heavy* impact and the amount that shows *no observable* impact are proportionally the same in Rex and Kappen.

However, *severe* impacts in Kappen Cave are more extensive and are the result of excavation pits that were left uncovered. Of special note are the doubled circles shown on the Kappen Cave VIAM. The circles represent prehistoric artifacts that have been left in the cave but are not in their original context.

FIGURE 6

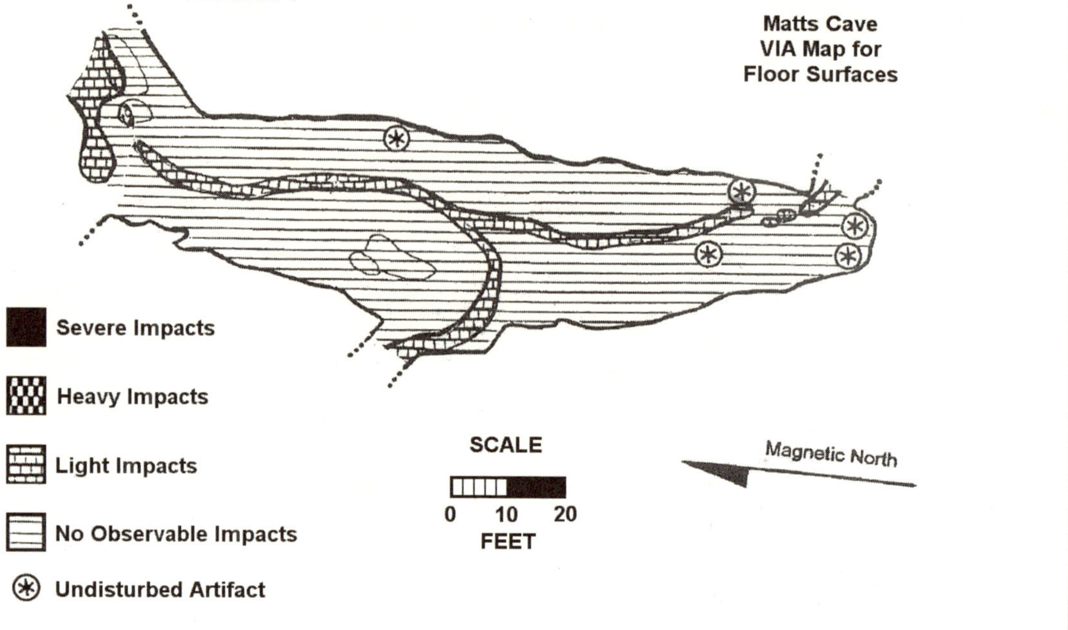

Matts Cave. Shown in Figure 6, Matts Cave, like Kappen and Rex, is relatively difficult to access, but unlike the other two caves, the entrance to Matts is much more obscure. Cumulative visitation to Matt's Cave is only about 10 visitors. There are no severe or heavy impacts within the cave. All visitors remained on the *lightly* impacted trail, which traverses the middle of the passage. Of interest are the circled asterisks, which represent prehistoric artifacts that are in their original context and have not been touched in modern times.

Comparison. Study and comparison of the VIAMs in Figures 4, 5, and 6 give a graphic impression of resource conditions within each cave. Assuming visitor use patterns are related to resource conditions, it is possible for the cave manager to predict the conditions of other caves in the region from the documented impacts of these three caves. These three VIAMs were selected to represent conditions that are expected within all known caves in this region.

Visitor impact mapping can also be used for long-term monitoring.

Visitor Impact Mapping for Long-Term Monitoring

Visitor impact mapping is also useful for long-term monitoring. Figure 7 presents a VIPM for the west portion of SP Cave. Point numbers and descriptors have been omitted for clarity. Points are coded according to impact over a 4-year period (from June 1991 to July 1995). Points marked with a *Happy Face* remained unchanged. Points marked with a *Sad Face* became more severely impacted. About half of the points reinspected in 1995 had become more severely impacted. Most damage involved breakage of additional stalactites and smearing of mud onto previously pristine flowstone.

Another example for long-term monitoring with visitor impact mapping shows a floor that consists mostly of pristine mud-cracked surfaces, located in the east portion of SP Cave. Figure 8 shows the 1991 VIAM for floor surfaces. Figure 9 shows the remapped portion as it appeared in 1995, four years later. Close study of the two maps reveals that a few trails became wider and more severely impacted. Figure 10 combines the two maps to

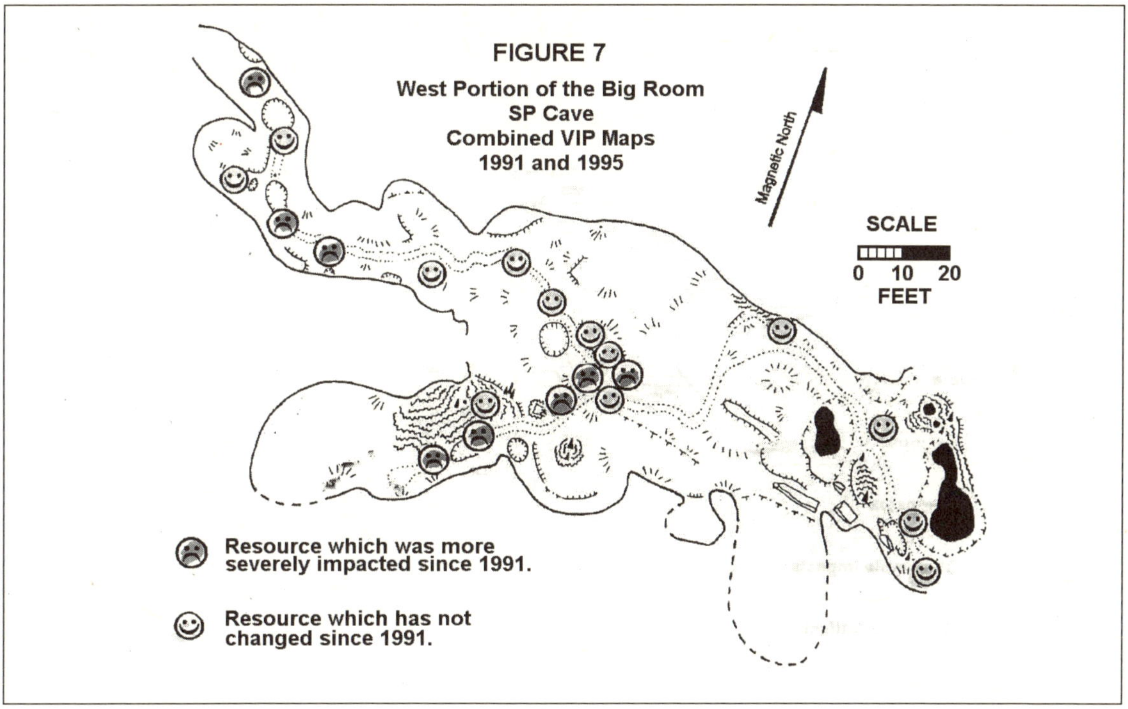

Hans Bodenhamer

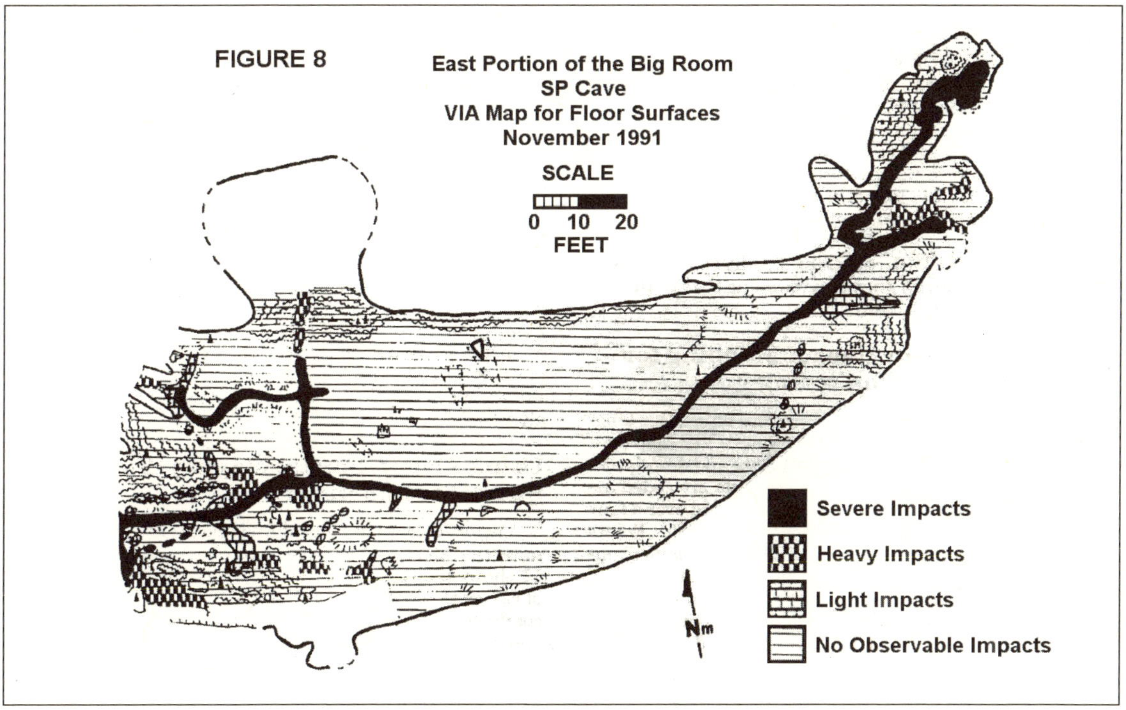

Hans Bodenhamer

graphically show changes in floor surfaces. Table 3 presents the calculated change in area for each class of impact. The information presented in Figure 10 and Table 3 may provide the cave manager with a tangible and quantitative look at resource changes over time.

Visitor Impact Mapping for Short-Term Monitoring

Visitor impact mapping is also used to determine the effects of one-time, special events. Figure 11 shows VIAMs for floor surfaces of a portion of El

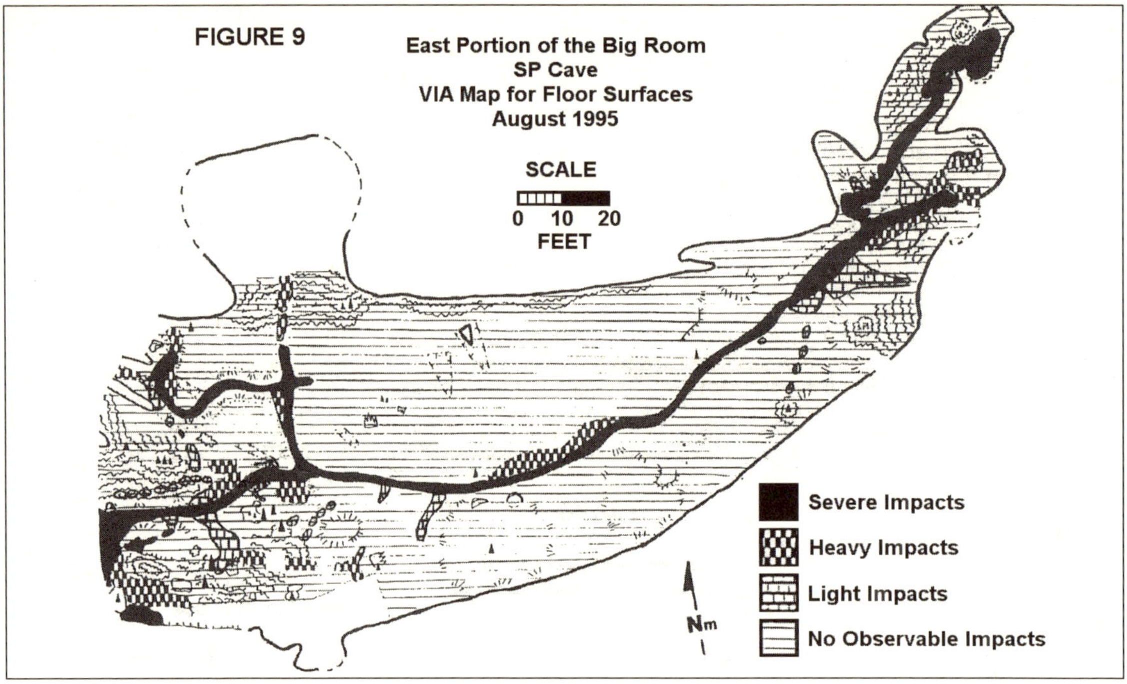

Hans Bodenhamer

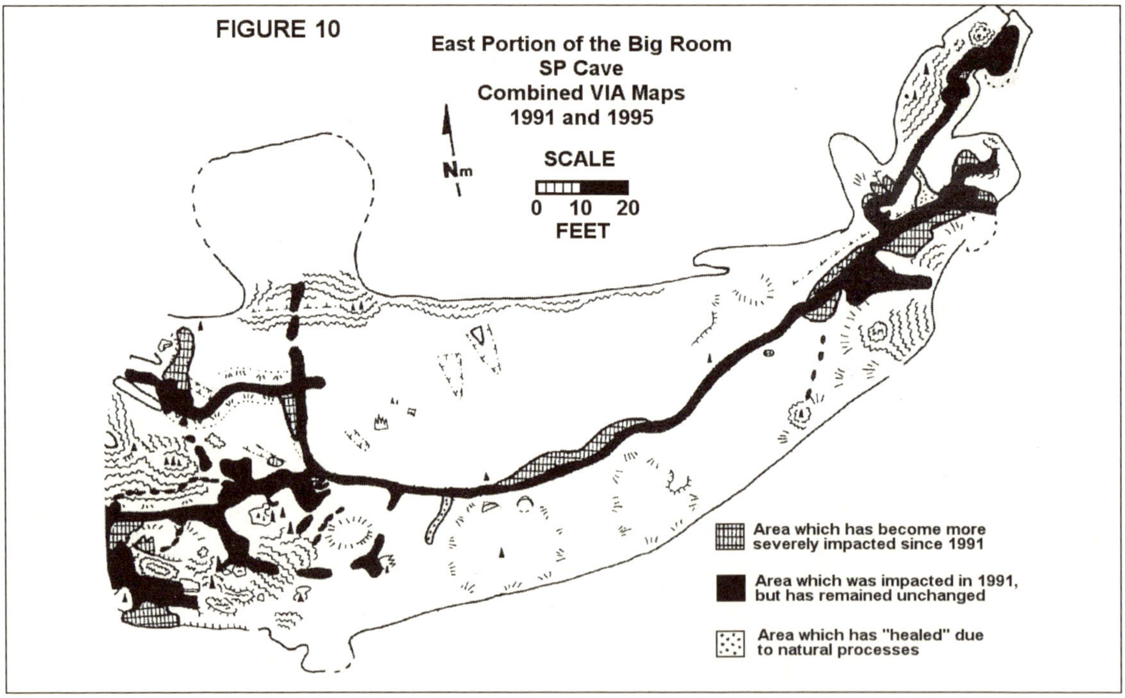

Hans Bodenhamer

Cap Cave before and after a rescue practice. Study and comparison of the VIAMs shows that the amount of *severely* impacted area increased. This increase is a result of rescuers hauling a stretcher-bound victim up a small pit. The rescuers walked further out onto the lip of the pit than previous visitors had, and the stretcher scraped and broke off chunks of bedrock as it was dragged over the lip.

Interestingly, photomonitoring was also repeated before and after the rescue practice. Photomonitoring detected *more* impacts than visitor impact mapping. Impacts detected with photomonitoring mostly include small mud smears on previously pristine bedrock walls.

Table 3
East Portion of the Big Room
SP Cave
VIA Map Surface Areas

	November 1991		August 1995		Change	
Impact Class	Floor Area (sq ft)	Percent of Total Area	Floor Area (sq ft)	Percent of Total Area	Floor Area (sq ft)	Percent of Total Area
No Impacts	9,696	87.7%	9,497	85.4%	-199	-1.8%
Light Impacts	211.5	2.0%	274.5	2.5%	+63	+0.5%
Heavy Impacts	304.5	2.7%	368.5	3.3%	+64	+0.6%
Severe Impacts	904.0	8.1%	976.0	8.8%	+72	+0.7%
Totals	11,116	100%	11,116	100%	—	—

Visitor Impact Mapping Versus Photomonitoring

In comparison, both visitor impact mapping and photomonitoring have advantages and disadvantages:

- Visitor impact mapping seems to be better at detecting and quantifying large scale or isolated impacts in expansive areas.
- Photomonitoring seems better at detecting detailed changes in small areas.

For example, a large area of pristine floor surface may be better monitored with visitor impact mapping. Trails and off-trail footprints throughout the area are located on the map and most changes are easy to detect.

However, an articulated skeleton of a squirrel may be better monitored with a photograph. Visitor impact mapping can record that visitors have stepped on or near the skeleton, but in order to determine if visitors are

Visitor impact mapping is also used to determine the effects of one-time, special events.

FIGURE 11

El Cap Cave Upper Entrance
VIA Map for Floor Surfaces

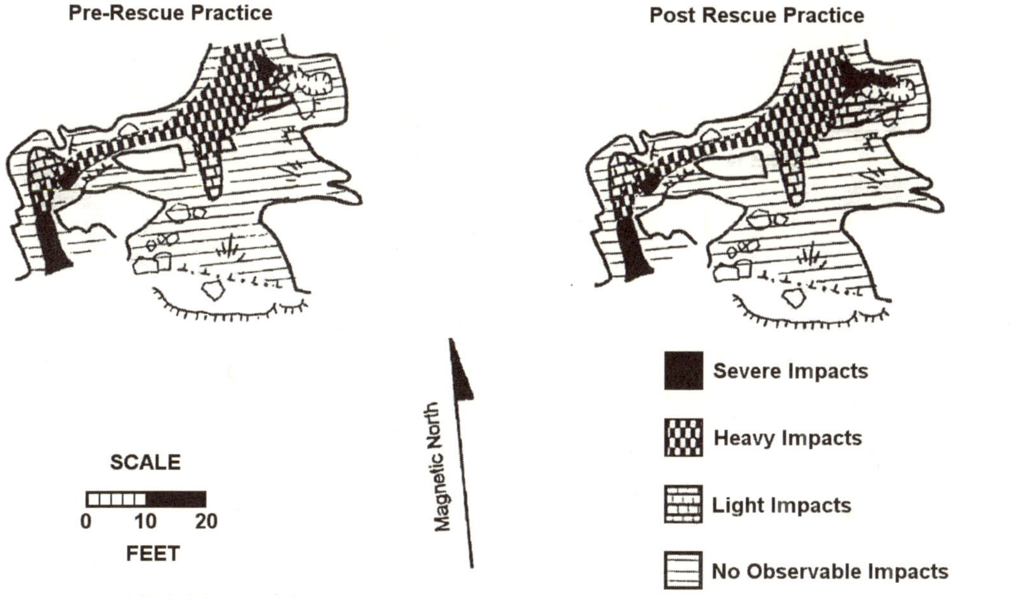

Hans Bodenhamer

actually picking up and moving individual bones, a photo is clearly more informative. Prudent cave managers should probably develop a coordinated monitoring system using both visitor impact mapping and photomonitoring to best fit their resource management needs.

Time and Impact

When people first view visitor impact maps, often their impression is that the process must be very tedious and time consuming. In practice, it usually takes about the same amount of time to establish visitor impact mapping as it does to establish a good photo transect.

When people first view visitor impact maps, often their impression is that the process must be very tedious and time consuming. In practice, it usually takes about the same amount of time to establish visitor impact mapping as it does to establish a good photo transect. However, compared to traditional photomonitoring with tripod, compass, and plumb line, repeating the visitor impact mapping usually goes more quickly. (See traditional photomonitoring versus streamlined photomonitoring with permanent stations, page 207.)

It does seem to take a lot of time to develop detailed base maps. Good visitor impact mapping requires a scale of about 2.5 meters = 1 centimeter (20 feet = 1 inch), and detail should be such that an isolated footprint can be located and drawn in an area with a diameter of about 1 meter (3 feet).

Fortunately there are a lot of cave mapping enthusiasts, and these cavers are likely to volunteer for a project that would generate a detailed base map. Perhaps cave cartographers would also be interested in developing visitor impact maps.

When doing any sort of cave mapping, it is extremely important to drastically limit additional impacts. (See survey, page 187.) Whether developing detailed base maps or establishing and repeating visitor impact mapping, it is best to stay on well-established, designated trails.

In order to see impacts and fragile resources that are off the trail, use a very bright light. Hand-held, self-contained spotlights that are suitable for visitor impact mapping are available through various outdoor sports retailers.

If there are off-trail areas that cannot be seen, even with a bright light, there is a simple, impact-free solution—don't map them. In almost all cases, the type of monitoring information that can be generated by staying on trails will be more than adequate to detect resource changes and direct management decisions.

Cited Reference

Bodenhamer H. 1995. Monitoring human-caused changes with visitor impact mapping. In: Rea GT, editor. *Proceedings of the 1995 National Cave Management Symposium: Spring Mill State Park, Mitchell, Indiana, October 25-28, 1995.* Indianapolis (IN): Indiana Karst Conservancy. p 28-37.

Section B—Developing Cave Management Programs

Photographs as Cave Management Tools

Val Hildreth-Werker

Take nothing but conservation-wise photos.
Leave nothing but careful footprints on established trails.
Kill nothing but time.

Cave softly ... and erase your trace.

Photographs provide visual records that document, educate, and entertain. Cave pictures are made for various purposes—personal enjoyment, historic documentation, resource inventory, salon entry, restoration before-and-after, and so on. Any photographic image may someday become an important element in cave management decisions.

Photographs can have significant influence on cave conservation. We refine and change caving ethics as we learn more about the detrimental impacts humans cause in cave environments. Photos can support protection of critters, microbes, bones, minerals, speleothems, cultural remains, geologic features, and cave systems—photographers can create support materials for all sorts of cave resource conservation.

Photographers have the opportunity and responsibility to communicate updated stewardship standards—through our in-cave behavior and reflected in the resulting images. Photos that portray outdated modes should systematically include explanation statements describing improved ethics.

This chapter defines photodocumentation, photoinventory, and photomonitoring. Simplify and enhance cave project photography with formulas for effective photos, strategies for file organization, and systems for efficient archiving.

This chapter defines photodocumentation, photoinventory, and photomonitoring to support resource management.

Photodocumentation, Photoinventory, and Photomonitoring

Three categories of cave photography support resource management decisions.

- *Photodocumentation* is the photographic recording of events, projects, procedures, expeditions, evidence, locations, or scientific examples. Photodocumentation includes organizing the photos to tell a coherent visual story. Documentation or inventory photos should always include an object or a person to indicate scale.

- *Photoinventory* is much like a picture inventory of valuable household items for insurance purposes. Photoinventory is an organized, labeled collection of pictures recording the valued resources and features of a cave or karst system.

- *Photomonitoring* is based on established photo stations that enable the same photographs to be accurately repeated over time. Photomonitoring sequences may document and detect visitor impact, vandalism, speleothem growth and decline, water level fluctuation, trail conditions, and other anthropogenic or natural impacts.

Most importantly, write the date and place the photo was made on every cave photograph. Immediately record the date and location on each slide mount, on the back of each print, or electronically file the information with each digital image.

Photographs that record or document events, projects, procedures, expeditions, evidence, locations, and scientific data examples are called photodocumentation.

Photodocumentation

Some cavers take a lot of pictures. We use images to record expeditions, projects, procedures, events, locations, evidence, and scientific samples. Picture collections become photodocumentation when the images are properly labeled, organized, and captioned to tell a coherent story. Photographs made for scientific documentation should include a measure or an object for scale reference.

Documenting Evidence or Stages of an Event

- Basic documentation is sometimes communicated in a single image. (See photo that shows before-and-after gypsum cleaning in one shot, page 420.)
- A series of two images can document before-and-after—the conditions before and results after an event, a project, or a time interval. (See Southwinds photo pair, page 207.)
- A series of three or more images can describe a process or tell a story. (See Lower Cave stalactite repair series, page 468.)

Rules of Three for Better Documentation Photos

- Shoot at least three images at each site:
  - Shoot from a distance or use a wide-angle lens to make long shots that record an overview of the area.
  - Move in (or zoom in) for medium shots and include people involved in activities.
  - Make close-up photos to record details.
- Shoot three angles of each image area—move the camera to higher and lower positions and find interesting angles that best communicate information.
- Bracket exposures and make notes about the camera settings for later comparative analysis. Shoot at least three bracketed exposures on important documentation shots.
- Compose the image before you shoot. Improve your photos by using the rule of thirds (Figure 1). Mentally divide your frame into thirds (nine sections)—composition is often improved by placing the subject at one of the four points where the lines intersect.

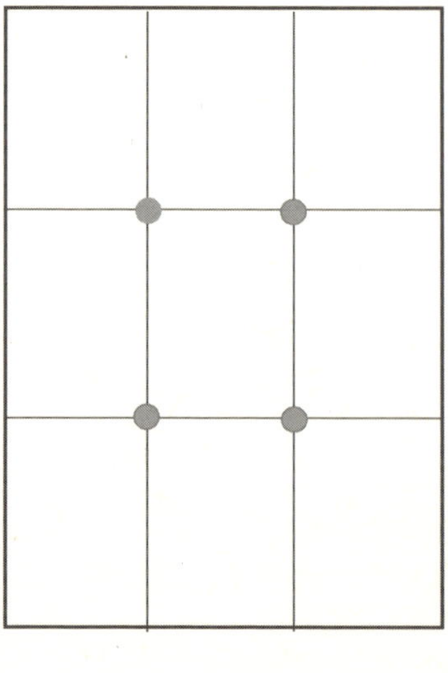

Figure 1. Rule of Thirds (left). To enhance photographic composition, mentally divide the frame into nine equal portions. Place the subject at one of the four intersection points. (See Figures 3, 4, and 5, page 205.)

Figure 2. Bull's Eye (below). Deliberately center the composition only when centering will enhance the visual impact of an image. (See Figures 6 and 7, page 205.)

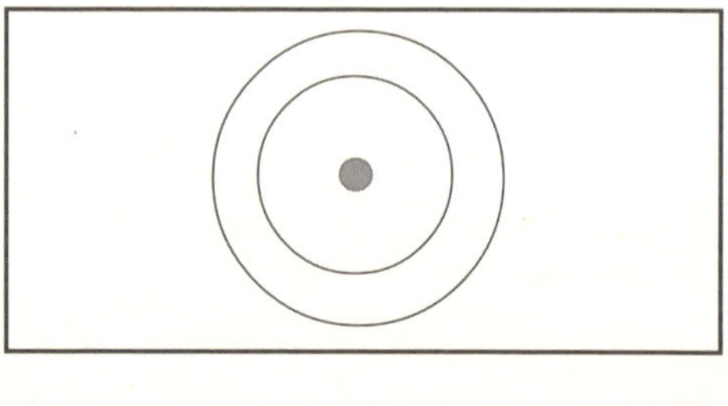

Photos © Val Hildreth-Werker

Figure 3. The white helmet in this image is positioned on an imaginary intersection for rule of thirds composition. The trail in this image, though unmarked, is an obviously compacted pathway that carries the viewer's eye down from the helmet, then back up into the passage.

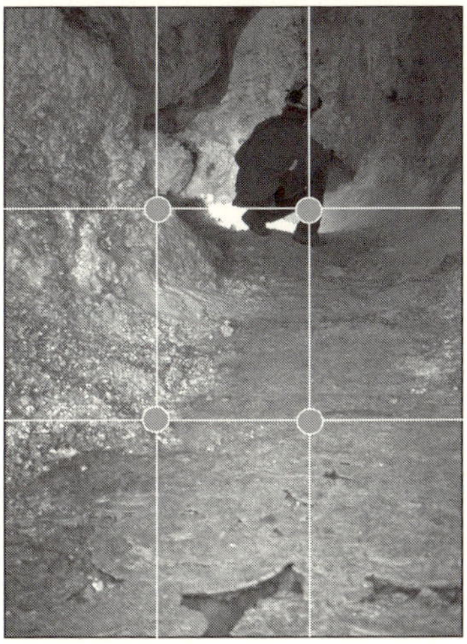

Figure 4. The caver is positioned in an imagined rule of thirds intersection. The viewer's eye is drawn from the silhouetted caver down the discolored path. The hole in the flowstone at the bottom of the trail creates more appeal by prompting the eye to move up through the image again.

Figure 5. Visualizing the rule of thirds, each caver is positioned on an opposing intersection point. The eye is drawn between the white circle on the foreground T-shirt and the caver sitting in the background. This interesting composition tends to hold the viewer's attention.

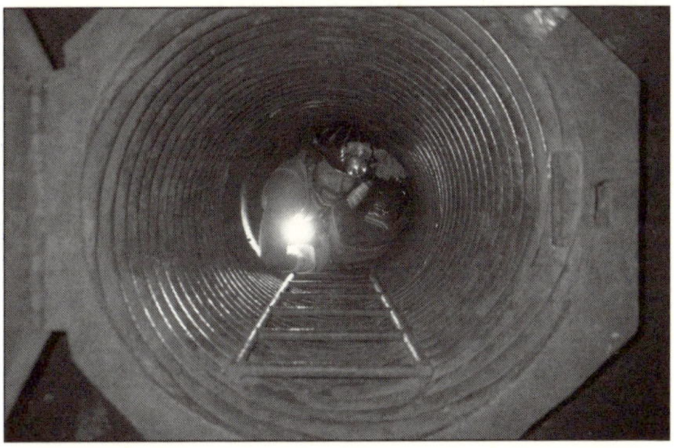

Figure 6. A caver is centered in the old culvert entrance to Lechuguilla cave. The rings of the galvanized culvert give a literal bull's eye impression to this image while the off-center flash adds interest to the composition.

Figure 7. The cave cricket is intentionally centered in the frame of white calcite. Note the slightly off-center nylon cap covering a photomonitoring station, adding visual interest to the image.

- Only use a bull's eye or centered composition when you deliberately choose to center the subject, not because you forgot to compose the image. Mentally compose the image three other different ways before exposing a bull's eye shot (Figure 2).

Document Participants in Action

- Make simple shots of human activity. Organize photo gear so action shots can be made quickly:
  - With a normal, mid-focal-length lens, shoot 2–3 meters (6–10 feet) from the participants.
  - Shoot two frames—the first candid, the second posed.
  - Add a slaved flash at the side only if convenient, conservation-wise, and quick.

Reference Scale for Scientific Documentation

- Photos made for scientific documentation should include a measure or an object for scale reference.

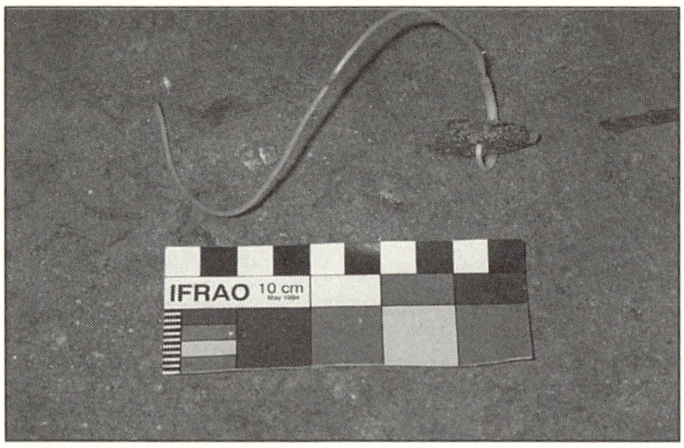

Figure 8. The IFRAO color scale, developed by the International Federation of Rock Art Organizations, is ideal for cave photographers. The fig seed, inadvertently left by a caver at a lunch stop, sprouted in the cave's dark silence. (See page 10 of color section.)

- The International Federation of Rock Art Organizations (IFRAO) has designed a color scale for photographic documentation of archaeological resources (Bednarik 1991; Bednarik and Seshadri 1995). The IFRAO scale is an excellent tool for cave photographers (Figure 8). On this useful Web site, <colca.mapaspects.org/methods/meth_graphics/IFRAOse.pdf >, a PDF file of the color scale can be downloaded and printed directly onto photographic-quality paper. The digital and archival benefits of color matching with the IFRAO scale are described described by Bednarik on another Web site <http://www.cesmap.it/ifrao/scale.html>.
- A small nonreflective ruler is adequate for scale only.
- Or use a common object of recognizable scale to indicate the approximate dimensions of the subject matter.

Photoinventory

The term *photoinventory* describes the process of photographing specific cave features or passages using a simple, somewhat repeatable, stand-and-shoot method. The photographer stands in an easy-to-remember, obviously discernible trail location and shoots the speleothem, feature, or scene.

Usually, nothing is left in the cave to identify stand-and-shoot locations. Over time, stand-and-shoot spots are memorized with the aid of notes, or locations are matched to previous photos. This technique yields approximate repeat shots. Even if a location marker is placed in the cave, repetition of the shot will not be exact.

Like a picture inventory of valuable household items organized for insurance purposes, photoinventory images should provide simple, readily accessible information.

Photoinventory images should be individually labeled, organized, and archivally stored as a collection of pictures that record the valuable resources and features of a cave or karst system. Photoinventory collections can benefit and support resource management decisions.

Tips for Photoinventory

- Good lighting greatly enhances images that are to serve as visual records for identification and inventory. Use composition and lighting to emphasize the subject. A simple, repeatable lighting solution works well for photoinventory images:
  - Shoot with a camera-to-subject distance of 2–5 meters (6–15 feet).
  - Use two flashes. Fire one flash from the camera. Slave the second, off-camera flash. For consistency, the slaved flash can be held to one side at arm's length.

Figure 9 (1999) and **Figure 10** (2002). Nest of cave pearls at the bottom of Boulder Falls in Lechuguilla Cave. These two images show no change during a three-year period. Because several pearls once disappeared from this nest, stand-and-shoot photographs are made periodically at this site and filed as photoinventory images.

Photos © Val Hildreth-Werker

Figure 11 (April 8, 1994) and **Figure 12** (August 8, 1997). The entrance to Southwinds in Lechuguilla was very white and pristine on the day after it was discovered in 1994. Three years and about 150 cavers later, the entryway had become dusted to a dull, brown-colored appearance. Cavers should move gently through this area to avoid kicking the soft rock flour into airborne particles that discolor the features. (See page 6 of color section.)

- Choose camera position, image composition, and lighting techniques that can be approximately repeated if future comparative photographic analysis is necessary.

Photomonitoring

Photomonitoring sequences yield time-based information for both qualitative and quantitative comparison of change, growth, and impact. Images are created from identical camera locations monthly, annually, or once every few years.

Photomonitoring systems can facilitate consistent documentation of human impacts and natural fluctuations in cave environments. Comparative images made from established photo stations provide monitoring data that pinpoint changes.

Environmental photomonitoring is a management and research tool for documenting, evaluating, and protecting resource values. Management decisions are supported with tangible, time-based evidence of changes in cave environments. Habitat fluctuation, visitor impact, natural growth and decline, water levels, trail conditions, cultural sites, paleontological remains, and other resource conditions should be photographed. Repeat photos can be made from the same point again and again, creating a time-lapse sequence that shows change or stability over time.

Photomonitoring images that are carefully labeled and archived become visual documentation tools for immediate, intermediate, and long-term cave management.

Photomonitoring systems enable consistent documentation of human impacts and natural fluctuations in cave environments.

Traditional Photomonitoring Methods

Historic methods of environmental photomonitoring are labor-intensive and time-consuming. Tedious documentation is required to achieve rigorous duplication of tripod placements, heights, angles, and camera settings (Uhl 1981). The repetitious use of levels, plumb-lines, measurements, and compass readings further complicates traditional methods of cave photomonitoring, often renders inaccurate results, and increases impact to fragile resources.

Streamlined Photomonitoring Installations

During the 1990s, Werker and Hildreth-Werker (1994, 1996) began developing systems for permanent photomonitoring installations to ease the

Figure 13. Traditional environmental photomonitoring techniques are cumbersome and time-consuming. With traditional systems it is difficult to avoid causing repetitive impacts to areas near the photo stations.

Figure 14. The streamlined photomonitoring system developed by Jim Werker and Val Hildreth-Werker is installed in caves throughout the western U.S.

time burden and redundant cave impact of traditional photomonitoring methods. Installing a permanent photomonitoring system enables quick, accurate, and repeatable images every time a photo station is used.

Stainless steel sleeves are installed in the floor, wall, or ceiling of a cave passage to achieve consistent image alignment, angle, and camera-to-subject distance. A specially designed mount accepts the camera and plugs into each sleeve. The same image area is reliably duplicated during subsequent photographic sessions—today, tomorrow, or 50 years from now.

With permanent photomonitoring installations and the specially tooled monorod, images made during subsequent photo sessions are readily compared on a light table, computer screen, or a projection system. Multiple projectors, multi-image dissolves, or digital comparisons can dynamically display temporal change and anthropogenic or natural impact.

Photomonitoring programs supply data for resource protection decisions and provide documentary photos for storytelling and interpretive activities.

Contemporary digital technology enables faster, more consistent photographic documentation. Advances in electronic imaging make it possible to compare visual data through layers, graphs, and statistical analyses. Innovative photographic technology and GIS applications will open future avenues for quantitative analysis of photomonitoring data.

Tips for Photomonitoring

- Selection criteria for a photomonitoring camera (digital or film-based) includes excellent lens quality, cave-friendly size and shape, adequate flash characteristics, sufficient exposure settings, and appropriate mounting positions.
- A camera with auto-focus, automatic settings, and in-camera flash makes photomonitoring sessions much easier, faster, and more consistently repeatable.
- Images exposed with only the camera flash make the most easily repeated type of photo-monitoring picture. Photograph subject matter located close to a photomonitoring station using only an in-camera flash (no additional external flashes). The in-camera flash yields a flat-looking image, but renders data that can be easily compared in multiple photos. However, multiple flashes are sometimes necessary to adequately light an image area.
- Choose identifiable positions for photo stations. Photopoints should be easily located during subsequent shooting sessions.

Figure 15 (1995) and **Figure 16** (2003, trail flagging added). These two photomonitoring images graphically show the success of management strategies to minimize new impacts along a trail leading to the New Mexico Room in Carlsbad Cavern.

- In some caves, photopoints are installed at systematic intervals to enhance the monitoring objectives. In other caves, randomly located photopoints are more useful.
- If there is no appropriate durable surface for installing a photomonitoring sleeve, choose a suitable stand-and-shoot location and include this shot to augment the chronological sequence of photomonitoring data. Stand-and-shoot photos are also useful for recording close-up details of subjects. For efficient identification, clearly labeled images indicate stand-and-shoot. (See photoinventory, page 206.)
- In addition to thorough written descriptions of each photo station, note the camera settings and flash techniques. Augment documentation with a photograph of each station area clearly showing the photopoint location and flash positions in relation to other features at the site.
- Tie photo stations to the existing cave survey and add the photomonitoring stations to maps of the cave.
- If using station markers in a cave environment, the recommended materials are stainless steel, nylon, or Mylar®. (See survey markers, page 185 and page 189; also see cave-safe materials, page 167.)
- Photographic transparencies (35-millimeter slides) combined with digital images work best for photomonitoring projects. Fine grain films, typically ISO 100 or slower allow for enhanced resolution of small details but often require more flash power for proper exposure in caves. Kodachrome® films, when properly processed by approved labs and stored correctly, promise archival stability for several decades.
- Kodak® E-100S, a professional transparency film that records accurate colors with flash photography, is this author's first choice for

In addition to thorough written descriptions of each photo station, note the camera settings and flash techniques.

> **Permanent Photomonitoring Installations**
> **Val Hildreth-Werker and Jim C. Werker**
>
> Permanent photomonitoring systems can streamline environmental documenting processes. Designed by Jim Werker and Val Hildreth-Werker for cave installations, the streamlined system is efficient for any environmental photomonitoring application.
>
> A small stainless steel sleeve is installed at each station. Each sleeve is covered with an unobtrusive nylon cap to prevent debris from entering. The camera is mounted on a specially designed monorod that fits the sleeves to assure accurate alignment and efficient operation. The systems include simple methods for archiving and retrieving both analogue and digital photographic data (Werker and Hildreth-Werker 1994, 1996).
>
> Both Jim and Val are long-time cavers and have backgrounds in research and development.
>
> Val has been photographing caves since the late 1970s and brings professional experience in medical and commercial photography, animation, multi-image, and video production. The tedium, time consumption, and redundant cave impact inherent in traditional photomonitoring led to the development of the streamlined method. The idea of installing permanent photomonitoring systems in caves simmered until

photomonitoring film. For slide developing, use Kodak "Q" Lab® services (photo labs that are inspected and certified for quality control in film processing). With archival processing and proper storage, Kodak E-100S retains color balance and quality.

- Proper processing and storage enhance the archival potential of any film.
- Prints (black-and-white or color) should not be used a the only media for photomonitoring because processing and printing variations for negatives can be extreme.
- Video photomonitoring has not proved successful because it is more efficient to compare individual, high-resolution images.
- For increased archival protection, scan photomonitoring images, (transparencies or negatives) and store the files on digital media. Some CD-ROM manufacturers claim 100-year storage stability. Ideally, archive digital images with at least 2,500 pixels on the long axis. In terms of common digital camera formats, a five megapixel file is adequate (note that a five megapixel camera file will make an 18 megabyte TIFF file). However, storage in JPEG format compressed to no less than 10 percent of the original file size is adequate for archival purposes (10:1 compression). Many software packages offer some degree of control over the level of JPEG compression—always choose the lowest degree of compression to achieve the least loss in original image data. Even this will result in substantially smaller file size than a TIFF. In the digital world bigger is always better, and larger image files will give better capability to resolve small details (Dave Bunnell, personal communication 2003).
- Immediately after processing or downloading, record the date and location on all photographs.
- Label all slides and place in archival, polypropylene sleeves for preservation. Polypropylene slide pages work well. (Polyvinyl pages are not recommended because they emit gases that rapidly deteriorate photographic emulsions.) Sequential slides or digital images from each photo station are paged or filed to facilitate easy retrieval, viewing, and comparison.
- Store complete photo sets in at least two separate locations. Update all sets with future photographs. If photomonitoring images are digital, then archive high-quality prints separately. Store electronic files on CD-ROM, DVD-ROM or other commonly accessible media.
- Photos typically grow more valuable over the decades and centuries. Long-term archiving and retrievability are important factors. Ideally, store photomonitoring images in public archiving institutions that accept cave photos. Find archiving facilities through cave societies, libraries, museums, and other public institutions.

Organizing and Archiving Cave Photographs

Cave photographs should be organized for efficient retrieval. Otherwise, photographic information is of little benefit to resource managers. Conservation management decisions and law enforcement cases are frequently supported with photographic documentation, but a cave manager has to be able to access the photos.

Changes in cave environments; habitat fluctuations; water levels and conditions; effects of moisture variations; trail maintenance status; progress of survey, construction, or restoration projects; growth and decline of speleothems; documentation of geologic features; condition of cultural materials or paleontological sites; visitor impact; vandalism; gate status; entrance vegetation; and other resource information should be photographically recorded and archived.

Every photo that is labeled with at least the place and date the image was created has potential future use for cave or karst management decisions. If labeled with the location and date, comparative photos created over time are invaluable for law enforcement situations.

However, if those potentially useful photos cannot be found when a need arises, then their usefulness is certainly diminished. Some type of efficient organization is obviously important. Photo filing systems vary as much as the photographers themselves. Most resource management offices already have photo filing systems in place. If a filing system is too complicated and time-comsuming to catalog and store images, then revamp and simplify the system so it is easy and user-friendly.

Label Photographs

The date is the most overlooked detail on many photographic images. If necessary, other information can be reconstructed from an accurate date through comparison with field notes, journals, datasheets, or other photographs.

When photos come back from the processing lab, *immediately* write a tiny date on slide mounts or in a corner of prints. When downloading digital images, include (at least) the date and cave name with the file. When time permits, make complete labels that include the following information:

- Date the image was photographed
- Name of the cave (or code if name information is especially sensitive)
- Location in the cave (or name of the passage or survey station)
- Name of the photographer (and assistants)
- Address, phone, e-mail, contact information for photographer
- Names of the people appearing in a photograph
- Name of the landowner or managing agency
- Caption explaining the photo
- File number for personal image retrieval system
- Copyright (optional)
- Request for photo credit'

Copyrighting photographic work is optional. Copyrighted material may complicate future use of simple documentation photos. Copyrights may make photographs less serviceable to the managing agency, organization, or landowner. Regardless of copyright decision, request for photo byline or photo credit may be included on the label.

Organizing and Archiving—Tools That Work

Streamline labeling and retrieval systems and minimize organizational time with the following suggestions:

- The ink in most Pilot® ball point pens will not smear on plastic slide mounts and resin-coated photographic papers.
- Ink-jet or laser printer labels work well and look great, but when time is short, write directly on the slide mount or photo print (front border or back corner) and save tedious label making for later.
- A photographic loupe aids in quickly viewing images on a light table.
- All photographs should be labeled and archivally sleeved for preservation in polypropylene slide or print pages. Avoid pages made of other materials—PVC, polyvinyl, vinyl, sticky pages, and so on—these products emit gases that rapidly deteriorate photographic materials. Today, most archival pages are made of polypropylene and the packaging will indicate archival quality.
- For prints, choose a photo-safe, archival album with slots for the negatives

Val described the concept to Jim. He designed and created hardware.

Jim, a mechanical engineer for Sandia National Laboratories since the 1960s, brings decades of professional experience in tooling design, engineering, and fabrication. For 30 years, he worked at the National Labs and the Underground Nuclear Test Site in Nevada.

Jim and Val collaborated to develop the complete photomonitoring system. Installation of their first system began in 1993. From designing the prototype through the evolutionary adjustments of fabricating, archiving, and digitizing, they created an easily repeatable means for chronological environmental documentation with photographs.

The couple also developed a specialized system using infrared photographic technology for monitoring the Mexican free-tailed bat population at Carlsbad Caverns National Park. They photomonitored the colony annually and collected population data during a 10-year study (Route and others 1999, 2005).

The Werkers have installed permanent photomonitoring systems in several caves of the western U.S.; caves in the Guadalupe Mountain region of New Mexico; caves of Arizona, Texas, Utah, Nevada, and Oregon; caves managed by the National Park Service, the Bureau of Land Management, the USDA Forest Service; and caves managed by private land owners. Patent is pending.

unless you already have a well-established negative filing system.

- For slide storage, archival clam-shell boxes made of polypropylene with ring binders inside work well for keeping dust away from slide pages. Archival slide pages and ring binder boxes are worth the investment—purchase through large photographic suppliers. These products greatly facilitate image organization and retrieval.

Photography Can Promote Cave Stewardship

"She's a photographer ... she can do anything she wants." That's what some people believe. On several occasions, I've overheard people say that about me. But, when it comes to caves, I wholeheartedly disagree with this sentiment. Wielding an official-looking camera rig does not give license for go-anywhere-do-anything ethics in caves.

The longing to get a shot should never overpower the importance of reducing impact to the cave. Before acting, first think about protecting the resources—the cave should come first.

Conservation Ethics for Cave Photographers

Promote leave-no-trace attitudes by staying on trails and adopting caver minimum-impact ethics. If there are no marked trails, look for the most impacted path and choose to stay on it.

Always appoint an assistant to be in charge of spotting the photographer. It is easy for cave photographers to get preoccupied and forget to watch out for their feet, backs, elbows, helmets, lenses, and equipment. Assign a conservation-minded caver who will speak up when the photographer needs caution or correction.

Check with cave managers and speleologists for proper protocol before entering sensitive areas. Preplan restoration gear and erase your trace as you move through the cave. Become aware of state-of-the-art conservation ethics and insist that models, cavers, and assistants cave softly. Shoot for no new impact. (See the minimum impact code for cave photographers, page 215—remember to eat over a zippie bag, use nonmarring boots, wear lint-free clothes, and pack clean flowstone shoes.)

In the spirit of leave-no-trace ethics, cave photographers are developing ways to restore along the way. Some cave photo teams now carry a little restoration kit with simple tools.

Erase Your Trace

For decades, cavers promoted a follow-the-footsteps-of-others ethic by placing feet inside the first set of footprints to avoid causing new impact. But, more people are caving, and ethics are changing. Pathways of single footprints have expanded into multiple scars across cave floors.

Today, rather than retracing former impact, cavers are developing gentle ways to prevent and erase impacts. Even the traditional caver motto has changed—*Take nothing but pictures. Leave nothing but footprints. Kill nothing but time.* Many are now augmenting the adage to reflect new ethics—*Take nothing but conservation-wise photos. Leave nothing but careful footprints on established trails. Kill nothing but time. Cave softly ... and erase your trace.*

If trails are not marked, choose routes across durable surfaces. Avoid creating new footprints and causing new damage to cave passages.

Immediately remove or restore any inadvertent foot or hand imprints if the restoration activities will not cause additional harm. Erase left-over footprints so they cannot tempt others to follow.

In the spirit of leave-no-trace ethics, cave photographers are developing ways to restore along the way. Some cave photo teams now carry a little restoration kit with simple tools. Use only clean, contaminant-free equipment. When feasible, get water from the immediate cave passage for restoration tasks. Use supplies in a single area to avoid cross-contamination. With just a few lightweight tools, many types of restoration and repair can be accomplished (Hildreth-Werker and Werker 1997, 1999; also see chapters in the restoration sections of this volume):

- Small nylon-bristled brush or a new, soft toothbrush for muddy prints
- Packable spray bottle for water
- Absorbent sponge to blot up gunk
- Two or three clean zip-closure plastic bags
- Stainless steel or plastic tweezers to remove small debris
- Nonlatex, powder-free surgical gloves
- Light flowstone shoes

Any cave photographer can support and promote conservation ethics through their images.

If setting up a shot is going to be extremely time-consuming, some photo assistants and models might be willing to work on restoring a small area while the rest of the team participates in shot preparation. By planning and appointing an experienced restoration team leader, simple tasks—certain trail maintenance, flowstone cleanup, or small speleothem repairs—can be quickly accomplished.

Photos Can Help Protect, Preserve, and Perpetuate

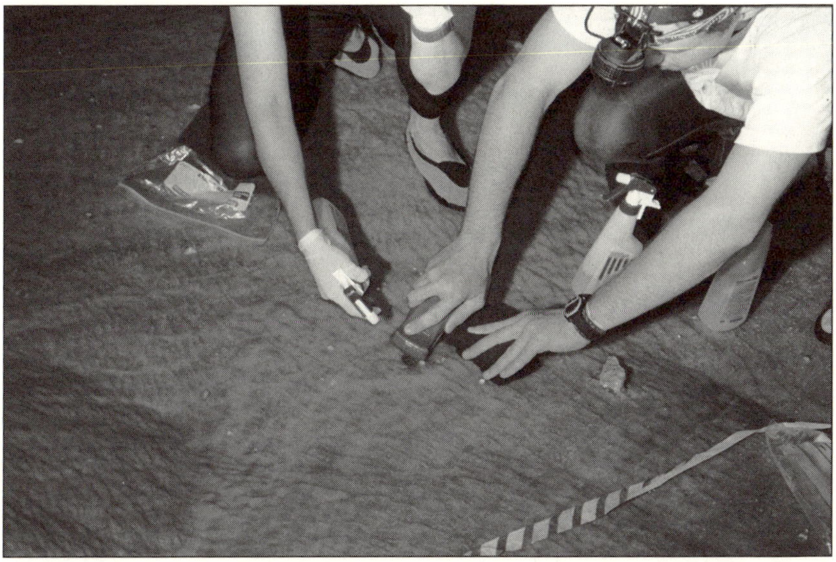

Figure 17. Two cavers work as a team cleaning flow-stone with sponges, water in spray bottles, and light brush strokes. By today's standards, both cavers should be wearing nonlatex, powder-free surgical gloves to help minimize new impacts. Cavers should also check the soles of flowstone shoes to confirm they are non-marking—be especially careful to test aqua socks that have dark-colored soles.

Photographs document our journey through the history of cave exploration and conservation. Each image freezes a single moment in speleological time. Pictures become silent mirrors, chronicling subterranean systems, human ethics, and environmental impacts.

Cave photographs should illustrate good caving practices. People tend to repeat or emulate what is seen in pictures—blunders, as well as new techniques, can make bold statements in photographs. Photographers can create images that focus other cavers toward low-impact methods and thoughtful resource stewardship. Any cave photographer can support and promote conservation ethics through their images.

Those who make and show photographic work of caves have an intense responsibility to communicate the best possible conservation messages.

Any published photograph, exhibition image, or salon entry that depicts outmoded caving practices should include a disclaimer, warning, or explanation. Insist that a statement describing better practice is boldly displayed or printed with any photograph that illustrates questionable behaviors (Figure 17).

Use captions to educate. If photos showing cave damage are used for instruction, or if photos with outdated practices are used for historical documentation, clarify the intention in the caption or context. Through careful wording, cave photographers can encourage low-impact caving ethics and champion current best practices in cave conservation.

Use captions to educate. Any published photograph, exhibition image, or salon entry that depicts outmoded caving practices should include a disclaimer, warning, or explanation.

Cited References

Bednarik RG. 1991. The IFRAO Standard Scale. *Rock Art Research* 8:78-8.

Bednarik RG, Seshadri K. 1995. Digital colour re-constitution in rock art photography. *Rock Art Research* 12:42-51.

Hildreth-Werker V, Werker JC. 1997. Lechuguilla restoration: Techniques learned in the Southwest. *NSS News* 55(5):107-113.

Hildreth-Werker V, Werker JC. 1999. Restoration, trail designation, and microbial preservation in Lechuguilla Cave. In: Stitt RR, editor. *Proceedings National Cave Management Symposium 1997: October 7-10 in Bellingham, Washington, and Chilliwack and Vancouver Island, B.C., Canada.* National Cave Management Symposium Steering Committee. p 181-189.

Route B, Bemis T, Roemer D, Hildreth-Werker V, Werker J. 1999. Methods for monitoring large colonies of Mexican free-tailed bats. In: Stitt RR, editor. *Proceedings National Cave Management Symposium 1997: October 7-10 in Bellingham, Washington, and Chilliwack and Vancouver Island, B.C., Canada.* National Cave Management Symposium Steering Committee. p 159-168.

Uhl PJ. 1981. Photomonitoring as a management tool. In: Mylorie JE, editor. *First International Cave Management Proceedings, Murray State University*. p 123-128.

Werker JC, Hildreth V. 1994. Streamlined photomonitoring. *NSS News* 52(2):66.

Werker J, Hildreth-Werker V. 1996. New improved system for photomonitoring. In: Rea GT, editor. *Proceedings of the 1995 National Cave Management Symposium: Spring Mill State Park, Mitchell, Indiana, October 25-28, 1995.* Indianapolis (IN): Indiana Karst Conservancy. p 311-312.

Figure 18. Photographers, flashes, and assistants remain within the trails in Lake Louise, Lechuguilla Cave, New Mexico.

Minimum-Impact Code of Ethics for Cave Photographers

Val Hildreth-Werker and Jim C. Werker

The goal of this code of ethics is to encourage cave photographers to minimize human impacts to caves. As we learn more about cave environments, we are in a continuing process of evaluating and redefining caver ethics.

These guidelines for low-impact cave photography come from the experiences and thoughtful contributions of many cavers. Think safety—take care of yourself and your team.

Think stewardship—avoid impacting cave resources. Take care of aesthetic, cultural, paleontological, geological, hydrological, mineralogical, meteorological, biological, as well as microbial resources. Move gently and photograph softly.

- Packs, vertical gear, boots, aqua socks, gloves, helmets, and lint-free clothing should be freshly washed to avoid transfer of mud, dust, and microbes from other environments. Photo gear and tripods should also be cleaned before each cave trip.
- Use only nonmarring/nonmarking soles. Whether light or dark colored, many boot soles will leave marks on calcite and other cave surfaces—test by striking soles across untreated limestone or concrete. No heavy waffle stompers. Traditional black lug soles will leave marks on cave surfaces. Be careful with blond rubber soles—chunks from soft soles tend to break off and leave debris in cave passages.
- Use padded camera cases rather than hard-edged boxes when feasible.
- Minimize photo gear—smaller, lighter, more compact.
- Always ask to be spotted when photographing fragile areas. Spot the feet, too.
- Instruct models to refuse if the photographer suggests a move that is too close to delicate speleothems. The model usually has the best vantage point. Appoint an assistant to spot each model.
- Avoid disturbing bats and other cave-dwelling creatures.
- Take nothing from caves. Removal of natural or historical objects is unethical and illegal unless you have a collection permit for authorized research. (Contemporary trash usually should be removed.)
- Wear gloves. Rather than grabbing handholds along the trail, use a gloved knuckle for balance where possible.
- Pack in clean gloves for use in gloves-off areas and in pristine passages. Powder-free, nonlatex surgical gloves are recommended. Fresh surgical gloves also help keep photo gear cleaner.
- Carry clean flowstone shoes—lightweight, lint-free, and nonmarking. Use water-sports booties, smooth-soled athletic shoes, aqua socks, or nylon slippers. Include a plastic bag to contain dirty boots. Do not use bare feet or socks—lint, skin flakes, blood, and body oils should not be left on cave surfaces.
- Insist that all team members change to clean garments when entering pristine areas to make photographs.
- Avoid contamination. Know which special-attention areas require clean clothes, shoes, and gear. Do not enter restricted areas wearing general caving attire. Keep these areas pristine.
- Avoid isolated pools. Tens of thousands of skin flakes and fragments

The goal of this code of ethics is to encourage cave photographers to minimize negative human impacts to caves.

Photographers are not exempt— they should adopt minimum-impact ethics.

of debris fall from each of our bodies every hour.

- Don’t comb or brush hair in caves. Try to avoid scratching.
- Don’t smoke or use tobacco in caves. Smoke and fumes can kill bats, invertebrates, and other cave-dwelling animals.
- Remove all human wastes. Contain and carry urine, feces, spit, and vomit out of caves.
- Eat over a plastic bag. Carry out all crumbs and debris. Don’t eat on the move.
- Stay on established trails. Sit within the pathway and be careful not to set packs outside the trail. Keep tripods within the trail. Always look for and use the most impacted areas whether traveling or stopping. Where trails are not obvious, stay within the most compacted or durable zones.
- Don’t lean on walls, ceilings, or speleothems. Don’t sit on formations. Avoid trampling floor deposits. When traveling through the cave, use small points of contact for balance instead of dirty open palms.
- Rather than retracing footprints and handprints, gently erase and restore.
- Photographers are not exempt—adopt minimum-impact ethics. Stay on trails. Stay away from sensitive or off-limits areas. Look carefully, move gently, don’t kick up dust, and avoid anything you don’t have to touch.

Section B—Developing Cave Management Programs

Using the Law to Protect Caves: A Review of Options

George N. Huppert

The legal system of the United States provides many avenues to protect caves. It seems paradoxical that, with all of the laws covering vandalism, theft, and trespass, caves still need special additional protection. Why do caves need specific protection laws?

Those who enforce the law are sometimes misinformed and believe that caves have little value. Groups of cavers, conservation societies, and scientists work diligently to persuade the legal system to recognize the substantial values associated with caves. However, in pragmatic terms, applying and enforcing laws for the effective conservation of caves, karst areas, and unique associated biota can be difficult at best.

Often, the biggest obstacle is convincing the enforcing authorities that caves are, in fact, worth the application of these protective laws. Fortunately, over the past three decades there has been a gradual improvement in this attitude. Success has been slow but productive. Many federal and state agencies, and to a much lesser degree local agencies, have accepted the wisdom of protecting caves. Still, enforcement remains spotty and can be quite variable, even within the same agency.

Laws are not static entities, but changeable records. Modifications can be developed and legislated. Court decisions can create precedents that may change the interpretation and application of the law. Usually the changes are helpful, clarifying some phraseology in the text, or including aspects of caves or karst that were omitted in the original law.

Unfortunately, modifications can also be detrimental, even to the point of requiring repeal. Keeping current with this body of legislation is necessary in order to make the best use of it. Fortunately, the Internet has made this once formidable task easier. Many federal and state laws are readily available through search engines on the World Wide Web.

This chapter is a compendium including a brief history of cave laws, descriptions of existing state and federal laws that may be useful for cave protection, and a list of Web sites for finding further information.

This chapter is a compendium including a brief history of cave laws, descriptions of existing state and federal laws that may be useful for cave protection, and a list of websites for finding further information.

Background

The first significant compilation of cave laws was produced by Louise Power (1974) in her role as Chairman of the Legislation Subcommittee of the Conservation Committee of the National Speleological Society. Power's effort provided a handbook to guide enactment of cave protection laws in the states. Next, an article and list of state cave protection laws by Stitt (1976) was published in the proceedings of the first cave management symposium held in the United States in 1975.

These two publications started a series of articles that follow a similar format: Huppert and Wheeler (1982), Bradshaw (1982), and Anonymous (1995). These earlier articles reflect the emphasis of the times—primarily to formulate and pass cave protection acts. Later articles, such as Huppert (1995), LaMoreaux and others (1997), and Lera (2002) provide greater detail.

The Puerto Rico Act specifically protects karst as well as the caves found within karst terrains, whether on public or private land.

State Laws

During the past 100 years, many states have enacted laws to protect caves. Colorado passed the first cave protection law in 1883, followed by Wyoming and South Dakota. All three states later revoked the first statutes because they were vague and difficult to enforce. (Also see states, page 220.) A number of other states enacted laws early in the 20th century that have since been repealed or superseded by superior legislation.

With the exception of Oklahoma, Maine, and Hawaii, (plus Colorado–see sidebar, page 220) all cave protection acts presently in force were passed between 1976 and 1993 (Huppert and Wheeler 1986, 1992). There are now 24 states with specific cave protection acts. Puerto Rico and the Cherokee Nation also have similar legislation (Huppert 1995). (Some states have resource protection laws that incidentally mention cave resources. Such laws may have been passed in different years than those listed below.)

There is great variation in the application and effectiveness of cave protection laws. Apparently, the drafters of the acts gradually learned from earlier efforts, because the later laws tend to be the best constructed. The following specific cave protection laws are listed by Huppert (1995) and Lera (2002).

Alabama 1988
Arizona 1977 Effective 1978.
Arkansas 1989
California 1976 Effective 1977.
Colorado 2004
Florida 1980
Georgia 1977
Hawaii 2002
Idaho 1982
Illinois 1985 Effective 1986.
Indiana 1983
Kentucky 1988
Maine 2001
Maryland 1978
Missouri 1980 Effective 1981. The state also has a 1959 show-cave inspection law.
Montana 1993
New Mexico 1981
North Carolina 1987
Ohio 1988 Effective 1989.
Oklahoma 1967
Pennsylvania 1990 Effective 1991.
Tennessee 1991
Texas 1979 A 1977 law was repealed, and then it was enacted into a different part of the statutes. Texas also has a law against littering in caves and another law protecting cave owners from liability.
Virginia 1979
West Virginia 1977

A Model Law: 1995 Puerto Rico Act

An encouraging event occurred on August 21, 1999, with the enactment of Law 292 for the Protection and Conservation of Karst Physiography in Puerto Rico (A Vale, written communication by facsimile on April 23, 2000). This Act specifically protects karst as well as the caves found within karst terrains, whether on public or private land. Protection also includes formations, minerals, flora, fauna, and all natural materials. The sale of such cave items is specifically forbidden. Law 292 is truly enlightened legislation and could be used as a model for similar laws elsewhere.

Unfortunately most state cave protection acts only protect caves on state lands. The few that cover caves on private lands usually require the landowner's permission to press enforcement. This is often difficult to obtain. Landowners frequently place no real value on caves and may not want to be bothered with the inherent problems and expenses of legal actions. An additional obstacle to enforcement may be that the violators are neighbors, friends, or relatives who were invited onto the property.

Protecting Archaeological Sites

Examples of states that protect cave resources as part of archaeological, paleontological, historical, or natural sites include Nevada, Vermont, Washington, and Wisconsin. The Nevada statute, passed in 1959, is perhaps the strongest of these. Based on archaeological evidence, the requirements of an archaeological protection act may well offer full protection for a cave and perhaps a portion of the surrounding surface landscape. Therefore, finding such assets in a cave may be the best route to protecting a cave in some states.

Examples of states that protect cave resources as part of archaeological, paleontological, historical, or natural sites include Nevada, Vermont, Washington, and Wisconsin.

Other Legal Avenues

Additional legal routes can protect caves. Most states have endangered species acts that cover a surprising number of animals and plants. For example, in 2001 Texas passed a law specifically protecting bats. In other states, the protection of biota requires protection of the habitat, which may include caves.

Many states also have their own wilderness acts, usually listed by state name, such as the Arkansas Wilderness Act of 1984. Caves may fall under the definition of wilderness in some states. Pennsylvania has a law that provides for an assistance program to help landowners who suffer damage from sinkholes. Tennessee requires stricter codes and oversight when a sewage disposal site is constructed on karst. Kentucky has several laws that are specifically meant to protect caves, karst, and karst groundwater in addition to the state cave protection law (LaMoreaux and others 1997). Other states should have similar laws.

Many states have provided some level of legal oversight for significant caves by including them in their system of state parks, wildlife, or natural areas. Texas and Missouri have probably protected the greatest number of caves in this manner.

Protection Through Tourism: Kartchner Caverns

A recently developed show cave in Arizona, Kartchner Caverns, is especially worthy of mention (Hose and Pisarowicz 1999). Kartchner was discovered in the early 1970s and its location was kept secret for years. Four years after discovery, the cave was revealed to the landowners and the state government in order to protect the site from potential encroachment by community, commercial, or industrial development.

It was decided that careful development as a public show cave would best assure the permanent protection of Kartchner Caverns. A decade passed while negotiations ensued to raise money for the purchase of the property and to transfer the cave to the state. Then it took another decade to plan for development, raise funding, and open the cave to the public in November 1999. Kartchner is perhaps the most studied cave in the United States in terms of resource conservation and show cave development. Many millions of dollars were expended in the effort to implement best current practices in the conservation and development of Kartchner Caverns. (See Kartchner Caverns, page 141.)

It was decided that careful development as a public show cave would best assure the permanent protection of Kartchner Caverns.

State Cave Laws
Thomas Lera

Twenty-eight states have cave protection laws. Texas has an additional law that specifically protects bats.

The definition of a cave varies widely by state and ranges from "historic site," as defined in Vermont, to Kentucky's definition of "...any naturally occurring void, cavity, recess, or system of interconnecting passages beneath the surface of the Earth containing a black zone including natural subterranean water and drainage systems, but not including any mine, tunnel, aqueduct, or other man-made excavation, which is large enough to permit a person to enter."

Definitions of vandalism include removing materials found in caves, killing or removing plant and animal life, breaking or tampering with doors or gates, and other destructive or damaging acts.

Penalties for vandalism are classified in every state as a misdemeanor ranging from Class A (Class 1) to Class E (Class 5).

The penalty is either criminal or civil and ranges from $50 to $2,500 and may include up to one-year imprisonment.

In most states, a subsequent violation either increases the penalty or the offense becomes a felony.

(Continued on next page)

Undeveloped Caves on Public Land

Commercially undeveloped caves located on state lands may well enjoy greater *de facto* protection than their *de jure* protection status, simply by neglect and remaining unknown to the public.

Often, overworked personnel are tempted to ignore such nontourist caves. Nevertheless, this method does not provide viable protection once the location of a cave becomes known.

Fortunately, protection of such caves is improving. More state-level managers are becoming aware of conservation management practices through the National Cave and Karst Management Symposia, the associated published proceedings, and the efforts of many cave conservation organizations such as the National Speleological Society and the American Cave Conservation Association. It is increasingly common to find access to these caves controlled by gates and permit systems.

States Needing Cave Laws

Reviewing specific state laws that protect caves raises the question of whether there are other states where such acts would provide benefit. It may be important to note that some states have few, if any caves.

Much of the land in the western United States is under federal jurisdiction, and should, at least in theory, be protected by federal statute (see federal laws below). On the other hand, cave protection laws that apply on private land would be beneficial in states where most caves are on privately owned land.

Huppert (1999) lists the following states as those most needing specific laws to protect caves.

- Alaska
- South Dakota
- Minnesota
- Iowa
- Colorado (See sidebar.)
- Utah
- Wyoming
- New York

People are discovering significant caves on remote possessions of the United States. The caves on the island of Guam have recently received intense scientific attention and they appear to deserve strong protection. As more caves are discovered, the need for adequate legislation in the states listed above will become more apparent and, in fact, it might be necessary to add states to the list. A more detailed account of state cave protection acts was published in *American Caves,* a journal of the American Cave Conservation Association (Anonymous 1995).

Federal Laws

Numerous federal laws have been passed to protect caves or their contents. Many of these laws apply to all lands in the United States and its territories. Others are more restrictive in their application.

Federal Cave Resources Protection Act of 1988

A controversial law is the Federal Cave Resources Protection Act of 1988 (P.L. 100-691). The significance of this law lies more in its precedence than its content. The Act requires many federal land managers to take into consideration and plan for all cave resources under their jurisdiction. Prior to 1988, many known caves could be ignored. The law has some drawbacks, however.

For example, it does not apply to all federal lands and it only applies to *significant* caves. There is considerable argument about what is actually meant by the term *significant*. See Appendix 1 for full text of the Federal Cave Resources Protection Act of 1988 (FCRPA).

National Park Service Act of 1916

Some of the nation's most outstanding caves are protected under the National Park Service Act of 1916 (P.L. 64-235; Vol. 39 U.S. Statutes, page 535). Examples of national parks and monuments where caves or karst are the main theme on display include sites across the U.S.

- Mammoth Cave National Park in Kentucky
- Carlsbad Caverns National Park in New Mexico
- Oregon Caves National Monument in Oregon
- Wind Cave National Park and Jewel Cave National Monument, both in South Dakota
- Timpanogos Cave National Monument in Utah

Caves are often protected within the confines of larger national parks. Examples include parks in several regions.

- Grand Canyon National Park in Arizona
- Great Smoky Mountains National Park in North Carolina/Tennessee
- Hawaii Volcanoes National Park in Hawaii
- Lehman Caves of Great Basin National Park in Nevada

When searching for legislative information, it is important to note that many national parks were mandated by their own enabling legislation.

The Antiquities Act

The full name of The Antiquities Act is the National Monuments Act, An Act for the Preservation of American Antiquities of 1906 (c. 3060, 34 stat. 225; 16 U.S.C. 431 *et seq*.). It allows for presidential mandate to establish national monuments. In 1908, President Theodore Roosevelt designated Jewel Cave as the first national monument.

The Wilderness Act of 1964

The Wilderness Act of 1964 (P.L. 88–577) has helped conserve caves in designated wilderness areas such as the Bob Marshall Wilderness Area in Montana. Much effort has been expended in attempts to designate several caves as wilderness areas on their own merits. Some of the unsuccessful tries are described in articles by Huppert and Wheeler (1986, 1992), Bishop and Huppert (1990), and Huppert (1993). Most other wilderness areas have had their own formative acts passed by Congress. These are usually titled by the name of the area, such as the Carlsbad Caverns Wilderness Act of 1978.

The Wild and Scenic Rivers Act of 1968

This law (P.L. 90–542) was enacted by Congress and allowed for the protection of caves along some of the nation's most beautiful rivers. Among the most important river areas for cave conservation are the Buffalo National River in Arkansas and the Ozarks National Scenic River in Missouri.

The National Environmental Policy Act of 1969

An important federal environmental law (1970) (P.L. 91–190, 83 Stat. 852 42 U.S.C. 4321 *et seq*.), the National Environmental Policy Act (NEPA) embodies significant legislation. Passed by Congress in late 1969, and

(Continued)

On August 4, 2004, Colorado became the twenty-eighth state to provide caves with legal protection from vandals and trespassers through a new Colorado Cave Protection Act.

The penalty is a Class 2 Misdemeanor with the maximum penalty of $2,500 and possible revocation of the individual's driver's license.

Those convicted of property damage will also be required to personally make repairs or provide appropriate compensation to the owner.

Damages include: defacing cave passages; breaking or removing speleothems; breaking or harming cave gates and locks; removing any cave resource including animal and plant life, paleontological deposits, and cultural materials.

Some of the nation's most outstanding caves are protected under the National Park Service Act of 1916.

signed by President Nixon on January 1, 1970, NEPA requires federal agencies to consider the consequences of their actions when environmental issues are involved.

An environmental impact statement (EIS) must be written by a federal agency (or by the responsible governmental unit) for any project that would use federal funds to support an action that might significantly and adversely affect the quality of the environment. The agency must also pursue mitigating actions to reduce adverse impacts when projects are implemented. The EIS process has conserved, or at least reduced, the impact of management decisions on numerous caves. Many federally managed caves or karst regions have an associated EIS.

The Endangered Species Act of 1973

The Endangered Species Act (P.L. 93–205) provides a mantle of federal protection for endangered species which is usually more effective than the protection that state endangered species statutes can offer. As mentioned above with state laws, the protection of a species may well require the broader protection of the habitat which may overlap political boundaries. Numerous animals use caves and karst as habitat for all or part of their life cycle. In Texas, the protection and management of more than 200 caves come under the provisions of this federal law and similar state laws (Texas Speleological Survey, unpublished data 2002).

The Archaeological Resources Protection Act of 1979

This Act (P.L. 96–95) gave protection to prehistoric sites on federal lands. Early inhabitants of the present United States often used caves as storage sites or as permanent or temporary domiciles. Wisely applied, the Archaeological Resources Protection Act could give significant protection to other cave assets. Unfortunately, this Act is not applied as widely as it could be on federally managed lands.

It is rare for a law to be enacted to protect a single cave. Perhaps the most significant such law is the Lechuguilla Cave Protection Act of 1993.

Lechuguilla Cave Protection Act of 1993

It is rare for a law to be enacted to protect a single cave. Perhaps the most significant such law is the Lechuguilla Cave Protection Act of 1993 (P.L. 103–169). The cave, which extends 1,567 feet deep and more than 115 miles in length, is located within the boundaries of Carlsbad Caverns National Park and Carlsbad Caverns Wilderness Area.

In a legal sense, Lechuguilla Cave is perhaps the most protected cave in the country. It is directly protected by The National Park Service Organic Act of 1916. (The legislation that specifically enabled the park is found in Ch. 279, 46, Stat. 272, 1930 and also found in Title 16, 407–407c. and P.L. 88–249.)

However, Carlsbad Caverns National Park was initially protected as a national monument by Executive Order No. 1679 (October 25, 1923–43 Stat. 1929), signed by President Calvin Coolidge. President Coolidge twice expanded the monument through Executive Orders (No. 3984 in 1924 and No. 4870 in 1928), followed by President Herbert Hoover who expanded the boundaries of the National Park by Executive Order (5370).

Both Carlsbad Caverns and Lechuguilla Cave are protected under the Federal Cave Resources Protection Act of 1988. In addition, they are protected by the Federal Wilderness Act of 1964 and the Carlsbad Caverns Wilderness Area; the latter was designated under the National Parks and Recreation Act of 1978 (P.L. 95–625). Additional federal laws may offer other specific protections.

As exploration continues, Lechuguilla Cave may extend beyond the rigid boundaries of the park and wilderness area. Entities with mining operations on adjacent lands are required to assure strict protection of the cave if any passage is found outside the present protection zone.

National Cave and Karst Research Institute Act of 1998

A law that could have significant impact is the National Cave and Karst Research Institute Act of 1998 (P.L. 105–325). The Act mandated that the National Park Service establish the Institute, stipulated that the Institute will be located in the vicinity of Carlsbad Caverns National Park in New Mexico (but not inside park boundaries), and that the Institute cannot spend federal funds without a match of nonfederal funds.

The main purposes of the Institute are to facilitate speleological research, to enhance public education, and to promote environmentally sound cave and karst management. The Institute is authorized to carry out its objectives internationally as well as nationally. The concept has been approved, initial organization has been implemented, and the Institute has entered the initial stages of permanent staffing.

The main purposes of the National Cave and Karst Research Institute are to facilitate speleological research, to enhance public education, and to promote environmentally sound cave and karst management.

National Parks Omnibus Act

Another significant law passed in 1998 is the National Parks Omnibus Act (P.L. 105–391). This Act gives park managers another tool to protect resources in the parks and allows the withholding of information on the location and nature of resources in national parks (DL Pate, personal communication on February 14, 2000). Section 207 of the Act, dealing with confidentiality of information, provides managers a way to protect resources from freedom of information requests. The law states that resources which are:

> *...endangered, threatened, rare, or commercially valuable, of mineral or paleontological objects within units of the National Park System, or of objects of cultural patrimony within units of the National Park System, may be withheld from the public in response to a request under section 552 of Title 5, United States Code, [unless an exemption is granted by the Secretary of Interior].*

Other Federal Cave Laws

A list follows, though probably not exhaustive, of other federal laws that could possibly have a role in cave and karst protection (Huppert 1995).

- Eastern Wilderness Act
- Endangered American Wilderness Act
- Forest and Rangeland Renewable Resources Planning Act
- Historic Sites Act
- Multiple Use Sustained Yield Act
- National Forest Management Act
- National Historic Preservation Acts (two, enacted in 1966 and 1976)
- National Parks and Recreation Act
- National Wildlife Refuge System Administration Act
- Native American Graves Protection and Repatriation Act
- Reservoir Salvage Act

LaMoreaux and others include most of the following federal laws in a list they prepared in 1997.

- Clean Air Act
- Clean Water Act
- Coastal Zone Management Act
- Comprehensive Environmental Response, Compensation, and Liability Act (CERCLA)
- Disaster Relief Act
- Energy Policy Act
- Federal Emergency Management Act (FEMA)
- Federal Food, Drug, and Cosmetic Act (FFDCA)

Complete sets of federal and state laws can be found in any major law school library. Technology has provided another route for accessing state and federal laws through the World Wide Web.

- Federal Insecticide, Fungicide, and Rodenticide Act
- Forest Land Management Planning Act
- Global Climate Change Protection Act
- Global Change Research Act
- Magnuson Fisheries Act
- Marine Plastics Pollution Research and Control Act
- Marine Protection, Research, and Sanctuaries Act
- National Earthquake Hazards Reduction Act
- National Parks Resource Protection Act
- Ocean Dumping Ban Act
- Oil Pollution Act
- Renewable Natural Resources Planning Act
- Resources Conservation Recovery Act (RCRA)
- Safe Drinking Water Act (SDWA)
- Shore Protection Act
- Soil and Water Conservation Act
- Toxic Substances Control Act
- Weather Service Modernization Act

The application of some of the above laws to caves and karst may seem a bit of a stretch to some readers. However, LaMoreaux and others (1997) point out that all of the above laws can in some way be used to protect groundwater and thus may involve karst.

Local or Regional Ordinances and Regulations

Among the countless city and county ordinances in the United States, local governments often have regulations in force that protect surface and ground waters or rare biota. They also include codes or statutes that establish parks, prehistoric or historic sites, or other unique phenomena. It is well beyond the scope of this book to present even a small collection of such samples. Conservationists should become familiar with such local regulations, especially where they may be used to protect caves and karst.

Two unique situations should be noted. The Edwards Aquifer in Texas and the Floridan Aquifer in Florida are two of the most heavily used and studied karst aquifers in the country. State laws and regional zoning protect both of these important aquifers that provide domestic and irrigation water for millions of people. Refer to LaMoreaux and others (1997) for a detailed description of the legal circumstances related to these two aquifers and several other unique situations.

On the local level, zoning ordinances and construction codes should be checked to see if they might be useful in protecting karst. An example of a very complete local law is the 2001 revision of the San Marcos (Texas) Edwards Aquifer Protection Regulations. It requires detailed surveys of caves, sinkholes, and other karst features; provides for protection of particularly sensitive features; requires buffer zones around those features and streams; and establishes limits on impervious cover in the surrounding area (City of San Marcos 2001). This ordinance serves to minimize pollution reaching the karst groundwater system and could stand as a model for other karst areas. There are undoubtedly similar but lesser known regulations in other parts of the country.

How to Find the Laws

Complete sets of federal and state laws can be found in any major law school library. Helpful individuals at most law libraries will guide your search or answer your questions. However, there are a limited number of law collections, and many cavers dealing with the legal ramifications of

NSS Cave Vandalism Deterrence Reward Commission

Thomas Lera

The National Speleological Society (NSS) offers a reward for information that leads to the legal conviction of cave vandals anywhere in the United States. (See commission sidebar page 110.)

The NSS will consider a reward of $250 to $1,000 for information leading to a local, state, or federal conviction of any person (or persons) for acts of cave vandalism.

- Breaking, breaking off, cracking, carving upon, writing on, burning, or otherwise marking upon, removing or in any manner destroying, disturbing, defacing, marring, or harming the surfaces of any cave or any natural material which may be found therein, whether attached or broken, including speleothems, speleogens, and sedimentary deposits.
- Breaking, forcing, tampering with, or otherwise disturbing a lock, gate, door, or other obstruction designed to control or prevent access to any cave.
- Dumping, littering, disposing of, or otherwise placing any refuse, garbage, dead animals, sewage, or toxic substances harmful to cave life or humans in any cave or

(Continued on next page)

cave conservation do not live near one. Technology has provided another route for accessing state and federal laws through the World Wide Web (list of suggested sites below). However, local laws are often inaccessible through this vehicle.

Law Libraries

Probably the easiest way to find a particular act in a law library is by its popular name. For example, with the title Federal Cave Resources Protection Act of 1988 (FCRPA) in hand, you can go to *Shepard's Acts and Cases by Popular Name* for the complete citation.

The latest edition is continuously updated with bimonthly cumulative supplements to keep it current within 60–90 days. *Shepard's Acts* is a 3-volume set that lists state and federal laws alphabetically by their common name and provides their complete legal citation. For example, the FCRPA is P.L. (public law) 100–691. This means the 100th U.S. Congress passed it and 691 is the chronological number assigned to the final bill, representing the order in which the president signed it.

The complete reference for FCRPA is as follows:

- U.S.C. (United States Code)
- 1988 (year of code edition at time of passage)
- Title 16 (this is conservation law)
- 4301 (first section of the law)
- *et seq.* (Latin for "and the following" sections)

If listed as "USCA," ("A" stands for annotated), the act is published by the West Group. If the listing is cited as "USCS," ("S" stands for Service), it is published by LEXIS Law Publishing. The official federal government citation is "U.S.C." However, the two annotated versions are used more frequently in legal research. They both supply more informative footnotes, case references, notes on the law's history, and cross-references. State laws have somewhat similar citations.

Usually state cave protection laws are in the volumes titled as follows:

- Conservation
- Environment
- Natural Resources
- A related or similar reference

Internet and World Wide Web

Fortunately, the World Wide Web has made the task of finding laws considerably easier. Some useful Web sites and their contents are listed here.

- National Speleological Society
  On the National Speleological Society homepage, click on "Conservation" or use the site search engine to find cave laws. Federal and state cave protection laws are listed.
  <http://www.caves.org/>

- Find Law
  This site directs you to a complete listing of federal and state legislation.
  <http://www.FindLaw.com/>

- Many city and some state ordinances are found on this site.
  <http://www.spl.org/govpubs/municode.html>

- Library of Congress Legislative Search Engine
  Full text of proposed bills, legislation and the Congressional Record are found here.
  <http://thomas.loc.gov/>

sinkhole.
(Continued)

- Removing, killing, harming or otherwise disturbing any naturally occurring organisms within any cave.

- Excavating, removing, destroying, injuring, defacing, or in any manner disturbing any burial grounds, historic or prehistoric resources, archaeological or paleontological site or any part thereof, including relics, inscriptions, saltpeter workings, fossils, bones, remains of historical human activity, or any other such features which may be found in any cave.

Application for this reward must be made within three months of such conviction, To notify or apply, contact:

NSS Cave Vandalism Deterrence Reward Commission
2813 Cave Avenue
Huntsville AL 35810

For more information, see the NSS Web site: <http://www.caves.org/> and click on Cave Conservation.

During the past 100 years, many states have enacted laws to protect caves.

- Government Printing Office
  On this site from the Government Printing Office, find Regulations by keyword in the *Federal Register and Code of Federal Regulations*. The site also allows individuals to search for government documents by subject and by depository libraries holding them.
  <http://www.access.gpo.gov/su_docs/>

- Legal Online
  This privately produced site provides the laws of all fifty states. Some of the states also include bills presently being considered in their legislatures.
  <http://www.LegalOnline.com/statute2.htm>

Other useful law school Web sites are:

- **<http://www.WashLaw.edu/>**
- **<http://www.Law.Cornell.edu/>**

Individual states also maintain online access to their government legal services including a listing of their laws. Gaining proficiency in using various search engines is very useful in extending the breadth of a Web search. Often, one search engine may contain useful information not found in another.

DogPile is a particularly helpful engine, which actually searches multiple other engines:

- **<http://www.DogPile.com/>**

The Web addresses listed above were accurate at the time of this book's publication. However, addresses may change over time.

A wide variety of laws at all levels of government may be useful in the protection of caves and karst.

Conclusion

A wide variety of laws at all levels of government may be useful in the protection of caves and karst. It is the job of concerned individuals to become educated about pertinent legislation, codes, and regulations that apply to their particular situations.

Legislative education is an ongoing process because the legal system, at all levels, is constantly changing. If conservationists are not vigilant, there is always the chance an important law for cave protection will be negatively modified, rendered ineffective, or even repealed. Environmental conservation problems never seem to go away. The forces of ignorance or aggressive land development are often ready to compromise or destroy spelean assets.

Acknowledgements
The Editors gratefully acknowledge Betty J. Wheeler and Thomas Lera for reviewing and updating this chapter after Dr. Huppert's untimely passing.

Cited References
Anonymous, 1995. An evaluation of state cave laws. *American Caves* 8(1):18-19.

Bishop SJ, Huppert GN. 1990. Taking wilderness underground. In: Lime DW, editor. *Managing America's Enduring Wilderness Resource: Proceedings of the Conference, Minneapolis, Minnesota, September 11-17, 1989*. St. Paul (MN): Tourism Center, Minnesota Extension Service and Minnesota Agricultural Experiment Station, University of Minnesota. p 367-370.

Bradshaw E. 1982. Cave laws of the United States. In: Wilson RC, Lewis JJ, editors. National Cave Management Symposia: *Proceedings [for 1978 and 1980]*. Oregon City: Pygmy Dwarf Press. p 214-227. (Presented at the 1980 symposium.)

City of San Marcos. 2001. *Edwards Aquifer Protection Regulations San Marcos City Code Chapter 94, Article 6*. <http://www.ci.san-marcos.tx.us>. (Access Directory, find Edwards Aquifer Protection Regulations.) Draft accessed 2003 Jan 1. [The final approved version is available at <http://www.caves.org/section/ccms/bat2k/index.htm>. Accessed 2003 Jan 31.]

Hose LD, Pisarowicz JA. 1999. Kartchner Caverns: Research Symposium. *Journal of Cave and Karst Studies* 61(2):38-126. [This is a special issue with 13 articles devoted to scientific studies of Kartchner Caverns.]

Huppert GN. 1993. Underground wilderness: The time is right. In: Foster DL, Foster DG, Snow MN, Snow RK, editors. *1991 National Cave Management Symposium Proceedings: Bowling Green, Kentucky, October 23-26, 1991*. Horse Cave (KY): American Cave Conservation Association. p 263-266. [Also listed as 1991.]

Huppert GN. 1995. Legal protection for caves in the United States. *Environmental Geology* 26(2):121-123.

Huppert GN. 1999. State cave protection laws…where next? *NSS News* 57(4):103-104.

Huppert GN, Wheeler BJ. 1982. State legislation concerning the protection of caves. In: Wilson RC, Lewis JJ, editors. *National Cave Management Symposia: Proceedings [for 1978 and 1980]*. Oregon City: Pygmy Dwarf Press. p 45-47. (Presented at the 1978 symposium.)

Huppert GN, Wheeler BJ. 1986. Underground wilderness: Can the concept work? In: Lucas RC, editor. *Proceedings—National Wilderness Research Conference: Current Research, Fort Collins, CO, July 23-26, 1985*. Ogden (UT): USDA Forest Service, Intermountain Forest and Range Experiment Station. General Technical Report INT 212. p 516-522.

Huppert GN, Wheeler BJ. 1992. The case for underground wilderness. In: Krumpe E, editor. *Designation and Management of Park and Wilderness Reserves: Official Proceedings of the Symposium held during the 4th World Wilderness Congress 1987*. Moscow (ID): College of Forestry, Wildlife and Range Sciences. p 1-7.

LaMoreaux PE, Powell WJ, LeGrand HE. 1997. Environmental and legal aspects of karst areas. *Environmental Geology* 29(1-2):23-36.

Lera T. 2002. *A Summary of Legislation and Organizations Involved in the Preservation of Caves and Bats*. <http://www.caves.org/section/ccms/bat2k/index.htm>. Accessed 2003 Jan 4. [Editors' note: Many laws associated with cave protection are available on this site.]

Power L. 1974. *A Handbook on Cave Conservation Legislation*. Huntsville (AL): Conservation Committee of the National Speleological Society, unpaged.

Stitt RR. 1976. State cave protection laws and their enforcement. In: Aley T, Rhodes D, editors. *National Cave Management Symposium Proceedings: Albuquerque, New Mexico, October 6-10, 1975*. Albuquerque (NM): Speleobooks. p 91-97. [Cave Research Foundation also in listing.]

Additional Reading

Juberthie C. 1992. *Underground Habitats and Their Protection*. Council of Europe: Strasbourg. [Editors' note: details of existing cave protection legislation in Europe.]

Section B—Developing Cave Management Programs

Management Tools for Supporting Conservation Ethics

Dale L. Pate

Caves are unique, fragile environments that can provide educational, scientific, and recreational opportunities for casual visitors, cavers, and researchers. These activities, when not properly managed, can seriously impact and even destroy the very features that make these opportunities possible. In many caves, these features are nonrenewable. Once lost, they are gone forever. Active participation of management is essential in conveying conservation messages to all who enter the cave. Several important management tools may help people develop conservation ethics:

- Education and research
- Written guidelines
- Monitoring of cave use
- Access control

Active participation of management is essential in conveying conservation messages to all who enter the cave.

Education and Scientific Research

Conservation education should target three groups of people:

- Management and employees (including volunteers)
- Recreational cavers
- The general public

Ideally, education begins with management. Owners and managers need to establish conservation objectives based on thorough understanding of their cave resources—the fragility of underground resources as well as surface impacts that may affect the cave system. The manager must assure that everyone who enters the caves, for whatever reasons, is adequately informed about conservation ethics and protection guidelines. This goes for employees, cavers, and sightseers.

Speleological research is a significant component of many conservation-directed cave management decisions. Surface development activities can affect caves in numerous ways. Many show caves have been impacted because owners or managers did not understand the relationship between surface features and the cave below.

The study of infiltration routes into Carlsbad Caverns is a prime example of how scientific results can influence management decisions. Before planning any construction, management should know the nature and location of the cave, its surrounding rock, and how water and other substances may infiltrate the cave. Some caves are located in limestone terrains known as karst, while others are found in lava flows, and yet others in a variety of different rock types. Regardless of the substrata, should buildings, sewer lines, gas tanks, and other manmade structures be placed near or over a cave? Or should they be located some distance away? Existing sewer lines, parking lots, and other structures may be negatively impacting the cave below. Scientific studies can aid management in deciding the appropriate placement for manmade structures.

Knowledge about the nature of the cave is essential for making appropriate conservation decisions. Whether the resource is being developed as a show cave or whether it is a wild cave that is open to recreational visits, some features may be extremely vulnerable or fragile. Every cave is different, and certain areas may need special attention. For example—cave floors may be covered with unobtrusive, fragile speleothems that need special protection; or the presence of mineral sites may limit activities; or protection of biota may require specific precautions.

Whether the resource is being developed as a show cave or whether it is a wild cave that is open to recreational visits, some features may be extremely vulnerable or fragile.

In many caves, floors have been trampled because paths were not clearly delineated during the original exploration. Subsequent visitors crushed features underfoot—but, not maliciously—they simply did not know that their actions were impacting the cave. Many minerals and other unique floor features have been lost this way.

Minerals make up much of the visual beauty of caves and are found on the floors, walls, and ceilings. In fragile areas, even the way a person moves through the passage can mean the difference between intact speleothems and broken ones. A careful, informed person can move through delicate areas with little or no new impact.

Cave life is often small and hard to see in caves. Invertebrates are easily crushed underfoot. In the low-nutrient environments of many caves, biological populations are small and vulnerable to environmental changes. Education about cave biology is essential for increasing the protection of cave habitats. In recent years, scientific studies have shown that caves harbor unique and potentially beneficial microbes. Microbial studies have shown that there are amazing and diverse ecosystems in caves, especially in isolated pools. Waste leaking from sewer lines and buildings on the surface may infiltrate the cave and destroy cave-dwelling populations. The management policies for caves are changing to reflect and support new knowledge about cave ecosystems.

Establishing Guidelines

Trust is a key element in cave conservation. Written guidelines can help develop trust between management and stakeholders. Formatted in clear

NPS Photo

Figure 1. This aerial view shows the National Park Service infrastructure built directly over Carlsbad Cavern. The cave entrance is in the lower central part of the photo.

and easy-to-understand language, guidelines can spell out exactly what is expected. The document should be reviewed with each person who enters the cave. Guidelines also provide tangible standards for evaluating the actions of those who enter the cave and for eliminating confusion about expectations.

Lechuguilla Cave, located in Carlsbad Caverns National Park, is an excellent example of how guidelines can be essential to the protection of the cave. In 1992, management began developing an 8-page document titled *Guidelines for Entering Lechuguilla Cave*, which later became an appendix of the park's *1995 Cave and Karst Management Plan.*

Figure 2. Beginning in 1992, the development of Guidelines for Entering Lechuguilla Cave has helped to cultivate a conservation ethic within the caving and research communities that work in Lechuguilla.

Guidelines for Entering Lechuguilla Cave

- The reason behind developing guidelines for entering Lechuguilla Cave is to allow limited access for scientific research, survey when in association with exploration, and NPS management related trips while impacting the cave as little as possible. Of primary importance are (1) impacts to the cave and (2) the safety of all who enter.

BEFORE ENTERING THE CAVE

- Everyone *must* sign a permit.

- Expedition leaders are ultimately responsible for the personnel on their expedition. Expedition leaders should do their best to recruit cavers who are willing to follow the guidelines that have been established. Before assigning anyone to the Far East [branch of Lechuguilla Cave], an expedition leader should be reasonably sure that individual is fully prepared for such a trip.

- Every team entering the cave will have one designated team leader. Team leaders have tremendous responsibilities. They are responsible for the safety of their team and for the actions of their team members. If a team member is acting in an unsafe manner or not being careful and actually causing more damage to the cave, then it is the team leader's responsibility to correct that person's actions. If problems persist, then the team leader must abort the trip and have the team leave the cave. A team leader should gear team activities to the least experienced member of the team. This pertains to speed of travel as well as climbing leads. A team should also stay together unless an emergency requires different actions.

- Everyone entering the cave is responsible for their actions while in the cave. They are also responsible for reporting to the team leader acts by other team members that are unsafe or damaging to the cave. The overall goal is to allow limited access to the cave while minimizing all impacts. It is everyone's responsibility to assure that Lechuguilla Cave remains as pristine as possible and that each team member is very safety aware while in the cave.

- Clothing, boots and caving gear should be clean before entering the cave to minimize the introduction of foreign bacteria, molds, and fungus into this unique ecosystem.

- BOOTS MUST HAVE NON-MARKING SOLES. If you are in doubt, scrape boot over white floor or limestone rocks. Marking boots will definitely leave marks.

- Using a frame pack in the cave can be a problem and a major nuisance. It is recommended that these NOT be used..

- Electric lights are a must. No carbide lights or open-flame lights will be allowed in Lechuguilla. One of the reasons for this is safety, because of the flammability of the massive sulfur deposits found in the cave.

1995 Cave and Karst Management Plan - Carlsbad Caverns National Park

The guideline document defines the conservation ethics for Lechuguilla Cave and allows exploration and scientific research to continue.

The guidelines were initiated as a result of communication problems between management staff and those working in the cave. Now, everyone who enters Lechuguilla must first attend an orientation that includes discussion of all the items listed in these guidelines.

These guidelines have proven to be an essential tool for cave management and have helped to cultivate a conservation ethic among the caving and research communities. The document includes the following items:

- Lists reasons for developing the guidelines
- Assigns responsibilities to expedition leaders, trip leaders, and individuals
- States what is expected from everyone before they enter the cave, when traveling through the cave, and while the teams are working in the cave
- Lists locations of campsites and resource protection zones
- Discusses climbing and the use of bolts
- Defines survey standards
- Describes expectations for exploration of new areas
- Explains other issues that are critical to the protection of the cave

Over the years, these guidelines have been refined and changed. Hard lessons have been learned by management as well as cavers and scientists. The ultimate goal of the park has been to allow limited access to Lechuguilla Cave while protecting the extremely fragile features. The guideline document defines the conservation ethics for Lechuguilla Cave and allows exploration and scientific research to continue. (See Appendix 5 for complete text of Guidelines for Entering Lechuguilla Cave, page 549.)

Observations and Monitoring

Written guidelines lay the foundation for good working relationships, and it is essential to follow up on that foundation by monitoring the actions of those entering the cave. In addition to resource monitoring projects, both short-term and long-term observations can aid in building conservation ethics.

Short–Term

By taking an active role, the manager can ensure that conservation ethics and the protection of cave features are high priorities for everyone who visits the cave. By establishing guidelines, management clearly defines expectations for behavior in the cave.

By visiting areas that teams or employees have been to, visual observations like damaged speleothems or unnecessary footprints can be recorded. Through first-hand observation of new impacts, the manager can tell if a conservation ethic has been established. Any deviations from the guidelines are noted and dealt with on a case-by-case basis.

Figure 3. Inappropriate footsteps found off the delineated trail can have lasting impacts.

NPS Photo Val Hildreth-Werker

Long–Term

It is valuable to retain records of impacts to cave features and the extent of impacts that occur over a long period of time. Permanent photomonitoring points and visitor impact maps are two methods for documentation.

By making repeat photographs of an area, visual records can be preserved and analyzed over a long period of time. (See photomonitoring, page 207.) Impact mapping may be more time-consuming and

NPS Photo Val Hildreth-Werker

Figure 4. When permanent photomonitoring points are established, cave management may monitor impact trends and speleothem changes over long periods of time. Installed by Jim Werker and Val Hildreth-Werker, this photomonitoring station at Leaning Tower of Lechuguilla, provides visual documentation to aid management decisions for Lechuguilla Cave, New Mexico. (See page 6 of color section.)

may show different levels of detail and change. (See impact mapping, page 193.) These valuable management tools can record impacts, show changes over time, and indicate how management decisions can affect conservation ethics and protection of cave resources.

Permanent photomonitoring points and visitor impact maps are two methods for documentation.

Controlling Access

Controlling access to caves is sometimes an important management objective for supporting conservation ethics and preserving cave features. Access control may involve the use of several types of management tools:

- Developing access policies
- Providing guided trips into certain areas
- Gating the entrances to caves or areas within caves
- Denying access to sensitive resources

Access Policies

Carlsbad Caverns National Park enforces access policies that allow visitors a wide range of cave experiences while protecting cave resources. The main paved trail through Carlsbad Cavern is open to the general public. Guided trips are offered to areas in Carlsbad Cavern, Spider Cave, Slaughter Canyon Cave, and Ogle Cave. Experienced cavers can apply for permits to visit eight other caves on the park. Other caves, including Lechuguilla Cave, are open only for approved research.

Designing a gate is not to be taken lightly—all ramifications and potential modifications should be thoroughly evaluated before proceeding.

Guided Trips

Guided cave trips encourage resource protection and are excellent educational opportunities for small groups. A guide is able to point out the fragility of the cave and discuss conservation measures. With a guide watching over the visitors on tour, resource protection and safety issues can be addressed.

NPS Photo Dale L. Pate

Figure 5. Stan Allison stands at the entrance gate to Slaughter Canyon Cave. Built a number of years ago, this gate is not considered bat-friendly. A new gate in the future will correct this problem.

Gates

Though intrusive in a number of ways, gating cave entrances or passages often enables management to control access. Designing a gate is not to be taken lightly—all ramifications and potential modifications should be thoroughly evaluated before proceeding. Impacts created by a gate should be weighed against impacts the cave will suffer if it does not have that gate. (See cave gates, page 147.)

Discretion and Secrecy

Controlling knowledge about certain caves or certain cave areas is a method sometimes used to protect caves. Obviously, if fewer people know about a cave, then fewer will visit. Fewer visits cause less impact. However, secrecy is not always the best answer. A manager must weigh the need for discretion and secrecy with other factors to best protect resources.

Deny Access

The cave manager also has the option of using another tool. Access can be denied. A cave or an area can be closed to a single individual, to certain groups, or to all visitors. It may become necessary to deny access to stop unwarranted impacts created by individuals or groups. When there are no other options for ensuring ethical behavior, denying access will at least halt the damage. Denying access will get the message across when nothing else works.

Summary

To protect cave systems and preserve their fragile resources, all users need to adopt conservation ethics. Since every action can have adverse effects on a cave, it is essential that the owner or manager take an active role in developing and encouraging conservation-directed behaviors.

It is important for owners and managers to learn and foresee how impacts will result from their management decisions. For example, management should evaluate the range of potential impacts before allowing cave access to work crews, restoration groups, research teams, or recreational visitors.

Education and research programs, guidelines, observation, and access control are excellent management tools for establishing cave conservation ethics. An effective cave management program using the right conservation tools may mean the difference between a cave remaining relatively pristine and a cave becoming trashed.

To protect cave systems and preserve their fragile resources, a conservation ethic needs to be developed among all users.

Additional Reading

Alexander EC. 1991. The evolving relationship between Mammoth Cave National Park and its hydrogeologic symbionts. In: Foster DL, editor. *1991 National Cave Management Symposium Proceedings: Bowling Green, Kentucky, October 23-26, 1991.* Horse Cave (KY): American Cave Conservation Association. p 11-56.

Baichtal JF, Swanston DN, Archie AF. 1995. An ecologically-based approach to karst and cave resource management. In: Rea GT, editor. *Proceedings of the 1995 National Cave Management Symposium: Spring Mill State Park, Mitchell, Indiana, October 25-28, 1995.* Indianapolis (IN): Indiana Karst Conservancy. p 10-27.

Bodenhamer H. 1995. Monitoring human-caused changes with visitor impact mapping. In: Rea GT, editor. *Proceedings of the 1995 National Cave Management Symposium: Spring Mill State Park, Mitchell, Indiana, October 25-28, 1995.* Indianapolis (IN): Indiana Karst Conservancy. p 28-37.

Buecher RH. 1995. Footprints, routes, and trails. In: Rea GT, editor. *Proceedings of the 1995 National Cave Management Symposium: Spring Mill State Park, Mitchell, Indiana, October 25-28, 1995.* Indianapolis (IN): Indiana Karst Conservancy. p 47-50.

Despain J. 1993. The creation and implementation of restrictive cave management policies: Three examples from Sequoia and Kings Canyon National Parks. In: Pate DL, editor. *Proceedings of the 1993 National Cave Management Symposium: Carlsbad, New Mexico, October 27-30, 1993.* [Huntsville (AL)]: National Cave Management Symposium Steering Committee. p 175-183.

Goodbar JR. 1995. Cave permits and permitting systems. In: Rea GT, editor. *Proceedings of the 1995 National Cave Management Symposium: Spring Mill State Park, Mitchell, Indiana, October 25-28, 1995.* Indianapolis (IN): Indiana Karst Conservancy. p 99-107.

Gookin J. 1991. Secrecy and discretion as cave management tools. *NSS News* 55(4):114-115.

Gookin J. (1995). How cavers learn ethics. In: Rea GT, editor. *Proceedings of the 1995 National Cave Management Symposium: Spring Mill*

State Park, Mitchell, Indiana, October 25-28, 1995. Indianapolis (IN): Indiana Karst Conservancy. p 108-110.

Gurnee R. (1993). Cave management is control of access. In: Pate DL, editor. *Proceedings of the 1993 National Cave Management Symposium: Carlsbad, New Mexico, October 27-30, 1993.* [Huntsville (AL)]: National Cave Management Symposium Steering Committee. p 218-221.

Pflugh DC. 1993. Resource-responsible professional caver education programs on public lands. In: Pate DL, editor. *Proceedings of the 1993 National Cave Management Symposium: Carlsbad, New Mexico, October 27-30, 1993.* [Huntsville (AL)]: National Cave Management Symposium Steering Committee. p 96-102.

NPS Photo Val Hildreth-Werker

Figure 6. The Visitor Center at Carlsbad Caverns National Park is located above Bat Cave Draw. The black path from the Visitor Center leads to the cave's historic entrance. On the horizon are the Guadalupe Mountains.

Section B—Developing Cave Management Programs

Are We Managing Caves or Conflicts?

Jerry L. Trout

Many individuals with deep interests in cave resources have pursued their passions by becoming involved in cave management, either as a career or as an avocation—and have done so as a result of the desire to preserve and protect underground resources. But are we really managing caves? Caves don't need us—they manage themselves without interference from people.

When no one enters or impacts a cave, it needs no human assistance. However, human entry may cause obvious negative impacts and surface activities may inadvertently affect cave systems. The human actions that impact caves, karst terrains, and ground water quality need to be managed.

Cave managers are often immersed in managing people with tenuous, opposing, and complex social values. Rather than managing resources, we seem to be more in the business of managing human relationships, disputes, and conflicts.

Rather than managing resources, we seem to be more in the business of managing human relationships, disputes, and conflicts.

In outdoor recreation management, conflict is typically defined as "goal interference attributed to another's behavior" (Jacob and Schreyer 1980). The goal interference model refers almost entirely to conflicts of interest—however, conflicts also arise from misunderstandings and value-based opinions. For example, distrust results from disputes like, "I understood the area was off-limits to recreational caving" or sentiments like, "cave photographers compromise conservation ethics to get the perfect picture."

Sources of Conflict

Burton (1990) believes that only value-based disagreements truly qualify as conflict, while Amy (1987) describes three general sources of conflict:

- Misunderstandings—conflicts arise when groups are differently informed about an issue.
- Interests—people want to use the same resource for different activities.
- Values—disputants differ in deeply rooted beliefs about what is right.

Conflicts tend to flare between recreation users and other natural resource stakeholders. Clashes between recreation interests and advocates of commodity land uses are common. For example, mining companies and mining-dependent communities may oppose plans to designate an area as wilderness or a cave as significant because these designations typically carry restrictions on mining activities.

A variant on this theme involves conflict between recreation interests and environmental organizations. For example, outfitter-guides that advocate publication of cave locations are opposed by government agencies as well as NSS-affiliated organizations that believe such listings threaten the protection of cave resources.

There are also conflicts between resource managers and their constituents. Managers usually view themselves as stewards of the land or mediators

Conflicts are often more effectively addressed if the focus is on "managing" rather than "resolving."

between competing interests—however, stakeholders left with unfavorable solutions to shared-use problems tend to view managers as culpable individuals who hold competing interests. Conflicts can also occur within agencies as recreation managers disagree with colleagues who are responsible for commodities or environmental protection (Brunson 1998).

Conflicts are often more effectively addressed if the focus is on "managing" rather than "resolving." Perceived restrictions associated with increased cave usage tend to cause simmering disagreements that intensify into contentious conflict. That's when management of cave resources shifts to management of human relationships. Recreation use conflicts sometimes cannot be entirely resolved since they are rooted in fundamental differences over how one should experience the natural environment.

Trout (2003) further discusses management of cave use conflicts. Conflicting interests typically attempt to tilt the balance of a dispute in their favor. Lincoln (1990) describes a continuum of strategies that intensify with the degree of conflict: inaction, negotiation, facilitation, mediation, arbitration, administrative appeal, judicial appeal, legislative appeal, nonviolent civil disobedience, or violence. Public agencies often succeed in managing conflict by focusing on the objective of balancing the needs and interests of multiple constituencies. Agencies increasingly promote collaborative decision-making processes that allow the conflicting interest groups to join in crafting solutions that will minimize clashes.

Education and information campaigns should be among the first strategies for managing conflict. For example, outreach programs to cave user groups teach proper etiquette, such as confining travel to a single path through a cave. Informational approaches are often coupled with improved enforcement of existing rules (Trout 2003). Strategies for managing resources by managing relationships are described in various chapters of this manual.

Cited References

Amy D. 1987. *The Politics of Environmental Mediation*. New York: Columbia University Press.

Brunson MW. 1998. Conflict among user groups: An overview of major issues and opportunities. In: *Forest Service Recreation Short Course*. Utah State University.

Burton J. 1990. *Conflict: Resolution and Prevention*. New York: St. Martin's Press.

Jacob GR, Schreyer R. 1980. Conflict in outdoor recreation: a theoretical perspective. *Journal of Leisure Research* 12:368-380.

Lincoln WF. 1990. *Dispute Resolution—A Handbook for Land Use Planners and Resource Managers*. Eugene (OR): University of Oregon, Bureau of Governmental Research and Service.

Trout JL. 2003. Conflict among user groups. In: Rea GT, editor. *Proceedings of the 15th National Cave Management Symposium: Tucson, Arizona, October 16-19, 2001*. Coronado National Forest: USDA-Forest Service. p 44-49.

Section B—Developing Cave Management Programs

Model Ethics System For Resolving Cave Conservation Dilemmas

John M. Wilson

The exploration of ethical theory in this chapter provides guidelines and methods to assist cavers in determining whether actions are ethical. An ethical act is one that should be performed, and this discussion seeks to explain how to make ethical decisions in certain caving and cave related situations. Ethical acts are also compatible with the stated highest value and its component values.

Model, Methods, and Guidelines

Discovering what is ethical can be arduous and risky. Effective use of these six guidelines resolves most ethical dilemmas:

- Use appropriate methods.
- Clarify primary values.
- Obtain relevant knowledge.
- Create effective solutions.
- Avoid factual and logical error.
- Facilitate compliance.

Methodology is this chapter's primary focus, but five other guidelines are also described.

Methodology is this chapter's primary focus, but the other five guidelines are also described and further study is recommended.

Classic ethical methods are described and unified in the Appropriate Methods Ethics System. This system is a model for cavers who face tough conservation decisions. As the difficulty of the dilemma increases, methods that require more time, research, calculation, and evaluation are applied. These four methods (rules, contracts, objective methods, and combined objective and intuitive methods) are based on principles from several major schools of Western ethical philosophy. Although proponents of each school of ethical philosophy may apply their model to all situations, most models are easy to use within only certain situations.

The Appropriate Methods Ethics System helps cave restorers and conservationists who need to resolve dilemmas, but who lack the time or inclination to review ethics philosophy. This system offers methods, not a list of what is right and wrong. These methods can be used to help cavers analyze ethical conflicts and identify those in which inappropriate methodology is a factor in an unresolved problem. A thorough discussion of compliance is beyond the scope of this chapter; however, the degree of compliance with ethical proposals can be improved by using appropriate methodology to achieve ethical results.

Appropriate Methods Ethics System

The Appropriate Methods Ethics System uses a sequence of four methods, or levels, to evaluate conservation proposals and determine whether an act

The Appropriate Methods Ethics System uses a sequence of four methods, or levels, to evaluate conservation proposals and determine whether an act is ethical

is ethical. Each system level should be used until it leads to an acceptable solution or until it fails; however, any of the methods, except the first, can be used to resolve any ethical problem. Level one uses only established rules, laws, and procedures; therefore, it is not applicable when a new solution is required to resolve a problem. The levels are proposed in a sequence that suggests using the least expensive and easiest first. If a method is inadequate, one uses the next level. Using the appropriate method usually offers better solutions, and it saves time and effort.
When there is no proposal, one must be developed. The proposal could be to do nothing. Developing creative proposals to solve ethical dilemmas is often an essential component of the conflict resolution process.

The First Level — Using What Is Known

First-level or normal decisions require actions that are compatible with appropriate practices and customary relevant laws, rules, and regulations. One reviews a proposed act for compliance with customary practices and applicable laws. The first level goal is to find an appropriate solution from the present knowledge base. This contrasts with the goal of the other three levels, which is to create new solutions.

We make most of our decisions using the first-level method, because doing what we know tends to be intuitive, easy, and quick. Rather than inventing the correct solution for every new situation, people do what is legal and expected based on law and custom. First-level methods are the standard operating procedure. First-level methods may be in the form of a written law, an informal custom, or a combination. The advantages of using the first-level methods include reasonable evaluation cost and less thought. In this level, people apply their knowledge to predictable situations. Successful results developed using higher levels are added to the first level, and they become part of the public repertoire of standard ethical behaviors

Case Study 1 — An Easy Example

Submitted by the author

Several cavers discover acts of vandalism in one of their favorite caves. Fifteen stalagmites and stalactites are broken and left on the ground. The group decides to repair the damage.

Analysis. The group will use the most appropriate means to make the repairs, as explained in this book. The cave owner understands what is involved and wants the restoration performed. No laws prohibit any part of the planned restoration. The expense is mostly in labor willingly volunteered, but other expenses are being paid by a friend for whom the amount is nominal.

The decision to restore the cave in this situation is deceptively easy because there are few conflicting interests and no ethical conflicts. There may be better alternatives for using the resources that will be committed to this cave restoration; however, because the emotional impetus to act is not likely to be transferable, these alternatives probably would not benefit from diversion to another project.

Rules and Transition Problems

Harm or an injustice may be caused when laws and procedures are applied inappropriately. Some of the worst ethical behavior has occurred in the name of law, tradition, and authority. The expressions "I was doing what I was told," "I was doing things the way they have always been done," or "We have rules and cannot make an exception" typify rule-obedience inflexibility. The transition from standard to other levels of ethical reasoning can be difficult.

This first level of ethical procedures is based on rules, not the consequence of an act. Most people recognize that if bad things are happening, and if the present rules are not helping, something more should be done. Sometimes people are unable to react because of their belief or faith in the law. There are three reasons people comply with laws and rules regardless of the consequences:

- Some people do not care about the outcome or are unconcerned with ethical behavior.
- Some people believe in the absolute authority or absolute correctness of a rule or law in general. Ethicists refer to this school of thought as *deontological*, from the Greek word *deon*, which means, "that which is binding." The concept of duty or obligation is crucial for deontologists. For them, murder, lying, and vandalizing a cave are always wrong, and indirect consequences of an act are not usually important. For example, mitigating circumstances such as lying to a cave vandal about the location of a cave is not acceptable.
- Some people may have inaccurate information or make a logical error.

This first level of ethical procedures is based on rules, not the consequence of an act. Most people recognize that if bad things are happening, and if the present rules are not helping, something more should be done.

Case Study 2 — A Lost Cave

Submitted by the author

A cave photographer has documented his love of caves with many outstanding photographs. One cave in Virginia had a section of rare, delicate mineral formations that he had captured on film with award-winning photographs. Many years ago, this cave was beginning to suffer vandalism and the local cavers wanted to place a gate near the entrance to keep out the vandals.

The photographer vigorously opposed the gating of his favorite cave. His argument was simple: cave gates are ugly and unnatural, and they destroy the aesthetic value of the cave, and people should not change caves. The photographer was unable to prevent the gate, but delayed its installation, and vandals destroyed the mineral formations before the gate was installed.

Analysis. When cave vandalism occurred in the caves he loved, the photographer was most upset. The photographer is unwilling to use another method to resolve the problem, even when it is clear that his ways are failing.

He is comfortable using only current laws and beliefs. When asked how he would stop the vandalism, his suggestion was to use long jail terms for cave vandals. He was deeply influenced by the psychological impact of cave gates. He is a "rules" type person. He believed that his obligation is to do no wrong. If each person followed the law, a gate would not be necessary. His arguments are internally consistent. He believes that right is absolute and right conduct does not depend on the circumstances or consequences, only the intent to do right matters.

This belief is referred to as *ethical absolutism*, and it is part of the Natural Rights school of philosophy. This philosophy says that as long as he acted rightly, he is not responsible for failed results. Critics of Natural Rights would contend that he is also committing the *natural fallacy* error by using the natural state of the cave to assert how it should be protected.

Changing Methods

The Appropriate Methods Ethics System applies the best ethics methods to a situation. It suggests using a higher-level method when the present methods produce unsatisfactory results, such as when two acts or their consequences conflict with each other or with higher values, cause unfairness, injustice, or unnecessary harm to someone or something. For example, property right laws offer a legal justification for cave owners to sell the mineral formations from their cave although this conflicts with natural resource conservation values.

The Second Level — Contracts

At the second level, it may be necessary to relinquish a law, rule, or agreement and replace it with a new contract. This level requires people to develop contracts, usually through negotiation. When used appropriately, negotiating compares valid competing interests fairly, and it can be quick and effective. Many people are familiar with using contracts, and they can reach a reasonable agreement through negotiation with minimal time and effort.

One way to evaluate the fairness of a contract is to imagine that you will have to live with the consequences of the solution, but that you do not know which party you will be after the solution is adopted. That is, have each participant in the conflict pretend that they occupy the role of the other party and must accept the solution.

At the second level, it may be necessary to relinquish a law, rule, or agreement and replace it with a new contract. This level requires people to develop contracts, usually through negotiation.

In second-level negotiations, justice is often the major goal of the contract. When justice is the primary goal, the goal should be made explicit, but the parties may agree to other primary values. If no primary value is stated, self-interest is the implied primary value when people are negotiating.

Level two is a simple version of overlapping consensus, which was developed by John Rawls (1971). In its most advanced form, competing interests are compared objectively, with each advocate unaware of which course of action would eventually be beneficial to him. The crucial component of the system of overlapping consensus of basic rights is the Veil of Secrecy. Philosophers call variations of this negotiating approach *contractarian*. Most common decisions are resolved with level-one procedures and rules. When level-one methods are inadequate, one may expect to resolve most remaining problems at level two. In most cases, it is simpler to achieve a contract than to meet the fact-finding and calculation standards required in level three, the next level, so most people will attempt to exhaust contract techniques before using level three.

Case Study 3 — The Terminator
Submitted by Diana E. Northup

This kind of conflict has arisen in the exploration of New Mexico's Lechuguilla Cave. The explorers find a new area with big exploration potential *and* big scientific potential. Because the cave's microbial community is potentially easily affected by human visitation, scientists ask that an area be designated off limits until it can be studied. Such study can last from one to five years, a duration that may be perceived as an eternity to an explorer with a hot lead.

The scientists are called "terminators," because they terminate exploration in certain areas of the cave. If scientists do not work with the explorers, then explorers may not reveal wonderful new areas for science. The situation is like walking a tightrope for both parties.

Analysis. Lechuguilla cavers are developing a set of rules for resolving the sometimes conflicting interests of science and exploration. They appear to have successfully developed contracts to balance the gains. Knowledge and pleasure are gained from exploration while scientific limitations delay some gratification to both the explorers and the scientists.

Case Study 4 — What Do You Get from Discovering a Cave?
Submitted by Douglas M. Medville

Caver group A discovers a virgin cave that contains attractive speleothems. The group explores the cave, practicing leave-no-trace ethics. But not being surveyors, they do not map the cave. A few weeks later, caver group B independently finds the cave. Thinking it is a virgin cave, group B surveys it, leaving what they feel to be unobtrusive survey stations: a carbide dot here, a metal tag at a station junction there.

A few days later, caver group A returns to the cave and sees the evidence of group B's survey. Group A feels that whoever did the survey has diminished the attractiveness of the cave. Group A does not want group B to re-enter the cave. Group A seals the entrance, preventing group B from continuing their survey.

Through a third party, groups A and B learn of each other's existence and communicate. Group B wishes to have the cave's entrance unsealed so they can continue their survey and promises to remove all traces of their first survey if it offends group A.

Group A is unwilling to unblock the entrance, citing a desire to protect the cave from further visitation, both by group B and other cavers whom they feel would diminish the cave's attractiveness. In addition, they express a concern that group B's photographs and map, if circulated in the caving community, will increase the visitation rate. The cave is on public land and there are no private landowner considerations.

Neither group owns the cave. Each group found the cave independently. Both groups feel that they have a stake in the conservation of the cave. Group A wants to see the cave protected, even to the point of making it impossible (or very difficult) for others to enter the cave. Group B believes that another group should not assume access control of a cave that they do not own, especially since group B found the cave on their own and are engaged in a survey. Finally, group A feels that they have explored the entire cave, while group B, through their survey, knows of passages that group A did not see and plans to survey these passages.

Analysis. What should be done? Who is correct? This type of problem is ideal for level-two methods. Part of the difficulty is that each of these groups believes it has some right to or ownership of the cave.

Many cavers believe that certain ownership rights such as exploration or access control belong to the discoverer. Property owners may not agree. Group A is claiming rights to the cave that are beyond those usually given to cave discoverers in the United States.

A facilitator should be able to help these two groups understand that the best interest is served by developing a mutually beneficial proposal. Such solutions might involve each group in the future management of the cave. All will benefit from an approach that offers the legal owner (the government) something in the future, such as their speleological expertise. The cave owner may benefit from these services enough to enter into a contract and grant certain privileges to either or both of these groups. Like the discovery of gold in a new place, it is difficult to control the exploration of a cave.

Many cavers believe that certain ownership rights such as exploration or access control, belong to the discoverer. Property owners may not agree.

Case Study 5 — Mud Filled Caves
Submitted by the author

In Alaska, a karst island with extensive old-growth forest offers an excellent source of timber, which is being harvested using clear-cut techniques. The island receives more than 250 centimeters (about 100 inches) of rain a year. Once an area is logged, erosion removes much of the topsoil and fills caves with mud and other debris. The lumber company agrees to leave a 0.5-hectare (1.2-acre) uncut buffer around each cave entrance.

Analysis. This plan was not successful because a 0.5-hectare (1.2-acre) stand of trees was too small to withstand high winds, and most of the trees in these buffer areas were blown down. Furthermore, the island forests do not seem to recover from clear-cut lumbering. The combination of poor soil, erosion, and wind has prevented the regeneration of the forest. Logging on this island provides a one-time economic gain.

This agreement was reached through negotiation, which supposedly considered the interests of all parties. Employment was stated as an important factor when making the case for continued logging. With this type of negotiation, there are few clear and immediate losers. The primary loss is of the common forest environment. It is difficult to convince people with direct economic interests that exploiting commonly held resources will cause indirect harm and economic loss in the long run. This type of conflict is better resolved using the consequential methods.

When no one articulates an interest, level-two methods are less successful. By focusing on articulated interests, level-two methods may not find other wrongs until subsequent events expose the inadequacies. People not yet born, other species, and commonly-held resources like caves, air, and oceans are examples of unrepresented constituencies.

The limitations of level-two contracts occur when negotiated solutions equally protect the interests of people close to the situation but not the interests of more distant people. Factors unconsidered by the current negotiators can weaken a contract. This is particularly true when common resources are allocated. Not all well-intentioned, negotiated resolutions have desirable consequences.

The Third Level—Consequences Using Objective Methods

This level requires adding the advantages and subtracting the disadvantages of an act's consequences. It involves considering present and future benefits and costs and comparing factors using a common denominator. Comparing apples to oranges requires the use of a common denominator, such as the level of sweetness or the monetary value of each fruit. This method relies on objective calculation. Cost benefit analysis and risk assessment are examples of appropriate methodologies for this level, and success is determined when a standard is achieved.

For *consequential* ethical methods, one compares the overall values of different actions in terms of their compatibility with the primary value. (For examples, refer to the section below on consequential highest values.) Benefit is measured in terms of how an act contributes to the attainment of the primary value, and accurate measurement of relevant data is important. This method can be important in allocating scarce resources. Wise, appropriate, and effective use of resources is in society's interest, and it contributes to the creation of additional resources and conservation of irreplaceable natural assets.

This level requires adding the advantages and subtracting the disadvantages of an act's consequences. It involves considering present and future benefits and costs, and comparing factors using a common denominator

Exercise 1

How would one devise a common denominator to enable a comparison of the caving experience in a cave with economic development of the land above it?

Level three's objectivity requires that values be compared with a standard unit of measurement. Different ethicists have created terms to use for comparing value. *Utilitarianism*, for example, employs "units of utility" as the common denominator. While this may help people with negative emotional associations with the concept of money as an ethical tool, creating a new standard will cause additional work in establishing relative value. The development of world markets has resulted in structures that regularly calculate values for exchanging or insuring almost everything, thus reducing the amount of work in assigning common denominators. By using an established practice, people making level-three calculations have less work. They must still include relevant factors, and they may have to assign some values that have no market, such as the value of free time. If the thought of using money as a common denominator for ethical calculations is not appealing, other common denominators are possible.

Within the cave conservation community, there may be agreement on a goal but not on the ways to accomplish the goal. Using level-three methods to evaluate the effectiveness of differing cave-protection decisions can unite people behind a cave conservation solution.

After applying all level-three methods, if the problem remains unresolved, if the value is impossible to measure accurately, or if significant incommensurable values occur, one should use the level-four methods. One should not expect to use level four often, and returning to level two might be better. Another possibility is that level-three methods are successful, but there is a compliance or marketing problem.

After applying all level-three methods, if the problem remains unresolved, if the value is impossible to measure accurately, or if significant incommensurable values occur, one should use the level-four methods.

Case Study 6 — Cave Modification and the Scientist

Submitted by the author

A meticulous scientist who studies caves has worked many years for the geologic survey of a midwestern state. He established an elaborate and effective meteorological experiment in Ice-water Cave that included a number of water level recording locations. Ice-water Cave floods often, and it is a significant hydrological information source. The information from his study advanced the hydrological understanding of the area and offered a way to make accurate predictions of dangerous high water levels in the cave.

This scientist is quiet and shy. He readily volunteers his services and talents for opportunities to advance the study of caves. While friendly, he is no politician and not particularly effective at swaying group opinion. He appears uncomfortable when conflicts arise.

In one-to-one situations, he is personable. He worked effectively with the cave owner and established a relationship that allowed him and his grotto to explore, map, and study Ice-water Cave. When a European scientist approached him with a request to measure spelean growth in the midwestern U.S. over an extended period, Ice-water Cave seemed a good choice with its large, active speleothems that appeared to be old. An inconspicuous place was available for a test boring. The 6-millimeter (about 0.25-inch diameter) bore could be studied, and most of the drilled core would be returned once it was analyzed. Plans were made to implement the drilling.

The grotto officially manages Ice-water Cave. The grotto chairman was concerned that this drilling was vandalism, and he made it clear that he thought taking this formation core was wrong. In his mind, right and wrong are easy to distinguish because breaking or marring cave formations is spelled out as vandalism in state cave law and by cave organizations.

The drilling was performed, and the core was studied and returned to its original place. The chairman was incensed that the grotto now had a cave vandal as a member. He persisted with character attacks on the scientist, who was bewildered, and after several months the shy scientist withdrew from the grotto. Since he was no longer a part of the management of Ice-water Cave, he removed his scientific equipment from the cave.

Analysis. Drilling holes in speleothems is cave vandalism by most measures, but this type of scientific study is permitted by most cave laws in the United States. The ethical exception claimed by the scientist is that the intent of this small modification will potentially accomplish a net good, even if no new insight or knowledge is gained—a level-three argument. Most people acknowledge the value of scientific research.

The principal parties cited no established law or custom that resolved this issue to the satisfaction of both parties. Current state cave laws may resolve this issue in favor of the scientific research. Because the chairman and at least a few others were unconvinced, level-three methods are used here as an example of how this conflict is easy to resolve.

What is lost by taking the sample? The physical losses are a small amount of calcite, a damaged formation, wear and tear on the cave, along with the expense of performing the study. To allow quantitative comparisons, the small financial loss can be calculated. The removal of the mineral core does not harm the biological habitat. The aesthetic loss is small because the drilled area is inconspicuous (even if multiplied by the loss all subsequent visitors would experience). The scientist might claim that is unlikely that most people visiting this cave would experience measurable loss. Because the cave is rarely visited, the total loss to people is relatively small, maybe even zero. The gain to society may be training the researcher, or some significant new information may be obtained. How much this one piece of information may help is difficult to determine; however, it is likely to be greater than zero.

Case Study 7 - Caving in the Wilderness

Submitted by Douglas M. Medville

Cavers are engaged in the exploration, survey, and study of caves in wilderness areas in USDA National Forests. Some of the caves contain streams flowing down pits. To facilitate safe exploration and survey, the cavers set bolts at the drops to keep the ropes and cavers out of the water. Citing a concern for keeping wilderness areas pristine, and having seen scars and bolts left in cliffs by rock climbers, the Forest Service holds public meetings on a proposed rule that would forbid bolting in wilderness areas. Members of the caving community attend the meetings and oppose the proposed rule, citing safety concerns and the value of bolting in limited circumstances to conduct scientific studies, carry out resource inventories, and survey the caves. They point out that the bolts are invisible to the general public, but they are willing to forego using power equipment to place bolts in wilderness areas.

Following the public meetings, the Forest Service promulgates a rule in the Federal Register prohibiting the placement of bolts in wilderness areas in USDA National Forests, citing conservation ethics and a desire to keep such areas as natural as possible. While their major concern is the placement of rock climbing bolts on outdoor cliffs, they make it illegal for cavers to place bolts in caves too.

Is the Forest Service correct in its rule making? Does keeping caves as natural as possible outweigh the benefits of placing bolts in caves?

How important is this naturalism, especially when the bolting furthers exploration and scientific studies that may benefit cave conservation and management?

Rule-based procedures may fail to accomplish what is best.

Analysis. Medville describes the classic example of *rule obedience* by a *rule-based authority*. The Forest Service followed its procedures correctly and did not come up with a good solution for cavers and land managers. Rule-based procedures may fail to do what is right. Resolution of this conflict appears to have failed using level-two methods.

Maintaining wilderness areas as natural may be a primary value, instead of a goal. If the Forest Service considers some artificial objects as acceptable in wilderness areas because they contribute to accomplishing higher values such as safety, long-term preservation, or gains in knowledge, then this value can be resolved. The best results in this type of problem require that a solution resolve the problem for both sides. One possible solution would be to recognize that climbing aids in caves perform a different function than they do on the surface and that the improved safety might help maintain the wilderness. For example, the rescue of an injured caver in a wilderness area could result in greater damage and the introduction of unnatural items to the wilderness and cave. Once a new proposal is developed, the advantages and disadvantages can be calculated using level-three methods. The natural fallacy may be a factor in this case, as might be conflicting primary values or inappropriate primary values.

The Fourth Level—Consequences Using Objective and Intuitive Methods

Dilemmas may require using both objective and intuitive ethics methods. At level four, all of the consequences of an act, including dissimilar values and incompatible measurements are considered. The situation may determine the weights assigned to each variable. This level should be used when the solution attained using the third level is unacceptable, because the results conflict with the highest value.

Unlike with the third level, some factors have no common denominator and must be compared subjectively. Third-level information and conclusions may be used, but the information from such research is tempered by invoking appropriate intuitive values.

The objective and intuitive level is particularly useful when some consequences are difficult to quantify; thus, consequences are measured by how an act contributes to the attainment of the highest value. The process of subjectively comparing incommensurable values is subject to unverifiable errors, because there may not be an objective standard. Such a discrepancy may be an acceptable risk in complicated problems.

Dilemmas may require using both objective and intuitive ethics methods. At level four, all of the consequences of an act are considered, including dissimilar values and incompatible measurements.

Exercise 2

A cave that has a history of use by recreational cavers has come under the control of a cave conservation organization. The cave was a significant habitat for endangered bats, but the bat population has declined over the years to almost nothing. Bat censuses have reported a few individuals of an endangered resident bat species in the cave.

Knowledgeable bat scientists have stated that if there were no cavers in the cave, the bats might return. Knowledgeable cavers have stated that it would be difficult to keep the recreational cavers out of the cave, and some of these recreational cavers have threatened to remove barriers that might block the entrance. There is ready access from the highway to the cave. But no one lives in the immediate area, and no one is available to guard the entrance.

Using level-four methods, develop a proposal for this conservation group—a proposal that is compatible with its charter. Their charter is similar to those of NSS cave conservancies, and it is compatible with the conservation policy of the NSS.

Beyond Methods — What is Important?

Some ethical dilemmas may be unresolved due to incompatible primary values. In such cases, people temporarily accept a truce. Ethics methods may achieve these results, but they will not resolve conflicts between incompatible primary values. While good methodology is important for resolving ethical dilemmas, it supplies only the *how to,* not the *why*. This discussion of primary values is only a starting point for understanding their use.

The Appropriate Methods Ethical System is most effective when the primary value is explicit; it does not predetermine a set of values. People using this system need to be familiar with primary values, and they need to know when to recognize them as factors in conflicts. Some ethics systems are associated with certain primary values. For example, for utilitarians, happiness or the greatest good for the greatest number is the primary value. Most other schools of thought propose primary values.

Because the reasons for conserving and restoring caves are likely to be similar to other types of environmental conservation, it is important for cave-interested people to explain how cave conservation applies to a broader environmental statement. The selection of what is most important may have a profound effect on the ethical solutions obtained. People have many different primary values, and sometimes a person or group may have more than one value or conflicting primary values. In other cases, new values become more important. The concept of cave rights provides an example of how primary values might change.

Do caves or other natural features have rights outside of human interests; that is, does a cave have a stake?

Cave Rights

Do caves or other natural features have rights outside of human interests; that is, does a cave have a stake? Reference to caves also refers to significant natural features in this section. While caves have great value and enhance human and other life, do they have the right to exist, and do they merit preservation and protection regardless of human interests? Do caves have rights that equal or exceed human rights?

Some philosophers maintain that the concept of cave rights is an error, because caves have legal protections but not human rights. Attributing that which is exclusively human to non-human animals or inanimate objects might be an error; however, any person, group, or institution may hold as true any set of primary values. Theoretically, an ethical system could maintain that caves have a stake and could have rights equivalent to those of humans. Some of such a system's requirements and considerations are presented as an example of factors involved when elevating a subsidiary value to a primary value.

There are two ways rights are bestowed on caves, by deity or nature, or by humans through institutions like governments. If a person believes that caves have rights equivalent to human rights granted by nature or deity, it is a religious belief. Therefore, this value is a matter of faith and is not subject to objective verification.

If caves had legal rights equal to those of people, these rights would need to be compatible with other highest law so that it could be rationalized. Although some people maintain that rights are inalienable, continuing enforcement is only as effective as the people and institutions that uphold them. Institutions can withdraw rights. Human rights or their equivalents are not granted to other living things or objects, and no serious effort exists to grant legal rights to caves; however, there is a movement to grant some limited human rights to certain animals.

The accepted approach to cave protection is to establish legal protections for caves and institute other measures such as education, active surveillance, and physical barriers. However, human conservation interests, not concern for caves because of intrinsic worth, generally motivate such protections.

People select primary values based on their beliefs. For these values to influence public policy or social behavior, others must adopt them. The following questions may help define one's beliefs about cave rights. What is the purpose of a cave? Why do caves need rights? Why should caves have rights as opposed to protections? Should people use caves for economic gain? Is cave modification and damage acceptable when rescuing a person from a cave? If so, how much is acceptable? Does the level of cave protection depend on the significance of the cave? What sacrifices should be made to protect caves? Should caving be banned in some caves? Who should pay for cave protection? Who represents the caves?

Exercise 3

Use level-three methods and develop two ethical proposals to govern the sale of speleothems by cave owners. The first proposal is to be compatible with individual freedom as a highest value, and the second is to be compatible with an environmental school.

Identification of the Primary Values

For exercises in identifying primary values, several terms need clarification. This section describes *primary value*. Also included are abbreviated statements for *consequential* and *nonconsequential highest values*.

Criteria for a Primary Value

- A primary value is a statement of what is most important. It describes what is most important to a species, group, or individual.
- A primary value has no components that are in conflict with each other. When there are conflicts between two components of a highest value, then at least one of them is a secondary value. If an ethical system has conflicting highest values, then solving some ethical problems may not be possible.
- A primary value helps clarify the purpose of ethical methods and offers a standard on which to evaluate ethical behavior.
- Primary values may be absolute or evaluated by their consequences.
- Primary values are in competition for acceptance.

A primary value statement has no components that are in conflict with each other.

Consequential Highest Values

- Acts should contribute to achieving the greatest good for the greatest number of people—promoted by utilitarians and many humanitarians.
- Acts should contribute to achieving the greatest good for humankind—promoted by some environmentalists.
- Acts should maximize freedoms of individuals—promoted by the natural rights school of thought.
- Equal justice is to be available to all people—promoted by some contractarians.
- Acts should promote the health of the ecosystem—promoted by Aldo Leopold (1970) and some environmentalists.

Nonconsequential Highest Values

- Nature determines moral acts and these acts have intrinsic value—Natural Law.
- God determines moral acts—most religious law is duty based and has different primary values. These systems cannot use levels three and four, and sometimes have more than one highest value.
- Acts should follow the Ten Commandments. Fletcher (1977) makes a consequential and plausible argument for agape (love) as an absolute highest Christian value. He brings Christian doctrine under a unified statement, which is expressed in the first two of the Ten Commandments.

Errors

Successful application of the Appropriate Methods Ethics System depends on avoiding error. Some logic texts identify logical errors and fallacies, but oversimplified, doctrinaire statements can also be a cause of error. Examples of *doctrinaire statements* are: job preservation is paramount; common resources are available for anyone who can take them; and cave owners can do anything they want with their caves.

Two common fallacies committed in cave and environmental conflicts merit detailed discussion. Following are some examples.

Error Example 1 — Denial

Denial may be in two forms:

- A person denies that there is a problem or an injustice or that wrong has occurred or will occur. An example of denial is when grotto leaders endorse the social ostracism of a caver at grotto functions because of trivial personality characteristics or differences in opinion.

- A person denies that there is a need to use an ethical method other than a rule-based authority to solve a problem. At one time, photographs depicting cavers without hard hats were prohibited in the NSS photo salon. Recently, photographs that depict a person touching a cave mineral formation have been banned from salons and publications. In general, these may be good rules; however, there are valid exceptions with good explanations. When a person maintains that there are no valid exceptions, denial may be a likely symptom.

Error Example 2 — The Natural Fallacy

Philosophers have sought to derive basic truths from nature. By applying fundamental truths to ethical dilemmas, they believe that they can deduce and justify ethical principles. While this method is essential for ethical systems to be compatible with natural law and to avoid fundamental errors, people sometimes mistake these natural laws for natural events.

Natural laws describe what is always true under certain circumstances. *Natural events* appear to be random occurrences that may or may not happen the same way at any time. Using nature to devise an ethical code involves the risk of using a random event as a moral principle.

Natural laws describe what is always true under certain circumstances, such as gravity, chemical reactions, human needs, and so on. *Natural events* appear to be random occurrences that may or may not happen the same way at any time, such as human wants. Using nature to devise an ethical code involves the risk of using a random event as a moral principle, the Natural Fallacy.

The statement "cave modification is wrong because the modifications are not natural" is an example that most philosophers would identify as a *natural fallacy*. The error is in using the natural state of the cave to assert that it has ethical standing that prevents an action. This natural fallacy is defined as deriving claims about what ought to be done from claims about what is the case. It assumes that the natural state of a cave is the best state.

To avoid the natural fallacy in establishing an ethical principle, one could show the event is the result of a natural law and not a chance event. To ethically maintain that a natural feature should not be changed or an organism should remain in its present environment, one could show that they have significant value beyond mere existence. The next case study demonstrates how this error can be a subtle part of more complex issues.

Case Study 8— Cave Restorationist Versus Cave Preservationist

Submitted by Joseph C. Douglas

One pitfall in contemporary cave resource protection is that conservation means different things to different people, and some of these meanings are in tension with each other. There may even be conflicting primary values within cave conservation. For some, biological protection of caves is the ultimate value, while others may stress the protection of geological features. Some cavers think primarily of aesthetic values when they undertake conservation work, and others focus on cultural or historical resource protection. For some, responsible multiple-use of caves is paramount, or recreational and educational values are emphasized. There is considerable overlap between the objectives of use and protection, and cooperation is possible (and necessary). However, when the emphasis is on underground preservation, or when preservation gives way to restoration, value-based conflict can appear.

Cave conservation, preservation, and restoration are important and need to be considered carefully before taking action. Restoration involving environmental modification may conflict with preservation. Aesthetic restoration is sometimes in tension with biological preservation and historical preservation. Removing old wood and even some types of trash from a cave improves the appearance of the space, but it may damage the cave's biota. Similarly, restoration projects that involve graffiti removal can damage cultural resources. While graffiti can be ugly, it may contain or overlay important prehistoric or historic resources. Awareness of the variety of important resources in caves, willingness to learn about them, and communication between speleologists with different values and expertise can help ameliorate conflict in the cave conservation community.

Summary

A model that employs four methods (rules, contracts, objective methods, and combined objective and intuitive methods) can be used to resolve cave-related ethical conflicts. By having four methods, one is more likely to find a solution to ethical dilemmas. Identification of primary values and avoiding fallacies and error are crucial factors for conflict resolution.

A model that employs four methods (rules, contracts, objective methods, and combined objective and intuitive methods) can be used to resolve cave-related ethical conflicts.

Cited References

Fletcher J. 1977. *Situation ethics: The new morality*. Philadelphia: The Westminster Press. 176 p.
Leopold A. 1970. *A Sand County Almanac: With Essays on Conservation from Round River*. New York: Ballantine Books. 295 p. [First published in 1949.]
Rawls J. 1971. *A Theory of Justice*. Cambridge (MA): Harvard University Press. 538 p.

Additional Reading

Barnes J, editor. 1995. *The Complete Works of Aristotle*. Princeton: Princeton University Press. 2256 p.
Bayles M, Henley K. 1989. *Right Conduct*. New York: Random House.
Elliot R. 1997. *Faking Nature: The Ethics of Environmental Restoration*. London and New York: Routledge. 192 p.
Graham G. 1988. *Contemporary Social Philosophy*. Oxford: Basil Blaskwell, Ltd.
Gregor M, editor. 1996. *Practical Philosophy by Immanuel Kant*. Cambridge University Press. 664 p.
Griffin J. 1986. *Well-being*. Oxford: Clarendon Press. 403 p.
Hare R. 1981. *Moral Thinking*. Oxford: Clarendon Press.
Hargrove E. 1989. *Foundations of Environmental Ethics*. Englewood Cliffs (NJ): Prentice Hall. 229 p.
Hargrove EC, editor. 1992. *The Animal Rights/Environmental Ethics Debate: The Environmental Perspective*. Albany (NY): SUNY Press. 273 p.
Hegel G, Knox TM, translator. 1967. *Hegel's Philosophy of Right*. New York: Oxford University Press. 377 p.
Kohlberg L. 1984. *The Psychology of Moral Development*. San Francisco: Harper & Row, Publishers.
Leena V. 1997. *The Intrinsic Value of Nature*. Amsterdam and Atlanta (GA): Rodolpi.
McClelland D, editor. 1982. *The Development of Social Maturity*. New York: Irvington Publishers. 225 p.
Mill JS. 1967 Reprint. *A System of Logic: Ratiocinative and Inductive*. London: Longmans, Green and Co, Ltd. 644 p.
Newton LH, Dillingham CK. 1993. *Watersheds: Classic Cases in Environmental Ethics*. Belmont (CA): Wadsworth. 249 p.
Norton BG. editor. 1986. *The Preservation of Species: The Value of Biological Diversity*. Princeton: Princeton University Press. 305 p.
Passmore J. 1974. *Man's Responsibility for Nature: Ecological Problems and Western Traditions*. New York: Scribner's. (Also see: 1980. 2nd ed. London: Duckworth.) 227 p.

Peacock K. 1996. *Living with the Earth: An Introduction to Environmental Philosophy*. Toronto: Harcourt Brace Canada.

Pojman LP. 1994. *Environmental Ethics: Readings in Theory and Application*. Boston: Jones & Bartlett. 560 p.

Schmidtz D, Willott E, editors. 2002. *Environmental Ethics: What Really Matters, What Really Works*. New York: Oxford University Press. 624 p.

Taylor PW. 1986. *Respect for Nature: A Theory of Environmental Ethics*. Princeton: Princeton University Press. 342 p.

Wenz PS. 2001. *Environmental Ethics Today*. New York: Oxford University Press. 326 p.

Wenz PS. 1988. *Environmental Justice*. Albany: State University of New York Press. 384 p.

Section C—Improving Caver Ethics

NSS Conservation and Preservation Policies

NSS Cave Conservation Policy

The National Speleological Society believes:

- Caves have unique scientific, recreational, and scenic values.
- These values are endangered by both carelessness and intentional vandalism.
- These values, once gone, cannot be recovered.
- The responsibility for protecting caves must be assumed by those who study and enjoy them.

Accordingly, the intention of the Society is to work for the preservation of caves with a realistic policy supported by effective programs for:

- Encouraging self-discipline among cavers.
- Education and research concerning the causes and prevention of cave damage.
- Special projects, including cooperation with other groups similarly dedicated to the conservation of natural areas.

Specifically:

- All contents of a cave—formations, life, and loose deposits—are significant for their enjoyment and interpretation.
- Caving parties should leave a cave as they find it.
- Cavers should provide means for the removal of waste.
- Cavers should limit marking to a few, small, removable signs as needed for surveys.
- Cavers should especially exercise extreme care not to accidentally break or soil formations, disturb life forms, or unnecessarily increase the number of disfiguring paths through an area.

Scientific collection is professional, selective, and minimal. The collecting of mineral or biological material for display purposes—including previously broken or dead specimens—is never justified, as it encourages others to collect and destroy the interest of the cave.

The Society encourages projects such as:

- Establishing cave preserves.
- Placing entrance gates where appropriate.
- Opposing the sale of speleothems.
- Supporting effective protective measures.
- Cleaning and restoring over-used caves.
- Cooperating with private cave owners by providing knowledge about their cave and assisting them in protecting their cave and property from damage during cave visits.
- Encouraging commercial cave owners to make use of their opportunity to aid the public in understanding caves and the importance of their conservation.

> **NSS Conservation Policy History**
>
> Editors' Note: Rane Curl reported that Al Amon, George Moore, and he wrote the NSS Conservation Policy in 1960.
> The NSS Board had asked Amon—who was chairman of the NSS Conservation Committee—for a "code of conservation ethics." The NSS Conservation Policy was the result.
> Curl added that he's as suprised as anyone that the 40-year-old policy has survived this long without revision (McClurg and McClurg 2003).
>
> McClurg D, McClurg J. 2003. A History of the San Francisco Bay Chapter, NSS #79, 1957 to present. In: Proffitt M, editor. *Range of Light, Realms of Darkness: A Guide for the 2003 NSS Convention*. Huntsville (AL): National Speleological Society. p 152.

Where there is reason to believe that publication of cave locations will lead to vandalism before adequate protection can be established, the Society will oppose such publication.

It is the duty of every Society member to:
Take personal responsibility for spreading a consciousness of the cave conservation problem to each potential user of caves. Without this, the beauty and value of our caves will not long remain with us.

Adopted by the National Speleological Society Board of Governors on December 28, 1960.

NSS Cave and Karst Preservation Policy

Caves and areas of karst and pseudokarst development are sensitive environments that often interact with surface and subsurface waters and ecosystems. They frequently harbor recreational, historical, and natural resources of considerable significance.

The National Speleological Society believes:
- All caves and cavernous areas are important.
- Cave wilderness, like surface wilderness, is a valuable resource that should be protected regardless of official designations or boundaries.
- Caves and cavernous areas, with their unique environment and development, may require management measures that are independent of geographic boundaries or designations established for the management of other surface or subsurface resources.

Accordingly, the Society endorses, supports, and advocates the implementation of the following precepts:
- No cave or cavernous area should be altered or modified without a full, balanced, and conservative study of the impact of such action, including input from knowledgeable persons specifically experienced in the exploration and scientific study of caves and cave resources.
- Cave resources should be protected by keeping wild caves wild and free from human manipulations and alterations that hamper the free play of natural forces, endanger cave and karst ecosystems, or diminish the pleasure of future visitors.
- Special efforts should be made to preserve the integrity of ecological and hydrologic systems within caves and cavernous terrains.

Where formal designation of Cave Wilderness is a useful tool in protecting this resource, the National Speleological Society will support such designation through actions of its Board of Governors.

Adopted by the National Speleological Society Board of Governors on November 15, 1994.

Section C—Improving Caver Ethics

When to Quit

James F. Baichtal

The cave was formed in a grayish-blue and white banded marble breccia with a red-to-pink matrix. There had been a long habitation by early humans, and a midden nearly blocked the entrance. Charcoal from torches was found on many of the ledges. Shells were stacked where children had played maybe thousands of years in the past. Bones of deer and other forest animals as well as birds, marine mammals, and fish were scattered in the passages.

The cave's entrance is approximately 30 meters (98 feet) from the beach and 20 meters (65 feet) above sea level. The cave formed when the present spring surfaced at a slightly higher level as part of an extensive karst groundwater system that flows from a lake nearly 1.3 kilometers (4,260 feet) away and 180 meters (600 feet) higher in elevation. Today, the main spring flows from just below the cave into the sea. At higher flows, a second spring rises from a pit at the base of a cliff and forms a substantial stream to the north.

The main large room of the cave has two levels into which many old phreatic passages come. We explored these passages which ended in breakdown or became too tight. One passage continued for over 60 meters (200 feet) into a maze of phrcatic tubes, a spongework. Lying on the floor, one could hear and feel the pulse of the huge stream in the passage beneath. On this particular day Steve and I were exploring and mapping the maze of tubes, hoping to find a way down into the main stream passage below. We worked up into one of the highest tubes, surveying as we went. We rounded a corner and the passage in front of us was quite decorated. Soda straws and stalactites hung from the ceiling, moonmilk coated the edges of the passage, and gravels on the floor contained several otter bones. The passage continued but we concluded our survey.

We decided not to disturb this passage. By careful negotiation we may have been able to work our way past the moonmilk and under the formations—but why?

This passage was trending away from the stream passage below and no air movement was detected in this portion of the cave. Exploration of this passage was not necessary for our purposes. We could have increased the overall surveyed length of the cave or found another entrance. Maybe the passage doubled back under us and down to the stream passage, but it was unlikely.

We left the passage unsurveyed and untouched. We made the conscious decision not to leave our tracks, not to risk harming the formations. Steve and I discussed our options and left the passage as we found it, just as it had been for thousands of years.

Steve and I discussed our options and left the passage as we found it, just as it had been for thousands of years.

Section C—Improving Caver Ethics

Slow Down

Paul Burger

Cavers do not like being told what to do or how to act. Explorers are generalized by die-hard conservationists as reckless and destructive. Conservationists are generalized by explorers as people who want to close caves. There is no way to dissuade the extremists of both camps.

For this discussion discard the extremists and concentrate on the middle, the average caver with only a minor interest in cave gating and heavy-handed management and just a passing desire to use mining techniques to get deeper into a cave.

The best advice for preserving a cave while doing exploration is to slow the heck down—not exclusively how fast cavers actually move through a cave, but how fast cavers make decisions.

The best advice for preserving a cave while doing exploration is to slow the heck down—not simply how fast cavers actually move through a cave, but how fast cavers make decisions.

First, address the actual speed issue. Moving quickly through a cave is more about how efficiently you move than how fast you actually try to move.

When most of us started out, every crawl and squeeze was awkward, every breakdown pile was an unstable shifting mass trying to crush us, and every climb was a daunting, holdless mudslope.

As we gained experience and got our "cave legs," we became more comfortable with these environments. Once we became more comfortable underground, we could move past these obstacles more efficiently and get to places faster.

It is important to make the distinction between speed and efficiency (exerted energy) because a great deal of damage is caused by people trying to keep up with someone who is traveling more smoothly through the cave. Keep an eye on your team, especially if you have newcomers or people who are tired at the end of a long trip. Running them into the ground might make you feel like a cave stud, but when one of the team crashes into a formation or stumbles and puts his muddy glove on a pristine wall, regret will (or at least should) follow. Trip leaders need to recognize these struggles and adjust the trip speed accordingly. Unless the cave is flooding or someone has a bomb strapped to his chest that will explode if you don't make it out of the cave by a certain time, there is no reason to push the team beyond their level of efficiency.

The speed of decision making also affects the condition of a cave. In the heat of exploration, we commonly opt to tackle problems head on. If we hit a dirt plug, we dig it out—if we hit a wall, we climb it—and if we find a delicate room with a lead going out the other side, we cross the floor and push it. Often it turns out that there was a different route to get to the same place.

Sometimes, it is best to slow down and check all of the passages around that dig or delicate room. Look at the passages you already know beyond that climb to see if there is a way up that doesn't require bolting. Decisions to move by using higher-impact pushing techniques should be made with as much information as possible.

There is no doubt that a good survey can help in finding routes around obstacles. While "survey as you go" works for most projects, it does not work all the time.

On the subject of survey and its close cousin cave inventory, I want to suggest what borders on the blasphemous. It is not necessary to survey and inventory every square inch of cave. There are times when only one person goes to the back of a delicate crawl or through a short, well-decorated loop. One caver causes far less impact than a whole survey team going through the same passage. (See survey it or not, page 188.)

Sometimes it is better for the cave if you just sketch the passage rather than survey it—but make sure it *does* appear on the map. If you leave it unsketched or sketch an open lead, someone will push the passage again and do the damage you hoped to avoid.

Common sense should also apply to inventorying cave passages that have already been mapped. Think about cave inventory (or photomonitoring, impact mapping, or whatever) and ask these questions.

- Is the information we're getting worth the impact?
- Do we have enough inventory around the delicate passage to characterize the area?
- Should we consider doing *no inventory* for this area?

For example, Jewel Cave has the world's "strategic reserve" of manganese and there is a problem with the manganese wandering around the cave on the bottom of cavers' boots. A complete inventory of the cave would result in more manganese being tracked into relatively pristine areas—so, only enough inventory is done on the previously surveyed passages to characterize the area.

Certainly, these ideas can be abused and some cavers will use them to justify wanton scooping. But it's impossible to stop the scoopers anyway. I just suggest these ideas as a way to slow down, look at what information you are trying to collect, and see if it is worth additional impact on the cave. These suggestions are based on my observations and no small number of my own mistakes. This commentary is merely intended as a starting point for discussion and seed for thought. If it slows down just one person and saves one single soda straw, then it was worth it.

© Val Hildreth-Werker

Figure 1. Fragile cornflakes—should we cross the thin, delicate rafts of this virgin passage? Does it go? Should we survey further? I was asking these questions when I stopped, reached forward to carefully place the slaved flash, gently put a glove in the foreground, and made this photo in a crawling passage beyond the Leaning Tower of Lechuguilla.

Section C—Improving Caver Ethics

Digging: An Apparent Contradiction in Cave Conservation

Rick Olson

On the surface, it seems inconsistent. The National Speleological Society, dedicated to cave exploration, study, and conservation, has both a Conservation and Management Section and a Digging Section. At the annual national convention, we have *blasting* workshops where people learn how to blow things to bits, and we have *restoration* workshops where people learn how to glue busted things back together.

Could anything be more contradictory? Apparently so, but actually not. Digs can ultimately lead to advances in cave conservation. The editors of this book asked me to explore the conservation issues associated with digging because I am neither a strong opponent nor an adamant proponent of excavating cave entrances or passages.

In the process of exploration, documentation, or study, humans can cause heavy impacts on cave resources. Accordingly, the National Speleological Society (NSS) calls on cavers to *leave the cave as they find it*. (See the NSS Cave Conservation Policy, page 253).

Figure 1. Rare gypsum speleothems in the Chandelier Ballroom of Lechuguilla Cave have become an icon for extraordinary cave resources. Had it not been for the entrance excavation, these and hundreds of other remarkable features would remain unknown.

© Val Hildreth-Werker

In other words, as we move through passages we should minimize impact—we should *cave softly*. Does this leave-no-trace philosophy indicate that digging out passages or entrances to allow human entry has only negative consequences for conservation of cave resources?

One test of the value of anything is to imagine taking it away, and then evaluate the consequences. If we had decided years ago that digging should not be done, then what would the consequences be? What would we lose?

- Lechuguilla Cave in Carlsbad Caverns National Park was ultimately entered via an extensive dig (Cahill 1991).
- Most of the grand finds in Jewel Cave National Monument are beyond a low spot named The Dugway (Conn and Conn 1977).
- In Mammoth Cave, the section known as New Discovery could only be accessed after a slab was reduced to rubble, and more recently a trunk passage with undisturbed artifacts left by Native American cavers was found after moving small breakdown (Olson 1999).
- La Cueva de las Barrancas, a pristine

cave now protected as a scientific laboratory, was discovered because air was sucking noisily into a fist-sized opening and a dig ensued (Hildreth-Werker 2001).

There are many other similar examples, but these are sufficient to make the point.

Would we choose not to know about these important finds? Would *not* knowing serve the best interest of these resources? My answer in both cases is no.

Recognizing that only virgin cave has a shot at remaining truly pristine, one possible management alternative is to secure the entrance of a new cave without exploring, documenting, or studying it. This might be a valid approach if closing off the entrance would actually protect a cave. But it can't. In the long run, simply gating an entrance or keeping it secret cannot provide the highest level of protection for caves.

Most caves cannot be protected by secrecy or closure because caves are vulnerable to the unintended side effects of surface land use and mineral extraction. Except in special circumstances, cave passage locations cannot be reliably predicted or remotely sensed. Even if we could determine where cave passages run without physically entering them, we would still be totally ignorant of the resources within.

The bottom line is that we cannot begin to manage resources we are unaware of, or do not understand. Paradoxically, digging has actually facilitated the conservation of some caves even if excavations are controversial.

What can new discoveries do for us? Caves offer tremendous scientific, aesthetic, and recreational values. Caves are like libraries. They have dark corridors with their own special smells, and stored within their walls are enormous amounts of information. In caves, we can learn lessons from previous human cultures and we can study life in earlier geologic eras. In caves we can discover previously unknown species and habitats. We may even learn new details about the nature of life itself and how life may exist on other planets.

Some may still conclude that it is simply better to leave virgin passages undiscovered. Others will conclude that not exploring is contrary to the human spirit. I conclude that careful, deliberate exploration can lead to advances in cave conservation.

© Val Hildreth-Werker and Allan Cobb

Figure 2. In Barrancas, a cave protected as a laboratory for speleological research, cavers have decided not to enter Yucca Flats. Instead of traversing the deep mud, they are looking for a less obvious route to explore the room. La Cueva de las Barrancas was discovered as a fist-sized opening whistling with airflow and the entrance was enlarged to allow human access. (See page 11 of color section.)

Cited References

Cahill T. 1991. Charting the splendors of Lechuguilla Cave. *National Geographic* 179(3):34-59.

Conn H, Conn JE. 1977. *The Jewel Cave Adventure: Fifty Miles of Discovery Under South Dakota*. Teaneck (NJ): Zephyrus Press. p 179.

Olson R. 1999. *Significant new discoveries in Mammoth Cave*. Memorandum to Mammoth Cave National Park Superintendent Ron Switzer. 1 p.

Hildreth-Werker V. 2001. Jim Werker and the Barrancas story: Vision of a cave laboratory. *NSS News* 59(3):67-69.

Section C—Improving Caver Ethics

Virgin Digs

Jim C. Werker

Hiking down a canyon, we heard a whishing sound off in the distance. We stopped to listen. It sounded like rushing water—but that was unlikely in the desert Southwest. Realizing it might be air, Mike Reid and I headed for the sound. We did find air—it was whistling into a fist-sized opening.

A promising air-sucking hole was in the bottom of a rugged canyon, not a likely place for a cave, so some say. But, caves are where you find 'em, so we began to think about a dig.

Before beginning the dig, we measured the hole. To maintain nature's original air exchange in this cave, we had to design a gate that could re-create the cave's natural quantity of airflow. Week after week, a few caving friends returned, enlarged the entrance, and kept quiet about the find. We installed a gate that was designed to allow approximately the same amount of airflow as the original hole. Before opening up any cave entrance, first consider a few important issues.

Think Through the Consequences

A variety of questions should be answered before initiating a cave dig. Who owns the land and what are the legal ramifications? What critters or plants might be impacted by a dig? Who needs to be contacted if bones or cultural materials are discovered? How might a dig in this location affect water flow—where will water go and how might debris change the flow path? Is water likely to be directed into the new opening? How will air exchange be measured and monitored? There are plenty of questions specific to each region and each site—evaluate the immediate as well as long-term consequences.

© Val Hildreth-Werker

Figure 1. Jim Werker pops through the Barrancas entrance gate. This solid gate was designed to replicate and maintain the natural air exchange that was measured at the cave before the fist-sized opening was enlarged.

Dig Checklist

Jim C. Werker and Val Hildreth-Werker

- Record the dimensions and calculate the area of the natural opening(s) at any dig site. If a gate is necessary, design it to replicate the original airflow. A solid gate or air lock closure may be required if no natural opening existed. The gate design might include fabricated apertures or ports to mimic smaller quantities of air exchange.
- Before initiating a dig, think through the consequences. Explore the potential legal, social, safety (and other) ramifications. Obtain permission from the landowner or land managing agency.
- What organisms might be impacted by a dig? Evaluate both flora and fauna.
- Who should be contacted if bones or cultural materials are discovered?

(Continued on next page)

(Continued)

- How might the dig affect water flow? Where will water go? Is flow likely to be directed through the new opening? How might debris change the flow path? Evaluate the immediate and long-term consequences.
- Is it wise to keep the dig secret? Think ahead and prearrange a plan of action to meet the challenges if the secret is unleashed.
- While constructing a gate or waiting for authority, a temporary closure may be necessary for security.
- On cave maps, use a symbol (perhaps a spade) to indicate entrances or connections made by significant digging projects (digs that require tools). By indicating dig sites and the digging techniques used, future researchers in cave archaeology, biology, hydrology, or geomorphology can seek information that may be important in the analysis and interpretation of cave features. In searchable cave survey databases, include information about connections made by digging.
- Dig safely and cave softly.

If a dig site is on federal, state, or county land, environmental regulatory procedures may be required. While constructing the gate or waiting for authority, a temporary closure may be needed for security. When the legalities and safety requirements are established and it is okay to proceed, measure the natural opening(s) of the cave before any digging begins.

Record the dimensions and calculate the area of the natural opening where air is entering and exiting the cave. Design the gate to allow for approximately the same rate of air exchange. If necessary, construct the closure and include fabricated apertures to create exchange that emulates the original airflow. Air that moves past the closed gate should mimic or replicate the original calculation.

The final gate design should either allow the original quantity of air to pass through—or the gate should in some way obstruct the new passage to simulate the original lack of airflow. If *no* natural air exchange flowed through the original breakdown or fractures and holes, then consider designing a solid gate or an air lock closure.

If encouraging bat habitation is among the objectives, research the habitat requirements for species in your area. Gate tolerance varies for different species. Check with Bat Conservation International, Inc. <http://www.batcon.org/>, and consult with bat specialists from universities, state agencies, and the US Fish and Wildlife Service. (See additional sources of information for cave gates, page 165.)

If there is time or inclination to gather additional baseline data before initiating the actual dig and deciding whether to build a gate, use meteorological instruments to record conditions inside and outside the entrance during a variety of weather conditions and seasons. Establish time intervals for monitoring and collect data sets before the gate is installed. Continue the monitoring schedules during and after gate construction and compare to the original data sets. Gate modifications may be necessary. When gates are built for virgin cave dig sites, they should be designed to mimic the original conditions and replicate airflow. (See cave gates, page 147.)

© Jim C. Werker

Figure 2. Airflow is monitored at the entrance of Barrancas, a cave now established as a speleological preserve and laboratory. Val Hildreth-Werker records meteorological data at the entrance gate.

Section C—Improving Caver Ethics

Do Cavers Need a Code of Conduct?

Val Hildreth-Werker and Jim C. Werker

Cavers don't like being told how to behave. But inappropriate caver behavior propagates considerable negative impact. More people are entering caves—flashlight spelunkers, weekend recreationists, enthusiastic students, keen scientists, and veteran explorers. Even casual observers find it obvious that cave environments are damaged by the normal activities of apparently well-meaning cavers.

Traditionally, experienced cavers mentor newcomers. Novices are taken underwing and trained in caving ethics. Ideally, rookies either become competent or quit caving. Attrition rates are high—many get their fill of the caver hierarchy, lose interest, and drop out along the way.

When new cavers stick with the system, they become increasingly involved in project caving, eventually take on leadership roles, and develop caving lifestyles. In some parts of the country, cadres of well-versed cavers maintain this influential peer-training system. But in other areas, there are simply too many inexperienced people who impulsively enter wild caves—who go caving, but receive no mentoring.

Over the past two decades, the speleological community has faced a two-fold challenge. More people are going into wild caves, and current best practices in caving ethics are rapidly changing. The mentors can't keep up. One-on-one training has become unrealistic. To do nothing is to invite more inappropriate impacts. If we want to protect cave systems from cave visitors, we need to expand our training arsenal and adopt new educational tactics.

Minimal Impact Codes and Codes of Ethics have been developed by speleological societies around the globe. A code of conduct is used to describe a set of principles and expectations for members of a particular group.

Caving Codes and Guidelines

Minimal impact codes and codes of ethics have been developed by speleological societies around the globe. A code of conduct is typically used to describe a set of principles and expectations for members of a particular group. Thus, a code of caving conduct contains clear guidelines and expectations for behavior in a cave.

American cave environments are diverse. Caves across the U.S. are categorized and managed very differently. Consequently, specialized codes that outline protective measures and describe expected behaviors have been developed for individual cave sites.

Clarifying Expectations for Lechuguilla

The National Park Service realizes the importance of clearly defining expectations for cavers who work in Lechuguilla Cave. In 1992, the cave resource management team at Carlsbad Caverns began developing a document titled *Guidelines for Entering Lechuguilla Cave* (Appendix 5, page 549).

All who enter Lechuguilla must first attend an orientation that includes guideline clarification and discussion. The guidelines became an appendix

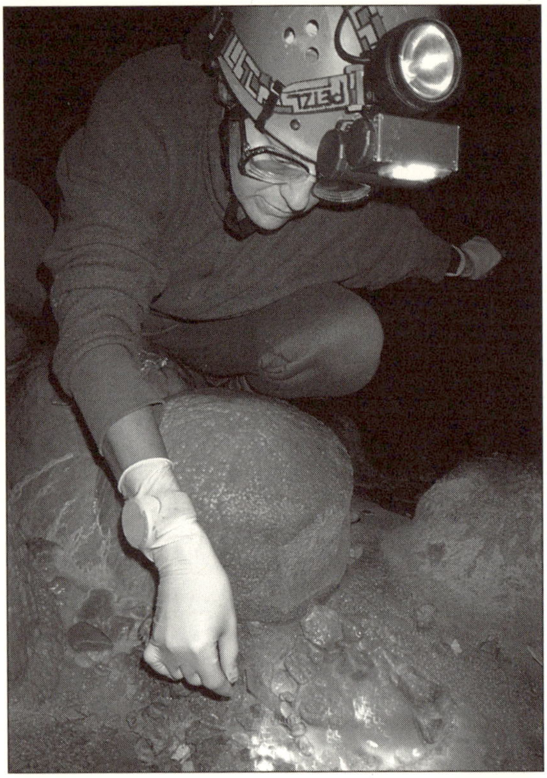

Figure 1. (top left). A specialized Minimum-Impact Caving Code was established to help protect La Cueva de las Barrancas as a speleological preserve for scientific research. In this photo, Penny Boston, attired in clean caving garments, is placing slides for long-term microbial studies in a small virgin pool in Barrancas. (See page 3 of color section.)

Figure 2. (top right). To reduce impact, Jim Werker remains within the designated drop route as he descends to Big Kiss in Barrancas. (See page 11 of color section.)

Figure 3. (middle right). Jim Werker is in the entrance crawl of Barrancas, installing bolts at the top of the first vertical drop. (See page 11 of color section.)

Figure 4. (bottom right). Dave Hamer carefully adjusts the rigging to help protect delicate features as he ascends to Eagles Aerie in Barrancas. (See page 11 of color section.)

Photos © Val Hildreth-Werker

of the park's *1995 Cave and Karst Management Plan*. (See establishing cave management guidelines, page 230).

The dig that opened Lechuguilla Cave occurred in 1986. In the early 1990s, Lechuguilla was a relatively young expedition cave and caver impacts were readily identified. By the mid-1990s, the impacts in Lechuguilla had great influence on improving caver conservation ethics.

For our own expeditions into Lechuguilla, we began to augment the park guidelines with a written code of conduct to further clarify behavior expectations. We updated our list of expectations frequently and it became an informal agreement with anyone who went into Lechuguilla with us. In 1996, we titled our document *Minimum-Impact Guidelines*.

Later, as acceptance increased, our title became *Minimum-Impact Code of Ethics* (Werker and Hildreth-Werker 1999). The draft continues to change as cave conservation parameters are better defined. Updates are essential to the success of this surprisingly effective educational tool.

Point out unsafe or damaging behavior. It is every caver's responsibility to ensure that cave environments remain as pristine as possible and that every team member is safe and aware of conservation ethics.

Minimum-Impact Code of Ethics

Eventually, we developed a simplified sheet of minimum-impact guidelines for general caving situations. *The Minimum-Impact Code of Ethics for Caving Groups* is designed for introductory caving trips, cave conservation workshops, and educational outreach presentations. The example on page 266 can be used as a template, but should be revised to fit specific caves or caving areas.

Specialized Codes and Protocols

Increasingly, we write specialized codes of conduct and protocols for specific cave systems. Barrancas, a virgin cave discovered in 1991, is now protected as a research laboratory for geomicrobiology and astrobiology (Boston 2001; Hildreth-Werker 2001). Abiding by the code established for Barrancas is essential to its continuing status as an underground speleological laboratory and prototype site for research in extraterrestrial life detection and contamination control. Thus, we included an agreement statement and signature line in the minimum-impact document (Werker and Hildreth-Werker 2003). The cave management plan requires that anyone who enters Barrancas must sign this agreement. (See Cave Implementation Schedule: La Cueva de las Barrancas, Appendix 7, page 567.)

Cave conservation practices differ widely across the United States as terrains and protection demands vary. When cavers travel outside of their local caving areas, they sometimes receive behavior guidelines that facilitate quick updates on the principals and protocols in various regions or in specific caves. Through a code of conduct, written emphasis is placed on the specialized conservation demands of particular caves or karst systems.

Cave conservation practices differ widely across the United States as terrains and protection demands vary. Behavior guidelines facilitate quick updates on regional protocols.

- Find the NSS Cave Conservation Policy and the NSS Cave and Karst Preservation Policy in this volume (page 253) and on the NSS Web site: <**http://www.caves.org/**>.

- Find links to a variety of minimum-impact and international caving codes on the NSS Conservation Division Web site: <**http://www.caves.org/conservation/**>.

- Or link to a page of international codes through the NSS Cave Conservation and Management Section: <**http://www.caves.org/section/ccms/caving_codes.htm**>.

Minimum-Impact Code of Ethics for Caving Groups

Val Hildreth-Werker and Jim C. Werker

The goal of this code of ethics is to encourage practices that minimize negative impacts to caves. As we learn more about cave environments, we evaluate and redefine caving conduct. These guidelines, compiled from the experiences and contributions of many cavers, describe low-impact caving techniques.

Avoid harming cave resources—aesthetic, cultural, paleontological, geological, hydrological, mineralogical, meteorological, biological, as well as microbial. Move gently and be good stewards. Think safety—take care of yourself and your team. Take care of the caves.

- Each caver should wear a helmet with a light attached. Each caver should carry water, food, a bottle for urine, and three sources of light with extra batteries and bulbs.
- All cave packs, vertical gear, boots, gloves, helmets, and lint-free clothing should be freshly washed to avoid transfer of mud, dust, and microbes from other cave environments.
- Use footwear with nonmarring/nonmarking soles. Whether light or dark colored, many soles do leave marks on calcite—test soles on untreated limestone or concrete. No heavy waffle stompers. Traditional black lugged soles will leave marks on cave surfaces. Be careful with blond rubber soles—chunks from soft soles tend to break off, leaving debris in cave passages.
- Use soft or padded cave packs. Avoid hard-edged boxes. Choose gear that is smaller, lighter, and more compact.

- Don't disturb bats or other cave-dwelling creatures. Watch for insects and avoid crushing them underfoot.
- Don't smoke or use tobacco in caves. Smoke and fumes can kill bats, invertebrates, and other cave-dwelling animals.

- Wear gloves. Check gloves for mud, dirt, and holes.
- Know which areas require clean clothes, shoes, and gear. Don't enter pristine areas with muddy or dusty garments and gear.
- Avoid isolated pools.
- Limit scratching of skin and hair. Tens of thousands of skin fragments and debris fall from each human body every hour. Never comb or brush hair in a cave. Reduce the input of organic carbons.

- Leave nothing in caves. Carry out trash. Do not mark on cave surfaces. Never mar a cave with graffiti.
- Remove all solid and liquid wastes. Carry an emergency pee bottle and burrito kit. Carry out all urine, feces, spit, vomit, and other waste.
- Avoid dropping crumbs and food particles. Eat over a plastic bag. Carry out crumbs and debris. Don't eat on the move.
- If you light a candle, catch the wax drips on a suitable base such as heavy foil.
- If carbide lamps are allowed, carry the spent carbide out of the cave in plastic bottles with threaded lids.

- Stay on established trails. Sit inside the trails. Keep packs and other items within the path. Choose the most impacted pathways.
- Move carefully and gently through the entire cave—avoid kicking up dust.
- Always spot each other in fragile areas. Especially watch heads, backs, hands, feet, and packs.
- Spot each other on climbs. Remember to maintain three points of contact.
- Touch as little as possible. Avoid leaning on walls, ceilings, or speleothems. Don't sit on formations. Look and avoid trampling floor deposits. When movement requires handholds, look first to avoid delicate features and use knuckles or fingertips for balance rather than dirty open palms.
- During survey and exploration, establish pathways on durable surfaces to minimize future impacts.

- Take nothing from caves. Removal of natural or historical objects is unethical and illegal unless you have a collection permit for authorized research. Check with cave managers, archaeologists, biologists, and historians before making decisions about large items or cultural materials. (Recently-deposited trash usually should be removed. Always carry extra plastic bags and use common sense safety precautions.)

- Point out unsafe or damaging behavior. It is every caver's responsibility to ensure that cave environments remain as pristine as possible and that every team member is safe and aware of conservation ethics. Cave softly … and leave no trace.

Cited References

Boston PJ. 2001. Microbes, minerals, and Mars: The NASA cave connection. *NSS News* 59(3):70.

Hildreth-Werker V. 2001. Jim Werker and the Barrancas story: Vision of a cave laboratory. *NSS News* 59(3):67-69.

Werker JC, Hildreth-Werker V. 1999. Restoration, trail designation, and microbial preservation in Lechuguilla Cave. In: Stitt RR, editor. *Proceedings of the 1997 National Cave and Karst Management Symposium: October 7-10 in Bellingham, Washington, and Chilliwack and Vancouver Island, B.C., Canada.* Huntsville [AL]: National Cave Management Symposium Steering Committee. p 181-189. [Also listed as 13th National Cave Management Symposium.]

Werker JC, Hildreth-Werker V. 2003. Cave management plan for an underground laboratory: La Cueva de las Barrancas, prototype site for Mars studies. In: Rea GT, editor. *Proceedings of the 15th National Cave Management Symposium: Tucson, Arizona, October 16-19, 2001.* Coronado National Forest: USDA Forest Service. p 77-87.

The Minimum-Impact Code of Ethics for Caving Groups can be used as a template, but should be revised to fit specific caves or caving areas.

Section C—Improving Caver Ethics

Packaging Human Waste

Val Hildreth-Werker

On long cave trips, it is essential to carry secure containers for urine and solid waste. Expedition cavers have devised systems to securely package human waste so it can be carried out of the cave.

This chapter gives instructions for a tried-and-true burrito kit that can be rolled up and tucked in the pack. The system is handy for unexpected emergencies as well as multi-day expeditions.

On long cave trips, it is essential to carry secure containers for urine and solid waste. Expedition cavers have devised systems to securely package human waste so it can be carried out of the cave.

Pee Bottles

Cavers should carry a container for urine on every trip. Even in cave passages that periodically flood and wash out, packing out all human waste is increasingly important.(See human waste, pages 35, 71, 125, and 276.)

Boldly mark the designated pee bottle and use a different shape than your drinking bottle—it's dark in there, and it's not fun to accidentally raise the wrong one to your lips. Make sure the lid provides a secure, leakproof fit.

Believe it or not, the flexible Platypus® bottles are quite durable for use as expedition pee bottles and they withstand multiple trips. (You can also purchase little Platy patches that really do work for repairing pinholes.) Some cavers need a small funnel to use with the Platy bottles.

Burrito Bags

This burrito wrapping has worked successfully in the field for over a decade and involves no foil, kitty litter, or granular chemicals. (Tiny bits of litter scatter and pieces of aluminum foil break off and disperse in the cave.) The protocol described here works equally well for caving, kayaking, climbing, or any other outdoor expedition. Make the light, compact packets ahead of time.

- **Turkey Baking/Basting Bag.** A medium-sized roasting bag or large turkey bag is best. It is strong and very light. This is by far the toughest, most durable, odor containing, and leak-proof bag we've tried.

- **Cheap Paper Towels.** Paper towels are more substantial in cave humidity and provide better grip than toilet paper. The less expensive paper towels have less lint. (Lint irritates skin.) Extra paper towels come in handy for nose blowing.

- **Reynolds® Plastic Wrap.** Alternate squares of plastic wrap between layers of paper towel to avoid the problem of the plastic sticking to itself. Most cavers use two or three paper towels per burrito. Reynolds plastic wrap works best—other wraps have failed. The extra expense will secure your burrito and your travel comfort. Plus, Reynolds offers those nice, strong, camouflaging colors. (Other wraps are almost okay, but they get too sticky and rip too easily.)
- **Antibacterial Wipes.** Add individual foil packs of antibacterial towelettes to burrito rolls. Or, take several sheets from a plastic

container of wipes and repackage them in a separate zippie. The wipes are also handy for sponge bathing and hand cleaning. (Those with suspected sensitivity or allergy to antibacterial and antimicrobial agents should avoid these products.)

- **Freezer Zips.** Fold and place a small freezer Ziploc® on the layered stack of alternating paper towels and plastic wrap squares before rolling it all together for transport. Add gallon-sized zippies for compact packaging of all burritos at the end of multi-day expeditions. Additional zippies will secure the entire parcel for travel and proper disposal.

Instructions

Roll the above items into individual, ready-to-use "kits" for transport. Use one or more squares of Reynolds plastic wrap to hold the stacked items all together in a rolled burrito shape. For easy access, package several burrito kits together inside a quart-sized zippie. The following bullet points detail a secure wrapping procedure:

- Open the turkey bag and make it into a bowl shape at your squat spot.
- Try to avoid mixing urine with the fecal matter. This combination creates excess methane and your baggie will get gassy, which could make a risky travel scenario.
- To cut the odor, spread an antibacterial wipe over the top before rolling up the burrito. For personal hygiene, use a clean antibacterial wipe on hands when finished.
- When done, twist and tie multiple knots in the top of the baking bag. Be sure to get the air out.
- Gather the Reynolds plastic wrap around the whole bundle and secure it in a small freezer zippie.
- Several used burritos will stack together nicely in a quart-sized zippie freezer bag.
- For travel, wrap the stack in several more zippies for a compact, secure trek. Some like to use a small waterproof dry bag or a sturdy Nalgene® bottle for transport. Place the package in a well-padded position inside your pack.

At the end of the expedition, wrap the disposable packet with several more opaque plastic bags and properly discard. Do not leave it in the house or truck. Do not flush. Do not drop down porta-potties. Do not put in open trash cans. Find a trash receptacle that will soon be emptied and carried away. On public lands where burritos are common practice, inquire about current disposal protocol.

At the end of the expedition, wrap the disposable packet with several more opaque plastic bags and properly discard. Do not leave it in the house or truck.

W.R. Elliott

Figure 3, page 39. The manure earthworm, *Eisenia foetida*, was identified in a polluted, midwestern show cave. Sewage infiltration brought large numbers of two earthworm species into the cave.

W.R. Elliott

Figure 4, page 39. Green and gray mold colonies, probably *Penicillium*, were found on a shelf in a sewage-polluted midwestern show cave.

Diana E. Northup

Figure 4b, page 63 (left). A bat carcass was found consumed by yellow mold on a cave wall near Bat Cave in Carlsbad Cavern, New Mexico.

Figure 4c, page 63 (below left). Human vomit covered with fungi was found along the visitor trail in Carlsbad Cavern, New Mexico.

Figure 4d, page 63 (below right). In the warm, humid environment of Lechuguilla Cave, New Mexico, caving ropes are sometimes covered with fungi.

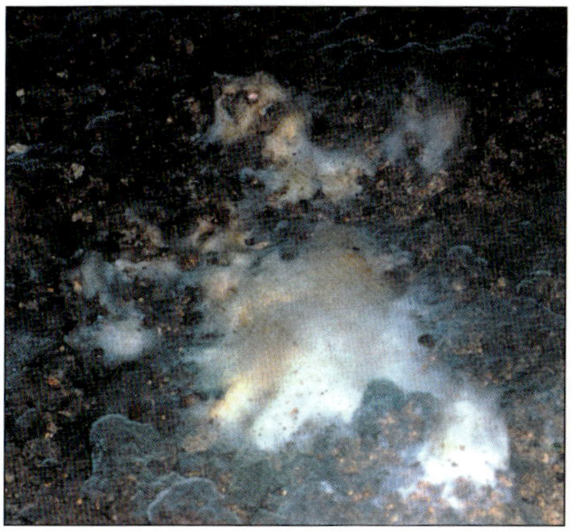

© K. Ingham and D.E. Northup

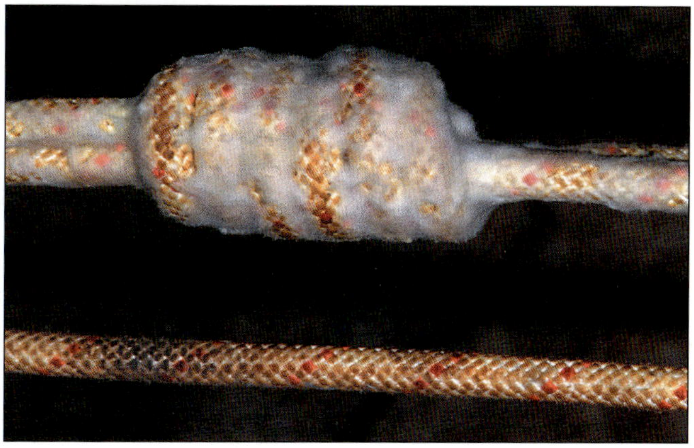

© Val Hildreth-Werker

Gus Frederick

Penelope J. Boston

Figure 1a, page 62 (above left). A moss garden grows near the entrance of Ape Cave (lava tube), Mt. St. Helens National Volcanic Monument, Washington.

Figure 1c, page 62 (above right). Brightly colored slime molds, unusual fungi, yeasts, and bacterial assemblages also grow in Four Windows Cave, El Malpais, New Mexico.

Figure 5a, page 64 (right). Actinomycetes on gypsum paste (pH 1.0) are visible in Cueva de Villa Luz, Mexico.

Figure 5b, page 64 (below left). Mike Spilde collects white moonmilk paste that has the consistency of Crisco® in Spider Cave, New Mexico.

Figure 6a, page 64 (right). Microbial mud mats called biovermiculations are on the walls of Cueva de Villa Luz, Mexico.

Figure 6d, page 64 (below right). Scientists study pool fingers in Carlsbad Cavern, New Mexico.

Louise D. Hose

© Kenneth Ingham

© Kenneth Ingham

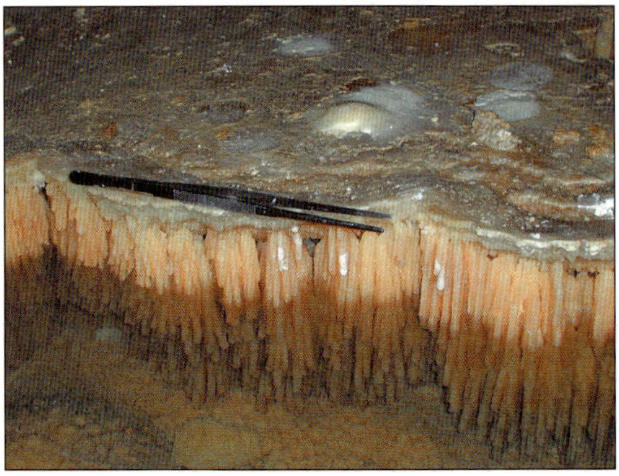

© Kenneth Ingham

Sherry Cady

David H. Jagnow

© Kenneth Ingham

Figure 7a, page 65 (above left). Colorful surface microbial mats thrive in hydrothermal springs at Yellowstone National Park, Wyoming.

Figure 7b, page 65 (above right). Conspicuous biofilms like this black iron mat in Coldwater Cave, Iowa, may contain significant microbial populations, extracellular materials like polysaccharide slimes and enzymes, and precipitated minerals like iron oxides.

Figure 7c, page 65 (middle left). Wearing protective respirators, Diana Northup and Penny Boston take oxidation reduction potential measurements across from Sulphur Spring in Cueva de Villa Luz, Mexico.

Figure 7d, page 65; also **Figure 1, page 264** (below left). Attired in clean clothes, Penny Boston carefully places a sterile glass slide in a virgin pool. The long-term study will remain *in situ* for several years before retrieval for laboratory microbial analysis. Invisible biofilms on flowstone and other moist cave surfaces are enough to support cave-adapted invertebrates in La Cueva de las Barrancas, New Mexico.

Figure 7e, page 65 (below right). Mike Spilde collects a sample from colorful encrustations of manganese and iron oxides that can contain a rich microbial diversity in Lechuguilla Cave, New Mexico.

© Val Hildreth-Werker

© Val Hildreth-Werker

Figure 1d, page 62. Ladybugs massing on an algae-covered boulder below the entrance to Carlsbad Cavern, New Mexico.

Figure 2d, page 62 (right). Fish live in a stream within Cueva de Villa Luz, Mexico.

Figure 3, page 49 (below left). Indiana bats (*Myotis sodalis*) cluster together on a cave ceiling.

Figure 5, page 51 (below right). In 2005 the Virginia big-eared bat (*Corynorhinus townsendii virginianus*) was designated the official state bat of Virginia.

Rick Olson

© Val Hildreth-Werker

Figure 6, page 85 (above left). This wood rat skull was found in a wood rat latrine. Near cave entrances, the nests and middens of wood rats often contain important paleontological remains.

Figure 8, page 87 (above right). This Pleistocene bat skeleton, discovered during a bat skull study, is extremely well-preserved in Land of Awes, Lechuguilla Cave, New Mexico.

© Val Hildreth-Werker

Figure 11, page 87 (left). Bat bones are clearly visible against the white calcite surface where they were found in Ogle Cave, Carlsbad Caverns National Park, New Mexico.

Rickard S. Toomey III

Figure 12, page 87 (left). These bat bones are preserved in small, dry, rimstone pools. Delicate bones are easily destroyed by careless traffic.

Figure 14, page 87 (below). Bat staining is clearly visible on this cave ceiling.

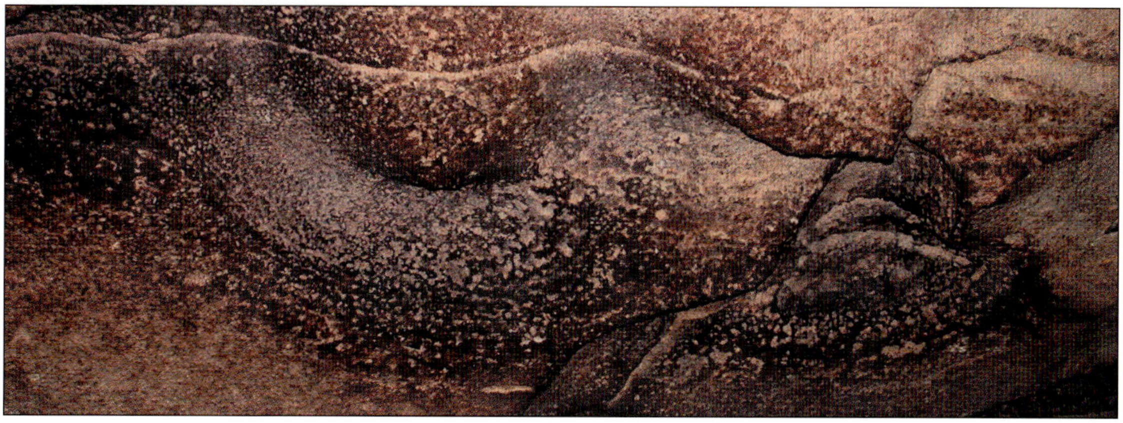
Rick Olson

© Val Hildreth-Werker

© Val Hildreth-Werker

© Val Hildreth-Werker

NPS Photo Val Hildreth-Werker

© Val Hildreth-Werker

© Val Hildreth-Werker

Figure 1, page 170 (above left). Aluminum carabiners were left on a traverse in Virgin Cave, New Mexico. After only six months in the cave, abundant pitting and corrosion occurred. Material strength was greatly reduced and some of the biner gates would no longer close. Stainless steel cara-biners are recommended for long-term use in cave environments.

Figure 2, page 171 (above right). In 1986, after digging open the entrance of Lechuguilla Cave, a galvanized culvert with a solid steel gate was installed. In a few years it was rusting and deteriorating.

Figure 3, page 171 (middle left). In 2002 the rusted entry culvert in Lechuguilla Cave was replaced with this special-design air lock and culvert made of stainless steel.

Figure 4, page 233 (middle right). Permanent photomonitoring points are established to help cave managers monitor impact trends and speleothem changes over time. Installed by Jim Werker and Val Hildreth-Werker at the Leaning Tower of Lechuguilla, this photomonitoring station provides visual documentation to aid management decisions for Lechuguilla Cave, New Mexico.

Figure 11, page 207 (left–April 8, 1994). The entrance to South-winds in Lechuguilla was very white and pristine on the day after it was discovered in 1994.

Figure 12, page 207 (right–August 8, 1997). Three years and about 150 cavers later, the Southwinds entryway had become dusted to a dull, brown-colored appearance. Cavers should move gently to avoid kicking the soft rock flour into airborne particles that discolor the features.

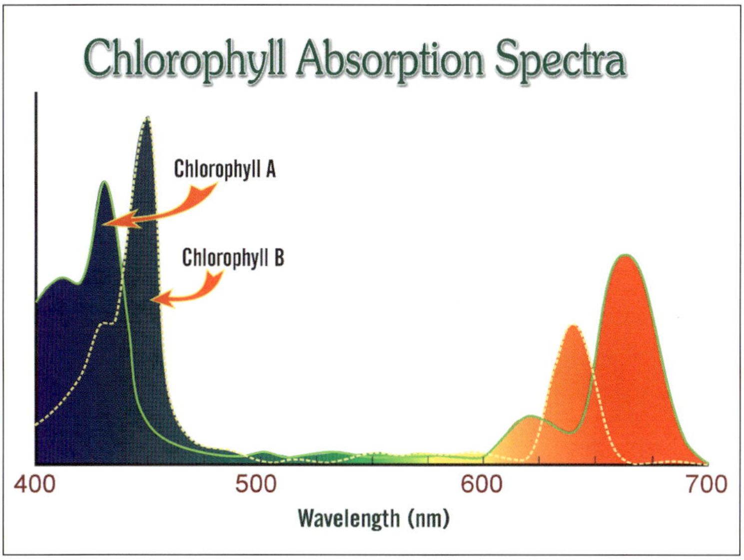

Gary Berdeaux

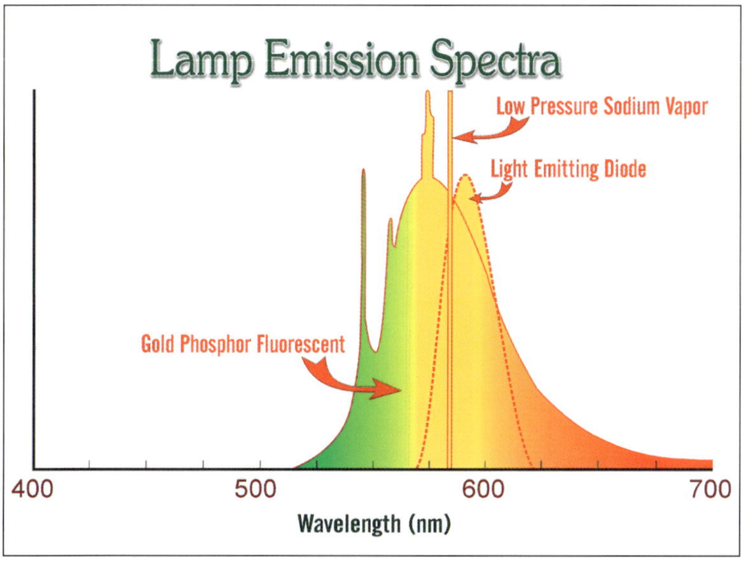

Gary Berdeaux

Figure 1, page 345. Compare the first diagram showing the absorption spectra of chlorophyll to the second diagram.

Figure 2, page 345. In the second diagram, the emission spectra of yellow lamps fit between the absorption peaks of chlorophyll. Yellow lamps support less lamp flora than white light sources. Gold-sleeved lamps are similar to gold-phosphor lamps.

Figure 3, page 346 (below left–before). This is the test area before yellow LEDs were installed in Frozen Niagra, Mammoth Cave, Kentucky. Lamp flora is an important consideration when designing lights for show caves. Algae, mosses, cyanobacteria, and other opportunistic plants are often induced by lights placed along cave tour trails. These photosynthetic species introduce major distortions to cave ecosystems.

Figure 4, page 346 (below right–after). After installing yellow LEDs, algae and mosses are no longer visible in the test area of Frozen Niagra, Mammoth Cave National Park. There is no regrowth of lamp flora after three years of illuminating this area with 595 nanometer LED lamps.

Rick Olson

Rick Olson

© Val Hildreth-Werker

© Val Hildreth-Werker

Figure 5, page 178 (above left). Weave flagging tape between rocks to define cave trails. This technique aids visibility and increases visitor compliance.

Figure 8, page 181 (above right). Double strips of continuous flagging confine impact to a limited pathway. This trail is flagged wider than 0.5 meter (18 inches) to facilitate safe travel and handholds along the slope.

Figure 11, page 181. On climbs, wider flagged areas accommodate handholds and various climbing styles.

Figure 12, page 181 (below left). To reduce further impact, Phyllis Hamer stands in a flagged posing area specified for portraits in the Chandelier Ballroom of Lechuguilla Cave. Unobtrusive orange flagging marks the trail and the photo zone.

Figure 15, page 182 (below left). Mark wider areas in trails as resting zones. Keep all gear inside the trails.

© Val Hildreth-Werker

© Val Hildreth-Werker

© Val Hildreth-Werker

Figure 5, page 192 (left). During survey and exploration, choose routes through new passages to preserve pristine qualities. Trail designation is typically assigned to survey teams.

Figure 9, page 181 (right). This photo shows excellent trail flagging technique using natural protrusions for tape tie-downs. Double lines of continuous delineation along the pathway help protect the larger flowstone area from being discolored and marred like the trail. Generally, 0.5 meter (18 inches) provides adequate trail width for efficient walking and pack hauling.

Figure 13, page 181 (left). Flagging tape hangs from the ceiling to define the route through this gypsum maze. The discolored blotches are handholds, which started as small, individual spots cuing cavers to limit impact to a single finger for balance rather than using a dirty open palm. To help reduce impact in pristine passages, encourage cavers to wear surgical gloves (and flowstone shoes where possible.)

Figure 14, page 181 (right). To prevent color-bleeding, carefully place the white side of flagging tape down against wet flowstone surfaces. The two-toned or striped flagging tapes are often completely white on one side.

Figure 18, page 182 (left). Use a whisk broom to gently comb away footsteps outside the trail. Leaving off-trail footsteps encourages more to follow.

Figure 19, page 182 (right). A caver is replacing the older, brittle blue/white flagging with red/white placed at the trail level through the unusual Boogins in La Cueva de las Barrancas, New Mexico.

© Val Hildreth-Werker

Figure 2, page 413 (above right). Gently lay an anti-microbial wipe over the mold patch to avoid distributing spores. Use a light disposable glove and invert it over the wipe after scooping up the patch of fungi. To protect against breathing mold spores, wear a surgical mask when cleaning up fungi.

Figure 3, page 414 (above). In some cave environments, colorful blotches of mold grow on spilled food particles. Developed by the International Federation of Rock Art Organizations (IFRAO), the color scale is ideal for cave photographers.

Figure 8, page 206 (middle right). The IFRAO scale is an excellent tool for cave documentation and close-up photos. Above the scale is a fig seed that sprouted in the cave's dark silence, inadvertently left by a caver during a lunch stop.

Figure 1, page 419 (right). Jim Werker uses light strokes with a stainless steel brush to clean gypsum surfaces. In this maze passage, Werker easily captures soiled gypsum particles in a small plastic bag. Always use a catchment to contain the scrapings.

© Val Hildreth-Werker

© Val Hildreth-Werker

© Val Hildreth-Werker

© Val Hildreth-Werker & Allan Cobb

© Val Hildreth-Werker

© Val Hildreth-Werker

© Val Hildreth-Werker

Figure 2, page 260 (above left). Instead of traversing and marring the drip-pocked mud of Yucca Flats, cavers are looking for a less obvious route to explore the cham-ber in La Cueva de las Barrancas, New Mexico. Discovered as a fist-sized opening whistling with air-flow, the entrance was enlarged to allow human access and the cave is now protected as a laboratory for speleological research. A specialized Cave Management Prescription and Minimum Impact Caving Code help protect Barrancas as a speleological preserve for geomicro-biology and other scientific studies.

Figure 2, page 264 (above right). To confine impact, Jim Werker remains within the scarred drop route as he descends to Big Kiss in Barrancas.

Figure 3, page 264 (middle left). Jim Werker is in the entrance crawl of Barrancas. To reduce future impact, he installs bolts at the top of the first vertical drop.

Figure 4, page 264 (left). Dave Hamer carefully adjusts the rigging to help protect delicate features as he ascends to Eagle's Aerie in Barrancas.

Figure 1, page 388 (above). In Pellucidar, cave drip water washed the muddy boot prints of careless cavers into pools. Jim Werker adapted a diaphragm hand-pump to filter sediment and circulate water back into the cave pools. Dave Hamer operated the pump. Water and silt were filtered through sponge material stuffed into a clean container.

Figure 2, page 388 (right). In the Pellucidar area of Lechuguilla Cave, New Mexico, Phyllis Hamer and Aimee Beveridge positioned sponge dams downslope to catch runoff water while restoring an area of flowstone and pearls.

Figure 3a, page 388 (below left–before). Plastic toothpicks, tweezers, and other small tools worked well for restoration in Pellucidar. Toothbrushes were used lightly on flowstone, but not on cave pearls.

Figure 3b, page 388 (below right–after). This small area of pearls in Pellucidar required several days of intensive cave restoration work.

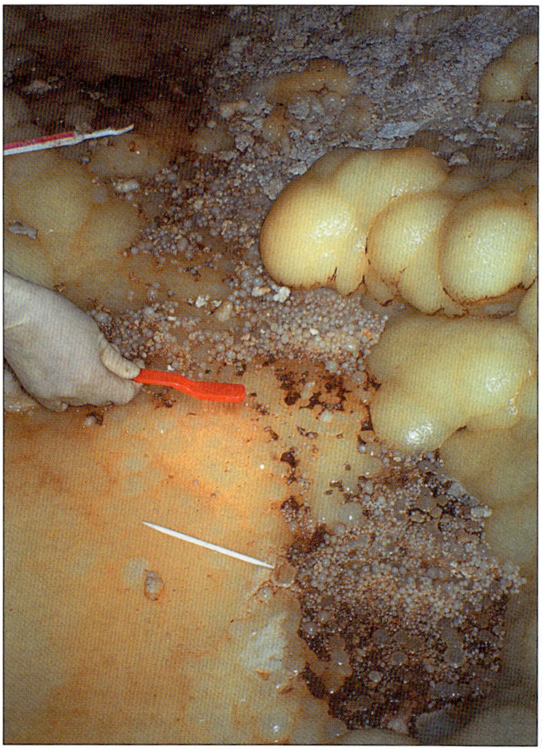

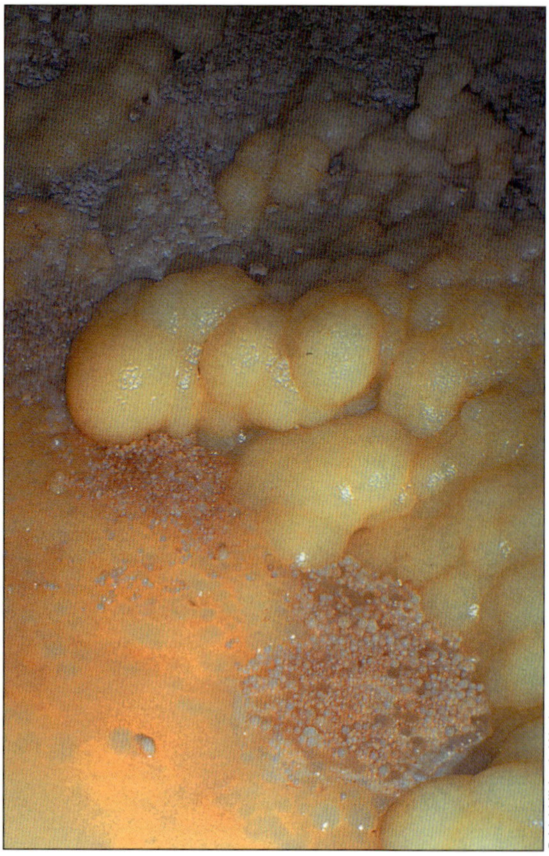

Figure 4, page 389 (above). Much of the restoration work in Pellucidar was done on rope. Justin Shaw rinsed his sponge into a dry bag secured with a cord.

Figure 5, page 389 (left). The Pellucidar restoration required clean clothing and careful technique. Justin Shaw is on rope. Allan Cobb is upslope. Phyllis Hamer is working on flowstone and pearls at the top of the image.

Figure 6a, page 389 (below left–before). Muddy boot tracks left in Pellucidar marred the entire area of flowstone. Silt settled on the bottoms of formerly pristine pools.

Figure 6b, page 389 (below right–after). Cavers used a large-volume diaphragm hand-pump to filter silt out of this Pellucidar pool.

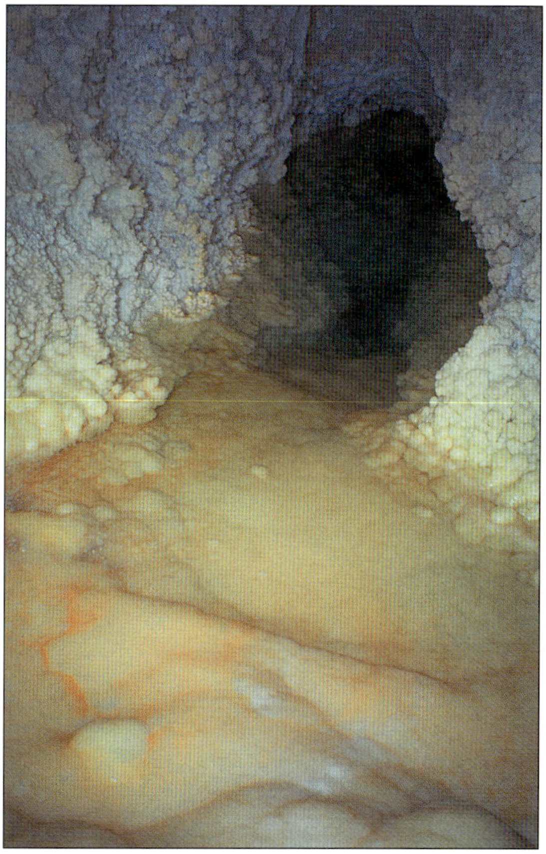

Figure 1, page 401 (left). Remove muddy boot prints from flowstone before new layers of calcite make the imprints permanent.

Figure 2, page 399 (right). Use pressurized sprayers to achieve a more efficient spray for some types of cave restoration. Be cautious—test small areas and avoid damaging cave resources. Manual-pump sprayers are available in a variety of sizes.

Figure 2, page 401 (middle right). Use a spray bottle of water and a soft, closed-cell sponge to blot up muddy prints from flowstone surfaces. Wear nonlatex, powder-free exam or surgical gloves.

Figure 9, page 404 (below left). Solid-colored flagging tape will discolor wet flowstone. Use white-backed flagging instead of solid colors on wet calcite surfaces.

Figure 10, page 404 (below right). Orange flagging tape faded onto this wet flowstone. Fortunately, the discoloration was not calcited over and the orange color was easily sponged away.

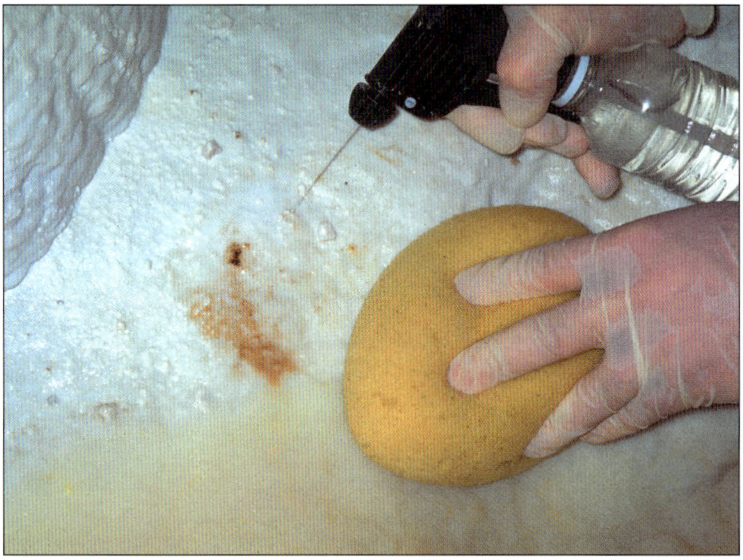

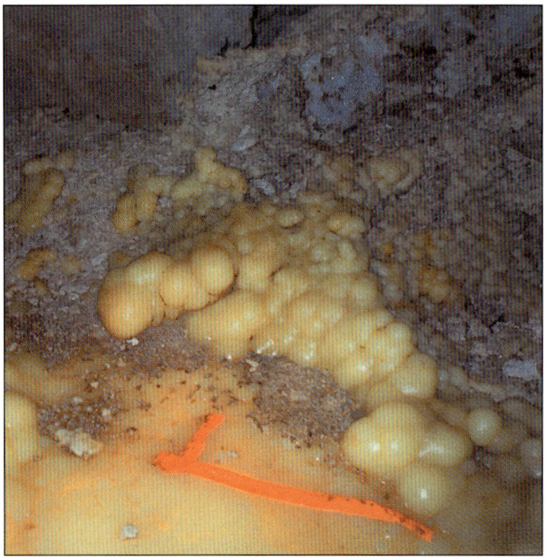

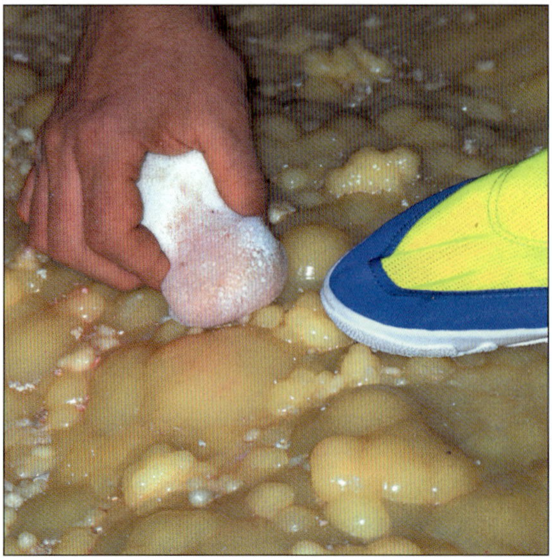

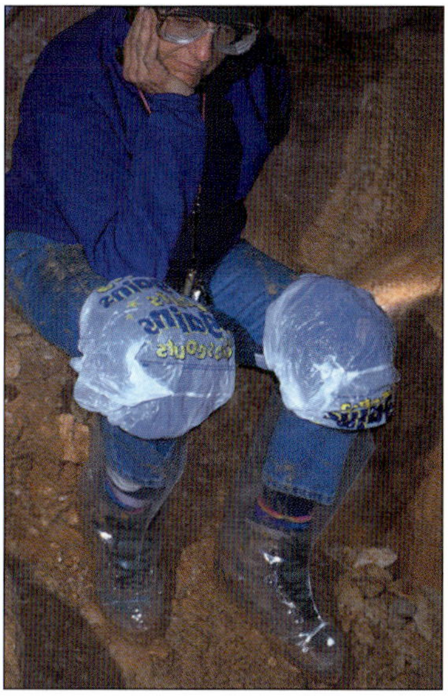

Figure 1, page 432 (left). Non-marking aqua socks with light soles make good footwear for many pristine cave surfaces. This is a transition zone for stepping out of boots and into flowstone shoes.

Figure 3, page 428 (right). When entering a muddy chamber in La Cueva de las Barrancas, Penny Boston wears boot covers and bags over her kneepads to avoid tracking mud through the cave.

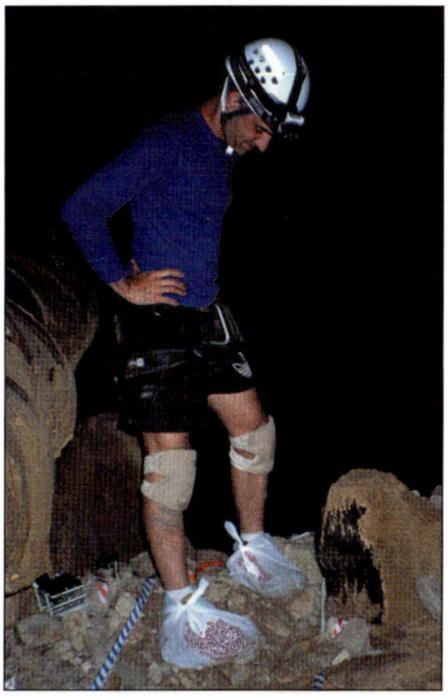

Figure 3, page 432 (left). Plastic shopping bags often make adequate boot covers for short distances if cavers walk gently. David Joaquim is checking a science and monitoring station on a muddy trail in La Cueva de las Barrancas. Beside the trail, test tubes remain *in situ,* allowing cultures to grow in the stable cave environment for months before removing for analysis.

Figure 4, page 432 (right). In a pristine cave area with intermittent mud, Harley Shaw slips hazmat boot covers over his flowstone shoes to avoid spreading mud onto clean flowstone surfaces.

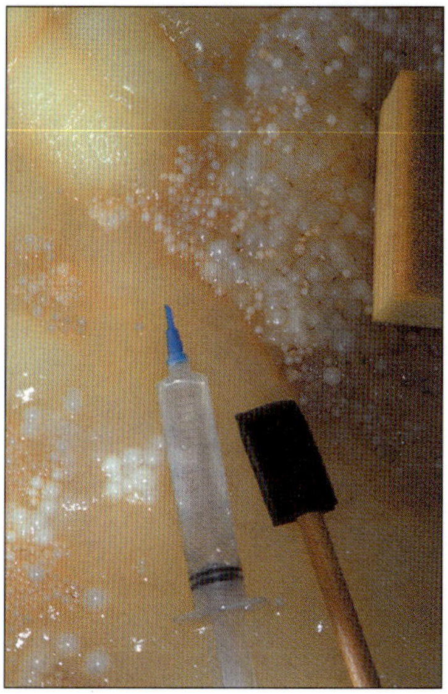

Figure 5, page 424 (left). Dressed in a Tyvek® suit, surgical gloves, and flowstone shoes, Gosia Allison-Kosior gently restores pearls in Pearlsian Gulf, Lechuguilla Cave.

Figure 6, page 424 (right). Both before and after are apparent in this cave pearl image. The left side of the flowstone has been cleaned, and the right side still has particles of mud. Syringes, foam clean-room brushes with plastic handles, and sponges are the tools for tedious cave pearl restoration. The foam brushes work like magnets to attract muck from the pearl nests. Avoid scrubbing cave pearls because abrasive action may disrupt microbial communities.

Figure 1, page 424 (left). This delicate aragonite probably cannot be restored. Any attempt to remove the dirt particles would likely create more damage. Cavers should move gently and avoid dropping dust onto this speleothem in Wind Hicks Cave, New Mexico.

Figure 2, page 424 (right). Imagine how easily gypsum flowers can be destroyed when cavers kick up dust or fail to look before placing a hand or foot.

Figure 3, page 424 (middle right). Doug Feakes was one of the first cavers to view this wall of virgin aragonite in Southwinds, Lechuguilla Cave, New Mexico. Perhaps the speleothem-covered wall can be preserved in this pristine state by moving gently and following trail protocols.

Figure 3, page 438 (below left). In a New Mexico Cave, the Candle Table was pinned and epoxied back into its original position. Use the proper cave-safe materials for speleothem repair—stainless steel pins and archival, museum-grade epoxy.

Figure 4, page 424 (below right). To mitigate careless impacts, muddy imprints were tediously removed from the cave pearl nests in Pearlsian Gulf, Lechuguilla Cave. The dark smear near the center of this image is the marring footstep of a perhaps well-intentioned but ill-informed caver.

Section C—Improving Caver Ethics

Health and Hygiene Related to Cave Conservation

Stephen R. Mosberg, M.D

Although a complete discussion of wilderness medicine or caving-related medical issues is beyond the scope of this book, several broad health categories influence good practices in cave conservation.

- A variety of factors affect caving performance—heat stressors, hydration status, nutrition, and general fitness.
- Hygiene practices help maintain health—human waste disposal, sanitation at camp, and water quality.
- It is important to be aware of several illnesses that may be contracted in cave environments.

Hypothermia

Hypothermia is a common topic in caving literature and is described in first aid and rescue publications. The definition of *hypothermia* is straightforward. *Hypo* means "low" and *thermia* means "heat." Many factors can lead to hypothermia or low body heat, but the underlying mechanism is always the same—heat is lost faster than the body can produce it.

Several factors may predispose humans to hypothermia—nutrition, hydration, inactivity, body *habitus* (lack of body fat), individual variance, alcohol use, and illness.

Hypothermia can be divided into three general categories.

- *Acute hypothermia* occurs suddenly—for example, falling into a stream (sometimes called immersion hypothermia).
- *Subacute hypothermia* occurs over an intermediate period of time (hours)—and would be more common on caving trips and expeditions.
- *Chronic hypothermia* occurs over a longer period of time (days, weeks, or months)—for example, a person living in an unheated room.

Many published lists detail the signs of hypothermia and correlate the symptoms with body temperatures. The accuracy of these lists is often questioned, but the details are not as important as understanding the general trend.

Many factors can lead to hypothermia or low body heat, but the underlying mechanism is always the same—heat is lost faster than the body can produce it.

Signs and Symptoms of Hypothermia

- As body temperature falls, shivering begins and coordination decreases.
- Fine movements become more difficult and eventually impossible.
- Walking becomes difficult and *ataxia* (stumbling) ensues.
- The ability to reason decreases. Decision making becomes difficult and poor judgment ensues.
- As body temperature falls, the victim becomes *stuporous* (sleepy).
- With progressive decrease in body temperature, shivering ceases and death from irregular heartbeat is not far behind.

It is easy to understand how lack of coordination and poor judgment might impact the cave environment. Hypothermic cavers tend to become clumsy and make bad decisions. Preventing and treating hypothermia is discussed in great detail in other publications. (See sources for additional information at the end of this chapter.) Appropriate clothing, adequate fuel intake (food), proper equipment, detailed planning, and awareness of your own status as well as that of your companions will go a long way toward preventing hypothermic conditions.

It is easy to understand how lack of coordination and poor judgment might impact the cave environment. Hypothermic cavers tend to become clumsy and make bad decisions.

Early Warning—Shivering

Shivering is the earliest warning sign of hypothermia. It is the body's first response to falling body temperature. The intense muscle activity of shivering generates heat by increasing heat production up to five times over the rate produced when one is simply sitting (Steinman and Hayward 1995). But heat production is not without a price: calories from food or body stores (fuel sources) are quickly consumed; muscles are quickly fatigued; and fine movements are difficult if not impossible. Don't ignore shivering hands or chattering teeth.

When you feel cold and before shivering starts, you should begin trying to rctain body heat or rewarming by other measures.

- **Put on a hat.** A huge amount of body heat is lost through the head and neck. Up to 70% of total body heat production can be lost through the head when the temperature is -15°C (5°F) (Bowman 1995).

- **Change to dry clothes**. At least put dry clothes next to the skin. Bodies perspire constantly and the moisture is transferred to clothing.
  - Some fabrics act like evaporative coolers. Cotton absorbs moisture, thus cooling the body by evaporation. Cotton also has high thermal conductivity and allows heat to flow from the body faster.
  - Nylon absorbs less moisture than cotton, but also retains enough water to cool by evaporation and has poor wicking qualities.
  - Wool is an excellent insulator, but holds moisture (not as much as cotton, but more than nylon, acrylic, and polypropylene) and becomes much heavier when wet.
  - Acrylic and polypropylene fabrics absorb minimal water, wick moisture away from skin, and allow body heat to dry garments.
  - Of all of these, polypropylene has the lowest thermal conductivity and highest insulating value while maintaining a high wicking ability.

- **Eat something.** Simple sugars (for example, candy) will provide the quickest energy. However, over the long term, high fat content or high calorie count foods will serve to refuel and reheat the body more efficiently than other food choices.

- **Warm up.** Use one of the traditional caver techniques to increase warmth.
  - Dig out your emergency body-sized plastic bag, crawl in, and put your carbide lamp or candle under the bag to warm up quickly.
  - Sit shoulder-to-shoulder with two warm cavers for heat transfer.
  - Skin-to-skin contact inside a sleeping bag is one of the quickest emergency procedures for restoring body temperature.
  - Two or more bodies clothed in polyester inside a perspiration-wicking bag may warm up even faster.

Hyperthermia

Hyperthermia, or heat stress, is the often forgotten, heat-related illness in caving. The definition is again easily understood from the word itself. *Hyper* means "high"—thus, hyperthermia is a state of "high heat."

Many medical and first aid texts make a point of differentiating between heat exhaustion and heat stroke. For our purposes, the distinction is not important. Because of overlap in signs and symptoms, the two conditions cannot be correlated well with different temperatures.

Several factors can contribute to heat-related illnesses—fatigue, lack of sleep, lack of acclimatization, drug use, obesity, sustained exertion, and dehydration.

Symptoms of heat-related illnesses can be easily overlooked.

- Headache
- Confusion
- Drowsiness
- Nausea
- Lack of coordination
- Loss of consciousness—a late symptom
- Cessation of sweating—a very late and ominous sign of hyperthermia

Again, it should be apparent that anything affecting the individual's ability to reason or function in a coordinated manner will adversely affect the cave environment.

Hyperthermia, or heat stress, is the often forgotten, heat-related illness in caving. The definition is easily understood from the word itself. *Hyper* means "high"—thus, hyperthermia is a state of "high heat."

Medical Emergency—Change in Level of Consciousness

With hyperthermia, there is one very important symptom to remember—*is there a change in the level of consciousness?*

This change can range from confusion to loss of consciousness. If there is an alteration in the level of consciousness, the person has a more severe heat-related illness. This is a true emergency and requires immediate attention (cooling) to prevent permanent brain damage or death.

Rehydrate and Lower Body Temperature

Treatment of heat illness centers on rehydrating the patient and lowering body temperature.

- An efficient way to lower body temperature is to wet the body with water or whatever is available (even urine will do) and then fan the person. (Since this method depends on evaporation, it will be less efficient in the 100% humidity found in many cave environments.)
- Immersion in cool water, previously thought to be contraindicated, is also a very effective cooling method and should be used when other measures fail to lower the body temperature (Dickinson 1995).
- For information on rehydrating a victim, see hydration status below.
- Again, decreased mental alertness is a very serious sign—anything from confusion to unconsciousness should be considered a medical emergency.

Prevention and Acclimatization

Prevention of heat illness requires awareness. Pay attention to signs and symptoms. Consciously monitor yourself and your companions. Plan for heat conditions and give attention to hydration status.

Preparation can help you avoid heat problems. Acclimatization in hot environments may take 7–10 days. However, exercise in a warm environment (1–1.5 hours daily) may help prepare the body for a hot environment. For example, exercising in the warmest, middle part of the day rather than in cooler morning or evening hours may facilitate acclimatization to warmer temperatures. Such an exercise program should be undertaken cautiously. Consult your physician before initiating any exercise program.

Hydration Status

Dehydration is an abnormal depletion of body fluids. The state of dehydration can be associated with the temperature-related problems described above, or dehydration can develop without heat-related stressors, as in severe vomiting or diarrhea.

Humans lose about 1.5 liters (1.5 quarts) of water per day through insensible losses—normal fluid loss through breathing or slight perspiration. With heavy exertion in a hot or humid environment, a person can easily lose 1.5 liters or quarts every hour (Hubbard and others 1995). One of the highest recorded sweat losses is about 3.5 liters (3.75 quarts) *per hour* (Armstrong and others 1986).

Confusion and Stupor

Like heat-associated illnesses, dehydration can cause confusion and lack of coordination. Symptoms of dehydration can accompany and mimic body temperature problems. Confusion and lack of coordination can manifest with hypothermia, hyperthermia, high fever, dehydration, fatigue, medical illness, diabetes, medication effects, and other conditions.

Like heat-associated illnesses, dehydration can cause confusion and lack of coordination. Symptoms of dehydration can accompany and mimic body temperature problems.

Over-hydration (for lack of a better term) can also cause the same dangerous symptoms. Drinking a lot of water without accompanying salt intake may cause the body's sodium level to fall to dangerously low levels. In a cave environment, as in any extreme environment, the onset of confusion and the loss of coordination may have severe consequences for the individual, the group, and, consequently, for the cave.

A state of decreased mental awareness is sometimes called *stupor* which reminds me of *stupid*. Hence, you get cold, or overheated, or dehydrated … you get stupid. You get stupid … you get dead.

Don't Count on Thirst

What about thirst? Unfortunately, thirst may be absent. The sensation of thirst is not prominent until the degree of dehydration exceeds the kidney's ability to deal with the fluid depletion (Hubbard and others 1995). The absence of this early warning often contributes to accidental dehydration. However, voluntary restriction of fluid intake can also contribute to dehydration—for example, if the caver wants to conserve water or decrease urine output. Restricting fluid intake is not a wise choice.

Fluid Intake

Watch your fluid intake—drink enough water. Prevention of dehydration, whether primary or occurring with heat-related disorders, is accomplished by fluid intake.

Because of the absence of thirst, it is important to consume fluids at regular intervals. Endurance athletes often try to consume 250–500 milliliters (8–16 ounces) every 20–30 minutes. This is not practical for caving, but indicates that sometimes you should consider taking in 1–1.5 liters or quarts every hour or so, even when you're not thirsty.

Much has been written and hypothesized about the type of fluids that can prevent or correct dehydration.

- Alcoholic beverages do not help hydrate—in fact, alcohol causes one to become dehydrated—alcohol causes the kidneys to clear more water by inhibiting the body's natural antidiuretic hormones. (This is the real reason you make more trips to the restroom when you're drinking alcohol.)

- Caffeinated beverages are also a poor choice as they tend to increase urine output.
- Generally, large amounts of water leave the stomach faster than small amounts, but huge gulps can cause discomfort.
- Cold liquids leave the stomach faster than warm ones, but cold fluids may not be available in the cave environment.
- Many commercial sports drinks are available. Sport drink supplements are based on the theory that glucose (or sucrose) promotes stomach emptying; that glucose plus sodium promotes water absorption by the small intestine; that glucose promotes performance; and that sodium is needed to replace salt loses. Research indicates no difference between water and glucose-sodium solutions in terms of absorption or effectiveness in rehydration or effect on performance.
- My personal preference is Gatorade® with a little extra added. To 1 liter (1 quart) of Gatorade, add about 1 liter or quart of water (dilute to half-strength); then add 2 milliliters (1/2 teaspoon) sodium chloride (table salt) and approximately 2 milliliters (1/2 teaspoon) potassium chloride (salt substitute). This combination reduces muscle cramps. However, beverages containing potassium (salt substitute) should not be used to *treat* heatstroke because the body's potassium may already be elevated.

Nutrition

A well-balanced, nutritionally sound diet is a *sine qua non* of good performance in any physical undertaking. Caving is no exception.

A well-balanced, nutritionally sound diet is a *sine qua non* of good performance in any physical undertaking. Caving is no exception.

One who is calorically deficient will not have the necessary stamina for prolonged exertion. Therefore, someone on a severely calorie-restricted diet should approach caving with caution.

Persons adhering to a vegetarian diet should pay special attention to the balance of protein they are taking in—it takes special care and effort to maintain proper proportions of different proteins in a vegan diet. This caveat should not be taken as prohibition of vegetarian diets. There is no problem with a properly balanced diet—vegan or otherwise.

Carbohydrate loading, consuming large quantities of complex carbohydrates 12–24 hours before exertion, has been practiced for a long time by endurance runners and other athletes and may be of some benefit.

It may also be wise to add extra salt to food for 1–2 days before heavy exertion, especially if one has been on a sodium-restricted diet. (If sodium restriction was prescribed by a physician or dietitian, definitely consult with them before adding salt). Muscle cramps are likely to result from inadequate sodium intake before or during heavy exertion. Increasing sodium intake in anticipation of (or in response to) exertion and sweating may help avoid muscle cramps, muscle weakness, and the consequent decrease in coordination that can create new impacts in caves.

Physical Conditioning

Before engaging in any physically exerting activity, one should prepare with a reasonable conditioning program. Generally, 60–90 minutes of aerobic exercise per day for several weeks is a good start. Aerobic exercise raises the heart rate. The exact nature, extent, intensity, and duration is left to the discretion of the reader with the warning that a physician should be consulted before beginning any exercise program—especially if you have chronic or serious medical conditions. Such conditions would include—but are not limited to—heart disease, diabetes, seizure disorders, hypertension, obesity, and thyroid disorders.

Failure to prepare for exertion will result in inevitable decreases in endurance and coordination, which can easily result in new impacts to cave resources. Decreases in performance abilities will predispose cavers to accidents with the ensuing rescues and the resultant impacts on the cave environment.

We humans literally leave piles of microscopic matter in our wake. We shed all the time—in the house, in the truck, on the trail, and inside caves.

Personal Hygiene

Personal hygiene may seem an unusual subject in a book about caving, but it can have great significance. Humans are walking bacterial farms—each body is covered from head to toe with bacteria. Especially high concentrations are found around the mouth and perianal areas. I know some readers are thinking, "Not me!" But no one is exempt from this biological fact.

Each of us sheds millions of skin cells every day. Dust mites that live in carpets and mattresses thrive on tiny flakes of skin and debris. We also shed countless ectoparasites and other microbes that live on our skin and hair. We humans literally leave piles of microscopic matter in our wake. We shed all the time—in the house, in the truck, on the trail, and inside caves.

A simple shower with antibacterial soap may reduce bacteria on a caver's skin, but you can never completely eliminate the bacteria. Laundering caving clothes, packs, vertical gear, and other equipment between cave trips will decrease the chance of introducing contaminates to a cave or causing cross-contamination between various cave environments.

Many enteric infections—those affecting the gastrointestinal tract—are spread by the simple failure to wash hands. Hand-washing is not practical in a cave. But it is easy to carry a small bottle of disinfecting-gel hand cleaner or a small supply of antimicrobial wipes.

Human Waste Disposal

The disposal of human waste is a subject that many people avoid because of the unpleasant connotations. Most of us are accustomed to simply pushing the lever, and flush, it's gone … out of sight, out of mind. The absence of plumbing or even pit toilets in most caves mandates our consideration of this subject.

The absence of plumbing or even pit toilets in most caves mandates our consideration of this subject.

In caves with flowing water or those subject to flooding, it has been common practice to dispose of liquid wastes wherever it was convenient on the assumption that it would be diluted and "flushed away." The cavalier implementation of this practice should be reconsidered. Liquid wastes not deposited directly into flowing water may remain for a long time where they are left, thus altering the chemistry of the cave environment, perhaps to the detriment of those organisms that live there.

Wastes deposited into flowing water produce their own problems, apart from the effects on the chemistry of the water and the organisms that may live there—you may be washing or walking in that same water. Furthermore, that water is probably somebody's drinking water (maybe yours).

Solid wastes in "wet" caves are often, unfortunately, treated the same way. This cavalier attitude has led to the degradation of the environment around many alpine climbing routes.

Microorganisms can move through soil—bacteria can penetrate to about 30 meters (100 feet) in sandy soil and viruses can travel about 90 meters (300 feet) laterally (Backer 1995).

Some sources recommend that feces be spread on rocks in the sun to be desiccated (Backer 1995). The USDA Forest Service recommends that feces be buried 20–30 centimeters (8–12 inches) deep, off trail, and 60 meters (200 feet) from water sources (United States Department of Agriculture-Forest Service 2003; Leave No Trace®, Inc. 2003). Obviously, neither of these options is viable in cave environments.

Every effort should be made to remove solid wastes from caves—no other option exists. This subject has been discussed jokingly as well as in serious debates for a long time. The simplest and most reliable method remains to defecate directly into a zip-closure plastic bag, and double-bag it into another sealable bag for aesthetics and safety. The usefulness of additional bags and desiccating or deodorizing agents (for example, powdered bleach or baking soda) is left to the individual to decide. Some cavers have found that "oven baking bags" are reliable containers. Commercially available dry bags have also been suggested as an outer container for added security. (See burrito bag instructions, page 269; also see human waste, pages 35, 71, and 125.)

Preventing Cave-Related Infectious and Environmental Diseases

A complete discussion of cave-related illnesses is beyond the scope of this chapter. For detailed information, check with the Centers for Disease Control and Prevention <http://www.cdc.gov/> and other sources listed at the end of this chapter. A few of the illnesses to consider are rabies, histoplasmosis, leptospirosis, and exposure to radon gas.

Rabies preexposure vaccine is recommended for anyone who will be around large numbers of bats.

Rabies

Preexposure prophylaxis is available for rabies and consists of three simple injections—a relatively painless and relatively safe procedure. Rabies preexposure vaccine is recommended for anyone who will be around large numbers of bats.

If rabies prophylaxis is obtained, it is important to periodically get blood *titers* checked, booster shots when appropriate, and postexposure vaccine if one's skin is broken by a bat bite (MMWR 1999).

Although there are reports of persons developing clinical rabies without an obvious bat bite, aerosol transmission of rabies virus has not been conclusively shown to occur (Gibbons 2002). Exposure to rabies can be treated with postexposure prophylaxis.

Exposure may be a bite (but since bats are usually small, the bite usually is not noticed) or contamination of an open wound or mucous membrane by bat saliva or nerve tissue (brain matter). Consequently, postexposure rabies prophylaxis should be considered when direct contact between a human and a bat has occurred—unless the exposed person can be absolutely certain a bite, scratch, or mucous membrane exposure did not occur.

Postexposure prophylaxis consists of an injection of immune globulin (antibodies) and a series of five vaccination shots (MMWR, 1999). Appropriate postexposure prophylaxis is widely available in the U.S., but some other countries use different vaccines and regimens.

Caution should be exercised if seeking postexposure prophylaxis overseas. The most current information is readily available from the Centers for Disease Control and Prevention <http://www.cdc.gov/>. Unfortunately, *rabies encephalitis* (rabies virus infection of the brain) is a zero-tolerance disease—essentially anyone who develops the clinical syndrome of rabies will die.

Unfortunately, rabies encephalitis (rabies virus infection of the brain) is a zero-tolerance disease—essentially anyone who develops the clinical syndrome of rabies will die.

Histoplasmosis

There is no prevention for *histoplasmosis* other than avoiding dust contaminated with bat guano or using a HEPA filter respirator. Interestingly, it appears that this fungus grows in soil contaminated with bat or bird guano but not in the guano itself.

Histoplasmosis is fairly common in the eastern United States, especially in the Mississippi and Ohio River areas. Many persons living in these areas already have some antibodies to histoplasmosis. Persons residing outside these endemic areas (those areas where the fungus is common) may be more susceptible to infection or prone to more severe disease, but this is only a generalization.

There is no prevention for histoplasmosis other than avoiding dust contaminated with bat guano or using a HEPA filter respirator.

Leptospirosis

Leptospirosis is an infection caused by a corkscrew-shaped bacterium carried in the urine of some animals (including bats) and it can infect humans when they wade in water contaminated by animal urine. This infectious disease is easily treated with antibiotics (such as doxycycline).

Radon

The presence of radon gas in caves and its clinical significance for humans is still a matter of some debate. Radon gas is a *carcinogen* (a substance that causes cancer) but it appears to work synergistically with smoking. Exposure to radon gas in the home, where the exposure is long-term and continuous, appears to be a greater risk than the relatively short-term exposure one might encounter on a caving trip, even a multi-day expedition.

Other Illnesses

Other infectious diseases which have been discussed in relation to caving include hantavirus, blastomycosis, coccidioidomycosis, and plague. For additional details about these illnesses, refer to the sources for additional information at the end of this chapter.

- *Hantavirus* can cause hantavirus pulmonary syndrome, but epidemiological factors (that is, where the deer mice live) make it an unlikely risk for cavers.
- *Blastomycosis* is a fungal infection found in the southeastern U.S. and is sometimes spread by digging in contaminated soil. Using respirators when digging to find cave passages may be helpful.
- *Coccidioidomycosis* is a fungal infection found in the southwestern U.S. (also called Valley Fever or San Joaquin Fever).
- *Bubonic Plague* is endemic though uncommon in southwestern parts of the country and is caused by the bite of an infected flea.

Disclaimer

Every situation and every individual is different and must be evaluated on its own.

No treatment should be started or carried out based solely on this or any other protocol.

No medication, treatment, or medical device should be administered or applied without complete training in, and a thorough understanding of the uses, limitations, and possible complications of said treatment or device.

The opinions expressed herein are solely those of the author and do not represent the views, opinions, treatment protocols, or standards of any other individual, group of individuals, medical group or organization, hospital, other organization, the NSS, or any section thereof or any of its other internal organizations.

Cited References

Advisory Committee on Immunization Practices. 1999. Human rabies prevention, United States, recommendations of the Advisory Committee on Immunization Practices (ACIP). *Morbidity and Mortality Weekly Report* (MMWR) 199948(RR-1):3-13.

Armstrong LE, Hubbard RW, DeLuca JP, Christensen EL. 1987. Heat acclimatization during summer running in the northeastern United States. *Medical Science Sports and Exercise* 19(2):131-136.

Backer HD. 1995. Field water disinfection. In: Auerbach PS, editor. *Wilderness Medicine: Management of Wilderness and Environmental Emergencies*. p 1086.

Bowman WD. 1995. Wilderness survival. In: Auerbach PS, editor. *Wilderness Medicine: Management of Wilderness and Environmental Emergencies*. p 368.

Centers for Disease Control and Prevention (CDC). Internet <http://www.cdc.gov/>. Accessed 2003 Feb 5. [The Centers for Disease Control and Prevention (CDC) is recognized as the lead federal agency for protecting the health and safety of people at home and abroad, and providing credible information to enhance health decisions.]

Gibbons RV. 2002. Cryptogenic rabies, bats, and the question of aerosol transmission. *Annals of Emergency Medicine* 39(5):528-536.

Hubbard AW, Gaftin SL, Squire DL. 1995. Heat-related illnesses. In: Auerbach PS, editor. *Wilderness Medicine: Management of Wilderness and Environmental Emergencies*. p 175, 198.

Leave No Trace®, Inc. (LNT). 2003. Internet <http://www.lnt.org/TeachingLNT/LNTEnglish.html>. Accessed 2003 Feb 8.

Steinman AM, Hayward JS. 1995. Cold water immersion. In: Auerbach PS, editor. *Wilderness Medicine: Management of Wilderness and Environmental Emergencies*. p 109.

United States Department of Agriculture Forest Service, Mt. Adams Climbing Information. 2003. Web site < http://www.fs.fed.us/gpnf/wilderness/climbing-mount-adams.htm>. Accessed 2003 Feb 8.

Additional Reading

Auerbach PS. 2001. *Wilderness Medicine*. 4th edition. St.Louis (MO): Mosby. 1910 p.

Auerbach PS. 2003. *Medicine for the Outdoors: The Essential Guide to Emergency Medical Procedures and First Aid*. 4th edition. New York: Lyons Press. 512 p.

Forgey WW. 2000. *Wilderness Medicine: Beyond First Aid*. 5th edition. Guilford (CT): Globe Pequot Press. 246 p.

Forgey WW, editor. 2001. *Practice Guidelines for Wilderness Emergency Care*. 2nd edition. The Wilderness Medical Society. Guilford (CT): Globe Pequot Press. 109 p.

Hempel JC, Fregeau-Conover A, editors. 2001. *On Call: A Complete Reference for Cave Rescue*. Huntsville (AL): National Speleological Society. 384 p.

Sources for Additional Information

Medical Section of the National Speleological Society
Web: <**http://www.caves.org/section/medical/**>

Centers for Disease Control and Prevention (CDC)
1600 Clifton Rd NE
Atlanta GA 30333
Phone: 404-639-3311
Travelers' Information Hotline Phone: 877-394-8747
Web: <**http://www.cdc.gov/travel/**>

International Association for Medical Assistance to Travelers (IAMAT)
417 Center St
Lewiston NY 14092
Phone: 716.754.4883
Web: <**http://www.cybermall.co.nz/NZ/IAMAT/**>

The Wilderness Medical Society
3595 East Fountain Blvd Suite A1
Colorado Springs CO 80910
Phone: 719-572-9255
Fax: 719-572-1514
Web: <**http://www.wms.org/**>

Section C—Improving Caver Ethics

Public Relations: An Essential Element in Cave Conservation

Jay R. Jorden

What if you cleaned up a cave and no one noticed? All those hours spent with a toothbrush and spray bottle, scrubbing mud off soiled formations and picking lint out of crevices. The sore elbows and knees from seemingly endless scouring of tenacious, multicolored graffiti spray painted across flowstone. Flagging trails in the pristine, unmapped section of the cave only to find a muddy footprint on the next trip where some careless explorer had stepped away from the designated zone to snap a photo. Picking up beer cans and broken bottles inside the entrance. Patiently and painstakingly reassembling a stalagmite smashed into small calcite fragments by an earlier adventurer who had rushed through the cave in a time when the need to preserve its beauty was not even on the radar screen.

After a hard day of restoration, you may find little sympathy from some of your friends who've been off enjoying themselves exploring in a new cave. They gently kid you with comments like, "Why would you want to work that hard for free?"

Visitors to your cave club just want to go caving and may need some prodding to go on a restoration trip. The officers don't see the need for having a conservation chairman. "Why do we need that level of bureaucracy?" they ask. Sometimes, the prospect of having everyone else care as much as you do about cave conservation can be a distant dream, indeed.

That's where public relations (PR) with its many powerful tools enters the picture.

© Jay Jorden

Figure 1. Labor Day 2002 restoration project at La Gruta del Palmito in Nuevo Leon, Mexico. Participants labor alone or in small groups within Bustamante Cave's dark recesses, armed only with spray bottles and sponges in their attempts to remove spray-painted graffiti from large columns and flowstone.

What is Public Relations?

Abraham Lincoln once remarked: "Public sentiment is everything. With public sentiment, nothing can fail. Without it, nothing can succeed."

The need for effective public relations is often unstated, but nevertheless crucial. In a one-sentence definition, PR attempts through a variety of methods to put the organization's best foot forward and shape positive public opinion, regardless of the circumstances.

That entails more than a passive approach to publicity—not only cooperating with the media in providing information, but also reaching out with a conservation message in the hope of generating publicity. In this regard, PR is considered the long-term program of molding public sentiment, with publicity more precisely being the nuts-and-bolts, day-to-day methods of helping to achieve goals.

The definition adopted by the Public Relations Society of America in 1988 is widely accepted: "Public relations helps an organization and its

© Jay Jorden

Figure 2. A reporter from a Monterrey, Mexico television station interviews Orion Knox (at left) of Austin, Texas. A cave restoration participant looks on during the Labor Day 2001 weekend cave cleanup inside Bustamante Cave, Nuevo Leon, Mexico.

publics adapt mutually to each other."

By this point, the question of "publicity or no publicity" should have been answered. A truism in media relations is that if reporters get no cooperation from the primary contact on a story, they will make telephone calls and follow all leads until they get it anyway. Often, that comes at the expense of the primary contact who has defaulted on the obligation to provide details and misses the opportunity to help mold the message.

The idea is that if a reporter can't get the story from you, he or she will get it from someone else—who may not be your friend or may not agree with your goals and objectives. Weighing in after the fact to try to straighten out the misconceptions or inaccuracies is either impossible or ineffective. So if someone in your conservation group says matter-of-factly, "I don't talk to the press and you shouldn't either," resist that suggestion because it is detrimental in the long run.

If someone in your conservation group says matter-of-factly, "I don't talk to the press and you shouldn't either," resist that suggestion because it is dctrimental in the long run.

An example of this dynamic is in organized cave exploration and conservation groups. For many years, the stated policy of such organizations has been to avoid generating interest in caving among individuals who had no preexisting desire to pursue the activity. The reason has been that caves are nonrenewable resources and population pressure can adversely affect them.

However, after someone has independently developed an interest either through books, television, or other media, the National Speleological Society (NSS) and other groups want to be the source for educating and training that person in proper technique, safety, conservation, and equipment. It's a fine line to walk, and one which involves a certain amount of outreach.

A network of NSS internal organizations scattered around the country and even in some international venues provides this passive public relations approach to outreach. It involves some media relations, but no deliberate program of publicity. Outdoor-oriented groups in other activities would likely view this approach as "hiding one's light under a bushel basket," in contrast to their more aggressive outreach programs for gaining membership.

Nonetheless, successful case studies in cave conservation abound. They include the discovery and subsequent preservation at Arizona's Kartchner Caverns and at Lechuguilla Cave on Carlsbad Caverns National Park in New Mexico; efforts by Austin, Texas-based Bat Conservation International to change public perception; and the largely unrecognized work by thousands of other conservationists around the world to ensure that nonrenewable cave resources, including biological, geological, and cultural are available for generations to come.

Public relations can be tough to get a handle on. In one respect, PR is so pervasive that it's like the air we breathe. What we say in a group setting, what we write in our newsletters, and who we try to enlist in our efforts to influence opinion are all part of the practice. How we conduct ourselves in the public domain influences how others perceive our activities.

Several questions need to be answered, including when to publicize cave conservation efforts and, if so, whether the effort will be passive or aggressive.

Cavers and others must be motivated to join the cause, to help complete important tasks, and to continue involvement when burnout approaches or other competing interests surface. It is important that cave conservationists vigorously advocate their worth, both to the volunteers who are perhaps their best allies and to members of the general public, whose support is greatly needed. Support must be earned in the highly competitive public arena where a wide range of causes and interests vie for sentiment and where the demands on leisure time have never been greater.

Through publicity, cave conservationists can attempt to create favorable public opinion. They can inform the general public of their goals and objectives, activities, and problems.

Every organization—and every cause—has PR problems, although they may not be immediately identified as such. For organized caving and cave conservation, the noncaving public's challenge in perceiving the work of conservationists and speleologists must be met. To many, cavers are those who go "crawling around in holes in the ground."

© Jay Jorden

Figure 3. Norma Yolanda Rosales de Santos, Presidente Municipal of Bustamante, Nuevo Leon, Mexico, is interviewed by a Monterrey television crew at a banquet in honor of La Gruta del Palmito restoration project participants in September 2001.

In one part of Mexico, cavers have had to convince local residents that they are not worshipping the devil in the pits they descend. And why save bats? That's a fair question from someone who might believe that bats inherently have rabies, fly into big hairdos, or suck human blood.

Competing interests are an issue for public relations practice. Within the organized caving community, issues of cave and bat conservation compete with exploration, the value of sport caving, cave ownership, and others. Through all of these, the National Speleological Society and other nonprofits have the task of motivating their own members and volunteers to identify and handle the important work before them.

A broader goal of PR for the volunteer, nonprofit realm in which cave conservation resides is to make people want to join the cause, donate their time, and provide financial and other resources needed to carry on the work. Cave conservationists working through nonprofits and organizations financed through public appeals must maintain a climate of high acceptance so that as needs grow and increased budgets are required, more funds can be acquired.

A broader goal of PR for the volunteer, nonprofit realm in which cave conservation resides is to make people want to join the cause, donate their time, and provide financial and other resources needed to carry on the work.

The PR Toolkit

The basic toolkit of public relations includes the following attitudes, actions, and tangible tools:

- Ambition and determination to influence public opinion in a favorable way about the conservation group and its causes.
- Access to phone directories and media lists such as *Editor and Publisher International Yearbook.*
- Business cards, introductory letter, or other means of helping to legitimize and document conservation purposes.
- Camera or access to a photographer.
- Computer and access to the Internet.
- Telephone.
- Conservation-related brochures, either in stock and available from the National Speleological Society, 2813 Cave Avenue, Huntsville, Alabama 35810-4431. Access some brochures through the NSS Web site <http://www.caves.org/brochure>. Or tailor your own brochures to specific needs, such as cave gating.
- A message, or position statement, distilling the goals and objectives of the conservation group.
- Background sheet or *backgrounder* summarizing the conservation group's position statement and the message it is trying to convey to the

public, including press contact(s) and other pertinent, *boilerplate* (standardized text) information on the organization and its members.
- Audiovisual aids, such as a PowerPoint® presentation, analog or digital slide shows, videos, pictorial manuals, Web site, and so on, to help deliver the conservation group's message.
- Lists of those who can serve as volunteer publicists in the conservation organization.

In proactive public relations it's advantageous for conservationists to have their press contacts develop working relationships with media representatives.

Community Relations

For cave conservationists and members of conservation task forces in the NSS and other related organizations, a great initial challenge exists. Volunteers reside in many different communities where conservation is not a major element of life. Here, an extensive community relations program is needed to make residents aware of the organization's existence and conservation goals. It's a good idea to have regional representatives who should pursue these tasks:

- Answer fully and knowledgeably any questions from local media representatives.
- Periodically weigh community attitudes toward cave conservation.
- Listen to and evaluate any community demands upon conservationists.
- Equip themselves to execute positive programs designed to interpret concepts to the community.

This provides a firm base from which any private or public disputes with community residents can be resolved fairly and efficiently. Officers and principals of conservation groups who are willing to become involved should be encouraged to speak out on public issues or participate in social developments.

Press Relations

In proactive public relations, it is advantageous for conservationists to have their press contacts develop working relationships with media representatives. There should be continuing formal and informal exchange of information between the organization and the media.

For these high-profile contacts, it's good to know and be known by editors and writers, who then feel free to call and ask for background on controversies and issues. Without this background, in emergency situations, writers and editors who lack a perspective on the organization's work end up writing only as accurately as they can from a limited vantage.

Generally, a strong relationship of trust between writer and source leads to quality news and information. However, a strong media relations program as described here requires the staffing and depth to answer immediately and well all inquiries from major press sources.

© Val Hildreth-Werker

Figure 4. Television journalist and helicopter pilot, Bob Martin, reports for CBS and KRQE News13 in Albuquerque, New Mexico. He does it all—flies the chopper, handles the lights and audio, shoots the video, and reports on camera—a perfect solution for conservation-directed media coverage in caves. For this influential story on science and conservation, there was one journalist, one belt of batteries for a video light, and three cavers.

When a conservationist is approached by a news reporter or representative of a media organization, the goal is to refer that person to the designated press officer or contact for the group. The press officer ideally should have some experience with the media. If none is available and the reporter is on deadline, here are some guidelines:

- Treat the reporter at least as

well as someone who is interested in joining the conservation group. Writers are in a position to help the cause with their work product. Even in those rare occasions where a reporter might plan for some reason to write a critical article about conservation, if treated with personal courtesy, he or she may end up with material that is less damaging—the reporter is likely to write more negatively if treated curtly or boorishly.

© Val Hildreth-Werker

Figure 5. A chopper hop to the cave's parking lot, three cavers, and one journalist—this was a terrific low-impact way to create a cave conservation message for television. Pictured are Diana Northup, Bob Martin, Val Hildreth-Werker, and Jim Werker on McKittrick Hill in New Mexico. (A fun caver note: We placed the cave permit on the windshield of the helicopter before we entered the cave.)

- If the query is by telephone, it is proper to ask for name, affiliation, and telephone number and then answer questions if no press officer has been designated. If the position exists, take a message and get the press officer to make the callback.

Ground Rules for Press Officers

- Ask for a business card from the reporter to verify affiliation.
- Request information on the scope of the article on which the writer is working.
- If not personally familiar with the publication or news outlet, ask the writer for details, including readership or viewership.
- Deal in concrete, truthful answers. Good writers will cross-check facts.
- If you can't answer a question, say so and offer to follow up later after researching it.
- Remember what is said to the writer. That information is used for later follow-up with the reporter.
- Ask for copies of the clips, if dealing with a print journalist, or videotape when dealing with broadcast representatives.

Fundamentals of Good Press Relations

- The people on the front lines of media contact in the conservation organization must be convinced that reporters are not out to use them. Reporters have a basic responsibility to be accurate, fair, and impartial—to present readers or viewers with a well-rounded, enlightening report.
- Define a clear message and stay on it with all comments.
- Dealings with the working press should be handled as if the media contact fully intends to continue cooperating for years to come.
- After establishing ground rules, shoot straight with reporters and you will be treated in kind.
- Print and on-air deadlines must be respected. Time is always a factor in editorial material. A wire service that once had the motto of "a deadline every minute" has in the technologically advanced new millennium refined that to "a deadline every millisecond." When writers call the NSS headquarters in Huntsville, Alabama, for information, they are usually working on a deadline. Television and radio representatives have critical deadlines, such as early-morning "drive time." Good PR meets these deadlines.
- Good writing in news releases is essential. There is much to be said for uncomplicated thought, simply stated. Beware of badly written, self-serving press releases.
- Try for imagination and freshness. Hold a press conference in a cave

or at an entrance? It might be a possibility.

- Also, avoid overestimating the impact of one little news story in one newspaper or in one region, or *quid pro quo* stories—those written in exchange for advertising, as in special sections. To change attitudes in a dramatic way, nationwide media exposure is required.
- There's no such thing as a free lunch—or free publicity. For nonprofits and volunteer conservationists, that means donated time and plenty of legwork to prepare releases, make the telephone calls, set up interviews, and follow up.

The end result of good PR is good publicity, channeled through the media at all levels—print and electronic.

History

The study of public relations has a long and distinguished history. Edward L. Bernays traces public relations from the beginning of time: "Men first communicated by signals, then by speech, then by writing. After writing, various types of mechanical devices were developed for conveying fact, thought, and meaning. Whenever and wherever there were such developments, they were also employed to express and mold opinion."

Bernays traces the course of PR from this origin in the Dark Ages, through the early American colonies of the 1600s, and finally through the "public be damned" philosophy of the latter part of the 19th century to the "public be informed" philosophy of the early 20th century. He believes that modern PR saw its inception between 1919 and 1929, and that between 1929 and 1941 it became an accepted and acknowledged part of the general business practice.

Bernays, in *The Engineering of Consent*, published in 1956, said: "Public relations is the attempt, by information, persuasion, and adjustment, to engineer public support for an activity, cause, movement or institution."

Perhaps the best cave-related case for PR is exemplified by bat conservation efforts. The early Romans believed bats to be creatures from the nether realms. The flying mammals were depicted in early drawings as inhabiting the underworld. They were threatening creatures, rumored to have supernatural powers, to be feared—and sometimes killed.

Now, of course, we know differently. But changing attitudes took years of slow, painstaking efforts by caver advocates and NSS members like Merlin Tuttle, founder of Bat Conservation International, to inform and educate the public. Through magazines, television, and other means, Tuttle and others have helped the public realize that bats have a beneficial place in ecology by eating many insects and pollinating many species of plants during nightly flights.

In retrospect, when using Bernays' definition of public relations as a guide, such efforts have been successful in engineering public support in the ongoing fight to save bats and their habitat.

Figure 6. This photograph of The Big Room in Carlsbad Cavern was made in 1993 during Tom Zannes' *Spirit of Exploration* video shoot for Carlsbad Caverns–Guadalupe Mountains Association. Cavers hauled in more than two miles of cable to power 40 movie lights placed in the Big Room. The short-term lighting mixture of daylight-balanced, incandescent, and fluorescent lights produced variations in color and enhanced the sense of depth in the video. This was the largest lighting job ever recorded in a cave for movie or video photography. Only a few large-chambered show caves in the world can handle this quantity of production gear without creating tremendous impacts on the cave environment.

Publics

The issue of publics is an important one for public relations work—picking the right strategies and methods are crucial to success. Publics are both external and internal—different strategies and methods are used to reach and influence them.

Organized caving members, boards of directors, staff, contributors, and internal organizations make up many of the internal constituencies which must be served through publications, Web sites, benefits of dues-paying membership, and so on.

On the outside, the picture becomes more complicated. The list of outside publics may include related groups such as the American Association for the Advancement of Science, community groups, government, and legislators.

> The issue of publics is an important one for public relations work—picking the right strategies and methods are crucial to success. Publics are both external and internal—different strategies and methods are used to reach and influence them.

Primary publics must be identified and programs developed for each. As an example, legislative and governmental publics are especially important to cave conservationists. They are increasingly affected by legislation regarding state and national parks and public lands, as well as hearings and decrees by legislative committees and state, federal, and local regulatory agencies. Aggressive public relations can be useful—efforts by the NSS were largely responsible for procuring the passage of the Federal Cave Resources Protection Act.

Vulnerable indeed is a group that does not seek out an able, sensitive listening post in Washington. Conservationists need to get advance signals of legislative matters that affect them. The purpose of public relations here is to give governmental representatives a frame of reference for evaluating and guiding their Washington activities. In state capitals, help in lobbying is crucial.

Usually, success in governmental public relations requires a series of activities that may extend through many months:

- Fact finding
- Interpretation of governmental actions to officers of the conservation group
- Counseling on management practices in light of developments
- Interpretation of conservationists' actions to legislators
- Advocacy of a position
- Using legislative connections as a springboard
- Support for action with government

Cave conservation has many parallels with the practice of environmental public relations. The general public in the United States only became aware of needs in the natural environment in the early 1960s and 1970s, although some forward-thinking people were concerned as far back as the 1940s, when the NSS was formed.

The emergence of an environmental ethic in this country as well as passage of far-reaching federal and state legislation and local laws protecting wildlife, governing air and water pollution, solid waste disposal, noise, land use, and hazardous substances have had an enormous impact on the activities of business and industry.

The parallels in cave conservation are numerous and include many broad environmental conservation concerns:

- Land use
- Habitat
- Vandalism
- Herbicides and pesticides
- Solid waste disposal
- Water contamination
- Other pollution issues

Audiences include governmental officials, scientific and academic groups,

Audiences include governmental officials, scientific and academic groups, citizens' groups, specialized and general media, community leaders, and managers of companies.

citizens' groups, specialized and general media, community leaders, and managers of companies. Printed booklets, films, and slide presentations usually are effective in telling an organization's environmental accomplishments to a general audience.

However, for scientists, governmental bodies, teachers, and environmentalists, many organizations rely on face-to-face meetings in small group settings or formal presentations at seminars and symposia. In those situations, information conveyed must be complete enough for a scientist to process. The door should be left open for two-way communication.

Besides seminars, symposia, and other scientific meetings, other forms of communication with the scientific community are through direct mail to well-targeted groups, articles in the semiprofessional press, personal visits to selected leaders, publication of materials in appropriate professional journals, letters to journal editors, and advertisements in such journals.

In dealing with the environmental press, strive to know the editor of the publication you've targeted and the work he or she has done. Be well-versed in the subject at hand and identify news angles that can help market the message you want to convey.

Steps for Environmental PR

- Marshal all of the available facts on the situation and prepare a *PR environmental inventory*.
- Establish an early-warning system so that the organization is not caught by surprise with restrictive legislation or other adverse developments.
- Prepare *position papers* (easy-to-use, in-depth background material) on all sides of the environmental issues.
- Establish and cultivate contacts with governmental, academic, environmental, and scientific groups.
- Set up a procedure to answer criticism immediately and forcefully.
- Explain and promote the positive environmental actions of the cave conservation group through every means available.

Establish Credibility

To establish the organization's reputation for credibility in environmental issues, successful communicators must adhere to the following guidelines:

- Demand strict accuracy in all materials for dissemination.
- Keep attuned to major developments in the conservation arena.
- Collect, analyze, and interpret available facts.
- Prepare environmental policy statements and other informational materials on matters projected to involve public attention.
- Prepare to challenge sensationalized attacks.
- Call on experts for help and assistance in communicating technical stories.
- Keep conservation support groups informed of environmental programs and plans.

Nuts-and-Bolts Methods

There are many ways to publicize cave conservation efforts and work toward influencing public opinion. Treatment of the subject here is not meant to be exhaustive.

Annual or Biennial Reports

Even 501(c)3 not-for-profit corporations, such as the NSS, produce reports containing financial information, mission statements, and summaries of activities for designated periods. They are used to target information for grant writing purposes as well as for legislative constituencies and other groups.

Audiovisual Tools

Slide shows, videos, and PowerPoint presentations are effective in communicating the cave conservation message with strong visual impact and clarity.

Backgrounders

The Society's Public Relations Committee uses a four-page background sheet in communications with journalists, prospective book authors, and others who contact it for information. Such *backgrounders* succinctly state the organization's purpose, history, organization, policies, activities, and services.

Brochures and Pamphlets

The National Speleological Society prints several four-color brochures that promote conservation of cave resources. Titles include *Fragile Underground, Bats,* and *Lava Tubes*. These brochures provide details on protecting nonrenewable cave resources and cave-dwelling animals. *Guide to Responsible Caving* is an NSS booklet designed to introduce people to conscientious caving practices. Brochures, pamphlets, and other related materials are distributed on request by the NSS office and can also be downloaded from the Society's Web site at <http://www.caves.org/>.

Such *backgrounders* succinctly state the organization's purpose, history, organization, policies, activities, and services.

Displays and Bulletin Boards

At meetings of related conservation groups, seminars, and other gatherings, static displays can be a useful way of carrying the organization's message. Photos of recent activities with accompanying captions, news releases, newspaper clippings, brief articles, and other information can be placed on upright multimedia boards.

A table-top, museum-quality display about the NSS is available for events. Plan to pay shipping costs, and find information about display availability on the NSS Web site <http://www.caves.org/> under conservation, or contact the NSS Office in Huntsville, Alabama.

News Releases

Preferably one page, these dispatches are prepared for faxing, mailing, or e-mailing to newspapers, radio and television stations, wire services, and other media outlets. A good release addresses the traditional *Five Ws and H* of journalism: *Who, What, Where, When, Why, and How.*

Preferably, the *lead* or first paragraph of the release will include the majority of these five. Writing releases is a craft, since it is likely the most efficient and cost-effective way for a conservation group to get its message out in wide distribution.

For maximum effect, it's best to identify a *news angle* or *peg* for an assignment desk. Think of it as trying to give an editor a reason why he or she should assign the release to a reporter for follow-up calls and potential development of an article or broadcast piece. A news release is always followed with telephone calls to complete the pitch and offer to provide more information.

Newsletters and House Publications

As simple as a front-and-back memorandum or as lavish as a glossy magazine, newsletters are important workhorses in PR. Publicity committees often launch campaigns within their pages. But newsletters should be more than mere propaganda vehicles. Journalistic standards such as impartiality—getting both sides of an argument—and fairness should not stop at the newsletter editor's door. Exhibiting such traits lends publications

For Immediate Release
September 13, 2002
Thomas Lera, NSS Administrative VP
2813 Cave Avenue
Huntsville AL 35810-4413
Phone: 256-852-1300
E-mail: avp@caves.org

Bat Stamp Series Issued by USPS

The United States Postal Service is issuing some very attractive American Bat stamps, featuring photographs from a National Speleological Society member.

The pane of 20 stamps consists of four different stamp designs featuring photographs of bats found in the continental United States: the red bat, the pallid bat, the spotted bat and the leaf-nosed bat. The photographs on the stamps and on the selvage were all taken by Dr. Merlin D. Tuttle, NSS Member #13020HM, of Bat Conservation International, Inc., based in Austin, Texas. The bats were released unharmed after being photographed.

Tom Lera, the NSS administrative vice president and a supporter in getting these stamps approved will be on hand for the ceremony at the Congress Avenue Bridge, which shelters a bat colony. More than 2,000 people were also expected to celebrate Austin's famous winged mammals at the sixth annual, "Freetail-Free-For-All," sponsored by Bat Conservation International to raise awareness about the city's unique urban wildlife and to help inspire young conservationists.

The Huntsville, Alabama-based Society is the nation's largest caving organization, composed of more than 12,000 members in more than 150 grottos or local chapters.

The Society, founded in 1941, promotes conservation and safe exploration through training programs. The NSS motto states, "Take nothing but pictures, leave nothing but footprints, kill nothing but time."

Figure 6. Typical press release. Preferably one page, these dispatches are prepared for faxing, mailing, or e-mailing to newspapers, radio and television stations, wire services, and other media outlets.

more credibility.

Policies

A written PR policy statement formalizes the conservation campaign's goals and objectives. Along with a policy statement for internal distribution, it is useful to develop such a document for public consumption.

The internal communication serves as a compass to keep the group's publicists on track with a unified message—the external policy is a distillation of that desired message. One example of an external statement is the NSS conservation policy:

> *Caves have unique scientific, recreational, and scenic values. These values are endangered by both carelessness and intentional vandalism. These values, once gone, cannot be recovered. The responsibility for protecting caves must be assumed by those who study and enjoy them.*
>
> *Accordingly, the intention of the Society is to work for the preservation of caves with a realistic policy supported by effective programs for: the encouragement of self-discipline among cavers; education and research concerning the causes and prevention of cave damage; and special projects, including cooperation with other groups similarly dedicated to the conservation of natural areas...*

The policy continues, urging every NSS member to take personal responsibility for communicating the cave conservation ethic to all cave users. (See page 253 for complete text of NSS Conservation Policy; also see NSS Preservation Policy, page 254.)

Speakers' Bureaus

Structured services for providing speakers on request can provide conservation volunteers another forum for discussing their views.

Conservation PR plans should contain a statement of the strategies envisioned in achieving organizational objectives. Effectiveness of publicity efforts can be hard to measure. Returns often cannot be related to raw numbers in terms of dollars spent or volunteer caver hours expended. Instead, the measure must be related to organizational objectives, especially those evolving to address the public's mood in terms of conservation awareness. Since public trends and moods are changeable, PR must be alert, flexible, and responsive to such challenges.

Additional Reading

Bernays EL. 1956. *The Engineering of Consent*. Norman (OK): University of Oklahoma Press. 246 p.

Cutlip SM, Center AH. 1978. *Effective Public Relations*. Englewood Cliffs (NJ): Prentice-Hall. 612 p.

Lesly P, editor. 1978. *Lesly's Public Relations Handbook*. Englewood Cliffs (NJ): Prentice-Hall. 663 p.

Reilly RT. 1981. *Public Relations in Action*. Englewood Cliffs (NJ): Prentice-Hall. 468 p.

Geocaching and Caves: Finding a Common Ground

Hazel A. Barton

The sport of geocaching is an example of a Public Relations opportunity for the caving community. Geocaching is an orienteering sport. The participants use GPS (global-positioning system) units to find a specific location where a cache has been hidden. Individuals, who collectively call themselves geocachers, have placed these caches in over 150 countries, including the U.S.

The geocachers post the location coordinates of their caches on geocaching Web sites. Examples of Web sites include: <http://www.geocaching.com/>; <http://www.navicache.com/>; <http://www.keenpeople.com/>; <http://www.terracachers.org/>. Other geocachers then use their GPSs to find the cache, usually a small plastic container or ammo can, which contains some kind of reward. The person who finds the cache may take one of these rewards, provided they replace it with an item of their own (hence the often eclectic "rewards"). They then sign the register and post comments about the cache on the Web site.

Geocaching has thousands of active participants all over the world. Geocachers pride themselves on how many locations they have found, or how many interesting geocaches they've placed. They have a motto that encourages participants to act in a conservation-minded manner—*Cache in, trash out.*

NSS Geocaching Committee

Since the inception of the NSS Geocaching Committee (a subcommittee of the NSS Conservation Committee), the aim has been to address the problem of placing geocaches in sensitive and potentially dangerous caves, and to prevent the posting of sensitive cave locations on the Web. We are using the same technique that cavers traditionally use to help preserve cave environments—education and outreach.

Geocaching is a rapidly growing sport, practiced by many outdoor enthusiasts and conservation-minded individuals. Many of these people simply lack an understanding of cave environments—they don't know the inherent risks and the delicate nature of cave ecosystems. The intent of many cachers is to share their excitement of caves by placing caches—not to deliberately damage caves or provoke cavers. It is important that we remain objective, and not get angry at anyone who expresses an interest in caves (remember, all cavers started out as noncavers).

We must continue to help geocachers understand the delicate and sensitive nature of caves. Many cavers take geocachers caving and explain why some cave environments do not constitute a good caching environment.

Equally important are the efforts to identify caves that may be conducive to caching, and to encourage the creation of virtual caches, which do not require placement of a permanent cache. We are also suggesting that geocachers use multi-caches for caves, which prevent the placing of actual cave locations on the Web. Indeed, many cavers who are becoming geocachers are working at the grassroots level to help educate other geocachers.

Despite all these positives, there have unfortunately been some negatives along the way. Tensions can get high regarding the placement of geocaches in caves, and the knee-jerk reaction of "not in our caves" does nothing to calm the situation or help cachers understand our concerns. In one case, out of vengeance against cavers who removed or destroyed caches, one cacher began placing caches in as many caves as he could find. This resulted in the placing of over a dozen cave locations on the Web, even going as far as to name one cache "Vengeance Cache." Luckily cooler heads prevailed and the cacher graciously removed these caches and the Web locations.

Cavers need to try and be accommodating to this new sport, just as others were to the growing sport of caving in the U.S. Geocachers are becoming organized and participation is increasing. In some areas, we will be forced to play in the sand pit together, so now is a good a time to start figuring out how.

At the initiation of the geocaching community, a caching "ombudsman" was created to mediate between cavers and cachers when cave-cache issues arise. The ombudsman can be contacted at <geocaching-issues@cavescience.com>. Members of the NSS Geocaching Committee are fulfilling this mediating role. In the future, we would like to generate a list of caver/cacher contacts in every state. The caver contacts will help educate geocachers interested in caves about safe caving and protecting cave environments.

Part 3
Restoration

Section A—Introducing Cave Restoration

Overview of Cave Restoration

Val Hildreth-Werker and Jim C. Werker

Human actions can negatively impact caves and their contents. Cave restoration efforts serve to remediate or repair damage caused by human carelessness, inadvertent actions, or intentional vandalism. Project planning for cave restoration often begins with thorough inventory and documentation of features, resources, and values.

Inventory and Project Planning

In-cave resource inventory involves the systematic recording and archiving of scientific and cultural information found within a cave. An inventory may include data on aesthetic values, archaeology, biology, climatology, cultural material, chemistry, geology, history, hydrology, microbial habitats, mineralogy, meteorology, paleontology, speleogenesis, and impacts caused by human visitation. (See resource inventory, page 19.)

When assessing a cave environment call on trained speleologists for assistance and consultation. Significant resources may be present, but visible only to the trained eye.

During inventory processes, evaluate damage and potential risks to resources. Clarify the purpose of restoration. Clearly defined goals backed by careful project planning and implementation should minimize alteration of resources during cave restoration efforts. Inform all restoration personnel about the objectives and cautionary measures for each project.

Cave restoration efforts serve to remediate or repair damage caused by human carelessness, inadvertent actions, or intentional vandalism.

Do No Harm

First among all restoration objectives is *do no harm*. Perform restoration tasks without damaging cave values. Sometimes deciding not to attempt restoration is the best choice. (See do no harm, page 329.)

Cave restoration should focus on one of three purposes. Decisions will usually emphasize one of the following values:

- Restoration to a former natural condition
- Restoration to a previous historic period
- General improvement of the current aesthetic state

In a show cave with decades of accumulated impact, partial restoration may be the best achievable option. Restoration to a former natural condition is more likely to be successful in a wild cave passage that is not as ecologically disturbed. (See biology, page 33.) In karst terrains where sinkholes have been used as trash dumping sites, cleanup projects may help mitigate pollution and improve groundwater quality. (See sinkholes, page 381.)

Volunteers

Caving groups volunteer thousands of hours annually to cave and karst projects. Cavers donate their time, labor, materials, and expertise for conservation efforts. On public and private lands across the United States, cavers provide significant volunteer value to cave and karst protection. (See volunteer value, page 321.) Grants, cost shares, partial reimbursements, cooperative agreements, and project contracts with caving groups may provide support for restoration projects. (See agreements, page 309.)

> **Why Call it Cave Restoration?**
> **Val Hildreth-Werker and Jim C. Werker**
>
> How should we describe or define cave restoration?
> Is it clean up?
> Remediation?
> Rehabilitation?
> Reclamation?
> Rejuvenation?
> Renovation?
> Refurbishment?
> Revivification?
> Repair?
>
> Cave restoration may be described with any of these words. Most cavers use the term cave restoration. In this manual, the nitty-gritty tactics are more important than semantics. This book offers the practical how-to.
> In this volume, we present techniques for restoration and methods for repair in separate sections. Taking no action is also emphasized as sometimes being the best common sense solution for the conservation of a particular cave system.
> One section targets restoration objectives that typically involve the restoration of cave passages.
> Another section concentrates on restoring specific speleothem types and pertains to both show caves and undeveloped caves.
> And the final part of this book describes techniques for speleothem repair.
> Chapters include information to acquaint cavers and cave managers with how to think about, plan, and perform
>
> *(Continued on next page)*

Categories of Cave Restoration

Cave restoration projects employ a variety of tasks, tools, and skills. Projects might include rubble and debris removal, sinkhole cleanout, historic marking evaluations, spray paint cleaning, lint picking, habitat restoration, or speleothem cleaning and repair.

Natural and cultural cave values can be damaged by overzealous restorationists; thus inventory, reconnaissance, and project planning are essential elements for any successful restoration. In wild or undeveloped caves that receive intermittent visitation, task areas and techniques are carefully defined and the number of cavers is usually limited to small, efficient teams with experienced restoration leaders. In caves that have high levels of visitation or historic commercial use, significant new negative impacts are less likely to occur from restoration efforts.

Habitat

Restoring entrance features, airflow, or hydrologic conditions may support rehabilitation of species and habitats (Aley 1989). (See gates, page 147; also see hydrology, page 121.) Old trash and woodpiles may be providing special habitat areas for cave species and are usually removed in stages to allow fauna to migrate to new areas and biofilms to recover. (See woodpiles, page 37 and page 72.)

Lint

Some restoration involves carefully removing lint and dust (garment fibers, epidermal matter, hair, and small particles of debris) from along tour trails where it accumulates, discolors formations, and provides habitat for opportunistic organisms (Jablonsky and others 1995). (See lint, page 351.)

Lamp Flora

Algae, mosses, and other opportunistic photosynthetic organisms often grow near electric lights in show caves. Historically, sodium hypochlorite solutions and other chemical agents have been used to control lamp flora; however they must be prudently applied at minimum concentration and rinsed carefully because cave biota can be indiscriminately killed. (See bleach, page 349.)

Today, manufactured chemicals are not recommended for cave cleaning tasks because fumes, residues, and byproducts are harmful to biota and cave systems. Anthropogenic agents typically produce toxic byproducts through outgassing and degradation. (See anthropogenic, page 57.) Recommendations for improved control of undesirable algae and microflora include advanced lighting technologies and wavelengths selected to inhibit photosynthesis. Recent experience demonstrates that lamp flora can be eliminated with properly designed, located, and controlled lighting. (See lamp flora, page 343.)

Large Projects

Large projects removing rubble, artificial fill, and debris usually require multiple project days scheduled over several years. It may require groups of cavers working multi-day sessions to restore speleothems covered with

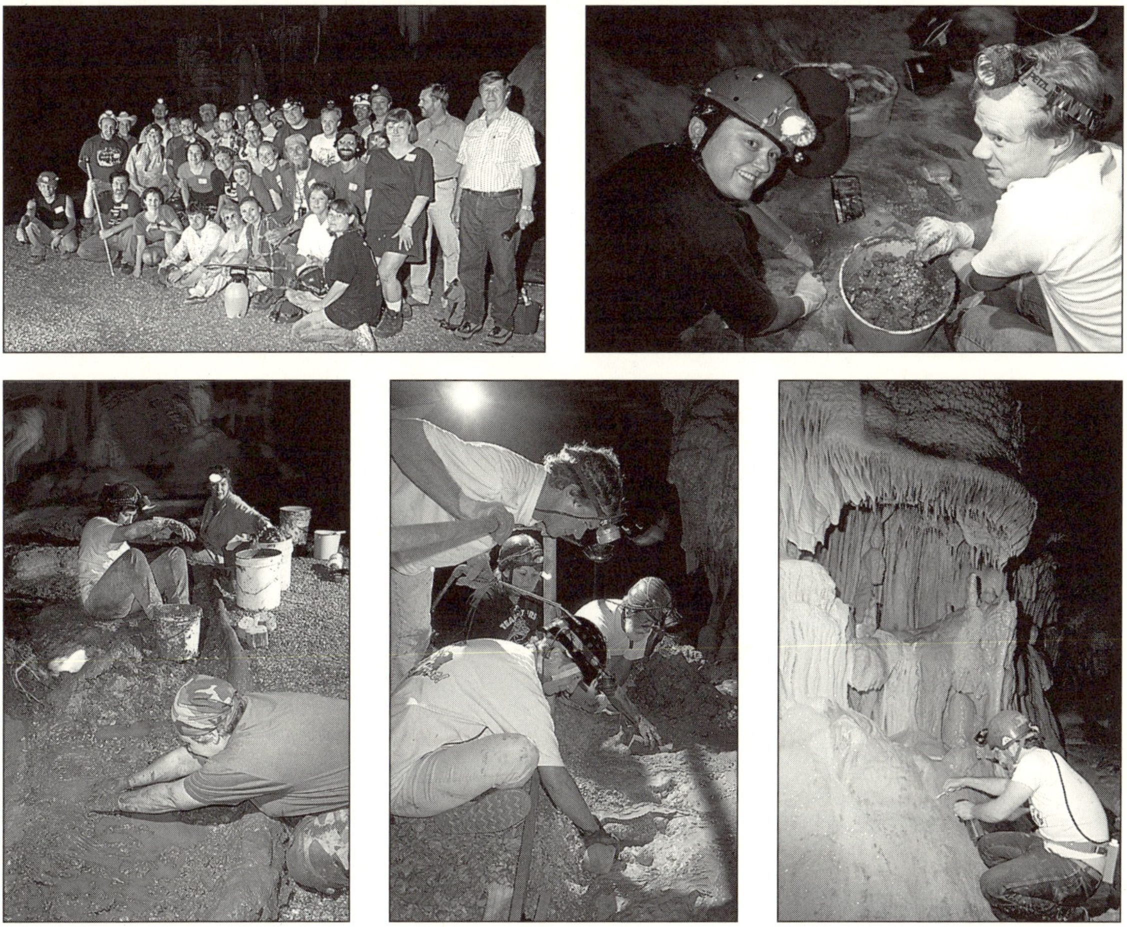

Photos © Val Hildreth-Werker

Figures 1-5. It takes many cavers to tackle the variety of restoration tasks during volunteer weekends at Cave Without a Name, Texas. Projects include digging fill material from show cave floors, sponging up muddy debris from rimstone, and cleaning delicate areas with toothbrushes.

excavation sediments or to remove construction debris and artificially deposited clastic sediments from show cave chambers. (See artificial fill, page 367.)

For example, in the late 1980s, tons of clay and rock were removed to restore natural floors at Wisconsin's Mystery Cave (Netherton 1993).

At Caverns of Sonora in Texas, 50 or more cavers gather for annual bucket brigade projects—teams remove tons of blast rock from trail construction, yet recognize and leave the natural breakdown in place (Veni 1998). (See bucket brigade, page 303.)

Labor-intensive restoration of Moondyne, a severely damaged former show cave in Australia, is an example of spectacular success in the restoration and repair of damaged tourist caves. After all the rubbish, infrastructure, and pathways were removed, the walls and speleothems were cleaned to appear pristine. The cave was restored to serve as a guided educational site (Bell 1993).

Another example of tremendous restoration success is Hidden River Cave, located in the town of Horse Cave, Kentucky. Because the town's sewage had been deposited in the cave, the American Cave Conservation Association came to the rescue, successfully restored the cave system, and subsequently developed the site into a productive karst education program and museum. (See Hidden River, page 331.)

(Continued)

conservation-oriented cave restoration. Many of us have reinvented the wheels of cave restoration technique each time we encountered an unfamiliar task.

The field-proven information presented in this manual is a starting point for new collaborations. Techniques will evolve as new ideas emerge. Research will advance informed decisions and methods. Contact the editors of this book or go to <http://www.caves.org/> with questions and improvements for cave restoration.

Small Projects

In contrast, delicate restoration projects on gypsum or cave pearls may require multiple project days or may only take a few hours of attention from one or two cavers. Some projects require minimal time and human resources. Even a single sponge stroke may prevent a muddy boot print from becoming permanently embedded in flowstone.

Restoration leaders evaluate the degree of previous impact in a cave passage, establish practical objectives, choose remediation techniques that will avoid creating new damage, and plan methods to protect the area from further damage.

Rock Art, Historic Writing, Mud Glyphs, and Graffiti

The history of an area may be literally written on the cave walls and should be documented and protected. Consultation with scientists, as well as historical and cultural preservation experts, is always appropriate. Document sites before erasing contemporary graffiti and removing trash. Be careful. Pictographs, petroglyphs, and historic signatures may be layered under contemporary graffiti. Historically important mud glyphs may be easily overlooked. (See rock art, page 99; see graffiti, page 333.)

Trash Removal

Approach trash removal carefully. Use common sense and safety protocols. Define and communicate specific safety precautions. It is wise to consult with historical preservation experts before planning trash and rubble removal projects. Hidden under the layers of recent contemporary articles may be items of historical importance.

Sensitive Area Protocols

Pristine environments that need restoration deserve careful strategies. When restoring recently discovered or rarely visited chambers, cavers follow stringent minimum impact guidelines. Some protected passages require special restoration techniques and changing to fresh, clean garments. Restoration in sensitive areas with suspected microbial significance might require more specialized clean-room techniques or sterile procedures (Hildreth-Werker and Werker 1999). (See microbiology, page 61; also see protocol, page 427.)

Trail Delineation

Trails through cave passages may be clearly marked to help confine visitor impact. Even in undeveloped caves, delineating trails on durable surfaces and previously compacted routes will aid in the preservation of sediments, small floor speleothems, and invertebrate populations. Flagging tape was first used to delineate pathways in 1970, Aguzou Cave, Aude, France (Cabrol 1997).

Unfortunately, if footprints are visible beyond the designated paths, others will tend to follow and trails will expand. Footprints outside of delineated trails may be erased to restore the natural appearance of clastic sediments and other surfaces. Use gentle combing motions with lightweight nylon brushes to erase footprints without stirring up dust. Camouflage deep footprints in rock flour or sand with natural sediments and soils taken from alongside the trail. Removing visible traces of human travel tends to discourage future damage (Hildreth-Werker and Werker 1997). (See trails, page 175.)

Some projects require toothbrushes or tweezers while others need shovels or power drills.

Well-Considered Decisions

Restoration leaders evaluate the degree of previous impact in a cave passage, establish practical objectives, choose remediation techniques to avoid creating new damage, and plan methods to protect the area from further harm.

During any cave project, unusual restoration concerns and decisions may crop up. For example, some cave surfaces are covered with thin, natural crusts that are easily scraped and marred by careless footsteps—can this damage be restored? Speleothems in lava caves differ from travertine formations—how do restoration techniques differ? Cleaning up carbide dumps can present some dangers—what is the protocol for safe removal?

Throughout this book, specific techniques are based on solid restoration concepts and research supported by common sense. Choosing to do nothing is sometimes the best decision. This manual offers guidelines—it is up to the reader to assimilate the information, create the options, and make wise choices for attempting cave restoration.

Cave Restoration Tools and Materials

Choose appropriate tools for various restoration tasks. Some projects require toothbrushes or tweezers while others need shovels or power drills. Be sure the tool fits the task. In show caves, industrial skills and equipment are necessary for accomplishing infrastructure-related tasks such as removal or remediation of wiring, guardrails, walkways, or PCB-filled transformers. It is wise to contact hazardous waste removal experts if toxic waste is encountered during cave or sinkhole restoration. When cleaning and repairing more delicate speleothems, gentle expertise and small tools are necessary.

Specialized materials are needed for long-term speleothem repair applications in cave systems. Recommended products include certain archival epoxies, museum-grade cyanoacrylate adhesives, and stainless steel support rods.

Choose Tools to Avoid Harming Speleothems

Scrub brushes with soft nylon bristles, soft absorbent sponges, surgical gloves, flexible plastic scrapers, buckets, hand-held sprayers, water, and human perseverance are effective tools for cleaning speleothems and many cave surfaces. (See flowstone, page 401.) Place catchments, towels, or sponge dams downslope to capture restoration runoff. All catchments should be removed from the cave and runoff water should be disposed of properly. (See runoff water, page 339 and page 396.)

Use Cave-Safe Materials

Use specialized materials for long-term speleothem repair applications in cave systems. Recommended products include certain archival epoxies, museum-grade cyanoacrylate adhesives, and stainless steel support rods. Degradation and outgassing characteristics are reduced and longevity characteristics are strengthened in these relatively cave-safe products. (See speleothem repair materials, page 445.) This handbook also presents repair strategies for draperies, soda straws, helictites, gypsum crust, flowstone, rimstone dams, and even large speleothems. (See repair techniques, page 455.) Poulter (1987) and others have developed similar methods in Australia.

Stay Up To Date

Research is needed to better define the safest materials for cave systems. Biochemists and materials engineers should confirm materials characteristics before products are used in cave environments for the construction of infrastructure or for repair and restoration (Spate and others 1998). (See cave-safe materials, page 167.) Future laboratory studies and field monitoring will facilitate scientific analyses of degradation, outgassing, and longevity for materials typically placed in cave systems.

As speleological research advances understanding of cave and karst ecological processes, restoration methods will continue to evolve and become less intrusive. Updated state-of-the-art restoration techniques will be developed based on new data. Cavers who perform restoration and repair are also defining low-impact caving guidelines. (See minimum-impact ethics, pages 263, 266.) If people will adopt improved caving ethics and learn to avoid unnecessary impacts to cave passages, then perhaps the need for restoration and repair will taper off.

Cited References

To accompany this restoration overview chapter, it was suggested we initiate the development of a comprehensive reference list to detailed information on cave restoration. The text of this overview includes numerous cross-references to restoration techniques in this volume. The additional reading list includes documents of historical significance as well as contemporary works that were not cited within the text of the chapter. Please contact the editors so we are sure to correct any omissions in future editions.

Aley T. 1989. Restoration and maintenance of natural cave microclimates. *NSS News* 47(2):39-40.

Bell P. 1993. Moondyne Cave: Development to guided adventure. *Cave Management in Australasia* 10:9-11.

Cabrol P. 1997. Protection of speleothems. In: Hill C, Forti P, editors. *Cave Minerals of the World.* 2nd edition. Huntsville (AL): National Speleological Society. p 294-309.

Hildreth-Werker V, Werker JC. 1997. Lechuguilla restoration: Techniques learned in the Southwest. *NSS News* 55(4):107-113.

Hildreth-Werker V, Werker JC. 1999. Restoration, trail designation, and microbial preservation in Lechuguilla Cave. In: Stitt RR, editor. *Proceedings of the 1997 National Cave Management Symposium: October 7-10 in Bellingham, Washington, and Chilliwack and Vancouver Island, B.C., Canada.* Huntsville [AL]: National Cave Management Symposium Steering Committee. p 181-189. [Also listed as 13th National Cave Management Symposium.]

Jablonsky P, Kraemer S, Yett B. 1995. Lint in caves. In: Pate DL, editor. *Proceedings of the 1993 National Cave Management Symposium: Carlsbad, New Mexico, October 27-30, 1993.* [Huntsville (AL)]: National Cave Management Symposium Steering Committee. p 73-81.

Netherton W. 1993. Mystery Cave trails: Past, present and future. In: Jorden JR, Obele RK, editors. *Proceedings of the 1989 National Cave Management Symposium: New Braunfels, Texas.* Austin (TX): Texas Cave Management Association; Texas Parks and Wildlife Department. p 141-147.

Poulter N. 1987. The restoration of the Jewel Casket, Yallingup Cave, W.A. *Australian Caver* 115:6-8.

Spate A, Hamilton-Smith E, Little L, Holland E. 1998. Best practice and tourist cave engineering. In: Smith DW, editor. *Cave and Karst Management in Australasia XII.* Carlton South, Victoria, Australia: Australasian Cave and Karst Management Asssociation. p 97-109. [*Cave and Karst Management in Australasia.* 12:97-109].

Veni, G. 1998. 1998 Sonora restoration project. *The Texas Caver* 43(6):87-89. Reprinted in *SpeleoDigest* 1999:301-302.

Additional Reading

Aley T. 1972. Control of unwanted plant growth in electrically lighted caves. *Caves and Karst* 14(5):33-35.

Aley T, Aley C, Rhodes D. 1985. Control of exotic plant growth in Carlsbad Caverns, New Mexico. In: Vandyke JE, editor. *Proceedings of the 1984 National Cave Management Symposium. Missouri Speleology* 25(1-4):159-171.

Bergthold L. 1997. Conservation, restoration, and low-impact caving: New approaches and attitudes. *NSS News* 55(3):72-73.

Bitting C. 1989. Water-based cave cleaning. *NSS News* 47(2):42-43.

Bonwick J, Ellis R, and Bonwick M. 1986. Cleaning, restoration, and redevelopment of show caves in Australia. In: *Proceedings Ninth*

International Congress of Speleology, vol 2.
Buecher RH. 1996. Footprints, routes, and trails. In: Rea GT, editor. *Proceedings of the 1995 National Cave Management Symposium: Spring Mill State Park, Mitchell, Indiana, October 25-28, 1995.* Indianapolis (IN): Indiana Karst Conservancy. p 47-50.
Crisman B. 1985. Cave restoration at Carlsbad Caverns National Park. *NSS News* 43(5):177-180.
Douglas JC. 1998. Discovering history in American caves. *NSS News* 56(9):261-63.
Douglas JC. 2001. Historic preservation at Hubbard's Cave: Inventory and management of cultural resources. In: Rea GT, editor. *Proceedings of the 14th National Cave and Karst Management Symposium: Chatanooga, Tennessee, October 19-22, 1999.* Chattanooga (TN): Southeastern Cave Conservancy. p 46-50.
Dunlap K. 1996. Inexpensive temperature monitoring in caves. *Proceedings of the 1995 National Cave Management Symposium: Spring Mill State Park, Mitchell, Indiana, October 25-28, 1995.* Indianapolis (IN): Indiana Karst Conservancy. p 76-87.
Elliott WR. 2000. Conservation of the North American cave and karst biota. In: Wilkens H, Culver DC, Humphreys WF, editors. *Subterranean Ecosystems.* Ecosystems of the World 30. Amsterdam: Elsevier. Chapter 34, p 665-689. Internet. <http://www.utexas.edu/tmm/sponsored_sites/biospeleology/>. Accessed 2006 May 6.
Faimon J, Zimak J, Stelcl J, Kubesova S. 2003. Environmentally acceptable effect of hydrogen peroxide on cave lamp-flora. *Environmental Pollution* 122(3):417-422.
Follet K. 1999. Graffiti control and removal: Lessons from an urban setting. *American Caves* Winter: 9-15.
Frith RS. 1976. Acid on–Graffiti off. *NSS News* 34(11): 203.
Fry J, Olsen R. 1999. Walkway development and construction relative to reducing visitor impact in the historic section of Mammoth Cave. In: Rea GT, editor. *Proceedings of the 1999 National Cave and Karst Management Symposium: Chattanooga, Tennessee, October 19-22, 1999.* Chattanooga (TN): Southeastern Cave Conservancy. p 64-72.
Grobbelaar J. 2000. Lithophytic algae: A major threat to the karst formation of show caves. *Journal of Applied Phycology* 12:309-315.
Hagan S, Sutton M. 1991. Cave restoration can cause damage. *NSS News* 49(6):170-171.
Hazslinszky T. 1985. *International Colloquium on Lamp Flora: 10-13 October 1984.* Budapest (HU): Magyar Karszt-és Barlangkutató Társulat. 155 p.
Hazslinszky T. 2000. *International Conference on Cave Lighting Proceedings.* Budapest (HU): Hungarian Speleological Society. 164 p.
Hildreth-Werker V, Werker JC. 2001. Restoration in Pellucidar. *NSS News* 59(3):76-77.
Horrocks RD. 1994. Artificial fill removal project: Timpanogos Cave System, Timpanogos Cave National Monument. *NSS News* 53(4):102-107.
Horrocks RD, Green DJ. 2000. Lehman lint liquidators. *NSS News* 58(3):72-73.
Hubbard DA Jr. 1995. Cave conservation and cave clean-ups: Not always one in the same. *NSS News* 53(2):31.
Irving D. 1989. Wind Cave Restoration Field Camp. *NSS News* 47(2):34-36.
Jablonsky PL. 1990. Lint is not limited to belly buttons alone. *NSS News* 48(5):117-119.
Jablonsky PL. 1991. Headlines! Lint accumulations are becoming more prevalent in caves. *American Caves* Fall:18.

Jablonsky PL. 1992. Implications of lint deposits in caves. *NSS News* 50(4):99-100.
Jameson RA. 1995. Zinc leaching from galvanized steel in Mystery Cave, Minnesota. In: Rea GT, editor. *Proceedings of the 1995 National Cave Management Symposium: Spring Mill State Park, Mitchell, Indiana, October 25-28, 1995*. Indianapolis (IN): Indiana Karst Conservancy. p 178-185.
Kohlwey K. 1992. 1991 Timpanogos Cave restoration project. *NSS News* 52(1):12-13.
Lemon L. 1976. Algae control and removal from cave formations. In: Aley T, Rhodes D, editors. *1976 National Cave Management Symposium Proceedings: Albuquerque, New Mexico, October 6-10, 1975*. Albuquerque (NM): Speleobooks. p 64-65.
Lewis JJ. 1993. The effects of cave restoration on some aquatic communities [abstract]. In: *1991 National Cave Management Symposium Proceedings: Bowling Green, Kentucky, October 23-26, 1991*. Horse Cave (KY): American Cave Conservation Association, Inc. p 346-350.
National Caving Association. 1999. *Cave Conservation Handbook.* Price, G. editor. London (UK): National Caving Association. [Softbound 1999. First published 1997 in ring bound format.]
National Outdoor Leadership School. 1998. *Leave No Trace Outdoor Skills and Ethics: Caving.* Tuohy L, editor; Austin A, illustrator. Boulder (CO): Leave No Trace, Inc. 20 p. Internet <http://www.lnt.org/LNTPublications/s&e.pdf.docs/S&ECaving.pdf>. Accessed 2003 Apr 5.
National Speleological Society, Inc. 2003. *A Guide to Responsible Caving.* Jones CJ and Anonymous, editors; Dale M, design. Huntsville (AL): National Speleological Society. 24 p. Internet < http://caves.org/brochure/grc.pdf>. Accessed 2003 Dec 5.
Meyer PJ, French JA. 1995. Cathedral Caverns: Techniques for restoration and cleaning. *NSS News* 53(7):186-193.
Meyer PJ, French JA. 1996. Crossings Cave: Test bed for cave restoration. *NSS News* 54(2-3):54-58.
Netherton W. 1993. Mystery Cave trails: Past, present and future. In: Jorden JR, Obele RK, editors. *Proceedings of the 1989 National Cave Management Symposium: New Braunfels, Texas.* Austin (TX): Texas Cave Management Association; Texas Parks and Wildlife Department. p 141-147.
Newbould RL. 1976. Steamcleaning of the Orient Cave, Jenolan. In: Hamilton-Smith E, editor. *Cave Management in Australia.* Broadway (AU): Australian Speleological Federation.
Northup DE, Reysenbach A-L, Pace NR. 1997. Microorganisms and speleothems. In: Hill C, Forti P, editors. *Cave Minerals of the World.* 2nd edition. Hunstville (AL): National Speleological Society. p 261-266.
Ohms M, Rogers R. 1999. Black Hills restoration camp. *NSS News* 57(4):99-100.
Olson, R. 2002. Control of lamp flora in Mammoth Cave National Park. In: Hazslinszky T, editor. *International Conference on Cave Lighting Proceedings*. Budapest (HU): Hungarian Speleological Society. p 131.
Poulter N. 1980. Restoration, stabilization and gating of the Christmas Star Extension of Crystal Cave (WI-62), Witchcliffe, W.A. In: *Proceedings 12th Conference, Australian Speleological Federation.* p 10-14.
Rajczy M, Buczko K, Hazslinszky T. 1997. Efforts for the restriction of lamp flora: Results and experiences between 1984 and 1996. In: Tóth E, Horváth R, editors. *Proceedings of the Research, Conservation, Management Conference Aggtelek, Hungary, 1-5 May, 1996.* Aggtelek (HU): Aggteleki National Park Directorate. p 54.
Rhodes D. 1976. Cave restoration, an impossible task? *NSS News*

34(11):200-203.
Rogers N. 1995. Mammoth Cave restoration field camp. *NSS News* 53(2):44-44.
Rohde K, Kerbo R. 1978. Cave restoration: An underview of correcting overviewing. In: Zuber R, Chester J, Gilbert S, Rhodes D, editors. *National Cave Management Symposium Proceedings: Big Sky, Montana, October 3-7, 1977.* Albuquerque (NM): Adobe Press. p 55-58.
Rohde K. 1982. Cave restoration and management. In: Wilson RC, Lewis JJ, editors. *National Cave Management Symposia: Proceedings [for Carlsbad, New Mexico 1978 and Mammoth Cave, Kentucky 1980]* Oregon City: Pygmy Dwarf Press. p 205-207.
Rohde K. 1989. Cave restoration begins at Wind Cave. *NSS News* 47(2):32-34.
Rogers N. 1995. Mammoth Cave restoration field camp. *NSS News* 53(2):44.
Schmitz SA. 1996. Lost Soldiers Cave restoration project. *NSS News* 54(2/3):59-63.
Schaper J. 1996. Techniques and volunteer issues in the restoration of Cathedral Cave. In: Rea GT, editor. *Proceedings of the 1995 National Cave Management Symposium: Spring Mill State Park, Mitchell, Indiana, October 25-28, 1995.* Indianapolis (IN): Indiana Karst Conservancy. p 279-281.
Spate A, Moses C. 1994. Impacts of high pressure cleaning: A case study at Jenolan. In: Henderson K, Bell P, editors. 1994. *Cave Management in Australasia X.* Carlton South, Victoria (AU): ACKMA. p 45-48. [*Proceedings of the Tenth ACKMA Conference, Rockhampton, Queensland.* ACKMA 10:45-48.]
Thornber N. 1953. Appendix A: Speleological code and ethics. In: Cullingford CHD, editor. *British Caving: An Introduction to Speleology.* London: Routledge and Kegan Paul, Ltd.
Tranel MJ. 1992. Changing to prevent change: Building a science and resource management program. *NSS News* 52(1):22-25.
Vale E. 1993. It's an open and shut cave: Plugging artificial entrances at Onondaga Cave State Park. In: Foster DL, Foster DG, Snow MN, Snow RK, editors. *1991 National Cave Management Symposium Proceedings: Bowling Green, Kentucky, October 23-26, 1991.* Horse Cave (KY): American Cave Conservation Association, Inc. p 129-131.
Veni G. 1995. The restoration of a vandalized show cave: Cave Without a Name, Texas. In: Pate DL, editor. *Proceedings of the 1993 National Cave Management Symposium: Carlsbad, New Mexico, October 27-30, 1993.* [Huntsville (AL)]: National Cave Management Symposium Steering Committee. p 65-72.
Veni G. 1996. Primer to speleothem repair. *NSS News* 54(2-3):64-67.
Veni G. 1997. Speleothems: Preservation, display, and restoration. In: Hill C, Forti P, editors. *Cave Minerals of the World.* 2nd edition. Hunstville (AL): National Speleological Society. p 301-309.
Vories KC, Throgmorton D, Harrington A, editors. 2004. *Proceedings of Bat Gate Design: A Technical Interactive Forum held in Austin, Texas, March 4-6, 2002.* Carbondale (IL): USDOI Office of Surface Mining, Alton, Illinois and the Coal Research Center, Southern Illinois University at Carbondale. 434 p. Proceedings available on the Internet <http://www.mcrcc.osmre.gov/bats/>. Accessed 2004 March 30.
Watson J, Hamilton-Smith E, Gillieson D, Kiernan K, editors. 1997. *Guidelines for Cave and Karst Protection.* Gland (SZ) and Cambridge (UK): IUCN International Union for Conservation of Nature and Natural Resources. 63 p.
Welbourn C. 1976. Biological consideration for cave cleanup projects. *NSS News* 34(11):204.

Werker JC. 1995. Formation repair. In: Pate DL, editor. *Proceedings of the 1993 National Cave Management Symposium: Carlsbad, New Mexico, October 27-30, 1993.* [Huntsville (AL)]: National Cave Management Symposium Steering Committee. p 61-64.

Werker JC. 1996. Formation repair. *NSS News* 54(2-3):65-67.

Werker JC. 1996. Proven techniques in repairing formations. In: Rea GT, editor. *Proceedings of the 1995 National Cave Management Symposium: Spring Mill State Park, Mitchell, Indiana, October 25-28, 1995.* Indianapolis (IN): Indiana Karst Conservancy. p 308-310.

Werker JC, Hildreth-Werker V. 1999. Restoration, trail designation, and micro-bial preservation in Lechuguilla Cave. In: Stitt RR, editor. *Proceedings of the 1997 National Cave and Karst Management Symposium: October 7-10 in Bellingham, Washington, and Chilliwack and Vancouver Island, B.C., Canada.* Huntsville [AL]: National Cave Management Symposium Steering Committee. p 181-189. [Also listed as 13th National Cave Management Symposium.]

Werker, JC. 2004. Characteristics of materials used in cave and mine gates. In: Vories KC, Throgmorton D, Harrington A, editors. *Proceedings of Bat Gate Design: A Technical Interactive Forum held in Austin, Texas, March 4-6, 2002.* Carbondale (IL): USDOI Office of Surface Mining, Alton, Illinois and the Coal Research Center, Southern Illinois University at Carbondale. 434 p. Proceedings available on the Internet <http://www.mcrcc.osmre.gov/bats/>. Accessed 2004 March 30.

Section B—Organizing Cave Projects

Bucket Brigade at Caverns of Sonora

George Veni

Opening a cave to the public often requires the enlargement of passages, which generates literally tons of rubble that sometimes get left in the cave. Caverns of Sonora, located near Sonora, Texas, is full of delicate, spectacular speleothems and has earned the reputation of being the world's most beautiful show cave.

Though the rubble left during development wasn't particularly noticeable, the owners wanted to return the cave to as natural a state as possible. Fourteen restoration projects between 1990 and 2005 have removed over 200 tons of rubble from the cave. Each annual project required 70 volunteers as described in the following excerpt from a project report (Veni, 1998).

Fourteen restoration projects between 1990 and 2005 have removed over 200 tons of rubble from the cave. Each annual project required 70 volunteers.

Buckets from Pit to Entrance

Shuttling the buckets from the pit to the entrance was done along a rotating chain of cavers. Twenty cavers dug in the pit, with one shift often working from morning until lunch, and a different ten working from after lunch until supper. The ten cavers produced enough rocks to keep everyone else busy moving the buckets of rubble out of the cave. The start of the haul line began with the "mules" pulling the buckets up the pit. Someone else off-loaded the buckets, and passed them to another caver who carried them down to a set of truck dollies where they were rolled off to the first set of stairs. Three cavers passed the buckets up the stairs and into the arms of 14 cavers lined through a narrow section of passage to where another set of dollies awaited. After another roll through a couple of passages, the buckets were then passed up three flights of stairs, two landings, and out of the entrance where they were emptied into a flatbed truck trailer.

On the following page are photos of a typical restoration project at Caverns of Sonora in 2002.

Cited Reference

Veni G. 1998. 1998 Sonora restoration project. *Texas Caver* 43(6):87-89. Reprinted in *Speleo Digest* 1999:301-302.

Figures 1-5. Removing construction rubble from Caverns of Sonora in Texas. The annual restoration event uses a rotating chain of 70 cavers. Photos show some of the steps in the process. Haul line pulling buckets up the pit (top left). Truck dollies rolling buckets to the stairs (top right). Chain of cavers passing buckets up the stairs (middle left). Passing buckets up a pit (bottom left). Passing buckets along one of several corridors (bottom right).

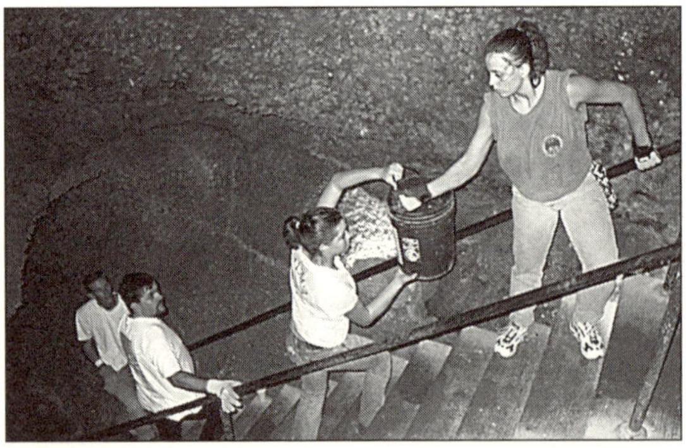

Photos George Veni

Section B—Organizing Cave Projects

Managing Cave Conservation Projects

Douglas M. Medville

Many cave conservation projects involve hundreds, if not thousands, of hours of effort. The work involved in organizing and managing cave projects is often overlooked or at least underestimated by enthusiastic volunteers. Management efforts can encompass writing proposals, obtaining funding, developing work plans, coordinating volunteer efforts, planning logistics, gathering materials, keeping financial records, reimbursing volunteers for expenses, reporting progress to the land manager or cave owner, managing interpersonal relationships among the volunteers, and reporting the final accounting for funds and hours expended. Project management is a serious effort in its own right.

The work involved in organizing and managing cave projects is often overlooked or at least underestimated by enthusiastic volunteers.

Scoping and Getting Started

Before undertaking any cave restoration project, determine the magnitude of the project. What is to be accomplished? Define the tasks required to achieve the project goals. Is the project a one-time effort involving a few people or is it a multi-week (or longer) project in one or more caves involving people from several groups, perhaps coming from different geographic areas, and requiring a variety of skills? Long-term, complex projects require more management time.

The project manager or leader should evaluate the objectives of the project and define specific skills that are required. If the project involves trash pickup, for example, special skills may not be required. On the other hand, specialized technical skills will be required if the project includes speleothem repair, vertical work, removal of hazardous materials, or other such tasks. From the beginning, the project leader should define tasks and delegate responsibilities.

Ownership and Proposals

This seems obvious, but before getting started, project personnel must identify the cave owner or manager, discuss the details of proposed work, and obtain the owner's or manager's concurrence before initiating any work.

If the cave is on public land, permission must be granted by the resource manager who is responsible for the cave. During initial interactions with the cave manager or private owner, explore the possibility of obtaining financial or other support for the conservation work. For example, the agency or owner may be able to provide housing or camping space, make a contribution for the purchase of tools and supplies, or arrange for special equipment. If the cave is on public land and special funding is available, the agency may be able to reimburse volunteers for travel expenses.

To secure this level of support, especially for caves on public lands, the project manager may need to prepare a written proposal that describes the work to be done and budget requirements. If the cave manager is willing to seek agency funding, cost share agreements will benefit all involved. Front-end paper work, proposals, work plans, and budgets, will be necessary if a cost share agreement is possible.

© Val Hildreth-Werker

Figure 1. These two cavers, Billy and June Hoffman, experienced in restoration, are cleaning a small area of trail at the bottom of a rope that ascends along a delicate climb.

After the agency accepts the proposal, a written agreement is prepared. This agreement specifies the work to be accomplished, the level of effort involved, and the responsibilities of both the restoration group and the agency. For example, the agency may agree to provide tools, open gates to allow access to the cave, and provide housing or other support services. The agency may also require the group to sign permits, waivers, or volunteer agreements. Some agencies will provide liability coverage or worker's compensation in case of injury. A Memorandum of Understanding, a Cooperative Management Agreement, or a Cost Share Challenge Agreement may be required to clarify the obligations of each party. To minimize the possibility of misunderstandings about the obligations of both parties, these issues should be thoroughly explored with the land manager before beginning the project. (See agreements, page 309.)

If the land-managing personnel have had no prior experience with cave restoration efforts, they may not be aware that caves on their district could benefit from conservation or restoration work. If an educational process is warranted, the restoration project leaders can start by introducing the Federal Cave Resources Protection Act of 1988, providing copies of existing cave cleanup and restoration agreements with other agencies, and presenting a slide show that describes the problem and remediation procedures. (See Appendix 1, Federal Cave Resources Protection Act of 1988, page 507.)

Conservation outreach brochures are available and can be downloaded from the NSS Web site < http://www.caves.org/brochure/>. The project leaders can also encourage the land manager or owner to enter the cave, see the site where the proposed work will be carried out, and observe the problems that will be corrected by restoration.

If the cave is on private property, a verbal understanding of what will be accomplished may suffice. Signed liability release forms can be provided to landowners and may help them be more amenable to allowing the restoration work. As with public land managers, the restoration project leaders should explore the possibility of the landowner providing some support. For example, the landowner might offer space for parking or camping on the property, or provide a place where cavers could clean up at the end of the day's work, or arrange a meal for the group.

For cave conservation and restoration projects, both on public and private land, an agreement describing the *rules of engagement* may be a good thing. This type of agreement typically defines: days and hours the work will take place; maximum number of people; rules of conduct; and vehicular access to the cave for transport of supplies and equipment.

Open communication between the owner or land manager and the conservation project leader reassures all involved that work is progressing. The owner can provide immediate feedback to the project manager and address small problems before they become irritants and bigger problems. For example, participants may need to park in a different place, gates may need to be left open (or closed), or the owner may need to receive notification (or give specific permission) before a team shows up to work.

Conservation outreach brochures are available and can be downloaded from the NSS Web site < http://www.caves.org/ brochure/>.

Specialized Skills

In scoping the project, the organizer needs to identify the skills required to get the job done. Cave restoration can include a variety of tasks: lint removal, spray paint removal, speleothem cleaning and repair, rubble and fill removal, gate installation, or restoring an entrance or passage to a former natural state. For example, if a show cave is to be converted to its original state, the project may require skills in removing electrical wiring, guardrails, walkways, PCB-filled transformers, and other infrastructure-related tasks. If the restoration objectives include repair of fragile speleothems, different expertise and tools are required.

Once the goals and expertise are identified, the organizer must find people with the right skills who can commit to the planned dates and times. If twelve cavers with specific skills are expected to be present on a prearranged weekend, then that number should show up. Always arrange backups who can be called at the last minute to fill in for no-shows.

Task Sizing and Allocation

For larger restoration projects, tasks may be handled by a number of different teams. For example, trash pickup, graffiti assessment, and graffiti removal might require three teams, each focusing on their specific goals. Delegate responsibilities and allow team leaders to work as autonomously as possible. Accepting responsibility for task completion, or team leader buy-in, is vital to success. If the project extends over multiple weekends or if multiple tasks are scheduled simultaneously, team leaders can arrange their schedules and resources as necessary.

Develop a list of objectives and scope tasks into achievable amounts. Designate people and equipment for each task. The slots left blank will lead to a clear list of needs. Find additional volunteers and resources to fit the open slots.

What Can I Do To Help?

During the course of a project, people will come forward and offer to help by giving time or providing equipment. Some will have specific things they want to do and others will just ask, "Where do you need me?" Good project managers refer to their needs list showing open slots for people, equipment, or tasks waiting to be started or continued.

Keep a list of people who offer help and note their skills and expertise. Some prefer to work outside the cave on the surface, some prefer to do administrative paperwork, some are good with computers, accounting, prefabricating parts, or welding, and some have equipment that can support the project. Keep an ear open.

The reverse is also true. If the project manager needs equipment, tools, or a skill, and it cannot be located, get the word out. Sometimes there are unexpected resources among volunteers who are already helping and the need can be met.

Infrastructure and Record Keeping

The project leader needs to set up an infrastructure that is appropriate to the project objectives. For weekend and weeklong efforts, this includes providing a place to camp, appropriate toilet facilities, a place to wash up bodies and gear after the day's work, a place to prepare meals, and if funds permit, food and fluids. If facilities and amenities are expected to be provided by land-managing agencies per a written or verbal agreement, make sure, in advance, that these will be available when needed. Communicate with the landowner or manager frequently, especially as the project dates approach. Volunteers don't want to show up on a restoration weekend and find out the owner has other plans and doesn't want to host a group of cavers.

NSS Conservation Task Forces

Editors' Note: NSS Conservation Task Forces are created to focus on local and regional issues. The Conservation Task Force (CTF) Coordinator stands ready to lend assistance with advice and networking to any group of conservation-oriented cavers who care about protecting the future of cave and karst resources. CTFs may be formed to address any cave or karst concern on public or private property. Whether a single cave or an entire region, a situation requiring secrecy or publicity—if you are involved in cave or karst conservation projects, your work may benefit from CTF designation. Visit <http://www.caves.org/> for information.

© Val Hildreth-Werker

Figure 2. Equipment for cave restoration comes from many sources. Tools are often supplied by volunteer cavers and by the landowner or managing organization. Good project managers assure that plenty of the right equipment is on hand.

If an agency is able to reimburse the restoration group for any expenses such as mileage, meals, lodging, or other costs, describe the terms in the written agreement. Assign a responsible person to keep track of expenses, hours contributed, and the overall volunteer value generated by the project. Recording the dollar value of volunteer effort enables the restoration group to receive recognition and any reimbursement detailed in agreements. (See volunteer value, page 321.) It also enables the project leader and the land manager to compare the effort delivered against the proposed activities and to compare the expenses incurred with those budgeted. For long-term, complex projects, especially where cost sharing is involved, financial management is an absolute necessity.

For an agency to establish a formal agreement with a restoration group, it may be necessary to work through a recognized entity such as the National Speleological Society (NSS). Agreements are typically arranged through an NSS internal organization such as a grotto, NSS Region, NSS Conservation Task Force, or NSS Conservancy. The NSS is recognized as a not-for-profit organization under Section 501(c)(3) of the Internal Revenue Code. This status may make it easier for the agency to contract with restoration groups that are affiliated with the NSS. (See task force sidebar, page 307.)

Project Documentation and Recognition

As the work proceeds, the project leadership should provide photographic and written documentation of progress. Before, during, and after pictures are effective in communicating the changes in the cave and the human effort committed to the project. Photos provide positive feedback for landowners, agencies, and cavers. Photographs can also demonstrate the groups' abilities when talking with other cave managers and owners about future projects. (See photodocumentation, page 204.)

A written record of progress should also be provided to the owner or agency (and is necessary for agency agreements or other contractual relationships). Cave managers must use documentation to justify projects to their agencies, to obtain incremental funding that is dependent on progress, and to obtain funding for future efforts. Cavers use project documentation to record volunteer value for income tax records, for reimbursements, and for grant applications to fund additional work.

Project leaders should also see that appropriate publicity and recognition is given at the end of the project, through grotto newsletter articles, *NSS News* articles, and newspaper or other media coverage. Be sure to show appreciation to all cooperators in the project: the landowner or managing agency, agency personnel or landowner family members, cave resource manager, project leaders, equipment lenders, resource providers, team leaders, as well as the cavers and other participants who did the work.

While many volunteers are motivated by the self-satisfaction of doing the restoration work, most people also hope for public recognition from peers. Although motivation varies significantly among cavers, it is nonetheless important to understand this component of the volunteer psyche and set up appropriate recognition. Publish the results in caving newsletters and include a group photo. Make sure everyone's name is spelled correctly. Other recognition ideas include certificates of appreciation, project T-shirts, and tickets for show caves.

After the restoration effort is complete and the volunteers have all gone home, a final report and financial accounting must be submitted for work conducted under a written contract or agreement. Completing the paperwork is an important, tedious, and often thankless task. Project management is serious business and the work is (almost) never done.

Section B—Organizing Cave Projects

Assistance Agreements: Partnerships for Cave Conservation

James R. Goodbar

Assistance agreements can be efficient, effective management tools for cave projects and are used in forming partnerships between agencies and caving groups. Sophisticated volunteer agreements establish affiliations between agencies and special interest groups for the benefit of public resources. As specialized legal instruments, assistance agreements address specific tasks or problems. Agreements may involve the exchange of money or provide for the mutual acceptance of services without the exchange of money.

Resource managers need to know what type of agreements are available, how to implement them, and when to use these documents. This chapter introduces situations where various types of assistance agreements are beneficial and discusses the authorities to authorize them.

This chapter introduces situations where various types of assistance agreements are beneficial and discusses the authorities to authorize them.

No matter what kind of assistance agreement is used, it is important for all parties to *clearly state the objectives*. Two types of agreements are discussed here. The first type is an agreement in which money is exchanged. The second involves no money.

Cost Sharing and Cooperative Agreements

There are two basic kinds of assistance agreements in which there is a transfer of money, property, or anything of value. They are the cost sharing agreement and the cooperative agreement. Cost sharing agreements are legal instruments used to reflect a relationship between the federal government and a state or local government or·other recipient when the principal purpose of the agreement is the transfer of money or other valuables to the recipient in order to accomplish a public purpose authorized by federal statute.

Cost Sharing Agreements

Categories of cost sharing agreements include the challenge grant and the challenge agreement. The challenge grant is used in cases where the federal government *is not* substantially involved, and the challenge agreement is used where substantial involvement *is* anticipated between the federal government and the recipient during performance of the contemplated activity.

Substantial involvement occurs in situations where the terms of the agreement indicate that the recipient, during performance of the contemplated activity, can expect federal agency collaboration or participation in the management of the project. Substantial involvement is specifically described by each federal agency. (For example, Bureau of Land Management describes substantial and nonsubstantial involvement in their Manual 1511, titled Assistance Agreements.)

These two types of agreements promote cost sharing projects by requiring the recipient to obtain or provide additional funding from nonfederal sources at a mutually agreed upon sharing ratio. The authority by which federal agencies can enter into these types of agreements is the Federal Grant and Cooperative Agreement Act of 1977 (P.L. 95–224), as amended by P.L. 97–258 (31 U.S.C. Chapter 63, 6301–6308). (More detailed procedures for processing these agreements can be found in the Office of Management and Budget Circular A–110.)

There are two basic kinds of assistance agreements in which there is a transfer of money, property, or anything of value. They are the cost sharing agreement and the cooperative agreement.

Contract or Cost Share. Cost sharing agreements do not require competitive procedures in order to award them. However, these agreements are not contracts and should not be viewed as sole source contracts or used as a way to circumvent the procurement process. If a more binding commitment is needed to ensure completion of the project and to establish more stringent remedies in case an action is not performed, then a contract should be used.

An Example of Cost Sharing. A cost sharing agreement might be used if a university pays the salary of one of their research professors to conduct biological studies in a specified area and manner and if the agency provides the necessary room, board, transportation, or other appropriate funding. The amount of agency involvement would determine if the cost sharing agreement is a challenge grant or a challenge agreement.

Cooperative Agreement

Cooperative agreements are similar to challenge agreements with one important exception. Cooperative agreements do not require the recipient to obtain additional funding from a nonfederal source. For example, a cooperative agreement may be used to provide funds to a caving organization for the assistance they provide in inventory, mapping, and preparing files of caves located in a specified area. (See examples of cooperative agreements below and in Appendix 4, page 541.)

Developing an Assistance Agreement

Generally, money is appropriated by Congress each fiscal year for cost sharing agreements. Before entering into cost sharing agreements, money must be made available to the agencies for this purpose. The development of an assistance agreement is quite simple. There are only three parts—purpose, authority, and duties and responsibilities.

Purpose

The purpose section should briefly describe the objectives of the agreement including any special emphasis or focus. If necessary, a list of objectives can be developed to more clearly outline the purpose.

Legislative Authority

The legislative authority for entering into an assistance agreement with a federal agency is cited in the authority section on agreement documents. There are several different authorities that authorize the use of assistance agreements. These authorities may be laws or executive orders and may vary depending upon the type of project entered into.

Always confirm the enabling legislation for each federal agency authorizing entry into assistance agreements. Listed here are authorities that may be appropriate:

- Probably the single most important enabling legislation that allows the use of assistance agreements is the Federal Grant and Cooperative Agreement Act. This Act authorizes three types of instruments:

contracts, cooperative agreements, and grants.

- Fish and Wildlife Conservation and Water Resources Development Coordination Act, as amended under P.L. 85–624, 16 U.S.C. 661.
- Federal Water Project Recreation Act, as amended under P.L. 89–72.
- Wild and Scenic Rivers Act, as amended under P.L. 90–542, 16 U.S.C. 1282.
- National Trails System Act, as amended under P.L. 90–543, 16 U.S.C. 1246(h).
- Endangered Species Act of 1973, as amended under 16 U.S.C. 1531.

Duties and Responsibilities

The duties and responsibilities section in agreement documents should clearly state the activities and products required of the cooperators and the agency. (For example, see the sample cooperative agreements below and in Appendix 4, page 541.)

Cooperative Management Agreements

A useful assistance agreement in which no funds are exchanged is the Cooperative Management Agreement (CMA). In most cases this type of agreement is tiered off of a broader Memorandum of Understanding (MOU). The MOU document defines cooperative relationships between two or more parties in their pursuit of mutual objectives.

The terms of the CMA should be written out in sufficient detail for all parties to clearly understand what is expected. Documents should outline the mutually agreed upon plan of action between the agency and the cooperators and identify the responsibilities and performance standards that apply to each of the participants.

A useful assistance agreement in which no funds are exchanged is the Cooperative Management Agreement (CMA). In most cases this type of agreement is tiered off of a broader Memorandum of Understanding (MOU). The MOU document defines cooperative relationships between two or more parties in their pursuit of mutual objectives.

Lost Cave

The first cooperative cave management agreement entered into by the Bureau of Land Management (BLM) was for Lost Cave in Carlsbad, New Mexico. This agreement was first signed in 1983 on a one-year trial basis. The cooperators were the BLM and the Pecos Valley Grotto. The agreement outlines the BLM's responsibilities of issuing permits, contacting the Grotto for trip leaders, and reviewing and approving the leader list submitted by the grotto chairman.

The Lost Cave Cooperative Management Agreement is very simple, but very specific. It spells out the responsibilities of each party, what should be in the required reports, and when these reports are due. Excellent cooperation is usually achieved with this type of agreement.

NSS Southwestern Region

The CMA between the New Mexico BLM and the Southwestern Region of the NSS (SWR) is an example of a more comprehensive document. This agreement is more flexible and outlines broader areas of cooperation and responsibilities. It also incorporates a Group Volunteer Service Agreement. This agreement can remain in effect for a period of five years until it needs to be reauthorized. When an authorized BLM volunteer project is conducted the participants need to sign only a participants list and not go through the additional paperwork of the Volunteer Service Agreement. This simplifies and expedites the use of volunteers. (See SWR CMA, page 313. Also see updated version of SWR CMA in Appendix 4, page 541.)

LOST CAVE COOPERATIVE AGREEMENT

The Roswell District of the Bureau of Land Management and the Pecos Valley Grotto, National Speleological Society, jointly agree to the following management plan for Lost Cave, located in Eddy County, New Mexico. This agreement can be cancelled by either party at any time by written notification.

The BLM shall be responsible for issuance of recreational, educational, and scientific permits for cave visitation to Lost Cave and will contact the designated Grotto representative(s) for leaders for trips. Additionally, the BLM shall review and approve the leader list submitted by the Grotto Chairman and include the approved leaders in the BLM Volunteer Program.

The Grotto shall clean up and maintain the cave, the cave gate, access within the cave, and lead permitted trips for the BLM. All recreational, educational, and scientific trips shall have a valid BLM cave permit. However, work trips shall not require permits, but a record of the work effort and participants shall be maintained. A BLM-approved trip leader shall accompany all trips, work or otherwise, to the cave.

A semi-annual report of activities at the cave shall be given to the BLM on March 15 (for October to March) and September 15 (for April to September). This report shall list the trips led during the period, the level of work effort during the period, and the materials donated or consumed.

The Grotto Chairman, or his/her designate, shall be responsible for adherence to this agreement and shall obtain consultation and concurrence from the BLM prior to any modification to the cave.

This document was originally approved in March 1983. It was signed by: R. Thomas Dillon, Jr., Chairman, Pecos Valley Grotto, National Speleological Society on March 18, 1983; and by G. Ben Korlei, Area Manager, Carlsbad Resource Area, Roswell District, New Mexico, Bureau of Land Management on March 16, 1983.

Most importantly, be specific in written agreements. Define the responsibilities of each party and what is expected of the cooperators.

Summary

If monetary assistance is provided, use either a cost sharing agreement or a cooperative agreement:

- If the cooperator plans on obtaining additional funds provided by a nonfederal agency, use a cost sharing agreement:
  - If the cost sharing agreement is one in which the agency will not be substantially involved, use a challenge grant.
  - If the agency will have substantial involvement in the project, use a challenge agreement.
- If money is exchanged but no additional nonfederal funds are obtained, use a cooperative agreement.
- When no money or anything of value is to be exchanged, use a cooperative management agreement.

Most importantly, be specific in written agreements. Define the responsibilities of each party and what is expected of the cooperators. In times of tight budgets and minimal personnel, partnerships with various user groups can make a critical difference in accomplishing cave and karst resource management goals.

COOPERATIVE
MANAGEMENT AGREEMENT

BETWEEN

NEW MEXICO
BUREAU OF LAND MANAGEMENT

AND

SOUTHWESTERN REGION OF THE
NATIONAL SPELEOLOGICAL SOCIETY

AGREEMENT NUMBER BLM-MOU-NM-2002-005

I. INTRODUCTION

A. PURPOSE

l) This Cooperative Management Agreement (CMA) is entered into between the New Mexico Bureau of Land Management (BLM), an agency of the Department of the Interior and the Southwestern Region of the National Speleological Society (SWR). The Agreement will provide for a more direct, efficient and cost effective means of cave resource management on public lands in New Mexico by providing an opportunity for the SWR to participate in cave management.

2) This CMA will allow for more comprehensive cave management activities and initiate a long-term cooperative cave management effort with the SWR to contribute to the protection and preservation of cave resources on public lands in New Mexico. As part of this agreement the SWR will participate in a Participating Management Team (PMT) as defined in section V of this Agreement.

3) This Agreement addresses general management actions related to cave resources. Site-specific projects will be addressed in separate Volunteer Service Agreements or other appropriate documents. The terms of this agreement will be applicable to all BLM/SWR site-specific projects.

4) The BLM authorized Officers are encouraged to authorize projects, agreements and activities under their jurisdiction as outlined in the Memorandum of Understanding between the U.S. Department of the Interior, Bureau of Land Management, the National Speleological Society and the Cave Research Foundation, agreement number WO204, dated June 11, 1984.

B. OBJECTIVE

Through a cooperative effort, the BLM and the SWR will work together to maintain and improve cave resources within BLM New Mexico public lands, seek and use the skills, knowledge and expertise in SWR to plan, develop and implement cave management and conservation efforts with BLM New Mexico and use volunteers and volunteer labor under a Volunteer Services Agreement or appropriate documents.

C. AUTHORITY

1) This Agreement is made under the following authorities:
The Federal Land Policy and Management Act of 1976
43 U.S.C. 1701 (1976)
31 U.S.C. 6301-6308 (1982)
Federal Cave Resources Protection Act of 1988

2) Direction and guidance for Cooperative Management Agreements for cave resources management is set forth in the Memorandum of Understanding between the United States Department of the Interior, Bureau of Land Management, the National Speleological Society and the Cave Research Foundation (1984).

D. BENEFITS

1) The SWR, through formal or informal interactions, will play an increased role in the management of caves located on BLM New Mexico public lands. This role would allow BLM and SWR to combine expertise and knowledge to the benefit of the cave resource. The CMA will enable the SWR to help develop and implement cave management policy, plans, and regulations.

II. DEFINITIONS

Assistance: Information, advice, and service given to the BLM by the SWR. This assistance will be utilized if applicable in preparing documentation and policy. The BLM may decline assistance if it conflicts with the management of caves on public land.

BLM New Mexico: All public lands within New Mexico under BLM administration and federal subsurface lands in Oklahoma and Texas.

Cave Digs: An attempt to modify an entrance or passage to allow further exploration. Individuals involved in a dig are not covered under this agreement or any volunteer agreement. NEPA documentation and BLM authorization must be completed prior to any dig taking place in caves within the public lands in New Mexico.

Cave Management Policies: Documents which create policy or set precedent, signed by BLM management are legal administrative documents. Congressional acts, Interior policies, Codes of Federal Regulations, and other policies are considered cave management policies if they relate to caves.

Collections: Any item of natural or cultural makeup taken from caves is considered a collection. In order to collect any item from a cave, a collection permit must be issued by the Field Office in which the cave is located.

Cooperative Management Agreement: A long-term agreement between parties for joint management of specific resources. The CMA delineates each party's role in the management of areas. There can be a commitment by each party to absorb part of the cost of managing an area, but there is no transfer of funds involved. They cannot be used to authorize non competitive contracting with the cooperator.

Copyright: All publications, films, slides, videos, artistic or similar endeavors, resulting under a Volunteer Service Agreement, as specifically contracted for, will become property of the United States, and such will be in the public domain and not subject to copyright laws.

Development: Assistance in the creation of an idea, plan, policies or regulations, etc.

Planning Efforts: Resource Management Plans, individual management plans, Environmental Impact Statements, Environmental Assessments, and other written documents in the BLM planning process. For further information see National Environmental Policy Act Handbook H–1790-1.

Participating Management Team: A work group consisting of members from the BLM, SWR, and other organizations concerned about the cave resource. The Southwestern Region representation on the PMT will be headed by the Chairman of the Southwestern Region. He or she will appoint SWR members to the PMT. All Southwestern Region PMT members must be SWR members and NSS members in good standing. BLM representatives will be designated by BLM management. This agreement does not prohibit BLM or the SWR from entering into other agreements with other parties.

Project: A planned undertaking which may encompass phases of work including preparation of plans, specifications, estimates, acquisition and actual construction, treatment, rehabilitation, restoration, or maintenance.

Project Leader or Official in Charge of Activity: Directs or oversees the work of the volunteer(s), provides any necessary instructions and guidance, and ensures that the volunteer(s) are aware of potential hazards. Reports all accidents, injuries and property damage to the appropriate officials.

Public Lands: Lands managed by the BLM. These lands are also termed BLM-administered lands.

Volunteer(s): Persons who contribute their services to the BLM and receive no wages or salary for their work, although Volunteer(s) may be reimbursed by the BLM for all or a portion of their incidental expenses.

III. TERMS OF AGREEMENT

A. This Agreement shall be in effect for five years from the date signed. This agreement may be renewed or extended for an additional five years (not to exceed a total of 5 years) by mutual written agreement of all parties to this agreement.

B. All BLM/SWR activities covered under this Agreement will be consistent with BLM's management decisions in land-use and activity plans.

C. The BLM or the SWR may cancel this agreement at any time provided 30-day written notification is given to the other party. The grounds for cancellation shall be addressed in the notification.

D. The cooperating parties recognize that this Agreement cannot be used to grant the SWR or its members the right of exclusive use to any cave(s) located on public lands.

E. No actions are to be taken by any member of the SWR which deal with direct regulatory or law enforcement activities.

F. This Agreement shall not obligate the BLM to expend funds in excess of appropriations authorized by law, budgeted and/or approved. Activities under this agreement are dependent on the availability of adequate BLM personnel to monitor and supervise such activities. Should it be necessary to reimburse the SWR for any work that will be ordered, a contract or assistance agreement will be awarded before work is started.

G. Volunteer Services Agreements, formulated in accordance with BLM Manual Section 1114, are to be used when the authorized officer for the BLM determines that cooperating personnel need to be:

1) protected under the Tort Claims Act
2) subject to Workmen's Compensation
3) identified as responsible for completing specific tasks or projects.
4) reimbursed for expenses incidental to volunteer work
5) assigned to complete administrative functions for the BLM

SWR Personnel working on BLM cave program projects will be covered by the BLM Volunteer Service Agreement. Such Agreements will be renewed to correspond with the renewal/extension of the CMA.

IV. AREAS OF RESPONSIBILITY

This section describes the general and specific areas of responsibility of both the SWR and BLM.

A. SWR Responsibilities

1) Assure that actions taken by any person's working under this Agreement are consistent with safe caving and Leave-No-Trace practices.

2) The project leader shall give an orientation program on safety, responsibilities, hazards, required equipment, fragile or delicate areas, emergency procedures and gate combinations to all individuals prior to entering a cave, including projects covered by the volunteer service agreement.

3) Submit to BLM a written progress report for ongoing projects within ten (10) days after a work session. This would include session objectives, work accomplished, work to be accomplished in the next session, and an accounting of volunteer hours by individual. A final written report will be submitted to the BLM within 30 days following completion of any project. This would include project objectives, results, and recommendations, as well as a final accounting of volunteer hours.

4) Report all accidents, injuries and property damage within 24 hours after an incident in a written incident report, submitted to the appropriate BLM office.

5) Notify the BLM Field Office in charge of a cave at least seven days in advance of any work being initiated under this agreement.

6) Ensure that SWR members are aware that they are financially responsible for loss and damage to their personal equipment and BLM equipment loaned to them for use while in volunteer status, (excluding normal wear and tear) and that SWR members are to provide their own vehicles, food, and personal caving equipment, except when provided by BLM through a Volunteer Service Agreement.

B. BLM RESPONSIBILITIES

The BLM will actively assist the SWR to the extent that work priorities, budget and manpower constraints allow, with achievement of mutually agreed upon tasks. BLM responsibilities include, but are not limited to, the following:

1) Prepare Volunteer Service Agreements and Project Plans for each project.

2) Supply any necessary information such as lock combinations, previous research and inventory results, maps, slides, photographs for presentation or photo copying. BLM may also provide equipment and supplies as available for major projects.

3) Provide personnel necessary for such activities as project coordination, support, and advice, if not constrained by personnel or budget.

4) BLM will provide an opportunity for the SWR to participate in

cave resource related project planning, activity planning and resource management planning in the initial stages (scoping) as early as possible, consistent with Federal regulations.

5) Provide work space in BLM offices, when available, for SWR members working under this Agreement, including access to cave files, aerial photos and maps.

6) BLM may reimburse SWR personnel for incidental expenses when set forth by mutual agreement in project-specific Volunteer Service Agreements, Challenge Cost Share or Cooperative Agreements.

V. ACTIVITIES COVERED BY AGREEMENT

This Agreement allows the SWR to become involved in the following general cave management activities:

A. Participating Management Team (PMT)—A work group consisting of members from New Mexico BLM offices, SWR and other cave oriented groups will meet semi-annually, or as needed, to discuss accomplishments under this Agreement. The PMT will develop work plans on a timely basis which identify volunteer projects and other cooperative ventures. Through the PMT the SWR is afforded the opportunity to provide technical assistance to the BLM in developing cave management policy program direction. The PMT will be notified in a timely manner, of national, state or district mandates or changes in policies and regulations which may affect cave management policy and procedures.

B. Cave Monitoring—Cave monitoring involves such activities as establishing photo points, maintaining photo files, monitoring visitor use and periodically changing locks or lock combinations on gates. Cave monitoring may also involve the installation and maintenance of cave registers, including restocking the registers and forwarding completed pages to BLM.

C. Cave Inventory—The SWR may assist the BLM in conducting inventories of caves and their contents. Included in the inventories will be such items as cave flora and fauna, geology, mineralogy, archaeologic or paleontologic material, rare or delicate speleothems and hazards to cavers. Cave inventories will be conducted using the BLM Cave Inventory and Classification System format.

D. Exploration—The SWR may conduct exploration in known caves as well as assist in locating previously unknown caves on public land. Cave digs will be authorized under BLM permits. NEPA documentation will be completed prior to authorization of the action. Because of the risks associated with cave digs, volunteers cannot be authorized under a Volunteer Service Agreement. A report will be given to the BLM by the SWR on any passage or cave discovered with details and locations of any notable features found. If any artifacts, cultural, or paleontological resources or remains are discovered, they must remain in place and BLM must be immediately notified.

E. Mapping—The SWR may conduct mapping work in caves on public land. If done as a volunteer project, the SWR will supply BLM with a copy of any cave map produced or updated as a result of the mapping activity, along with a copy of the mapping notes. Cave maps will be provided to the BLM on reproducible media. Cave entrance locations will be given to the BLM for caves mapped under a Volunteer Service Agreement with as much accuracy as possible. GPS coordinates may be used along with topographic maps. All cave maps will be considered confidential and the use of these maps and locations will be treated as proprietary information for internal BLM use only. The copyright of the map remains with the SWR, (mapper/cartographer) unless the mapping was accomplished while under a Volunteer Service Agreement with the BLM.

F. Cave Restoration—Cleanup or restoration projects may be conducted as needed in caves specified by BLM or upon SWR initiative. This may include, but is not limited to, speleothem restoration and cleaning, trash removal, dismantling unnecessary rock cairns or flagging tape and removal of modern graffiti.

G. Cave Conservation—The SWR may assist the BLM in performing conservation activities such as designing, installing, repairing or removing cave gates and fences, marking routes in caves, as well as conservation and Leave-No-Trace training of new cavers and the general public.

H. Interpretation and Education—SWR may assist BLM or give programs independently to local civic or school groups on such subjects as cave exploration, education, conservation, bats or geology. The SWR may also assist the BLM in developing environmental education programs and materials such as handouts, brochures and slide shows. Members of the SWR may be asked by the BLM to act as trip leaders. The SWR has the right to refuse any trip leader request.

I. Research—SWR may conduct or arrange for scientific research in BLM caves through appropriate BLM authorizations. A copy of any findings, reports and/or publications resulting from such research will be forwarded to the BLM within six months of the end of the project.

J. Other Projects—SWR may be provided an opportunity to assist BLM in the performance of other management duties or projects as approved by BLM.

VI. PROPRIETARY INFORMATION

A. Information specifically requested by BLM from members of the SWR through work as volunteers under a Volunteer Service Agreement with the BLM belongs to the Federal Government and shall be exempt from the Freedom of Information Act in accordance with the Federal Cave Resources Protection Act of 1988.

B. Information gathered by the SWR other than under a Volunteer Service Agreement belongs to the SWR. Any information held by the SWR but loaned to BLM to aid in resource management decision making will be treated as propriety information and shall be exempt from the Freedom of Information Act in accordance with the Federal Cave Resources protection Act of 1988.

VII. PARTICIPANTS

The participants to this CMA are the Southwestern Region of the National Speleological Society and the New Mexico Bureau of Land Management

This document was approved in January 2002. It was signed by: Bob Rodgers, Chairman, Southwestern Region of the National Speleological Society on December 8, 2001; and Michelle Chavez, State Director, Bureau of Land Management, New Mexico State Office on January 16, 2002.

Section B—Organizing Cave Projects

Documenting Volunteer Value

Val Hildreth-Werker and Jim C. Werker

Cavers give tremendous amounts of time, labor, and expertise. They spend from their personal incomes to participate in survey and exploration as well as cave protection and management. The caving community is recognized for providing significant volunteer contributions to cave and karst resources on both public and private lands. Some federal officials believe the caving community contributes more consistant volunteer time and expertise than any other land-using group in the United States. But we need to document the value of caver contributions. How do we calculate Volunteer Value?

Volunteer caver hours include in-cave work, preparation time, documentation time, and some travel. Why should cavers learn to record Volunteer Value? By defining and calculating volunteer value, the caving community can quantify caver contributions of stewardship, time, and expertise. Systems for documenting volunteer value are described in this chapter.

Caver Volunteer Value represents extensive economic benefits to private, county, state, and federal lands. By assigning dollar values and adding up the sums, the data can be used in grant application processes, cave management proposals, and karst protection efforts. We should therefore record Volunteer Value for all cave projects—survey, cartography, inventory, science, administration, restoration, cleanup, and conservation.

By defining and calculating volunteer value, the caving community can quantify caver contributions of stewardship, time, and expertise. Caver Volunteer Value represents extensive economic benefits to private, county, state, and federal lands.

History of Volunteer Value: Fee Demo Compromise Achieved

NSS action opposing the federal Fee Demonstration Program eventually led to the development of the NSS Volunteer Value Program. The U.S. Congress, supported by the Recreation Coalition, initiated Fee Demo in the early 1990s, suggesting that federal agencies design programs to collect more fees for recreational opportunities on public lands. Unexpectedly, cave programs from Region 3 of the USDA Forest Service (southwestern U.S., specifically Arizona and New Mexico) were submitted and selected as Fee Demo projects. Through persistent efforts of NSS members and Forest Service officials, the USDA Forest Service Region 3 Cave Fee Demonstration was dropped and cavers agreed to help monitor and maintain the caves through increased volunteerism and development of a prototype Volunteer Value Program (Jagnow 1998).

The proposed Cave Fee Demo was shown to lack cost effectiveness and to potentially put nonrenewable underground resources at great risk. Yes, the cave programs need more funding. Yes, the Fee Demo project "promised" that most of the funds would be used for cave programs—at least during the first three years of the demo—but in reality, the collected fees become federal funds that can eventually be pulled from the programs that generate them. (For example, only the proceeds from donation boxes located in visitor centers on federal lands go back, 100 percent, into on-site programs.)

Volunteer Value: Banking Our Contributions to Caves and Karst
Bern Szukalski

Cavers encompass a broad range of expertise and skills that are applied directly to caves and karst through many types of projects. This combined volunteer effort represents considerable value in terms of the time and expertise donated by cavers toward the protection, exploration, scientific investigation, conservation, and preservation of caves.

More so than any other resource-use group, cavers freely contribute their time and skills toward the betterment and understanding of speleological resources.

These volunteer efforts represent significant value to caves and karst on private, state, and federal lands.

The Volunteer Value program was established to document the volunteer investment of NSS cavers, and to assign a monetary value to that investment.

Why is this important?

(Continued on next page)

While up to 80 percent of the money collected through Fee Demo programs may, through a proposal and selection process, go back into site development, there is no guarantee where the Fee Demo funds will be used. (For current information, search the Web for Fee Demo. For current updates, historic background, development of the Fee Demo Program, and links, see <http://www.wildwilderness.org/>. Use the Wayback Machine on this comprehensive site to access federal and Recreation Coalition documents from the 1990s.)

Another major objection for inclusion of caves in Fee Demo is advertising and promotion. Participation in the Fee Demo Program requires that marketing strategies be initiated to advertise the fee site (or, in this case, the cave adventure) as a recreational resource. Marketing wild cave trips is reminiscent of wild river trip promotions in the Grand Canyon that led to increased visitation and liability concerns, ultimately resulting in virtual monopolization by outfitter-guide services. Most cave resources are nonrenewable and cannot be replaced. We have all seen cave resources destroyed through ignorance, inadvertence, or overuse. Because of marketing requirements, Fee Demo can be interpreted as being contrary to The Federal Cave Resources Protection Act that was passed by Congress in 1988 to protect caves and their contents. The prospect of charging fees does not justify marketing or promoting undeveloped caves as recreational resources.

Persistent resistance to cave fees convinced the Forest Service of cavers' convictions to take care of the caves. NSS President at that time, Fred Wefer, appointed Ray Keeler to chair a temporary NSS Ad Hoc Cave Fees Committee. Cavers from Arizona, New Mexico, and Texas served on the committee. Concerned interaction with the Forest Service proved NSS commitment to find better solutions than Fee Demo could offer. Cavers offered increased volunteer assistance for improved cave management, and together, the NSS and the USDA Forest Service developed a Volunteer Value program.

Volunteer Value Defined: Agreement Signed

USDA Forest Service Region 3 agreed to set aside cave fee demonstration projects while cavers agreed to give significant in-kind volunteer contributions. Cavers agreed to contribute at least $100,000 annually in Volunteer Value to the Region 3 cave programs. Work projects were initiated and cavers from across the U.S. traveled to participate.

But, there was no agreement in place to determine how Volunteer Value would be calculated or tallied. Jim Werker and Val Hildreth-Werker, at that time Co-Chairmen for NSS Resource Protection and Preservation, collected data from previous cost share agreements, volunteer projects, NSS Ad Hoc Cave Fees Committee e-mail discussions, USDA Forest Service (USDA-FS) volunteer programs, and federal agency guidelines.

Jim Werker presented a draft proposal to the Forest Service at the 1998 NSS Convention in Sewanee, Tennessee. Jim Miller, Dispersed Recreation Program Manager for the USDA-FS in Washington, D.C., suggested additions and edits. The document was reviewed and approved by Jim Miller; Jerry Trout, USDA-FS National and Region 3 Coordinator/Cave Resources; Fred Wefer, NSS President; and Doug Medville, NSS Administrative Vice President. The agreement defines the monetary value of volunteer efforts. (See volunteer value agreement, page 323.) It serves as an addendum to the national Memorandum of Understanding between the USDA-FS and the NSS. The agreement establishes the foundation for developing the NSS Volunteer Value Program. Cavers are increasingly documenting their Volunteer Value for all cave projects on private, state, or federal land, worldwide.

Volunteer Value Agreement Between the USDA Forest Service and the National Speleological Society

This document serves to define Volunteer Value and represents an agreement between the USDA Forest Service and the National Speleological Society. Cavers are providing volunteer services to Forest Service cave programs in lieu of Fee Demonstration Projects being applied to caves.

Assigning monetary values to volunteer efforts for caves managed by the Forest Service will benefit all concerned. The purpose of this agreement is to define and clarify a Volunteer Value system that will be consistently implemented in all Forest Service cave programs. By applying this system for calculating Volunteer Value, confusion and disagreement can be avoided. This agreement has been designed with input from NSS cavers, nonaffiliated cavers, and Forest Service personnel.

Through this Volunteer Value agreement for cave programs, in-kind contributions from volunteer organizations, nonaffiliated volunteers, and NSS volunteers will be recorded and calculated as follows.

1. "Salaries" and/or volunteer hours:

A. Valued at the GS5/1 hourly rate for standard volunteer arrangements or minimum $10.00 per hour.

B. Volunteer hours include:

- Travel time to the cave and/or work site, portal to portal, round trip, not to exceed work time. (Refer to 1.C.)
- Time spent in the cave and/or work site actively involved in the project.

C. Volunteer hours do not include:

- Time spent in activities such as sleeping, eating, and preparing food.
- Time spent for recreational caving. (Recreational caving should be done on other trips.)
- Travel time from home that exceeds the total time spent in the cave and/or work site actively involved in the project. (For example, if travel time totals 12 hours, and 8 hours of active work time at the site was accomplished, then 8 hours work time and 8 hours travel time is counted for that individual when calculating Volunteer Value.)

D. When tasks require specialized professional expertise, value for "salaries" and/or hours should be assigned according to equivalent GS ratings or other definable rates.

2. Travel:

A. Valued per mile at the government rate for business, to be adjusted as the government rate changes.

B. Mileage should be listed to and from the site. (Mileage for all volunteers in one vehicle is listed once.)

- All mileage directly related to reaching the work site, portal to portal, round trip. (Mileage from home to the work site and return).
- All mileage for each vehicle. (Not for each person.)

(Continued on page 324)

(Continued)

By calculating and documenting contributions we can quantify the benefits cavers provide to public land managers and private landowners.

This in turn can be used for grant acquisitions, cave and karst management proposals, public policy issues, and federal land management decisions.

Volunteer Value is easy to document and record. At the NSS Conservation Web site <www.caves.org/committee/conservation/> you'll find additional details on the Volunteer Value program, as well as downloadable forms.

It only takes a few minutes to fill these out, and document your contributions for the benefit of all. Include documenting your volunteer efforts as part of every trip and activity, and spread the word to support the Volunteer Value program.

3. **Supplies:**
   A. All monies spent on supplies used during the project. If reimbursed, this is not counted as Volunteer Value.
   B. All monies spent on supplies for preparation and documentation for the project. If reimbursed, this is not counted as Volunteer Value.
4. **Equipment:**
   A. Valued at $150.00 per caver for basic caving gear such as helmet and light. (Calculated one time each 3 year period or $50.00 per year.)
   B. All monies spent on caving equipment such as ropes, climbing gear, cameras, survey equipment, etc. If reimbursed, this is not counted as Volunteer Value. (Re-list as equipment is replaced.)
5. **Preparation and documentation:**
   A. Time spent in organizing and preparing the project.
   B. Time spent completing documentation of project.

This document was approved on August 6, 1998, and signed by: Jim Miller, FS National Dispersed Recreation Program Manager; Jerry Trout, FS National & Region 3 Coordinator/Cave Resources; Fred Wefer, NSS President; and Doug Medville, NSS Administrative Vice President.

Volunteer Value Worksheets for All Cave Projects

In order to calculate and make tangible this asset we call Volunteer Value, we created three simple forms for recording caver contributions. These generic worksheets are designed to be filled out in the field, electronically, or online, and can be used for documenting in-kind volunteer labor, expertise, and contributions for any cave project. General values can be assigned, or specific individual values can be inserted for specialized skills.

Use Volunteer Value forms to document all cave projects—*survey, cartography, inventory, science, administration, restoration, maintenance, conservation*. Record in-cave work as well as preparation and documentation time. The calculation sheets are facilitating design of a national database to handle caver volunteer value data for inclusion in NSS Biennial Reports, funding and grant applications, and conservation-management issues. Template worksheets and type-in sheets are included in this volume (pages 325–328). The forms can also be accessed through the NSS Web site <http://www.caves.org/conservation/>. Click on Volunteer Value.

Cited Reference

Jagnow D. 1998. History of USFS cave fee demo. *NSS News* 56(3):64-66.

Additional Information and Downloadable Forms

NSS Web site home page: <http://www.caves.org/>
Click on Cave Conservation, then click on Volunteer Value.
Or go to: <http://www.caves.org/committee/conservation/> and find Volunteer Value Worksheets.

Project Title ______________ **Agency** ______________ **Project Date(s)** ______________

*Optional work column should be tallied from all WORKSHEETS(s)#2 & recorded here at end of project if individuals have more travel time than work time. Include month / dates / year

VOLUNTEER VALUE REPORT WORKSHEET #1 – SIGN-IN SHEET: PARTICIPANTS, TRAVEL, AND SUPPLIES

NAME	City	State	Driver D	Mileage Roundtrip (Record once per vehicle)	Rider R	Travel Total time round trip	*Work Total time all project work	Date Arrive	Date Depart	Supplies List items for project and record total $ Don't include if reimbursed by agency/owner	Cost	Comments, notes, phone, e-mail, etc.
1.												
2.												
3.												
4.												
5.												
6.												
7.												
8.												
9.												
10.												
11.												
12.												
13.												
14.												
15.												
16.												
17.												
18.												
19.												
20.												
VV 1	Optional: Use this space to record four column totals.			Total		Total	Total				Total	NSS

Figure 1. Volunteer Value Report Worksheet # 1

Project Title ______________________ **Agency** ______________________ **Project Date(s)** ______________________
(Indicate task leaders in comment space.) Include month / dates / year

VOLUNTEER VALUE REPORT WORKSHEET #2 – TASKS AND WORK HOURS
For easier record keeping, note volunteer hours on your permits, then transfer information from each permit to a copy of this form.

Date ☞	
Project or Cave Name	
Volunteer Names (Volunteer Work Hours may include travel time from and to camp, either here or on SHEET #1.)	**Hours worked**
1.	
2.	
3.	
4.	
5.	
6.	
7.	
8.	
9.	
10.	
11.	
12.	
Project Preparation & Documentation Note names and hours spent.	Record total doc and prep time in this space.
TOTAL work hours in column at right ☞	Total hours on this page.

Notes. Trip report. Or describe project. Include observations, comments, and suggestions.

Supplies and expenses not already listed This space is for additional supplies/equipment not listed elsewhere, e.g., copies, postage, etc. **Don't include if expenditures will be reimbursed by agency or owner.**	**Cost**

Task Categories (circle if applicable): bats, biology, gates, geology, graffiti, rock art or historical signature analysis, impact mapping, meteorology, microbiology, mineralogy, preparation and documentation, restoration, research, photo documentation, photo inventory, photomonitoring, resource inventory, rigging, speleothem repair, survey and cartography, trails, other (describe):

NSS VV 2

Figure 2. Volunteer Value Report Worksheet # 2

Project Title ____________ **Agency** ____________ **Project Date(s)** ____________
Include month / dates / year

VOLUNTEER VALUE REPORT WORKSHEET #3 – PROJECT TOTALS

Total Number of Volunteers for this Project ____________
Total Number of Participants

Project Coordinator(s) ____________
Name, phone number, e-mail address

Total Volunteer Vehicle Mileage Add up total mileage from Sign-In WORKSHEET #1 for this project. Multiply mileage by 32.5¢ or currently designated government rate.	Total Miles		32.5¢ or Current Rate	**Total Volunteer Value for Mileage** Mileage **x** 32.5 cents per mile or current rate. **Value in Dollars**
Total Volunteer Travel Time Add up total hours from Travel column on Sign-In WORKSHEET #1. ROUNDTRIP Travel Time FROM HOME is recorded for EACH Volunteer. Multiply travel time by $10.00/hour or current GS 5/1 rate.	Total Travel Time		$10.00 or GS 5/1 rate (if you don't know, use $10)	**Total Volunteer Value for Travel Time** Travel Time **x** $10.00 or GS 5/1 rate. **Value in Dollars**
Total Volunteer Work Time Transfer and add up total hours from Task WORKSHEET(s). Volunteer work hours include travel from and to camp. Multiply work time by $10.00/hour, or current GS 5/1 rate, or special equivalent expertise rates. If special rates apply, please indicate the rates beside specific names on WORKSHEET(s) #2.	Total Hours Worked	Show calculation SUBTOTAL work hours multiplied by $10.00 rate	Show calculation SUBTOTAL work hours multiplied by other $ rate	**Total Volunteer Value for Project Work** TOTAL of the two Subtotal spaces. **Value in Dollars**
Total Cost of Volunteer Supplies for Project Transfer supply totals from other WORKSHEET(s).	Total Cost of Supplies on Form(s) #1		Total Cost of Supplies on Form(s) #2	**Total Volunteer Value for Cost of Supplies** TOTAL the Cost of Supplies.
TOTAL VOLUNTEER VALUE FOR THIS PROJECT ☞ **Add the dollar amounts from the last column and record your total in-kind volunteer contribution for this project.**				**Total Volunteer Value for Project** Add the four total dollar amounts from column above. Record grand total here. $

VV 3 NSS

Figure 3. Volunteer Value Report Worksheet # 3

VOLUNTEER VALUE REPORT

USFS Lincoln NF 29 September – 2 October 2000

VOLUNTEER VALUE REPORT WORKSHEET #1 – SIGN-IN SHEET: PARTICIPANTS, ROUND-TRIP TRAVEL, AND SUPPLIES

Name	City	St	Driver	Mileage	Travel	Project*	Arrive	Depart	Supply Cost
Jim Werker	Albq	NM	Driver	700	12 hrs	28 hrs	9/29/00	10/02/00	$12 batts monitor
Val H-W	Albq	NM			12 hrs	28 hrs	9/29/00	10/02/00	$32 batts&photo
Dave Hamer	Tucson	AZ	Driver	1030	18 hrs	20 hrs	9/29/00	10/01/00	$ 4 batts
Phyllis Hamer	Tucson	AZ			18 hrs	20 hrs	9/29/00	10/01/00	$ 4 batts
Allan Cobb	Austin	TX	Driver	1110	20 hrs	20 hrs	9/29/00	10/01/00	$ 4 batts
Aimee Beveridge	Austin	TX			20 hrs	20 hrs	9/29/00	10/01/00	$ 4 batts
Stan Allison	Carlsbd	NM	Driver	100	4 hrs	8 hrs	10/01/00	10/02/00	$ 4 batts
Gosia K-A	Carlsbd	NM			4 hrs	8 hrs	10/01/00	10/02/00	$18 batts&photo
Totals				**2940**	**108 hrs**	**152 hrs**			**$82**

***Project**: Indicate work time in this category. May include travel time from camp to project site and back to camp

VOLUNTEER VALUE REPORT WORKSHEET #2 – TASKS & WORK HOURS

Name	Hours	$ Cost	Explain
Val Hildreth- Werker	6	$ 5.00	preparation and documentation for trip report, photo labels
Jim Werker	5	$ 17.00	copies, mailing, faxes, sewing and materials
Allan Cobb	2		documentation
Totals	**13**	**$22.00**	

Task Categories (underline if applicable):
bats, biology, gates, geology, graffiti, rock art or historical signature analysis, impact mapping, meteorology, microbiology, mineralogy, preparation and documentation, restoration, research, photo documentation, photo inventory, photomonitoring, resource inventory, rigging, speleothem repair, survey and cartography, trails
other (describe):

VOLUNTEER VALUE REPORT WORKSHEET #3 – PROJECT TOTALS

Total Volunteer Vehicle Mileage	**Total Mileage** 2940	**X Rate** @ 32.5¢	**Total Volunteer Value Mileage** $ 955.00
Total Volunteer Travel Time	**Total Travel Time** 108 hrs	**X Rate** @ $10.00	**Total Volunteer Value Travel Time** $ 1080.00
Total Volunteer Work Time	**Total Project Hrs** 13 + 152	**Calculate Hours** 165 @ $10.00	**Total Volunteer Value Proj Work** $ 1650.00
Total Cost of Volunteer Supplies for Project Transfer supply totals	**Worksheet #1** $82.00	**Worksheet #2** $22.00	**Total Volunteer Value Supply $** $ 104.00

TOTAL VOLUNTEER VALUE FOR THIS PROJECT

Total Volunteer Value for Project

Add dollar amounts from four categories in last column.

Record grand total for in-kind volunteer contribution here. ☞ **$ 3789.00**

Figure 4. Volunteer Value Report compiled from the three worksheets.

Section C—Restoring Cave Passages

Do No Harm

Val Hildreth-Werker

Primum non nocere.
First, do no harm.

All-important in cave restoration is the notion of leaving things better than we found them—*while avoiding* new harm to biota, habitat, or other cave resources.

When considering a project—assess, plan, and avoid creating new problems. Thoroughly evaluate the restoration tasks. Carefully design the workload and crew assignments. Conscientiously avoid causing new restoration dilemmas in the cave passages. When the project is complete, clean up any leftover mess. These are simple, universal precepts.

Can we actually achieve *no harm* when we're in a cave sopping, scrubbing, scraping, and disrupting the microcosm?

The words, *no harm*, have a meaningful and memorable ring. It is a lofty, idealistic objective.

The potential of causing no new harm is relative to the damage that has already been done, the number of visitors that have already tromped through, the fragility or durability of the resources, and the historical stewardship of a passage. Previous anthropogenic alterations within a cave will influence restoration goals and no harm objectives. Project planning includes defining what natural or historical period to target restoration along a continuum of ecosystems, anthropological impacts, and aesthetics.

Many cave values are nonrenewable or irreplaceable. Some cave species and some speleothems are fragile. Subterranean ecosystems, sometimes hanging in delicate balance, may be easily disturbed. Cultural materials may be forever lost. Harm happens regardless of our best conservation-directed intentions. In many cave systems, *no harm* is literally impossible if humans enter.

More realistically, we strive to minimize our harm—and we should recognize that whatever we do in a cave may damage the resources in some way we have not yet anticipated.

In caves, causing no new harm may sometimes mean doing nothing to remediate horrible impacts. Other times, we decide to attempt restoration with sincere and well-considered efforts toward minimizing new damage to cave values.

With our inadequacies acknowledged, maybe we should choose an axiom like *minimize harm,* but instead we use the admonition *do no harm*. It is a strong statement—it reminds us to reach for the depths of ideal caving ethics and informed restoration decisions.

Cave safely ... restore softly ... and do no harm.

Many cave resources are nonrenewable or irreplaceable. Some cave species and some speleothems are fragile. Subterranean ecosystems, sometimes hanging in delicate balance, may be easily disturbed.

Section C—Restoring Cave Passages

Hidden River Cave: A Karst Reclamation Success Story

Michael Ray Taylor

On September 7, 1867, while en route to Mammoth Cave, Kentucky, naturalist John Muir stopped at lesser-known Horse Cave. In his book, *A Thousand-Mile Walk to the Gulf*, Muir wrote of the place:

> *It seems a noble gateway to ... the birthplace of springs and fountains and the dark treasuries of the mineral kingdom. This cave lies in a village of the same name which it supplies with an abundance of cold water. Cold air issues from its fern-clad lips. In hot weather crowds of people sit about in the shade of the trees that guard it. This magnificent fan is capable of cooling everybody in town at once.*

Within 50 years of Muir's visit, Horse Cave had become one of America's most notorious examples of groundwater pollution.

In the 1920s and 30s, raw sewage—thick with spoiled whey from a local creamery—was piped directly into the underground stream. The "magnificent fan" spread the stench for miles, making the town virtually uninhabitable every summer for 6 decades. Chromium and other chemicals from a metal-plating factory further polluted the abundance of cold water.

In 1943 the cavern, by then renamed Hidden River Cave, was closed to the public as a health hazard—at a time when cavers had spotted tantalizing unexplored passages leading from the known trails.

In 1987, after nearly 50 years of closure, the American Cave Conservation Association (ACCA) located its national headquarters on the Hidden River Cave grounds and began efforts to restore and reopen the cave. The ACCA worked with local politicians and residents to get a new regional waste treatment system constructed in order to stop the discharge of sewage into Hidden River Cave's groundwater recharge basin.

ACCA Collection

Figure 1. Entrance to Hidden River Cave in the town of Horse Cave, Kentucky.

In 1987, after nearly 50 years of closure, the American Cave Conservation Association (ACCA) located its national headquarters on the Hidden River Cave grounds and began efforts to restore and reopen the cave.

As the cave began to recover, opportunities for exploration opened up. On September 8, 1991, David Foster, Executive Director of the American Cave Conservation Association, began survey trips utilizing volunteers from the Cave Research Foundation (CRF) and the Central Kentucky Karst Coalition.

"This cave may be longer than Mammoth," said an excited Foster after that first trip, echoing the sentiments of legendary caver Jim Dyer who explored but never surveyed the cave. Once beyond the restored scenic entrance, survey crews, organized by CRF volunteer Mike Yocum, followed a 15–meter (50–foot) diameter trunk passage lined with massive mud banks. For 600 meters (2,000 feet) they skirted the remains of an old wooden tourist trail.

"It was a gloomy, dark place," recalled caver James Wells. "Chunks of rotting wood poked up from the mud. They crumbled at the slightest touch. The walls seemed to suck up light. But the cave was amazingly clean. We crossed a place the old-timers used to call The Cottage Cheese Swim, because of all the floating whey. I didn't see a trace of white in the water. The air smelled fine."

The "cottage cheese" had vanished thanks to two new waste treatment plants, a $15 million project that diverted sewage and whey from the cave. Volunteer crews removed decades of refuse that had blocked the entrance area. The town of Horse Cave (it kept its name) supported Foster's efforts by joining in the construction of a $2 million museum devoted to caves and cave conservation.

The American Cave and Karst Center opened in July 1992, and has since been visited by nearly 100,000 people. The Center uses Hidden River's unique history as an educational tool for teaching about the problems of living on karst terrains.

Throughout the 1990s, Yocum's survey crews mapped over 11 kilometers (7 miles) of new passage in Hidden River Cave, and survey trips continue into the 21st century. Using software that aligns the map of Hidden River Cave with satellite photographic data, a recent study was able to locate contemporary sources of groundwater pollution entering the cave stream. While the river of whey and horrible stench are distant memories, the fight to preserve karst waters is ongoing, and continues to improve with technology.

The surveyed length of Hidden River Cave remains only a fraction of that of Mammoth, but its now-clean passages are likely to beckon surveyors for many years to come.

Section C—Restoring Cave Passages

Cave Graffiti: The Writing is On the Wall

James R. Goodbar and Val Hildreth-Werker

We've probably all seen it, and some of us have even tried to remove it. Viewed as unsightly and unnecessary, the ugly writing on cave walls is called graffiti. But hidden among the contemporary spray paint may be culturally significant pictographs, petroglyphs, mud glyphs, historic writing, and other markings that should be preserved. Bilbo and Bilbo describe rock art and historic writing in another chapter in this volume (page 99).

Whether one graffito or many graffiti, some are old and some are recent. Some were applied with a candle or carbide lamp. Other types are graffiti nightmares in black enamel paint or the reflective hot pink day-glow stuff. Regardless of how the marks were applied on cave walls, ceilings, and floors, there are basic guidelines for evaluating, documenting, and preserving or removing it. The techniques described in this chapter were developed through firsthand experience over the past quarter century.

Cave environments vary tremendously from region to region—cave conditions in humid areas like the eastern U.S. are very different from the cave environments found in the arid western states. What works well in Kentucky may be ineffective in New Mexico. Methods that work in limestone caves may cause irreparable damage to gypsum caves, and vice versa. Restoration techniques are sometimes developed to fit the needs of specific sites. No technique can be recommended for all caves. Each cave must be evaluated on its own merits and restoration methods must be selected that will best serve each particular cave system.

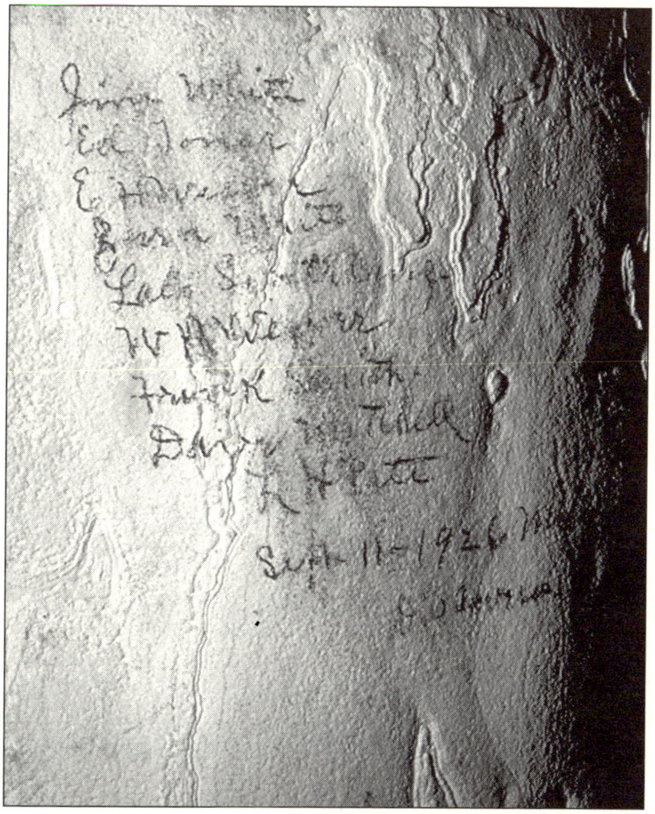

© Val Hildreth-Werker

Figure 1. Historic names are lined up on a speleothem. The top name is Jim White, the explorer who conducted the first tours of Carlsbad Cavern. These historic signatures are protected by the National Park Service and remain intact within Carlsbad Cavern.

What Is Cave Graffiti?

The word graffiti, strictly speaking, is a plural form. Graffito is the singular. In accordance with current usage, the term *graffiti* is used in this chapter—cavers often find multiple incidents of scribbles, signatures, or drawings on cave walls. Collectively, these are called graffiti. (See graffiti sidebar, page 334).

In caves, graffiti can be quite old and provide evidence of historical use by early visitors. It comes in all shapes and sizes and in a variety of media. Graffiti can be drawn or written with ocher, lampblack, or spray paints. It is found scratched into soft surfaces. Arrows, words, names, dates, cartoons, and pictograph replicas are marked on cave walls with crayon, lipstick, nail polish, and permanent marking pens.

Graffiti, no matter how bad it looks, is a rich part of human history. Before considering removal of any graffiti, a careful historical and cultural evaluation should be conducted. During an inventory of the cave's artifacts and features, expert archaeologists or historians should provide consultation.

Sources for local expertise may be found by contacting the State Historical Preservation Office (SHPO). (See consult with experts, page 116.) The SHPO personnel are helpful and willing to network expertise and resources. The historian or archaeologist asked to evaluate the cultural material should be knowledgeable in cave resources.

There are also archaeologists, historians, and trained volunteers in the NSS who are good resources for these tasks. Check with the NSS American Spelean History Association <http://www.caves.org/section/asha/>.

Following consultation, if the graffiti is deemed to have insignificant historical value, the cave managers may decide on documentation and removal. However, markings deemed historically or culturally significant should remain as they are and should not be removed, touched, or harmed in any way—historic signatures and historic or prehistoric cave art is easily damaged. (See rock art and historic writing, page 99).

It is always wise to seek permission from the landowner or cave manager and request assistance in determining whether it is appropriate to remove graffiti and what portions should be left on the cave walls to preserve historical or cultural data.

Laws and Regulations

Laws and regulations also provide helpful guidelines for determining when to preserve graffiti. The Archaeological Resources Protection Act of 1979 sets 100 years for the protection of cultural remains. However, 50 years is generally accepted in the field of archaeology as a date signifying historical importance. (See 50-year "rule", page 341; historic writing, page 110.)

Within the federal government, each agency has specific regulations that can be cited in the prosecution of known offenders. Generally, those regulations address the Federal Cave Resources Protection Act of 1988 or the destruction of government property. Many states also have cave protection laws that prohibit making new graffiti in caves. (See cave laws, page 217.)

Beyond laws and prosecution, consider the other consequences. If historical or cultural value is in question—no matter how vague the lingering doubts—markings can always be removed later. If removed before the value is determined, it's gone forever.

Document Content, Style, and Media

Some of the most common media for cave markings include incised or etched material, paint applied by hand, torch mark, candle smoked carbon, carbide soot, pencil, and spray paint. There are poorly understood chronological parameters for each. For example, spray paint is post World War II, and the carbide era is between 1900 and 1970. The same is true of prehistoric natural pigments (Joe Douglas, personal communication 2003). While photographs and drawings record content and style, field notes should thoroughly document the site as well as the media and give insight to the authenticity and importance of markings.

Always photograph, evaluate, and thoroughly document the content,

Graffiti Versus Graffito

Editors' Note: The word graffiti is listed in most contemporary dictionaries as the plural form of graffito.

Graffito is defined as *an* inscription or drawing made on a public surface such as a wall, or on a natural or cultural surface.

Thus, the plural form, *graffiti*, indicates several inscriptions or drawings and should be used with plural verbs.

In this volume, if we were following the dictates of perfect English usage, sentences would read: "Graffiti *are* painted along the walls of Good Grammar Cave. One brightly colored graffito *is* inscribed at the remote end of the short passage and *reads* 'Have No Fear, Graffiti Rule.'"

However, after much debate among reviewers, we decided to go with common current usage in the pages of this book and we wrote graffiti sentences to reflect the way cavers actually talk about graffiti.

Cavers tend to say, "New graffiti is covering authentic historic signatures in Current Lingo Cave."

Photos © Val Hildreth-Werker

Figure 2. Historic signatures have been vandalized with contemporary graffiti carved into the walls of Black Cave, New Mexico.

Figure 3. Historic signatures, including Robert Nymeyer's 1934 inscription, are preserved in Endless Cave, New Mexico.

Photos © Val Hildreth-Werker

Figure 4. Armed with a stiff brush, a spray bottle of water, and elbow grease, this caver is prepared to eradicate the spray painted arrow marring a cave wall.

Figure 5. Scrubbing with stiff nylon brushes and water eliminates most carbide stains.

BLM Photos

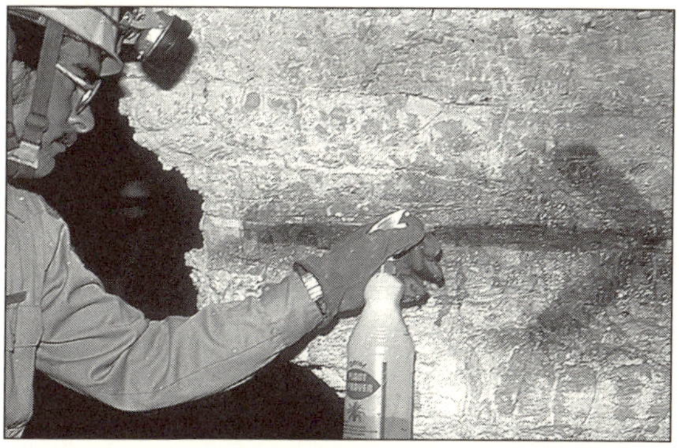

Figure 6 (before) and **Figure 7** (after). Buzz Hummel used a spray bottle, stainless steel brush, perseverance, and time to make most of this obnoxious arrow vanish.

Photos Joe Orman

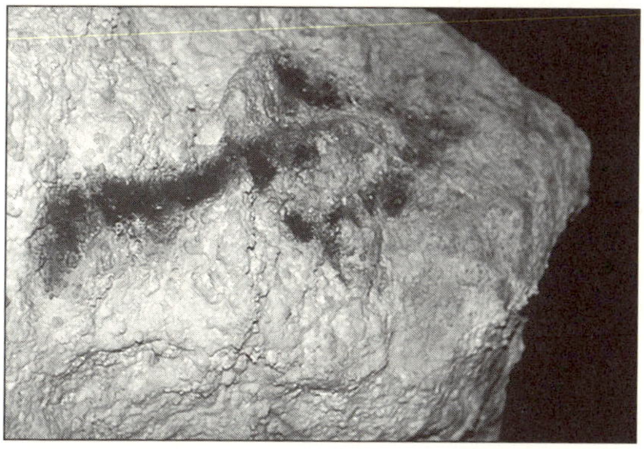

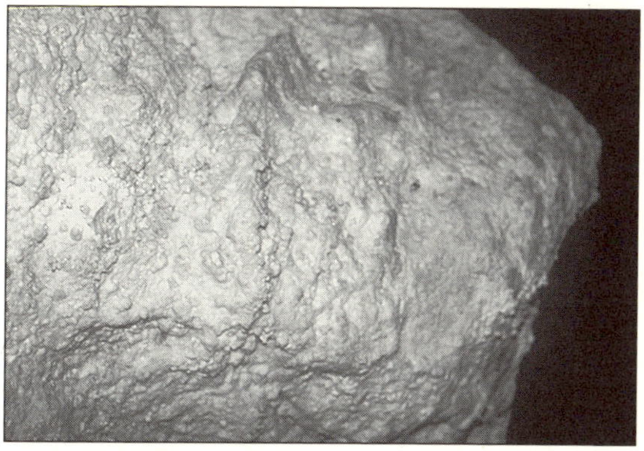

Figure 8 (before) and **Figure 9** (after). After two hours of scrubbing, the big black spray painted arrow was gone.

Photos © Val Hildreth-Werker

Figure 10 (before), **Figure 11** (during), and **Figure 12** (after). This caver, Chris Rollo, disgusted by the graffiti arrow spray painted on a cave ceiling, spent the better part of a day scrubbing it away with water and a stainless steel brush.

style, and media of graffiti before deciding whether to remove it. Keep in mind that evidence written in contemporary graffiti may facilitate the identification and prosecution of vandals. Use photodocumentation of offenses to pursue legal prosecution, enhance conservation education, and improve cave management. (See NSS Cave Vandalism Deterrence Award Commission, page 110 and page 224.)

Photographs and documentation of media also provide a small measure of protection against loss of historical or cultural data through accidental removal or ignorance. (See photodocumentation, page 204.)

> **Does Graffiti Beget Graffiti?**
> **James R. Goodbar**
>
> Perhaps the presence of graffiti invites or encourages new markings.
>
> Some sites have layer upon layer. Contemporary markings often cover important historical or cultural resources.
>
> Restorationists try to preserve historic markings. However contemporary graffiti is generally viewed as vandalism. Usually considered visually obtrusive, contemporary media may be detrimental to historical and cultural markings as well as cave biota. (See graffiti and vandalism, page 110.)

Tools and Techniques

If it is determined that contemporary graffiti should be cleaned off, the type of media and substrate may indicate how difficult it will be to remove the unwanted marks. Enamel spray paint seems to be the worst offender and requires extensive scrubbing, but carbide black can often be sprayed clean with water.

Judicious application of water may help where rough rock with small cracks and declivities make it difficult for brushes to reach. Hard work, perseverance, and specialized tools are the key factors to success. Delicate formations such as helictites, soda straws, and fragile gypsum formations require special care and treatment.

Scrub Brushes

Most graffiti removal efforts start with scrub brushes, water, and lots of elbow grease. Some brushes cause more harm than good, so it is important to select tools that minimize new problems. Even the softest brush can cause some damage to cave surfaces as well as biota. Removing the slightly weathered exterior surface of the wall may speed up erosion of newly exposed surfaces.

Cavers should weigh biologic, geologic, and aesthetic factors to determine whether the costs are acceptable. It is important to use new, clean brushes for restoration—be careful to *never introduce materials from home projects* into the cave. Make sure people wear safety glasses and other personal protection devices.

Nylon or Stainless Steel Brushes. Clean nylon and stainless steel brushes are relatively safe for cave use. Be careful not to create bristle grooves or scratch marks. Always assign an experienced graffiti scrubber to test tools. Both nylon and stainless brushes leave few environmentally damaging residues or bristles behind. However, because some bristles will fall out and paint flecks will fly, always spread plastic under the work areas and gather the debris at the end of the day. (See restoration brushes, page 419.)

- Stiff, nylon-bristled brushes cause minimal damage to cave surfaces.
- Stainless steel brushes, when used with a light touch, work well for some surfaces. Be aware that any metal brush, stainless steel included, will leave black marks on some surfaces.

Avoid Brass, Steel, or Natural Fiber Bristles. As stated above, nylon and stainless steel brushes are generally best for cave use. Stray bristles made of nylon or stainless steel are less offensive to cave environments than bristles made of other materials.

Even with catchments and careful scanning of the restoration area, some bristles are likely to escape and remain in the cave.

- Brass brushes leave a fine metal sheen on formations and rock.
- Steel wire bristles will leave black marks and break off and oxidize (rust), discoloring the surrounding cave surfaces. Deteriorating steel adds ferric hydroxide and ferric oxide to the cave ecosystem.
- Natural fiber brushes leave bristles behind that can provide nutrients for molds, mildews, and fungi. Natural fibers may disrupt a cave's ecological balance, providing new food sources for biota and microbiota.

Rotary Brushes, Grinding, and Sanding. Rotary brushes on electric drills are an option, but they have a tendency to scatter paint flakes and bristles over a large area. Rotary brushing or sanding can remove a lot of rock or flowstone in a very short time and this method should be used only if deemed appropriate. For this technique, assign a careful, gentle operator—someone with a good light source, exceptional close-up vision, and patient attention to detail. Always plan ahead and prepare adequate catchment procedures.

Scrub Gently. In scrubbing cave surfaces, be careful to avoid removing layers of mineral. The layers uncovered may not be the same color as the layers removed. Scrubbing away mineral layers may result in a well-defined clean area in the shape of the letters just removed.

Water Sprayers

When it is appropriate to remove contemporary graffiti from cave walls, water can help loosen the media and clear paint flecks from the scrubbing area. Clean, chlorine-free water is the best solution. It is usually not detrimental to cave biota and does not harm most speleothems. It is safe for human use, inexpensive, and readily available. (See water sources, page 393.)

For arrows and contemporary markings with the lampblack from carbide lanterns, it is important to note that water alone will usually clean off the black marks. Brushing is rarely necessary on carbide marks. (See carbide removal, page 411.)

Only use new, clean products in cave environments. Always avoid introducing household cleaning chemicals, herbicides, pesticides, or other human-manufactured chemicals. (See anthropogenic chemicals, page 57.)

Hand-Held Spray Bottles. Hand-held squirt bottles are easy to purchase, use, and carry into the cave. They are easily refilled. Adjustable nozzles can be used to spray wide areas or shoot a stream to loosen and remove debris such as lampblack and flakes of paint. (See pressurized water, page 397.)

Torch Marks
Joseph C. Douglas

When recording cultural resources in caves, look beyond graffiti to all types of wall, ceiling, and floor markings.

Torch marks, tally marks, and other markings may have historic importance. Torch marks are particularly important in that they can be culturally diagnostic in certain areas of the U.S.

For example, scattered torch stroke marks made by bundles of river cane are evidence of prehistoric visitation. But large single carbon torch marks might be either historic or prehistoric in origin.

If a graffiti removal project takes out the scatter of carbon marks, evidence is lost. In many dozens of caves these torch marks are the only clues into the prehistory of the site.

Some bristles will fall out. Paint flecks will fly. Always spread plastic under the work areas and gather the debris at the end of the day.

Garden Sprayers. Garden sprayers, in a variety of sizes up to 5 gallons, are useful tools for some graffiti projects. They commonly include a pump for creating pressure within the container. The resulting water stream is stronger and more continuous than the spray produced from a hand-held squirt bottle. Again, designate new garden sprayers for cave restoration work (used ones may contain insecticides or other chemicals that can wipe out cave communities). Sprayers may also be used to clean mud from speleothems and trails.

Bladder Bags. A close cousin to the garden sprayer is the bladder bag, or backpack pump used for fire fighting. The bladder bag consists of a 5-gallon rubberized canvas bag with shoulder straps. Water is squirted through a hand-held trombone pump attached to the bottom of the bag. Bladder bags are used where moderate pressures and high volumes are required.

Gravity Fed Water Delivery Systems. Gravity fed systems are often efficient, but first consult local cave scientists and evaluate the ecosystem before introducing significant quantities of water into a cave system. Runoff water should always be controlled. (See cautions described in restoration runoff, page 339; also see capture runoff, page 396.) If deemed appropriate, a new garden hose with a trigger nozzle is run from a water supply at the entrance. Depending on the gradient, a substantial head of water is produced and special caution must be exercised to avoid damaging delicate areas. Check that all couplings and gaskets are in good condition to prevent leakage.

Consult with scientists, cave owners, land managers, and knowledgeable cavers before cleaning, and select the least damaging water source for any cave restoration project.

Avoid Delicate Speleothems. Be careful when using high-pressure water devices in areas of delicate speleothems. The pressure combined with a larger volume of water can easily break or damage fragile cave formations.

Protect Invertebrates and Biofilms. While it is efficient to use large quantities of high-pressure water to help remove graffiti, there are drawbacks. Cave habitats may become flooded. Also pressurized spray can harm invertebrates and blast away microbiota. Thin biofilms of microscopic organisms living in moist or wet areas should not be sprayed or scrubbed. Pressure blasting can damage communities of microorganisms—microflora and microfauna are often impossible to see with the naked eye. (See biofilms, page 68.)

Where Should We Get Water?

For cave graffiti projects, it is generally acceptable to use clean, fresh water with no chlorine and commercial chemicals. For some cave systems, distilled water may be an option, if minimal amounts are applied with short contact duration.

No single recommendation is best for *all* caves. In caves with active streams and annual flooding, water from the cave may be the best choice for graffiti cleaning. However, be careful not to use water from isolated cave pools that are slowly refilled over geologic time. Carefully avoid cross-contamination within a cave—transporting cave water from an isolated passage to a different chamber may destroy local indigenous microbial populations.

Consult with scientists, cave owners, land managers, and knowledgeable cavers before cleaning, and select the least damaging water source for any cave restoration project. (Water sources for restoration projects are discussed in depth, page 393.)

Always Catch Restoration Runoff Water

Always collect the runoff water from graffiti efforts, regardless of where the project is located in the cave. Never allow runoff to contaminate cave pools, streams, or water sources. Use large sponges, lint-free towels, environmental remediation "pigs", or shop vacuums to soak up or contain restoration water. Vacuuming devices must be used cautiously to assure that natural loose materials and biota are not sucked away. (See shop vacs, page 358.)

When restoration water is scarce, runoff is sometimes strained and filtered back into a bucket for reuse. However, be careful—used water may contain flecks of paint or lampblack that should not be left in the cave. (See filtering restoration water, page 415.)

Runoff water may also contain lint and debris that should not be washed elsewhere in the cave. Be very cautious about deciding to introduce large quantities of water with hoses or pressure washers. The runoff is difficult to contain and pressurized methods generally cause graffiti byproducts, lint, and loose sediments to be redeposited elsewhere. (See restoration runoff, page 396.)

Weigh the Ecological Costs

Scrubbing layer upon layer of contemporary graffiti made by cave vandals is frustrating, irritating work. Understandably, innovative volunteers want to speed up the process with mechanical devices. However, assist from powered tools is often inappropriate for cave habitats.

On the other hand, a few caves that are heavily trafficked party sites are also playpens for graffiti vandals. The natural ecosystems and surface textures in these caves have already been severely altered. Often, the damage has occurred over many decades. Carefully evaluate the environmental costs before embarking on projects that will significantly change cave surfaces and diminish populations of the current biological communities.

Several authors and reviewers of this book have tried the following methods, realize the pitfalls associated with each, and hope that advanced technology will introduce new, less destructive alternatives.

Avoid Using Heat on Cave Walls

Application of direct heat inside caves is *not* recommended. Heat damages limestone surfaces—the rock heats up and spalls off along with the paint.

Heat should never be used on gypsum or speleothems. Direct application of heat to gypsum drives off water. Direct heat changes transparent selenite to white powder and destroys the crystals. Though it seems an easy solution, propane torches and other surface heating devices are not appropriate for cave walls and toxic fumes may harm cave-dwelling communities. (See fumes, page 40; also see logistics, page 155.)

Avoid Using Commercial Chemicals in Caves

Historically, cavers have tried everything from oven cleaners, hot solvents, and acids, to citrus-based cleaners, soybean products, and biodegradable magic pastes. Some products will remove paint, but should not be used on cave walls—commercial chemicals and acids can damage cave life as well as the people applying them. The fumes are harmful to humans and wildlife in the enclosed spaces of caves. Natural airflow carries toxins throughout the cave, disrupting or harming bat colonies and other fauna. (See materials and toxins, page 171.)

Some products that claim to be environmentally safe simply do not work and labels often overstate their benefits. When analyzed in the laboratory, products commonly do not live up to their advertised claims of environ-

Portable Carbon Dioxide Powered Sprayer

Val Hildreth-Werker

Noble Stidham of the Lubbock Area Grotto developed a high-pressure, low-volume pump in the late 1980s and published drawings for building the unit (Stidham 1989). The apparatus works much like a waterpick device.

The system produces a fine jet of water at approximately 250–300 psi (1,700–2,100 kilopascal). The water is pressurized by a positive displacement piston pump powered by a small CO_2 bottle. Water is pulled from a container via a siphon hose. Any 4–liter (1-gallon) plastic jug works well.

The portable pressure sprayer device is very water-efficient and can be used in some intricate and delicate areas without causing damage. It works well on lamp-black and soot from carbide lights or candles but has little effect on spray paint.

Stidham, N. 1989. A CO_2-powered graffiti remover. *NSS News* 47(2):43-46.

mentally safe content. Even if a product is truly biodegradable, it may provide an unusual food source for biota that depend on the cave's natural ecosystem. (See anthropogenic chemicals, page 57.)

It is always best to avoid changing the natural processes that sustain life within caves. Some forms of cave life are able to recolonize a restored cave, but not all. (For example, certain microbes may not exist beyond the pool where they currently live. If the ecosystem of that pool is destroyed by chemical contamination, that colony of organisms may be gone forever.) Usually, the safest substance to use is *water*. (See sources for restoration water, page 393.)

Dental Picks for Gentle Tasks

Use dental tools to gently pick and scrape off graffiti that is painted on difficult surfaces. This effective technique works well on gypsum bedrock and soft formations. It is labor-intensive but the results can be impressive.

Use dental tools to gently pick and scrape off graffiti painted on difficult surfaces. This effective technique works well on gypsum bedrock and soft formations.

Pat Jablonsky and Andrea Goodbar used stainless steel dental tools to remove day-glow hot pink names from gypsum wall crusts in Endless Cave, New Mexico. Names were etched off with dental picks and the dry paint chips were collected from the gypsum with a hand-held, battery-powered, dust-buster vacuum. The color of the gypsum underneath was a sparkling white compared to the light tan of the surrounding gypsum crust.

For a more natural look, they took crushed gypsum from around the trail

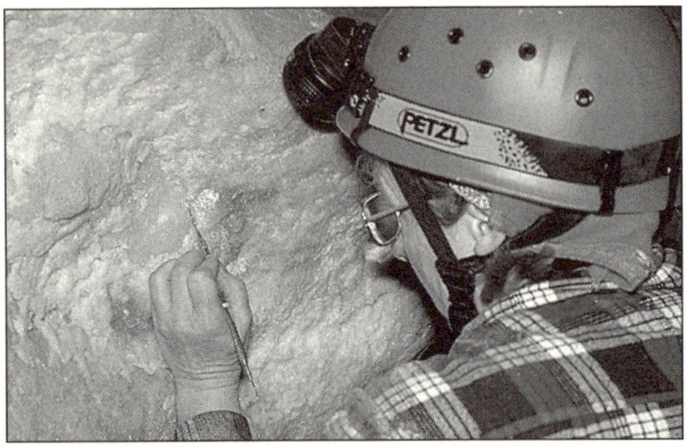

Figure 13. Pat Jablonsky uses a dental pick to etch off the graffiti spray painted on gypsum crust in Endless Cave, New Mexico.

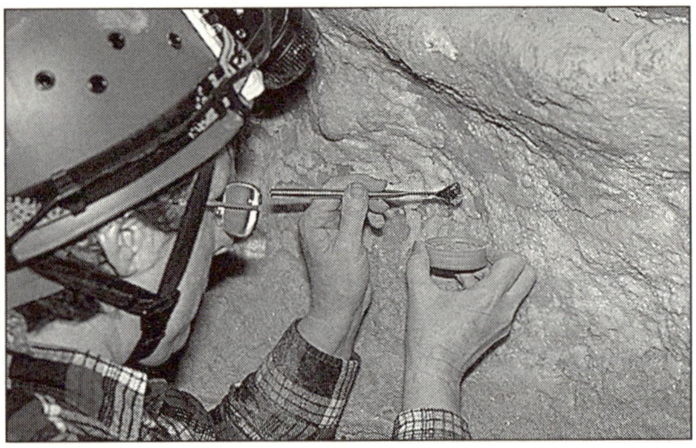

Figure 14. A make-up brush and tan-colored gypsum powder serve to camouflage the white sparkling crust where spray paint was removed. Application of the gypsum dust helps match the restored site to surrounding surfaces.

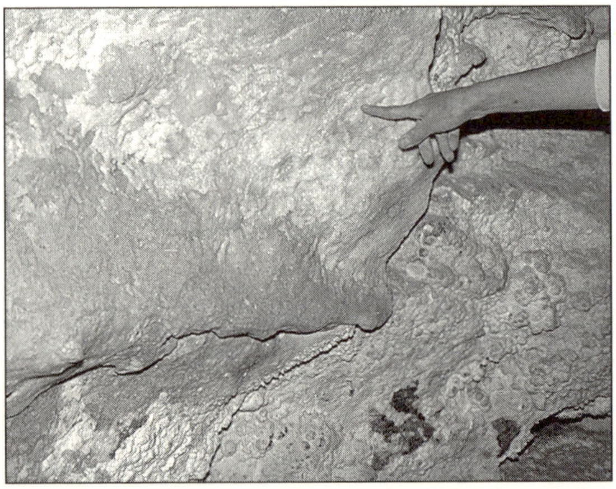

Photos James R. Goodbar

Figures 15 (before) and **Figure 16** (after). These two photos display the before and after of tedious dental pick techniques for restoring graffiti-covered gypsum surfaces.

and used a make-up brush to apply the gypsum powder to the restored area. They achieved a well-matched wall color and a cave passage that no longer invites new graffiti.

What About Camouflaging Cave Graffiti?

When all else fails, cover it up. Small amounts of dirt or mud from the cave floor will generally blend with the color of a cave wall if there are naturally occurring muddy surfaces. This technique can be used as a temporary measure until more volunteer time is found or a better method is developed. A lightly applied natural poultice sometimes loosens the medium for easier removal during the next restoration effort. (See Crocketts Cave, page 116.)

Camouflaging techniques are not recommended for speleothems. Dirt or mud will not stay on active formations. The crystalline micropores will become clogged. Camouflaging soil is almost always a different color than the speleothem and generally looks as bad as the graffiti.

The Summary and Conclusion Is Simple

Contemporary cave graffiti may be removed only after professional site analysis and careful documentation. If there are no issues involving historical or cultural preservation, low-impact removal techniques may mitigate the negative impacts of graffiti. Cave biota, habitat, and all other system conditions or concerns should always be carefully evaluated before planning restoration projects.

Nylon-bristled brushes and water are the safest tools for graffiti projects. Vigorous scrubbing with stiff nylon brushes along with judicious application of water generates the least negative impact on caves. More aggressive techniques increase adverse impacts to the cave and its ecosystem.

Additional Reading

There is an obvious lack of published material on both the techniques and philosophies concerning graffiti and cultural markings in caves. The speleological community needs continued research and publication on these significant conservation and preservation issues.

Bilbo B, Bilbo M. 1995. Management implications of historic writing and rock art in caves. In: Pate DL, editor. *Proceedings of the 1993 National Cave Management Symposium: Carlsbad, New Mexico, October 27-30, 1993*. [Huntsville (AL)]: National Cave Management Symposium Steering Committee. p 233-264.

Bitting C. 1989. Water-based cave cleaning. *NSS News* 47(2):42.

Douglas JC. 1998. Discovering history in American caves. NSS News 56(9):261-63.

Douglas JC, Smith MO, Simek JF. 2000. Identification and analysis of a Civil War soldier's name in South Carthage Cave, Tennessee. *Tennessee Caver* Spring: 7-13.

Douglas JC. 2001. Historic preservation at Hubbard's Cave: Inventory and management of cultural resources. In: Rea GT, editor. *Proceedings of the 14th National Cave and Karst Management Symposium: Chatanooga, Tennessee, October 19-22, 1999*. Chattanooga (TN): Southeastern Cave Conservancy. p 46-50.

Follet K. 1999. Graffiti control and removal: Lessons from an urban setting. *American Caves* Winter: 9-15.

Fifty Year "Rule"

Joseph C. Douglas

The 50-year "rule" is a guideline that may be useful but is not definitive. That is, there may be 40-year-old graffiti that deserves protection.

The 50-year "rule" is somewhat arbitrary. It came out of the federal government's need for a cut-off date for preservation of historic structures.

Then it was applied (or misapplied?) to markings in caves. Yes, 50 years is needed for National Register status. But it is not a hard and fast rule elsewhere.

Some of the work in contemporary history presents the dissenting position, especially Marion Smith's article (1991) on the 1968 exploration of Bull Cave, which he insists is historically important for a number of reasons despite being only 25 years past.

Smith MO. 1991. The survey of Bull Cave,Tennessee: Unpublished records of the 1960s explorers. *Journal of Spelean History* 25(1):1-10.

Frith RS. 1976. Acid on—Graffiti off. *NSS News* 34(11):203.
Gillieson D. 1996. *Caves: Processes, Development, and Management.* Cambridge (MA): Blackwell Publishers. p 237-267.
Hagan S, Sutton M. 1991. Cave restoration can cause damage. *NSS News* 49(6):170-171.
Hubbard DA Jr. 1995. Cave conservation and cave cleanups: Not always one in the same. *NSS News* 53(2):31.
Newbould R. 1976. Steam cleaning of the Orient Cave, Jenolan. *Cave Management in Australia, Proceedings of the First Australian Conference on Cave Tourism*. Broadway: Australian Speleological Foundation. p 87-90.
Rhodes D. 1976. Cave restoration techniques. *NSS News* 34(11):202.
Smith MO. 1991. The survey of Bull Cave,Tennessee: Unpublished records of the 1960s explorers. *Journal of Spelean History* 25(1):1-10.
Smith MO. 2003. Civil War inscriptions in caves [abstract]. *Journal of Cave and Karst Studies* 65(3):177. [Unpublished manuscript presented at 2003 NSS Convention in Porterville, California.]
Spate A, Moses C. 1994. Impacts of high pressure cleaning: A case study at Jenolan. *Cave Management in Australasia X*. Carlton South, Victoria (AU): ACKMA. p 45-48.
Stidham, N. 1989. A CO_2-powered graffiti remover. *NSS News* 47(2):43-46.
Welbourn C. 1976. Biological consideration for cave cleanup projects. *NSS News* 34(11):204.

Section C—Restoring Cave Passages

Control of Lamp Flora in Developed Caves

Rick Olson

Lighting in developed caves should be of sufficient intensity to allow safe passage for people and visitor appreciation of the cave. Maximizing the opportunity for dark adaptation of the human eye is crucial to achieving both of these goals because, if glare is prevented, human sensitivity to light can increase by a factor of 10,000.

Shielding light fixtures so that the light source is blocked from view is the single most important step toward maximizing dark adaptation. Using diffuse rather than spot lighting is probably second in importance since a bright spot has almost the same effect as a light shining directly into our eyes. What would be considered low light intensity in an office or home is considered quite adequate for the appreciation of a cave where shadow and light combine to highlight passage features.

High lighting intensity in caves is undesirable for both cave conservation and energy conservation. Incidence of graffiti on passage walls tends to be higher in brightly lit areas, presumably because the inscriptions will be visible and people can see well enough to write.

Heat from high intensity lighting can dry areas that would have remained moist. As the temperature rises from inefficient lighting, relative humidity drops. Adapted to humidity near saturation, cave life is highly vulnerable to desiccation compared to similar surface species, and mineral deposition may also be altered. For these reasons alone, energy efficiency in cave lighting is very important.

Another major consideration in cave lighting design is the growth of cyanobacteria (formerly known as blue-green algae), algae, mosses, and other plants induced by the lights we place in cave passages. Collectively, these photosynthetic organisms growing near lights in caves are called *lamp flora* in Europe (or *lampenflora*). The term lamp flora has been adopted in the United States.

A major consideration in cave lighting design is the growth of cyanobacteria (formerly known as blue-green algae), algae, mosses, and other plants induced by the lights we place in cave passages. Collectively, these photosynthetic organisms growing near lights in caves are called lamp flora.

Surprisingly, Barr (1967, page 182) plus Aley and Aley (1992) have pointed out that natural populations of algae and cyanobacteria have been found in small numbers within caves. However, other than in the twilight zones of entrances, the great quantities of lamp flora found along tour trails must be considered opportunistic in cave ecosystems. Historically, lamp flora have been considered largely a nuisance, but the gross enhancement of these photosynthetic species and the measures used to control them constitute a major distortion to cave ecosystems.

Opportunists

Moist areas in electrically lit cave passages are especially prone to the growth of lamp flora. These growths displace the natural microflora of bacteria, fungi, and natural algae associated with moist, dark surfaces in caves. Such distortion of the cave ecosystem is often severe in passages decorated with calcite speleothems.

The metabolism of bacteria associated with calcite speleothems may facilitate mineral deposition (Northup and others 1997) and their presence is possibly crucial to maintenance of the highly valued luster of these decorations. Therefore, restoration of natural microbial populations is desirable from both ecological and aesthetic viewpoints. Preventing the growth of lamp flora via non-toxic means, such as lighting design and technology, has generally been identified as the preferred method.

In Europe, the high level of cave impact caused by lamp flora has long been realized. In 1984, the first International Colloquium on Lamp Flora was held in Budapest, Hungary, and progress has been documented in subsequent meetings (Hazslinszky 1985, 2000).

Algae in Caves
Diana E. Northup

Algae are photosynthetic, single-celled plants that usually rely on sunlight to make their living. How-ever, several algal species are adapted to the low light levels of cave entrances, and may produce chlorophyll in darkness.

Some algae even grow *chemoorganotrophically* in total darkness (some algae use chemical nutrient sources instead of sunlight).

Green algae (Chlorophyta) and diatoms (Bacillario phyceae) have been extensively reported from caves. *Exotic* (nonnative, introduced) algal species drift into caves on air currents, enter in dripping water from surface sources, and are carried in by cave visitors.

Lighted formations in show caves often display the growth of abundant lamp flora (which may consist of some combination of algae, cyanobacteria, or mosses). In show caves, human traffic carries in algae cells on boots and clothing, and if there is a little light available, the humid cave atmosphere and organic nutrients may provide happy habitats for algae.

Electric lights in show caves stimulate the luxuriant growth of algae—however, in cave passages that are constantly dark, nonnative algae cells simply fail to grow and over time they die off.

Historic Lamp Flora Control

Historically, lamp flora have been killed or removed by a variety of means including herbicides in Lehman Caves and steam in Mammoth Cave, Kentucky (Harris 1981). More recently, Grobbelaar (2000) recommended the use of atrazine among other treatments for control of lamp flora.

At Kartchner Caverns in Arizona, a 12% solution of hydrogen peroxide was used with limited success. Growths returned in a month compared with three to six months for areas treated with diluted bleach (RS Toomey III, personal communication 2002). However, Aley (1976) found even 30% hydrogen peroxide to be ineffective, and such high concentrations are dangerous. (See hydrogen peroxide, page 349.)

For many years, these unwanted growths have been controlled through periodic spraying of 5% sodium hypochlorite, which is household bleach (Aley and Aley 1992) and in Hungarian caves 5% formaldehyde has been used successfully (Rajczy and others 1997).

However, use of manufactured chemicals can create ecological, aesthetic, and safety issues. (See anthropogenic chemicals, page 57.) Even when protective clothing is worn, burns to skin can occur from bleach. On cave surfaces, bleach indiscriminately kills everything contacted but does not necessarily remove the dead biomass. (See chlorine bleach, page 349.) The result is an unsightly layer of bleached organic material that may become calcified under some geochemical conditions. Also, cave-adapted invertebrates are easily killed by bleach. The unintended side effects can be minimized by thoroughly rinsing treated sites and carefully containing runoff.

Lighting Design

Indirect lighting, especially the use of aisle lights to reflect illumination off the trail surface, has been effectively used to reduce the intensity of light on undecorated passage surfaces in Mammoth Cave. Handrail-mounted lamps directed onto the trail surface have been used to advantage in Baradla Cave, Hungary. This lighting approach by itself, however, does not adequately display areas decorated with secondary cave minerals high above the trail surface.

One strategy for limiting growth of photosynthetic organisms is to provide illumination with wavelengths that are minimally absorbed by photosynthetic pigments. The use of green light to limit growth was recommended by Harris (1981), but the poor aesthetics of this alternative was unacceptable. The use of green light was shown to limit algal growth at Cango Caves, South Africa, but was also rejected there for aesthetic reasons (Oostthuizen 1981). Grobbelaar (2000) again recommended the use of green light at Cango Caves to prevent lamp flora.

A minimum impact alternative to consider is the use of electric lanterns. No lamp flora problems have been reported from caves lit with hand lanterns, and many other development-related impacts could be avoided. However, the logistics of keeping hundreds to thousands of lanterns in operating condition for caves with high visitation are daunting.

Variants of chlorophyll found in higher plants, mosses, algae, and cyanobacteria primarily absorb red and blue light (Lee 1989, page 18; Fogg and others 1973, pages 53, 146, 149, 150), and do not strongly absorb yellow light (Figure 1.) Lamps emitting light in the yellow to green range were recommended by Imprescia and Muzi (1984).

Most colored lights in developed caves will produce highly unnatural scenes, but since cave walls are often tinted with yellow earth tones by iron and organic acids, the visual impact of yellow light is minimal. In Mammoth Cave, visitors seldom comment on the yellow lights in place, and these comments are generally neutral.

If some relief from monochrome light is desired, then aisle lights can be directed onto the trail. The intensity of light reflected off the trail surface onto cave walls can easily be kept below the threshold where lamp flora will grow.

Cyanobacteria are more versatile than algae, as they can utilize the full

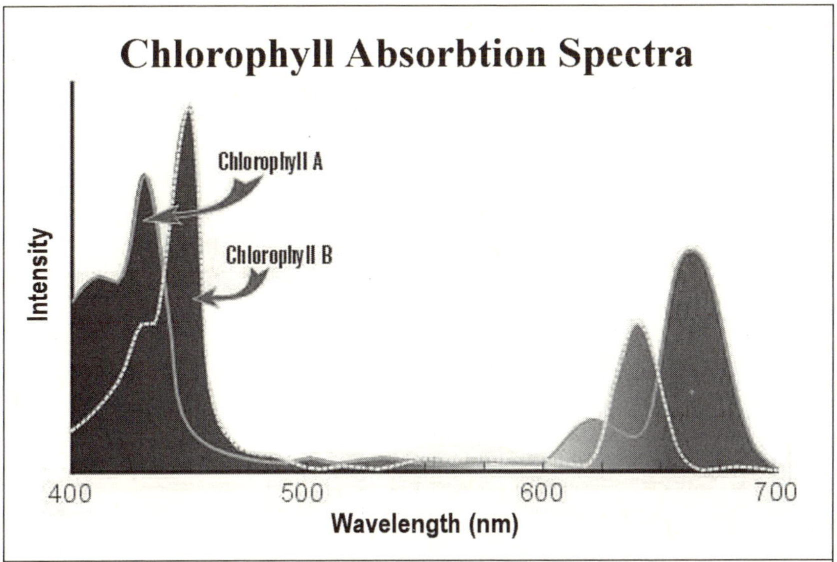

Figure 1. Absorption spectra of chlorophyll. (See page 7 of color section.)

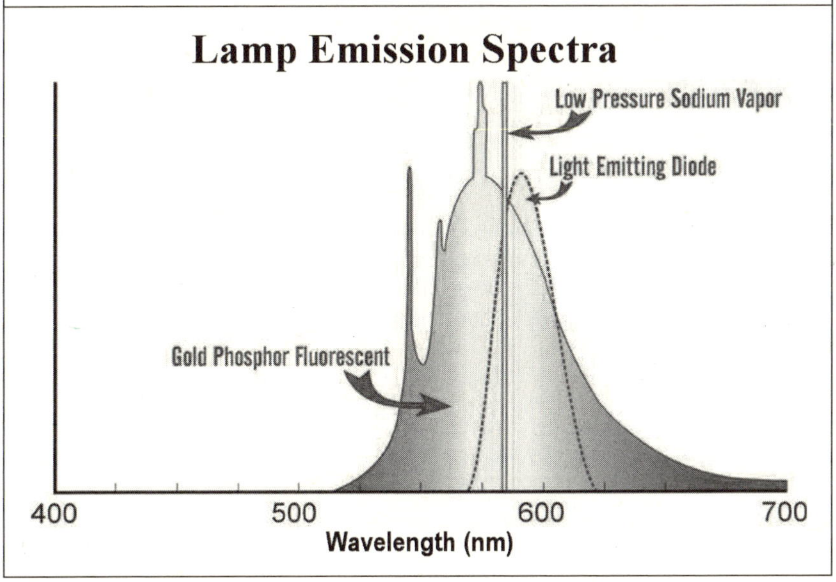

Figure 2. Emission spectra of yellow lamps. Gold–sleeved lamps are very similar to gold–phosphor lamps. (See page 7 of color section.)

Gary C. Berdeaux, Diamond Caverns, Kentucky

spectrum of light. In particular, a pigment known as phycocyanin allows them to absorb yellow. This of course makes growth more difficult to inhibit, but light supplied in a narrow spectral band limits photosynthetic efficiency since only one of two separate photosystems is engaged (Fogg and others 1973, page 53).

Types of yellow lamps include gold–phosphor fluorescent tubes, gold–sleeved fluorescent tubes, low- and high-pressure sodium–vapor lamps, and based arrays of yellow light emitting diodes (LEDs) (P Weinreb, personal communication 2000). Emission spectra for these lamps (Figure 2) largely fit in between the absorption peaks of chlorophyll shown in Figure 1. Therefore, the light from these lamps will support less lamp flora than white light sources.

Of these, low-pressure sodium–vapor lamps and ultra-bright LEDs have narrow emission bands that are well suited to this purpose (the width of LED emission bands is variable). Unfortunately, both types of sodium–vapor lamps attain full brightness slowly, and repeated starts greatly shorten lamp life. Moreover, Kartalis and Mais (2001) found in Alistrati Cave, Greece, that low-pressure sodium–vapor lamps support lamp flora if light intensity is very high, likely due to the phycocyanin pigments discussed earlier.

LED lamps are not as bright, but switch on instantly, are highly energy efficient, emit very little heat, no noise, and bulb life is phenomenal (100,000 hours or about 10 years continuous usage).

Test Results at Frozen Niagra

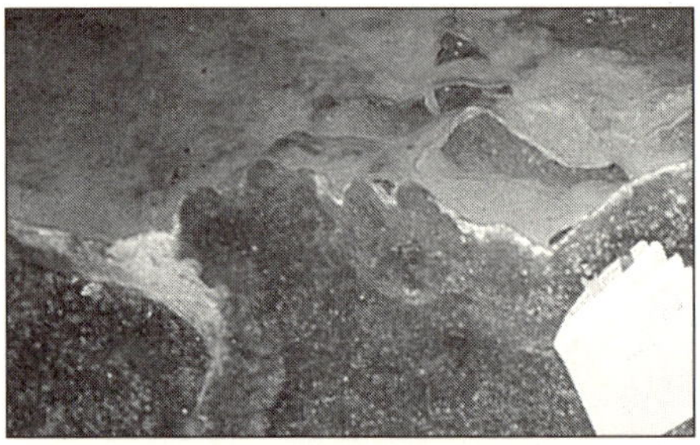

Figure 3 (before). Growth of lamp flora. Algae, mosses, cyanobacteria, and other opportunistic plants are often induced by lights placed along show cave trails. These photosynthetic species introduce major distortions to cave ecosystems.

Figure 4 (after). Test area in Frozen Niagra, Mammoth Cave National Park. There is no regrowth of lamp flora after three years of illuminating this area with 595 nanometer LED lamps. (See page 7 of color section for both photos.)

Photos Rick Olson

In a test at Mammoth Cave's Frozen Niagara section, problem areas were photographed, cleaned with bleach, and both gold–phosphor fluorescent and yellow LED lamps were installed at discrete locations. As experimental controls, some of the documented problem areas were left with existing white lamps.

After 2.5 years of daily tours, the white lamps supported growths as expected. The gold–phosphor fluorescent lamps also supported some growth of algae, moss, and cyanobacteria. This has two apparent causes. First, the emission spectrum of the gold–phosphor fluorescent lamps extends into the absorption area of chlorophyll; and second, the phosphor coating inside the tube became thin on the ends, which allowed white light through.

The only wavelength that did not support lamp flora was 595 nanometers in the yellow range. The 595 nanometer yellow LED lamps successfully prevented lamp flora growth in an area that consistently had heavy infestations in the past (Olson 2002). The light intensity on a nearby rimstone dam was 50 lux (4.6 foot–candles), which is twice the maximum intensity of white light recommended by Aley and Aley (1992) for prevention of lamp flora. Aley and others (1985) found that in alcoves, algae would grow with light as dim as 17 lux (1.6 foot–candles).

Mais and others (2001) of the Department of Karst and Caves at the Museum of Natural History in Vienna, Austria, discussed the possibility of using LED lamps in the show cave of Alland. They were encouraged by the positive results in Mammoth Cave, and will pursue this avenue. LED lamps with a slightly lighter yellow hue at 592 nanometers are now available, but have not yet been tested.

Conclusion

In developed passages, the goal is to provide safe appreciation of the cave with minimal ecological and aesthetic impacts. This is especially difficult in areas prone to lamp flora. We must provide enough light for people to maintain balance, see the trails and low ceilings, and for staff to interpret these highly valued resources. Diffuse, general lighting can be provided by aisle lamps directed onto the trail surface, such as those recently installed in Baradla Cave, Hungary, and in Mammoth Cave, Kentucky.

Feature lights in areas prone to lamp flora can be wavelength limited. Given that cyanobacteria can use yellow light at 595 nanometers, there is apparently an intensity threshold above which growth will be supported. However, it appears possible to adequately light the cave below such a threshold.

Coincidentally (and fortunately), the human eye is far more sensitive to yellow light than to the red and blue wavelengths which fuel photosynthesis. Whatever approaches are used to solve these cave management problems, we must keep our eyes open to new possibilities for reducing impacts of human use on cave ecosystems.

The only wavelength that did not support lamp flora was 595 nanometers in the yellow range. The 595 nanometer yellow LED lamps successfully prevented lamp flora growth in an area that consistently had heavy infestations in the past (Olson 2002).

Cited References

Aley T. 1976. *Recommendations for the control of plant growth in Blanchard Springs Caverns*. Ozark Underground Laboratory Contract Report to Ozark-St. Francis National Forest Under PO 334-10-76. 44 p.

Aley T, Aley C. 1992. Resource protection studies associated with relighting of a portion of the historic tour route, Mammoth Cave, Kentucky. In: *Report to Mammoth Cave National Park*. p 11, 20, 34.

Aley T, Aley C, Rhodes D. 1985. Control of exotic plant growth in Carlsbad Caverns, New Mexico. *Proceedings of the 1984 National Cave Management Symposium. Missouri Speleology* 25(1-4):159-171.

Barr TC Jr. 1967. Ecological studies in the Mammoth Cave system of Kentucky. *International Journal of Speleology* 3:147-204.

Fogg G, Stewart W, Fay P, Walsby A. 1973. *The Blue Green Algae*. London: Academic Press. 459 p.

Grobbelaar J. 2000. Lithophytic algae: A major threat to the karst formation of show caves. *Journal of Applied Phycology* 12:309-315.

Harris D. 1981. *The Control of Nuisance Plant Growth in Mammoth Cave, Kentucky*. Unpublished report to Mammoth Cave National Park. p 20.

Hazslinszky T. 1985. *International Colloquium on Lamp Flora: 10-13 October 1984*. Budapest (HU): Magyar Karszt-és Barlangkutató Társulat. 155 p.

Hazslinszky T. 2000. *International Conference on Cave Lighting Proceedings*. Budapest (HU): Hungarian Speleological Society. 164 p.

Imprescia U, Muzi F. 1985. Light sources and the flora growth in caves illuminated for touristic purposes. In: Hazslinszky T, editor. *International Colloquium on Lamp Flora: 10-13 October 1984*. Budapest (HU): Magyar Karszt-és Barlangkutató Társulat. p 103-113.

Kartalis N, Mais K. 2001. The influence of illumination in the Alistrati Cave, Serron, NE Greece. In: Hazslinszky T, editor. *International Conference on Cave Lighting Proceedings, Held in Budapest, November 2000.* Budapest (HU): Hungarian Speleological Society. p 65.

Lee RE. 1989. *Phycology*. 2nd edition. Cambridge: Cambridge University Press. 645 p.

Mais K, Pavuza R, Cech P. 2001. Preliminary investigations of the "Lampenflora" in the show cave of Alland (Austria). In: Hazslinszky T, editor. *International Conference on Cave Lighting Proceedings, Held in Budapest, Nov 2000*. Budapest (HU): Hungarian Speleological Society. p 121.
Northup DE, Reysenbach A-L, and Pace, NR. 1997. Microorganisms and speloethems. In: Hill C, Forti P, editors. *Cave Minerals of the World*. 2nd edition. Hunstville (AL): National Speleological Society. p 261.
Olson, R. 2002. Control of lamp flora in Mammoth Cave National Park. In: Hazslinszky T, editor. *International Conference on Cave Lighting Proceedings*. Budapest (HU): Hungarian Speleological Society. p 131.
Oostthuizen, H. 1981. The blue-green algae in the Cango Cave System. In: Mylroie J, editor. *Proceedings 1st International Cave Management Symposium*. Murray (KY): Murray State University. p 153.
Rajczy M, Buczko K, Hazslinszky T. 1997. Efforts for the restriction of lamp flora: Results and experiences between 1984 and 1996. In: Tóth E, Horváth R, editors. *Proceedings of the "Research, Conservation, Management" Conference Aggtelek, Hungary, 1-5 May, 1996*. Aggtelek (HU): Aggteleki National Park Directorate. p 54.

Additional Reading

Sheps L. 1972. The effects of photoperiod and some microenvironmental factors on plant growth in Lehman Cave, Nevada. *Bulletin of the National Speleological Society* 34(1):15-25.
Dobat K. 1999. Flore de la lumiere artificielle (Lampenflora—maladie verte). [Flora in artificial light.] In: Juberthie C and Decu V, editors. *Encyclopaedia Biospeloeogica*. p. 1325-1335.

Sources for Additional Information

Rickard S Toomey III
E-mail: <rick.toomey@wku.edu>

Cave Resources Manager, Kartchner Caverns State Park
PO Box 1849
Benson AZ 85602
E-mail: <manager@pr.state.az.us>

Peter Weinreb
Vice President, Light Southwest Inc.
6509 Poza Rica NW
Albuquerque NM 87107
Web site: <http://www.LightSouthwest.com/>

Section C—Restoring Cave Passages

To Bleach or Not to Bleach: Algae Control in Show Caves

Penelope J. Boston

Algal growth around lights, the infamous scourge of show caves, is a persistent problem. The use of very strong cleaning agents like chlorine bleach appeals to our desire to get rid of contaminating organisms like algae once and for all.

Alas—this strategy doesn't work well because the organisms grow so quickly when given the right conditions. The only way to really correct the algae problem is to change those fundamental conditions. Periodically dousing algae with chlorine bleach only kills some of the cells for a little while—the colony quickly grows back from the cells that are not killed during bleaching episodes. Algal colonies may also grow from new cells brought in on the shoes and clothes of visitors.

> Periodically dousing algae with chlorine bleach only kills some of the cells for a little while—the colony quickly grows back from the cells that are not killed during bleaching episodes.

Chlorine bleach (sodium hypochlorite, which is regular household bleach) is a contact poison and can indiscriminately kill microorganisms and small invertebrates, even when diluted with water. Chlorine bleach also emits highly poisonous chlorine gas and leaves an oxidizing residue, even when it is dry. The gaseous chlorine is particularly dangerous to humans and bats (indeed to all animals) in closed environments with limited air circulation—like small cave passages. There are very few situations in undeveloped caves where use of chlorine bleach is justified. Sewage contamination or exotic microbial invasions that cover extensive areas may qualify as extraordinary exceptions and cautious, prudent use may be appropriate.

Alternate Control Methods—Wavelength Selection

In developed caves, algae management through alternative control methods is preferred. Test results to date indicate that the growth of lamp flora can be prevented through wavelength selection. Lighting with LED lamps (light emitting diodes) that emit *nonphotosynthetically active* wavelengths (for example, 595 nanometers in the yellow region) can eliminate the challenge of algae and other lamp flora in show caves. (Olson 2002; Olson 2006, page 343 in this volume).

The suggestion of using *organism-inhibiting* wavelengths such as ultraviolet (UV—wavelengths that are shorter than about 360 nanometers) has been tried and may create more harm than good. UV requires extended contact times, presents serious human safety hazards, and can easily harm or kill nontarget biota (Rick Olson, personal communication 2003).

Select appropriate wavelengths that fail to support algal photosynthesis and unwanted algae will not grow in show caves.

Hydrogen Peroxide—A Better Option?

If cleaning of algal growth is necessary, a better option than chlorine bleach may be hydrogen peroxide (Grobbelaar 2000). Reports of effectiveness in using hydrogen peroxide (H_2O_2) for removing algae are mixed.

However, the primary failure of most contact disinfectants is lack of adequate *exposure time* to the disinfecting agent. Contact time is *critical* with all microbial disinfecting agents (Sand 1991; Wingender and others 1998).

A good strategy is to apply a small amount of the agent and allow it to remain in contact with the algae for a significant period of time—for example, at least 30 minutes with a 10% solution of hydrogen peroxide. This works better than a large amount applied for a short period of time. Additionally, it is much easier to make certain the area is thoroughly rinsed and that all residual disinfecting agent is removed from the site.

A good strategy is to apply a small amount of the agent and allow it to remain in contact with the algae for a significant period of time—for example, at least 30 minutes with a 10% solution of hydrogen peroxide.

When hydrogen peroxide comes in contact with surfaces, it oxidizes organic materials, much like chlorine bleach does. In sufficiently high concentrations, hydrogen peroxide is a contact poison for microbes. However, unlike chlorine bleach, the breakdown products of H_2O_2 are relatively harmless. Hydrogen peroxide quickly breaks down to water and oxygen, compounds that are less likely to harm cave microorganisms and are not toxic to humans. Even with the more benign bleaching characteristics of hydrogen peroxide, care should be taken to prevent it from coming in direct contact with any potential native microbial habitats (pools, muds, or speleothems).

Household hydrogen peroxide is sold as a 3% solution in water, but can be purchased from chemical supply companies at up to a 30% concentration. This more concentrated solution is a potential danger to the user and must be treated with caution as per the label and the MSDS (Material Safety Data Sheet) that is available from NIOSH (National Institute of Occupational Health and Safety) online at <http://www.cdc.gov/niosh/npg/npg.html>.

Summary

To summarize, chlorine bleach is *not* the best choice for controlling algae in show caves. Other disinfecting agents with less harmful byproducts are preferred for cleaning algal growth.

For long-term control of algae, the best solutions will be found through advanced lighting technologies and wavelength selection. Selection of appropriate wavelengths that fail to support algal photosynthesis can prevent unwanted algae in show caves.

Cited References

Grobbelaar JU. 2000. Lithophytic algae: a major threat to the karst formation of show caves. *Journal of Applied Phycology* 12:309-315.

Olson R. 2002. Control of lamp flora in Mammoth Cave National Park. In: Hazslinszky T, editor. *International Conference on Cave Lighting Proceedings*. Budapest (HU): Hungarian Speleological Society. p 131.

Olson R. 2006. Control of lamp flora in developed caves. In: Hildreth-Werker V and Werker JC, editors. *Cave Conservation and Restoration*. Huntsville (AL): National Speleological Society. p 343-348.

Sand KE. 1991. Effectiveness of commercial germicide products against the ropy slime producing lactic acid bacteria. *Journal of Food Protection* 54(8):632-4.

Wingender J, Griebe T, Flemming H-C. 1998. The role of biofilm and slime formation and the effect of chlorine, monochloramine, hydrogen peroxide, and silver on *Pseudomonas aeruginosa*. *Water Supply* 16(3/4):207-12.

Additional Reading

Faimon J, Zimak J, Stelcl J, Kubesova S. 2003. Environmentally acceptable effect of hydrogen peroxide on cave lamp-flora. *Environmental Pollution* 122(3):417-422.

Section C—Restoring Cave Passages

Cave Lint and Dust Removal Projects

Rodney D. Horrocks and Marc Ohms

The cumulative nature and impact of lint and dust problems in caves has only recently been recognized, researched, and addressed. Originally noticed in show caves, lint and dust accumulations have also been observed in heavily used wild caves.

Lint is not merely a benign aesthetic nuisance in caves. Lint and dust removal projects benefit caves in several ways:

- Restore natural conditions.
- Prevent unnatural speleothem dissolution.
- Remove artificial food sources.
- Eliminate unnatural odors.
- Restore visual scenes.

Lint is not merely a benign aesthetic nuisance in caves. Lint and dust removal projects benefit caves in several ways.

Lint consists of foreign materials shed from human visitors—fibers, hair, dander, and other debris that accumulates along heavily used routes (Figure 1).

Dust originates from trail construction projects, trampling of unpaved trails, and material brought into the cave by visitors.

Lint and dust accumulations can be removed with hand-held brushes, vacuums, or washing, depending on the substrates involved. Not only have methods been developed to restore natural conditions in caves, but techniques have also been developed to slow the accumulation and facilitate cleaning of these unnatural deposits.

Attempts to restore the aesthetic appeal of developed caves occurred as early as 1960 in Jenolan Caves in Australia (Newbould 1976). Lint was recognized as an unsightly foreign material as early as 1969 at Timpanogos Cave (Horrocks 2000). However, it was not until 1977 that lint was specifically targeted as a detrimental foreign substance in caves and was identified as a major component of the dirty-appearing passages in many developed caves (Roth 1987). Test cleaning was conducted at Carlsbad Caverns in 1977 and 1986, and the first "cave lint camps" were organized in 1988 (Jablonsky 1992a). This chapter emphasizes lint and dust removal in developed caves. However, the same techniques can be used in wild caves.

NPS Photo Jon Jasper

Figure 1. Close-up of collected lint and hair in Timpanogos Cave National Monument, Utah.

Reasons for Conducting Lint and Dust Removal Projects

Developed caves can only be partially restored to natural conditions. Cave managers must compromise between restoring caves to pristine states and causing additional negative impact. Good cave managers must recognize when restoration and cleaning projects need to be halted. Cave managers or project leaders should stop the work, rethink the situation, and possibly retrain the restoration team before additional damage is caused.

Because the impacts from lint and dust are cumulative, periodic restoration efforts are usually effective toward maintaining natural conditions in show caves.

Removing unnatural lint and dust can prevent foreign materials from being cemented onto the surfaces of actively growing speleothems. The incorporation of introduced materials into secondary deposits can cause permanent discoloration and unnatural textures. In fact, accumulations of lint and dust may partially or completely obscure the natural coloration and sheen of cave surfaces.

Lint has been shown to cause the dissolution of some speleothems (Jablonsky and others 1994). Lint may act as a condensation site for undersaturated cave water, and can trap carbon dioxide in or underneath the lint. Lint may also attract bacteria, yeast, molds, or algae that secrete or excrete acids that can dissolve the rock substrate (Crowle 1993a, 1993b). The accumulation of lint in humid cave environments may also result in unnatural musty odors (Moore 1993).

Many caves are extremely low-energy environments with limited food sources. Natural fiber lint, hair, and skin (dander) can represent a significant artificial food source and may provide a nutrient base for microbial life (Moore and Jesser 1995). These food sources can shift the balance of nutrients that sustain cave-adapted microorganisms (Michie 1997). Biota, such as microbes, mites, beetles, and spiders have been observed in lint accumulations (Jablonsky 1992a, 1992b). Whether the organisms are native to the cave or accidentals carried into the cave by humans, populations may artificially balloon as a result of artificial food sources. (See cave microbiology, page 61).

Figure 2. This collection of lint, trash, and chunks of blond boot sole was gathered in the Chandelier Maze, Lechuguilla Cave, Carlsbad Caverns National Park, New Mexico.

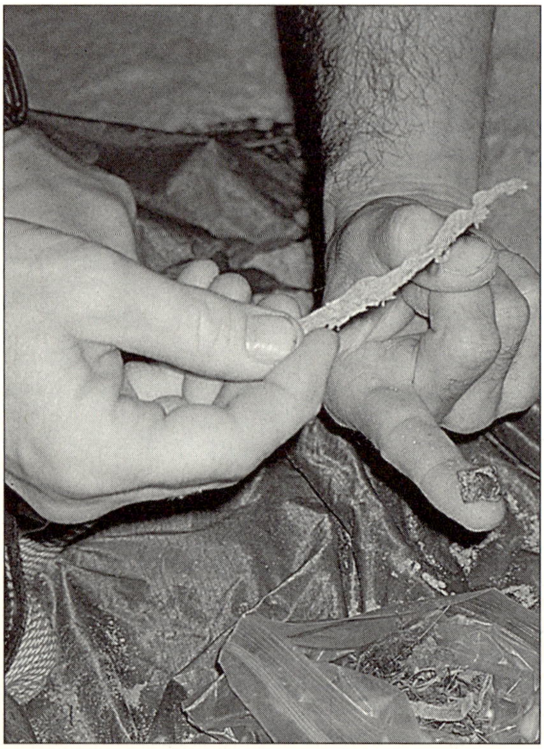

© Val Hildreth-Werker

Nature of Lint and Dust

Developed caves may have a dusty and dirty appearance caused by a combination of factors. In order of importance, they are listed here:

- Construction-generated dust
- Dust from tours traveling over unpaved trails
- Dust tracked into caves on shoes or brought in on clothes
- Lint accumulations
- Other materials shed from humans
- Natural dust

Human visitors shed a plethora of other materials into caves (Jablonsky 1995). Lint and dust removal efforts should address most of these foreign materials:

- Hair
- Dander
- Skin flakes and debris
- Mites
- Microbes
- Shoe rubber
- Pet animal fur

Lint is composed primarily of fibers from natural and synthetic fabrics. Natural cellulose-based fibers include cotton, hemp, and linen. Natural protein-based fibers include keratin, wool, alpaca, mohair, silk, and cashmere. Synthetic fibers include oil-based synthetic polymers (Michie 1997). Both natural and synthetic fibers are usually present. Old deposits of lint in caves may be almost entirely synthetic fibers. Natural fibers are more brittle and tend to shed more over time but they are often eliminated by deterioration or consumed by molds, fungi, or bacteria, resulting in concentrations of synthetic fiber lint (Jablonsky 1995).

Dust is composed of silica and fine clay mineral particles ranging from 0.1 micron to 100 microns in size.

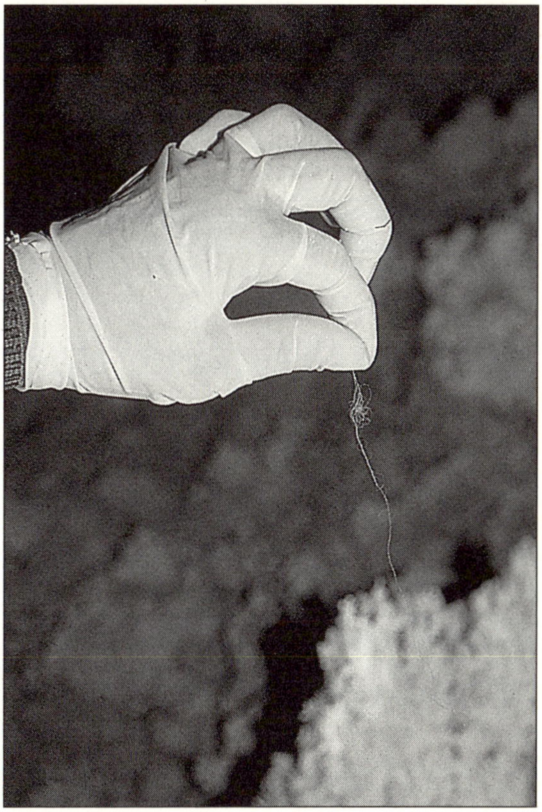

Figure 3. This strand of hairy lint was found off trail near Lake Chandalar, Lechuguilla Cave, Carlsbad Caverns National Park, New Mexico.

Recognizing Lint and Dust Accumulations

Due to the gradual accumulations of lint and dust in show caves, long-term employees may not recognize the subtle color changes and worsening conditions associated with these deposits. These conditions are noticeable only when the walls near trails are viewed close-up with a bright flashlight. Older deposits appear as a grayish mass made up of finely ground lint and dust. Black lights are useful in locating lint concentrations because many natural fibers are treated with optical brighteners that reflect long-wave ultraviolet (UV) light. When viewed under a black light, these fibers appear bluish-white against dark cave surfaces.

Accumulation of Lint

Lint accumulation is influenced by many different factors:

- Number of cave visitors
- Gravity
- Air movement
- Cave wall contours and textures
- Trail design
- Heat
- Humidity

Generally, more visitors lead to greater deposits of lint. Mammoth Cave, Kentucky, and Carlsbad Cavern, New Mexico, generate tremendous accumulations of lint in short periods of time. Observations in Lehman Caves, Nevada, showed noticeable lint accumulating one year after cleaning (Horrocks and Green 2000).

Lint accumulates faster at interpretive stops, especially those where visitors sit down. More lint is produced by a combination of longer times spent at sit-down spots and the abrasive nature of clothes rubbing against benches. Lint fibers also get caught in shoe soles and are tracked throughout tour routes.

Most lint initially falls directly to the floor and is then moved laterally to the edge of the trail by foot traffic. Lint also tends to concentrate on the lower portion of walls immediately adjacent to trails. Trail design and construction materials can influence how and where lint accumulates.

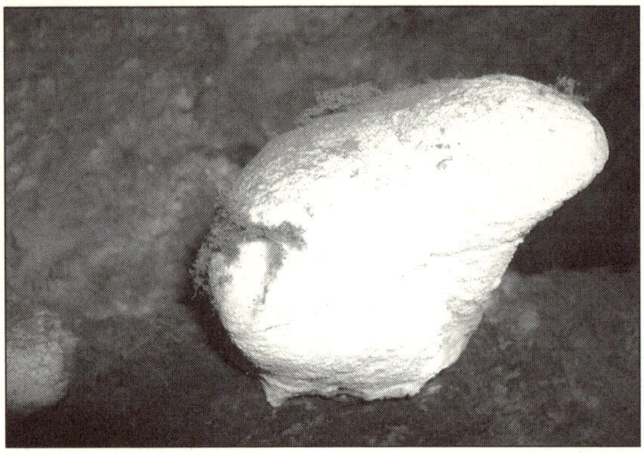
NPS Photo Jon Jasper

Figure 4. Along developed cave tour routes, lint is carried on air currents and will collect and mass along certain ledges. This example from Timpano-gos Cave National Monument shows thick lint deposits on a ledge.

- Trails with curbs 20- to 60-centimeters (8- to 24-inches) high, capture a large percentage of lint falling from visitors' clothes (Jablonsky 1995).
- Trails with grates and underlying tarps capture a larger percentage of lint. Grated trail surfaces must be periodically cleaned because the open spaces become choked with lint, dust, and other debris.

Airflow determines where lint accumulates. Over time, trampled fibers on the trail become airborne and are carried by air movement and redeposited on cave surfaces away from the trails. Large quantities of lint may be deposited on ledges high above a trail or on ceiling formations (Figure 4). Fibers less than 3 microns in size will become suspended immediately, without falling to the trail. These fibers descended at a rate of 1 meter (3 feet) every 9 minutes in laboratory conditions (Moser 1988).

Manmade entrances and tunnels may artificially increase cave airflow and compound lint problems. Lint accumulations tend to concentrate on flat or sloping shelves. Lint also accumulates on any rough surface, especially on stalactites, popcorn, helictites, spiked frostwork, or on low ceilings. In dryer areas, accumulations will often ball up into "lint bunnies."

Heat generated from lights and visitors can cause distribution, or redistribution, of lint near trails. Movement of people also causes airflow disturbances and may redistribute lint away from the trails.

High humidity in caves may cause older deposits of lint to have a higher percentage of synthetic than natural fibers. Due to the hydrophilic characteristics of natural fibers, they absorb moisture and break down into simple starches that either decompose or become a nutrient source for biota (Kraemer and Yett 1992).

Accumulation of Dust

Although dust occurs naturally in most caves, the accumulation of natural dust away from entrances is minute and collects over long periods of time (Ohms 2002). Very strong cave winds near entrances may suck material in from the outside. Dust can also become a problem during trail construction activities such as blasting and digging.

> Dust can remain suspended in the air for hours or even days. Heat rising from visitors can disperse dust widely over a short period of time, causing air velocities above a tour.

Unlike lint, dust can remain suspended in the air for hours or even days (Michie 1997). Due to air currents, dust particles less than 0.25 microns may take long periods to settle out of the air.

Heat rising from visitors can disperse dust widely over a short period of time, causing air velocities of 1 meter per second (3 feet per second) above a tour. These currents can carry fairly heavy particles up to the ceiling and then outward away from the party.

The heaviest particles fall quickly to the floor and the smaller particles become distributed evenly throughout the air (Michie 1997). The dust can become incorporated into formations where active deposition of secondary minerals is occurring. At Ngilgi Cave in Australia, every ten million visitors cause the deposition of a 30-micron layer that is as opaque as a layer of paint and completely hides the original rock and speleothem colors (Michie 1997).

In wild caves, cavers who wear dirty clothes or have dirty gear (especially dried mud) can shed 10,000 to 100,000 times as much dust as cavers who have clean clothes and gear. As these cavers start climbing over irregular terrain, the amount of dust released from fabric greatly increases.

One hundred dirty cavers may spread as much dust as 1,000,000 visitors with relatively clean clothes on a tour trail (this does not include lint) (Michie 1997). Cavers with clean clothes will only spread the dust that is already present. Caving group size is important, as small groups of cavers stir up less dust than larger groups.

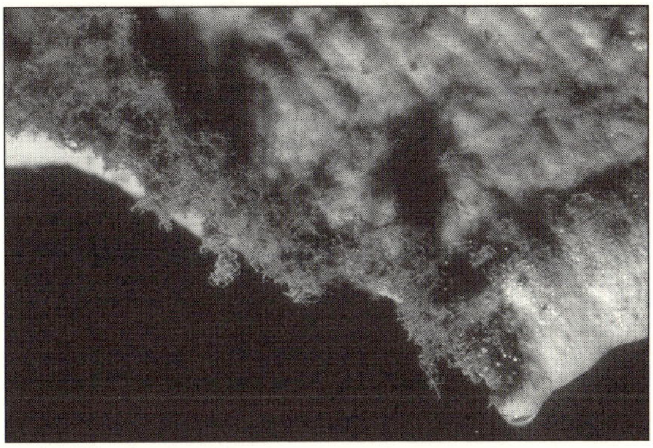
NPS Photo Rod Horrocks

Figure 5. Lint can accumulate on formations along developed tour routes, as shown in this photo from Timpanogos Cave National Monument, Utah.

Lint Camps

Lint camps are a great tool for cleaning sections of a cave thoroughly and systematically. Volunteers are often willing to travel long distances and return year after year for on-going projects. Cavers who are conservation minded, experienced, and love the resources are excellent lint volunteers.

Running a successful lint camp requires a fair amount of preliminary work. The following details should be worked out before the project starts:

- Identify areas to be cleaned and techniques that will be used.
- Schedule far in advance and advertise the event widely.
- Find experienced supervisors and team leaders.
- Secure nearby lodging or camping for the participants.
- Put together an introductory training outline for the first morning of the camp.
- Purchase supplies.

Several types of perks might be offered to the volunteers:

- Create a food plan and find a camp cook to prepare meals.
- Design a project T-shirt, certificate, and a reception with the cave management and staff.
- Arrange for evening cave trips or other perks for participants. To accommodate everyone, provide easy trips and difficult trips. Cave trips can be for recreation, survey, or photography.

Establish a ratio of 1 supervisor or team leader to every 5 or 6 lint pickers. The supervisors rove among their workers, answering questions and providing assistance.

In addition to the perks, the following suggestions will help keep enthusiasm high among the volunteers:

- If you are using local volunteers, weekend projects are ideal. Keep lint camps under three days to prevent burnout.
- If using volunteers from other regions, weeklong camps work best.
- Plan at least two different types of projects (for example, brushing and vacuuming lint or washing dust).
- Switch duties or work areas.

Lint camps should start with a short introduction, a demonstration, and a short field trip into the cave. Spend only one or two total hours with these introductory activities. The participants will mostly learn "on the job" by doing the work or watching veterans (Kraemer and others 1997):

- The introduction should cover rules, safety issues, goals, schedules, and perks.
- Follow the introduction with a demonstration of the techniques that will be used.
- Take volunteers into a relatively clean area away from the trail and closely examine cave surfaces with flashlights, so that they can see what the areas near the trail *should* look like.

- Finally, take them into a dusty area along the trail to show not only dust, but also what new lint versus old lint looks like.

Establish a ratio of 1 supervisor or team leader to every 5 or 6 lint pickers. The supervisors rove among their workers, answering questions and providing assistance. If too few supervisors are appointed, the lint pickers will spread out and miss areas.

It is common for workers to find coins, hair accessories, light bulbs, tickets, sunflower hulls, gum, trash, and other miscellaneous items. Remind participants that artifacts of historical importance may be encountered.

Cultural artifacts include any manufactured item that is over 50 years old. (See cultural and historical preservation, page 99; see 50-year rule, page 341.)

Objects found should be reported to the supervisor and work should be stopped in that area until the situation is evaluated. The state of preservation, historical significance, and age can help determine whether items should be collected or discarded. Before anything historically significant is removed, the date, name of discoverer, condition, and location should be recorded:

- If a cultural item is worthy of collection, bag it with a complete label.
- Number each item bag separately.
- Mark locations on a large-scale map.
- Record these data at the time of collection, even if the value is not immediately obvious. Important location data cannot be recreated at a later date.
- Photodocumentation may also be beneficial.

Figure 6. Removal of calcifying lint from wet flowstone along the visitor route in Jewel Cave National Monument, South Dakota, required the use of small tweezers and plastic toothpicks.

© Val Hildreth-Werker

Equipment and Tools

Lint and dust removal projects need the following types of equipment (Jablonsky 1993; Kramer and others 1995; Horrocks 1998):

- Detailed map of the area to be cleaned to use for documenting the sites
- Camera and film or digital photography equipment to document the sites before, during, and after the project is completed
- Hand-held polyester paintbrushes for gathering lint and dust, 2.5- to 10-centimeter width (1- to 4-inch)
- Plastic-bristled scrubbing brushes for wet areas
- Quart- and gallon-sized zip-closure plastic bags for collecting and storing lint and dust
- Hand-held spray bottles
- Lint-free rags and sponges for catching runoff from sprayers
- Plastic buckets, containers 4-liter to 20-liter-sized (1-gallon to 5-gallon-sized)
- Foam-cell collector brushes to collect lint off delicate surfaces
- Tweezers for delicate areas and in between rocks
- Small dust pans to brush lint and dust into
- Plastic cups to brush lint and dust into

Specialty equipment may also be necessary for successful lint projects:

- HEPA filter vacuum cleaner and bags (High Efficiency Particulate Air, high-filtration, wet-and-dry-tank style)

- Extension cords
- Flowstone shoes or aqua socks for walking on flowstone surfaces (see footwear, page 431)
- Backpack sprayer units with fine-mist nozzles
- UV black light to spot lint accumulations
- Respirators

Each worker should have the following personal gear. Always check to find out if the project will provide masks and other safety equipment:

- Helmet
- Helmet-mounted light (brighter is better and LED headlamps are especially effective)
- Knee pads and elbow pads
- Sturdy nonmarking boots
- Work gloves
- Surgical gloves (use powder-free, nonlatex gloves—some people are allergic to latex)
- Goggles (depending on job)
- Ear plugs (depending on job)
- HEPA filter respirator masks (depending on job)

NPS Photo Jon Jasper

Figure 7. Cami Pulham uses the static properties of a polyester brush to pick up lint in Timpanogos Cave National Monument, Utah.

Safety Measures

For lint and dust removal projects, establish the following safety precautions. When minors (under 18) are participating as volunteers, their parents or guardians should sign a parental consent form. Adult volunteers may need to sign liability release forms.

HEPA filter respirators as well as protective gloves and clothing should be worn in dusty environments or where guano (wood rat, mouse, bat, bird, cricket) or amberat (dried wood rat urine) may be disturbed (often found near cave entrances).

Dust masks may be ineffective against very small particles—you must choose the proper HEPA filters for respirators to ensure protection. Instruct personnel not to rub their eyes or face and to wash hands thoroughly before eating.

If vacuum cleaners are used, ear protection should be provided. Likewise, goggles should be available and required for some tasks.

During the removal of construction-generated dust, the most common safety hazard is overfilled buckets. Overfilling can shorten the life of buckets and cause spills. It can also result in strained muscles or back injuries.

Buckets should be marked inside with a fill line that is one-third to two-thirds above the bottom of the bucket and supervisors should not allow workers to exceed that mark. The weight should be geared to the smallest member of the crew or the individual in the most awkward position for carrying a bucket. If possible, individuals in assembly line crews should be of similar size. Workers need to be reminded how to lift buckets without causing back strain. Lids may be used to prevent spills.

Cleaning Techniques

The most common tool for removing lint is a polyester-bristled paintbrush (Figure 7). A swirling or dabbing motion with a polyester brush creates static electricity that attracts the lint (Horrocks and Green 2000). Since the electrical charge in lint is eventually lost, sweeping or vacuuming may

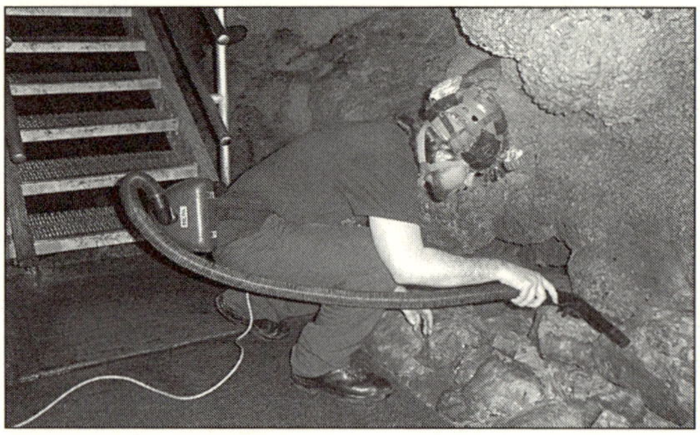
NPS Photo Rene Ohms

Figure 8. Ryan Brown uses a hip-mounted HEPA vacuum to pick up lint along the side of a developed trail in Jewel Cave National Monument, South Dakota.

work better on old deposits.

Bristles on polyester brushes do not break as often as those on other brushes. Before using any brush, check to see how well the bristles are attached. Once a brush starts to lose bristles, discard it.

Lint picked up with brushes can be deposited in zip-closure baggies by placing the brush in the bag and squeezing the bag shut with the other hand as the brush is pulled out. This method works best when there is more lint than dust or when rough surfaces are encountered. If a lot of dust is mixed in with the lint, gently sweep the particles into plastic baggies, cups, or small dustpans. The baggies will mold to the cave surface and allow brushing directly into the bag without spilling material. Sweeping should be confined to small areas and care should be taken to prevent lint and dust from becoming airborne.

If delicate surfaces are encountered, tweezers or foam-cell collecting brushes work well. Lint pickers at Carlsbad Cavern have successfully used foam brushes (P Jablonsky, personal communication 2000).

Vacuums are effective in picking up lint and dust. HEPA vacuums filter out the smallest particles and do not reintroduce particulate debris back into the cave air. A variety of attachments can be used to clean various surfaces (Figure 8). However, HEPA filter vacuums are expensive, noisy, and pick up everything.

For dry lint and dust, non-HEPA vacuums should be avoided, since they allow dust particles to escape and circulate back into the cave. At Jewel Cave, non-HEPA vacuums (wet-vacs and shop-vacs) were successfully used to clean the tarps underneath catwalks where the lint was wet and no dust was generated (Ohms, personal communication 2000).

Flowstone, bedrock, calcified sediments, or trails with curbs are ideal surfaces for water-based methods. Theoretically, water should be applied as a final stage of cleaning, after the lint has been collected—however, water-based methods should be avoided if there are natural loose sediments under the material being cleaned. (See pressurized water, page 397.)

Figure 9. A caver vacuums the lint tarp underneath an open-grate stairway in Jewel Cave National Monument.

Although pressure washing can be effective (Newbould 1976; Jablonsky 1990, 1991; Schmitz 1996), this method introduces the disadvantage of washing everything away, including natural dust, delicate speleothems, and biota, and redepositing it elsewhere in the cave (Ressler 1988).

Pressure washing uses large quantities of water and runoff is difficult to control. Wet-vacs or shop-vacs should be used to suck up excess water generated by backpack sprayers and large hand-held sprayers. Sponges and lint-free rags can soak up small amounts of runoff and buckets can be used to capture runoff if spraying overhead. It is always important to capture restoration runoff and not simply transfer the problem somewhere that is out of sight and out of mind. (See runoff water, page 396.)

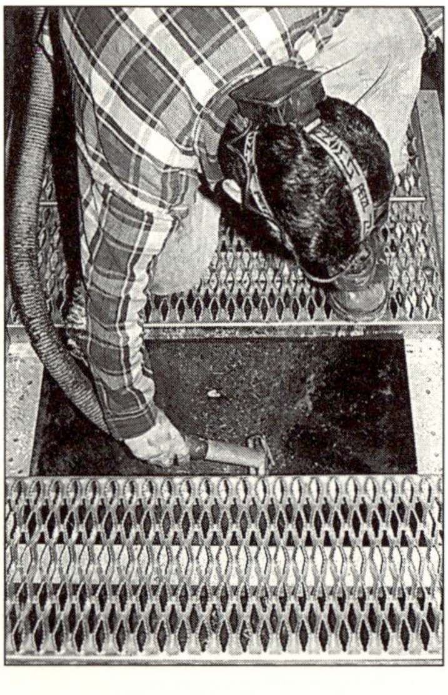
NPS Photo Rene Ohms

Do not use water from cave pools for large spraying operations that may lower water levels and harm cave life. (See water sources for restoration, page 393.) Avoid spraying water in areas where excess water will drain into cave pools. Runoff containing lint and dust will pollute pools and artificially increase water levels.

Be aware that fresh tap water may contain chemicals such as chlorine or fluoride. If tap water is allowed to stand overnight some of the chemical component may dissipate. Before using water, determine the pH. Water that is acidic will have a slightly corrosive effect which could damage delicate formations. If the water is too basic it may leave behind mineral deposits. (See restoration water, page 394–395.)

Be extremely cautious spraying water around lights and other components of the electrical system. It is wise to shut off the power while using water to reduce electrical risks and avoid exploding the light bulbs.

Try dry methods before using water to clean an area. In wet areas, the lint will tend to lump together and form balls or mounds. A HEPA filter vacuum cleaner may be effective in this situation but paintbrushes are worthless. A stout bathroom scrubbing brush or tweezers will also remove wet lint. Once the larger lint accumulations have been removed, water techniques may work for the final touch.

Special Issues

Before any work is done, it is important to shoot precleaning photographs of the area. Take close-ups and overviews. Remember to include an object for scale in photodocumentation shots. Pictures of the cleaning crew in action will document the historical aspects of the project. Post-cleaning pictures finish the photodocumentation. For the postcleaning photographs, recreate the same photograph as the precleaning pictures. (See photodocumentation, page 204–207.)

If the hand-held-brush method is used to remove lint and dust, lint from specific areas can be collected and weighed with a sensitive scale. Project leaders can then determine where lint is accumulating, target future cleaning areas, and devise methods to slow the deposition or trap the lint (Horrocks and Green 2000). If a lot of dust is being swept up with the lint, it can be separated with a polyester brush and weighed (or both lint and dust can be weighed together).

Techniques for removing lint must be adjusted to the surfaces being cleaned. Delicate formations (including some frostwork) can sometimes be cleaned by spraying very light mists of water on the lint and dust accumulations and then catching the resulting runoff in tarps (Horrocks and Green 2000). Acidic water could dissolve the crystals of delicate speleothems and must not be used for this type of cleaning.

Some gypsum surfaces can be cleaned. Crusts can often be cleaned with hand-held paintbrushes or HEPA vacuum cleaners. A water-based cleaning technique was successfully used on gypsum in the Snowball Dining Room of Mammoth Cave (Aley 1989). (See gypsum restoration, page 419.)

Trash from many cave sites is deposited in the local landfill. Dispose of lint and dust debris in the same way.

Preventing or Slowing Lint and Dust Accumulations

Curbs with heights of 20 to 60 centimeters (8 to 24 inches) help contain lint and dust, thereby reducing accumulations beyond the trails (Jablonsky 1995; Fry and Olson 1999). Curb configurations depend on the number of visitors, characteristics of the cave, trail design, and trail construction materials. Rough-surfaced curbs have been shown to catch lint better than smooth ones (Jablonsky 1995).

Trails with curbs must be cleaned regularly—otherwise the lint is ground up by foot traffic and becomes airborne (Figure 10). Trails should not be swept with a broom because sweeping reintroduces lint into the air, creates dust, and leaves bristles behind. HEPA filter vacuums should be

Figure 10. An example of a lint curb along a developed cave trail in Mammoth Cave National Park, Kentucky, and the associated "lint bunnies" that collect along its edges.

Rick Olson

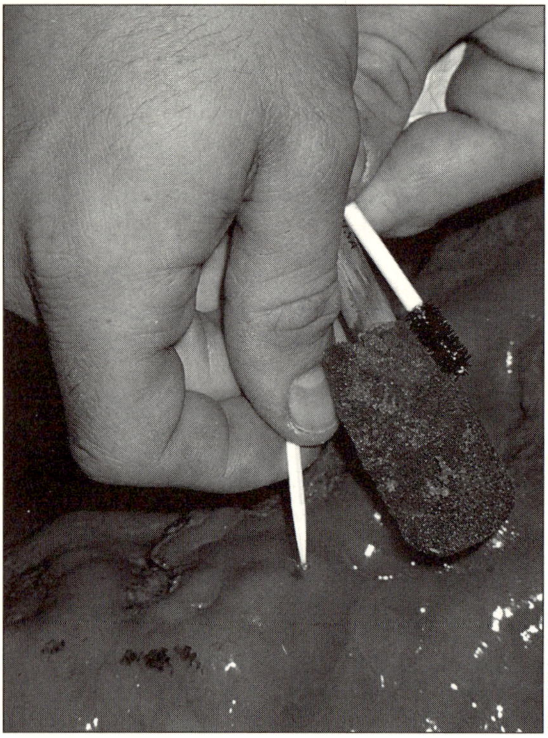

© Val Hildreth-Werk

Figure 11. In Jewel Cave National Monument, South Dakota, lint can quickly calcify into wet flowstone surfaces. Plastic toothpicks, tiny stiff-bristled brushes, and foam clean-room brushes are the tools of choice for this lint-picking chore.

used to clean trails with curbs. In caves without electricity, a plastic broom may be somewhat effective if short, slow strokes are used. If spraying techniques are used on trails, all waste water should be collected and removed from the cave.

If visitors are walking directly on unpaved cave floors, sediments may become airborne. Cement trails or raised platforms with catchment tarps underneath are often successful in reducing lint.

Walkways, bridges, and platforms made of metal grating with underlying tarps capture lint and tracked-in dirt (Figure 9). Generally, grates near an entrance are cleaned annually, and those farther in the cave are cleaned every other year. Tarps should be made of nylon and designed to be easily removed for cleaning. They should be taken outside the cave and thoroughly washed every 3 to 4 years.

Where public seating is provided, the seating surface should be smooth and nonabrasive to reduce lint deposits.

When construction is taking place within a cave, tarps can be hung in a passage to contain and block dust. To contain dust as it is being created, hoods may be built over equipment or mist spraying may be used in conjunction with HEPA filter vacuums. Periodic vacuuming during the project will also help control construction dust.

When an artificial entrance or gate is designed, airflow and dust influx should be considered. If no entrance existed before (that is, if it is a new, entirely artificial entrance) the closure should be airtight or should re-create the original airflow. If an entrance has been artificially enlarged, the gate or door should have air spaces that equal the size of the natural aperture in order to maintain natural airflow. (See air exchange, page 262.)

Several successful techniques have been developed to prevent or slow lint accumulations. Curbs, grates, and catchment tarps allow continuous lint collection and easy periodic removal (Jablonsky and others 1994). Other techniques have been developed to prevent lint fall, including misting visitors before they enter the cave. (See Kartchner Caverns, page 141.) Finally, restricting, monitoring, and training cavers can greatly reduce dust accumulations along major routes in wild caves.

Cited References

Aley T. 1989. Restoration and maintenance of natural cave microclimates. *NSS News* 47(2):39-40.

Crowle F. 1993a. *Lint-associated Rock Deterioration in Caves*. Unpublished manuscript, cave management files, Wind Cave National Park, May 10, 1993. 6 p.

Crowle F. 1993b. *Importance of Microbes in Lint-associated Cave Mineral Erosion*. Unpublished manuscript, cave management files, Wind Cave National Park, June 10, 1993. 4 p.

Fry J, Olsen R. 1999. Walkway development and construction relative to reducing visitor impact in the historic section of Mammoth Cave. In: Rea GT, editor. *Proceedings of the 1999 National Cave and Karst Management Symposium: Chattanooga, Tennessee, October 19-22, 1999*. Chattanooga (TN): Southeastern Cave Conservancy. p 64-72.

Horrocks RD. 1998. *Lint Camp Orientation Outline*. Great Basin National Park, Cave Management files. 2 p.

Horrocks RD. 2000. *The History of Cave Management at Timpanogos*

Cave National Monument. Unpublished manuscript, Timpanogos Cave National Monument, cave management files. 3 p.

Horrocks RD, Green DJ. 2000. Lehman lint liquidators. *NSS News* 58(3):72-73.

Jablonsky PL. 1990. Lint is not limited to belly buttons alone. *NSS News* 48(5):117-119.

Jablonsky PL. 1991. Headlines! Lint accumulations are becoming more prevalent in caves. *American Caves* Fall:18.

Jablonsky PL. 1992a. Implications of lint deposits in caves. *NSS News* 50(4):99-100.

Jablonsky PL. 1992b. *Develop Preventive Measures for Future Accumulations of Cave Lint*. Unpublished Semi-Annual Report, U.S. Department of the Interior, National Park Service, Rocky Mountain Regional Office, November 2, 1992; Second Year Extension Proposal, March 15, and October 22, 1993; and final report June 1, 1994.

Jablonsky PL. 1993. *Investigator's Annual Report*. Unpublished report, U.S. Department of the Interior, National Park Service, Rocky Mountain Regional Office. March 23, 1999.

Jablonsky PL. 1995. Lint in caves. In: *Proceedings of the 1993 National Cave Management Symposium: Carlsbad, New Mexico, October 27-30, 1993*. [Huntsville (AL)]: National Cave Management Symposium Steering Committee. p 73-81.

Jablonsky P, Kraemer S, Yett B. 1994. *Research Topic: Develop Preventive Measures for Future Accumulations of Cave Lint*. Final report submitted to Wind Cave National Park.

Kraemer S, Yett B. 1992. Semi-annual Report, Research Topic: Develop Preventive Measures for Future Accumulations of Cave Lint. Wind Cave National Park, Cave Management files. 11 p.

Kraemer S, Yett B, Petty K. 1995. *Jewel and Wind Cave Lint Camp*. Wind Cave National Park, Cave Management files. 2 p.

Kraemer S, Yett B, Petty K. 1997. *Lint Camp Report, 1997*. Wind Cave National Park, Cave Management files. 2 p.

Michie NA. 1997. *An Investigation of the Climate, Carbon Dioxide, and Dust in Jenolan Caves*, N.S.W., Ph.D. Thesis. School of Earth Sciences, Macquarie University, Australia. 298 p.

Moore J. 1993. *Survey of the Biota and Trophic Interactions within Wind Cave and Jewel Cave, South Dakota*. Progress Report, Wind Cave National Park, 11/29/93. 35 p.

Moore J, Jesser R. 1995. *Survey of the Biota and Trophic Interactions within Wind Cave and Jewel Cave, South Dakota*. Final report, Wind Cave National Park, Cave Management files. 8 p.

Moser P. 1988. All the real dirt on dust. *The Exploratorium Quarterly* 12(1):15-19.

Newbould RL. 1976. Steam cleaning of Orient Cave, NSW, Australia. *Jenolan Caves Historical and Preservation Society Occasional Paper No 1*. p 5-15.

Ohms M. 2002. *Dust Deposition Along the Candlelight Tour, Final Report*. Wind Cave National Park, Cave Management files. 4 p.

Ressler D. 1988. *Thoughts About Lint—The Great Fuzz Battle Begins*. Cave Management files, Wind Cave National Park.

Roth J. 1987. *Preliminary Survey of Lint Accumulation in Carlsbad Caverns*. Unpublished manuscript, January 7, 1987. Carlsbad Caverns National Park Library. 3 p.

Schmitz SA. 1996. Lost Soldiers Cave restoration project. *NSS News* 54(2/3): 59-63.

Additional Reading

Ohms M, Rogers R. 1999. Black Hills restoration camp. *NSS News* 57(4):99-100.

Pickard R. 1993. Particulate matter accumulations in the Aberombie Cave. *Labyrinth, The Journal of the University of Technology.* Sydney Speleological Society, no 42. p 15-30.

Spate A, Moses C. 1994. Impacts of high pressure cleaning: A case study at Jenolan. In: [1998] *Cave Management in Australasia X.* Carlton South, Victoria (AU): ACKMA. p 45-48. [*Proceedings of the Tenth ACKMA Conference, Auckland, New Zealand.* ACKMA 10:45-48.] [Cave Management in Australasia 10:45-48.]

Tinsley J, Miller K, Johnson R. 1989. Monitoring and sedimentology of selected caves at Lava Beds National Monument. Abstract in: *Cave Research Foundation 1991 Annual Report*. p 31.

Went WF. 1970. Measuring cave air movements with condensation nuclei. *National Speleological Society Bulletin* 32(1):1-9.

Guidelines for Trash and Rubble Cleanup Projects

George Veni

Sometimes caves and sinkholes become trash dumps. Large rubble piles from trail building are often left in show caves. Other chapters discuss detailed aspects of removing artificial fill (page 367) and cleaning out sinkholes (page 381). While much of the restoration and rubble removal at these sites is intuitive and obvious, below are some guidelines to minimize surprises and ensure the work goes smoothly.

Sometimes caves and sinkholes become trash dumps. Here are some guidelines to minimize surprises and ensure the work goes smoothly.

Trash Pits and Sinkholes

Trash is sometimes dumped in sinkholes or cave entrances. Working inside the cave may not be necessary, so access and removal is relatively simple. Following are some important items to consider for restoration work that involves trash removal.

Gloves and shots. Everyone should wear thick leather gloves and be current on their tetanus shots. While many underplay the threat of tetanus, by the time the symptoms appear it can be fatal. Check with your doctor for other immunizations you should consider.

Recycling. Decide in advance if it is feasible to recycle the glass, aluminum, and other materials removed from the cave or sinkhole. If so, prepare areas and containers for sorting and temporary storage, and be sure to line up a recycling facility that will accept the materials. If possible, sort the recyclables as they are dug out of the pile—otherwise assign enough people to sort them at the top.

Disposal. Arrange in advance how you will dispose of the materials. You may need to take them to a public landfill and pay a small fee. Contact the landfill first. Some only accept certain types of trash and have vehicle or load requirements. See if a public agency can be involved to waive the fee and perhaps provide a dump truck. If the trash volume is small, it could be divided between the cavers and set out with their weekly trash pickup.

Interview the owner. Find out what the owner knows about the age, content, and extent of the trash. It will prove important in planning the restoration project and for the considerations listed below.

How old is it? In many states, human-produced materials more than 50 years old are considered archaeological or historical artifacts (cultural artifacts). Their handling and disposal may be regulated. Contact your State Historical Preservation Office (SHPO) for advice if you suspect a dump could hold historical material. If possible, find an archaeologist to monitor the excavation for rare or unusual items worthy of preservation that most people wouldn't recognize. (See 50-year rule, page 341; also see SHPO page 116 and page 334.)

Is it toxic? Use common sense. Avoid physical contact with potential hazardous materials and fumes. If you're not sure, leave it alone, even if you must cancel the restoration project, and call a county or state inspector to assess and remove it. For recent dumps where containers of hazardous materials are more likely to be intact, use additional leak-proof containers for transport. Hazardous materials should not be disposed of with the nonhazardous trash—make disposal arrangements with the appropriate facilities. With most old trash dumps, bottles and cans have long ago leaked their chemicals into the groundwater. However, toxic substances in car batteries and other hazardous materials may still present hazards.

Rubble Piles

Many privately owned show caves are first developed on a shoe-string budget. Unfortunately, as trails are cut, the rubble is dumped in piles or down pits and out-of-the-way passages. Special considerations in rubble removal are covered in the following section. Three primary concerns are essential when removing trail rubble:

Many privately owned show caves are first developed on a shoe-string budget. Unfortunately, as trails are cut, the rubble is dumped in piles or down pits and out-of-the-way passages.

- Exercise caution to protect the cave from additional damage.
- Learn from the owner if electrical or other utility lines may be hidden in the rubble.
- Decide with the owner how to sort and what to do with whole or broken speleothems that may be found.

How to Remove Large Volumes of Materials

Regardless of whether it is trash or rubble, following are some guidelines for removing large volumes of materials.

Limit the number of diggers. Cavers love to dig, but this craving must be curbed. In projects where a lot of material needs to be removed, especially if the project is conducted in a cave or down a deep sinkhole or pit, many cavers will be needed to haul away the material produced by only a few diggers.

Cavers love to dig, but this craving must be curbed. In projects where a lot of material needs to be removed, many cavers will be needed to haul away the material produced by only a few diggers.

Buckets. For most projects, hauling debris in 20-liter (5-gallon) plastic buckets is the easiest method for transporting materials. The size of the bucket limits weight to a reasonable amount. Buckets are sturdy, hold large and small items, do not leak, and are convenient to stack, load, and transport. At least 30 buckets are needed for small projects and over 100 may be needed for large projects. At any given time, many of the buckets are in transit in or out of the cave or are waiting to be filled. Of course, plans should be made in advance for removing large or awkward items that will not fit in buckets.

Partly fill the buckets. Depending on the weight of items loaded, it may be important to fill buckets only half-way. Full buckets are often too heavy for most cavers to carry. The half-way fill also works if the material is likely to spill out, or if the weight or contents are more likely to break the buckets.

Don't repack buckets. Once a bucket is filled, do not empty or repack it until it is ready to be dumped. Repacking results in extra effort, less efficient use of people, spillage throughout the cave, and more cleanup. Wheelbarrows have been used in some projects, but they can create messes in the cave. Also, time is lost loading and unloading wheelbarrows, and they are likely to damage the cave if the passages are narrow.

Pass, do not carry buckets. Carrying buckets is tiring, inefficient, and creates a lot of potential for injury to cavers and caves. The most efficient way of moving buckets over irregular terrain, like stairways in a show cave or uneven ground in a sinkhole, is to pass them, in assembly-line fashion. Place cavers about 2 to 3 paces apart. This requires a few more people, but is easier on the people and the cave, results in faster removal of the buckets, and produces far less confusion than cavers running all over the place.

Dollies. Upright or truck dollies should be used in flat-floored passages. Half-filled buckets can be stacked 3 or 4 high and wheeled quickly out of the cave or to an area where people will pass them onward. If passages are narrow, a prearranged location should be established for the dollies to pass each other.

Dump trucks. All removed material should be dumped into a truck and carried off for proper disposal at a landfill. If only rubble and dirt are removed, a flat-bed trailer may suffice to carry the trail rubble to a field where it can be deposited and blended with the landscape.

Heavy equipment. Some excavations may be beyond what can be feasibly accomplished by human hands and buckets. Some types of work will require heavy equipment such as backhoes, skid loaders ("bobcats"), and excavators. People with specific skills and training in heavy equipment should operate it. For safety reasons—unless required for other specialized tasks—no one else should be allowed in the excavation area. Slopes (and sometimes even level surfaces) in and near sinkholes may be unstable under the weight of heavy equipment. Extreme caution is necessary. The equipment should not be fueled in the sinkhole unless there is no alternative—always take precautions to prevent spills of fuel, oil, or hydraulic fluid. To prevent damage to the natural sinkhole walls and floor, equipment operators must be careful not to dig too deeply in any single movement. Some fill can be left and later removed more carefully by hand.

Some excavations may be beyond what can be feasibly accomplished by human hands and buckets. Some types of work will require heavy equipment.

Section C—Restoring Cave Passages

Removing Artificial Fill from Developed Caves

Rodney D. Horrocks and John E. Roth

Artificial fill changes and potentially impacts the natural conditions of caves in undesirable ways. In developed caves, introduced fill results from disturbing natural sediments, dumping trail construction or blasting debris, or constructing artificial structures. Through a series of mitigating steps, artificially introduced debris or materials can be removed and natural features can be restored. It is important to differentiate natural cave sediments from human-caused debris and to define goals for restoring natural conditions in a cave. Safety issues, tools, and techniques for removing artificial fill are discussed in this chapter.

In developed caves, introduced fill results from disturbing natural sediments, dumping trail construction or blasting debris, or constructing artificial structures.

Reasons for Conducting Artificial Fill Removal Projects

Artificial fill has been introduced since caves were first developed for visitors, but it was not until the late 1970s that show cave managers recognized the need to restore natural conditions by removing unnatural materials (Rohde and Kerbo 1978). The accepted practice, described by Irving (1989), is to "remove all unnecessary, unnatural materials from the cave, but leave natural materials in the cave."

Cave managers remove artificial fill to restore any combination of natural conditions—airflow, water flow, speleothem growth or dissolution rates, biotic communities, or visual scenes. Fill remediation projects may also remove harmful materials from and restore natural conditions to the cave environment.

Artificial fill can disrupt the volume and speed of air flowing into or out of a cave. The presence of fill, excavated pits and passages, or the altering of cave entrances can lead to unnatural drying out or wetting of passages. These alterations can even cause unnatural freezing or thawing deep in caves. Where feasible, the volume and configuration of an area may be restored to the original state before the fill was introduced. On another hand, borrow pits that were dug to obtain trail construction materials may be refilled with artificial fill, as long as that fill is native to the cave or is similar to the fill that was excavated, and the locations are carefully documented.

Artificial fill can alter temperatures to the extent that bats will abandon some roosts. Likewise, bats may be attracted by temperature changes where there are no historical or paleontological records of their presence. In such cases, the value of restoring natural conditions should be carefully weighed with other factors such as the species present, its vulnerability, and the availability of suitable, alternate roosting sites.

Natural hydrologic flow paths can be disrupted by the presence of artificial fill, trails, or dams. These obstructions can entirely change biotic communities within a cave, as well as the flow of nutrients within the cave ecosystem, especially if formerly dry passages are flooded. Blockages may also change the nature and amounts of speleothem deposition. Speleothem growth rates can be unnaturally disrupted when covered with artificial fill. This may lead to the cementation of the fill, which adds difficulties and dilemmas to the fill removal process.

By re-exposing a natural substrate that was covered with artificial fill, biotic communities may be restored when original surfaces or food sources are again accessible. Disturbance of some types of natural fill may increase habitat for microbes, egg-laying crickets, and egg predators.

Sometimes, the reason to restore a room or passage is largely aesthetic. Restoration projects may enhance part of a tour route and add interesting visual scenes where none existed since the initial impact (Figure 1).

Various impacts may occur from construction materials that are harmful to the cave, its biota, or to people. The following materials are major targets for fill removal projects:

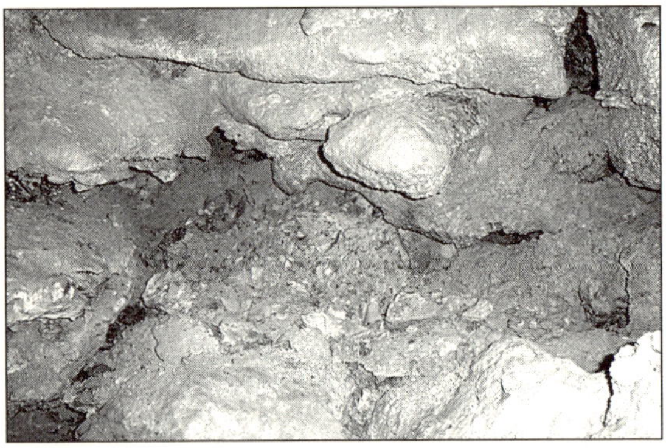

NPS Photo Matt Reece

NPS Photo Bonnie Curnock

Figure 1a (before, at left) and **Figure 1b** (after, at right). Removal of artificial fill at Lenas Arbor along the Natural Entrance Tour Route in Wind Cave National Park. Fill in the center of the photo was removed. Note that dust on the surfaces of the surrounding rock is also gone.

- Galvanized steel, which may leach zinc into cave waters and impact cave biota (Jameson 1995)
- Epoxy-covered trail chips
- Asphalt (carcinogenic aromatic hydrocarbons)
- Aluminum railings, which may leach aluminum into the cave in the presence of acidic water
- Batteries
- Old transformers, especially any containing PCB's (*polychlorinated biphenyls*)
- Control panels with mercury switches

Before removing potentially toxic materials, consult Material Safety Data Sheets (MSDS) for specific information about the safe handling of each substance. (See MSDS, page 70 and page 172.)

Types of Artificial Fill

Artificial fill is differentiated from natural fill as material added to or internally disturbed by human activity in a cave.

The goal during any artificial fill removal project should be to avoid disturbing any natural or *in situ* fill. At least two experienced people should evaluate each potential fill area. If everyone involved cannot agree that the material was the result of human activity, then the site should be left until the nature of the fill is better understood.

Artificial fill usually results from one of four activities—blasting; trail projects; utility line projects; or construction projects. Typically, these projects were originally instigated to facilitate the visiting public by aiding access, providing services, or artificially enhancing visual scenes. Occasionally, these projects were conducted to create level areas for equipment or to allow returning guides to bypass tour routes.

Blast debris may result from entrance enlargement, tunnel construction; ceiling expansion; platform construction; trail leveling, widening, or deepening. In the past, gravel-sized to very-large angular blocks of rock were sometimes dumped in the nearest convenient spot so that laborers would not have to haul them out of the cave. In some cases, tiny chips of rock reveal that blast debris was dumped into a convenient hole. These activities usually generated large quantities of artificial fill that may partially block existing passages and greatly reduce their volume and aesthetic appeal.

Trail construction and utility line projects in show caves produce varying volumes of debris, including silt, sand, clay, gravel, or broken calcite. During construction, this material was often stuffed into every available pit or alcove along a trail. When flowstone or other speleothems are found among this type of debris, the material should stay in the cave and not be dumped outside where collecting might be encouraged. Materials that belong in the cave may be stored in a dead-end side passage (so it does not affect natural airflow), and certainly away from visitors' sight and reach.

During show cave development, trail construction and utility line projects produced varying volumes of debris, including silt, sand, clay, gravel, or broken calcite. This material was often stuffed into every available pit or alcove along a trail.

If possible, speleothems should be repaired and reattached in their original locations. (See speleothem repair and questions, page 441.) Broken speleothems should never be used as interpretive "touching stones" since this type of activity may lead to additional touching when visitors wonder how the other speleothems feel. Touching stones may also encourage collecting and vandalism.

Trail construction projects or artificial structures in developed caves can disrupt natural conditions. Sometimes, minerals in nonnative gravels will cause potential leaching or deposition problems, or even provide artificial food sources for microbial life. Structures such as rock walls, concrete pads, artificial dams, asphalt trails, soil cements, or gates are often found in developed caves. Artificial dams can reduce habitat for some species while creating new habitat for others, both of which disturb cave ecosystems. An improperly built gate can reduce or increase airflow, impede or increase natural biota, and change air temperatures and relative humidity.

Calcium hydroxide in Portland cement is very soluble and is easily dissolved and quickly redeposited as "soda straws" or "flowstone" below concrete structures. These deposits grow at an accelerated pace compared to normal calcium carbonate speleothems. Managers should attempt to keep concrete trails in the lowest part of a passage or in areas where undersaturated water does not drip or run onto the pathway. If possible, visitor trails should not be built under actively dripping stalactites where new speleothems are apt to develop on the constructed path. If concrete structures are built in cave passages, use high tensile strength cement with low amounts of calcium hydroxide. This type of concrete will last longer in cave environments. (See concrete, page 169; also see concrete products, page 476.)

Recognizing *In Situ* Deposits

Because it is often very difficult to recognize in-place, or *in situ,* deposits, artificial fill removal projects should be preceded by testing, and work crews should be supervised closely. As in archaeological excavations, test pits should be dug in the suspected fill to determine what has been dumped, what the natural *in situ* material looks like, what tools are going to be needed, and what issues may be encountered. Even a quick analysis of the types and history of naturally laid material in the cave is helpful in identifying artificial fill and from where in the cave the material was excavated.

The presence of speleothems (such as popcorn lines, moonmilk deposits, cave rims, and the orientation of coralloids) or atmospheric corrosion sites can provide information for identifying areas where fill did not naturally exist. Natural fill is often marked by a change in color. The presence of

NPS Photo Rod Horrocks

Figure 2. An over zealous cave restoration worker dug into these *in situ* (in place) sediments in the Assembly Room in Wind Cave National Park. Notice the laminations in the natural cave fill, which in this case are multicolored and readily visible.

speleogens (sculpted bedrock that is sometimes dissolved underneath natural fill), or a change in wall rock color *may* indicate that natural fill has been artificially removed, although such sediment can be removed by natural process as well. A number of criteria can indicate if the deposits are *in situ*—layering, compaction, voids, broken surfaces, tool scratching, calcite deposition, organics, and historical artifacts.

Normally, fine-scale layering is a sure sign of *in situ* deposits. However, artificial layering can be created by maintenance workers spreading down alternating layers of sand or gravel to prepare an area for trail construction. Coarser layering can be caused by dumped materials that came from multiple sites. In such instances, coarser blasting material may be overlain by finer materials that were dumped on top, creating artificial layering. However, if fine laminations, cross bedding, or pebble imbrications are present, the layers are natural (Figure 2). Natural sediments may show varying degrees of calcification and may create weakly cemented deposits.

Compaction is another good indicator of *in situ* materials. Fine-grained water-lain materials usually have very little air space between the grains. However, compaction can also result when a deposit has been walked on for extended periods of time. Tests pits will reveal that such compaction is less than 15 centimeters (6 inches) deep. Voids, especially those near natural walls, are often good indicators that material has been re-excavated and dumped.

Pack rat nests can indicate when fill removal efforts have reached the natural floors. Sometimes the organics have decomposed from these nests, leaving only bones to indicate their former existence and the presence of *in situ* materials below.

Undisturbed *in situ* rocks in caves are often covered with a thin veneer of calcite or other materials—however, fresh fractures on rocks commonly indicate the presence of artificial fill. Carbonate rocks broken during blasting and other construction activities often have a sparkly surface with small amounts of calcite dust. However, blast debris may be covered with a thin film of mud, which may be calcified and mask fractures or tool scratches.

Scratches are sometimes found on rocks that have been excavated with tools. Thus, scratches, especially randomly oriented ones, are good indicators of artificial fill. Since the calcite dust produced by scratches is easily dissolved, prehistoric scratches are often not visible.

The easiest fill restoration projects involve the removal of artificial fill dumped on flowstone, calcified sediment, crusts, or stalagmites where identification of the material is more obvious. However, watch for sediments that may be deposited on calcite surfaces, mimicking artificial fill conditions. Other calcite deposits can also indicate *in situ* materials—examples include floor coatings, wall coatings on rocks, and coralloids such as popcorn.

Metal tools should not be used to remove fill when approaching these surfaces because the soft calcite is easily scratched. Plastic tools such as putty knives, hand brooms, and gardening hand shovels work well, depending on the material encountered (Netherton 1993). However, sometimes a thin layer of partly cemented fill or blasting mud can only be removed with dense-bristled metal brushes, which may cause minor scratching. Only clean, rust-free, nonpainted stainless steel and new nylon brushes should be used in cave projects—old ones tend to shed steel, rust, or nylon into the cave. (See restoration tools, page 406.)

Lumber or burned wood may indicate the presence of artificially dumped debris. Look for squared ends and saw or axe marks on such pieces. Creosote or pressure-treated wood usually should be removed. Be careful not to remove cultural resources without the assistance of an archaeologist.

Removal of rotting wood piles may cause population crashes among cave-dwelling species. Check wood for fauna and consult a biologist if cave-dwelling species are found. (See wood piles, page 37.) Wood may have to be removed in stages to mitigate impact to cave biota. Wood should not be removed if it will disintegrate during transport and thus provide an unnatural nutrient windfall (Lewis 1993). Buckets or bags may be used if this is a concern.

The presence of recent cultural items usually indicates the presence of artificial fill. However, burrowing animals may take items into *in situ* materials. These objects may also fall into cracks between rocks or be washed in during flooding. Many states consider cultural objects more than 50 years old as having potential archaeological significance and an archaeologist should be consulted on the disposition of these items (more information is included below).

The presence of construction debris is usually a good indication that the fill has been disturbed. Such debris may include pea gravel, sand, cement splatter, or trash. If the gravel and sand is a rock type not native to the cave and is not in a stream deposit, it is usually safe to assume it has been dumped at the site. Sharp, angular gravel or boulders with some calcite dust usually indicate blasting. Blocks of breccia are sometimes hard to distinguish from concrete, but usually will have some amount of calcified mud, as opposed to mud-free concrete surfaces.

Certain natural processes may produce materials that resemble artificial fill. If the project is near an entrance, piles of frost-wedged wall rock or thin layers that have spalled off speleothems may resemble artificial fill. Speleothems desiccating in domes, humidity changes on gypsum layers, or atmospheric corrosion of ceilings may leave piles of debris on the floor that resemble artificial fill. Additionally, flood events may wash material into a cave. The presence of charcoal in these flood deposits may help in determining what is natural and what is not.

Special Issues

It is very useful for cave restorers to include or at least have access to experts with engineering, archaeological, biological, and paleontological experience to address the many special issues encountered in artificial fill removal projects. Special concerns include cultural artifacts and graffiti, broken speleothems, cement-covered speleothems, recemented fill, destabilization, utility lines, and artificial lakes.

It is very useful for cave restorers to include or at least have access to experts with engineering, archaeological, biological, and paleontological experience to address the many special issues encountered in artificial fill removal projects.

It is common to find cultural items mixed in with artificial fill. A qualified archaeologist should determine their preservation, age, historical significance, and whether they should be collected or discarded. Cultural artifacts include any manufactured item that is over 50 years old. (See cultural resources, page 110; also see 50-year rule, page 341.)

If a qualified archaeologist directs that certain cultural items are worthy of collection, they should be bagged with a completed artifact label. Mark locations on a large-scale map and record the depth items were found below the fill surface. Number and bag each item separately. Record these data at the time of collection, even if the value is not immediately obvious—this important information cannot be accurately re-created at a later date. The presence of datable artifacts may aid in determining when overlying layers of fill were dumped.

Pieces of flowstone, rocks with popcorn, or wall coating mixed in with artificial fill are often found and should remain in the cave. These materials may be hand-picked from the artificial fill if they are in large pieces or sieved out if in smaller pieces. If they are dumped outside, they may encourage collecting, both by individuals and commercial operators. Reattach identifiable speleothems where possible. Dump remaining materials in an area of the cave that is not accessible or visible to visitors, near their origin, and where they do not impact natural airflow or clog natural hydrological drains. Thoroughly document in-cave dumping sites.

Concrete was often applied directly to flowstone surfaces as sidewalks, platforms, and light shields. Concrete also can be found splattered on speleothems adjacent to paved trails or covering wire runs between transformers and light fixtures. Restoration should not be attempted if speleothems would be damaged to a degree greater than the benefits derived from restoring the natural surface. However, rapid deposition can cover over shallow scratches resulting from restoration. Concrete can often be removed with hammers, chisels, and dental picks. It may sometimes be popped off using a light blow to a chisel parallel to the contact between smooth flowstone and concrete. This technique cannot be used when bumpy surfaces, such as popcorn, are covered with concrete. Any chiseling should be done at an angle and not perpendicular to the surfaces being cleaned (Schaper 1995).

If a significant amount of time has elapsed since the artificial fill was dumped, dripping water may have cemented the fill together. If more than a thin film has accumulated, several questions should be asked and answered before proceeding:

- Is the artificial fill impeding natural airflow or hydrological drains?
- Is the fill preventing the growth of buried speleothems?
- Are there biological considerations?

If the answer is yes to any of these questions, seriously consider removing the fill. When buried speleothems are encountered below a drip point, dripping water may have already cemented the fill to those speleothems and the resulting conglomeration will have to be carefully removed.

Removal of artificial fill can destabilize large breakdown blocks. Carefully supervise volunteers if they dig underneath or around such rocks. If a large block is also artificial fill, it may be broken up before becoming unstable. A hammer drill with plug and feathers is a highly effective, low-impact technique for breaking up large blocks (Figure 3). Exercise care to reduce dust caused by hammer drills. Another technique involves filling drilled holes with expanding materials. Expanding materials need to be completely removed after they have broken up the rock.

NPS Photo Rod Horrocks

Figure 3. Use a hammer drill with plug and feathers to split large blocks of rock into smaller, safer chunks of rubble. This is a low-impact technique when workers take care to contain the dust that results from drilling.

Anytime artificial fill is removed from a developed cave, workers may encounter buried utility lines, especially electric cables, water lines, or phone lines. High-voltage cables can be life threatening. Maintenance employees or the original developers should be contacted before work begins. If an electrical cable is known to be buried in a certain area, turn off the power before the project begins. Test all continuous wires with an electromagnetic device to make certain they are not live, and then carefully remove the artificial fill. Once found, use hand tools to uncover the wire throughout its entire length. Lines can then be bundled together and lifted, or tied out of the way so they are not stepped on or hit with metal tools as the project progresses.

Test all continuous wires with an electromagnetic device to make certain they are not live, and then carefully remove the artificial fill. Once found, use hand tools to uncover the wire throughout its entire length.

If an artificial lake has been built in a developed cave, the lake can be slowly drained using small tubing. Or the water might be used for cleaning and restoring muddy or soiled areas, but not relatively pristine areas. If this water is safe to use for cleaning, collect the restoration run-off and do not allow it to run onto cave floors. Avoid relocating any problem to a new place. (See restoration runoff, page 339 and page 396.) Use lint-free rags or sponges to sop up the cleaning water. Exercise care to avoid causing unnatural erosion.

Equipment and Tools

Before an artificial fill removal projects starts, several items should be gathered or constructed:

- Permits (from agency/owner)
- Rocking screens (detailed information below)
- Large-scale cave maps
- Parental Consent Form (if youth volunteers are used)

Additional items that will be useful are listed here:

- Camera, film, and flash
- Wheelbarrows
- Buckets, 20-liter (5-gallon) with handles
- Powerful portable lights
- Extension cords
- Square-nose shovels
- Round-nose shovels
- Short-handled shovels
- Folding shovels
- Garden trowels (nonpainted metal and plastic)
- Push brooms
- Tarps

When final, detailed restoration is performed after the bulk of the artificial fill is removed, the following tools may be needed:

- HEPA filtered vacuum cleaner with special tools to reach into cracks
- Hand-held squirt bottles
- Paint brushes (2.5- to 5-centimeter width or 1-inch to 2-inch)
- Auto part brushes (to reach into small places)
- Toothbrushes
- Rags (lint-free) and sponges
- Handled scrub brushes

Some artificial fill removal projects may require specialty equipment:

- Breaker bars for moving large blocks
- Impact hammer for use with plug and feathers
- Plug and feathers for breaking up large blocks

- Powered wheel barrows for outside use only
- Hammers and chisels for removing concrete
- Dental tools for removing cement or loose fill in small spaces
- Quart-sized zip-closure bags for bagging artifacts—first consult with an archaeologist
- Artifact labels (cards for artifact bags)
- Permanent markers for labeling artifact bags

The most common safety hazard in cave fill removal is overfilled buckets. Big, heavy buckets can easily cause strained muscles, back injuries, or bone-bruised heels.

Each worker should have the following personal gear:

- Helmet
- Helmet-mounted light
- Leather gloves
- Knee pads and elbow pads
- Sturdy boots
- Change of clothes
- Goggles (depending on job)
- Ear plugs (depending on job)
- Dust masks (depending on job)

Safety Measures

Safety is the most important aspect of cave restoration. Planning can greatly increase the safety margin at artificial fill removal projects. When youth volunteers are participating, their parents should sign a parental consent form. Volunteers may need to sign a liability release form.

It is important to inspect the restoration site beforehand to look for potential safety hazards. At the excavation site, each worker should wear gloves, safety helmet, sturdy boots, and appropriate clothes (layered clothing for cold caves—layers can be removed if workers overheat).

There should be at least a 1-to-6 supervisor-to-worker ratio to properly oversee the progress of the excavation. The supervisors or team leaders are responsible for monitoring the safety of workers and assuring protection of cave surfaces.

There are a number of special techniques for accomplishing artificial fill removal tasks—research and testing, documentation, tallying fill totals, excavation, transporting materials, and locating buried speleothems.

- Make sure that no excavator is undermining sediment banks or large rocks.
- See that workers are wearing appropriate safety gear for individual tasks.
- Monitor whether workers are overtiring or overheating.
- Assure sufficient spacing between workers (so they are not swinging equipment and hitting each other or the walls).

Buckets and wheelbarrows should be operated by individuals physically capable of handling them. No one should be tossing rocks or tools. Many cave restorers are inexperienced cavers and need to be trained in safe and conservation-oriented caving and climbing techniques. Vertical or climbing skills are needed to reach many fill sites.

The most common safety hazard in cave fill removal is overfilled buckets. Big, heavy buckets can easily cause strained muscles, back injuries, or bone-bruised heels. Overfilling is hard on the buckets, and causes them to break or spill. Depending on the volunteers participating, buckets should be marked inside with a line only one-third to two-thirds above the bottom of the bucket and supervisors should be careful in seeing that workers do not exceed that mark. The weight should be geared to the smallest member of the crew or the individual in the most awkward position for carrying a bucket. If possible, individuals in assembly-line crews should be of similar size. Remind workers how to lift buckets to avoid causing back strain.

The second most common safety hazard is an overtired crew. The loss of coordination in tired workers can lead to accidents. Establish a workload maximum—for example, 3 hours in the morning and 3 hours in the afternoon with plenty of breaks in between and plenty of time to exit the cave at a safe speed. One way to detect tiredness and maintain high moral is for supervisors to banter with workers up and down the line and show interest in what they are accomplishing.

The third most common safety hazard is moving rocks that are too large. Blasted rock can be broken up into smaller pieces before removal. (See the section above on special issues.)

Other potential hazards include the following:

- Insufficient work space—the surrounding space needs to be cleared of people whenever overhead tools are used. Use picks and sledgehammers with fiberglass or plastic handles because wooden ones are more likely to splinter.
- Overheating—high temperatures and humidity may lead to overheating in some cave environments.
- Respiratory diseases—use filter masks in any dusty environment. Use HEPA filter respirators where guano (woodrat, mouse, bat, birds, crickets) or amberat (dried woodrat urine) is present.
- Venomous biota—exercise caution near entrances where disease-carrying mites, rattlesnakes, or the more venomous spiders (*Meta* sp., black widow, brown recluse, hobo) are present. Check skin for ticks or other mites after hiking to or from a remote cave.
- EMT response time—an EMT should accompany each group or be able to respond within 20 minutes.
- Unusual odors—if any unusual or strong odors are encountered, immediately instruct all participants to leave the artificial fill removal project. The source of that odor should be determined before the group is allowed back to the site.

Techniques

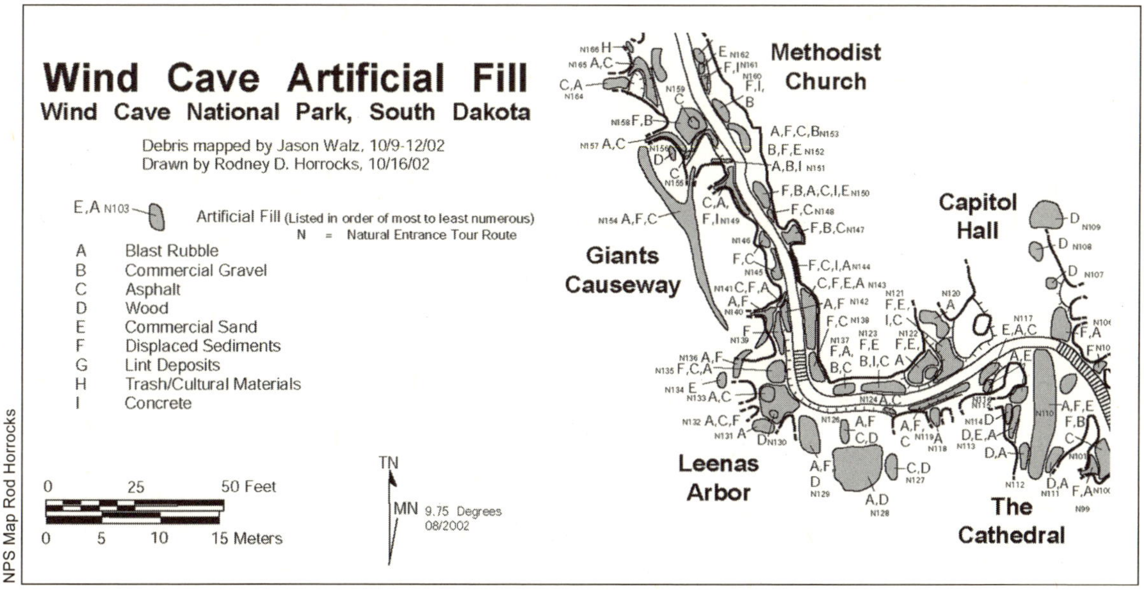

Figure 4: A section from the restoration map along the Natural Entrance Tour Route in Wind Cave National Park. All the before-and-after photos taken and all the cultural items collected during the restoration project were referenced to the unit numbers assigned to each deposit on this restoration map.

Cave Restoration Weight Log

A=asphalt, F=fill, L=lint, W=wood, M=mixed load

Date	Crew Member	Material	Weight	Date	Crew Member	Material	Weight

Cave Restoration Artifact Log

Tour Route: ___________ **Recorder:** ____________ **Project:** ____________

Artifact ID	Photo #	Date	Location	Description

Cave Restoration Photo Log

Tour Route: ___________ **Recorder:** ____________ **Project:** ____________

Date	Photo #	Site ID	Description	Orientation	Before/After

Figure 5: Log sheet examples for photos, artifacts, and weights. These tally logs are used during cave restoration projects at Wind Cave National Park.

There are a number of special techniques for accomplishing artificial fill removal tasks—research and testing, documentation, tallying fill totals, excavation, transporting materials, and locating buried speleothems.

A little research before a project begins can be very useful, particularly oral history interviews with anyone who remembers the restoration site prior to being filled (Netherton 1993). Before any artificial fill removal project begins, test the material to determine access issues, potential safety concerns, the extent of the deposit, the nature of the fill, what is *in situ* and what is not, the number of excavators and haulers needed, the tools required, and the techniques that will work for removing the debris.

It is important to document an artificial fill removal project before, during, and after with photographs. When taking pictures for before-and-after sequences, the camera should be mounted on a recoverable point (from a station that will never be removed during any stage of the project). For thorough documentation, photograph with black-and white-film as well as color and digital imaging techniques. Black-and-white film and photographic prints may last longer than color films, depending on the processing and storing techniques (See photo bullets, pages 209-211.) However, color images in caves will often give more information. A digital camera can be used, but the photographer should remember that the shelf life of digital storage products is largely unknown and files may have to be transferred to new media periodically. (See archival protection, page 210.)

If a project is completed in stages, it is good to document the extent of each period of excavation on a large-scale map (Figure 4).

Keeping track of the amount of fill removed is good for morale. If 20-liter (5-gallon) buckets are used, fill several to a predetermined level and weigh them all on a common-weight scale to arrive at an average weight for that particular type of fill. Individual bucket weights will vary depending on the material being moved and the void space in the buckets. The weights of these buckets may vary between about 14 to 23 kg (30 to 50 lbs). Once the average weight is determined, one person can keep track of the number of buckets dumped into a wheelbarrow or hauled out of the cave using a simple tally sheet (Figure 5). At the end of a work session, the number of ticks can be multiplied by the average bucket weight to obtain a good estimate of the amount of fill removed by a group.

Artificial fill removal projects should be closely supervised by experienced cave restorers. Expertise in archaeology and paleontology is beneficial. Project directors should be ready to bring in subject matter experts. Several potential disasters may occur if workers are not closely watched—avoid digging into *in situ* materials, destabilizing large breakdown blocks, damaging or not documenting historic artifacts, and spilling wheelbarrows or buckets in the cave.

Wheelbarrows should be operated only by the stronger workers. To prevent creating dust or spilling debris, wheelbarrows or buckets should not be emptied or transferred while inside a cave. Supervision should be available for each work crew, whether excavating, hauling, or screening.

Each artificial fill removal project will require different excavation techniques and tools, some of which will have to be developed specifically for individual projects. When developing these techniques, consider the following issues:

- Worker safety
- Minimizing dust and flying rock chips
- Proximity to speleothems
- Tool durability
- *In situ* materials

Workers should be able to adapt different tools to a specific project. If gravel-sized material is being removed, workers can get downhill from the area being excavated, put 20-liter (5-gallon) buckets between their legs, and simply use a garden trowel or a folding shovel to pull the fill into the bucket. Whenever the natural floor is approached, (see the section above on recognizing *in situ* deposits), the worker should switch to carefully remov-

When large groups of volunteers are available, an assembly line is the fastest method of moving large volumes of rubble. By spacing workers an arm's length apart, 20-liter (5-gallon) buckets can be passed out of an excavation, down passages to paved cave trails, and up or down stairs.

NPS Photo Rod Horrocks

Figure 6. Rocking screens are used to screen for items that should not be discarded with the artificial fill—for example, broken speleothems, fossils, and cultural materials. This project was located outside the Hansen Cave Entrance, Timpanogos Cave National Monument, Utah.

Ben Maxwell, Timpanogos Cave Natl. Mon.

Figure 7. A volunteer group hauls debris out of the cave in "assembly-line fashion." Crews of cavers have been known to pass up to one ton of debris in 15 minutes or less using this method at Timpanogos Cave National Monument, Utah.

ing the fill with hand tools, or use plastic tools if speleothems are present. If concrete walkways on artificial fill are being removed, they can be broken easily by tunneling underneath and using a sledge hammer to break the surface.

If historical artifacts, broken speleothems, or pieces of calcite are present in the fill, many of them can be hand picked during careful observation while excavating. If the fill is gravel size or smaller, the remaining fill should be screened so that small historical artifacts, fossils, and pieces of calcite are not dumped outside the cave—items that tend to encourage uninvited collecting. If delicate fossils are uncovered, they can be water screened with nonrocking, fine screens (1-millimeter mesh). Large rocking screens with double handles on one or two sides work well (Figure 6). Screens can be constructed out of lumber, PVC, or fiberglass, bolts, and window screen material.

When large groups of volunteers are available, an assembly line is the fastest method of moving large volumes of rubble. By spacing workers an arm's length apart, 20-liter (5-gallon) buckets can be passed out of an excavation, down passages to paved cave trails, and up or down stairs (Figure 7). If a longer distance must be crossed, the group can pass a collection of buckets, stack them up and then move the line forward, repeating the process until the paved trail is encountered. At Timpanogos Cave in Utah, volunteer crews were able to move up to a ton of fill out of an excavation site every 15 minutes using these methods (Horrocks 1994). Once the paved trail is reached, loose fill can be dumped into waiting wheelbarrows (if it doesn't create dust) and buckets can be stacked on dollies and rolled to the entrance.

It is also important for morale and efficient documentation to limit projects to a workable scale. Plan small enough projects so that workers can complete final cleaning of an area and see the results of their efforts. Projects should start at peripheries and work toward the established trails to prevent further impacts in restored areas.

It is also important for morale and efficient documentation to limit projects to a workable scale. Plan small enough projects so that workers can complete final cleaning of an area and see the results.

Before starting any excavation, check the ceiling for soda straws and stalactites to help predict the location of stalagmites (stalactites may not be currently dripping). Workers at each potential stalagmite location should be cautioned to watch for that speleothem. Likewise, flowstone on a wall or uphill from the fill can be used to predict the presence of buried flowstone surfaces. When excavating, it is always easiest to start on a flowstone or popcorn surface—use plastic tools and work along the flowstone, removing fill as you progress deeper into the deposit.

Dirty handrails and trails need to be cleaned during and after a fill removal project. Tools should be cleaned and sharpened if needed. Remove all restoration supplies that will rust or decompose in the cave. All other material stored for an ongoing fill removal project should be hidden from the sight of visitors on the developed trail. Clean mud off the equipment each night before leaving because some materials are difficult to remove from the tools the next day (Netherton 1993).

Summary

Through the mitigating steps described in this chapter, artificially introduced debris, rubble, and fill materials can be removed from the cave and natural cave features can be restored. Be certain the supervisors or team leaders can differentiate natural cave sediments from human-caused debris. In this volume, also see the brief list of trash and rubble guidelines describing safety, tools, and techniques (page 363).

Cited References

Horrocks RD. 1994. Artificial fill removal project: Timpanogos Cave System, Timpanogos Cave National Monument. *NSS News* 53(4):102-107.

Irving D. 1989. Wind Cave Restoration Field Camp. *NSS News* 47(2):34-36.

Jameson RA. 1995. Zinc leaching from galvanized steel in Mystery Cave, Minnesota. In: Rea GT, editor. *Proceedings of the 1995 National Cave Management Symposium: Spring Mill State Park, Mitchell, Indiana, October 25-28, 1995*. Indianapolis (IN): Indiana Karst Conservancy. p 178-185.

Lewis JJ. 1993. The effects of cave restoration on some aquatic communities [abstract]. In: *1991 National Cave Management Symposium Proceedings: Bowling Green, Kentucky, October 23-26, 1991*. Horse Cave (KY): American Cave Conservation Association, Inc. p 346-350.

Netherton W. 1993. Mystery Cave trails: Past, present and future. In: Jorden JR, Obele RK, editors. *Proceedings of the 1989 National Cave Management Symposium: New Braunfels, Texas*. Austin (TX): Texas Cave Management Association; Texas Parks and Wildlife Department. p 141-147.

Rohde K, Kerbo R. 1978. Cave restoration: An underview of correcting overviewing. In: Zuber R, editor. *National Cave Management Symposium proceedings: Big Sky, Montana, October 3-7, 1977*. Albuquerque (NM): Adobe Press. p 55-58.

Schaper J. 1995. Techniques and volunteer issues in the restoration of Cathedral Cave. In: Rea GT, editor. *Proceedings of the 1995 National Cave Management Symposium: Spring Mill State Park, Mitchell, Indiana, October 25-28, 1995*. Indianapolis (IN): Indiana Karst Conservancy. p 279-281.

Additional Reading

Benton JM. 1995. Marengo Cave second annual restoration weekend. *NSS News* 53(2):41-42.

Crisman B. 1985. Cave restoration at Carlsbad Caverns National Park. *NSS News* 43(5):177-180.

Crisman B. 1985. Pioneering cave restoration at Carlsbad Caverns. *Inroads* 47:40-43.

Kohlwey K. 1992. 1991 Timpanogos Cave restoration project. *NSS News* 52(1):12-13.

Meyer PJ, French JA. 1995. Cathedral Caverns: Techniques for restoration and cleaning. *NSS News* 53(7):186-193.

Meyer PJ, French JA. 1996. Crossings Cave: Test bed for cave restoration. *NSS News* 54(2-3):54-58.

Rhodes D. 1976. Cave restoration: a(n) (im)possible task. *NSS News* 34(11):200-202.

Rogers N. 1995. Mammoth Cave restoration field camp. *NSS News* 53(2):44-44.

Rohde K. 1989. Cave restoration begins at Wind Cave. *NSS News* 47(2):32-34.

Schmitz SA. 1996. Lost soldiers cave restoration project. *NSS News* 54(2-3):59-63.

Tranel MJ. 1992. Changing to prevent change: Building a science and resource management program. *NSS News* 52(1):22-25.

Vale E. 1991. It's an open and shut cave: Plugging artificial entrances at Onondaga Cave State Park. In: *1991 National Cave Management Symposium Proceedings: Bowling Green, Kentucky, October 23-26, 1991*. Horse Cave, KY: American Cave Conservation Association, Inc. p 129-131.

Veni G. 1994. The 3rd annual Caverns of Sonora restoration project. *The Texas Caver* 39(1):3-4.

Veni G. 1995. The 4th annual Caverns of Sonora restoration project. *The Texas Caver* 40(1):3-4.

Veni, G. 1998. 1998 Sonora restoration project. *The Texas Caver* 43(6):87-89. Reprinted in *SpeleoDigest* 1999:301-302.

Section C—Restoring Cave Passages

Sinkhole Cleanout Projects

Joseph H. Fagan and William D. Orndorff

Sinkholes serve as natural catchments where surface water enters the subsurface to recharge karst aquifers. Dumping trash in sinkholes leads to contamination of groundwater systems. Improper management practices result in polluting caves and water.

Over the decades, many landowners used caves and sinkholes as places to dispose of trash and sometimes even hazardous materials. Debris commonly found in sinkhole dumpsites can range from common household waste, old appliances and junk cars, to pesticide containers and dead animal carcasses. Debris from dumping frequently obstructs cave entrances that are found in sinkholes. As water flows through such an accumulation of waste, contaminants and pathogens wash into the cave environment and ultimately into drinking water supplies. Sinkhole cleanout projects are a practical way to protect caves and improve the quality of groundwater in karst aquifers.

Dumping trash in sinkholes leads to contamination of groundwater systems. Improper management practices result in polluting caves and water.

Some sinkholes tend to channel unfiltered pollutants into the subsurface more readily than others do. Sinkholes with cave entrances or ones receiving flow from active or intermittent streams are particularly vulnerable to contamination. These more sensitive sinkholes should receive a higher priority when planning and targeting cleanouts. Especially valuable are projects resulting in restoration and protection of habitat for stygobitic and troglobitic fauna in cave and karst systems. (See stygobites, page 36.) Sinkhole cleanout projects can significantly help efforts to preserve biological diversity in karst regions.

Landowner Involvement

Landowner involvement is the most important element in organizing a sinkhole cleanout. It is critical for the landowner to fully agree with project plans and grant legal permission before performing any work. Funding for sinkhole cleanouts usually depends on landowner participation.

Unless you or your organization is willing to absorb all the costs, the landowner will be a major player in cleaning up any sinkhole dump. Landowners can apply for funding through federal and state environmental cost share programs to pay for sinkhole cleanouts. Under some cost share programs, donated volunteer labor can apply toward required landowner contributions. Depending upon the specific details of a sinkhole cleanout project, and what cost share funding program is used, the required landowner contribution might range from 0 to 50 percent.

Landowner involvement is the most important element in organizing a sinkhole cleanout.

Government Assistance

Government programs at various levels can fund sinkhole cleanout projects. Federal programs such as the Environmental Quality Incentives Program (EQIP) and the Conservation Reserve Enhancement Program (CREP) may fund a cleanout if the sinkhole receives an active stream. Sinkholes with large drainage areas and significant intermittent stream flows into the groundwater may also qualify for funding under CREP. The

Federal Farm Bill provides funding for the CREP and EQIP programs through the US Department of Agriculture.

The US Fish and Wildlife Service funds and manages sinkhole and cave cleanout projects involving federally listed species with the Partners for Wildlife Program under the US Department of the Interior. In addition to paying for implementation of particular conservation practices, some programs provide an annual rental payment to landowners who continue to maintain conservation practices as specified under a contractual agreement. Cost share rental payments are usually contingent on the sinkhole being fenced off after the cleanout.

The US Fish and Wildlife Service funds and manages sinkhole and cave cleanout projects involving federally listed species with the Partners for Wildlife Program under the US Department of the Interior.

State Assistance

Some states sponsor cost share and tax credit programs to assist landowners with conservation practices such as sinkhole cleanouts. On July 1, 2002, the Virginia Agricultural Best Management Practices (BMPs) Cost Share Program, administered by the Virginia Department of Conservation and Recreation, adopted a Sinkhole Protection BMP. The Virginia Sinkhole Protection BMP (WQ–11) can pay up to 75 percent of the cost for a sinkhole cleanout with a maximum contribution by the state of $2,500.

Landowners in Virginia and elsewhere can also apply for financial assistance for conservation efforts through their local Soil and Water Conservation Office. Other agencies assisting landowners are the USDA Farm Service Agency and Natural Resources Conservation Service.

Cave Conservancies

In addition to government programs, private organizations sometimes fund sinkhole cleanout projects. Cave conservancies such as the Cave Conservancy of the Virginias, the West Virginia Cave Conservancy, the Michigan Karst Conservancy, and the Indiana Karst Conservancy have all supported sinkhole cleanouts. The Nature Conservancy of Tennessee and the Canaan Valley Institute are also involved with sinkhole cleanout work.

Much can be accomplished through cooperation between private organizations and government agencies. The Virginia Karst Program, administered under the Virginia Department of Conservation and Recreation Division of Natural Heritage, has coordinated many successful sinkhole cleanout projects using funds provided through outside sources, including the Cave Conservancy of the Virginias.

In addition to government programs, private organizations sometimes fund sinkhole cleanout projects.

Cleanout Costs

The cost of sinkhole cleanouts can range from a few hundred to many thousands of dollars. Most sinkhole projects involve coordination with excavation contractors who use heavy machinery such as backhoes, track hoes, and front-end loaders to perform the actual cleanout work. A sanitation service or similar waste hauler usually provides transportation of the removed materials to a nearby landfill.

Proper disposal of the debris removed from sinkhole dumpsites usually requires segregation of recyclable materials and the payment of landfill tipping fees. Some localities and regional public service authorities are willing to encourage sinkhole cleanout projects by waiving landfill tipping fees and may even provide heavy equipment if asked.

Contractors

Choice of a contractor should hinge on a company's equipment capabilities and experience as well as cost. A cheap, but unreliable contractor will cost a project far more in the long run than paying an experienced operator to do the job. Written contracts are advisable to prevent misunderstandings and to help ensure successful completion of sinkhole cleanout projects. Any contract should specify payment only for work actually performed. Be

Figure 1. Sinkhole dump in Giles County, Virginia. This site includes an old trailer without any sanitation connection. (They were straight-piping the bathroom onto the ground and into the sinkhole.)

Figure 2. County workers and caver volunteers watch the heavy-equipment operator performing cleanout work as part of a joint venture between the Virginia Department of Conservation and Recreation, the Cave Conservancy of the Virginias, and the VPI Cave Club Student Grotto of the NSS.

Photos Joey Fagan

Figure 3. VPI Cave Club volunteers at the sinkhole work site late in the day.

Figure 4. VPI Cave Club volunteers sort out recyclable items at the work site.

Figure 5. The cleanup is substantially complete with some debris still left to be removed from the site and transported to a landfill.

sure contractors can certify they are adequately insured—at least a half million dollars in general business liability and automobile liability coverage is usually required for contractors performing work on small government-funded projects. Employers are usually required to have workers compensation insurance as well as employer's liability insurance. Check state and local laws.

Erosion and Sedimentation Control

Implementation of proper erosion and sedimentation control practices is critical for every sinkhole project. Proper conservation practices will help protect the sinkhole, caves, and groundwater from further damage. Stormwater and sediment can pass directly into the karst aquifer from unstable slopes that lack appropriate vegetative ground cover. Sinkholes should have a vegetated buffer or filter strip around their perimeter to provide some filtration of water flowing into the subsurface. Plantings should include native species. Avoid the use of invasive species.

Choose products such as erosion control matting with an eye toward minimizing adverse impacts to wildlife—the plastic netting used in some erosion control materials can trap small animals. Professional natural resource conservationists such as those working for the Natural Resources Conservation Service, US Fish and Wildlife Service, and local Soil and Water Conservation Districts can usually recommend appropriate ground cover plantings and advise what erosion control practices are appropriate for a given site.

Sinkhole Protection

Fence building is an important part of sinkhole protection. Semi-skilled volunteer laborers can contribute significantly to sinkhole cleanout projects during fence construction. Most cost share programs require livestock exclusion from treated areas. Livestock ought to be fenced from any sinkholes that contain cave entrances or receive active or intermittent streams. Farm animals should be excluded from streams in karst regions whenever possible. Allowing livestock direct access to streams and sinkholes can result in higher levels of nitrates and pathogens in surface streams and groundwater. The same cost share programs used for sinkhole cleanouts will sometimes pay to develop alternate water sources for cattle and will fund livestock exclusion practices.

It is important to continue to check sinkholes after cleanouts are completed. Watch for indications of slope instability and erosion problems.

Monitoring Sinkholes

It is important to continue to check sinkholes after cleanouts are completed. Watch for indications of slope instability and erosion problems. Monitor the growth and stability of ground cover plantings. Place signs at the sinkhole to help discourage further dumping. Correct any problems as quickly as possible to put a stop to any resumption of dumping and to prevent associated environmental impacts.

Sinkhole News Stories

Sinkhole cleanout projects are newsworthy. People who understand the direct connections between sinkhole dumps and their drinking water supply are less likely to put waste in sinkholes in the first place. Tell news correspondents the important details about the project and the reason for the cleanout. Always give credit to the organizations involved in the project. Provide a prepared written news release to reduce the chances for tainted articles that contain misquoted statements and out-of-context casual remarks. A good news story, however, can be an invaluable tool for educating the public about the importance of protecting caves and groundwater quality in karst regions. (See public relations and press relations, page 284.)

Sinkhole cleanouts can help promote cave and karst protection. For media coverage and community collaboration, choose a sinkhole project that demonstrates karst contamination, groundwater pollution, and impacts to cave resources. Partnerships between conservation groups and government agencies can provide significant resources for performing sinkhole cleanouts.

To better promote conservation projects, place a strong emphasis on groundwater protection and cave habitat enhancement issues. Educating the public about the reasons for sinkhole cleanouts and the resulting benefits to water quality can create a larger group of stakeholders who will work to support cave and karst protection goals.

Additional Reading

Fagan JH, Orndorff W. 2003. Interagency cooperative sinkhole protection and karst remediation in Virginia. In: Rea GT, editor. *Proceedings of the 15th National Cave and Karst Management Symposium: Tucson, Arizona, October 16-19, 2001*. Coronado National Forest: USDA Forest Service. p 101-103.

Indiana Karst Conservancy. *Sinkholes, Groundwater, and Other Mysteries Below Your Feet in Southern Indiana* [brochure]. Internet <http://www.caves.org/conservancy/ikc/images/sinkhole.pdf>. Accessed 2006 May 4.

Veni G, DuChene H, editors. 2001. *Living with Karst: A Fragile Foundation.* AGI Environmental Awareness Series, no 4. Alexandria (VA): American Geological Institute. 64 p. Internet <http://www.agiweb.org/environment/publications/karst.html>, <http://www.agiweb.org/environment/publications/karst.pdf >. Accessed 2006 May 4.

Virginia Department of Conservation and Recreation. 2003 [2002]. Agricultural Sinkhole Protection WQ-11. In: *Virginia Agricultural Best Management Practice Manual.* Internet <http://192.206.31.57/agbmpman/wq-11.pdf>. Accessed 2006 May 4.

Zokaites C, editor. 1997. *Living on Karst: A Reference Guide for Landowners in Limestone Regions.* Richmond (VA): Cave Conservancy of the Virginias. 26 p. Internet <http://www.dcr.virginia.gov/dnh/livingonkarst.htm>. Accessed 2006 May 4.

Section D—Restoring Speleothems

Restoration in Pellucidar

Val Hildreth-Werker and Jim C. Werker

Pellucidar is a pristine, multi-level flowstone chamber in the Western Branch of Lechuguilla Cave in New Mexico. Cave pearls, subaqueous helictites, and other delicate features were threatened by muddy corrosion residue tracked through the area. Six cavers spent four days and more than 150 caver hours cleaning the pools, pearls, and flowstone of Pellucidar.

Since it is a sensitive area, Pellucidar required very clean and careful restoration techniques. Cavers wore powder-free, nonlatex surgical gloves throughout the project. Lightweight, white-soled flowstone shoes were spot cleaned often. In the Pellucidar area, we allowed only dust-free, lintless clothes—silky-feeling polyester, nylon, or Tyvek® suits worked well.

> Cave pearls, subaqueous helictites, and other delicate features were threatened by muddy corrosion residue tracked through Pellucidar.

Restoration garments were not used for travel in the cave. Dirty knee pads were secured inside zippie bags to serve as kneeling pads. Before entering the restoration area, cavers changed clothes and cleaned their helmets and lights with wet wipes.

Brushes, sponges, sprayers, and collapsible buckets were tethered in vertical areas. For all rope work in the restoration area, the team wore restoration clothes and flowstone shoes.

Constant dripping provided plenty of water at the site. For remediation of the muddied flowstone, we drew water with large 60-milliliter (60 cc) syringes and filled new, clean Platypus® flexible bottles.

Platypus containers are easy to fold and transport in cave packs. The threads on industrial hand-held spray nozzles fit Platypus threads. Industrial sprayer mechanisms have a plastic crisscross filter that fits the bottom of the spray tube to restrain crystals and sediments from blocking the tube and nozzle.

Mud and corrosion residues from old boot tracks left silt in three pools—about 7 square meters (80 square feet) of bottom area. Jim Werker adapted a diaphragm pump for removing silt from the pools. Pool water was suctioned up with a large-volume hand pump, then filtered through soft, spongy material that was stuffed into a cut-off water bottle (clean and new, of course). The water was then circulated back into the pools.

We scooped muck, cleaned flowstone, and restored four areas covered with cave pearls—about 5 square meters (50 square feet). Clean-room foam brushes worked like magnets to attract muck from around the pearls. Stainless steel tweezers and nylon toothpicks were handy. We collected grit with sparsely bristled brushes and swept the tidbits onto small, flat, nylon pan-scraping tools made for backpacking.

Orange flagging tape was removed from the pristine flowstone. Solid-colored tape will "bleed" on wet flowstone surfaces. White-backed flagging (for example, red/white candy-striped flagging tape) leaves no stain if the solid white side is in contact with flowstone.

The work in Pellucidar was an especially satisfying restoration experience. As the photos show, a good bit of rewarding work was accomplished.

Figure 1. A diaphragm hand pump was adapted by Jim Werker for removing silt from the Pellucidar pools. (Lechuguilla's drip water had washed muddy footsteps across the flowstone and into pools where the silt settled on the bottoms.) Pool water is filtered through sponge material stuffed into a clean, cut-off water bottle. Dave Hamer is operating the pump to filter and circulate water back into the pool. (See page 12 of color section.)

Figure 2. Sponge dams are positioned downslope as Phyllis Hamer and Aimee Beveridge restore an area of flowstone and pearls in Pellucidar. (See page 12 of color section.)

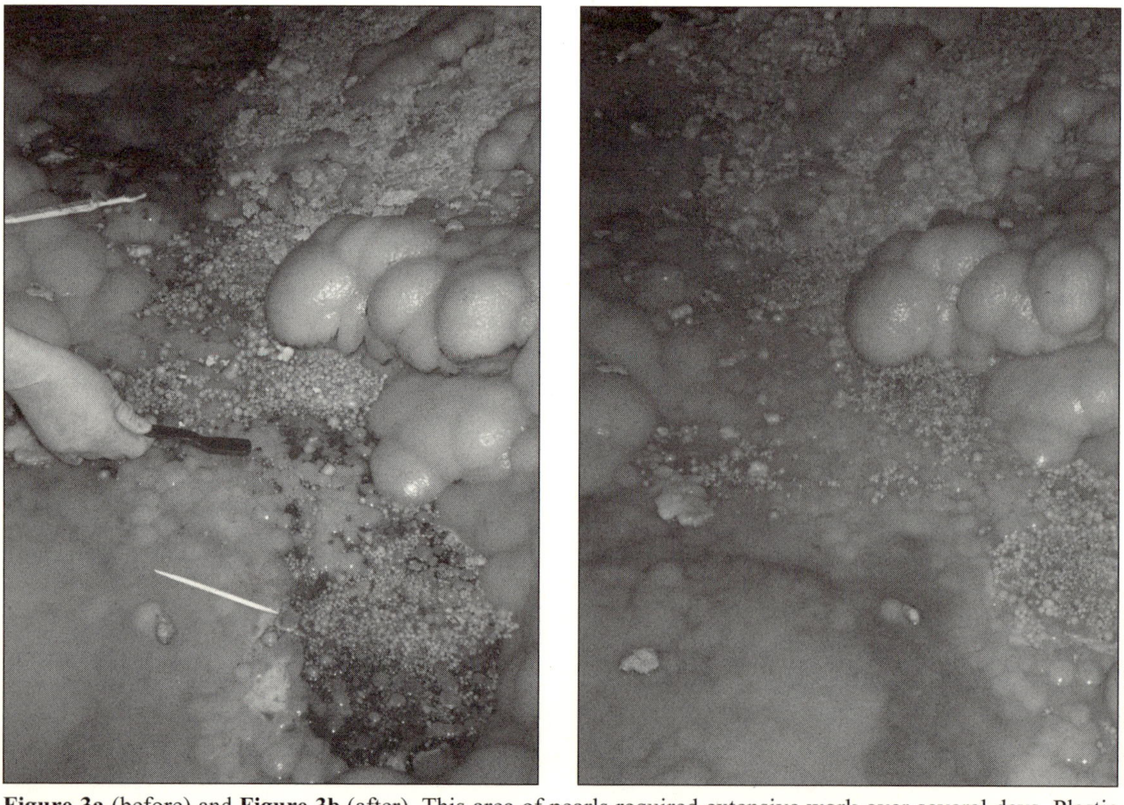

Photos © Val Hildreth-Werker

Figure 3a (before) and **Figure 3b** (after). This area of pearls required extensive work over several days. Plastic toothpicks, tweezers, and other small tools worked well for this tedious job. (See page 12 of color section.)

Figure 4. Justin Shaw is on rope rinsing his restoration sponge off into a small dry bag tethered with a cord. (See page 13 of color section.)

Figure 5 (left). The Pellucidar restoration required clean clothing and careful technique. Justin Shaw is on rope. Allan Cobb is upslope. Phyllis Hamer is working on flowstone and pearls at the top of the image. (See page 13 of color section.)

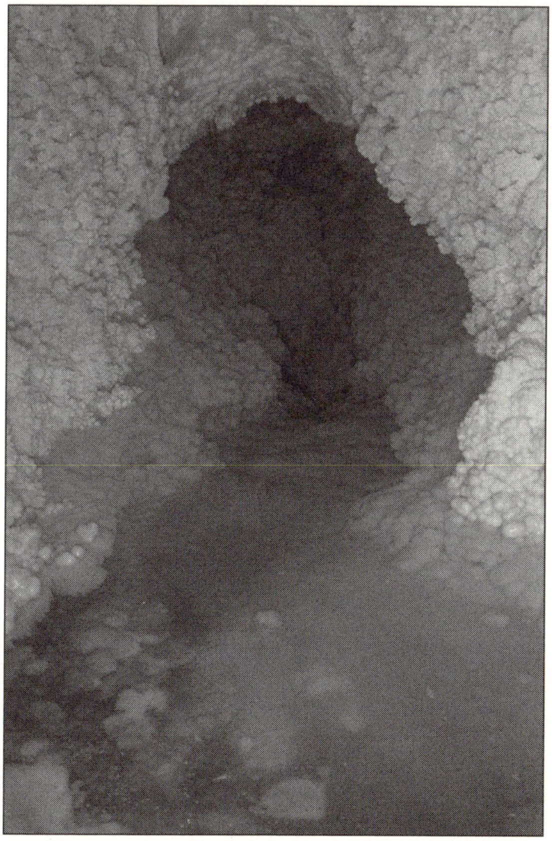

Photos © Val Hildreth-Werker

Figure 6a (before) and **Figure 6b** (after). Cavers used a large-volume diaphragm hand pump to filter silt (introduced by muddy boots) out of this Pellucidar pool in Lechuguilla Cave, New Mexico. (See page 13 of color section.)

Harms and Limits

Val Hildreth-Werker

In cave restoration, how do we reach that lofty objective of *do no harm*? First, do the reconnaissance work and look at the in-cave restoration or cleanup sites. Talk about the tasks, consult with others, bounce ideas around, and figure out how to best fix it, restore it, clean it, or remediate the damage. Establish prudent limits before starting actual hands-on project work.

During the assessment and planning process, define the spots that are ripe for harm, explore the potential limits of acceptable change, and prescribe boundaries for the restoration activities.

> How do we reach that lofty objective of *do no harm*? Establish prudent limits before starting actual hands-on project work.

Limit the objectives. Define reasonable goals and tasks to accomplish within the designated time frame. Limit the work areas to assure successful completion.

Know your own limits. Make judgment decisions regarding the achievable limits of restoration and the expertise of the crew. In some cases, doing nothing may be the best option. In other situations, a decision for no action may result in additional damage. Consult with managers or owners and speleologists before initiating any cave restoration effort.

Define limits of acceptable change. Carefully communicate what and how much. What should be left alone? How much remediation is acceptable? Define where to stop during restoration efforts. Explore new approaches, research current management practices, and find out how to apply prescribed techniques that are reasonably safe for caves before deciding how to clean or restore. Limits of acceptable change are usually established for recreational carrying capacities, but limits should also be determined for changes resulting from any human activity in caves.

Limit techniques and tools. State the objectives, make a plan, gather the tools, and if feasible, take a minimal crew of team leaders into the cave to try out the techniques before commencing larger restoration efforts. Make major project time more efficient by doing a planning trip. Train team leaders to keep an eye on specific concerns and areas prone to new damage.

Limit participation. Restoration team leaders are typically most effective when they are responsible for training and supervising only four to six cavers. Team leaders must watch out for potentially detrimental maneuvers as the cave restorers become increasingly enthusiastic or ambitious.

Limit the use of chemicals. Avoid using manufactured chemicals that are inappropriate or contrary to cave conservation standards. (See anthropogenic chemicals, page 57.)

Is It Natural or Introduced?
Val Hildreth-Werker

Before initiating cave restoration, explore these questions. Is impact caused by natural ecological processes? Or is damage human induced?

- If impact or change is a result of environmental phenomena, leave it alone and do not attempt restoration.
- If damage is obviously a result of human actions, restore it, repair it, or clean it up.

What if it is not possible to determine whether impact is caused by natural forces or by human influence? For example, when there are spots of mud splattered across otherwise pristine flowstone, did the blotches drop from the ceiling or were they tracked in by human feet?

If no clear answer is gained through observation or scientific information, there are three options for action.

- Leave the area alone and do nothing.
- Leave it alone and monitor the site to gather data for a future decision.
- Clean up a portion of the area and see if the impacting sediment, soil, or dilemma returns.

Review safety limits. Go over site-specific safety precautions with all participants.

Limit off-trail impact. It may be appropriate to limit travel routes by marking trails through the cave. Informative signs may also help direct appropriate behaviors. (See trails and signs, page 175.)

Establish compliance with limits. Some areas may be considered off limits during the restoration project. Team members should agree to stay away from designated protection or preservation zones. Establish monitoring projects to document the impact and compliance of future visitors.

Limit repeat damage. Discuss and brainstorm ideas with the restoration teams to help eliminate the need for future restoration. How can we encourage visitors to avoid repetitiously making the same mess we are cleaning up? Limit the replay of restoration.

Limit new impacts. Restoration projects provide surprisingly effective opportunities for promoting improved cave and karst stewardship. When properly promoted, a restoration project is an effective tool for educational outreach to user groups.

Section D—Restoring Speleothems

Sources for Cave Restoration Water

Val Hildreth-Werker and Penelope J. Boston

Water is the primary cleaning agent for most speleothem restoration. Many caves contain sources for restoration water within their subterranean passages—however, for some projects water must be brought in from outside sources. Where should the water come from? How can cavers determine the best water sources for specific cave projects?

This chapter proposes guidelines for choosing appropriate water sources to support cave restoration projects. Before any restoration or cleaning begins, examine the cave resources and think about how the different parts of the cave system fit together—the geology, hydrology, and biology. If scientists are actively engaged in study of the cave under consideration, consult with them. Avoid introducing chemical or microbial contaminants that might disrupt the natural ecology of a cave system.

This chapter proposes guidelines for choosing appropriate water sources to support cave restoration projects.

Common Sense Options

No single recommendation is best for all caves. There are no hard and fast rules, and common sense must ultimately lead decision processes.

In caves with active streams, restoration water may be readily available. However, if contamination results from episodic flooding, it may be harmful to transport stream water to pristine areas of the cave for restoration.

In caves where isolated pools fill over geologic time, water may be used in the immediate area if natural replenishment is sufficient. However, transporting water from isolated pools to other passages may harm or destroy localized indigenous microbial populations.

Water imported from surface streams and ponds is not recommended for cave restoration due to the abundance of chemical and microbial contaminants that would be introduced into a cave system.

The best option is clean, constantly replenished pool or drip water (when it is available). Fresh, pure cave water from a source that receives frequent drip replenishment is not likely to introduce new chemicals or microorganisms to the targeted area if the water is collected in the immediate vicinity of the restoration site.

Keep in mind—there may be unusual factors that prevent cave water from being a good source, such as leaking sewage pipes near the cave, rare or threatened species living in the proposed water source, or extreme differences in chemical or mineral composition.

When collecting water from isolated cave pools, take special care to avoid contaminating the water. Prevent the introduction of microbes from hands, clothes, boots, water bottles, and other sources.

In caves with insufficient pools or streams, capture drip water for restoration tasks. Do not abandon the collection devices and disrupt the amount of water available to the ecological community. Lack of water may adversely affect invertebrates and microbes living in the flow path.

The best option is clean, constantly replenished cave pool or drip water.

Plan methods to control, contain, and dispose of all runoff water that is generated during cave restoration projects. The categories below include restoration guidelines for choosing water sources, collecting in-cave water, and planning for disposal of the runoff.

Choosing a Water Source for Cave Restoration

We propose the following list of best practices. The list starts with sources that are generally recognized as the best options and works down to less desirable water sources. Choices near the top of the following list are not always available and decisions must be based on interdisciplinary assessments of ecosystem values.

We propose the following list of best practices. The list starts with sources that are generally recognized as the best options and works down to less desirable water sources.

When choosing a water source, determine the degree of difference between the imported water and the in-cave water. Will imported water contribute to abnormal mineralization or dissolution of calcite and related minerals at the restoration site? Is the restoration water chemically aggressive? It is wise to test potential restoration water sources for chemical composition and microbial contaminants. Water chemistry should closely resemble the *in situ* cave water, especially with respect to pH and calcium carbonate.

Recharge. Clean cave pools that receive constant recharge may provide excellent water sources for restoration.

Drips. Dripping speleothems (as well as other in-cave drip points) may produce sufficient water for restoration.

Cave Springs. Cave springs or cave streams may provide water sources for restoration. If water flows through a sinking stream, test for surface contaminants.

Surface Springs. Surface spring water is sometimes transported or pumped directly from the source to restoration sites in nearby caves.

Well Water. Well water might be used in a nearby cave for restoration.

Bottled Water. Bottled, filtered drinking water from a spring source or a city water supply has been carried into caves for restoration tasks when the water chemistry does not differ significantly from the water found in the cave.

Filtered Tap Water. Filtered tap water from a city water supply is a less desirable choice for cave restoration. Chemical additives in municipal water supplies are intended to control microbial populations. Before using tap water for restoration, it should stand in open containers for 48 hours to allow chlorine compounds to dissipate, which will partially mitigate the deleterious effects of chlorination and other chemical additives. (Tap water, if allowed to outgas, may be a good choice for restoration along heavily impacted cave routes.)

Distilled water may increase the potential for dissolution of speleothems.

Tanker Trucks. Municipalities have used tanker trucks to transport potable water to cave entrances for restoration projects. It may be necessary to allow chlorine and other outgassing chemicals to dissipate before using the water for cave restoration.

Distilled Water. Distilled water is hesitantly included toward the bottom of this priority list because it may increase the potential for dissolution of speleothems. Most distilled water is slightly acidic and undersaturated with calcium carbonate. Distilled water will be chemically aggressive on calcite. Abundant application and long contact times are not recommended if distilled water is used for cave restoration.

Deionized Water. Deionized water is similar to distilled and is not normally recommended for cave restoration. Deionized water is about pH 7 and it can be chemically aggressive against calcite speleothems. The pH for water saturated with calcium carbonate is between 8 and 8.2.

Filters. Portable backpacking filters might seem to be a promising idea—but they may clog rapidly and typically do not provide enough volume for most restoration projects.

Recycle. If restoration water is in short supply, recycle it. Allow the collected restoration runoff water to stand in containers so that sediment can settle to the bottom. Carefully remove the cleanest water from the top, strain or filter it, and reuse the water to continue restoration in the same area of the cave. (See filtering techniques, page 415.) Careful evaluation of traffic in the cave passage is important before attempting to recycle restoration water.

Contaminants. Do not import water from surface streams or ponds for cave restoration. The abundance of organisms, organics, and contaminants in surface streams and ponds preclude using these water sources for most cave restoration projects. As mentioned above, if water from a sinking stream flows through the cave, check it for surface contaminants before using for restoration.

Collecting Water from *In Situ* Cave Sources

Collect Nearby. Where possible, collect cave water for restoration in the immediate vicinity of the restoration site. Importing restoration water from a different area in the cave (even from an adjacent chamber) may introduce contaminants and new microbial populations. Scientists have found vastly differing microbial communities in pools only a few feet apart and in adjoining passages—it is important for cave restorers to avoid cross-contamination and preserve native populations.

Collect cave water for restoration in the immediate vicinity of the restoration site.

Keep it Clean. Collect cave pool water with new or sterile syringes, new basting bulbs, a siphon hose with on-off spigot, or a very clean, disinfected pitcher. Avoid reaching out over pools—remember that skin flakes and debris (organic carbons) as well as microorganisms from surface environments are constantly falling from human bodies. (See microbes, page 74 and page 276.)

Drip Sites. In caves with insufficient pools or streams, capture drip water to use for restoration. Set up buckets or funneled devices to collect restoration water between projects. Plastic or Mylar® sheeting positioned below multiple drip points may serve as adequate catchments—place funneled buckets at the low spots. Properly maintain drip collection sites and do not abandon the collection apparatus because invertebrates and microbes may be adversely impacted by reduced drip water in their habitat.

Capture and Dispose of Runoff Water

Define protocol to capture and dispose of restoration runoff water. Never allow runoff to contaminate cave features.

For each work site, define protocol to properly capture and dispose of restoration runoff water. Never allow runoff to contaminate pools, water sources, travertine surfaces, or other sensitive cave features. Use sponge dams and lint-free towels to capture runoff. Squeeze water and debris into buckets or other containers for transport and disposal. (See restoration runoff, page 339.)

Pack it Out. The best option is to transport restoration runoff out of the cave for proper disposal.

Designate Dump Spot. When disposal outside of the cave is not feasible, discard the runoff in designated safe locations in the cave that were scoped out during reconnaissance and preplanning. If runoff is not removed from the cave, it is typically disposed of in heavily traveled trails under rocks where the water can be absorbed. Take care to assure that discarded runoff is not draining into other passages that might be below a trail.

Plan Collection Strategies. The large quantities of runoff generated by pressurized spray devices require more aggressive collection strategies. Wet-dry vacuums and funneled tarps may facilitate the management of restoration runoff. Use vacuuming equipment cautiously and assure that the cave's naturally loose materials and organisms are not sucked away. (See shop-vacs, page 358.)

Summary

Use the best available source for restoration water. Base your choice on cave values, habitat, feasibility, and common sense. Collect all restoration runoff—the best option, when possible, is to dispose of runoff outside the cave. If water sources or runoff disposal options are unsuitable, postpone the restoration project.

Section D—Restoring Speleothems

Pressurized Water for Cave Restoration

Val Hildreth-Werker and Jim C. Werker

Pressurized water increases efficiency for many types of cave restoration. A little water pressure helps when cleaning muddy footprints and handholds, clearing layers of human-introduced soil, dust, or mud from trails and cave features, removing destructive contemporary graffiti, and washing away lint and muck.

Careful, informed reconnaissance and planning is most important for avoiding the pitfalls associated with pressurized methods. Direct pressure and flooding can harm biota, speleothems, minerals, paleo-remains, and cultural resources. Additional challenges are created by the abundant runoff water that results from high-pressure sprayers—all restoration runoff must be collected and properly discarded.

Plan cautiously. Direct pressure and flooding can harm biota, speleothems, minerals, paleo–remains, and cultural resources.

Do No Harm

Pressurized water may harm or destroy natural and cultural resources. First, carefully inspect the proposed restoration site and determine whether cave values will be put at risk by introducing pressure sprays. It is always wise to ask experts to assist in the preliminary evaluation phase of restoration projects.

Prudently determine the level of water pressure that is necessary for various tasks. The risk of damaging speleothems and cave surfaces increases exponentially as water pressure is increased. Exercise caution to avoid harming fragile speleothems, splattering nearby cave features, or destroying cave life with streams of high-pressure water and flooding. Define safe limits for each cave feature.

Wherever possible, water from within the cave is used for restoration because water imported from the surface may be chemically aggressive and cause damage to speleothems. Cave restorationists, speleologists, and cave managers are leaning toward lower, gentler pressure sprays while searching for less harmful methods. (See water sprayers, pages 337–338 and page 398.)

Initial studies at the Jenolan Caves, Australia, indicated that pressurized water was safe for cleaning durable speleothems (Bonwick and others 1986). Later, investigation with scanning electron microscopy found that the cave surfaces were suffering some damage (Spate and Moses 1993).

Be conservative. Use best practice guidelines:

- Minimize washing.
- Use low-pressure water.
- Control the runoff.

Figure 1. Industrial spray nozzles fit Platypus® flexible water bottles and many commercial drinking-water bottles. These products are easy to carry and durable enough for many cave restoration applications.

Water-Spray Devices

The following paragraphs start with the most gentle, reliable spray devices. The list progresses through apparatus that offer increased pressure and require greater skill and care to avoid harming cave values.

Hand-Held Spray Bottles. Industrial-quality hand-held spray bottles (available in packages of three at large home improvement centers) usually have a plastic crisscross screen filter that fits on the bottom of the suction tube and restrains crystals and sediments from blocking the spray mechanism. Threads on the industrial spray nozzles conveniently fit the top threads of Platypus® flexible canteens and some plastic bottled water containers.

Manual Pump Bottles. Increased pressure is achieved with spray bottles that have a small, manual pressurizing pump—available through industrial catalogues.

Large Pump Containers. Garden sprayers, up to 20-liter (5-gallon) capacity, should be designated for cave restoration only. Sprayers should be clean, free of chemical residues, and never used in the garden. Manually pump pressure into the container to create a mid-force pressurized spray.

Backpack Sprayers. Backpack sprayers or bladder bags provide moderate pressures and high volumes. Some are designed for fire fighting, and water squirts through a hand-held trombone pump attached to the bottom of the bag. Others are designed for various applications. (See caarbon dioxide powered sprayer, page 339.)

Gravity and Hoses. Gravity fed systems use a hose and siphon system originating from a higher elevation. Depending on the gradient, a substantial head of water may be produced that could cause considerable damage in caves. Watch for leaking couplings and gaskets on hoses that could cause damage to cave surfaces. Thorough monitoring of gravity fed systems is important for cave restoration projects.

Pipes and Hoses. If water pipes are installed in the passages of a show cave, hoses with spray nozzles may be appropriate for some restoration tasks. If the water is from a chlorinated source, consider carefully whether it will be used directly from the hose, or if logistics will include first allowing the chlorine to dissipate. (See water sources for restoration, page 394.)

Truck Transport. Pressure spray systems may use water that is transported to a cave entrance by a tanker truck. It may be necessary to allow chlorine and other outgassing chemicals to dissipate before using the water for cave restoration. Cavers have organized the logistics for community water trucks to deliver municipal water to cave entrances and to drop hoses down into cave passages through preexisting drill holes on the surface.

Runoff Control. Plan and test methods to capture, remove, and dispose of restoration runoff water. (See restoration runoff, page 396.)

Photos © Val Hildreth-Werker

Figure 2. Use pressurized sprayers to achieve a more efficient spray. When pressurized water is applied in cave restoration, cautiously test small areas to avoid damaging cave resources. Manual-pump sprayers are available in a variety of sizes. (See page 14 in color section.)

Figure 3. One caver is holding a sponge under the area while spraying with a pressurized bottle. Another caver has the bucket ready to capture runoff. Always be prepared to catch runoff when using any type of pressurized water for cave restoration.

Plan Carefully

Before initiating pressure-spray projects, always do a reconnaissance and decision-making trip into the cave. Carefully inspect the cave for biological, mineralogical, and cultural resources that might be destroyed by restoration efforts.

If pressurized water is deemed appropriate, take the time to plan well-considered protocols and equipment to collect and remove restoration water runoff. When using pressure sprays, constantly monitor the pressure. Always use the minimum pressure necessary for the task and avoid damaging cave features.

If pressurized water is deemed appropriate, take the time to plan well-considered protocols and equipment to collect and remove the restoration water runoff.

Cited References

Bonwick J, Ellis R, Bonwick M. 1986. Cleaning, restoration, and redevelopment of show caves in Australia. In: *Proceedings Ninth International Congress of Speleology*, vol 2.

Spate A, Moses C. 1994. Impacts of high pressure cleaning: A case study at Jenolan. In: Henderson K, Bell P, editors. 1994. *Cave Management in Australasia X*. Carlton South, Victoria (AU): ACKMA. p 45-48. [*Proceedings of the Tenth ACKMA Conference, Rockhampton, Queensland*. ACKMA 10:45-48.]

Section D—Restoring Speleothems

Flowstone and Dripstone Restoration

Val Hildreth-Werker and Jim C. Werker

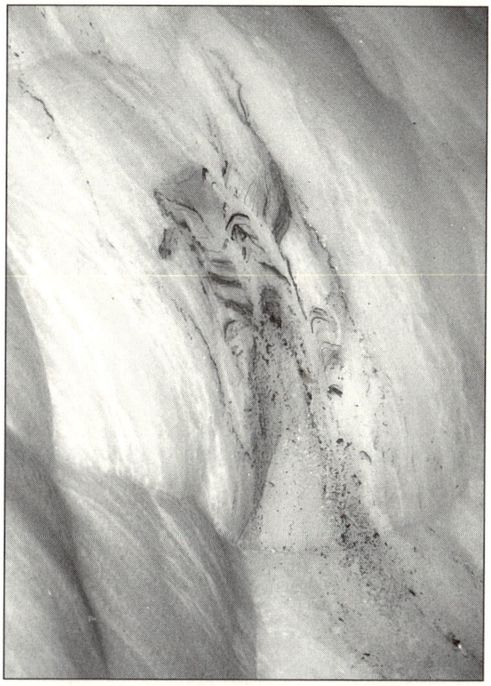

Figure 1. A muddy boot print mars white flowstone. Remove muddy marks from flowstone before a new layer of calcite covers and makes the imprints permanent.

It is easy to mar cave surfaces and leave unsightly scars on speleothems. With wise planning and careful execution, some cave features can be restored to resemble a former natural condition.

Muddy tracks, boot marks, hand prints, dust, lint, carbide, muck, scrapes, and other human impacts can cause discoloration and interruption of natural speleothem growth patterns. Ideally, restoration is performed before new calcite deposition seals these human impacts into speleothems. Some cave passages can be restored to display former aesthetic beauty. Some caves can be restored to a former ecological state.

This section addresses tools and techniques for cleaning and restoring durable speleothems. The general methods described here work well in most cave passages. Techniques focus on minimizing disturbance to cave-dwelling species and maintaining the natural ecological balance. Also described are specialized techniques for the restoration and preservation of pristine areas with emphasis on minimizing human disruption of the amazing diversity of subsurface microbes.

Always consider how cave life and other cave values may be impacted by restoration. Check with speleologists, scientists, and cave managers to define priorities and achieve balance between aesthetics, cultural significance, speleothem growth, microbial preservation, biological concerns, and mineralogical issues.

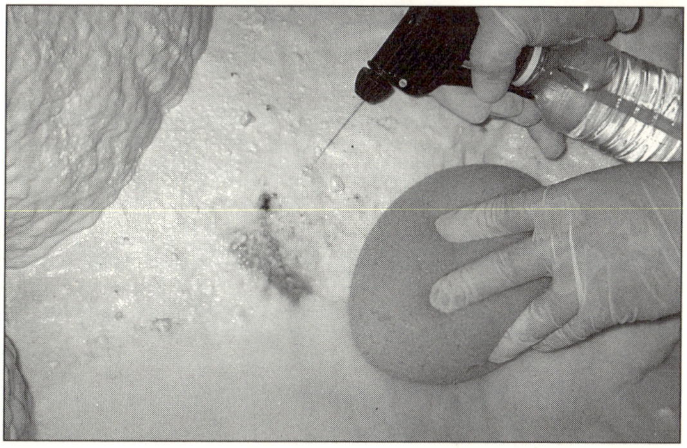

Photos © Val Hildreth-Werker

Figure 2. Use water in a spray bottle and a soft, closed-cell sponge to blot up muddy prints from flowstone surfaces. Wear nonlatex, powder-free gloves.

Flowstone and Dripstone

Spray and sponge. Gentle cleaning with water and blotting with a fresh sponge does not appear to harm most flowstone. It is always a good idea to try test spots first.

Use hand-held spray bottles and a soft sponge on fragile flowstone. Avoid wiping sponges across flowstone surfaces because the sponges will deteriorate rapidly and particles will be distributed in the cave (Figure 2).

Photos © Val Hildreth-Werker

Soft Brushes. Light brush strokes, sprayers, and sponges aid the restoration of muddied textured areas and "bathtub rings" on flowstone surfaces. Use a brush on flowstone only when necessary and make certain the bristles are not marring the cave surface by making dents, grooves, or scratches. Longer, relatively sparsely distributed nylon or plastic bristles tend to work best for most restoration (Figure 3).

Figure 3. Use light brush strokes, water sprayers, and sponges to clean textured flowstone surfaces. Be careful to test tools and avoid damaging soft surfaces.

Figure 4. Pick up bits of aragonite and broken pieces of cave popcorn to avoid continued crushing and grinding into flowstone surfaces.

Remove Chunks. In areas with fragmented aragonite, crunched popcorn, or other broken speleothems, the first task is to pick up the larger chunks that have not calcited over so crushing and grinding damage does not continue. Some breakage may be repaired. However, many pieces will likely be irreparable and should be gathered and stored in a relatively safe location in the cave, away from traffic, and near the area where they were retrieved (Figure 4).

Sediment. Layers of human-introduced mud and sediment require gentle digging, scraping, mopping, scrubbing, spraying, and sponging. (See Cave Without a Name photos, page 295.)

Spray and Runoff. Pressurized spray devices are sometimes appropriate for cave restoration—however, the large quantity of runoff water is difficult to capture and remove. Consider whether cave biota will be heavily impacted by direct pressure and flooding. (See pressurized water, page 397–399.)

Small Tools. Nylon toothpicks, stainless steel tweezers, or plastic dental tools will dislodge muck from small declivities (Figure 5).

Clean Clothes. For restoration in pristine flowstone-covered rooms, clean, lint-free clothing is recommended (Figure 6). Silky feeling polyester, nylon, or Tyvek suits work well. (See clean clothing, page 427; also see flowstone shoes, page 431 and surgical gloves, page 433.)

Capture Runoff. Capture all the restoration runoff water with sponge dams, lint-free towels, wet-dry vacuums, funneled tarps with buckets, and other catchment solutions (Figure 7). Squeeze water and debris from sponges and towels into buckets or other containers. Give special attention to planning the protocol for disposing of restoration runoff water. (See runoff, page 396.)

Collect Grit. As a final step, it may be necessary to rinse and sponge the area. Or it may be useful to perform a final light sweep to collect grit and particles that sponges miss (Figure 8).

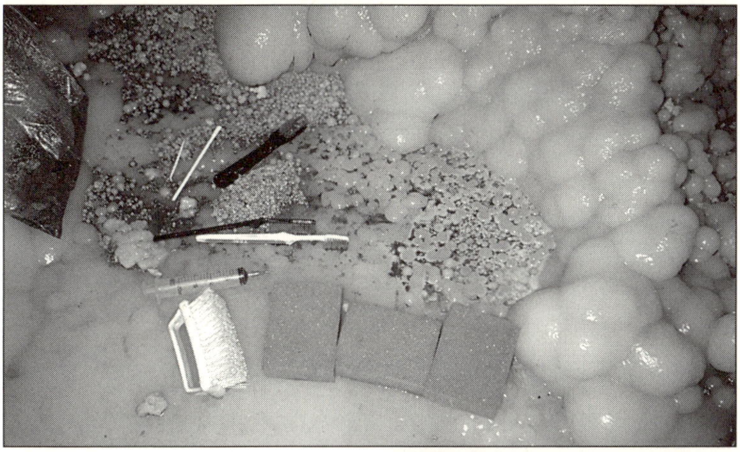

Figure 5. Flowstone restoration tools include sponges, spray bottles, brushes, nylon picks, tweezers, syringes, zip-closure bags, and sponge dams.

Figure 6. Cavers wear clean, lint-free polyester clothing (silky feeling fabrics attract less dirt), flowstone shoes, and surgical or exam gloves for restoration in pristine areas.

Figure 7. (below) Sponge dams help to stop and contain restoration runoff water.

Figure 8. (below) Wear nonlatex, powder-free exam gloves. Use a flat, nylon pan-scraping tool and a brush to collect final bits of grit left on flowstone surfaces.

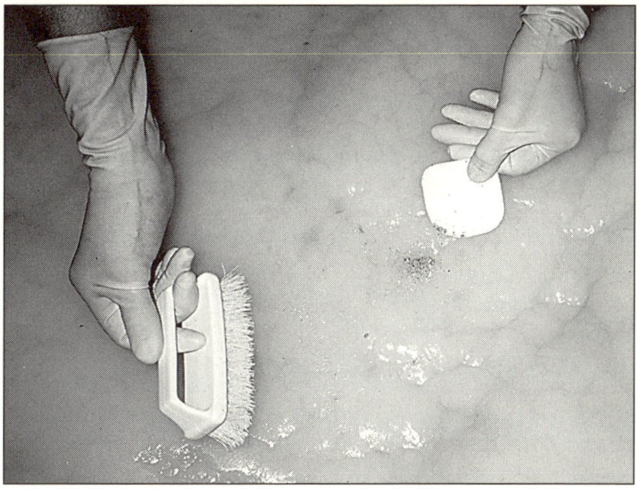

Tips for Flowstone and Dripstone

Sponge Prep. Thoroughly rinse sponges with fresh water before using for cave restoration. Some sponges are saturated with a soapy, softening, lubricating agent during the manufacturing process.

Water Rinse. Rinse and disinfect all supplies between uses. Conservative disinfecting is achieved by washing restoration supplies in fresh tap water, then allowing them to dry in the sun because ultraviolet rays kill many microorganisms.

Chemical Sterilization. Durable tools may be soaked in alcohol or chlorine bleach. Contact time is essential. Thorough rinsing is extremely important. Disinfecting agents must be thoroughly removed before taking supplies back into a cave—residual disinfectants can kill cave biota. Check with cave biologists before trying new ideas or products. (See anthropogenic chemicals, page 57; sterilizing and disinfecting, page 77.)

Use New. In pristine or isolated areas of some caves, spelean restorationists are using fresh, new, and sometimes sterilized supplies.

Autoclave. When extreme measures are required to avoid disturbing microbiota in sensitive areas, supplies can be autoclaved or steamed in a double boiler between uses. (See sterilizing, page 77.)

Pure Water. Use only clean, pure water for cave restoration, as advocated throughout this book. Do not use acids, manufactured cleaning agents, biodegradable products, or anthropogenic chemicals in cave systems. (See anthropogenic chemicals, page 57; also see water sources for restoration, page 393.)

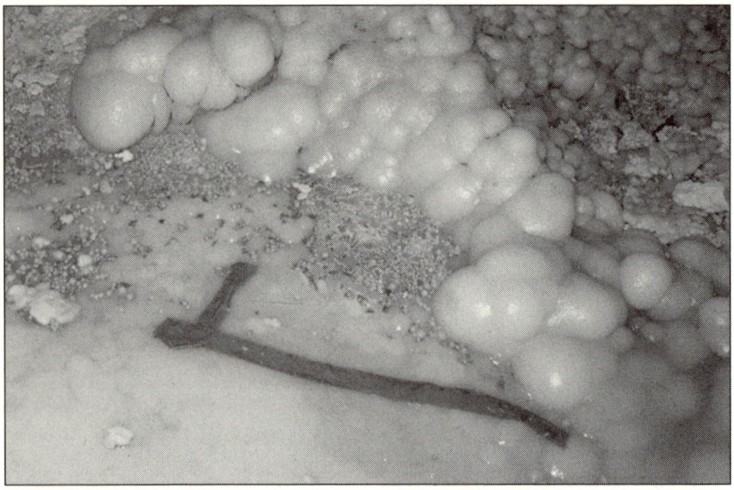

Figure 9. Solid colored flagging tape, such as this orange tape, will discolor *wet* flowstone. Use white-backed flagging instead. (See page 14 in color section.)

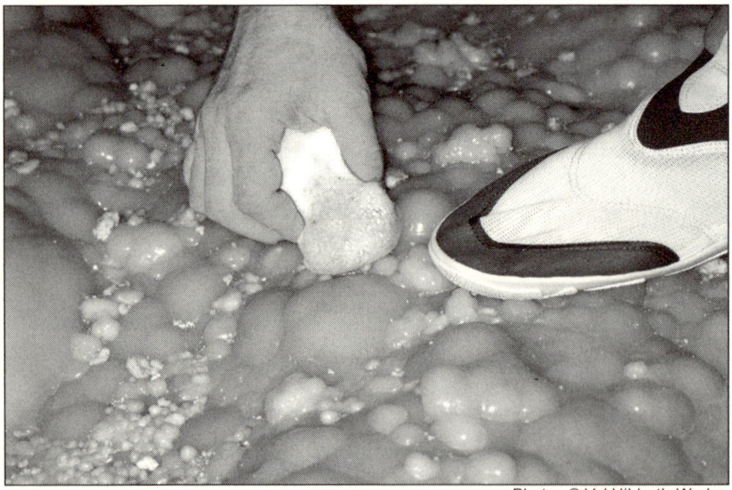

Photos © Val Hildreth-Werker

Figure 10. Orange flagging tape faded onto this wet flowstone. Fortunately, the discoloration was not incorporated into the calcite and the orange color was easily sponged away. (See page 14 in color section.)

Check pH. Acidic water is not used for typical flowstone restoration. It is wise to check the pH and avoid water that may be too aggressive on calcite surfaces. (See restoration water, page 395.)

Flagging Tape. Solid-colored flagging tape should not come in contact with wet flowstone surfaces because the colors bleed. When color from the flagging remains on flowstone, spray with water, blot the color away with a sponge, and use light brush strokes if necessary. Color bleeding can usually

Figure 11. To avoid color bleeding, use white-backed flagging on wet flowstone surfaces. Surveyors' flagging tape is best secured on small natural protrusions. Red/white striped flagging may indicate delicate zones outside of the trail-marking tape.

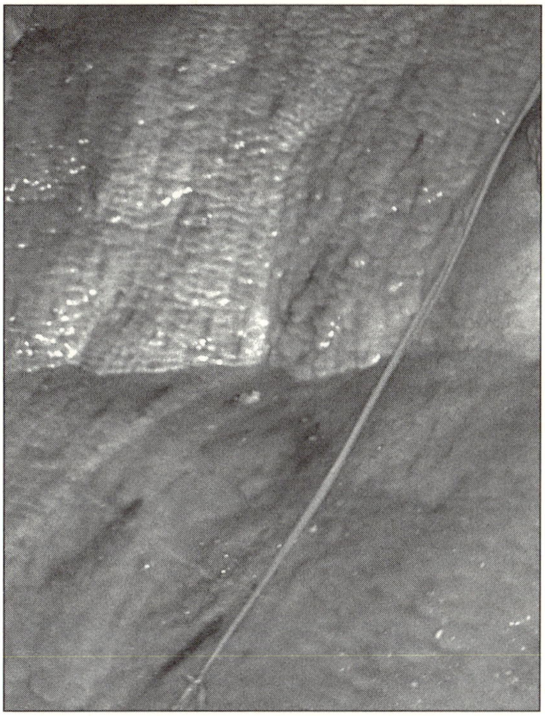

Figure 12. Black scuff marks made by boot soles can permanently disfigure flowstone. Lugged hiking soles made of traditional Vibram® left scars on this flowstone. Some contemporary soles are nonmarking. Test soles by striking across limestone or untreated concrete.

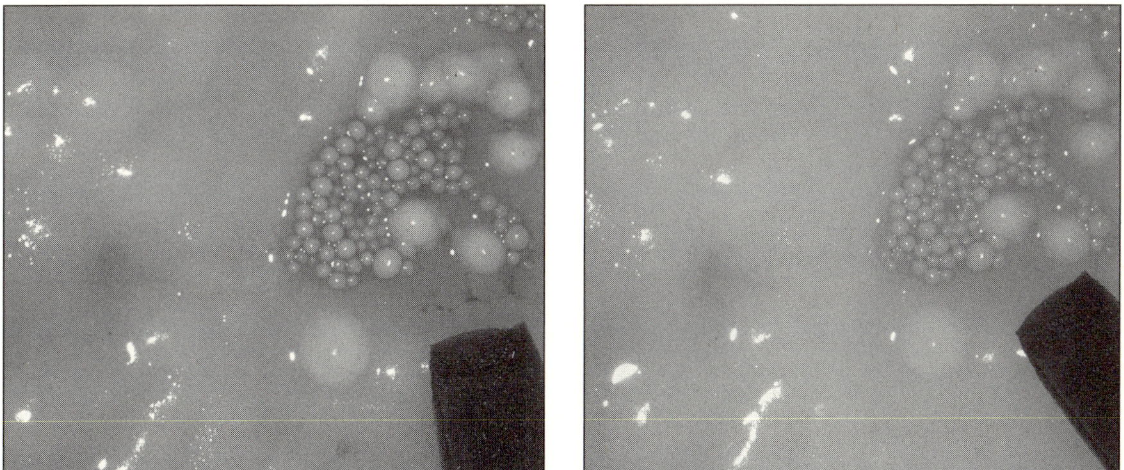

Photos © Val Hildreth-Werker

Figure 13a (before, at left) and **Figure 13b** (after, at right). The soft spongy material of a clean-room foam brush attracts soil particles and muck like a magnet. Plastic handles are preferred for cave restoration.

Photo © Val Hildreth-Werker

Figure 14. Transitions are best accomplished by stepping out of boots and into flowstone shoes already positioned in the clean area (or vice-versa).

be cleaned up unless the color has calcited into the flowstone surface (Figures 9 and 10). White-backed flagging does not bleed—for example, red/white candy-striped flagging can be used with the white side placed directly on flowstone surfaces and neatly secured on natural protrusions (Figure 11).

Soft Scrubbers. Black scuff marks, if not yet calcited into flowstone, can often be removed with an extremely soft sponge scrubby.

- Some restaurant-grade glassware sponges have a very soft scrubbing surface laminated on one side of ultra soft, tight-celled, foam material.
- Harsher dish scrubbers (marketed for nonstick pans) usually fail to remove scuffmarks and will scratch most flowstone surfaces.
- Flowstone may be irreparably damaged if cavers try to scrape calcited boot marks off.
- Eventually, calcite layers camouflage the marks—some irritating boot scuffs are best left alone (Figure 12).

Foam Brushes. Foam brushes with plastic handles are manufactured as clean-room supplies and for painting applications. The soft foam attracts muck from small declivities and speleothems (Figure 13a).

Failed Attempts. In attempt to avoid repeat restoration sessions, innovative techniques have been tried in some cave passages. *Be forewarned, these ideas have been abandoned*—rubber mats, long-term plastic sheeting, foot-bath stations—all these create extra problems, tend to become havens for fungi, and simply don't provide the expected benefit of protecting pristine flowstone areas. Simply changing from caving boots to flowstone shoes works much better (Figure 14). (See cave-friendly footwear, page 431.)

Restricted Access. Passages that are reclaimed or restored to a near pristine appearance may be marked off-limits or may be designated for flowstone shoes and surgical gloves. (See signs, page 183–185.)

Restoration Kit. To clean a solitary imprint of misplaced mud left on a speleothem, a little water and a small sponge may be adequate. Science, photo, and exploration teams sometimes carry a small restoration kit for correcting inadvertent impacts. (See compact restoration kit, page 212.)

Tools and Supplies for Common Cave Restoration Tasks

This is a general supply list that fits most cave projects involving flowstone or dripstone restoration. Larger, bulky items may be suitable for show caves. However, small, light, compact equipment is more adaptable to travel in wild caves. Only field-tested suggestions are included below.

Very Soft Sponges. Car wash sponges are adequate. Big, blue semi-tight-celled car washing sponges are designed to stay soft and absorbent, plus, the blue pieces are easier to pick up when sponges begin to deteriorate. The softest sponges do a better job of capturing grit and sediment—tight-celled sponges are more absorbent and act like muck magnets. A supply of new sponges should be available for every restoration trip.

Surgical Gloves. Powder-free, nonlatex surgical gloves are recommended for most cave restoration tasks. (See gloves, page 433–435.)

Hand-Held Spray Bottles. Threads on industrial hand-held spray nozzles fit the threads of flexible Platypus® bottles and some bottled drinking water containers. The bottom tube of the sprayer mechanism has a plastic crisscross filter that restrains crystals and sediments from blocking the tube and nozzle.

Garden Sprayers. Purchase one- to five-gallon capacity, new, clean, and free of chemical residues. Manually pump pressure into the container to create a mid-force pressurized spray. (For other spray devices see pressurized water, page 397–399.)

Buckets. For tedious travel in wild caves, use collapsible or folding buckets with handles.

Extra Sponges or Lint-Free Towels. For creating dams to catch restoration runoff water.

Brushes. Palm-fitting, hand-sized upholstery brushes with raised handles to prevent scraping knuckles against flowstone. Found in auto supply outlets.

Foam Brushes. Clean-room foam pads (or foam paint brushes on plastic handles). The soft foam attracts muck from small declivities and speleothems (Figure 13a).

Plastic Scrapers. Collect grit with sparsely bristled brushes (nylon bristles preferred) and sweep the tidbits onto flat, nylon pan-scraping tools made for backpacking.

Small Tools for Detail Work.
- Fresh, new toothbrushes.
- Soft bristled paint brushes.
- Whisk brooms with plastic or nylon bristles.
- Nylon toothpicks, hard plastic picks, and dental tools.
- Variety of tweezers (plastic and stainless steel).

Turkey Baster. Draw water with a new, clean turkey baster or a big 60-cc syringe and fill spray bottles.

Zippie Bags. Plastic zip-closure freezer bags, especially the one-gallon and two-gallon sizes.

Flagging Tape. Surveyor's flagging tape for trails and for special areas. On flowstone, use flagging tape that is white on one side (to prevent color bleeding.) Red and white striped is often used to indicate delicate or pristine areas. (See trails and flagging, pages 178–183 and page 404.)

Section D—Restoring Speleothems

Footstep Ethic

Val Hildreth-Werker

Leave nothing but careful footprints on established trails.

During the last decade of the 20th century, there was a gradual change in traditional caving ethics. In the past, when we found footsteps going into pristine territory, it was okay to put our boots right inside those imprints and follow the footprints to retrace the path.

But, as more people followed, too many feet quickly expanded the footstep routes into broad trails. The footstep ethic had to change. Footsteps should not be left to invite others to follow. Today, rather than retrace steps, we gently erase them. An off-trail footstep signals the need for restoration.

Figure 1. To limit new impacts, erase inappropriate footprints that lead off trail. Leftover footprints invite others to follow. Avoid making imprints—use proper footwear and stay on durable surfaces.

Erasing Footprints

- Remove individual muddy boot prints from flowstone with a sponge and spray bottle.

- Remove broadly impacting tracks with more aggressive flowstone restoration techniques. (See pressurized water, page 397; also see flowstone techniques, page 401.)

- Black scuff marks can sometimes be removed from smooth flowstone with the extremely soft scrubbing side of a restaurant grade sponge. (See scuff marks in flowstone tips, page 406.)

- On softer cave surfaces and cave soils, lightly comb away individual imprints with a small, nylon-bristled whisk broom. Take care not to stir up dust. (See trail maintenance, page 183.)

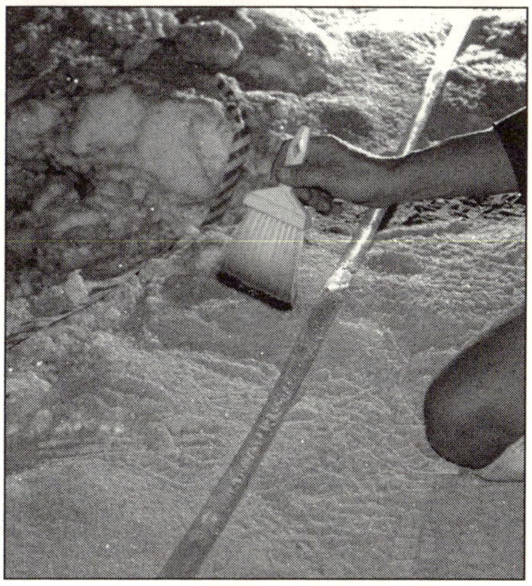

Figure 2. Use a lightweight whisk broom with nylon bristles to gently remove off-trail footsteps left in cave soils.

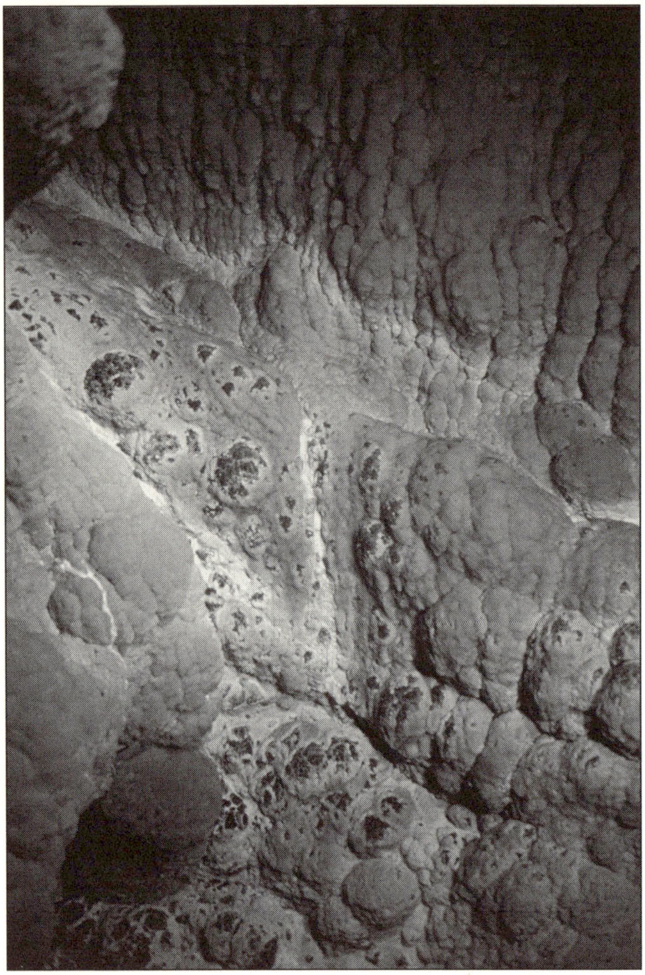

Figure 3. Two sets of boots created these scuff marks in a narrow passage covered with cave velvet. Two pairs of feet did this irreparable damage in what was virgin passage only minutes before. This photo was made in 1993. Today, ethics have changed and cavers typically use nonmarring flowstone shoes to survey and explore delicate virgin chambers.

Section D—Restoring Speleothems

Carbide and Soot

George Veni

Soot is among the easiest stains to clean from speleothems and cave walls. In the past, cave visitors in some regions typically made carbide soot marks on cave walls with their lamp flames. Be especially aware of avoiding harm to markings that have historical significance. Call in experts in archaeology and historic culture to help evaluate cave sites. (See the chapter on historic markings, page 99; also see contact experts, page 116 and page 334.)

Carbide Marks

If removal of carbide marks or arrows is deemed appropriate, slightly pressurized water will readily remove it. Carbide soot marks left on cave surfaces can usually be squirted with a spray bottle and they will disappear with little or no scrubbing. If gentle brushing is required, use nothing more than soft nylon bristles.

© Val Hildreth-Werker

Figure 1. Although not commonly used today, carbide lamps were once the mainstay of caver lighting. Take special care and wear gloves when removing old carbide dumps. Conscientious carbide cavers use a threaded plastic bottle to carry spent carbide out. Never dump spent carbide in a cave.

Spent Carbide

Removal of spent carbide requires more care. Wear gloves to avoid caustic burns. First remove, lift, or sweep up as much carbide as possible. Chisel away old carbide deposits that have formed hardened crusts. Be careful—a crust may seal the interior carbide, which can still be reactive after as much as 35 years (Veni 1994).

- Do not use a vacuum cleaner, even with a fine filter, because carbide dust may enter and corrode the motor and be ejected into the cave.
- If the carbide sits on sediment, scoop up and remove the sediment since it will be impregnated with the carbide—remove all sediment that smells of carbide.

Where carbide rests on rocks or speleothems, after sweeping or scooping up the majority, wash away any remaining carbide by isolating small areas for cleaning and dousing them one at a time with a relatively large volume of water to dilute the exothermic reaction and minimize corrosion of the underlying surface. Place sponges, towels, and other equipment immediately downslope to capture the runoff, which should be removed from the cave and properly disposed of along with the carbide and any contaminated sediment (Veni 1997).

Cited References

Veni G. 1994. *The 3rd Annual Caverns of Sonora Restoration Project*. The Texas Caver 39(1):3-4.

Veni G. 1997. Speleothems: Preservation, display, and restoration. In: Hill C, Forti P, editors. *Cave Minerals of the World*. 2nd edition. Hunstville (AL): National Speleological Society. p 301-309.

Section D—Restoring Speleothems

Mold Kits

Val Hildreth-Werker

In some caves, particularly environments with warmer temperatures, mold tends to grow on dropped food particles, candle wax, and any other debris left by careless cavers. The best practice is to clean it up when you find it. Otherwise, fungal spores may spread to other areas within a cave system.

The components of a simple, lightweight mold cleanup kit are listed here. Pack these items in a small zippie bag:

- Two prepackaged antimicrobial wipes
- One surgical glove or a disposable food service glove
- Surgical mask to protect the caver from breathing fungal spores (a very important item to help avoid uncomfortable, lengthy allergic reactions after cleaning up spots of mold)

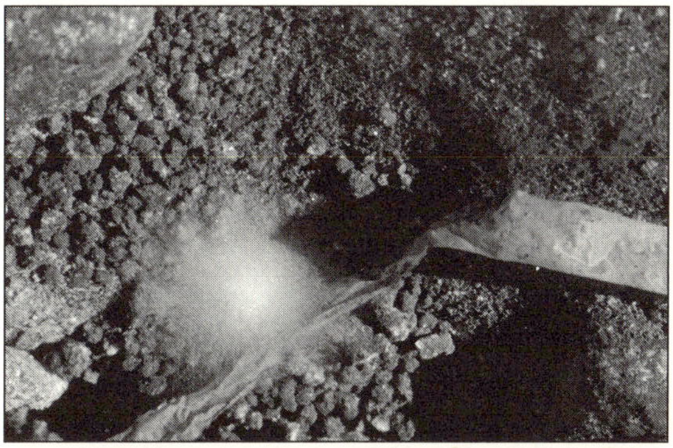

Figure 1. Dropped food particles provided the nutrients to fuel this ball of white fungi found in Lechuguilla Cave.

Gently Gather Fungi

On encountering a mold splotch in a cave passage, don the surgical mask and the glove from the mold kit. Open an antimicrobial wipe and gently lay it over the mold. Take care to avoid dispersing spores. (So-called antibacterial wipes may also provide antifungal benefit.)

With the gloved hand, use the wet-wipe to gather up the mold and some of the underlying substrate. Hold the bundled wipe, pull the glove off and draw it over the bundle in one motion to contain the entire unit and minimize distribution of spores.

Tie off the opening of the glove and secure it in a zip-closure bag (several zippies may be more secure).

Remove the packaged fungi from the cave and dispose of it promptly.

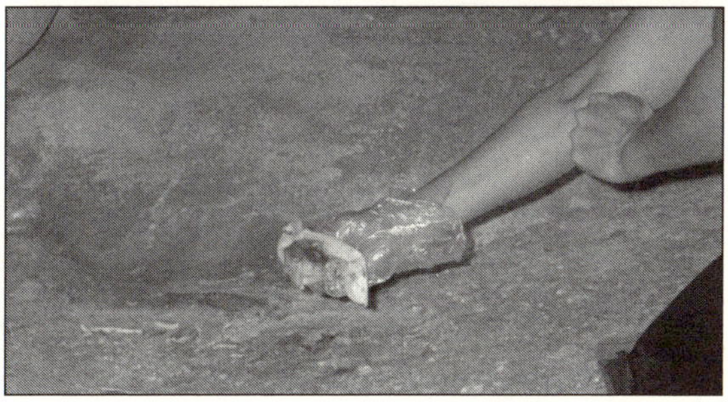

Photos © Val Hildreth-Werker

Figure 2. Gently lay an antimicrobial wipe over the mold patch to avoid distributing spores. Use a light disposable glove and invert it over the wipe after scooping up the patch of fungi. To protect against breathing mold spores, wear a surgical mask when cleaning up fungi. (See page 10 in color section.)

Avoid Dispersing Spores

Do not use any kind of spray on mold found in caves. Spray devices will only serve to disperse fungal spores. Spraying diluted bleach or any other substance potentially may spread the problem and extend the fungal bloom. Instead, gently gather the offending blotch with antimicrobial wipes, then package it and dispose of it properly.

In some cave environments, dropped food particles or candle wax left on cave surfaces will provide nutrient sources for fungi. We recommend using heavy foil or some other catchment under candles and eating over a plastic bag to catch crumbs before they hit the cave floor (Figure 4).

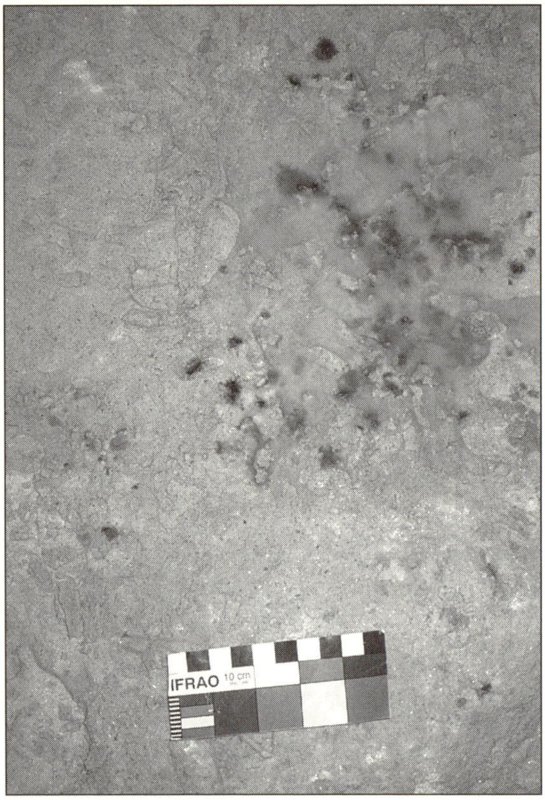

Figure 3. In some cave environments, colorful blotches of mold will grow on spilled food. Each little crumb will blossum into a fuzzy colony. (See page 10 in color section.)

Photos © Val Hildreth-Werker

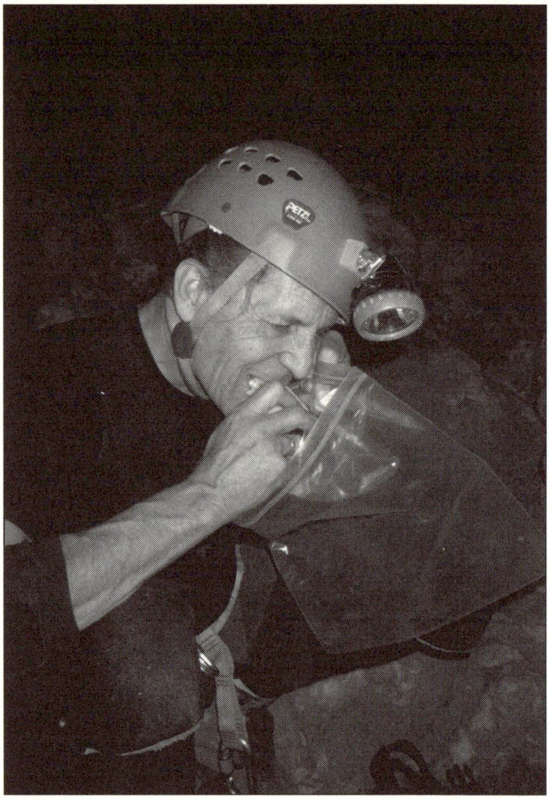

Figure 4. Cavers should eat over plastic bags to prevent scattering food particles. Packing out crumbs deprives opportunistic organisms of a free lunch.

Section D—Restoring Speleothems

Water Filtration and Cave Pool Restoration

Val Hildreth-Werker and Jim C. Werker

Simple filtering devices can help conserve cave water. For light flowstone restoration, filter the recovered runoff water and, if appropriate, use it again for more flowstone cleanup in the same area.

Filtering methods may also be used to restore silted cave pools. Muddy footprints sometimes wash into pretty drip pools alongside cave routes, which results in human-introduced silt covering pool bottoms. Muddy tracks can be pervasive—the resulting sediment and debris is unsightly and may be disruptive to native cave organisms. Water and sediment can be pumped out, and silt can be caught in a filtering device as the water is returned to its source. Field-tested techniques are described in this chapter.

Filtering Restoration Water

When restoration water is in short supply, simple recycling filters are put to work. Collect restoration runoff and debris with sponges or towels. Squeeze the collected runoff into containers, and allow time for the larger particles to settle.

If the water is not clear, strain it through the simple foam filter described below. Sediment will collect around the outer edges of the foam filter. If appropriate, use the clean water that passes through for another round of travertine restoration in the same area.

Always use filtered restoration water in the same area it was used the first time—never transport it to other sections within a cave.

Figure 1. To recycle cave restoration water, allow particles to settle out and then pour the cleanest water through a simple filtering device.

© Val Hildreth-Werker

Tips for Filtering

Place used restoration water in a bucket and allow the particles to settle to the bottom. Skim clean water off the top. If necessary, filter the water to reuse for restoration.

Settle. Allow particles and sediment from recovered restoration water to settle in buckets, plastic bottles, or zip-closure plastic bags propped upright.

Siphon. Dip or siphon the cleanest water off the top. Use a clean, disinfected turkey baster or 60-milliliter (60 cc) syringe to retrieve the clean water.

Strain. Strain or filter cloudy water for reuse in the same restoration area. Be sure to remove any flecks of paint, lint, carbide, and other human-introduced debris.

© Val Hildreth-Werker

Figure 2. One caver can operate a small hand-held pump while directing the water through a simple filtering device and back into the pool. The filter is made from a clean, inverted water bottle that is stuffed with soft sponge material.

Make a Simple Filter

For restoration in remote locations, water may need to trickle through a simple filter made of foam material compressed into a clean water bottle.

Step 1. Remove the bottom of a 20-ounce, 32-ounce, or 2-liter bottled-water container. (Water bottles with smooth, gently sloping top sections make the best funnel filters.)

Step 2. Remove the screw-top lid and stuff the bottle three-quarters full with very soft sponge or foam material. (Avoid treated foam products since they may add unwanted chemicals to the cave system.)

Step 3. Turned upside-down, the stuffed bottle performs as a filtering funnel.

Step 4. Silt particles and debris stop along the interface between the foam and the bottle. Clean water passes through the foam and is caught in a container to use again as recycled restoration water.

Pumping and Filtering Cave Pools

Is the sediment natural or is it introduced by humans? Silt caused by natural hydrological processes is usually left as is and no alteration is attempted. However, mud, silt, and sediment sometimes enter cave pools as a direct result of human activity—produced either by surface activity or by in-cave actions.

Due to human foot travel or activity, silt may collect on the bottoms of cave pools near trails. Removing water from puddles or pools is sometimes appropriate to mitigate sediment left by caver boots, to remediate visible mudflow caused by human activity, or to clean up debris introduced by other anthropogenic processes.

A small hand-held suction pump is sufficient for vacuuming undesirable silt out of puddles and small pools. A large-volume diaphragm pump may be more efficient for bigger pools that are marred by human-introduced sediments. As described above, a sponge stuffed into a cut-off water bottle usually provides sufficient filtration to improve the appearance. If cave biologists and management deem it appropriate, silt is filtered out and the water is circulated back into its source pool.

Figure 3. With this pool filtering device, a pair of cavers can work in tandem to direct the hose and operate the diaphragm pump. Pool water flows through a filter that captures particles and allows silt-free water to return to the pool. Both before and after are apparent in this photo. The dark area (top left) has not yet been cleaned.

© Val Hildreth-Werker

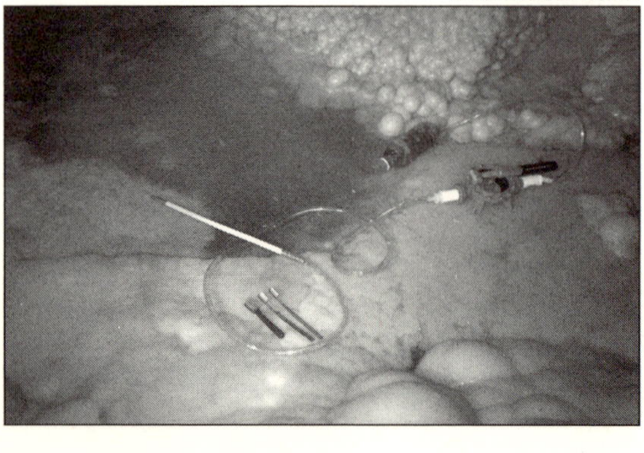

Tips for Restoring Cave Pools

Bathtub Rings. Use a sponge to remove bathtub rings that form along cave pool edges. The softest, most absorbent sponges work like dirt magnets when gently placed near the rings of sediment.

Dirt Magnets. Sponges made for paint application or clean-room sponge brushes also work as dirt magnets. For pristine cave applications, plastic handles may be better than wood.

Small Pools. For smaller pools and puddles, a

hand-held suction pump with a simple water bottle filter makes an adequate silt cleaning system.

Large Pools. Hand-operated diaphragm pumps are necessary for larger pools. Electric pumps may be more efficient for show caves. (If a generator is used for undeveloped caves, it should remain outside the entrance.)

Pump and Filter. Place a makeshift filtering funnel at the water outflow. The filter will capture debris and silt along the edges of the foam. (See filter making above.) For some pools, nylon or Tygon® tubing is attached to the pump. Two cavers can work together to operate the device, one pumping and filtering, one directing the hose. (See Pellucidar photos, page 388.)

Always use filtered restoration water in the same area it was used the first time—never transport it to other sections within a cave.

Return. When cleaning and filtering cave pools, return the water to the pool it came from.

Disinfect and Replace. Disinfect the pump and tubing before using the system in a different pool, and replace the filter with new, clean, soft, spongy material. (See disinfecting, page 77 and page 404.)

Summary

The simple water-recycling filters described in this chapter are cave-friendly and easy to carry in cave packs. The hand-operated pumps are surprisingly efficient.

It is best to remove captured sediments from the cave. When removal is not deemed necessary, find an appropriate location under a rock along the trail. Be certain there is no passage below where runoff will create problems.

Avoid cross-contamination issues. Use recycled restoration water only in the same area it came from. Always return recycled pool water back to its original source. Clean and disinfect equipment before using it in a different location. Use new filtering material for each new site.

Gypsum Cleaning

Jim C. Werker and Val Hildreth-Werker

Gypsum surfaces become soiled and discolored from contact with hands, dirty gloves, gear, garments, and boots. Clean and restore sulfate speleothems by gingerly removing the soiled crystals and layers. Be gentle with gypsum—use a light touch and minimize damage to the soft, crystalline surfaces.

Be gentle with gypsum—use a light touch and minimize damage to the soft, crystalline surfaces.

Techniques for Cleaning Gypsum

To restore most marred gypsum surfaces, carefully brush it with light sweeping motions to gently peel off layers of grime. Because a heavy hand will produce unsightly grooves, special care and light strokes are most important. Catch gypsum scrapings and debris in plastic bags, pans, or buckets to avoid leaving contaminants in the cave.

Cavers sometimes work on gypsum restoration in pairs. One holds a plastic trash bag under the restoration site to capture falling fragments and watches to help avoid irreparable damage. The other caver judiciously brushes the gypsum crust in repetitive sweeping motions. Frequent inspection of the restoration site with the aid of side lighting will reveal problems so the gypsum team can adjust their technique.

Depending on the location and the porosity of gypsum, different procedures may be indicated. Techniques are listed here, starting with tools for the softest porous gypsum and working down to methods that have been surprisingly effective on more dense gypsum surfaces.

© Val Hildreth-Werker

Figure 1. Jim Werker uses light strokes with a stainless steel brush to clean gypsum surfaces. In this maze passage, he easily captures scrapings in a small plastic bag. (See page 10 of color section.)

Grout Brushes. Tile grout brushes with stiff nylon or plastic bristles work well for cleaning soft gypsum surfaces.

Metal Brushes. Gypsum crusts of medium hardness often require metal brushes. Clean, new brushes with stainless steel bristles are necessary to avoid introducing rust, organic materials, and black marks. Other metal bristle materials typically leave black streaks on cave surfaces. (See stainless steel in cave-safe materials, page 167.)

Loose Bristles. Expect some bristles to break off or come loose from the brushes. Use catchments. Retrieve any stray bristles. Remove all debris from the cave.

Spray and Sponge. For cleaning dense, crystalline gypsum surfaces, a spray bottle and sponge may be

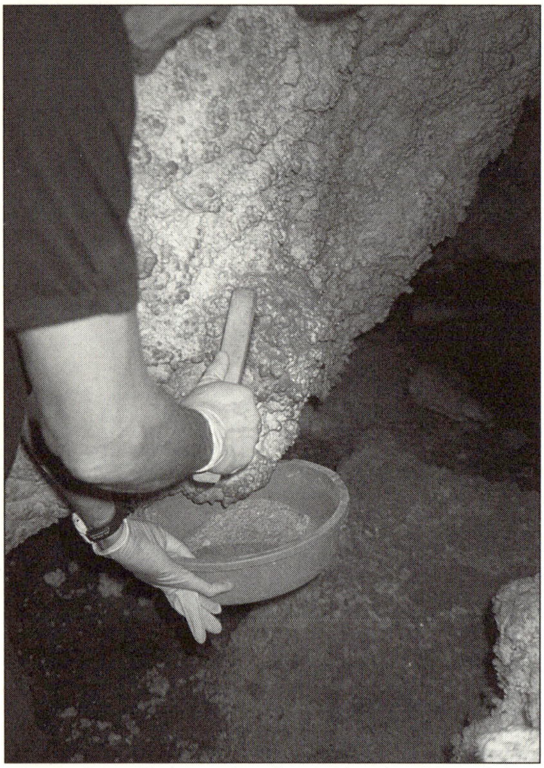

Figure 2. A tile grout brush or a brush with stainless bristles will clean gypsum surfaces. Catch all the scrapings and stray bristles. Always use gentle strokes.

Photos © Val Hildreth-Werker

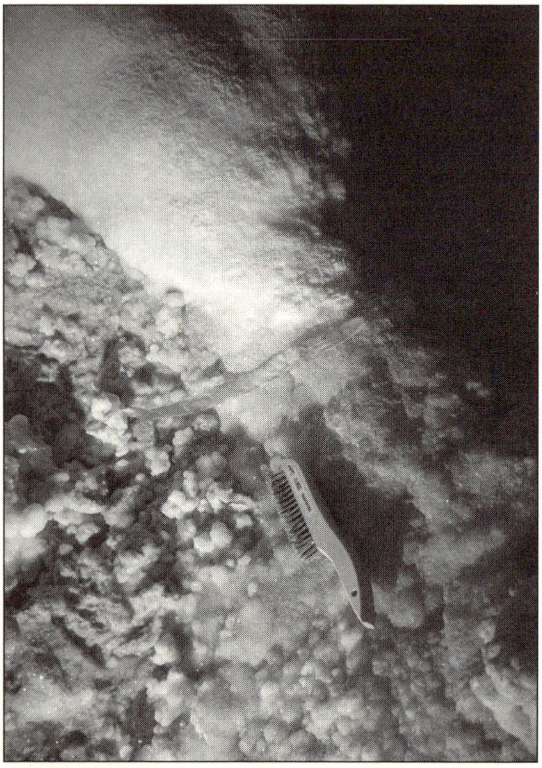

Figure 3 (before and after). Use a stainless steel brush to clean dense gypsum surfaces. The brush is sitting on soiled gypsum that is not yet restored.

effective. Spray water onto harder surfaces to loosen the soil, and then sponge up the grime.

Avoid Water. Water is ineffective on soft gypsum surfaces—applying water to porous gypsum only seems to suck any discoloration deeper below the surface.

Dental Picks. Use dental tools to pick off especially tenacious paint or soiling, fleck by tedious fleck. (See information on dental picks and gypsum photos in the graffiti chapter, page 340.)

Gypsum Tips and Precautions

Stainless Steel. Stainless steel brushes are generally preferred because stainless is more corrosion resistant than other metals, thus safer for cave environments. If stray bristles are inadvertently left in the cave, stainless will do less harm. Stainless bristles are specified in industrial supplier catalogues, but are difficult to find in local hardware stores.

Be Gentle. Avoid heavy-handed brushing that creates grooves. Use very light, sweeping motions and check the results often.

No Touching. Cavers can avoid creating big patches of tainted gypsum handholds by using one fingertip or a gloved knuckle as a tiny point of contact to assure balance. Better yet, cavers sometimes carry surgical gloves (nonlatex, powder-free) to wear through pristine passages. (See photos, page 421; also see caving gloves, page 433–435.)

Tools and Supplies for Gypsum

Stainless Steel Brushes. Lightly scrape away layers of grime.

Tile Grout Brushes. Stiff nylon bristles work on soft gypsum surfaces.

Dental Tools. Fleck off paint and debris, bit by tiny bit.

Goggles. Use eye protection.

Surgical Gloves. It is best to use nonlatex, powder-free exam or surgical gloves.

Large Plastic Trash Bags. Catch gypsum scrapings and stray bristles.

Figure 4a (before). The open palms of muddied gloves can make dirty blotches on sparkling gypsum surfaces. This large, dirty handhold started as a fingertip hold. Cavers should look first and touch only within small, carefully positioned soiled spots.

Figure 4b (before). A single finger or the point of a knuckle for a touch of balance may help reduce further damage to this cave passage.

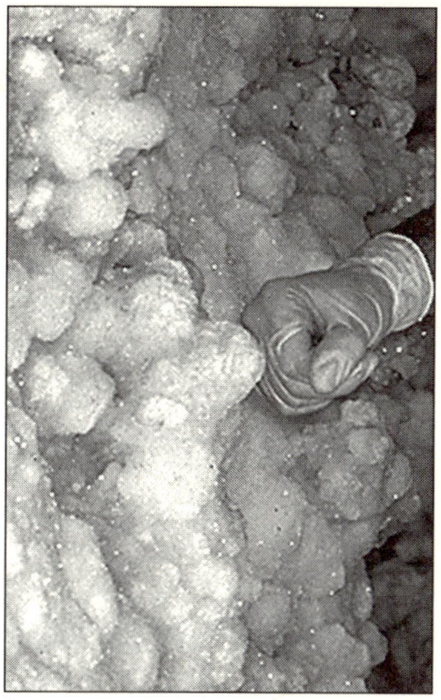

Figure 4c (after). After cleaning the gypsum handhold, cavers now keep a pair of surgical gloves handy to wear when traveling through the Chandelier Maze of Lechuguilla Cave. Here, the caver uses a gloved knuckle to help protect the restored gypsum passage.

Tarps, Plastic Sheeting, and Buckets. Large catchments may be necessary.

Spray Bottle and Sponge. On dense gypsum surfaces, try spraying water and sponging away the grimy marks.

Section D—Restoring Speleothems

Cave Pearl and Delicate Speleothem Restoration

Val Hildreth-Werker

Restoration of cave pearls and fragile speleothems can be a tedious labor of love. Keep in mind, most delicate speleothems cannot be cleaned or restored. It is best to act on foresight and implement precautionary measures to prevent impact in the first place.

When attempting to erase human impacts from temperamental speleothems, gentle techniques are employed that require small tools, excellent close-up vision, and an artistic, ultra-light touch. Several techniques to try on the various types of fragile speleothems are included below. Cave pearl restoration requires a different set of techniques and is included as a separate category in this chapter.

Restoration of cave pearls and fragile speleothems can be a tedious labor of love.

Restoring Delicate Speleothems

Restoration of fragile or delicate speleothems requires painstaking patience. Particular techniques are very specific to the site and to the individual who is actually executing a tedious task. Restoration of fragile areas involves using best practices to find a technique that works, depending on the character of environment, the type of speleothem, and the skill of the spelean restorationist.

Techniques listed here are for fragile, easily damaged speleothems such as coralloids, helictites, shelfstone, boxwork, soda straws, cave flowers, anthodites, frostwork, and so on. Restoration should not even be attempted on many fragile speleothem types—use good judgement.

Gloves. Always use surgical or exam gloves when working near fragile speleothems—nonlatex and powder-free.

Syringes. Syringes, in a variety of sizes, are used to vacuum dirt and particles from around the edges of delicate, wet speleothems. Or a light stream of water, pushed through a needle or tip, may wash particles from helictites, coralloids, boxwork, shelfstone, and other delicate travertine formations.

Water Misting. Popcorn and other coralloids are typically durable enough to mist with cave water for removing dust and lint. Soda straws may be able to withstand gentle misting from below. (Delicate soda straws may break if misted from the side.)

Foam Brushes. Clean-room foam brushes work like magnets to attract muck from moist coralloids, rimstone, sturdy heligmites, and shelfstone.

Paint Brushes. Soft-bristled paint brushes, synthetic-fiber feather dusters, and very soft toothbrushes are good dry-cleaning tools for more durable travertine and coralloid surfaces.

Figure 1. This delicate aragonite probably cannot be restored. Any attempt to remove the dirt particles is likely to create more damage. Cavers should move gently and avoid dropping dust onto this speleothem in Wind Hicks Cave, New Mexico.
(See page 16 in color section.)

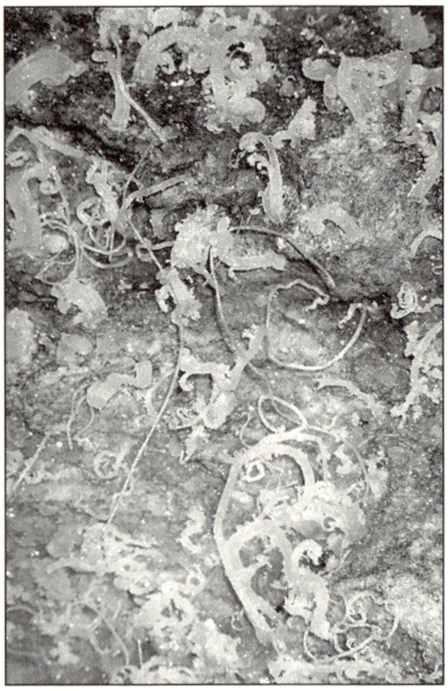

Figure 2. Imagine how easily gypsum flowers can be destroyed when cavers kick up dust or fail to look before placing a hand or foot.
(See page 16 in color section.)

Figure 3. Doug Feaks was one of the first cavers to view this wall of virgin aragonite in Southwinds, Lechuguilla Cave, New Mexico. By defining trails and following protocols, perhaps the speleothem-covered wall can be preserved in this pristine state.
(See page 16 in color section.)

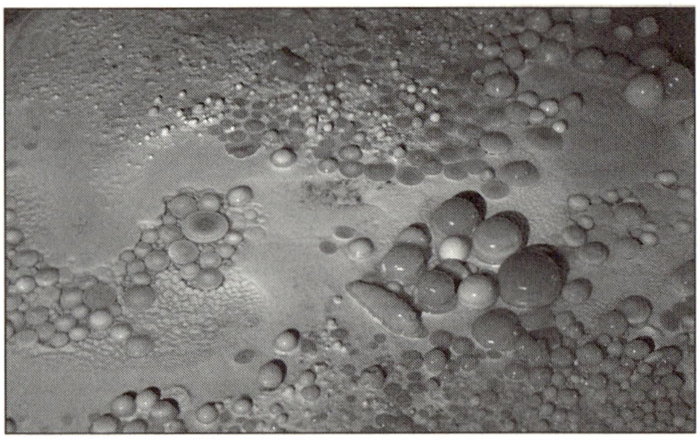

Figure 4. To mitigate careless impacts, muddy prints were tediously removed from the cave pearl nests of Pearlsian Gulf, Lechuguilla Cave, New Mexico. Note the dark footstep smear in the center of this image—the mark of another perhaps well-intentioned, yet uninformed caver.
(See page 16 in color section.)

Figure 5. Dressed in a Tyvek® suit, surgical gloves, and flowstone shoes, Gosia Allison-Kosior is restoring pearls in Pearlsian Gulf, Lechuguilla.
(See page 15 in color section.)

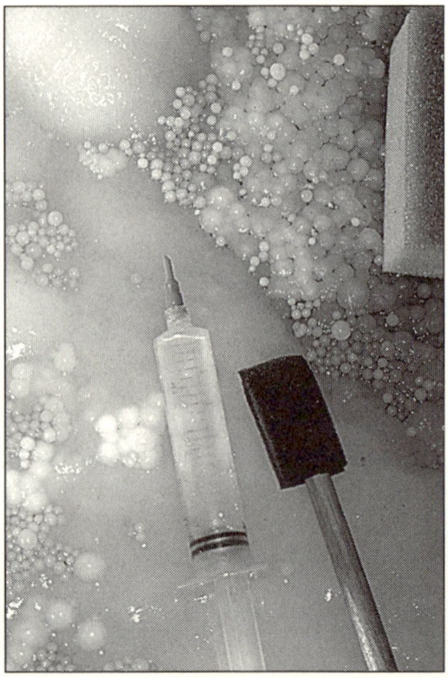

Figure 6. Syringes, foam clean-room brushes with plastic handles, and sponges are the tools for cave pearl restoration. Both before and after are apparent in this image—the left side of the flowstone has been cleaned, and the right side still has particles of mud.
(See page 15 in color section.)

Photos © Val Hildreth-Werker

Artist Brushes. Small, delicate artist brushes containing only a few bristles will sometimes remove particles of dust. A little static charge or a little moisture may help.

Tiny Dental Tools. Nylon toothpicks and stainless or plastic dental tools may facilitate removal of small particles.

Tweezers. Stainless steel and plastic tweezers are essential tools for many delicate restoration tasks. Tiny particles of misplaced debris may be gently lifted away with tweezers and a steady hand. Cleaning particles of debris from stout cave flowers or frostwork may require only a good pair of tweezers and gentle, steady patience—however, many speleothems are entirely too delicate to attempt any kind of restoration.

Air Blasts. Canned, compressed air may seem like a good idea—but canned air exerts too much pressure, breaks delicate speleothems, and causes lint and dust to become airborne.

Restoring Cave Pearls

Nests of cave pearls are sometimes inadvertently marred by muddy boot tracks. Though it is a time-consuming proposition, cave pearl restoration yields satisfying aesthetic rewards. However, extremely cautious techniques are recommended to avoid harming microbial communities that may live among the pearls or may even be partially responsible for the intriguing formation and luster of cave pearls. (See cave pearls, page 76.) Gentle techniques for restoring cave pearls are described here.

New and Sterile. Tools for cave pearls should be new and sterile if used in pristine areas. (If pearls are located in heavily trafficked areas, restoration tools should at least be clean and disinfected.)

Surgical or Exam Gloves. Nonlatex, powder-free gloves are essential for cave pearl restoration. Pearls located in isolated areas may require sterile gloves and sterile equipment. (See latex, page 433; also see sterilizing tools, page 77.)

Hypodermic Vacuum. Use syringes to vacuum dirt particles from underneath pearls and from the bathtub rings that may form along the rims of cave pearl nests. Needles and tips are useful in creating very gentle pressure for washing and for picking up particles.

Specialized Tips. A short length of hose or tubing slipped over the end of a syringe creates a soft, flexible suctioning device for the curvatures of pearl nests.

Save that Water. Use a sterile container to reserve the water drawn from pearl pools. (For example, the plastic bubble from sterile syringe containers often makes an adequate reservoir.) Allow particles to settle to the bottom and use a syringe to return the clean water to its original pool.

Muck Suckers. Clean-room foam brushes work like magnets to attract muck from around the pearls.

Detailed Pointy Work. Tweezers gently lift away tiny particles of debris.

Lint-Free Fabric or Paper. Small particles tend to cling to tweezers, but can be removed and captured for disposal by wiping the instrument across a lint-free surface.

Sponging. New, small, ultra-soft sponges will attract dirt from around the edges of pools. Thoroughly rinse new sponges before using. (See sponges for flowstone, pages 404–407.)

No Toothbrushes. Avoid using toothbrushes for two reasons—the surfaces of pearls and the linings of pearl nests may be soft and easily scratched. Also, brushing would disrupt any biofilm that may be partially responsible for cave pearl growth. (See cave pearls, page 76.)

Dress for the Occasion. Flowstone shoes, Tyvek® suits, clean lint-free clothing, wet-wipes to remove dirt from skin and helmets, kneepads or kneeling cushions sealed in zippie bags—all these guidelines for pristine areas are important when restoring cave pearls. (See full descriptions in pristine protocol, page 427–429; also see page 78.)

Tools for Cave Pearls

Sterile Supplies. Use sterile supplies if the restoration area is isolated or seldom visited.

Surgical Gloves. It is best to use nonlatex, powder-free exam or surgical gloves.

Flowstone Shoes. Change to light-soled flowstone shoes before entering the restoration area.

Clean, Lint-Free Garments. Use breathable Tyvek suits or clean silky fabrics.

Syringes. Carry a variety of sizes with needles, tips, and flexible tubing.

Sterile Container. Use for reserving cave pearl water so it can be returned to the pearl nest after any remaining sediment settles out.

Clean-Room Foam Brushes. The spongy foam material works like a gentle muck magnet when held against a rim of dirt.

Tweezers. Both stainless steel and plastic are handy.

Lint-Free Paper or Fabric. Deposit debris that sticks to tweezers or other tools onto a lint-free surface, then contain and carry it out.

New, Small Soft Sponges. Hold a sponge edge alongside bathtub rings to collect dirt. Rinse new sponges thoroughly before using near cave pearls.

Section D—Restoring Speleothems

Protocol for Pristine Passages

Val Hildreth-Werker and Jim C. Werker

Listed here are field proven recommendations for minimizing human impacts during caving activities. These procedures are used in pristine cave passages for exploration, survey, travel, research, restoration, and so on. The techniques described are effective both for working in and traveling through clean chambers where cavers want to avoid creating muddy tracks or introducing soils and nutrient sources.

Boot soles, flowstone shoe options, and glove choices are discussed in the two chapters following this one. Use this chapter as a starting point for developing specific minimum-impact standards and transition protocols that address the specific protection needs of various cave systems.

Clean Cavewear

To maintain underground chambers in a state more closely resembling the original virgin habitat, cavers need to prevent the human transfer of mud, cave soils, and organic carbons. Carefully plan the tasks, tools, and travel. Minimize contamination. The points below describe alternatives for cave gear and transition protocols.

Surgical Gloves. Use nonlatex, powder-free surgical gloves when tasks require direct contact with pristine cave surfaces. (See gloves, page 433.)

Plan Appropriate Footwear. Determine what combination of footwear is suitable for the tasks—boots, boot covers, aqua socks, neoprene booties, other types of flowstone shoes, or flowstone shoe covers. (See cave-friendly footwear, page 431.)

- Pull hazmat boots over thick-soled flowstone shoes to walk through soiled areas.
- Slip leather gloves over surgical gloves for a muddy crawl.
- Cover kneepads with lightweight plastic bags for short crawls, or waterproof nylon coverlets for longer crawls.
- Protect packs from short mud passages with plastic or nylon bags. Carry an extra small, light pack or bag for work in pristine chambers.

Wipe off Dust and Soils. Before entering a pristine area or changing into fresh garments, brush off cave dirt and use wet-wipes to clean dust and soil from skin, helmets, lights, other gear, and garments.

Remove Dirty Kneepads. A clean pair may be carried into the cave, or the dirty travel pair can be covered for

Figure 1. Penny Boston is wearing clean, lint-free garments and flowstone shoes for scientific studies in Pearlsian Gulf, Lechuguilla Cave, New Mexico.

© Val Hildreth-Werker

Figure 2. To collect water samples from a pristine cave pool, Penny Boston and Mike Spilde wear Tyvek suits, hair covers, surgical gloves, and flowstone shoes.

Figure 3. When entering a muddy chamber in La Cueva de las Barrancas, cavers wear covers on their boots and bags over their kneepads to avoid tracking mud throughout the cave.
(See page 15 in color section.)

Figure 4. This caver is in a transition zone (an area designated for changing from boots to flowstone shoes). The caver has one finger poised for balance on the adjacent rock, should he need it. His dirty kneepads will also be left with the boots.

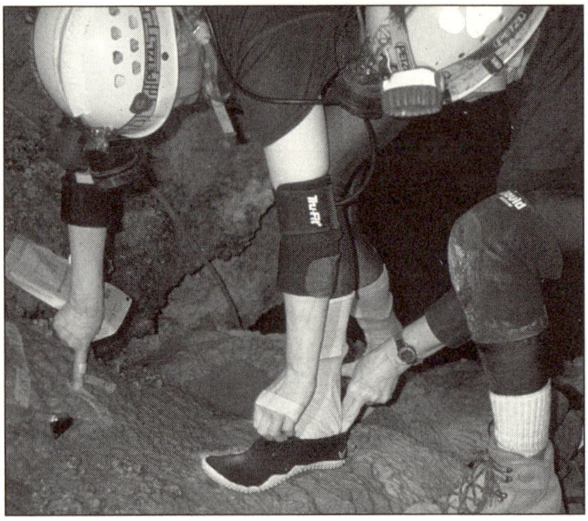

Figure 5. Only one caver needs to remove dirty kneepads and change to flowstone shoes so she can step off trail into a flowstone area to perform water testing at Liberty Bell in Lechuguilla Cave. A little help from a friend facilitates the transition.

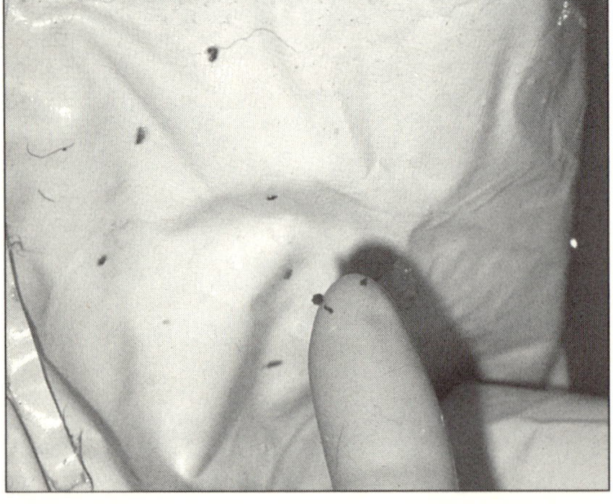

Figure 6. The dots are polypropylene lint balls from fuzzy caver garments. When laundering, separate polypro and nylon clothing from cottons to reduce pilling.

Figure 7. A nylon sheet is spread in the trail area and used as a staging area for scientific studies conducted in Spider Cave, Carlsbad Caverns National Park, New Mexico.

pristine tasks. Put each kneepad inside a one-gallon freezer zippie to contain dirt, lint, and debris—then use the contained kneepads as kneeling cushions (be certain the zip-closures will stay tightly sealed).

Change to Clean Clothes and Packs. Before entering clean passages, it may be appropriate to take off soiled travel clothes and change to coveralls, silkies, or Tyvek® suits. (Clean-room suits are available in paint supply outlets or through lab catalogues.) Depending on the conditions, it may work better to wear coveralls over the attire planned for the pristine passage and simply shed the coveralls before entering.

- For pristine areas, change to garments that are free of lint and cave dirt—silky feeling polyester, nylon, or clean-room suits made of Tyvek are lightweight and take up minimal room in cave packs.
- Various types of Tyvek will provide a light insulating layer for cooler temperatures, but for colder cave passages cavers may need warmer cave suits, layers, or advanced technology fabrics.
- Remember to include a clean pack or plastic sack for transporting gear into or through clean passages.

Hair Covers. Use surgical caps, loose nylon or lycra swim caps, or bandanas to contain hair and debris. Elastic surgical caps are made of breathable Tyvek and can be worn over or under a helmet to contain flecks of debris and strands of hair. Bandanas are available in a variety of fabrics. Lycra swim caps fit tightly and may need to be stretched out before wearing under a helmet.

Be Careful with Polypropylene Cavewear. Some fabrics create polypro lint balls that will fall off garments and scatter on cave surfaces. Retrieval of thousands of tiny lint balls is inefficient and irksome. To minimize lint dots, do not launder synthetic fabrics with cottons.

Protect the Flowstone. Spread out a plastic or nylon sheet as a staging area for clean items.

- Wipe off excess dirt and then move toward the staging area.
- Step out of soiled clothing onto the clean sheet.
- Stand or sit on the clean sheet to don gear for the pristine chamber.

Listed here are field proven recommendations for minimizing human impacts during caving activities. These procedures are used in pristine cave passages for exploration, survey, travel, research, restoration, and other caving activities.

Before entering a clean passage, it may be appropriate to take off soiled travel clothes and change to coveralls, silky garments, or Tyvek® suits.

Cave-Friendly Footwear

Val Hildreth-Werker

In caves, use footwear with nonmarring/nonmarking soles. Employ common sense and select comfortable, durable, safe footwear for various caving activities. From light-soled atheletic shoes to durable rubber boots—the variety of specialized footwear available today makes it easy to choose appropriate shoes for most any cave environment.

Boot Soles

Don't wear heavy waffle stompers in caves. Weight and bulk cause unnecessary travel damage and excessive trail compaction.

Avoid traditional black Vibram® soles—they definitely leave marks on cave surfaces. Advanced technology has afforded Vibram the ability to offer a variety of new soles—some are described as nonmarking. Check the soles before walking across sensitive cave surfaces. For the best test, strike the soles across a block of limestone before wearing them in a cave. (See scuff marks, page 405; also see soft scrubbers, page 406.)

Whether light or dark colored, many boot soles will leave scuff marks on calcite and other cave surfaces. Test all caving soles on untreated limestone or concrete.

Be careful with blond rubber soles—chunks from soft soles tend to break off and leave debris in cave passages.

Include an extra plastic bag to contain dirty boots through pristine areas where flowstone shoes are required.

Common sense should guide selection of durable, comfortable, safe footwear for various caving activities.

Flowstone Shoes

Carry clean, lint-free flowstone shoes with lightweight, nonmarking soles for entering pristine areas. Use water sports booties, smooth-soled athletic shoes, aqua socks, or nylon slippers. Carry a small sponge to spot-clean flowstone shoes while caving.

Choose flowstone shoes according to the nature of travel surfaces. Rough or rocky surfaces require durable flowstone soles for comfort and safety while flexible soles often give adequate protection and grip on slick, smooth surfaces.

Do not travel across cave surfaces with only bare feet or socks—lint, skin flakes, blood, body oils, and microbial residents living under toenails and in skinfolds should not be left on cave surfaces.

Between cave trips, flowstone shoes should be laundered in a washing machine with just a touch of mild detergent.

Include a plastic bag that is suitable for containing flowstone shoes that get soaked with water or soiled with cave muck.

Shoe Covers

When traveling through areas that have intermittent muddy floors, cover flowstone shoes with hazmat booties. Designed to slip over shoes, the hazmat boot covers can be donned over (or removed from) flowstone shoes

Figure 1. In transition zones, step out of boots and into flowstone shoes. Non-marking aqua socks with light soles make good footwear for many pristine cave surfaces. (See page 15 in color section.)

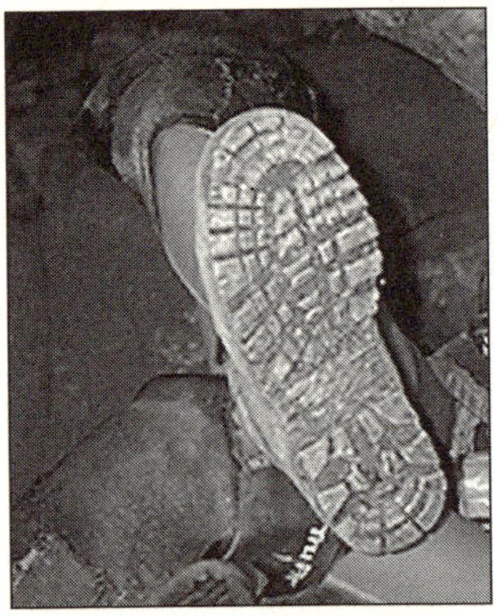

Figure 2. Use nonmarring boots for caving. Some boot soles leave tenacious scuff marks on flowstone. Test soles by striking across untreated limestone or concrete.

Figure 3. Plastic shopping bags often make adequate boot covers for short distances if cavers walk gently. David Joaquim is checking a science and monitoring station on a muddy trail in La Cueva de las Barrancas. Just outside of the trail, test tubes remain *in situ* to allow cultures to grow in the stable cave environment. (See page 15 in color section.)

Photos © Val Hildreth-Werker

Figure 4. In pristine areas with intermittent mud, Harley Shaw slips hazmat boot covers over his flowstone shoes to avoid spreading mud onto clean flowstone surfaces. (See page 15 in color section.)

while standing. Though durable enough to withstand mud-covered, rocky surfaces when worn over aqua socks, the hazmat covers tend to rip and tear when worn over normal caving boots.

Simple plastic bag boot cover may be adequate for brief zones of travel. Two-gallon Ziploc® freezer bags fit easily over many boot sizes and provide surprisingly durable protection and safety on most surfaces. (Hefty® bags with built-in zipping mechanisms speed up the transition process, making it easy to slip a foot in and secure the bag. However, with repeated use, the zippers stop working and it is wise to carry a pair of large elastics or Velcro® bands to secure the bags loosely around the ankles.)

In a pinch, or for very short distances, lightweight plastic shopping bags tied over boots provide temporary protection if the caver steps very gently.

When traveling through areas with intermittent muddy floors, cover flowstone shoes with hazmat booties.

Happy Feet Make Happy Cavers

From rock to mud to water, boots are made for caving. Choose wisely and keep your feet healthy and happy. Be kind to your soles for they bear you.

Practical Caving Gloves

Val Hildreth-Werker

For most travel in wild caves, gloved hands are viewed as more conservation-aware than bare hands. Every touch of bare skin leaves body oils, skin fragments, microorganisms, and debris on cave surfaces.

Glove Choices

Leather gloves are durable and create less lint than cotton and other natural fabrics. Leather gloves are appropriate for general caving in many environments, for vertical rope skills, and for tasks that require lifting and heavy labor. However, gloves made of durable synthetic materials wear well and produce minimal lint. In wet environments, greater protection may be necessary. Glove choice is based on the nature of travel and activity in specific cave systems.

Muddy hands and gloves leave ugly scars along travel routes in caves. Instead of grabbing a speleothem or a piece of wall with a big muddy palm, use the cleaner back of a knuckle or a fingertip as a small point of contact for balance. (See photos, page 421 and page 434.) Where something must be grasped as a secure hold, look for and use the most impacted spot instead of creating new muddy blobs. (Observe common sense safety measures—if a caver loses balance, it may be necessary to use the first-available durable protrusion as a quick handhold.)

To avoid making muddy scars in pristine passages, cavers carry lightweight synthetic gloves. Because oils and organics on our hands create damaging impacts, conscientious cavers pack extra gloves for quick access. In some caves, laminated signs that read *Gloves Off* mark locations where cavers are expected to remove their travel gloves and don their flowstone gloves (surgical, exam, vinyl, nitrile, disposable, or some other kind of clean handwear) (Figure 4). Synthetic gloves are good tools for cave protection and conservation.

> In caves, gloved hands are usually viewed as more conservation-aware than bare hands. Every touch of bare skin leaves body oils, skin fragments, microorganisms, and debris on cave surfaces.

Surgical Gloves

Nonlatex, powder-free surgical or exam gloves are recommended for most cave restoration tasks and for travel through pristine cave passages.

Surgical and exam gloves made of other materials—vinyl, nitrile, and other nonlatex products—are available through medical and lab suppliers, beauty and barber suppliers, food service outlets, and hardware stores.

Lightweight, plastic food service gloves are adequate for quick restoration jobs like small trash or mold removal—or for sponging up an occasional footprint on flowstone.

Holes and tears in lightweight surgical gloves can be a nuisance. Some cavers rarely rip their gloves while others go through several pairs a day. Heavier gloves made of nitrile or vinyl still allow dexterity, but are more durable and can be washed between trips. Only powder-free gloves should be used in caves and it is best to avoid latex products.

Figure 1. Reduce caver impacts created by muddy gloves. Avoid reaching up to touch cave ceilings and walls when a handhold is not necessary.

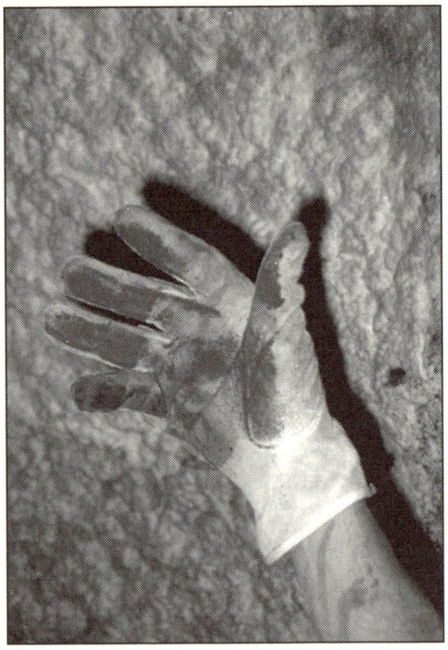

Figure 2. Muddy gloves leave ugly scars along cave routes

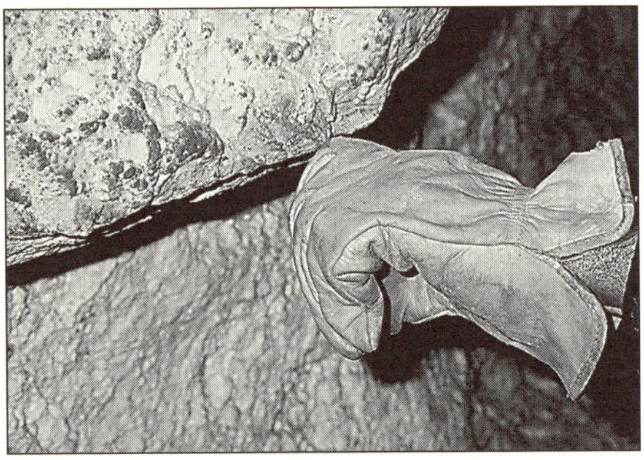

Figure 3. Instead of a dirty open palm, use a gloved knuckle for a single point of contact to help maintain balance.

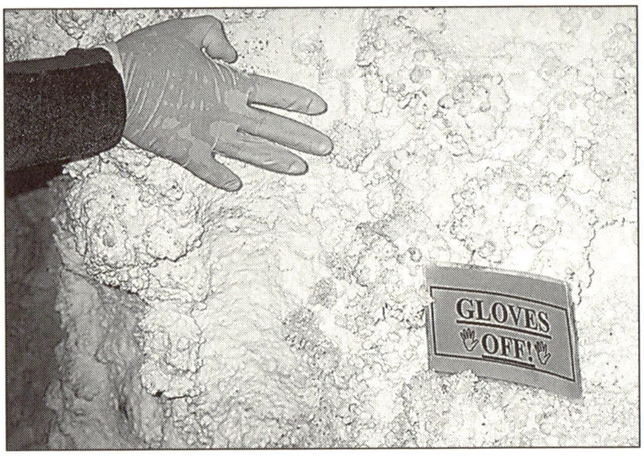

Figure 4. Along some routes, cavers are expected to remove their leather gloves and don a pair of nonlatex, powder-free surgical gloves to help keep pristine areas clean.

Glove choice is based on the nature of travel and activity in specific cave systems.

Avoid Latex Products

In caves, it is best to avoid contact with latex products. The term *latex* usually refers to natural rubber latex, the product manufactured from a milky fluid derived from rubber trees. Proteins flake off from natural rubber latex gloves. Protein particles remain in cave passages and provide a nutrient source for nonnative species introduced from surface systems. Latex proteins become fastened to the glove's lubricant powder and particles become airborne when the gloves are removed. Aerosolized particles then spread through the cave and to human mucus membranes.

The most common reaction to latex is *irritant contact dermatitis*—the development of dry, itchy, irritated areas on the skin, usually the hands. This reaction is caused by irritation from wearing gloves and by exposure to the lubricant powders added to them.

Allergic contact dermatitis results from the chemicals added to latex during harvesting, processing, or manufacturing. These chemicals can cause a skin rash similar to that of poison ivy.

Neither irritant contact dermatitis nor chemical sensitivity dermatitis will cause an immediate, life-threatening allergic reaction. (There are several types of synthetic rubber that are sometimes also referred to as "latex." The synthetic products do not release proteins that cause allergic reactions.)

However, people with suspected latex sensitivity (allergy) should avoid latex products because latex exposures at even very low levels *can* trigger severe, life-threatening allergic reactions in some sensitized individuals.

- Reactions usually begin within minutes of exposure, but can occur hours later and can produce various symptoms.
- More severe reactions may involve respiratory symptoms such as runny nose, sneezing, itchy eyes, scratchy throat, and asthma (difficult breathing, coughing spells, and wheezing).
- Rarely, anaphylactic shock may occur—an acute and critical condition that involves the respiratory system and requires immediate medical assistance.

Only powder-free gloves should be used in caves and it is best to avoid latex products.

For additional information about latex allergy, visit the National Institute for Occupational Safety and Health (NIOSH) and other safety sites at the following Web addresses.

- National Institute for Occupational Safety and Health
  Occupational Latex Allergies
  <**http://www.cdc.gov/niosh/topics/latex/**>

- Anesth.com
  Search for latex allergy
  <**http://www.anesth.com/lair.htm**>

- Family Village Home
  Latex Allergy
  <**http://www.familyvillage.wisc.edu/lib_latx.htm**>

- University of Rochester
  Guidelines for Reducing Employee Exposure to Latex
  <**http://www.safety.rochester.edu/ih/latex.html**>

Acknowledgements

Latex information for this chapter was collected in 2004 during phone interviews with Stephen R. Mosberg, MD, and Karmen Hopkins, MD. The editors gratefully acknowledge their contributions and reviews. We recommend researching the Web addresses listed above for updated information.

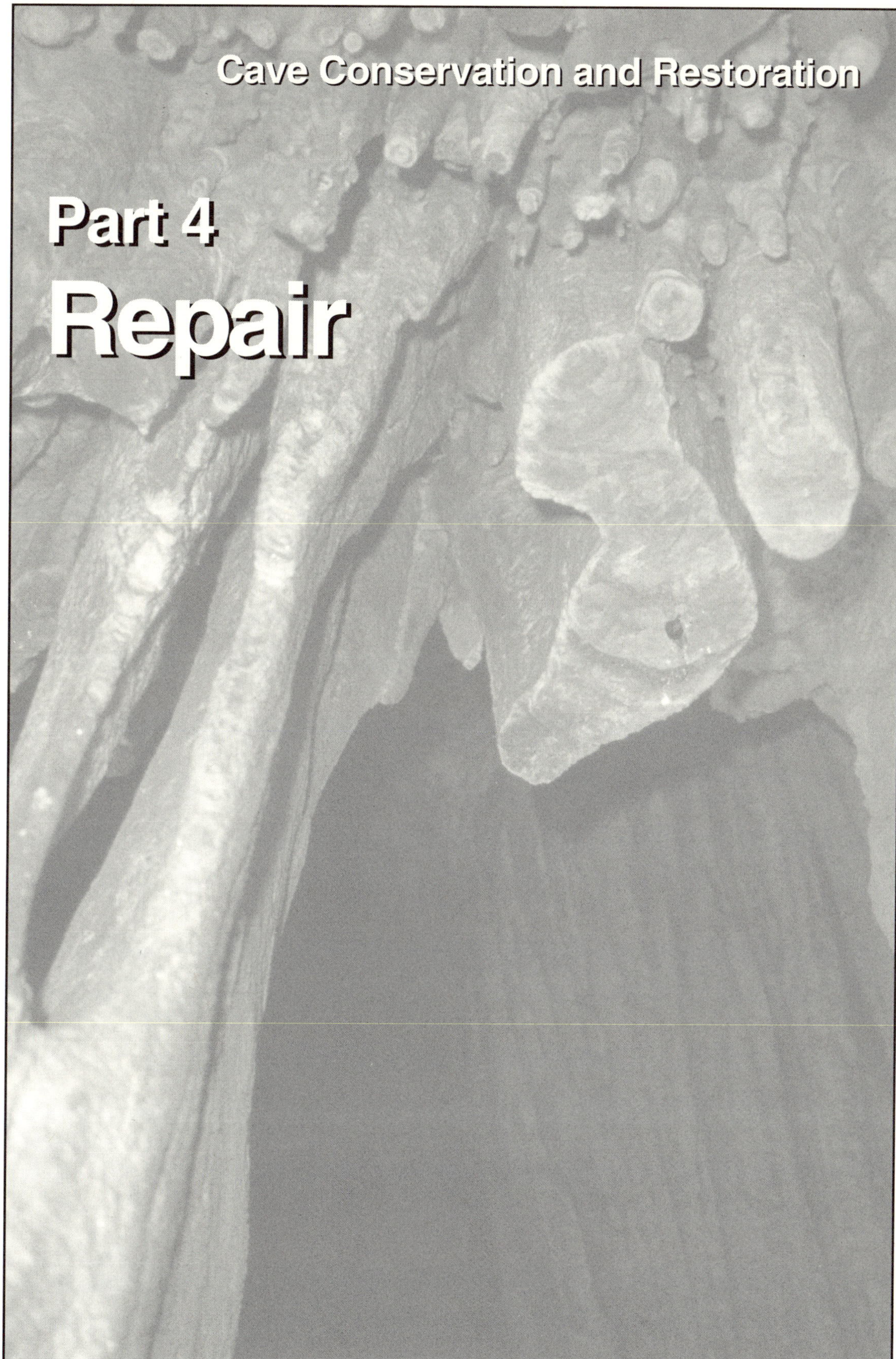

Cave Conservation and Restoration

Part 4
Repair

Section—A Introducing Speleothem Repair

Candle Table Revisited

Jim C. Werker

Vandals tried to steal the Candle Table from a well-known cave in New Mexico. The room of shelfstone and stalactites reminds cavers of nostalgic soda-fountain-style tables with narrow pedestals. These top-heavy "coke tables" are arranged around a very special shelfstone speleothem that once had a candle-like formation growing from ceiling drips positioned above the center of the table. Over time, changing water levels helped form this room of unusual spelean sculptures decorated with beautiful mineral deposits.

In the foreword of this volume, Ron Kerbo describes the day that vandals broke off the Candle Table and tried to haul it out of the cave. (See Candle Table, pages 3–5.) Ron and his friend, Joe, carefully carried the damaged speleothem along crawlways and returned it to its room of origin.

They reported the vandalism to Jerry Trout who was employed by the USDA Forest Service as the first officially titled cave specialist in a federal agency. Jerry believed the Candle Table could be repaired. He knew it had been forever damaged and would never be the same—but maybe the speleothem could again stand upright among the other calcite tables.

Like most cavers, Jerry is innovative and practical. He used glue and a sizeable nail that he had on hand—he thought it might hold the Candle Table long enough to give nature a chance to form an outer coat of healing deposits. The glue didn't work.

Jerry knew that I was knowledgeable about materials used in the underground tunnels at the Nevada Test Site. He asked me about adhesives that might hold better and I showed up with a selection of tools and supplies to offer a higher probability of success. That's how I got started with speleothem repair.

We fixed the Candle Table. In fact, we repaired it several times until the vandals gave up and quit kicking it over. They kicked or pried the shelfstone speleothem with enough force to bend the stainless steel stabilizing rod that we epoxied into the skinny base. The rod bent but the archival grade epoxy held the pin firmly and the Candle Table did not budge. The formation was precariously tilted on its pin, but was not gone.

For the final fix, we straightened it, used a larger diameter pin, and reinforced the outer crust at the base with a mixture of epoxy and pulverized rock. The Candle Table stands upright today, decades after the initial vandalizing break.

Word got around and I started getting calls to put other speleothems back together again. Better techniques began to evolve. In this section of the handbook, we offer methods from several cavers who have spent years developing expertise in formation repair. Proven techniques and current best practices are presented in the chapters of this section.

Figure 1. In 1961 Bob Trout admired the unusual Candle Table speleothem before the candle-shaped formation on top was stolen. Later, the Candle Table itself was repeatedly vandalized and repaired.

Jerry L. Trout

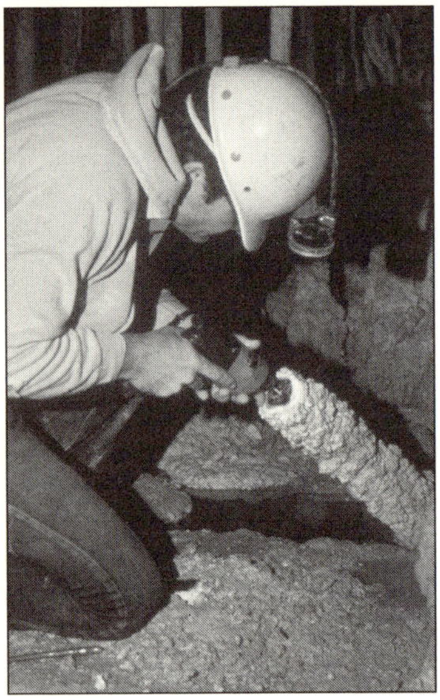

Jerry L. Trout Collection

Figure 2. Jerry Trout drilled the original hole for installing a stabilizing pin in the base of the Candle Table.

© Val Hildreth-Werker

Figure 3. The Candle Table was pinned and epoxied back in its original location.

© Val Hildreth-Werker

Figure 4. The outer crust of the base was reinforced with museum-grade epoxy mixed with calcite dust.

© Val Hildreth-Werker

Figure 5. Again, vandals tried to take the Candle Table. The stabilizing rod bent, but the cave-safe archival epoxy held the speleothem in place.

© Val Hildreth-Werker

Figure 6. After several episodes of vandalism, Jim Werker puts the finishing touches on yet another repair of the surviving Candle Table.

Section A—Introducing Speleothem Repair

Current Best Practices in Speleothem Repair

Val Hildreth-Werker

The following chapters present state-of-the art speleothem repair method and philosophy. The editors have reviewed techniques, evaluated concepts, and coordinated statements to describe the *current best practices* in repair techniques and cave-safe material choices.

Cave-safe is considered a relative term and is used only to describe the best of what is known. State-of-the-art techniques and materials will change rapidly as more is understood about cave environments.

Some materials that are considered relatively cave-safe today will likely lead to future improvements. Many practices of the past are now rejected because of unexpected, detrimental results (Hamilton-Smith and others 1998). The bottom line is clear—any human-introduced materials should be used sparingly.

Best practice is standard language in business management and people are sometimes suspicious of the terminology (Spate and others 1998). Why use the term *current best practice* for speleothem repair standards?

First, adding the word current in front of best practice reminds us to always keep our eyes, ears, and minds open for improved methods. Second, the term inherently awakens the essential process of questioning, "What *are* the current best conservation practices?" Thus, the phrase encourages research, evolution of ideas, and advances in methodology.

Current best practice in cave conservation and management is not an end product, but rather a conscious process of defining and enhancing standards.

There is one constant factor in cave conservation, restoration, and speleothem repair—mistakes lead to new methods. Never view today's best practices as prescriptions for all situations.

There is one constant factor in cave conservation, restoration, and speleothem repair—mistakes lead to new methods. Never view today's best practices as prescriptions for all situations.

Cited References

Hamilton-Smith E, McBeath R, Vavryn D. 1998. Best practice in visitor management. In: Smith DW, editor. *Cave and Karst Management in Australasia XII*. Carlton South, Victoria, Australia: Australasian Cave and Karst Management Association. p 85-96. [Also in: *Australasian Cave and Karst Management Association Journal*. 1997. 27:33-43.]

Spate A, Hamilton-Smith E, Little L, Holland E. 1998. Best practice and tourist cave engineering. In: Smith DW, editor. *Cave and Karst Management in Australasia XII*. Carlton South, Victoria, Australia: Australasian Cave and Karst Management Association. p 97-109.

Section A—Introducing Speleothem Repair

Speleothem Repair Planning and Documentation

Val Hildreth-Werker and Jim C. Werker

More people are visiting caves. Increased visitation generally results in more broken speleothems. Whether damaged through acts of ignorant vandalism, blatant carelessness, or inadvertent clumsiness, some of the busted formations can be repaired. As with any cave restoration project, before making decisions carefully consider the potential environmental impacts of repair materials, equipment requirements, and safety measures.

The first objective in speleothem repair is to avoid creating new problems—first, *do no harm.* Before starting a repair project, get permission from the landowner, manager, or agency. Go into the cave, look at the site, and evaluate the potential problems. Talk over the repair tasks and logistics, consult with others, bounce ideas around, and make certain the broken pieces fit tightly together—then make decisions about the repair technique to use.

Is it best to repair it or leave it alone? In some cases, doing nothing at all may be the best answer. Evaluate the potential repair from all angles and become aware of materials that are safest for long-term use in caves.

Materials that are considered reasonably safe for long-term cave applications are listed in the speleothem repair materials chapter (pages 445–450). Adhesives are notorious for introducing nasty substances into natural environments. Inexpensive adhesives may set up quickly and may appear to work okay, but the cheaper consumer glues and epoxies often have detrimental degradation characteristics, toxic outgassing, and rapid deterioration of chemical bonds. (See archival epoxies and adhesives, page 172 and page 446.)

Obtain manufacturer data sheets that detail the chemical properties of repair materials. Also check the federally regulated Material Safety Data Sheets (MSDS) that discuss hazards and the specific safety precautions for most products and materials. (See MSDS, page 70 and page 172.)

Know about the repair products you are using and ask for assistance in predicting how the product may react in cave environments. Consult with biologists, geologists, chemists, and other specialists working in speleological disciplines to ascertain the effects of various materials on specific cave environments. (See materials, page 167.)

> Before starting a repair project, get permission from the landowner, manager, or agency. Go into the cave, look at the site, and evaluate the potential problems.

Explore These Questions

Before getting to the nitty-gritty of the actual repair techniques, there are a few important questions and decisions to address.

Is it natural breakage? Most speleologists agree that speleothem pieces should be left alone if the break was caused by naturally occurring environmental changes. Repair the break if the damage was caused by human interference.

> **"If it ain't broke, don't fix it."**
> —Collier, Beard, French, Meyer, Veni, and Werker are all advocates of this popular axiom.
>
> **"If it's growing back on its own, think long and hard about repair."**
> —French and Meyer

Do the pieces actually fit each other? Most speleothem repair experts agree that orphan speleothem pieces should not be glued together to create a new formation out of unmatching broken pieces. Likewise, it usually is considered unethical to add speleothems where none were naturally placed by time and deposition.

What about fabricating missing chunks? It is often reasonable to restore an incomplete formation by filling in gaps with a mixture of epoxy and rock dust. However, it is usually considered inappropriate to fabricate or recreate totally missing speleothems.

What if calcite deposition prevents tight alignment? If new speleothem material has deposited on top of an old break, it may be inefficient or detrimental to attempt repair. However, this handbook offers several pinning and fabrication techniques that have worked in reconstructing speleothem joints disfigured by light calcite deposits.

Will the repaired speleothem be safe from future damage? Should it be repaired anyway—regardless of security factors or potential for additional vandalism?

Is a repair attempt worth the potential harm? What is the probability that more damage will be caused during the repair?

What are the safety considerations? Are scaffolding and fixtures needed to raise the formation into place? Should cavers or broken pieces be tied off for safety due to the location of the repair? Caution should prevail when repairing heavy speleothems. Evaluate location and weight before rehanging heavy stalactites. Consider liability issues before attempting repairs.

Cave dwellers? Will critters be affected by fumes from the adhesives or residue from the glues? Do bats use the site? Invertebrates? Other organisms?

Cleanup? Include planning for the logistics of cleanup after the project is complete. Remove all indications of the repair crew's presence.

Residual effects? Evaluate the potential residual effects of any proposed repair. Because all materials will eventually break down and the byproducts may be harmful, minimize the use of human-introduced materials as much as possible.

More Questions for the Pre-Planning Phase

Always plan in advance—*before* initiating the hands-on work. Reconnaissance trips set the stage for success and help prevent poor performance. Here are a few planning pointers for speleothem repair projects.

Where is the speleothem located? What are the logistics of reaching the repair site?

What materials are needed? What tools are required?

Scaffolding? Will the formation require fixtures for lifting and holding it in place until the epoxy cures?

Timing? If the repair is in a show cave with tours, timing may be important. Discuss scheduling with the cave manager so the administration agrees with and approves of the project.

Is the repair in an area with a lot of speleothems? Assign spotters to help the team avoid breaking more formations.

What type of adhesives will work for the specific repair? Is the area wet or dry? Remember, just because an area is dry now doesn't mean it will be in the future.

Personnel? Don't arrange for more help than is required to complete the job. Extra bodies create additional safety concerns.

Repairing a Stalagmite on a Slope in Slaughter Canyon Cave

Figure 1. Pieces of a stalagmite rest on the slope where this speleothem was broken in Slaughter Canyon Cave, Carlsbad Caverns National Park.

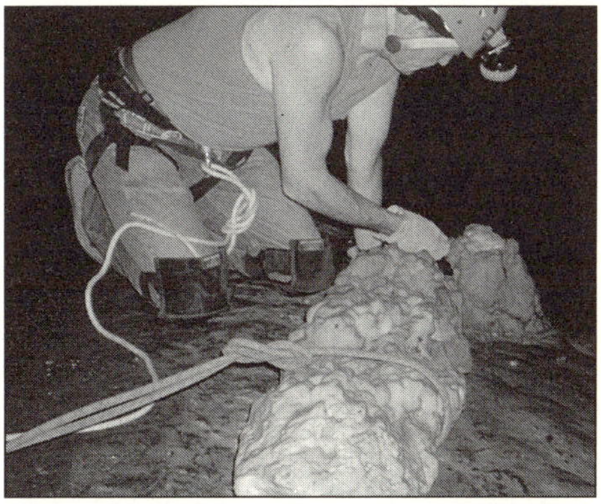

Figure 2. Since the repair is performed on a sloping flowstone surface, cavers and large broken pieces are secured with safety lines.

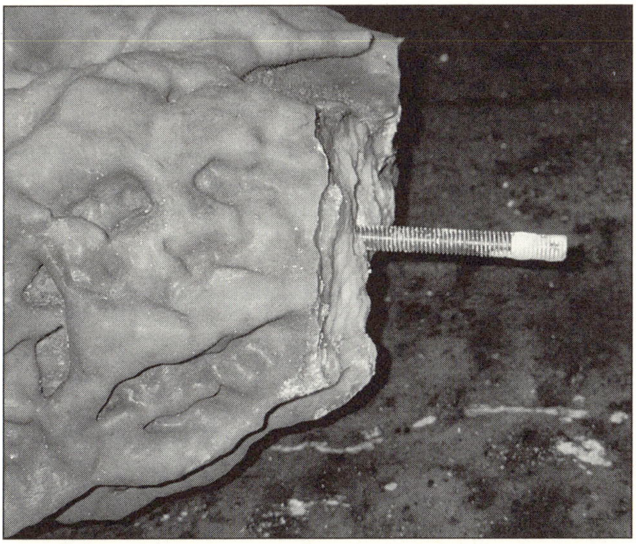

Figure 3. A stabilization pin made of stainless steel all-thread is epoxied into the center of the upper part of the broken stalagmite.

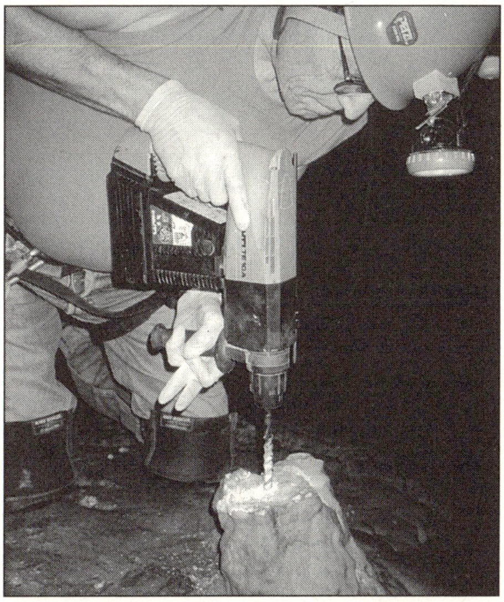

Figure 4. Jim Werker drills a larger hole in the stump to receive the stainless steel pin.

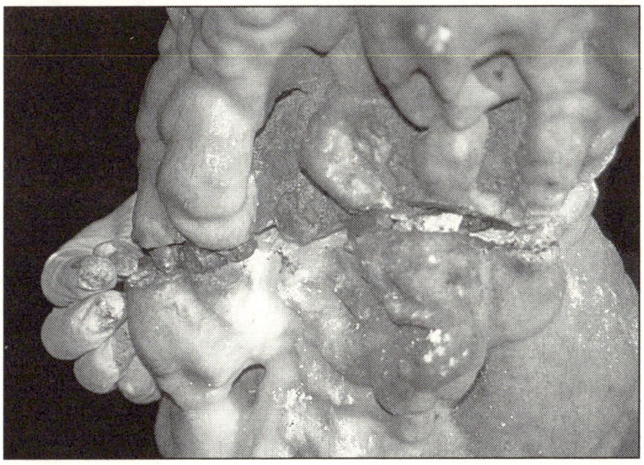

Figure 5. New calcite deposition on the break prevents solid mating of two pieces. Small rocks fill the gap between broken pieces and provide additional support.

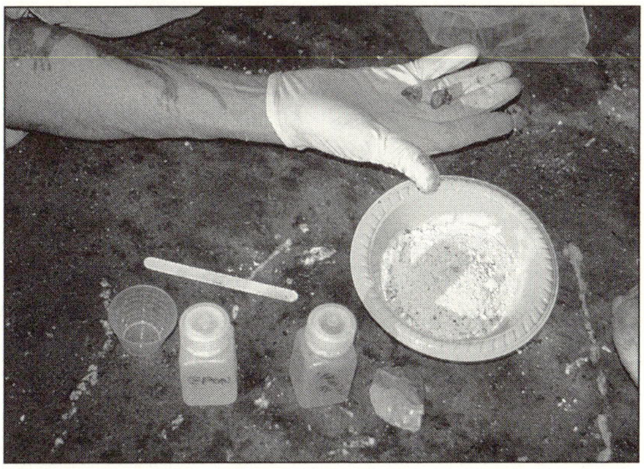

Figure 6. Pulverized rocks and archival epoxy form a thick, color-matched mixture for filling cracks and gaps in repaired speleothems.

Repairing a Stalagmite in Slaughter Canyon Cave *(continued)*

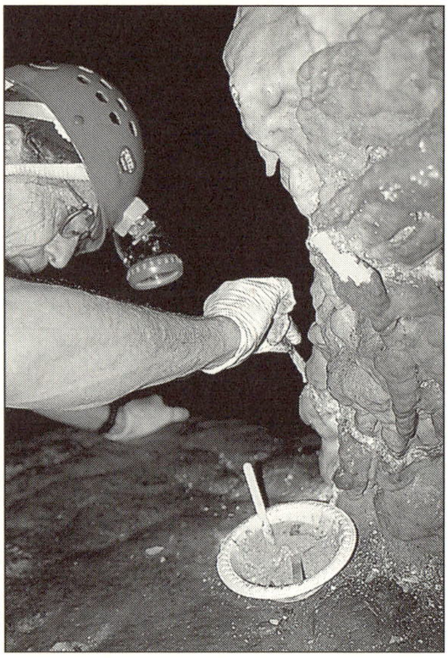

Photos © Val Hildreth-Werker

Figure 7. A thick mixture of rock dust and archival-grade epoxy is pushed in place with a craft stick to fill and sculpt the gap.

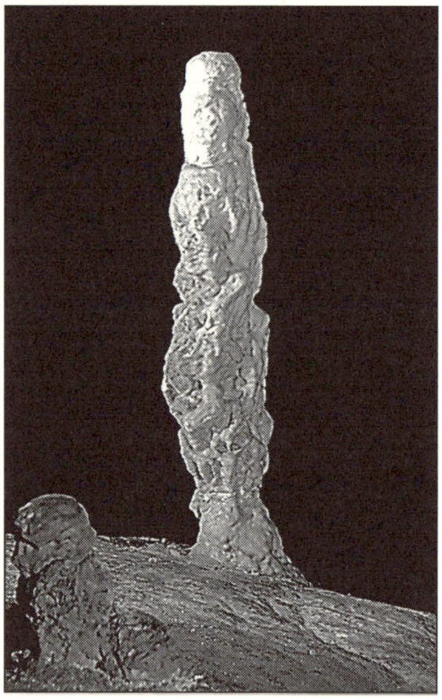

Figure 8. Repaired with stainless stabilization pins and archival epoxy, the reassembled stalagmite again stands tall on its base of sloping flowstone.

Document Repair Sites—Before and After

Use photos and written accounts to document speleothem repair projects. Make pictures of breakage and damage, repair procedures, personnel, and completed repair sites. Use photographs for historical documentation and file them with written reports that describe the methods and materials. Include mechanical drawings to enhance reports.

File the documentation in two or three separate places—historical information is useless if nobody can find it. Always provide a file for the landowner, manager, or agency. Submit a report to the local grotto or region files. Project reports may also be submitted to the NSS library for archiving. Why should repair projects be so thoroughly documented?

Before and after photographs are impressive historical documents that may be used for interpretation, research, educational outreach, protection, and funding opportunities.

Photograph repair procedures to report, document, and evaluate the techniques employed.

Documentation may help stewards identify the cause and make corrections if future abnormalities occur near repair sites. Reports may help in future planning for similar projects. Documentation can become a teaching aid for sharing repair how-to instructions.

Future cavers may be able to check on the longevity of repairs if a few details are documented. At least take pictures, record the date, the type of adhesive used, type of pin installed (if it has one), name of the cave and passage (if appropriate), and names of those who participated in the repair. (See photodocumentation, page 204.)

History is worthless if nobody can find it. File redundantly.

Figure 9. In Lower Cave of Carlsbad Cavern, Jim Werker used Epon 828 and Versamid 40 with a 3-inch 10-24 UNC stainless steel all-thread stabilizing pin to rehang this stalactite on June 5, 1999. Stan Allison, Cave Technician at Carlsbad Caverns National Park, assisted in the planning. (See additional photos of this repair, page 468).

Photos © Val Hildreth-Werker

Section A—Introducing Speleothem Repair

Speleothem Repair Materials

Jim C. Werker and Val Hildreth-Werker

The materials recommended in this book are relatively safe for long-term use in caves. Based on current best practices in caves, coupled with lab analysis and several decades of practical observation in subterranean environments, the materials listed in this section are generally considered safe for speleothem repair. Keep in mind that the term *cave-safe* is used to describe the best of contemporary understanding.

State-of-the-art techniques and materials for speleothem repair will change as knowledge advances. It is important to exchange information with others who repair speleothems and to make every effort to stay current.

Cave-Safe Epoxies and Adhesives

Cavers prefer to be out exploring caves, not repairing someone's speleothem accident or an act of vandalism. Cavers tend to hunt for new, sure-fire, quicker repair materials and techniques—but faster is not necessarily better.

Throughout this volume, only tested archival-grade epoxies and adhesives are recommended for use in cave environments. The epoxies, bonding agents, and quick glues found in neighborhood dime stores and variety outlets are certainly easier to obtain than the archival adhesives. However, mass-market products may create a lot more harm than good.

Most hardware store epoxies and quick-glues will break down rapidly and may add unwanted toxins and nutrient sources that will harm or destroy cave biota and habitat. Before choosing epoxies, hardeners, cyanoacrylate adhesives, solvents, or metal fixtures, research and understand the characteristics of the materials. (See cave-safe materials, page 172.) Don't trust the mighty marketing claims of wonder products.

All glues break down over time, but archival products are formulated to do less harm than those on the general market. In the future, technological advances will introduce better products that are safer for cave environments. Always carefully research and test products before using them for cave applications.

Any new product should be checked by research chemists and biologists who understand the specific cave environments where product use is proposed. Find out the physical characteristics and short-term effects of proposed products.

What are the components of the agent and how will the degradation and outgassing characteristics of the compounds affect cave-dwelling biota, ecosystems, chemistry, water quality, or the minerals of a cave system? Arrange for a cave-savvy chemist or a materials engineer to evaluate how the long-term degradation characteristics may interact with the naturally occurring chemicals and minerals of the cave.

Get the manufacturer's data sheets and the federally regulated Material Safety Data Sheet (MSDS). Understand the recommended safety precautions.

Throughout this volume, only tested archival-grade epoxies and adhesives are recommended for use in cave environments.

Units of Measurement

Editors' Note: In this chapter, dimensions for materials deviate from the standard metric/English format used elsewhere in this volume because construction materials are usually sold in English units in the United States.

Photos Jerry L. Trout

Figures 1a, 1b, and 1c. To reassemble this vandalized totem, Jim Werker used a simple epoxy and stack repair method with Epon 828 epoxy and Versamid 40 curing agent. This speleothem, repaired in the mid 1980s, still stands today in a New Mexico cave.

Wise product and material choices for cave environments are based on thorough research. Call on speleological consultants for input about any material that is considered for long-term underground use. Use educated common sense and constantly gather information about current best practices.

Sometimes referred to as archival epoxies and pure cyanoacrylate adhesives, the adhesive agents listed in this chapter have been used in subterranean applications over the past several decades and appear to be relatively safe for long-term installations in cave environments.

Epon® 828 With Versamid® or Epi-cure® 3234 (TETA)

- Epon 828 epoxy has been successful in underground environments and speleothem repair for decades.
- Epon 828 epoxy resin combined with Versamid 40 curing agent will bond dry surfaces, even in humid cave environments.
- For wet applications or speleothems with active dripping, Versamid 25 hardener will cure more efficiently. Epon 828 with Versamid 25 curing agent will bond underwater.
- The Epon family of archival adhesives will develop strong bonds with shear strengths up to 6,000 psi (41,370 kilopascal).
- Curing time can take 24–72 hours and sometimes longer in moist cave environments. Shrinkage is minimal. The bond is resistant to a broad range of chemicals.
- Mixing ratio is typically 1:1. Use one part Epon to one part Versamid (50:50 mix). However, for a more rapid drying time, use just a little more hardener for a 40:60 mix. Faster cure rates result in weaker bonds. Accelerating the drying time reduces the shear strength of the joint. A slower curing time typically increases the durability of the bond.
- Epon 828 and the Versamid hardeners were lab tested for long-term underground use at the U.S. Department of Energy Nevada Test Site and have proved successful for speleothem repair during several decades of use and observation.
- The combined epoxy and curing agent mixes to a creamy-white color and viscous consistency—when dry, it is colorless or a slightly shiny, translucent yellow.
- These products are available from the Shell® Chemical Company and through regional chemical or plastic product suppliers. However, the Versamid curing agents are increasingly difficult to locate and fabricators in the plastics industry are recommending a Shell replacement product, Epi-cure 3234 (TETA). Epi-cure 3234 is a highly concentrated curing agent with bonding properties and archival characteristics similar to Versamid.
- Epi-cure 3234 (TETA) is currently the recommended hardening agent for Epon 828 and is typically mixed in a 12:1 ratio, twelve parts Epon to one part TETA.

Hot Stuff® Super T and Special T

Fast-drying cyanoacrylate adhesives are useful for repairing soda straws, helictites, thin draperies, and other delicate speleothems or small applications in caves. Hot Stuff adhesives are industrial-strength products containing a very pure form of

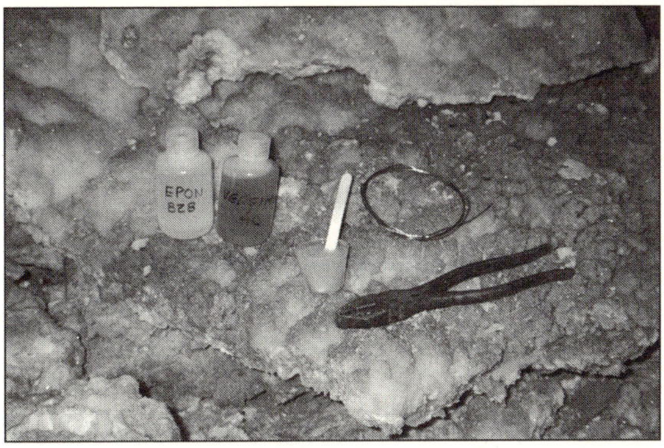

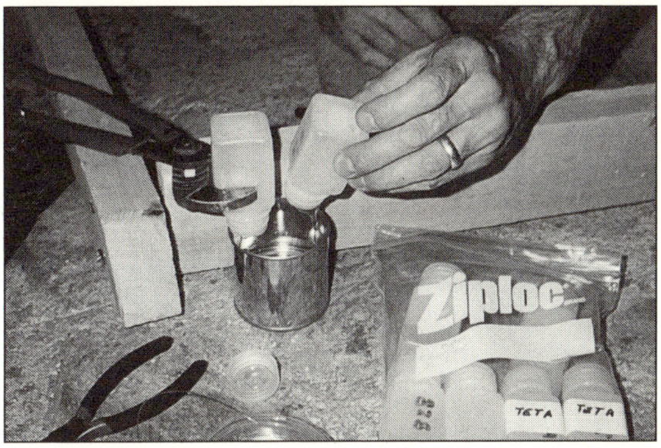

Figure 2 (Versamid) and **Figure 3** (TETA). Epon 828 epoxy mixed with Epi-cure 3234 (TETA) curing agent in a 12:1 ratio is a relatively safe adhesive for cave applications. The Epon and Epi-cure products are manufactured by the Shell Chemical Company and are available through regional chemical or plastic product suppliers. Versamid hardeners are also used with Epon 828. For speleothem repair, always have mixing cups, mixing sticks, and disposable gloves on hand. To ease handling the products in a cave, premeasure both the epoxy and the hardener into small containers.

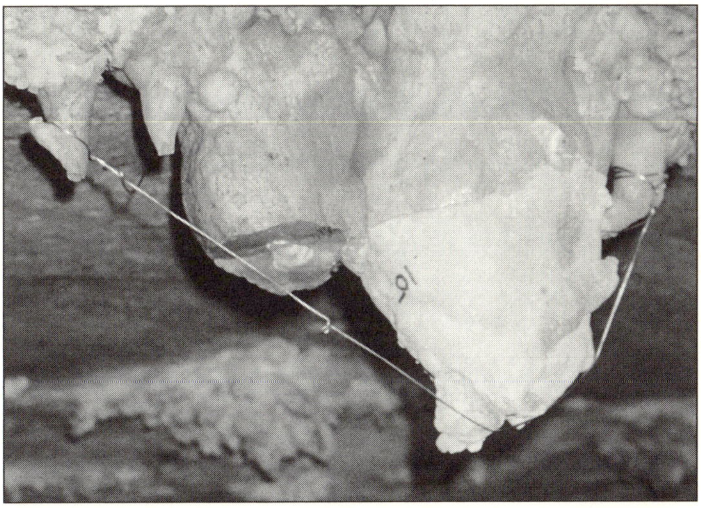

Photos © Val Hildreth-Werker

Figures 4a and **4b**. Hot Stuff Super T or Special T with NCF Mild Accelerator are relatively safe quick-setting adhesive products for cave applications. All Hot Stuff adhesives are industrial strength and contain a very pure form of cyanoacrylate that remains clear when it cures. Pieces are held in place with wire for curing (detail right).

cyanoacrylate that remains clear when it cures. Oil is not added to lengthen shelf life (as it is in many other instant glues).

- Hot Stuff Super T is often used in paleontological applications and model-building activities and works well for small repairs in caves.
- Hot Stuff Special T is a more viscous product that is formulated to fill in gaps and is extremely useful for cave applications.
- Both Hot Stuff cyanoacrylate adhesive products are clear and colorless when wet or dry.
- In most caves, Hot Stuff cures in 30–90 seconds but bonding can be accelerated with NCF Mild Accelerator (also a Hot Stuff product). Shear strength is weakened by accelerated curing times, but cyanoacrylate does not bond quickly in large quantities. One spritz from a spray pump bottle of NCF will cause chain reaction bonding.
- Hot Stuff Super T, Special T, and NCF Mild Accelerator can be purchased in model-building stores, quality woodworking shops, museum supply catalogues, and through sources on the Web.

Fast-drying cyanoacrylate adhesives are useful for repairing soda straws, helictites, thin draperies, and other delicate speleothems.

Stainless Steel Rods, Pins, and Wires

Chromium–nickel austenitic steels, commonly known as stainless steels, are more suitable than other products for speleothem repair. Stainless is tough and highly resistant to corrosion. Longevity is ten-fold that of mild steels. Stainless steels that are not austenitic will degrade more rapidly in most cave environments. Metals that work best underground include high austenitic stainless steels, particularly types 304L, 316L, and 321.

Chromium–nickel austenitic steels, commonly known as stainless steels, are more suitable than other products for speleothem repair. Stainless steel all-thread is the choice material for pin-stabilization of broken speleothems.

Stainless All-Thread

Stainless steel all-thread is the choice material for pin-stabilization of broken speleothems. The ridges of all-thread grab and adhere better than pins made from regular round-stock stainless. Most stainless all-thread is type 304. All-thread is stocked in assorted diameters and cut to the required lengths.

Installed as a stabilization rod for rehanging stalactites, all-thread will provide greater shear strength. If repairing naturally leaning stalagmites, the resulting shear load is better resisted with all-thread stabilization rods. Support pins made from regular round-stock stainless are usually successful for repairing stalagmites that have minimal shear load.

Consider the expansion characteristics of any material before using it to pin broken speleothems. Over time, corrosion will cause expansion, even when stabilization devices are epoxied inside of speleothems. Structural steel (also called mild or cold-rolled steel), plastic, fiberglass, wood, and nylon rods tend to expand and break apart the repaired cave formations when epoxied into the centers. Stainless is more corrosion resistant than most other materials. The corrosion-resistant characteristics of high-austenitic stainless steels minimize the potential for breakage from material expansion.

What about using other metals for pinning broken cave formations? Aluminum is not a good choice. (See characteristics of aluminum, page 169.) Some of the other alloyed stainless steels are very strong but their interaction with cave environments has not been studied.

Titanium is expensive and is sometimes alloyed with lead or other materials, but again, the interaction with cave environments is not known. There are other exotic metals and materials that might work, but they are expensive and no experience of use in caves is documented.

On the other hand, use of high-austenitic stainless steels in harsh environments has been documented for decades. Stainless materials are highly resistant, durable, and readily available.

Common sense should also play a major role in selecting materials for speleothem repair applications.

Stainless Steel Wire

Stainless wire makes sense from a theoretical materials standpoint and is a good choice for some speleothem repair jobs. In practice, stainless steel wire is difficult to bend and conform to the shapes of cave formations. However, if the wire is installed as a permanent material in a repaired speleothem, it may be worth the time and aggravation to use stainless wire.

Nevertheless, common sense should play a major role in selecting materials for speleothem repair applications. For example, copper wire bends easily, and while it should not be left on the outside of speleothems for long-term applications, it is extremely useful for temporarily holding epoxied seams together on repaired stalactites. Of course, copper wire should be removed as soon as the epoxy sets up, usually within a few days of application.

Stainless is the best all-around metal choice for long-term applications in caves.

What Size Support Pins?

Select pin diameter and length to fit the characteristics of the project. Determine the size of the pin according to the size and shape of the formation repair.

If you lack experience in bolt sizing, ask for help to decide the pin dimensions. Length depends on the application. We've used pins ranging in length from 2 inches up to 12 inches (about 5 to 30 centimeters). The smallest pin we've used is 1/16-inch diameter (1.6-millimeter). The largest is 5/8-inch diameter (1.5-centimeter).

If a speleothem is *not* likely to be vandalized again, use a short pin. The only real benefit a short pin provides is to prevent sliding from shear load while the epoxy cures. A longer pin provides additional strength.

When you have questions, seek consultation from cavers in construction trades, civil engineers, or those experienced in speleothem repair.

Repairs with Cement Products

Some cement products work well for some types of repairs in cave environments. Concrete is resistant to chemical and corrosive attack, has extremely good longevity characteristics, and introduces few toxins to most cave systems because the chemical composition resembles the natural composition of some rocks and cave passages. (See concrete, pages 169–170.)

If concrete products are chosen for speleothem repair projects, use high tensile strength cement with low-content calcium hydroxide. The calcium hydroxide in cement is very soluble—it will quickly dissolve, redeposit, and introduce new, unwanted "soda straws" or "flowstone" below or near cement-based structures or repairs. The concrete deposits grow at an accelerated pace compared to normal calcium carbonate speleothems. However, the concrete speleothem problem is reduced by using high tensile strength cement products that contain minimal amounts of calcium hydroxide. (See artificial fill removal, page 369; also see rimstone dam repair, page 476.)

Summary

Choose speleothem repair materials that are considered safe for cave environments. Archival-grade adhesives, stainless steel, and cement products with low-content calcium hydroxide are some of the best options. Contact trained interdisciplinary speleologists for review when other products or materials promise cave-safe qualities.

Choose speleothem repair materials that are considered safe for cave environments. Archival-grade adhesives, stainless steel, and cement products with low-content calcium hydroxide are some of the best options.

Choose Cave-Safe Speleothem Repair Materials

Jim C. Werker

Cave-Safe Archival Epoxy

- Epon® Resin 828 with Versamid® Curing Agent or approved equivalent.
  - Versamid® 40 for dry surfaces, even in humid conditions.
  - Versamid® 25 for wet, dripping, or underwater applications.
- Tetra® is currently replacing the Versamid products.
- Minimal harmful outgassing.
- Does not support growth of fungi.
- NOT just any ole' hardware store epoxy.

Pure Form of Fast-Acting Cyanoacrylate Adhesive

- Hot Stuff® Super T or Special T.
- Hot Stuff® NCF-Mild Accelerator (non-CFC).

Austenitic Stainless Steel Pins and Wires

- Stainless steel all-thread for stabilization rods and pins.
- Stainless steel wire.
- Other materials will expand and break speleothems. For example, steel pins will rust, expand, and discolor.

Cement

- Quik-crete® or quick-set cement.
- Form and sculpt by hand to match rimstone dams.

Use Cave-Safe Natural Materials for Coloring

- Grind cave rocks and mix with approved epoxy.
- Test soil or dust from the cave floor for color mix with epoxy.
- Try natural chalks to match unusual colors. Charcoals may work for matching dark colors.
- Avoid oil-based materials.

Test All Materials *In Situ*—Avoid Surprises

- Some products will grow fungi or introduce harmful nutrient sources for cave organisms.
- Test first. Some agents will become unstable in underground applications.

Section A—Introducing Speleothem Repair

Speleothem Repair Supplies and Equipment

Jim C. Werker

Based on some of the information describing cave-safe repair materials in the previous chapter, this is a general supply and equipment list for speleothem repair projects. Battery-operated power tools are required for many cave repairs. If electricity is available, standard power tools and industrial extension cords may be preferred. The tools and supplies included here provide a starting point for making site-specific lists. All items are not required for every job.

Equipment Tips

A battery-powered drill with a steel bit is sufficient to drill holes in medium-sized speleothems with diameters up to 4 inches (10 centimeters). For large speleothem repairs, the Hilti® battery-driven hammer drill is capable of drilling larger diameter holes to a greater depth to accommodate large stabilization rods. Take special care to avoid shattering the speleothem. The hammer function should not be used on small speleothems—a hammer drill will definitely shatter any small pieces.

Small battery-powered drills with steel or concrete bits work okay for drilling speleothems. Carbon steel bits last longer. However, regular steel, straight-shank twist drill bits are consistently more successful. When using the hammer function, it is important to use bits designated for hammer drills.

Have an assistant hold a large zip-closure bag under the drill to catch the drill dust. Use a can of compressed air to blow the dust out of the hole and into the zippie. There are two reasons to catch and retain speleothem drill dust and powder from drilling the holes for support pins.

- Drill dust is the best media to mix with epoxy for color-matching.
- Capturing the dust prevents it from becoming a powdery mess in the cave passage.

A hard-bristled toothbrush and pocketknife are handy for digging out compacted dust from damp drill holes. Some sites also need to be protected from repair operations with large drop cloths, nylon tarps, or plastic sheeting. Large plastic bags can be spread out as staging areas. However, in some caves, the plastic becomes very slick if the repair crew walks on it. Think, plan, and prepare for

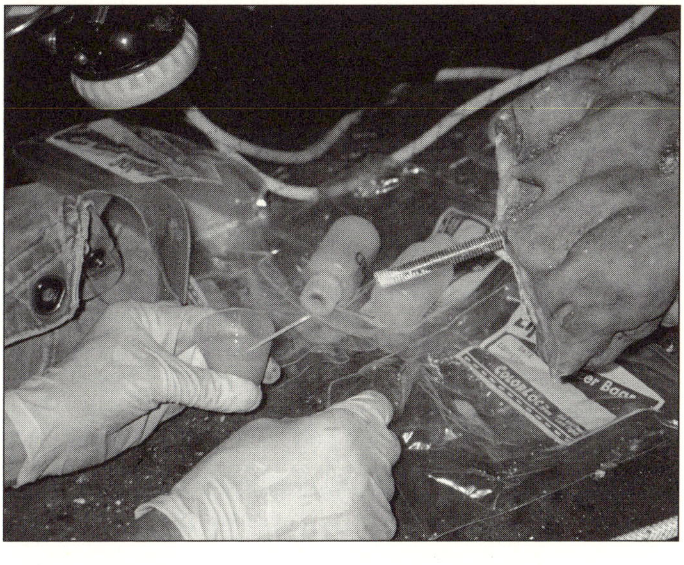

© Val Hildreth-Werker

Figure 1. If a speleothem repair needs increased stabilization and strength, use a stainless steel all-thread support pin. Epoxy the pin into the upper part of a broken speleothem. A larger hole in the mating section will easily accept the pin. Half-fill the larger hole with epoxy.

safety as well as cleanup procedures.

Equipment and Materials List

Adhesives

Approved archival-grade epoxy. Epon® 828 epoxy resin and Versamid® or Epi-cure® 3234 (TETA) curing agents are recommended. Use Versamid 40 catalyst for dry or humid applications or Versamid 25 catalyst for wet or dripping applications. Epi-cure 3234 is replacing the Versamid products. Mix Epon 828 with Epi-cure 3234 in varying ratios depending on the conditions. Standard mix is 12:1, twelve parts Epon 828 to one part Epi-cure 3234. Both Epon and Epi-cure are made by the Shell® Chemical Company.

All glues break down over time, but archival products are formulated to do less harm than those on the general market.

Approved museum-quality pure cyanoacrylate adhesive. Hot Stuff® Super T, Special T, NCF-Mild Accelerator, and a cyanoacrylate solvent are recommended.

Mixing cups for epoxy. Plastic specimen or medication cups provide measurement lines for quantities of resin and hardener.

Craft sticks for mixing and applying epoxy. Disposable dental tools also work well.

Paintbrush for applying epoxy. Usually 1/2-inch (12.5-millimeter) width works best.

Surgical gloves or vinyl gloves. Nonlatex, powder-free gloves are best.

Lint-free rags or disposable towels. Clean up as you go.

Small bottle of isopropyl (rubbing) alcohol. Use alcohol to clean and dry the mating ends of speleothems.

Easy-to-operate cigarette lighter. Avoid leaving dark residue on the speleothems if using a lighter to dry the mating ends. Do not overheat the speleothem pieces. Lightly dry only the broken surfaces—just enough for the adhesive to bond.

Mortar and pestle. Pulverize rocks for the epoxy mix. Or use a thick plastic or metal bowl with rounded bottom and a ball-peen hammer for crushing rock.

Drill and Pin

Stainless steel all-thread. Usually 3/16-inch or 1/4-inch (4- to 6-millimeter) diameter is sufficient for supporting small to medium speleothems.

The ridges of stainless all-thread grab and adhere better than pins made from regular round-stock stainless.

Hacksaw. Cut the stainless all-thread to useful lengths. Cut PVC pipes.

Magic Markers®. Thick, black indelible markers work best. Cut the tip off and position it in the center of two mating pieces to leave a guide mark for drilling.

Pencils. Make pencil tick marks across joint alignments to assure a quick fit after applying the adhesive.

Electric drill. Use straight-shank twist drill bits. Bits made of carbon-steel will last longer.

Hammer drill for large formation repair. Find bits suitable for hammer drills.

Dremel® tool with grinder and sander attachments. Use for texturing and finishing.

Canned air. Blow dust off surfaces and out of declivities.

Small sponge. Clean up drill dust.

Q-tips®. Clean up drill dust.

Small, soft paintbrush. Clean up drill dust.

Toothbrush. Clean up drill dust.

Drop cloths. Size depends on the repair job.

Catching and Saving Speleothem Drill Dust

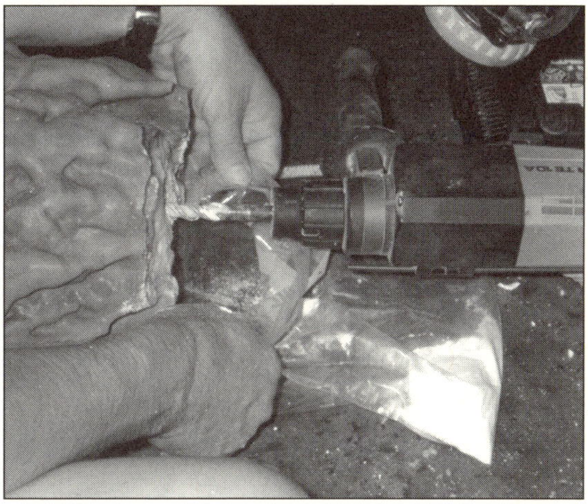

Figures 1 and **2**. Large trash bags will easily capture most drill dust. If the powder is needed for filling cracks, gaps, or drill holes, capture and save it in a Ziploc.

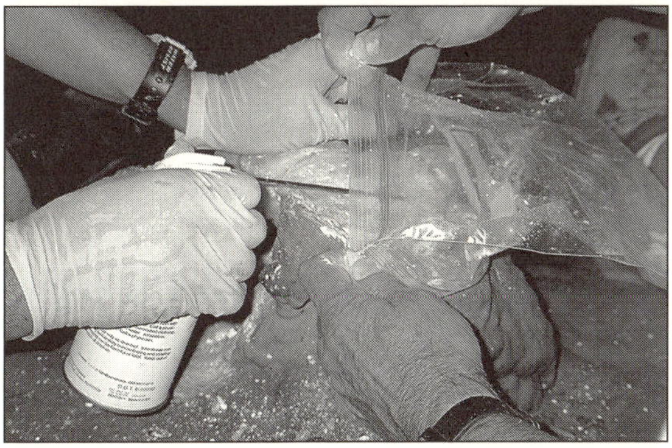

Photos ©Val Hildreth-Werker

Figure 3. Compressed air in a can is handy for removing drill dust if the dust can be captured in a plastic bag.

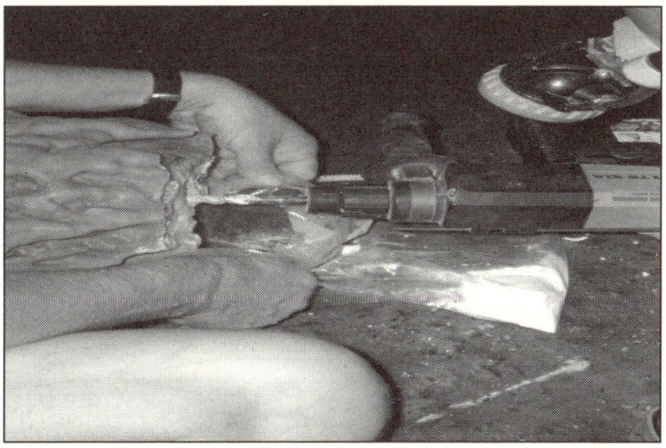

Figure 4. Toothbrushes are good tools to carry in speleothem repair kits. Use a toothbrush to clear the drill dust from large drill holes and to aid in the final cleanup of repair sites.

Plastic garbage bags. Use for trash, drill dust, and surface protection.
Large Ziploc® bag. Collect drill dust to use in epoxy and rock powder mixtures.
Goggles. Wear eye protection.

Support and Brace

Wire. Assorted gauges of stainless, copper, and bailing wire are handy.
Wire-cutters.
Pliers. Take both needle-nose and flat.
Small files. Both flat and round are useful.
Pocket knife or Leatherman®.
Inner tube. Use bicycle inner tube cut into various widths as large rubber bands.
Light bungee cords. Sometimes bungees are useful for pulling tension on the curing break joint of a leaning speleothem.
Rubber bands. Include a variety of widths and lengths.
PVC pipe. Determine diameter and length by size and shape of broken

Drill dust is the best media to mix with epoxy for color-matching. Capturing the dust prevents it from becoming a powdery mess in the cave passage.

speleothems. Assorted dimensions are useful.

Shims and wedges. Wood, metal, or plastic—various sizes from toothpicks to blocks.

Small jacks. Hydraulic jacks and scissors jacks work equally well.

Duct tape.

Assists and Safety

Structural steel, pipe, lumber, and chains. Build structures to move and support large cave formations. Select materials to make scaffolding, gantries, frames, and A-frames. (See specialized mechanical assist devices, pages 487–490.)

Clamping devices. Clamps are useful only for clamping boards or other bracing devices to mechanical assist apparatus. C-clamps, hose clamps, and other clamping devices should not be applied directly to speleothems.

Come-along and chain hoist.

Flagging tape. Delineate safety zones and speleothem repairs that need time to cure. Use *non*biodegradable flagging tape in caves.

Section B—Repairing Speleothems

General Techniques for Most Speleothem Repairs

Jim C. Werker and Val Hildreth-Werker

This section starts with basic techniques that apply to most speleothem repair situations, and then moves from simple repairs through more complex or specialized procedures. Topics in this part of the book describe principles for repairing stalagmites, columns, stalactites, draperies, flowstone, gypsum crusts, rimstone dams, helictites, frostwork, soda straws, and other secondary cave deposits.

This chapter covers basic techniques that apply to most speleothem repair situations. Subsequent chapters start with simple repairs and move through specialized procedures.

Mating Surfaces Must Be Clean and Dry

For proper bonding, the two matching speleothem surfaces must be absolutely clean and dry. Water in a spray bottle and a nylon-bristled brush will normally accomplish the cleaning task.

Cave passages are usually humid, speleothem surfaces are often wet, and residual moisture may inhibit bonding. Superficially dry the surfaces with a lint-free cloth and a can of compressed air (the kind designed to blast dust and dirt from computers). Then use rubbing alcohol to swab the mating surfaces. As the alcohol evaporates, the moisture will dry enough for the epoxy to bond the pieces.

Though most speleothem restorationists have tried heating speleothem breaks with torches to dry them out, heat is not recommended. The pieces tend to discolor and the heat often results in overdrying and spalling. If a broken speleothem is seasonally wet, schedule the repair during a drier time of the year unless new mineral deposits are quickly building on the broken joint.

Marking Pieces to Facilitate Alignment

When two mating pieces of speleothem are properly aligned, they snap or click into place with satisfying assurance, like the pieces of a three-dimensional puzzle. Once the click-point is located, place a few pencil tick marks across the break joint. The pencil ticks allow a quick matchup when the epoxy or cyanoacrylate adhesive is applied. Light pencil marks usually can be removed with a gloved finger.

Marking for Drill Alignment

What's the trick for drilling the holes so the all-thread support rod will fit properly and allow perfect alignment? Take a thick indelible felt-tip marker—black is the preferred color. Cut off part of the tip. Lay the black, inky chunk in about the center of the bottom section and click-fit the mating top section in place.

When the pieces are separated, a nice inky dot is visible on each section. Drill on those dots. Use the dots as guide points while assistants spot the

© Val Hildreth-Werker

Figure 1. Mark a tick across the break joint to create a means for quickly matching two mating pieces, especially when applying quick-setting cyanoacrylate adhesive. Pencil tick marks are preferred when the speleothem surface

drill operator to assure the drilling angles are aligned congruently for pinning the two pieces.

Always drill an oversized hole in the bottom section to allow fudge room for misalignment. The larger bottom hole provides space for the stabilization pin to shift so that a perfect click-fit can be achieved for the speleothem joint.

Tracing Shapes

When broken speleothem sections are too large or too awkward to lift into place for a test fit, use an alternate method to establish alignment. The following procedure is more tedious than the ink-dot method, and success is dependent on the speleothem shape.

Drill a hole for the support pin in the broken section. Over the broken and drilled end, place a piece of tracing paper, Mylar®, wax paper, plastic wrap, or Glad® Press-N-Seal®. Trace a line around the outer edge of the speleothem cross-section and mark the center of the drilled hole. Transfer the tracing media to the mating section, adjust the pattern for alignment, and drill a hole at the center mark.

This tracing method is also helpful when broken pieces are strewn throughout a cave. Tracings can be carried around the cave to match up with potential mating sections.

Bracing and Propping Devices

All kinds of devices can brace speleothems in place while the epoxy cures. Soft copper or aluminum wire offers excellent malleability for temporarily attaching sections of small speleothems while the epoxy dries. Soft elastics serve well for very small formation repair. Any combination of copper or aluminum wire, large rubber bands, dental elastics, inner tube, upholstery hooks, PVC pipes, boards, sand bags, and duct tape can be used to secure and support attachment points while the epoxy cures.

All of these materials should be removed from the cave as soon as the epoxy sets up. Only stainless steel products should be used for permanent installations. Some of the simpler propping ideas are introduced here, and

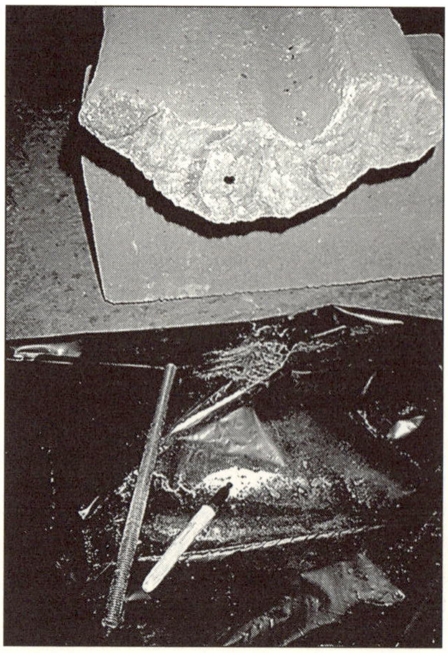

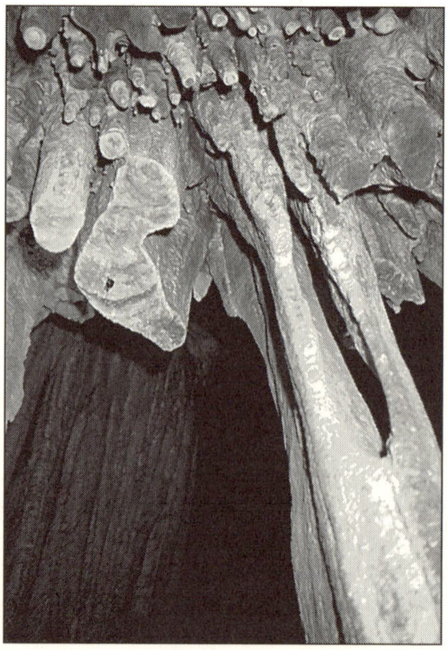

Photos © Val Hildreth-Werker

Figure 2 and **Figure 3**. Make sure the two mating pieces will click-fit. Cut off the tip of an inky felt-tip marker. Put the ink tip in the center of one broken speleothem piece. Click-fit the mating piece and the ink tip leaves matching dots on the two broken surfaces. The ink dots serve as guides for drilling positions.

more complex mechanical assists are included in another section of this manual (pages 487–494).

Use pipe, tubing, rocks, blocks, buckets, wedges, shims, jacks, tripods, and most anything else to prop, brace, or hold speleothems in place while the epoxy fully cures. Obviously, the tool needs to fit the job. Speleothems come in all sizes and shapes and so do propping devices. Typically, the archival grade epoxies will set in 24–72 hours, depending on temperature and humidity in the cave. Small speleothems repaired with quick-setting cyanoacrylate adhesives will be secure in 30–90 seconds and are often easily held in place by hand until the adhesive dries. A few field proven ideas are included here.

Rubber Bands. Large rubber bands, inner tubes with wire hooks, light bungee cords, and assorted types of wire are all useful for supporting smaller speleothem repairs while the epoxy sets.

Bungees. For top-heavy, tilted, or leaning stalagmites, installation of a support pin is usually necessary. Stalagmites can be braced with bungees or circles of inner tube arranged to pull in opposing directions. Keep a tight fit at the joint while the epoxy cures.

PVC Pipes. Use polyvinylchloride (PVC) pipe for bracing underneath stalactites while the epoxy sets. Cut job-specific lengths of PVC pipe to fit between the point of the stalactite and the cave floor or flowstone surface below. The drip end of many stalactites will fit nicely into the top end of the PVC pipe, and the rigid circumference of the pipe provides constant pressure to hold the speleothem in place. PVC diameters vary according to the job. Allow the PVC brace to remain in place until the epoxy cures thoroughly. Buckets, cans, wedges, and rocks are effective for propping underneath the PVC and forcing a tight fit on the epoxy joint. (See photos of PVC propping up repaired speleothems, page 466.)

Scissors Jack. For large stalactites, first establish and mark alignment of the mating pieces, install the support rod, apply epoxy to both sides, allow the epoxy to seep into pores and become tacky, then use a padded scissors jack or similar device to hold upward force against the stalactite while the epoxy cures. (See photos of Colossal Cave stalactite repair, page 504.)

Braces. Descriptions of a variety of other mechanical assists, leveraging devices, and specialized braces for delicate tasks are included toward the end of the speleothem repair section. (See mechanical assists, pages 487–494.)

Cave-Safe Repair Summary

Epon 828 with Epi-cure 3234 (TETA) or Versamid 40 curing agent (Versamid 25 for wet applications)
- Minimal harmful outgassing
- Does not support growth of fungi
- NOT just any hardware store epoxy

Stainless steel support pins and wire
- Austenitic stainless steel all-thread for pins
- Steel pins rust, expand, and discolor
- Stainless wire for permanent installations
- Test all materials before using

Only cave-safe natural materials for color-match.

Filling Cracks, Gaps, and Seams

Larger, less delicate speleothems—stalagmites, columns, stalactites, and draperies—may have gaps, cracks, fractures, or missing chunks that can be filled with a color-matched epoxy and rock dust mixture. Use pulverized cave rocks, powder from drill-holes, cave soil, clean crystalline limestone, or find an irreparably damaged speleothem near the repair site.

To achieve a workable consistency and match the speleothem color, use crushed and powdered media mixed with an approved epoxy. Apply multiple thin layers of the epoxy and rock dust mixture. Manually texture as each layer dries. Or use a Dremel® tool, file, or other implement to sculpt textures after the epoxy cures. Over time, active drips may further heal and cover the repaired seams and fracture sites.

Fill Holes and Cracks

- Mix archival epoxy with rock dust.
- Used crushed rock, drill dust, or soil from the cave.
- Match the texture after epoxy cures.
- Over time active drips may heal the gaps.

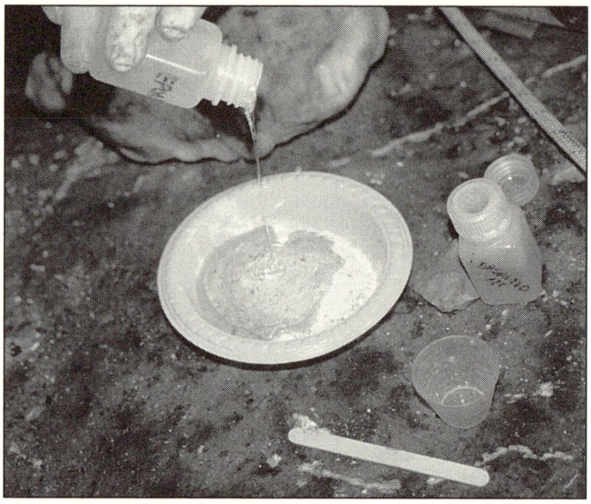

Figure 4. Mix archival-grade epoxy with crushed rock for filling cracks and seams.

Figure 5 (left). Pulverize bits of rock and mix with archival epoxy for coloring and filling cracks and seams.

Figure 6. Mix drill dust or crushed rock with an approved epoxy to create color and a workable consistency for filling gaps and cracks.

Figure 7. Press the epoxy and rock dust mixture into cracks and holes. Create texture as the epoxy is curing.

Photos © Val Hildreth-Werker

Color Matching

- Find matching rock dust color and mix with archival epoxy.
- Use crushed rock, drill dust, or soil from the cave.
- If natural chalks or charcoals are necessary, test them first.
- Take samples into sunlight to compare. Caving lights will misguide color vision.
- No synthetics, no oils—use only natural materials for coloring.

Natural Materials for Color-Matching

It is best to match speleothem color by using natural substances from the cave. Soil, rock, dust, or old broken pieces from the cave are easy to obtain, crush, and use as a natural aid for color matching. If no *in situ* materials are available, other natural materials such as chalk, charcoal, or minerals from outside the cave may work for color matches. However, first check out potential chemical interactions between materials and avoid unintended results. Always test small samples of materials and make inquiries with experts. Oil-based substances will introduce toxins to cave environments and should not be used for color matching.

Take samples outside in the daylight to compare the color of the cured epoxy mixture to the speleothem color. Cave lights are tricky. Color temperature variations can make rods and cones (the color sensors in human eyes) lie to the brain under cave light sources. Comparing color in daylight is the only way to achieve a close match.

Fabricating Parts and Pieces

While it may be appropriate to reconstruct sections to fill in missing speleothem pieces, faux speleothems should not be installed in caves, and totally absent speleothems should not be fabricated.

Current best practice philosophies in cave restoration frown on the idea of fashioning entire replacement parts for broken speleothems. Most cave managers agree that it is improper to go that far. If half of the formation is missing and the piece cannot be found, then it is better to just leave the diminished speleothem as is.

Fabrication techniques should be reserved for filling gaps, cracks, holes, and seams, not for creating fantasy speleothems or fake formations. (See fabricating missing pieces, page 473.)

Fabrication techniques should be reserved for filling gaps, cracks, holes, and seams, *not* for creating fantasy speleothems or fake formations.

Finishing Techniques

Rather than introducing finishing products into cave environments, use texturing tools and natural materials on the epoxied seams and filled cracks to mimic the appearance of natural textures. Technical ideas and basic principles are listed here.

Apply Epoxy in Thin Layer

- Do not apply epoxy all the way to the edges of speleothem joints—leave space for the epoxy to spread when the pieces are pressed together. Use an extremely thin layer of epoxy and try to avoid oozing at the seams.
- When ooze happens, remove it with lint-free rags or paper towels, popsicle sticks, surgical gloves, and other disposable tools. Veni (1997) suggests that a rag draped over a knife blade will remove early ooze from along a repair joint.
- After the epoxy dries, clean up any hardened ooze with files and sandpaper.

Finishing Textures

- Texture the repair to mimic the natural surface by adding and sculpting successive layers with an epoxy and rock dust mixture. Each layer

Photos © Val Hildreth-Werker

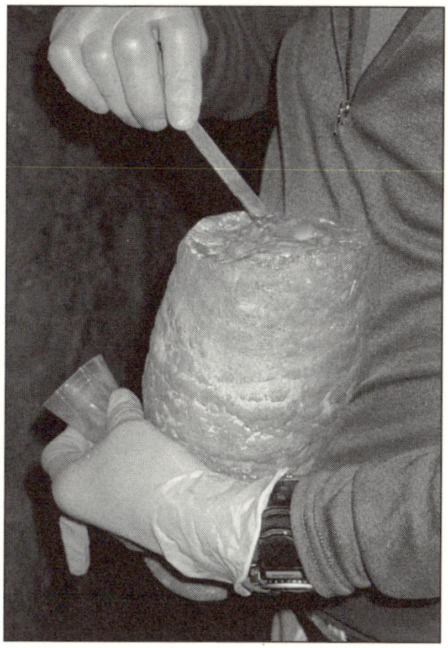

Figure 8. Apply epoxy in an extremely thin layer on speleothem break joints. Avoid spreading it out to the edges of the broken surface. If epoxy does ooze at a crack, be prepared to mop the seam with lint-free rags, mixing sticks, and other disposable tools.

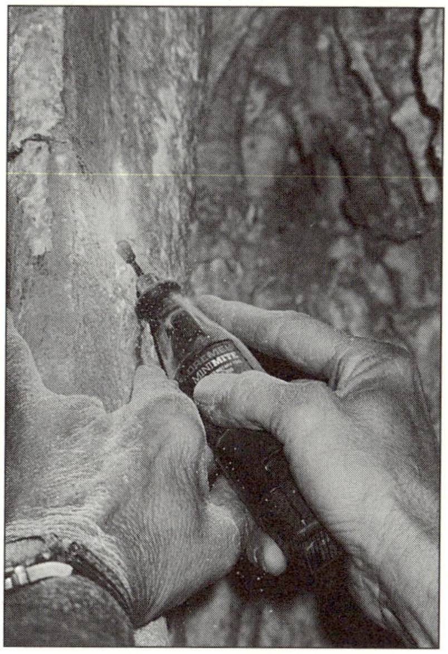

Figure 9. Try to blend, camouflage, or conceal the repair joints. Texture epoxied seams with a Dremel tool, sand paper, dental picks, drill dust, soil, or sand.

must dry before another is applied. To diminish the shiny look of dried epoxy, dust or sand is added on top as the final layer cures. Experiment with files, dental tools, utensils, and other texturing tools.
- Use dental tools to texture the epoxy before it completely cures.
- After the epoxy hardens, use a Dremel tool with a small grinder to create a matching texture and mitigate the shiny, wet appearance of cured epoxy.

Texture the repair to mimic the natural surface by adding and sculpting successive layers with an epoxy and rock dust mixture.

Avoid Paints and Surface Finishes
- Paints and other surface finishes will introduce toxins and should never be applied to speleothems.
- More information on the detrimental effects of paints and finishes is available in this volume (page 173).

Additional Reading

Meyer PJ, French JA. 1995. Cathedral Caverns: Techniques for restoration and cleaning. *NSS News* 53(7):186-193.

Meyer PJ, French JA. 1996. Crossings Cave: Test bed for cave restoration. *NSS News* 54(2-3):54-58.

Veni G. 1995. The restoration of a vandalized show cave: Cave Without a Name, Texas. In: Pate DL, editor. *Proceedings of the 1993 National Cave Management Symposium: Carlsbad, New Mexico, October 27-30, 1993*. [Huntsville (AL)]: National Cave Management Symposium Steering Committee. p 65-72.

Veni G. 1996. Primer to speleothem repair. *NSS News* 54(2-3):64-67.

Veni G. 1997. Speleothems: Preservation, display, and restoration. In: Hill C, Forti P, editors. *Cave Minerals of the World*. 2nd edition. Hunstville (AL): National Speleological Society. p 301-309.

Werker JC. 1995. Formation repair. In: Pate DL, editor. *Proceedings of the 1993 National Cave Management Symposium: Carlsbad, New Mexico, October 27-30, 1993*. [Huntsville (AL)]: National Cave Management Symposium Steering Committee. p 61-64.

Werker JC. 1996. Formation repair. *NSS News* 54(2-3):65-67.

Werker JC. 1996. Proven techniques in repairing formations. In: Rea GT, editor. *Proceedings of the 1995 National Cave Management Symposium: Spring Mill State Park, Mitchell, Indiana, October 25-28, 1995*. Indianapolis (IN): Indiana Karst Conservancy. p 308-310.

Werker JC, Hildreth-Werker V. 2004. Stalactite repair: Large speleothem hangs again in Colossal Cave, Arizona. *NSS News* 62(4):98.

Section B—Repairing Speleothems

Stalagmite Repair

Jim C. Werker

Most stalagmites, even totems, can be repaired if the mating pieces are available. Methods vary from simple stacking procedures to more complex repair techniques using epoxy, stabilization pins, and bracing devices. Little evidence of damage is apparent after most repairs are completed.

Stack It and Hope It Heals

Sometimes a really simple repair strategy works for stalagmites. If water is actively dripping on broken speleothems, fit the pieces back in place and allow mineral deposition and time to take course.

Epoxy and Stack

At repair sites with no healing drips, or in locations where repairs may require reinforcement, clean the broken joints and use an approved archival epoxy such as Epon® Resin 828 with a Versamid® or Epi-cure® 3234 curing agent. (See cave-safe epoxies and adhesives, pages 445–447.)

Starting from the center, apply a thin layer of epoxy on the broken base or stump of the stalagmite. To avoid oozing along the seam, do not spread the epoxy all the way to the edges. Fit the upper section into place, and allow several days for thorough curing. For multiple breaks, sometimes several

Stalagmite Repair: Stack to Heal

- Simply restack the pieces.
- Make sure each joint mates and conforms properly.
- Active drips may heal the break lines.

Stalagmite Repair: Epoxy and Stack

- First, stack the pieces and confirm the fit.
- Clean and dry the break joints.
- Apply a very thin layer of archival epoxy to the base.
- Align and stack.
- Brace if necessary.
- Allow time for the epoxy to set and dry.

Stalagmite Repair—Epoxy and Stack

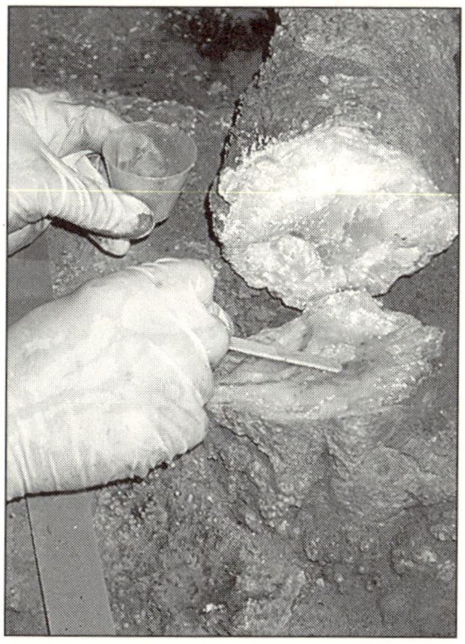

Figure 1. Spread a thin layer of archival epoxy on the clean, dry surface of the stalagmite stump.

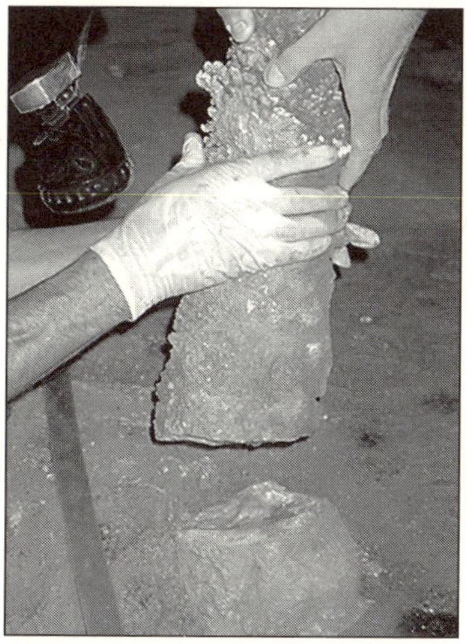

Figure 2. Click-fit the upper piece onto the base and allow several days for the epoxy to thoroughly cure.

Figure 3. The epoxy has set and the repair is complete. The stalagmite stands again.

Photos © Val Hildreth-Werker

> **Mating Speleothem Pieces**
> **John A. French and Paul J. Meyer**
>
> The owner of a small commercial cave once described a simple, but questionable method of formation repair. On finding a broken-off *orphan* stalagmite, he would simply jam the end into a mud bank where a small hole indicated a steady drip from the ceiling.
>
> While the purist will wince at this parody of serious renovation, it illustrates a problem. It is not always possible to find a stalagmite stump that matches the upper part of the formation.
>
> On the other hand, when a match can be found, a simple set-in-place method can be the easiest type of repair.
>
> The mating parts are often easy to identify by size, shape, color, and surface texture. When placed one on the other, and adjusted, there is often a satisfying tactile feedback when the two surfaces seem to snap together.
>
> The match is obvious when the broken edges conform properly and the alignment looks and feels right.
>
> Color of the two sections may vary if they have been separated for a long time, but the surface textures—knobby, smooth, rough, pitted, and so on—should be in good agreement.

broken stalagmite sections can be restacked during one session (Figures 1, 2, and 3).

Stalagmite Repair—Pin and Epoxy

When a broken stalagmite is in a high-traffic or precarious area where it is likely to be damaged again, drill a center hole in the upper and lower mating sections and use an archival epoxy to secure a stainless steel stabilization pin. For tilted, leaning, or top-heavy stalagmite breaks, reinforce repairs with stainless all-thread rods. (See repair photo sequence of a stalagmite on a slope, pages 443–444.)

Make certain the broken pieces fit and align properly, clean and dry the break joints, and mark center guide dots for drilling. (For techniques on marking the pieces to facilitate alignment, see page 455.) In the upper section of the broken stalagmite, drill a hole the diameter of the stainless all-thread pin. Clean out the drill dust with canned air and a small brush. Apply epoxy to the pin, insert the pin in the drill hole, and allow curing to begin while drilling the lower section.

Drill a slightly oversized hole in the lower section, remove the drill dust, and half fill the hole with epoxy. Apply a thin layer of epoxy to the broken joint of the lower piece, but avoid spreading it all the way to the edges or goo will ooze from the seam and create cleanup challenges. Fit the upper and lower pieces together and brace if necessary until the epoxy cures. Add final texturing to match or mimic the original surfaces of the stalagmite.

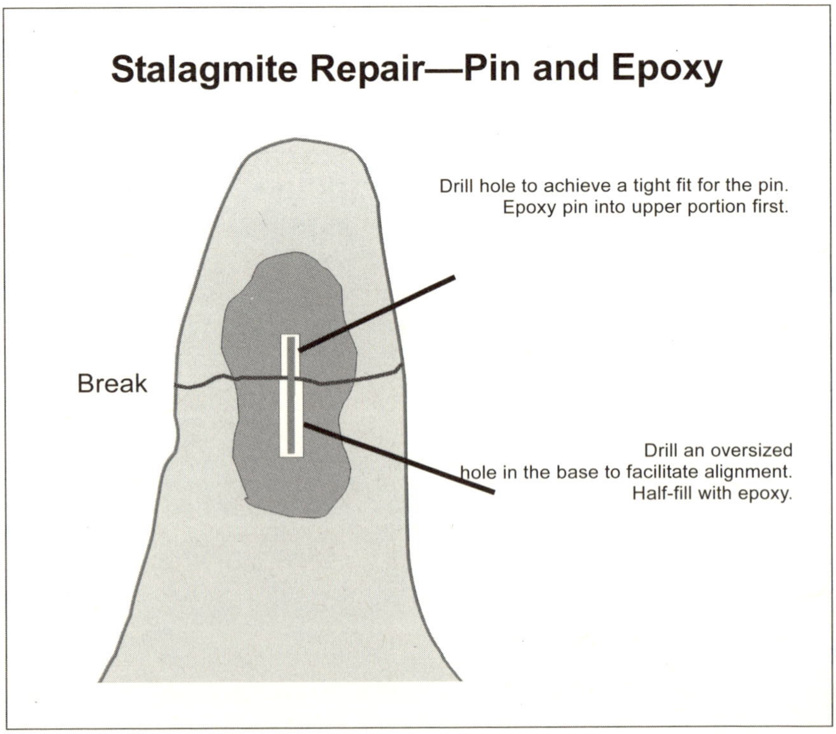

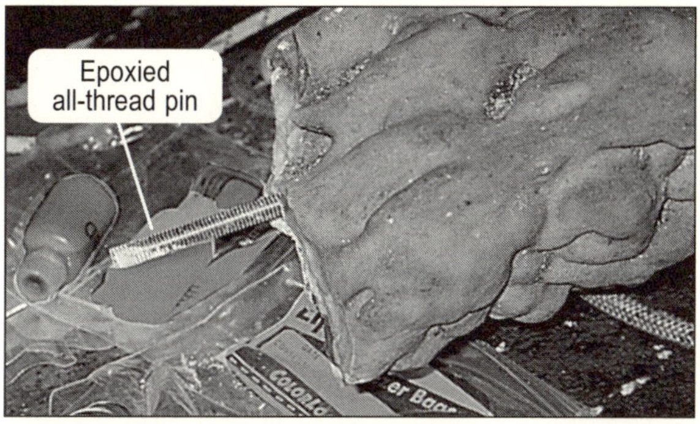

Figure 4. After drilling a hole in the center of the upper portion of a broken stalagmite, epoxy a stainless steel all-thread stabilization pin in place with Epon 828 and Epi-cure 3234 (TETA) curing agent.

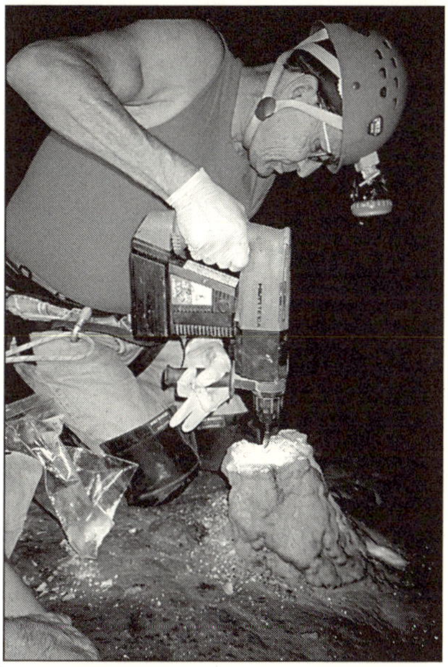

Figure 5. While allowing a little time for the epoxy in the upper piece to begin to set, drill a larger hole in the base of the stalagmite.

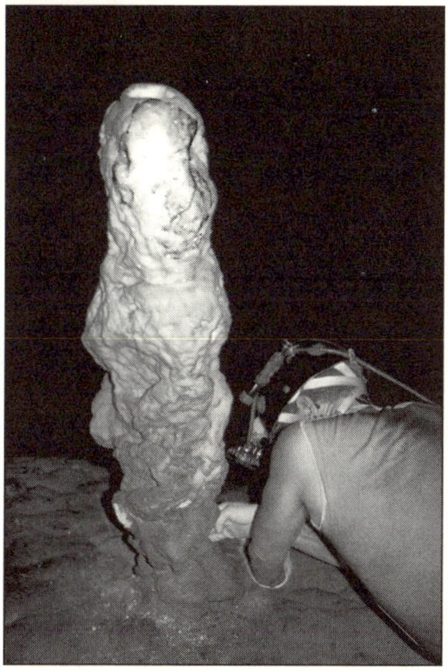

Figure 6. Half fill the larger bottom drill hole with cave-safe epoxy. Align the top piece and check the fit. Ideally, a tight click-fit achieves a tidy seam. If necessary, use an epoxy and rock dust mixture to fill cracks and gaps. (See stalagmite repair photo sequence made in Slaughter Canyon Cave, page 443.)

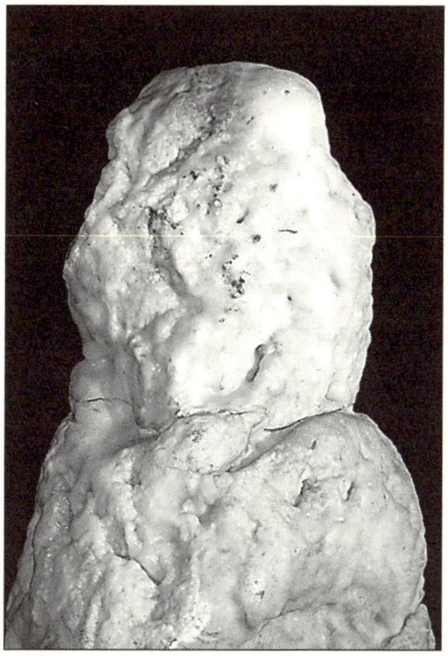

Figure 7. The top joint of this repair made a tight click-fit.

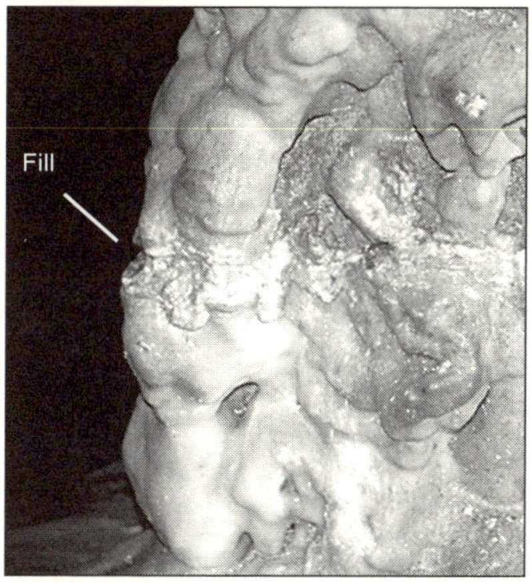

Figure 8. The bottom joint of the stalagmite had a large gap that was filled with an epoxy and rock dust mixture.

Photos © Val Hildreth-Werker

Stalagmite Repair: Pin and Epoxy

- Match the pieces and adjust for a sure fit.
- Cut part of the tip off a black indelible felt-tip marker.
- Place the marker tip in the center of the broken base.
- Align and set the top onto the base and check the fit—black ink marks will remain on both pieces to guide the drill.
- Drill the top piece to create a tight fit for the stainless all-thread.
- Clean drill dust from the hole and epoxy the all-thread into the upper section.
- Drill a slightly larger hole in the base to facilitate alignment of the two pieces.
- Clean drill dust out and half fill the bottom hole with epoxy.
- Fit the top piece that already has the all-thread epoxied in place.
- Adjust until the mating pieces align in a click-fit.
- Fill the gaps and seams with a color-matched epoxy and rock dust mixture.
- Brace if necessary and allow time for epoxy to dry.
- Sculpt the epoxy surfaces that are visible to match or mimic textures of the original speleothem.

Section B—Repairing Speleothems

Stalactite Repair

Jim C. Werker

Most stalactite repairs require a support pin since gravity is working against the epoxy. After confirming alignment of the broken pieces, clean and dry the break joints, then make inky guide dots for drilling the holes that will receive the pin. Some speleothems also need alignment tick marks across the break joint. (See cave-safe epoxies and adhesives, pages 445–447; also see marking pieces to facilitate alignment, page 455.)

Stalactite Repair: Epoxy and Prop

- Apply archival epoxy (or cyanoacrylate adhesive for smaller speleothems).
- Brace to hold a tight-fitting joint while adhesive dries.

General Stalactite Procedure

In the upper section of the stalactite, drill a hole that matches the diameter of the stainless all-thread pin. Clean drill dust from the hole, apply epoxy to the pin, insert the pin in the drill hole, and allow curing to begin while drilling the lower section. (Some stalactite repairs will need 24 hours for the pin to begin to dry in the upper section.)

Drill a slightly oversized hole in the lower section, remove the drill dust, and half fill the hole with an archival epoxy such as Epon® 828 with Epi-cure® 3234. Also apply a thin layer of epoxy to the broken joint of the lower piece, but avoid spreading the glue all the way to the edges. Fit the upper and lower pieces together and brace, prop, or wire with the methods described in illustrations on the following pages.

Tiny Stal Repairs

Tiny broken stalactites are easy to repair with a pure form of cyanoacrylate adhesive such as Hot Stuff® Super T. (See cave-safe epoxies and adhesives, page 446.) Check to make sure the pieces are clean and will snap-fit. Then apply a dot of Super T to the broken piece, attach to its mate, and hold in place for 60–90 seconds. For soda straws, apply two or three miniscule dots of Super T around the edge to leave the central canal open. (See fragile speleothem repair, page 485.)

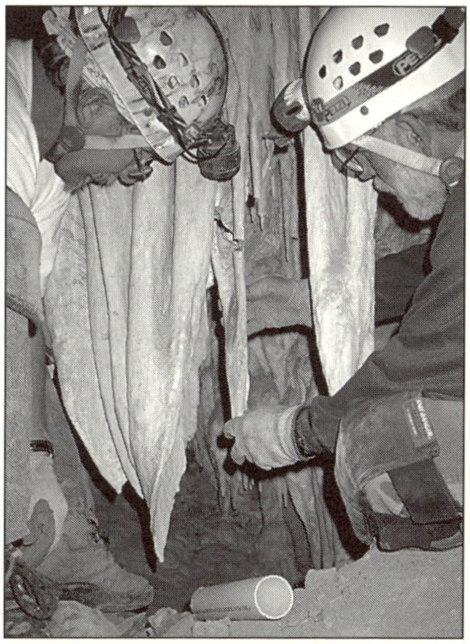

© Val Hildreth-Werker

Figure 1. When stalactites are too thin to install a support pin, and when the broken section is not too heavy, try repairing the joint with an archival epoxy. Use bracing devices to assure proper mating as the adhesive cures.

Alternate Technique

Sometimes, an approved cyanoacrylate adhesive is used in conjunction with epoxy, even on speleothems that require a support pin. Applied correctly, the quick-set glue can help hold the speleothem in place until the epoxy cures.

For example, after half filling the lower section drill hole with epoxy, spread a thin layer of epoxy in a circle outside of the hole, and then apply drops of Super T quick-setting adhesive on the inner circle adjacent to the drill hole and around the outer inside edge of the broken joint. (If a pin is required, overfilling the bottom drill hole with epoxy will result in overflow into the inner circle and will negate the effectiveness of the cyanoacrylate adhesive.)

Quickly align the two broken pieces and manually hold in place until the quick-set adhesive dries. Minor bracing may be beneficial. Allow 24–72 hours for the epoxy to cure before touching the speleothem.

Stalactite Repair—Epoxy and Prop

Figure 2. Brace with upward pressure and allow time for the epoxy to cure thoroughly. Steve Wilsey holds the joint in place while Sheridan Stone props the stal with a PVC pipe.

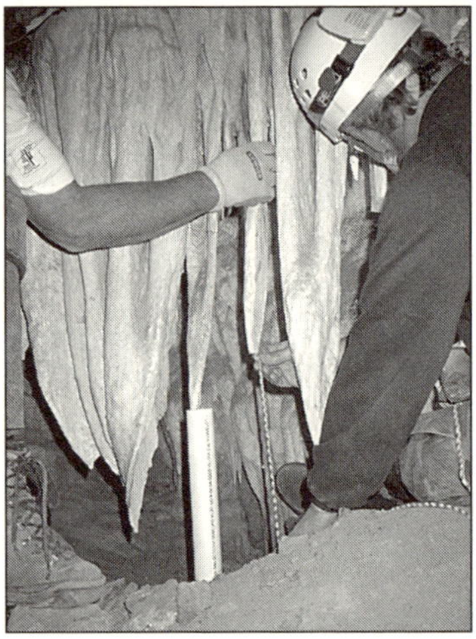

Figure 3. Measure the length needed for proper bracing and cut PVC pipe to size.

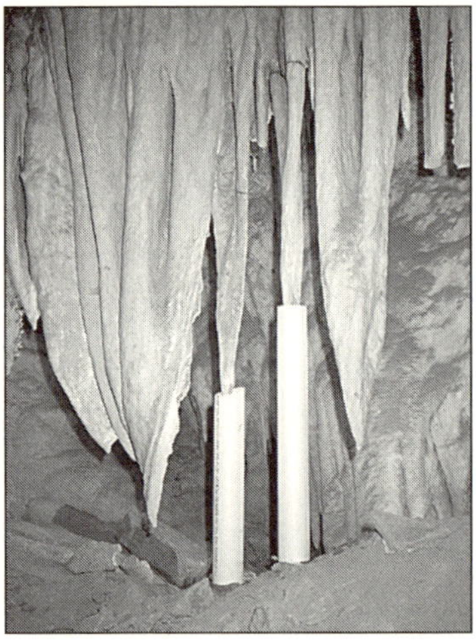

Figure 4. Two custom-cut PVC pipes brace stalactite repairs in Box Cave, Arizona.

Figure 5. Steve Smith and Jim Werker first check to make sure the broken speleothem properly click-fits and mates.

Figure 6. After using the ink-dot marking technique, Jim Werker drills a hole for the support pin.

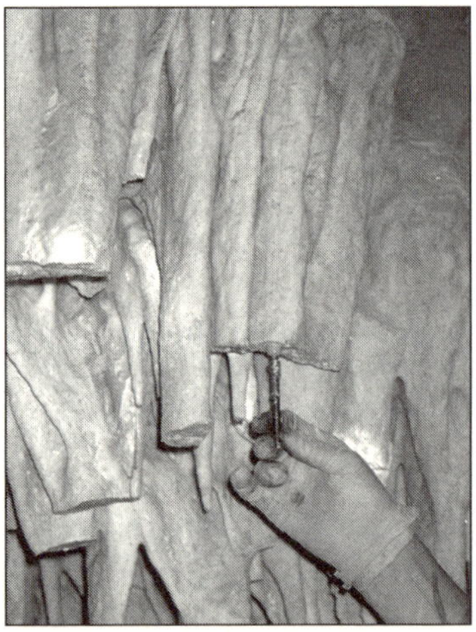

Figure 7. Stainless steel all-thread is epoxied into the drill hole to provide stabilization for the stalactite repair.

Photos © Val Hildreth-Werker

Figure 8. A larger hole is drilled in the lower piece.

Stalactite Repair—Pin, Epoxy, and Prop

Stalactite Repair: Pin, Epoxy, and Prop

- Drill and install support pin.
- Use archival epoxy and stainles steel all-thread pin.
- Prop or brace to hold a tight-fitting joint while epoxy dries.
- Remove propping devices after epoxy cures.

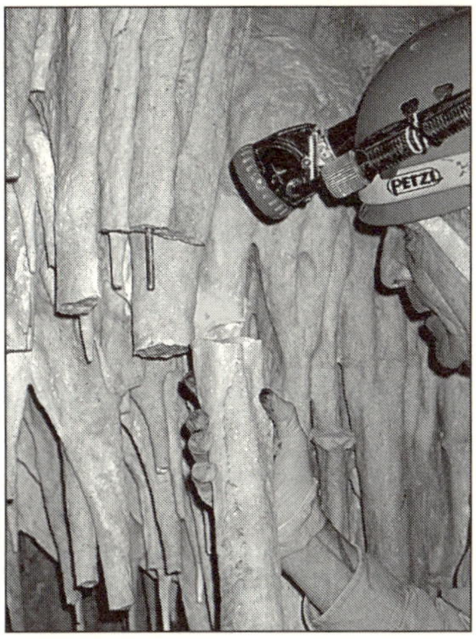

Figure 9. The lower drill hole is half filled with epoxy and positioned to receive the pin and reattach to the mating piece.

Figure 10. The repaired stalactite is supported with a PVC brace.

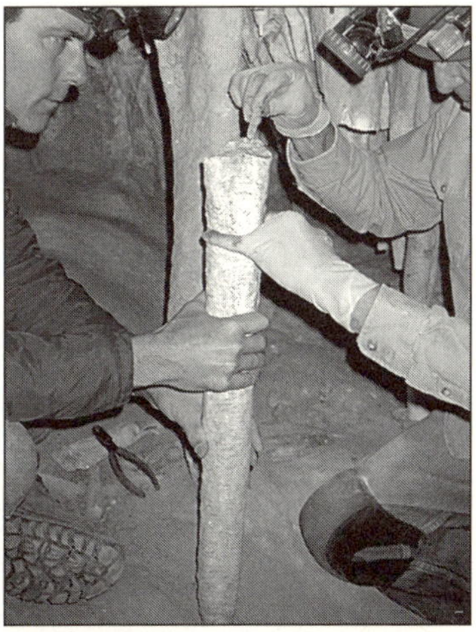

Figure 11. Steve Smith and Jim Werker prepare a second stalactite for reassembly.

Figure 12. Stalactites were pinned, epoxied, and braced with PVC during a speleothem repair workshop at Box Cave, Arizona. Epoxy was allowed several days to cure before cavers returned to remove the PVC.

Photos © Val Hildreth-Werker

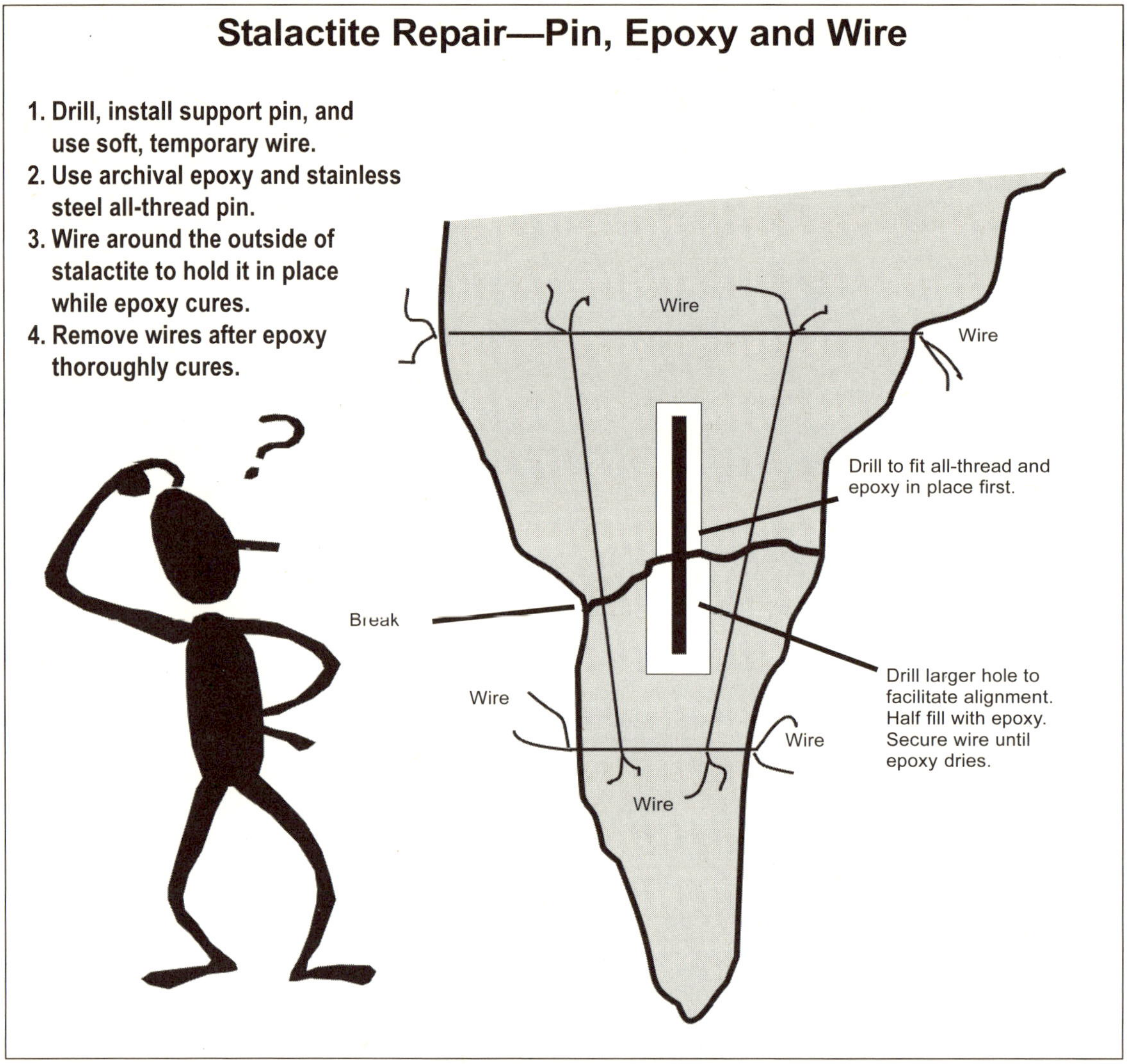

Figure 14. Drill the hole for a support pin in the top stal section that remains attached to the ceiling. Jim Werker is drilling to prepare a site for repair in Lower Cave, Carlsbad Caverns National Park, NM.

Figure 15. Reattach the stalactite with archival epoxy and a stainless stabilization pin. Support the repair with external wire. Clean up the epoxy ooze and drips.

Figure 16. Allow epoxy to cure thoroughly before removing the external wire support.

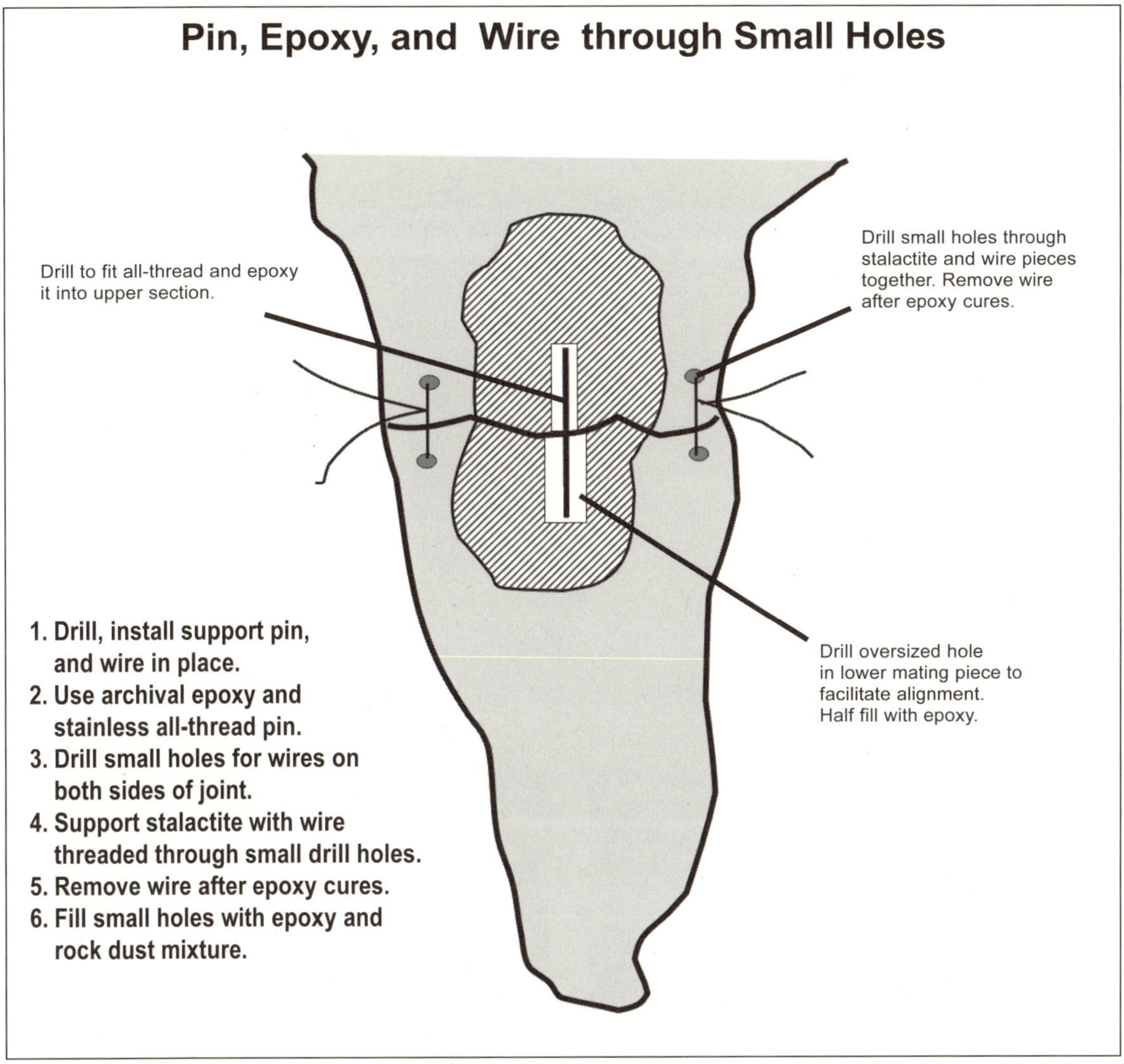

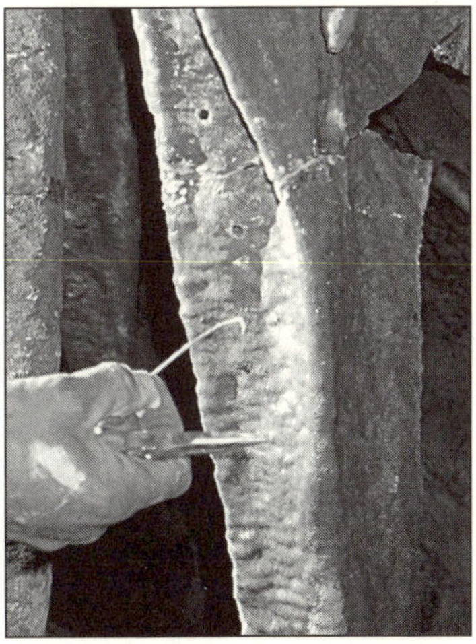

Figure 18. Install a stabilization pin if necessary and drill small holes above and below the break joint to accept supporting wires.

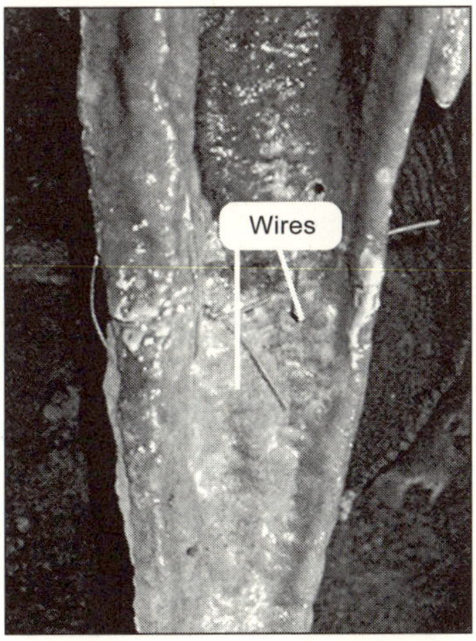

Figure 19. Thread wires through the small holes to hold the epoxied click-fit in place. Add bracing below the stalactite.

Figure 20. After the epoxy cures, remove support wires and bracing. Fill the holes with a color-matched epoxy and rock dust mixture.

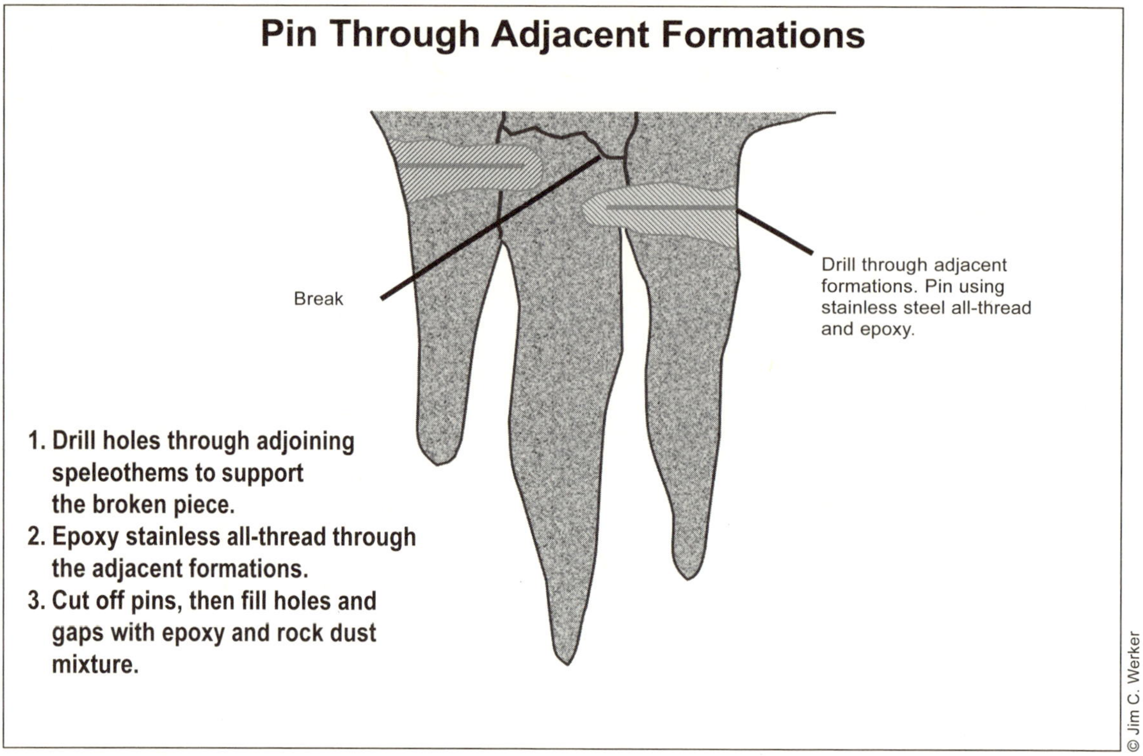

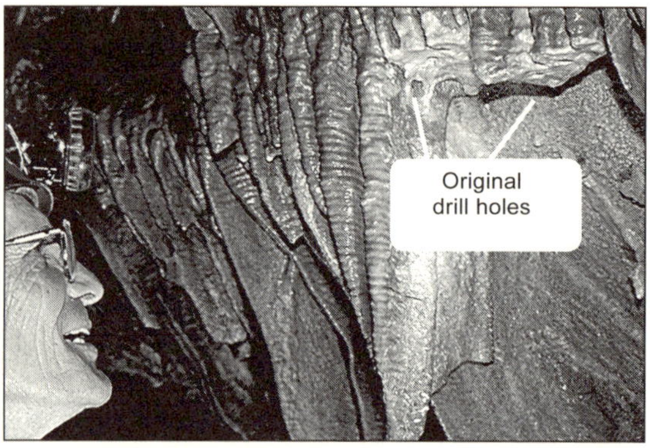

Figure 22. Visible in this photo are the original drill holes made for internal stabilization pins that were used the first time the formation was repaired.

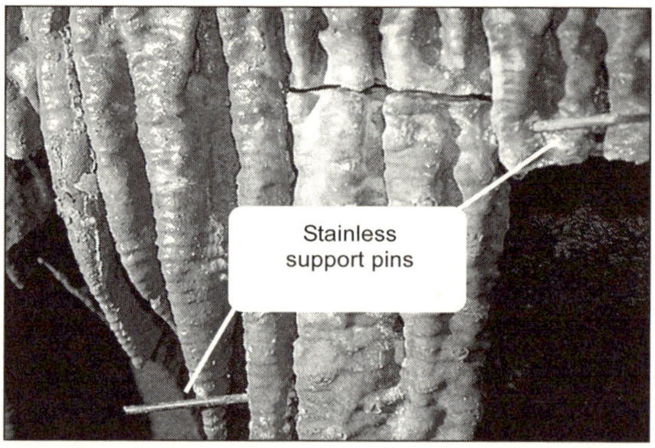

Figure 23. After vandals struck a second time, we reinforced the repair with stainless support pins installed through adjacent speleothems.

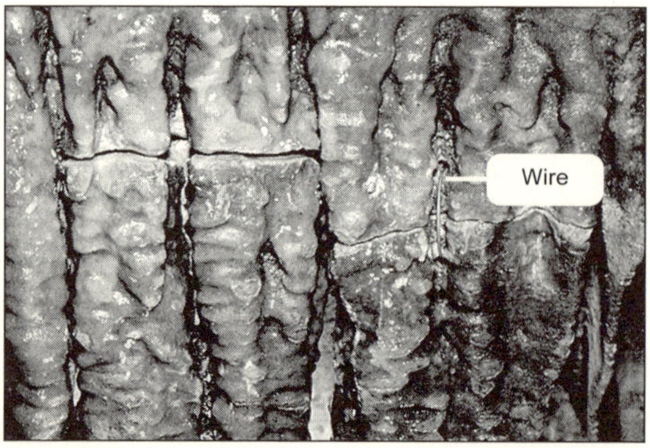

Figure 24. To provide support for another broken stal in the group, we threaded stainless wire through small holes drilled above and below the break. Pins were cut off after the epoxy dried.

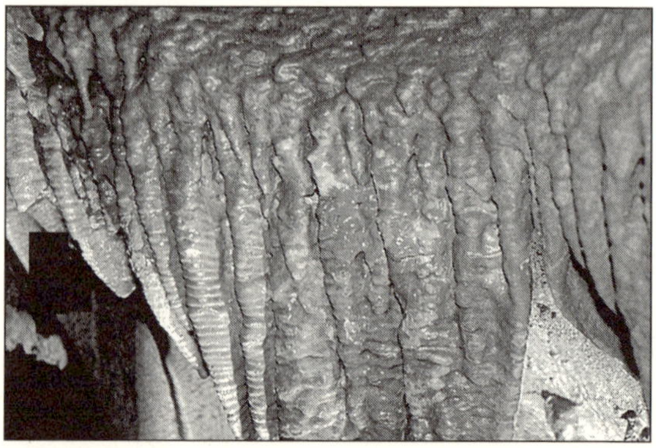

Figure 25. Because these stalactites receive intermittent mud flow, we partially filled the holes and gaps with a mud mixture. Time and new deposition will continue the healing process.

Section B—Repairing Speleothems

Column Repair

Jim C. Werker

Unique challenges accompany column repair. A creative combination of the methods described in the other repair chapters will likely apply. Advance planning is essential for making the last piece of the column slip into place like a perfectly adjusted three-dimensional puzzle.

Plan the Column Repair Sequence

- Plan the repair sequence so the final piece is shaped to easily slide into the last slot.
- Stick with super-thin epoxy joints all along the column repair and hopefully, there will be enough space remaining for the final section to drop into place.
- If one or more support pins are necessary, simplify positioning by installing smooth round-stock stainless instead of all-thread.

Does the Repair Need a Slip-Fit Pin?

If the final section requires a stabilizing rod, design a slip-fit pin and install it in the joining piece that is already in place above the last slot. Then drill a larger-than-normal hole in the final piece so the slip-fit pin has a better chance of falling into place. Finally, half fill the pin-receiving hole with epoxy, slide the final section into position, and hope for a good click-fit.

Column Repair: Slip-fit Pin and Thin Epoxy

- Match and mark all column pieces to facilitate fit.
- Carefully plan and deliberately choose the final piece.
- Apply super-thin epoxy and work at getting perfect joint alignments.
- If needed, design a slip-fit support pin out of smooth, round-stock stainless and install in the section directly above the final piece.
- Drill an oversized hole in the final piece to accommodate alignment of the slip-fit pin and half fill the hole with epoxy.
- Slide the last piece into place and position for a click-fit.

© Val Hildreth-Werker

Figure 1. Column repair is tedious. Use ultra-thin epoxy joints and plan an easy fit for the final piece. A slip-fit support pin made of round-stock stainless is an option for securing the final piece.

Section B—Repairing Speleothems

Drapery Repair

Jim C. Werker and Val Hildreth-Werker

Gather the broken fragments of drapery and piece them together like a jigsaw puzzle before beginning the actual reattachment with adhesives. A flat, dry surface aids in the matching process. It may be practical to transport the drapery pieces outside of the cave.

For pieces that go together with that satisfying snap-fit feel, make a pencil tick mark across the fracture for alignment, apply Hot Stuff® Super T along one edge, spray NCF Mild Accelerator on the matching edge, then quickly match the ticks and click-fit the edges. Along weight-bearing joints, larger pieces may also need a thin layer of archival epoxy. (See cave-safe adhesives and epoxies, pages 445–447.)

Fabricating Missing Pieces

Missing pieces can be fabricated from rock dust and epoxy. After mating and adhering the available pieces of the drapery, fill in the missing sections with a thick mixture of approved epoxy and crushed or powdered rocks. (Cave rocks or irreparably damaged pieces of speleothem might be pulverized for the mixture.)

Place masking tape or clear packing tape on one side of the gap. Spoon the epoxy fill into place. Add texture as the epoxy thickens and cures. Create surface texture and diminish the epoxy shine with a Dremel® tool after the epoxy is completely dry. Epoxy mixtures change color during the curing process—speleothem color is best matched by fabricating pieces in daylight conditions.

To reattach the completed drapery to the ceiling, install stainless steel stabilization pins or stainless wire supports and epoxy the parts into position. Permanent wire supports provide added security for rehanging drapery formations. It is important to use stainless steel wire in long-term applications. (See stainless pins, rods, and wires, pages 448–449.) Drill small holes for the wire through the portions of speleothem still attached to the ceiling and through matching positions on the pieced-together drapery.

Use a Dremel tool to grind small grooves between the holes for hiding the permanent stainless support wires. Make grooves on the front and back of the drapery break joint.

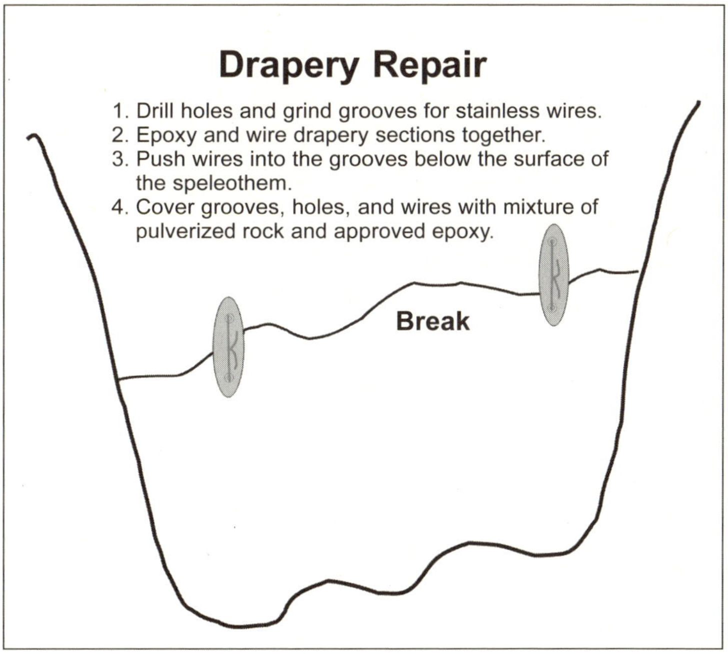

© Jim C. Werker

Drapery Repair Fit, Epoxy, Wire to Rehang, and Texture

- Fit pieces of drapery together and make tick marks across the joints.
- Use archival cyanoacrylate adhesive and epoxy.
- Fill in holes and missing pieces with an epoxy and rock dust mixture (may need to create forms to hold the epoxy mixture).
- Drill holes for stainless steel wire to permanently support reattachment.
- Grind grooves to receive the wires.
- Cover wires and fill grooves with epoxy and rock dust mixture.
- Texture with files and a Dremel tool.

Drapery Repair—Fit, Epoxy, Wire to Rehang, and Texture

Figure 2. During a week-long drapery repair project at Cave Without a Name in Texas, we pieced together over 300 fragments of broken drapery. We used both Hot Stuff Super T cyanoacrylate adhesive and Epon 828 epoxy with Versamid 40 hardener to adhere fragments of various sizes.

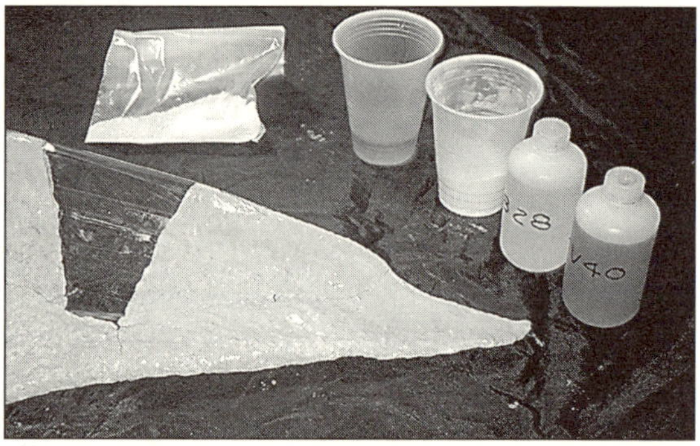

Figure 3. To fabricate the sections that were never found, we filled in gaps on the reconstructed drapery with a color-matched epoxy and rock dust mixture. To make a form for the epoxy, clear packaging tape was attached on one side and up along the edge.

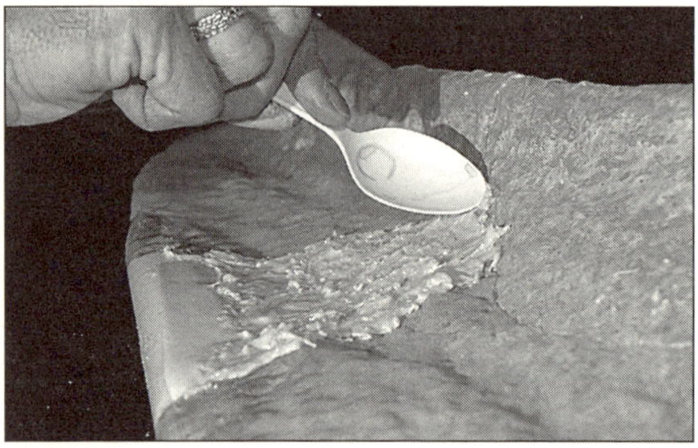

Figure 4. After allowing the epoxy to set up for several hours, we used a plastic spoon to create a base texture in the thickened epoxy mixture.

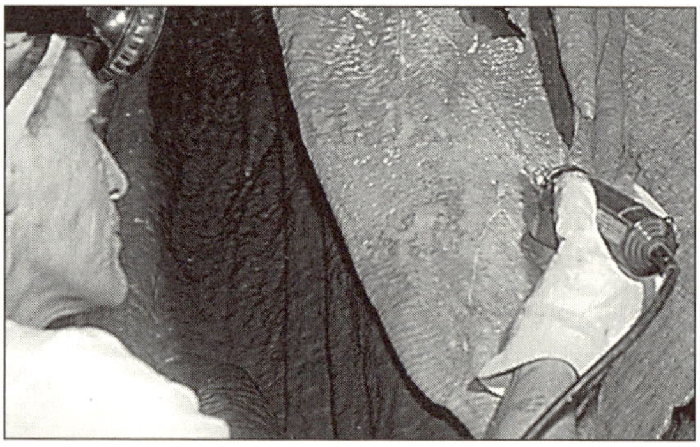

Figure 5. This drapery was too thin for internal support pins. We drilled small holes to accept permanent stainless wire that reinforced the epoxy reattachment joint. Jim Werker used a Dremel tool to grind grooves for the wires. Holes and grooves were filled with a color-matched rock dust and epoxy mixture. The Dremel was also used to create final texturing.

Photos © Val Hildreth-Werker

Figure 6. The repaired drapery again hangs in Cave Without a Name. Time and new mineral deposition will improve the color match.

Apply a thin coat of epoxy to one surface of the joint, fit to the tick marks, and tightly secure all support wires to hold the speleothem in place. Then push wires into the grooves below the speleothem surface and fill the slots with an epoxy and drill dust mixture. (See filling cracks and holes, pages 457–458.)

After the epoxy cures, smooth and texture the filled grooves with a Dremel tool and files. Though drapery repair is time consuming and each painstaking step must be scheduled separately, the completed repair is rewarding.

Section B—Repairing Speleothems

Rimstone and Travertine Dam Rebuilding

Jerry L. Trout

Rebuilding rimstone and travertine dams, when properly orchestrated, provides instant aesthetic improvement and is among the most visually enhancing and rewarding of all formation repairs. When water fills the pockets, pond, pool, or lake behind the rimstone dams, it often appears as if the entire scenic gallery is restored. Rimstone dam repairs sometimes achieve lasting benefits. When the rebuilt dams refill, populations of natural cave organisms may replenish, deposition or corrosion patterns may reestablish, and the presence of water may help deter new breakage.

Rimstone and travertine dams, easily trampled underfoot, are among the most frequently damaged speleothems. Dam configurations have tremendous variations.

- From paper-thin to several inches of thickness
- From only a narrow line a few millimeters high to several feet tall.
- From thick but fragile to extremely delicate.

Dams or gours are not difficult to repair, except when they are very thin or very tall. Dam repair is best accomplished with mortar-mix cement. Varying the amount of lime in cement will create a stiff, workable consistency, depending on the moisture present and the inherent height or thickness of the damaged area. (See concrete, page 179; also see Portland cement, page 369.)

Ace Hardware® products like Fast Concrete Patch® or Quick Plug® are good options. Similar products with differing brand names are found at hardware stores and construction supply outlets.

Rimstone Dam Repair

- Clean all mud and debris from rimstone.
- Hand sculpt the quick-setting cement (10–30 minute set time).
- Mold and texture it with gloved hands and tools.
- Repaired areas should hold water.

Photos © Val Hildreth-Werker

Figure 1. Tall, winding rimstone dams border a large, intermittent pool in Hidden Cave, New Mexico.

Figure 2. Quick-setting cement products are the recommended materials for reconstructing and sculpting broken rimstone dams.

All cements will cure slowly in caves with cool temperatures and high humidity. The lower the temperature and the higher the humidity, the longer cement materials will take to set and cure.

Materials

Mortar Mix. Use mortar mix when the thickness of the dam needs to be at least 1/2 inch (1.25 centimeters). The height of the repaired section should be no more than 8 times its thickness. Add increased amounts of lime for taller or thinner repairs. More lime (calcium oxide or calcium hydroxide) makes the cement mix set quicker. There is no particular formula for the amount of lime to add—however, if more than 10% is needed for the right consistency, a different repair material should be chosen. (See low-content calcium hydroxide, page 170 and page 369.)

In caves where temperatures range from 50–65° F (10–18° C), mortar mix will take approximately 1–4 hours to set, and 3–4 days to cure. Lower temperatures will lengthen both the set and cure times—warmer temperatures will shorten those times. Set and cure times are also extended if cave humidity exceeds 95%.

Experiment with coloring agents from within the cave—cave soil, crushed cave rocks, drill dust, or powdered material from irreparably damaged speleothems.

Fast Concrete Patch. This product can be purchased from Ace Hardware. Similar products are available elsewhere. Use for repairs when the rimstone dam thickness is less than 1/2 inch (1.25 centimeters) or the height is more than 8 inches (20 centimeters). In caves, Fast Concrete Patch will typically take about 20 minutes to set and 1 hour to cure.

Quick Plug Hydraulic Cement. This product is available through Ace Hardware. Use a quick hydraulic cement if the repair site is wet (also works for underwater repairs). Quick Plug has successfully served in rebuilding dams as thin as 1/8 inch (3 millimeters) and 7 inches tall (18 centimeters). In caves, Quick Plug typically takes about 4 minutes to set and about 45 minutes to cure.

Coloring Agents. Obtain coloring agents for concrete from almost any hardware store or construction supply outlet. Read the labels and avoid coloring agents containing ingredients that may be toxic to cave-dwelling organisms. Choose natural products made of materials that will not harm cave ecosystems.

Experiment with coloring agents from within the cave—cave soil, crushed cave rocks, drill dust, or powdered material from irreparably damaged speleothems. Rimstone colors have been successfully matched by simply adding dust or dirt found close to the repair site. Colored chalk composed only of natural ingredients also works well. Color-matching is best achieved by taking small samples outside into the daylight. It is important to note that most concretes cure to a lighter shade after drying.

When the repair site is damp, wet, intermittently full of water, or underwater, the precipitation of new speleothem material will help the repaired rimstone begin to match the speleothem color naturally, given enough time. In some cases, when the area is very wet or if there is water movement, natural color matching will occur within a few months.

Tools for Rimstone and Travertine Repair

Wire brushes. Use new stainless steel brushes (no brass or other metals that rust or mar).

Paintbrushes. Use new brushes with nylon bristles.

Rimstone Dam Repair

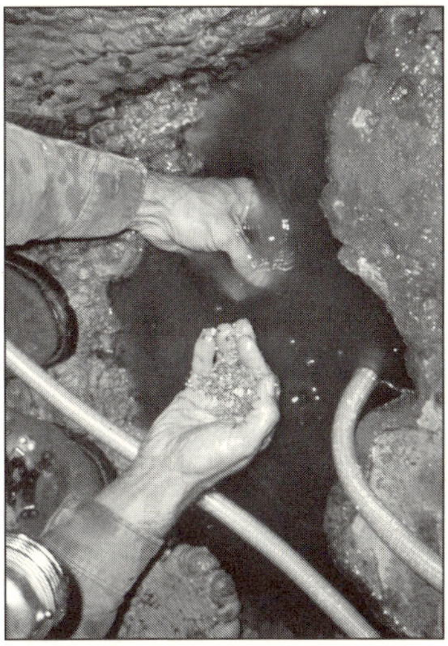

Figure 3. Thoroughly clean the repair area. Remove mud and debris before attempting to reconstruct rimstone dams with cement products.

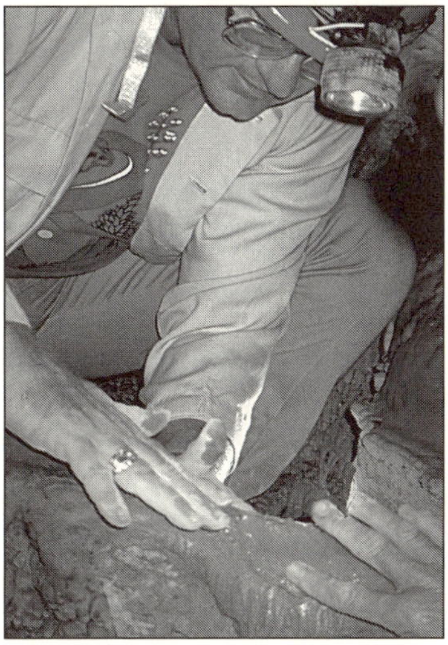

Figure 4. Jerry Trout uses quick-setting cement to manually sculpt missing pieces of rimstone.

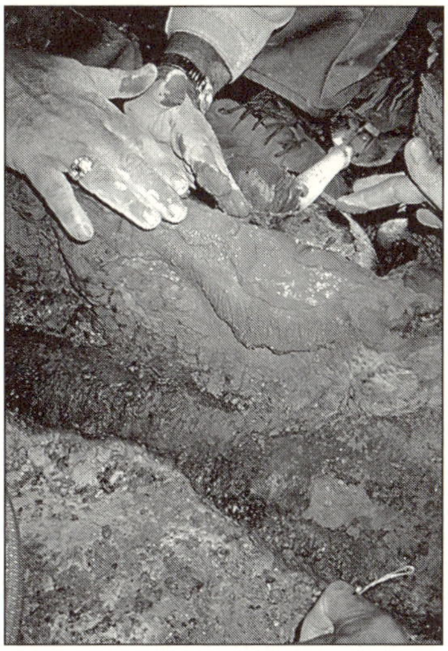

Figure 5. Use trowels, brushes, wires, and sculpting tools to shape the cement and match the texture of reconstructed sections to the original. Wear disposable vinyl gloves to help protect cavers' hands.

Figure 6a. These three photos show the progression of rimstone repair in Happy Jack Cave, Arizona. First, clean the repair site. All surfaces should be as wet as possible before applying patching materials.

Figure 6b. Next, use gloved hands and various tools to sculpt cement and fill in missing sections of the rimstone dams.

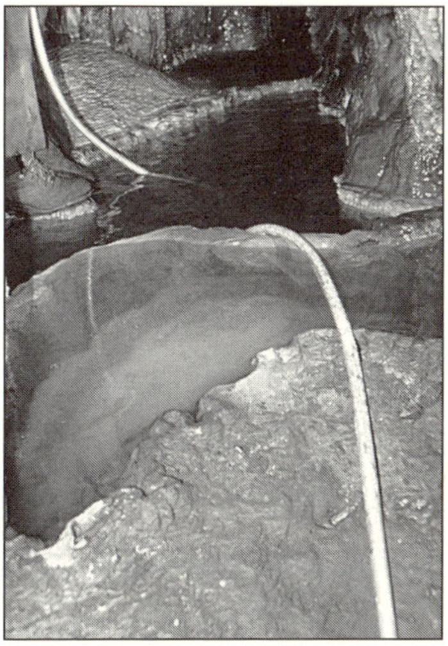

Figure 6c. Ths view shows the completed rimstone dam. Repaired rimstone dams should retain water like the originals.

Toothbrushes. New toothbrushes are best. Designate them for repair.
Trowels. Both small and large are handy.
Sponges. The soft, car-wash type works best.
Wire. Stainless steel is best, but other types are okay for temporary use.
Paper clips. Use for texturing.
Water. Bring water if the cave is very dry.
Duct tape. Make forms for rimstone cement.
Buckets and containers. Use for mixing.

Stirring paddles or wooden spoons. Use for mixing.
Surgical gloves. Nonlatex, powder-free gloves are best.
Goggles. Wear eye protection.
Drop cloth. Protect cave surfaces when mixing and coloring.
Thick garbage bags. Transport all mixing materials in bags.

Methods and Techniques

Surface Preparation

Remove all loose materials with wire brushes, paintbrushes, sponges, and water. Surfaces should be clean of dirt and loose debris. All surfaces prepared to receive cement repairs should be as wet as possible before applying the patching materials.

Surfaces should be clean of dirt and loose debris. All surfaces prepared to receive cement repairs should be as wet as possible before applying the patching materials.

Mixing

Mix all cement patch materials in a clean container with a 3:1 ratio—that is, 3 parts powder to 1 part water. Mixtures are typically adjusted according to environmental conditions. The consistency should be very thick, much like ready-mix concrete. In other words, the mixture should be as thick as possible, yet remain workable.

If possible, compare color matching outside the cave, in the sunlight. If it is not possible to take the materials or samples out of the cave, use a very bright white light—a broad beam, full spectrum light is preferred. Remember that most concretes will cure to a lighter shade after they are completely dry.

Application

Using gloved hands, trowels, and brushes, apply and shape the cement as required to fill in the missing pieces. Take care to match the thickness and shape of the original dam. Where large pieces are missing, mimic the patterns of the original dam by matching the texture and design of the remaining natural sections. If possible, use photographs of the dam that were made before the damage occurred.

A length of stiff tape—duct tape for example—is sometimes used to give stability to the repair while adding layers of cement. Remove the tape as soon as the concrete begins to cure.

If the dam repair is thin or several inches high, let each layer set up before adding subsequent layers. In some cases, the repair will take several sessions—allow each layer to set before adding the next, but all layers need to be added before the foundation layers completely cure. We have had success in reconstructing dams up to 4.5 feet tall (1.3 meters) and less than 1 inch thick (2.5 centimeters).

Do the final texturing with gloved hands and a variety of tools such as small paintbrushes, toothbrushes, sponges, wire, paper clips, and so on.

Clean up

Remove all patching material debris, tools, equipment, and packs in spill proof bags. Wash skin and tools immediately. Large amounts of water and soap are required for thorough cleaning of tools used in cement jobs.

Section B—Repairing Speleothems

Flowstone Repair

Jim C. Werker and Val Hildreth-Werker

Several different methods presented in this manual may be employed during flowstone repair. Use an approved epoxy or cyanoacrylate adhesive on simple broken pieces. If flowstone chunks are absent, patch and fill in with either an epoxy and rock dust mixture or quick-setting cement. (See fabricating missing pieces, page 473; concrete patch methods, page 476–478.)

If foundation sediment under the flowstone is missing, use a filler material, sediment from nearby in the cave passage, or cement with low-content calcium hydroxide. Then repair the broken flowstone pieces on top.

Refer to other chapters in the speleothem repair section of this volume and adapt the techniques to fit specific flowstone repair projects.

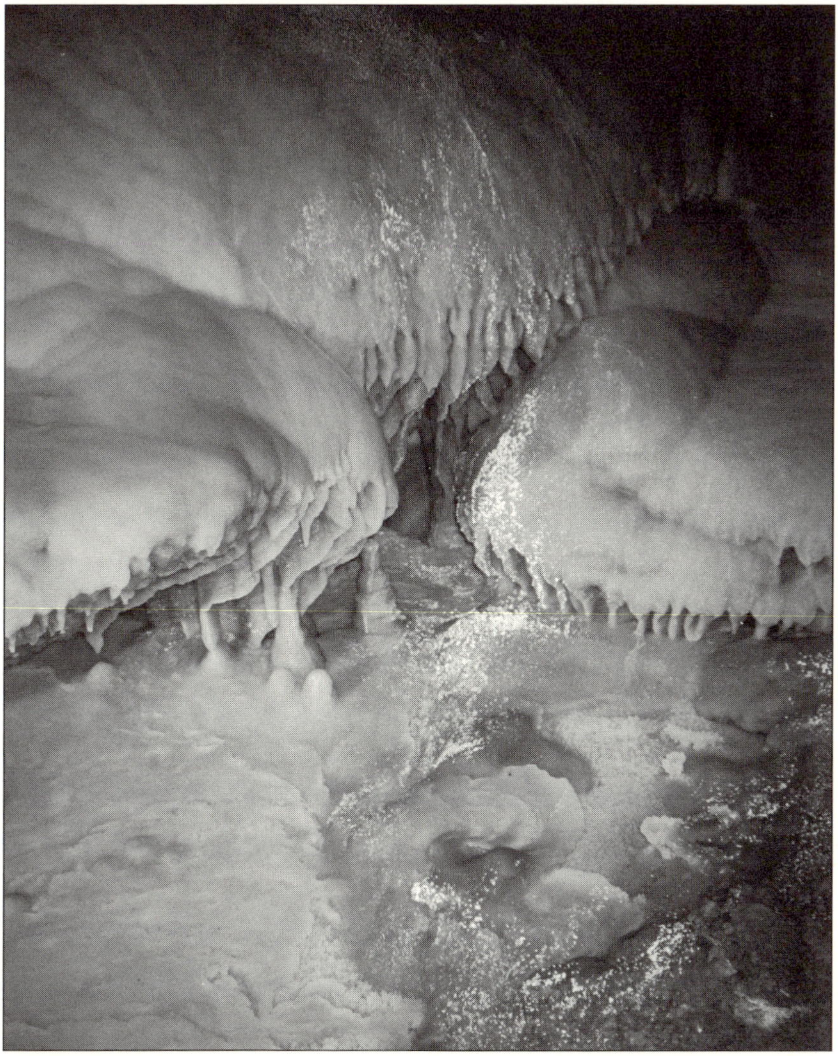

Figure 1. Flowstone areas are well worth protecting from unnecessary foot traffic to avoid the need for repair. Flowstone surfaces may harbor interesting microbial communities and invertebrate populations, especially in moist areas. Beautiful flowstone features also have significant aesthetic value in cave systems.

Section B—Repairing Speleothems

Gypsum Repair

Jim C. Werker and Val Hildreth-Werker

Depending on the texture and porosity, broken sulfate speleothems can sometimes be repaired. An approved cyanoacrylate adhesive may work on very small pieces. If extremely delicate technique is employed, even gypsum flowers may be good candidates for repair.

Larger repairs on damaged gypsum crusts usually require bracing or propping while an archival epoxy cures.

Depending on the texture and porosity, broken sulfate speleothems can sometimes be repaired.

Gypsum Crusts

In Floyd Collins Crystal Cave, Mammoth Cave National Park, Kentucky, gypsum crusts hang suspended from bedrock walls. Vandals used sledge hammers to break off chunks of gypsum weighing up to 70 pounds (32 kilograms) as well as numerous other speleothems. Mammoth Cave National Park invited us to develop techniques for rehanging large pieces of gypsum crust.

To suspend crusts from bedrock walls, mate the broken crust to intact edges and locate a position in the bedrock for a supporting rod. Set the broken piece aside. At the matching point on the bedrock wall, drill a hole and use epoxy to secure a stainless steel all-thread supporting rod. Use an archival epoxy to install the support rod approximately perpendicular to the wall. (See cave-safe epoxies, page 445–447.)

Drill through the broken piece of crust at the matching location. Lift the crust onto the rod, position the edges, and hang it somewhat like a picture on a wall. The stainless rod acts as a straight hook or pin to support the broken piece. Allow time for the epoxy to cure.

Cut off the rod even with the outer surface of the crust and cover the drill hole with a mixture of epoxy and gypsum dust. Reinforce the mating edges with a color-matched mixture gypsum powder and epoxy. Color match and texture the holes, cracks, and seams to complete the repair.

Figure 1a (before) and **Figure 1b** (after). These two photos show images of a vandalized gypsum crust repair in Floyd Collins Cave, Mammoth Cave National Park, Kentucky.

Gypsum Crust Repair

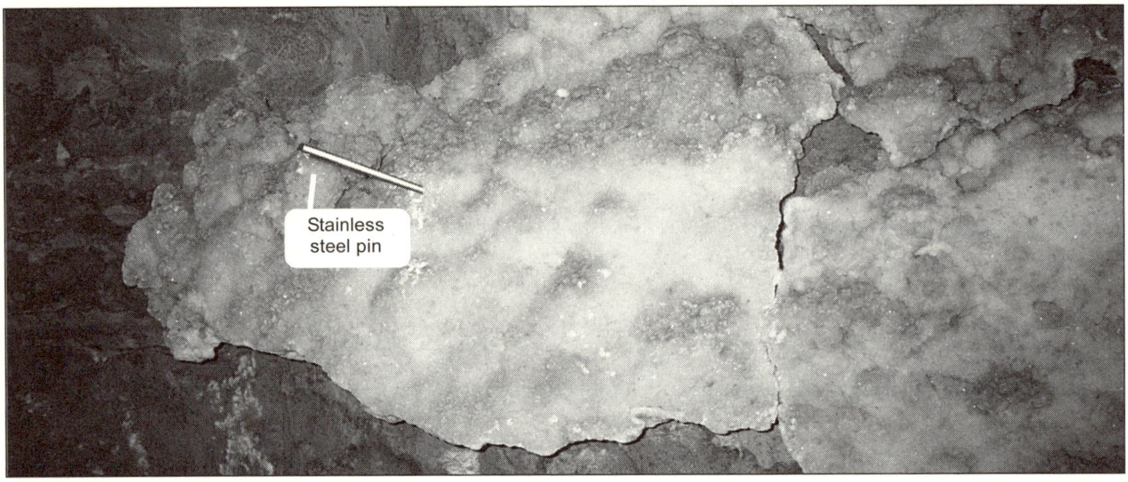

Figure 2. To repair gypsum crust, use an archival epoxy and install a stainless steel pin in the wall. With the pin acting as a hook, rehang the gypsum crust.

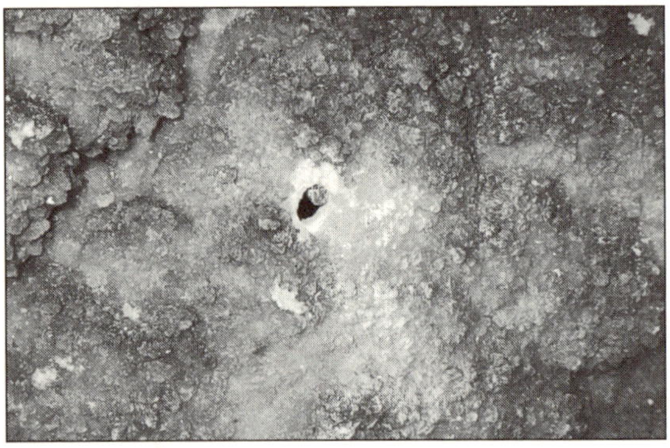

Figure 3a. This photo shows the drill hole in the gypsum crust with the pin cut off. The pin is cut off even with the surface of the crust.

Figure 3b. Fill the drill hole with a color-matched epoxy and crushed gypsum mixture.

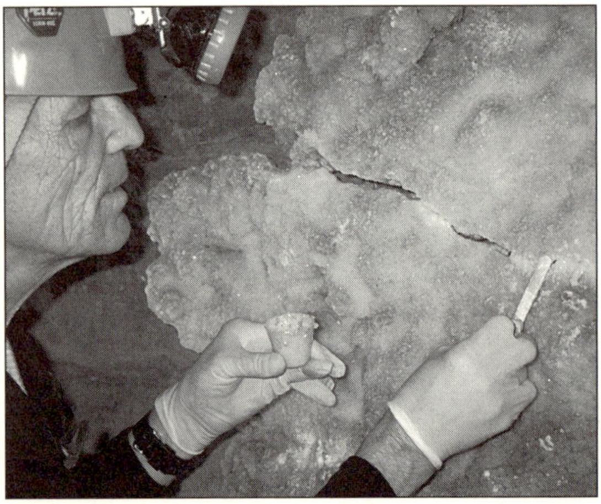

Figure 4. Jim Werker fills seams and gaps with a thick mixture of epoxy and gypsum dust.

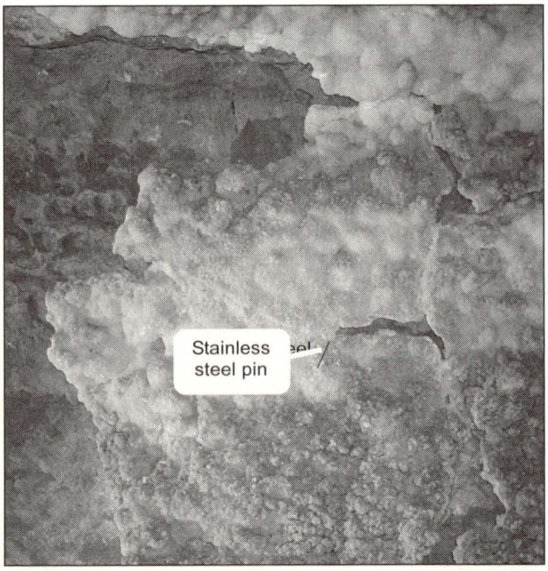

Figure 5. A second gypsum crust, weighing 70 pounds (32 kilograms) is reattached using the same perpendicular pin technique.

Photos © Val Hildreth-Werker

Section B—Repairing Speleothems

Lava Formation Repair

Jim C. Werker

Features in lava tube caves can be repaired with the same techniques and materials described in the various chapters on speleothem repair.

If lava cannot be pulverized to the appropriate powdery consistency for adding color to the epoxy mix, then charcoal or natural chalks may work as coloring agents.

© Dave Bunnell

Figure 1: An unknown caver observes a tubular lava stalactite.

© Gus Frederick

Figure 2: A large ice formation in Raufarholshellír, an Icelandic lava tube cave.

© Dave Bunnell

Figure 3: Elizabeth Rousseau beneath lavacicles.

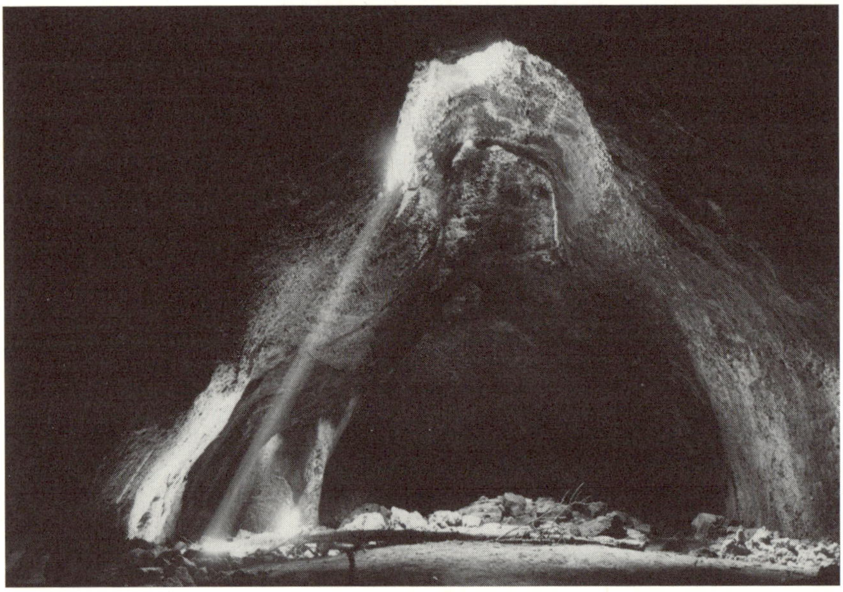

© Gus Frederick

Figure 4: A series of three "Hornito Holes" in this central Oregon tube provides a perfect skylight, as well as convenient hole to throw things into.

© Dave Bunnell

Figure 5: Ric Elhard with "shark's teeth" type lavacicles.

Section B—Repairing Speleothems

Fragile Speleothem Repair

Jim C. Werker and Val Hildreth-Werker

Delicate, damaged speleothems are often surrounded by intact, fragile features. The location of the repair site, the human realities of working among delicate formations, and the prevention of additional harm are important considerations for tedious repair tasks. Use common sense—many potential repairs in delicate areas should not be attempted. A relatively cave-safe, pure form of cyanoacrylate adhesive is typically used for fragile speleothem repairs. (See cave-safe adhesives, pages 445–447.)

Soda Straws. Place two to four miniscule drops of an approved cyanoacrylate adhesive (for example, Hot Stuff® Super T) around the central canal and hold the fragile soda straw in place for 30–90 seconds until the adhesive sets. When the repair is in an awkward location, apply the cyanoacrylate, align the joint for the snap-fit, connect the pieces, and shoot one quick spritz of accelerator on the joint (for example, Hot Stuff NCF Mild Accelerator).

Thin Draperies and Bacon Rind. Use cyanoacrylate adhesive along the broken face of one piece. Match the pieces and hold in place until the quick-set adhesive dries. For broken pieces that are too heavy for the fast-acting adhesive, see the introduction to the drapery repair chapter (page 473).

Thin, Delicate Travertine Dams. If the pieces are available, use cyanoacrylate adhesive along only one face of each break and carefully replace each fragment.

Helictites and Heligmites. Apply a few dots of a cyanoacrylate adhesive along the outer rim of the break. Leave capillary canals free of adhesive to give speleothems the chance for continued growth.

Popcorn, Coralloids, and Folia. For each delicate fragment of coralloid, tower coral, popcorn, folia, U-loops, and so on, find its original location with the proper snap-fit. Apply an approved cyanoacrylate or epoxy, depending on the size and shape of the speleothem. Hold in place until the adhesive sets. (See tripod speleothem repair system, page 491–494.)

Frostwork and Other Delicate, Lacey Structures. Cyanoacrylate adhesives like Hot Stuff Super T or the more viscous Special T usually provide the only answers for repairing fragile speleothem structures. Where breaks happen, dirt and soil usually also cause damage. For tips on cleaning delicate features, see restoration techniques (page 423–425). Remember, it may be better to leave the site as is and do nothing.

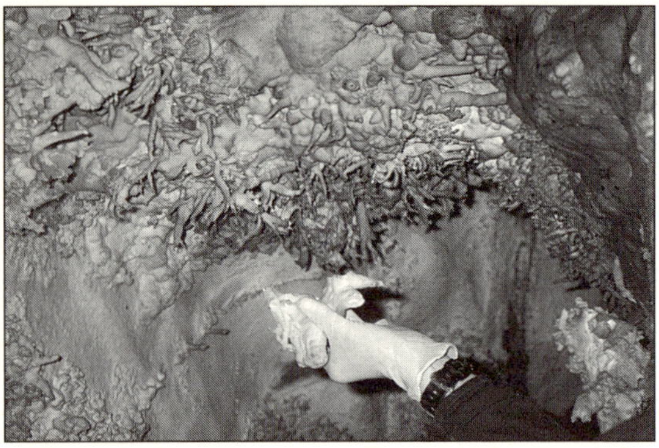

Figure 1. The empty scar where this speleothem was broken off is apparent on the helictite-covered wall. See the repair in Figure 2.

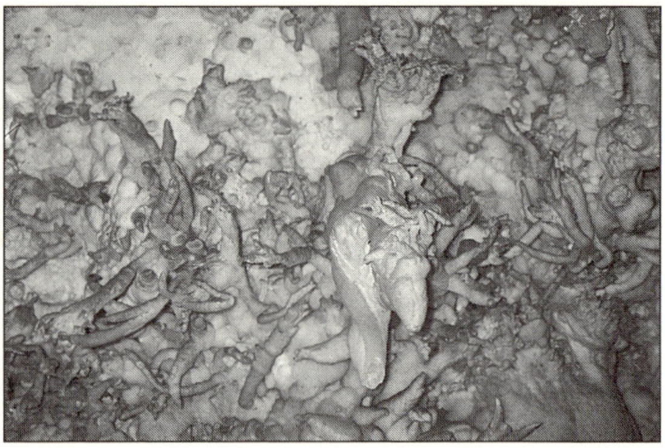

Figure 2. The forked speleothem was returned to its original location on the wall of delicate helictites. One broken tip of the repaired speleothem is still missing.

Figure 3. Delicate gypsum speleothems that adorn some cave passages are easily damaged. These are often impossible to repair.

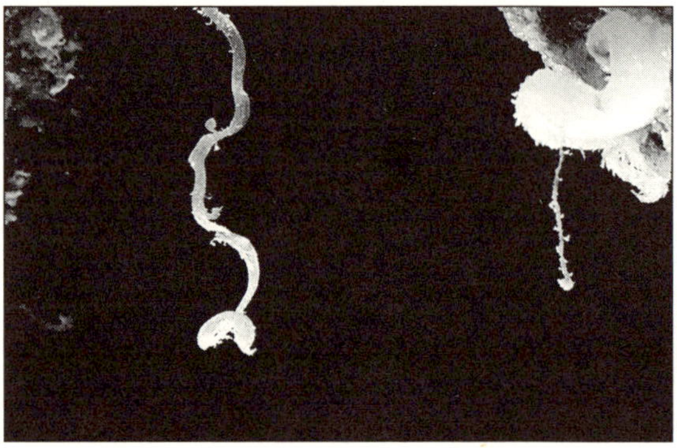

Photos © Val Hildreth-Werker

Figure 4. Move gently and spot each other through areas decorated with delicate features. Cave softly to avoid soiling or destroying fragile speleothems.

Section C—Specialized Mechanical Assists

Apparatus for Large Speleothem Repairs

Jim C. Werker

Large or tall stalagmites may require gantries to aid in lifting speleothem pieces into place (Figure 1). Sizable stalactites may require a scissors jack or other jacking mechanisms that apply upward pressure while the epoxy cures (Figure 4). Scaffolding and mechanical assist devices may be adapted for safe use in some cave environments. (See Colossal Cave repair, page 503–506.)

Specialized Mechanical Assists

Sometimes, a speleothem shape or location does not lend itself to typical bracing aids. More specialized mechanical aids are occasionally necessary. This section describes a variety of specialized bracing and lifting devices (page 489 and pages 491–494). Simpler bracing and propping devices are discussed in an earlier chapter (pages 456–457).

Figure 2. The gantry held the top part of the speleothem in place while Jim Werker applied epoxy. Notice the rock placed for safety in the gap while Jim reached in between the two pieces to spread the epoxy. (See other Endless photos, page 497.)

Photos Noble Stidham

Figure 1. A caver fell against and broke this 9-foot tall (2.75-meter) stalagmite, a totem that stands in the Grand Canyon of Endless Cave, New Mexico. Jim Goodbar is pictured with the overhead gantry system that was erected to raise the broken top into position for repair.

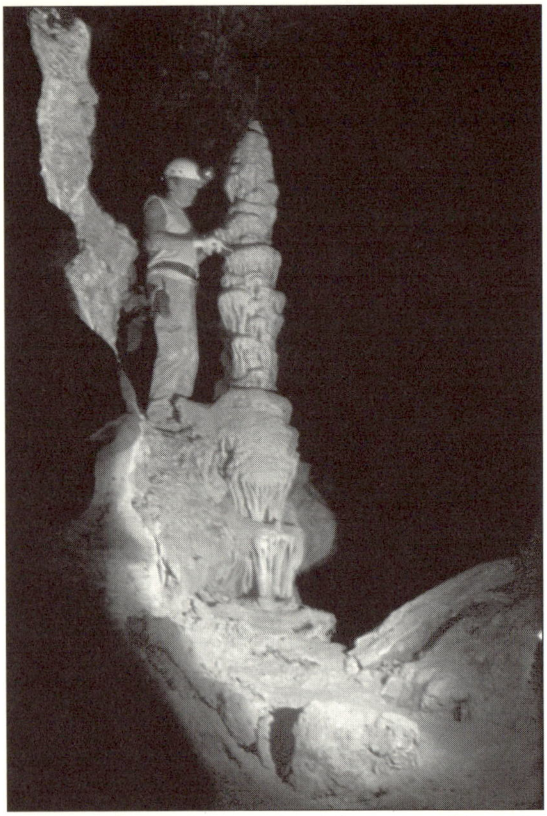

Figure 3. Jim Werker completed the repair with touch-up around the break joint of this tall, stately stalagmite in the Grand Canyon of Endless Cave, New Mexico.

Photos © Val Hildreth-Werker

Figure 4. Jim Werker constructed a wood support with a simple jacking mechanism to apply upward pressure to a large stalactite repair in Colossal Cave, Arizona. The two broken pieces in the foreground weigh a total of about 225 pounds (102 kilograms). (See other Colossal photos, page 504.)

Section C—Specialized Mechanical Assists

Assist Devices for Medium Speleothem Repairs

George Veni

Engineers are trained to come up with elegant, efficient solutions. It took Dan Hogenauer about 2 minutes to set up this system during a vandalism repair event at Cave Without a Name, Texas. Dan created a lever.

Use a Lever to Stabilize

Place a board set under the stalactite and set a rock on the opposite end of the board to hold the speleothem in position while the epoxy sets up and cures (Figure 1, next page). There are a couple of key concepts for this type of assist device:

- The fulcrum should be at about the same elevation as the lower tip of the stalactite when set in place.
- The weight at the opposite end of the board is set far from the fulcrum when more pressure is needed, or closer to the fulcrum when pressure needs to be reduced.

For repairs close to the floor, a fulcrum can be made by stacking rocks or available materials (gravel, buckets, boxes, and other items). For repairs more than 3 feet (1 meter) above the floor, scaffolding can be built to support a fulcrum at higher elevations (Veni 1995).

While this leverage method is quick and effective, it has limitations. The tip of the stalactite must not be fragile and the technique is easier to apply if the tip is rounded or flat. Regardless of the shape, the tip should be padded and protected.

If the break is horizontal, the upward pressure provided by leverage works well for supporting a stalactite while the epoxy cures. If the break is angled, support may be needed alongside the stalactite to prevent it from slipping out of place when upward pressure is applied.

Cited Reference

Veni G. 1995. The restoration of a vandalized show cave: Cave Without a Name, Texas. In: Pate DL, editor. *Proceedings of the 1993 National Cave Management Symposium: Carlsbad, New Mexico, October 27-30, 1993*. [Huntsville (AL)]: National Cave Management Symposium Steering Committee. p 65-72.

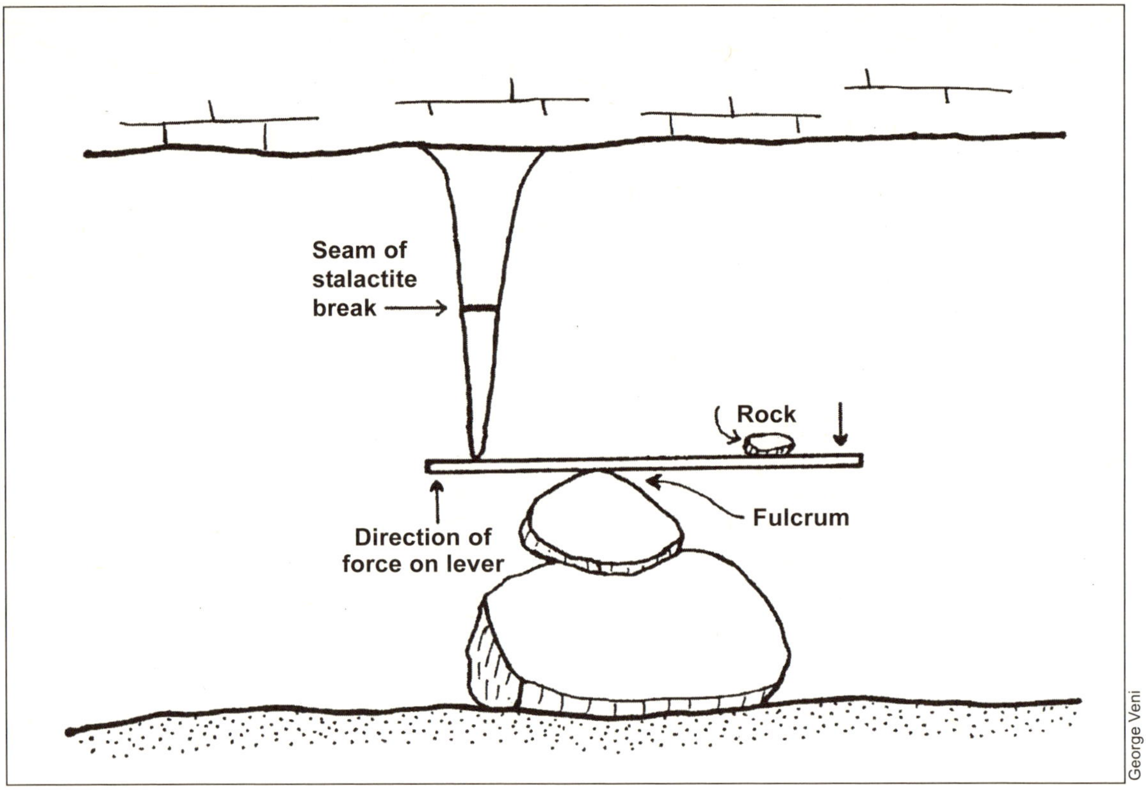

Figure 1. This diagram shows the lever method. Build a lever to stabilize broken stalactite pieces while the epoxy dries along the seam.

Section C—Specialized Mechanical Assists

The Tripod Speleothem Repair System

Al Collier

To secure the speleothem while an approved adhesive or epoxy dries, a simple system of multiple tripods, cross bars, and counterweights is employed. Using tripods to support speleothem repair is not a new idea.

Through years of trial and error, I have developed a number of innovative stabilizing concepts based on tripods. While any photographic tripod with a periscoping center leg will work, the Slik® 1200-G tripod is ideal because it is designed to be adaptable and works well with aluminum crossbars and counterweights.

A camera plate sits on a fulcrum and provides flexibility when attaching assorted crossbars and adjustable pads. Counterbalance is easily achieved by filling small laundry bags with rocks, which are typically found along the trail (when possible, use rocks found inside the cave).

The variations on this technique and its applications are endless. This section illustrates simple tripod systems for supporting repaired speleothems while the epoxy cures.

To secure the speleothem while an adhesive dries, employ a simple system of multiple tripods, cross bars, and counterweights.

Crawlway Stalactite and Drapery Repair

Place a short tripod with the crossbar centered on the fulcrum under the stalactite and apply the appropriate counterweight (Figure 2). To protect the surface of the speleothem, pad the end of the crossbar that supports the stalactite. Place the counterbalance on the opposite end of the crossbar.

Position the broken piece on the padded end and manually align the apparatus directly underneath the intact remnant of the stalactite. Apply epoxy to the lower break joint and then mate the two pieces by gently allowing the counterweight to apply pressure.

Once the epoxy has set sufficiently, slowly elevate the counterweight to release the upward pressure on the stalactite, and then remove the entire fulcrum assembly.

Tall Passage Stalactite and Drapery Repair

This technique works for repairing stalactites in passageways up to 8 feet (approximately 2.5 meters) high. Extend the center leg of the tripod vertically from the lower fulcrum and reverse the leg 180 degrees from standard (Figure 4).

Install an adjustable mini tripod head of compatible size on the extended end to create an upper fulcrum. Then install a shorter aluminum crossbar on the upper fulcrum. (Add extra weight to the center of the tripod to help stabilize it while making above-the-head repairs.) Place the tripod with the crossbar centered on the fulcrum under the stalactite with appropriate counterweight.

Pad the end of the crossbar that supports the stalactite to protect the surface of the speleothem.

Tom Dotter

Figure 1. For speleothem repair, a simple system achieves great flexibility. Al Collier uses tripods, camera plates, cross bars, and adjustable pads, with rock-filled bags for counterbalance.

Figure 2. In this crawlway, Al Collier places a short tripod with the crossbar centered on the fulcrum under the speleothem and applies the appropriate counterweight to reattach the stalactite.

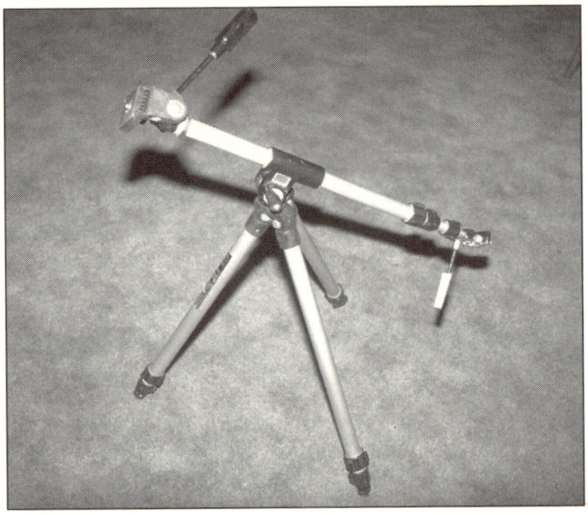

Figure 3. Use a basic photographic tripod that has a reversible periscoping center leg. Adapt the tripod to use with camera plates, adjustable pads, and rock-filled counterbalance bags.

Figure 4. For overhead repairs in tall passages, extend the center tripod leg vertically from the lower fulcrum. Create an upper fulcrum with a mini tripod head and short padded crossbar.

Figure 5. Al Collier repairs a short stalactite tip with Hot Stuff Super T cyanoacrylate adhesive, a spring-loaded tube, and a plumb bob.

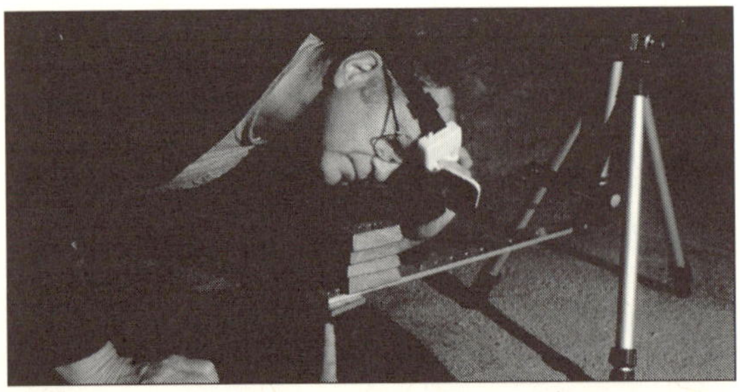

Figure 6. Reverse the tripod system and apply downward pressure instead of overhead bracing.

Figure 7. Use a similar technique for stalactite repairs in inaccessible locations.

Photos Tom Dotter

Put the counterweight on the opposite end of the crossbar. Position the broken stalactite section on the padded end and manually align it directly underneath the intact remnant overhead. Apply epoxy to the break joint of the broken piece. Then mate the two pieces by gently allowing the counterweight to apply pressure.

Once the epoxy has set sufficiently, elevate the counterweight to slowly release the upward pressure on the stalactite. Complete the repair by removing the entire fulcrum assembly.

Inaccessible Location Stalactite Repair

Use this method to repair stalactites in areas of the cave that are not readily accessible (such as under ledges.) Extend the center leg of the tripod horizontally from the main fulcrum and position it approximately 90 degrees from standard (Figure 7).

Install an adjustable mini tripod head of compatible size on the extended end, which creates a second fulcrum. Next, install a small aluminum pad on the second fulcrum.

Place the aluminum pad under the stalactite with appropriate counterbalance applied to the opposite end at the position of the primary tripod head. (Extra weight added to the center of the tripod helps to stabilize it while making this type of repair.) Place the broken piece on the padded end and manually align it directly underneath the intact remnant of the stalactite. Apply epoxy to the surface of the broken piece and mate the two pieces by gently allowing the counterweight to apply pressure.

Once the epoxy has set sufficiently, the counterweight is slowly elevated to gently release the upward pressure on the stalactite.

> The variations on these techniques and their applications are endless.

Standard Helictite Repair

Support helictite repairs in much the same way as stalactites, with a few minor modifications. Extend the center leg of the tripod horizontally from the main fulcrum and position it approximately 90 degrees from standard.

Install an adjustable mini tripod head of compatible size on the extended end, which creates a second fulcrum. Then install a small aluminum crossbar on the second fulcrum. The crossbar has an aluminum pad on one end, which is used to protect the speleothem.

Place the aluminum pad against the helictite and apply appropriate counterweight to the opposite end at the position of the primary tripod head (only for the purpose of balancing). (Extra weight added to the center of the tripod helps to stabilize it while making this type of repair.)

Place the broken piece near the padded end and manually align it directly in front of the intact remnant of the helictite. Apply epoxy to the surface of the broken piece and mate the two pieces by hand. Use a small bungee connected to one end of the aluminum crossbar to create tension on the opposite, padded end. Then gently place the padded end against the epoxied speleothem.

Once the epoxy has set sufficiently, manually shift the bungee end of the crossbar to release the gentle pressure from the helictite.

Stalagmite and Multi-break Stalactite Repairs

Use this technique to repair stalactites that have been broken into multiple pieces. It is also effective for small stalagmite repairs. Adjust the tripod to the appropriate height for the size of the speleothem being repaired (Figure 6).

The center leg is 180 degrees from normal with the large camera head pointing straight down. Attach a long aluminum crossbar to the camera head. At the opposite end, attach an aluminum pad.

Align the various parts of the speleothem directly underneath the pad. Apply epoxy to one of the broken surfaces. Match to the corresponding part of the speleothem. Place the pad on the upper piece. Then place appropriate weight on the bar and apply pressure to the speleothem to assure a tight epoxy joint.

Once the epoxy has set sufficiently, remove the weight from the bar and gently lift the pad from the speleothem. Repeat this process for each of the broken sections.

> We can protect and restore caves for future generations or we can make them into mud holes devoid of speleothems and life. The choices and the responsibilities belong to all of us.

Repairing Short Tips on Stalactites

Repair short stalactite tips in just minutes with a spring-loaded tube and a plumb bob (Figure 5). First, compare the alignment of the broken tip to the stalactite for proper fit. Then, move the tip down and add a drop of adhesive. Reattach the tip to the stalactite and hold in place with the spring-loaded tube and use the plumb bob to maintain vertical pressure until the epoxy dries. Finally, remove the tube very slowly.

Conclusion

Cave softly—don't break it and it won't need to be fixed. We can clean, renovate, refurbish, and repair some speleothems. We can come close to making some look pristine. However, chipping a name into flowstone or removing broken speleothems from a cave causes damage that can never be reversed. We can protect and restore cave environments for future generations or we can make them into mud-filled holes devoid of speleothems and life. The choices and the responsibilities belong to all of us.

Beware of Claims and Labels

Val Hildreth-Werker

Various products may be available that are safe and satisfactory for cave systems, but before using any material, thoroughly check it out for site-specific interactions and implications.

Research the characteristics of proposed materials and use discretion when introducing new products.

Watch out for environmentally friendly claims—unfortunately, statements on labels sometimes have minimal foundation in veracity.

Many environmental or biodegradable products degrade rapidly and provide new nutrient sources for cave-dwelling organisms. (See anthropogenic chemicals, pages 57–58.)

The byproducts of degradation may interact with naturally occurring substances in the cave and result in unnecessary damage.

Before causing harm, check out the idea or product with scientists and trained spelean experts. If a product seems to receive approval, avoid surprises and first test a small sample in the cave passage.

Success Stories and Blunders

Editors' Note: It took a lot of failures for speleothem restorationists to develop the methods presented in this book. (That's why we call them proven methods.) We originally decided to write this book so others would be able to start with this information and avoid our blunders, bad judgements, and failures.

We encourage you to build on our successes and devise new simple strategies to benefit future speleothem repairs.

Section D—Success Stories and Blunders

Endless Surprises

Jim C. Werker

A reputable chemical company read a newspaper article about cave repair projects. The article mentioned that I was using Epon® 828 to repair speleothems. A representative called and encouraged me to try their competing product. Claiming their epoxy resin was inert, colorless, and had the same archival characteristics as the Epon, the guy sent me a sample. I made the mistake of believing he knew what he was talking about and, rather than first testing a small quantity of the product, I jumped in and used the new, unproven epoxy to repair a stalagmite in Endless Cave, New Mexico.

Several weeks after the repair, I returned to Endless Cave to find a lovely lavender ring around the break joint of the speleothem. Looking down to the floor, I was further incensed at finding a ring of dead cave crickets around the base of the stalagmite.

Immediately, I started digging out the purple epoxy and rescued the repair and the future cricket population by replacing the purple goo with Epon. We do learn from embarrassing *faux pas.*

The moral of this story is … test first to avoid toxic surprises.

Learn from unfortunate *faux pas.* Test first and avoid toxic surprises.

Photos © Val Hildreth-Werker

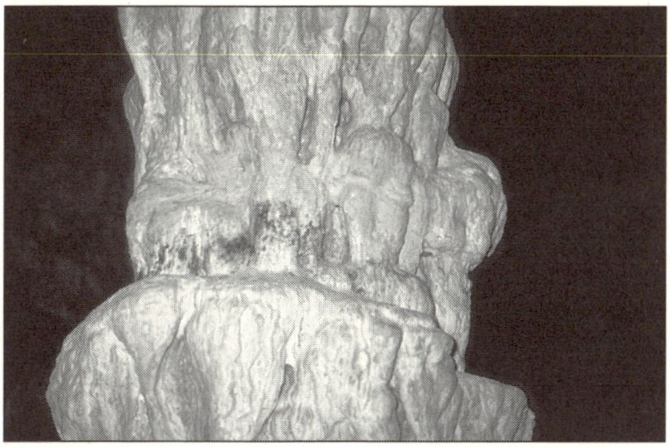

Figure 1. An untested, unproven epoxy grew a purple ring of fungi within a few weeks of application on this stalagmite repair in Endless Cave, New Mexico.

Figure 2. Jim Werker in a "Guano Happens" T-shirt working on the purple ring left by a new epoxy. He used Epon 828 and Versamid 40 to replace the unsuccessful adhesive.

Section D—Success Stories and Blunders

"Three R" Story

John A. French and Paul J. Meyer

All too often, repair, refurbishment, and restoration is completed only to discover new vandalism has undone many hours of underground labor. If the cave is too popular, too accessible, or unprotected, the speleothem repairers may well find their efforts better spent elsewhere. However, there are exceptions to all rules.

Several cavers took on the challenge of repairing one room of the much-visited, much-vandalized Crossings Cave in Alabama. Conditions in the cave varied from muddied formations to pristine rooms protected by a long, tedious crawlway; historic saltpeter diggings; and one room of former exceptional beauty. Damage to this room included extensive soda straw mutilation; dozens of stalagmite and stalactite clubbings; drapery, column, and chandelier breakage; and several large "Christmas tree" speleothems that were tipped over.

Unfortunately, no early photographs were located so that the extent of the damage could be fully appreciated. Cavers in the area often considered this cave as a "write-off," but a practical use for the cave evolved.

When we took on this restoration task, the initial intent was to fine-tune our repair skills. After several trips, we wondered if a more ambitious objective might be in order. We faced a classic question, "Would our repairs simply challenge the vandals to repeat the former breakage?" Maybe it was possible to launch a modest educational effort that would be appreciated by visitors and allow the cave to continue its slow recovery.

We eventually made several hundred repairs, cleaned floors and formations, and restored rimstone pools. Finally, we designated paths and viewing locations by marking off sections with flagging tape and laminated paper signs. (Both the flagging tape and the paper proved to be appetizing to indigenous pack rats and had to be replaced with more durable materials.)

Our rewards were not long in coming. Now there are frequent trips of school classes, scouts, church groups, and others. Thus, Crossings Cave has become a classroom that hopefully will lead to eventual appreciation of other caves in the area. Feedback from school teachers was relayed through local residents, "Will the formations grow back?"

Other cavers have gone out of their way to see this project for themselves. Maybe best of all, several students are planning to pick up where we left off and expand the repair, restoration, and refurbishment of a cave environment that was once considered hopeless.

Another goal may now be considered. If the current level of voluntary protection continues, this cave could become an underground laboratory for experimentation in the "Three Rs"... repair, restoration, and refurbishment. There is plenty of work remaining for volunteers.

No Sacrificial Caves

Jim C. Werker

The term *sacrificial cave* is a misnomer. Cavers should sincerely argue against the idea of further tainting a damaged cave by referring to it as a sacrificial site—as if it is undeserving of respect and reserved for cave vandals. Surprising restoration efforts have proven successful in some of the most unlikely caves.

Our rewards were not long in coming. Now there are frequent trips of school classes, scouts, church groups, and others. Thus, Crossings Cave has become a classroom that hopefully will lead to eventual appreciation of other caves in the area.

Section D—Success Stories and Blunders

Resurrection of The Flutestone

Jonathan B. Beard

In 1979, the cave room containing the Flutestone was discovered when a group of cavers made a brief dunk through a bathtub-sized pool under a low ceiling into virgin cave. The terminal room of the newly-discovered cave contained massive flowstone and a 10.5-foot (3-meter) stalagmite that resembled a giant flute, a narrow totem that was naturally slanted a few degrees.

Within ten years of this discovery, the Flutestone was pushed over by vandals who also broke a number of smaller speleothems in the cave. The

The totem repair project culminated in the attachment of the final top section. Once again, the Flutestone stood tall at its bizarre angle.

Jonathan B. Beard

Figure 1. The completed Flutestone repair. Caver Bill Heim (6-feet, 7-inches tall) is standing beside the restored totem.

> Ozark Highlands Grotto members were disheartened to find that vandals had made their way into the Flutestone Room and pushed the stalagmite over again. We plan to gate the cave and repair the Flutestone again.

long, slender stalagmite tumbled into more than 23 pieces.

In 2001, members of Ozark Highlands Grotto began a 4-month task of reassembly. First, we carefully carried all retrievable pieces out of the 4,109 foot (1,252 meter) cave.

Using a garden hose, we cleaned the pieces, and then allowed them to dry. In my garage (known as "The Calcite Clinic"), we realigned the pieces like some bizarre three-dimensional jigsaw puzzle. Archived photos of the Flutestone *in situ* provided instruction for the reconstruction.

Once realigned, we made aligned drill holes and groups of pieces were reattached using 3M® two-part epoxy paste with sections of stainless steel all-thread. After 23 pieces were reattached into five "master pieces", they were wrapped in plastic bubble wrap, slipped into heavy-duty plastic sacks and loaded into my vehicle.

Meanwhile, we prepared the base where the Flutestone once stood. We drilled a hole through the thin flowstone floor that covered underlying clay. We excavated the clay and filled the void with concrete. We attached the base "master piece" to the floor using an 18-inch (0.5-meter) piece of stainless all-thread and epoxy.

On a subsequent visit, we attached the next "master piece" to the base piece with all-thread and epoxy. It was allowed to set before continuing the reconstruction. In March of 2002, the project culminated in the attachment of the final master piece. Once again, the Flutestone stood tall at its bizarre angle.

Later that year, after the maternal gray bat colony had left the cave, Ozark Highlands Grotto members were disheartened to find that vandals had made their way into the Flutestone Room and pushed the stalagmite over again. We plan to gate the cave and repair the Flutestone again.

Section D—Success Stories and Blunders

Colossal Hanging

Jim C. Werker and Val Hildreth-Werker

A beautiful, 6-foot long (1.8-meter) stalactite was positioned at the bottom of a stone stairway built by the 1930s Civilian Conservation Corps in Colossal Cave Mountain Park near Tucson, Arizona. It hung in quiet splendor until 1995 when a visiting woman tripped on the stairs and fell into the huge speleothem.

The woman was not injured, but two large pieces crashed to the cave floor. The bottom piece weighed about 40 pounds (18 kilograms). The larger, upper piece weighed 185 pounds (84 kilograms). The two broken hunks were placed off-trail and remained in silent sadness for the next eight years.

In the spring of 2003, we received an invitation to assess the damage and made plans to repair the 225-pound (approximately 102-kilogram) stalactite. We designed and constructed sturdy support structures to assure visitor protection during the restoration of the large speleothem. (See Colossal repair photos, pages 504–505.)

With the assistance of several Arizona cavers, we reattached both broken pieces of the stalactite.

- Stainless steel all-thread support pins and Shell Epon 828® epoxy with Shell Epi-cure® 3234Tetra (TETA) curing agent provided the main support for both pieces of the sizeable stalactite.
- A wood support and metal jack system raised the larger 185-pound (approximately 84-kilogram) piece to a tight fit and the support apparatus remained in place for several weeks to allow time for thorough curing.
- Because the visitor trail runs alongside this sizeable stalactite, we reinforced the repair by installing additional stainless steel pins up through the formation and into the ceiling.
- Threading stainless wire between small holes drilled completely through both sides of the upper repair provided additional reinforcement for the joint.
- Between the holes, we carved grooves to receive the wire ends and then concealed all holes and grooves with a rock dust and epoxy mixture.
- After pinning the lower 40-pound (approximately 18-kilogram) piece into position, wood blocks and shims provided support while the epoxy cured.
- Touch-up and additional color-matching work completed our successful repair of the long, heavy stalactite.

Much appreciation is extended to the Arizona cavers who helped in this multi-episode repair over several weeks and to the gracious staff of Colossal Cave Mountain Park for their hospitality and support.

Do no harm ... cave softly and eliminate the need for speleothem repair.

A beautiful, long stalactite hung in quiet splendor until a visitor tripped on the stairs and fell into the huge speleothem.

Repair of 225 Pound Stalactite at Colossal Cave

Figure 1 (before). This sequence shows the steps in the reattachment of a 225-pound (102-kilogram) stalactite in Colossal Cave, Arizona.

Figure 2 (after). Jim Werker (in back), Ian Joaquim, and David Joaquim stand beside the completed repair of the long, heavy stalactite.

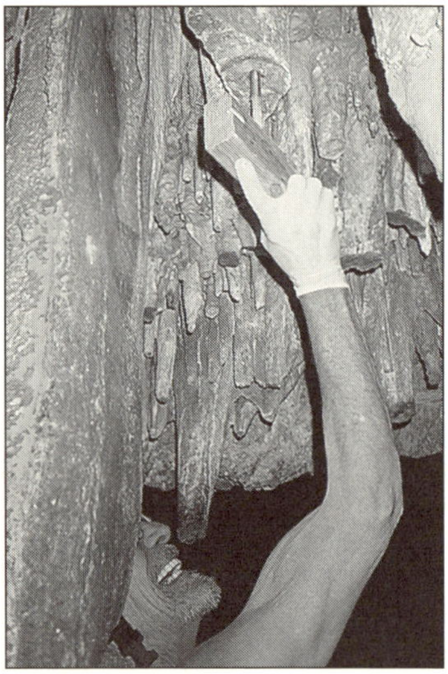

Figure 3. Jim Werker uses Shell Epon 828 epoxy with Epi-cure 3234 (TETA) catalyst to install a stainless steel all-thread stabilization rod into the upper break.

Figure 4. Dave Hamer drills a slightly larger hole to receive the upper stabilization pin in the large section of the broken stalactite. David Joaquim and Jim Werker assist by holding the 185-pound (84 kilogram) piece of speleothem.

Figure 5. Dennis Hoburg and others position the large stalactite while Jim Werker adjusts the screw jack support structure.

Photos ©Val Hildreth-Werker

Figure 6a (before) and **Figure 6b** (midway through repair). These two photos show the reattachment of the upper piece to the ceiling. In the second image, the support jack is cordoned off for safety and left in place for several weeks to assure proper curing of the archival epoxy.

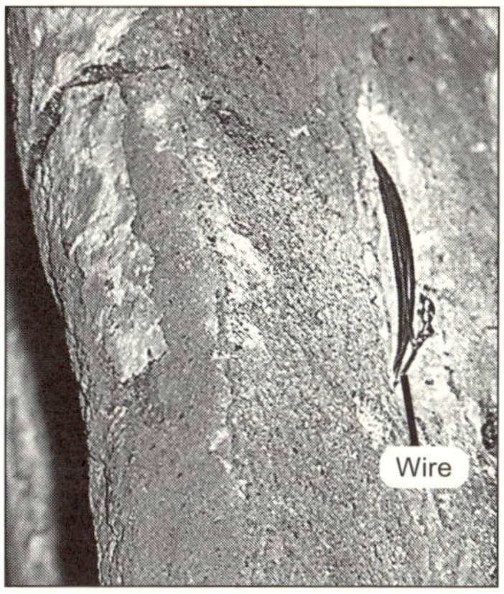

Figure 7. To reinforce the pin and epoxy repair, stainless steel wire is threaded through small drill holes above and below the upper break joint. Alongside the wire, there are additional stainless pins installed at an angle up through the break joint and into the ceiling.

Photos ©Val Hildreth-Werker

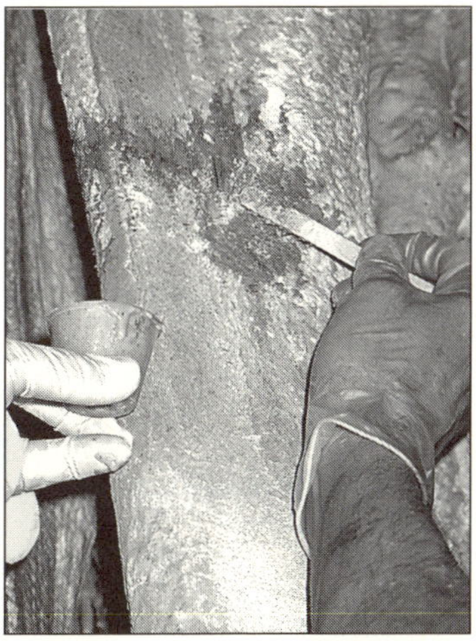

Figure 8. A mixture of rock dust and archival epoxy conceals the drill holes, grooves, and wire supports.

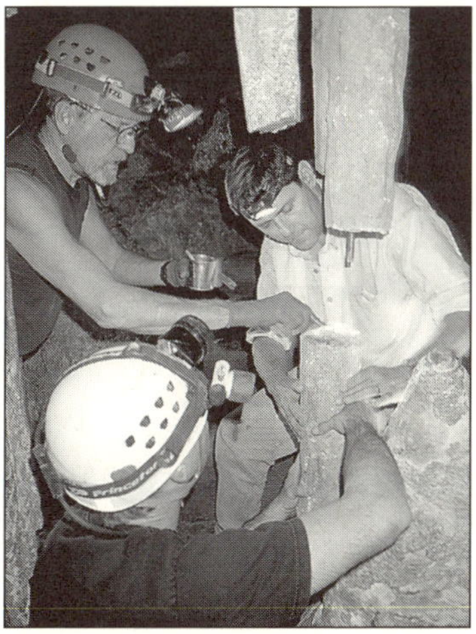

Figure 9. After allowing several weeks for the upper piece to thoroughly cure, the lower piece of stalactite is pinned and epoxied into position.

Figure 10. Wood blocks are placed under the lower piece of the completed stalactite. The lower epoxy seam and the drill dust is cleaned up. Bracing supports for the lower break will be removed after the epoxy cures.

Cave Conservation and Restoration

Appendixes

Appendix 1

Federal Cave Resources Protection Act of 1988

FEDERAL CAVE RESOURCES PROTECTION

16 U.S.C. CHAPTER 63 SECTIONS 4301–4309

Public Law 100-691; November 18, 1988; 102 Stat. 4546

TITLE 16 – CONSERVATION
CHAPTER 63 – FEDERAL CAVE RESOURCES PROTECTION

BE IT ENACTED BY THE SENATE AND THE HOUSE OF REPRESENTATIVES OF THE UNITED STATES OF AMERICA IN CONGRESS ASSEMBLED,

SHORT TITLE.

This Act may be referred to as the "Federal Cave Resources Protection Act of 1988, (FCRPA)."

Section 4301. FINDINGS, PURPOSES, AND POLICY.

a) FINDINGS.
The Congress finds and declares that:
(1) significant caves on Federal lands are an invaluable and irreplaceable part of the Nation's natural heritage; and
(2) in some instances, these significant caves are threatened due to improper use, increased recreational demand, urban spread, and a lack of specific statutory protection.

(b) PURPOSES.
The purposes of this Act are:
(1) to secure, protect, and preserve significant caves on Federal lands for the perpetual use, enjoyment, and benefit of all people; and
(2) to foster increased cooperation and exchange of information between governmental authorities and those who utilize caves located on Federal lands for scientific, education, or recreational purposes.

(c) POLICY.
It is the policy of the United States that Federal lands be managed in a manner which protects and maintains, to the extent practical, significant caves.

Section 4302. DEFINITIONS.

For purposes of this Act:

(1) CAVE.
The term "cave" means any naturally occurring void, cavity, recess, or system of interconnected passages which occurs beneath the surface of the earth or within a cliff or ledge (including any cave resource therein, but not including any vug, mine, tunnel, aqueduct, or other manmade excavation) and which is large enough to permit an individual to enter, whether or not the entrance is naturally formed or manmade. Such term shall include any natural pit, sinkhole, or other feature which is an extension of the entrance.

(2) FEDERAL LANDS.
The term "Federal lands" means lands the fee title to which is owned by the United States and administered by the Secretary of Agriculture or the Secretary of the Interior.

(3) INDIAN LANDS.
The term "Indian lands" means lands of Indian tribes or Indian individuals which are either held in trust by the United States for the benefit of an Indian tribe or subject to a restriction against alienation imposed by the United States.

(4) INDIAN TRIBE.
The term "Indian tribe" means any Indian tribe, band, nation, or other organized group or community of Indians, including any Alaska Native village or regional or village corporation as defined in, or established pursuant to, the Alaska Native Claims Settlement Act (43 U.S.C. 1601 et seq.).

(5) CAVE RESOURCE.
The term "cave resource" includes any material or substance occurring naturally in caves on Federal lands, such as animal life, plant life, paleontological deposits, sediments, minerals, speleogens, and speleothems.

(6) SECRETARY.
The term "Secretary" means the Secretary of Agriculture or the Secretary of the Interior, as appropriate.

(7) SPELEOTHEM.
The term "speleothem" means any natural mineral formation or deposit occurring in a cave or lava tube, including but not limited

to any stalactite, stalagmite, helictite, cave flower, flowstone, concretion, drapery, rimstone, or formation of clay or mud.

(8) SPELEOGEN.

The term "speleogen" means relief features on the walls, ceiling, and floor of any cave or lava tube which are part of the surrounding bedrock, including but not limited to anastomoses, scallops, meander niches, petromorphs and rock pendants in solution caves and similar features unique to volcanic caves.

Section 4303. MANAGEMENT ACTIONS.

(a) REGULATIONS.

Not later than nine months after the date of the enactment of this Act, the Secretary shall issue such regulations as he deems necessary to achieve the purposes of the Act. Regulations shall include, but not be limited to, criteria for the identification of significant caves. The Secretaries shall cooperate and consult with one another in preparation of the regulations. To the extent practical regulations promulgated by the respective Secretaries should be similar.

(b) IN GENERAL.

The Secretary shall take such actions as may be necessary to further the purposes of this Act. These actions shall include (but not be limited to):

(1) identification of significant caves on federal lands;

(A) The Secretary shall prepare an initial list of significant caves for lands under his jurisdiction not later than one year after the publication of final regulations using the significance criteria defined in such regulations. Such a list shall be developed after consultation with appropriate private sector interests, including cavers.

(B) The initial list of significant caves shall be updated periodically, after consultation with appropriate private sector interests, including cavers. The Secretary shall prescribe by policy or regulation the requirements and process by which the initial list will be updated, including management measures to assure that caves under consideration for the list are protected during the period of consideration. Each cave recommended to the Secretary by interested groups for possible inclusion on the list of significant caves shall be considered by the Secretary according to the requirements prescribed pursuant to this paragraph and shall be added to the list if the Secretary determines that the cave meets the criteria for significance as defined by the regulations.

(2) regulation or restriction of use of significant caves, as appropriate;

(3) entering into volunteer management agreements with persons of the scientific and recreational caving community; and

(4) appointment of appropriate advisory committees.

(c) PLANNING AND PUBLIC PARTICIPATION.

The Secretary shall:

(1) ensure that significant caves are considered in the preparation or implementation of any land management plan if the preparation or revision of the plan began after the enactment of this Act;

(2) foster communication, cooperation, and exchange of information between land managers, those who utilize caves, and the public.

Section 4304. CONFIDENTIALITY OF INFORMATION CONCERNING NATURE AND LOCATION OF SIGNIFICANT CAVES.

(a) IN GENERAL.

Information concerning the specific location of any significant cave may not be made available to the public under section 552 of title 5, United States Code, unless the Secretary determines that disclosure of such information would further the purposes of this Act and would not create a substantial risk of harm, theft, or destruction of such cave.

(b) EXCEPTIONS.

Notwithstanding subsection (a), the Secretary may make available information regarding significant caves upon the written request by Federal and state governmental agencies or bona fide educational and research institutions. Any such written request shall, at a minimum:

(1) describe the specific site or area for which information is sought;

(2) explain the purpose for which such information is sought; and

(3) include assurances satisfactory to the Secretary that adequate measures are being taken to protect the confidentiality of such information and to ensure the protection of the significant cave from destruction by vandalism and unauthorized use.

Section 4305. COLLECTION AND REMOVAL FROM FEDERAL CAVES.

(a) PERMIT.

The Secretary is authorized to issue permits for the collection and removal of cave resources under such terms and conditions as the Secretary may impose, including the posting of bonds to insure compliance with the provisions of any permit.

(1) Any permit issued pursuant to this section shall include information concerning the time, scope, location, and specific purpose of the proposed collection, removal or associated activity, and the manner in which such collection, removal, or associated activity is to be performed must be provided.

(2) The Secretary may issue a permit pursuant to this subsection only it he determines that the proposed collection or removal activities are consistent with the purposes of this

Act and with other applicable provisions of law.

(b) REVOCATION OF PERMIT.
Any permit issued under this section shall be revoked by the Secretary upon a determination by the Secretary that the permittee has violated any provision of this Act, or has failed to comply with any other condition upon which the permit was issued. Any such permit shall be revoked by the Secretary upon assessment of a civil penalty against the permittee pursuant to section 4307 or upon the permittee's conviction under section 4306 of this Act. The Secretary may refuse to issue a permit under this section to any person who has violated any provision of this Act or who has failed to comply with any condition of a prior permit.

(c) TRANSFERABILITY OF PERMITS.
Permits issued under this act are not transferable.

(d) CAVE RESOURCES LOCATED ON INDIAN LANDS.
- (1) (A) Upon application by an Indian tribe, the Secretary is authorized to delegate to the tribe all authority of the Secretary under this section with respect to issuing and enforcing permits for the collection or removal of any cave resource located on the affected Indian lands.
  - (B) In the case of any permit issued by the Secretary for the collection or removal of any cave resource, or to carry out activities associated with such collection or removal, from any cave resource located on Indian lands (other than permits issued pursuant to subparagraph (A), the permit may be issued only after obtaining the consent of the Indian or Indian tribe owning or having jurisdiction over such lands. The permit shall include such reasonable terms and conditions as may be requested by such Indian or Indian tribe.
- (2) If the Secretary determines that issuance of a permit pursuant to this section may result in harm to, or destruction of, any religious or cultural site, the Secretary, prior to issuing such permit, shall notify any Indian tribe which may consider the site as having significant religious or cultural importance. Such notice shall not be deemed a disclosure to the public for purposes of section 4304.
- (3) A permit shall not be required under this section for the collection or removal of any cave resource located on Indian lands or activities associated with such collection, by the Indian or Indian tribe owning or having jurisdiction over such lands.

(e) EFFECT OF PERMIT.
No action specifically authorized by a permit under this section shall be treated as a violation of section 4306.

Section 4306. PROHIBITED ACTS AND CRIMINAL PENALTIES.

(a) PROHIBITED ACTS.
- (1) Any person who, without prior authorization from the Secretary, knowingly destroys, disturbs, defaces, mars, alters,

removes or harms any significant cave or alters the free movement of any animal or plant life into or out of any significant cave located on Federal lands, or enters a significant cave with the intention of committing any act described in this paragraph shall be punished in accordance with subsection (b).

(2) Any person who possesses, consumes, sells, barters or exchanges, or offers for sale, barter or exchange, any cave resource from a significant cave with knowledge or reason to know that such resource was removed from a significant cave located on Federal lands shall be punished in accordance with subsection (b).

(3) Any person who counsels, procures, solicits, or employs any other person to violate any provisions of this subsection shall be punished in accordance with subsection (b).

(4) Nothing in this section shall be deemed applicable to any person who was in lawful possession of a cave resource from a significant cave prior to the date of enactment of this Act, November 18, 1988.

(b) PUNISHMENT.

The punishment for violating any provision of subsection (a) shall be imprisonment of not more than one year or a fine in accordance with the applicable provisions of title 18 of the United States Code, or both. In the case of a second or subsequent violation, the punishment shall be imprisonment of not more than 3 years or a fine in accordance with the applicable provisions of title 18 of the United States Code, or both.

Section 4307. CIVIL PENALTIES.

(a) ASSESSMENT.

(1) The Secretary may issue an order assessing a civil penalty against any person who violates any prohibition contained in this Act, any regulation promulgated pursuant to this Act, or any permit issued under this Act. Before issuing such an order, the Secretary shall provide such person written notice and the opportunity to request a hearing on the record within 30 days. Each violation shall be a separate offense, even if such violations occurred at the same time.

(2) The amount of such civil penalty shall be determined by the Secretary taking into account appropriate factors, including:

(A) the seriousness of the violation;

(B) the economic benefit (if any) resulting from the violation;

(C) any history of such violations; and

(D) such other matters as the Secretary deems appropriate. The maximum fine permissible under this section is $10,000.

(b) JUDICIAL REVIEW.

Any person aggrieved by an assessment of a civil penalty under this section may file a petition for judicial review of such assessment with the United States District Court for the District of Columbia or for the district in which the violation occurred. Such a petition shall be filed within the 30-day period beginning on the date the order assessing the civil penalty was issued.

(c) COLLECTION.

If any person fails to pay an assessment of a civil penalty:

(1) within 30 days after the order was issued under subsection (a), or

(2) if the order is appealed within such 30-day period, within 10 days after the court has entered a final judgment in favor of the Secretary under subsection (b), the Secretary shall notify the Attorney General and the Attorney General shall bring a civil action in an appropriate United States district court to recover the amount of penalty assessed (plus costs, attorneys' fees, and interest at currently prevailing rates from the date the order was issued or the date of such final judgment, as the case may be). In such an action, the validity, amount, and appropriateness of such penalty shall not be subject to review.

(d) SUBPOENAS.

The Secretary may issue subpoenas in connection with proceedings under this subsection compelling the attendance and testimony of witnesses and subpoenas duces tecum, and may request the Attorney General to bring an action to enforce any subpoena under this section. The district courts shall have jurisdiction to enforce such subpoena and impose sanctions.

Section 4308. MISCELLANEOUS PROVISIONS.

(a) AUTHORIZATION.

There are authorized to be appropriated $100,000 to carry out the purposes of this Act.

(b) EFFECT ON LAND MANAGEMENT PLANS.

Nothing in this act shall require the amendment or revision of any land management plan, the preparation of which began prior to November 18,1988 [enactment date of this Act].

(c) FUND.

Any money collected by the United States as permit fees for collection and removal of cave resources; received by the United States as a result of the forfeiture of a bond or other security by a permittee who does not comply with the requirements of such permit issued under section 4306; or collected by the United States by way of civil penalties or criminal fines for violations of this Act shall be placed in a special fund in the Treasury. Such moneys shall be available for obligation or expenditure (to the extent provided for in advance in appropriation Acts) as determined by the Secretary for the improved management, benefit, repair, or restoration of significant caves located on Federal lands.

(d) Nothing in this act shall be deemed to affect the full operation of the mining and mineral leasing laws of the United States, or otherwise affect valid existing rights.

Section 4309. SAVINGS PROVISIONS.

(a) WATER.

Nothing in this Act shall be construed as authorizing the appropriation of water by any Federal, State, or local agency, Indian tribe, or any other entity or individual. Nor shall any provision of this Act:

(1) affect the rights or jurisdiction of the United States, the States, Indian tribes, or other entities over water of any river or stream or over any groundwater resource;

(2) alter, amend, repeal, interpret, modify, or be in conflict with any interstate compact made by the States; or

(3) alter or establish the respective rights of States, the United States, Indian tribes, or any person with respect to any water or water-related right.

(b) FISH AND WILDLIFE.

Nothing in this Act shall be construed as affecting the jurisdiction or responsibilities of the States with respect to fish and wildlife.

Appendix 2

Cave and Karst Information Source Listings

- **Speleological Sources**
- **Federal Agencies**
- **Biological Sources**
- **Cave Mapping and Inventory Sources**
- **Geological Sources**
- **Archaeological Sources**

Speleological Information Sources

National Speleological Society
2813 Cave Avenue
Huntsville AL 35810-4431 U.S.A.
Telephone: 256-852-1300
E-mail: nss@caves.org
Web: <http://www.caves.org>
Look for NSS Sections in:

- Biology
- Conservation and Management
- Geology and Geography
- Survey and Cartography

American Cave Conservation Association
American Cave and Karst Center
PO Box 409
Horse Cave KY 42749
Telephone: 270-786-1466
Fax: 270-786-1467
Web: <http://www.cavern.org>

Cave Research Foundation
PO Box 343
Wenona IL 61377
Web: <http://www.cave-research.org>

Center for Cave and Karst Studies
Department of Geography and Geology
Western Kentucky University
Bowling Green KY 42101-3576
Telephone: 502-745-4555
Web: <http://caveandkarst.wku.edu/>

Karst Waters Institute
PO Box 537
Charles Town WV 25414
Telephone: 304-725-1211
Web: <http://www.karstwaters.org>

National Cave & Karst Research Institute
1400 University Drive
Carlsbad NM 88220
Telephone: 505-887-2759
Fax: 505-887-3051
Web: <http://www2.nature.nps.gov/nckri/>

Federal Agencies

Bureau of Land Management
Office of Public Affairs
1849 C Street NW Room 6-LS
Washington DC 20240
Web: <http://www.blm.gov/nhp/>

National Park Service
NPS Geologic Resources Division
PO Box 25287
Denver CO 80225-0287
Web: <http://www.aqd.nps.gov/>

USDA Forest Service (Headquarters)
PO Box 96090 (RHWR)
201 14th Street SW
Washington DC 20090-6090
Web: <http://www.fs.fed.us/>

US Fish and Wildlife Service
(Headquarters)
1849 C Street NW
Washington DC 20240
Web: <http://www.fws.gov>

Biological Information Sources

Karst Waters Institute
PO Box 537
Charles Town WV 25414
Telephone: 304-725-1211
Web: <http://www.karstwaters.org>

Bat Conservation International, Inc.
PO Box 162603
Austin TX 78716
Telephone: 512-327-9721
Web: <http://www.batcon.org>

Illinois Natural History Survey
(Biospeleology)
607 East Peabody Drive
Champaign IL 61820
Telephone: 217-333-6880
Fax: 217-333-4949
Web: <http://www.inhs.uiuc.edu/~sjtaylor/cave/biospeleol.html>

Missouri Department of Conservation
Administrative Office
PO Box 180
(zip code for PO Box is 65102)
2901 West Truman Boulevard
Jefferson City MO 65109
Telephone: 573-751-4115
Fax: 573-751-4467
Web: <http://www.conservation.state.mo.us>

Cave Mapping and Inventory Project Information Sources for Computer Software

National Speleological Society
2813 Cave Avenue
Huntsville AL 35810-4413
Telephone: 256-852-1300
E-mail: <nss@caves.org>
Web: <http://www.caves.org>

Cave Research Foundation
PO Box 343
Wenona IL 61377
Web: <http://www.cave-research.org>

Carlsbad Caverns National Park
Cave Resources Office
3225 National Parks Highway
Carlsbad NM 88220

Wind Cave National Park
Route 1 Box 190-WCNP
Hot Springs SD 57747-9430

Geological Information Sources

American Geological Institute
4220 King Street
Alexandria VA 22302
Telephone: 703-379-2480
Web: <http://www.agiweb.org>

US Geological Survey
Earth Science Information Centers
(Geological and Topographic Maps)

Anchorage, Alaska-USGS/ESIC
4230 University Drive Room 101
Anchorage AK 99508-4664
Telephone: 907-786-7011
Fax: 907-786-7050
E-mail: <gfdurocher@usgs.gov>
Hours: Monday through Friday 8:30 AM to 4:30 PM (Alaska Time)

Denver, Colorado-ESIC
Box 25286 Building 810
Denver Federal Center
Denver CO 80225
Telephone: 303-202-4200
Fax: 303-202-4188
E-mail: <infoservices@usgs.gov>
Hours: Monday through Friday 8:00 AM to 4:00 PM (Mountain Time)

USGS Information Services
(Map and Book Sales)
Box 25286
Denver Federal Center
Denver CO 80225
Telephone: 303-202-4700
Fax: 303-202-4693
Hours: Monday through Friday 8:00 AM to 4:00 PM (Mountain Time)

USGS Information Services
(Open-File Report Sales)
Box 25286
Denver Federal Center
Denver CO 80225
Telephone: 303-202-4700
Fax: 303-202-4188
Hours: Monday through Friday 8:00 AM to 4:00 PM (Mountain Time)

Menlo Park, California-USGS/ESIC
Building 3-MS 532 Room 3128
345 Middlefield Road
Menlo Park CA 94025-3591
Telephone: 650-329-4309
Fax: 650-329-5130
E-mail: <wmcesic@usgs.gov>
Hours: Monday through Friday 8:00 AM to 4:00 PM (Mountain Time)

Reston, Virginia-USGS/ESIC
507 National Center
Reston VA 20192
Telephone: 703-648-5953
Fax: 703-648-5548
TDD: 703-648-4119
E-mail: <ask@usgs.gov>
Hours: Monday through Friday 9:00 AM to 5:00 PM (Eastern Time)

Rolla, MO-USGS/ESIC
1400 Independence Road MS 231
Rolla MO 65401-2602
Telephone: 573-308-3500
Fax: 573-308-3615
E-mail: <mcmcesic@usgs.gov>
Hours: Monday through Friday 7:45 AM to 4:15 PM (Central Time)

Sioux Falls, SD-USGS/ESIC
EROS Data Center
Sioux Falls SD 57198-0001
Telephone: 605-594-6151
Fax: 605-594-6589
TDD: 605-594-6933
E-mail: <custserv@edcmail.cr.usgs.gov>
Hours: Monday through Friday 8:00 AM to 4:00 PM (Central Time)

Spokane, WA-USGS/ESIC
US Post Office Building Room 135
904 West Riverside Avenue
Spokane WA 99201
Telephone: 509-368-3130
Fax: 509-368-3194
TDD: 509-368-3133
E-mail: <esnfic@usgs.gov>
Hours: Monday through Friday 8:30 AM to 4:00 PM (Pacific Time)

Washington, D.C.-USGS/ESIC
US Department of the Interior
1849 C Street NW Room 2650
Washington DC 20240
Telephone: 202-208-4047
E-mail: <ask@usgs.gov>
Hours: Monday through Friday 8:30 AM to 4:00 PM (Eastern Time)

State Geological Surveys

Alabama
Geological Survey of Alabama
PO Box 869999
420 Hackberry Lane
Tuscaloosa AL 35486-6999
Telephone: 205-349-2852
Fax: 205-349-2861
Web: <http://www.gsa.state.al.us/>

Alaska
Alaska Division of Geological and Geophysical Surveys
794 University Avenue Suite 200
Fairbanks AK 99709-3645
Telephone: 907-451-5000
Fax: 907-451-5050
Web: <http://wwwdggs.dnr.state.ak.us/>

Arizona
Arizona Geological Survey
416 West Congress Street, Suite 100
Tucson AZ 85701
Telephone: 520-770-3500
Fax: 520-770-3505
Web: <http://www.azgs.state.az.us/>

Arkansas
Arkansas Geological Commission
Vardelle Parham Geology Center
3815 West Roosevelt Road
Little Rock AR 72204
Telephone: 501-296-1877
Fax: 501-663-7360
Web: <http://www.state.ar.us/agc/agc.htm>

California
California Division of Mines and Geology
801 K Street MS 12-01
Sacramento CA 95814-3531
Telephone: 916-445-1825
Fax: 916-445-5718
Web: <http://www.consrv.ca.gov/dmg/>

Colorado
Colorado Geological Survey
1313 Sherman Street Room 715
Denver CO 80203
Telephone: 303-866-2611
Fax: 303-866-2461
Web: <http://www.dnr.state.co.us/geosurvey/>

Connecticut
Connecticut Geological and Natural History Survey
Environmental and Geographic Information Center
Department of Environmental Protection
79 Elm Street
Hartford CT 06106-5127
Telephone: 860-424-3540
Fax: 860-421-4058
Web: <http://dep.state.ct.us/>

Delaware
Delaware Geological Survey
University of Delaware
Newark, DE 19716-7501
Telephone: 302-831-2833
Fax: 302-831-3579
Web: <http://www.udel.edu/dgs/dgs.html>

Florida
Florida Geological Survey
903 West Tennessee Street
Tallahassee FL 32304-7700
Telephone: 850-488-4191
Fax: 850-488-8086
Web: <http://www.dep.state.fl.us/geo/>

Georgia
Georgia Geologic Survey Branch
19 Martin Luther King Jr Dr SW Rm 400
Atlanta GA 30334
Telephone: 404-656-3214
Fax: 404-657-8379
Web: <http://www.ganet.org/dnr/environ/branches/geosurv/>

Hawaii
Department of Land and Natural Resources
Commission on Water Resource Management
PO Box 621
1151 Punchbowl Street
Kalanimoku Building Room 227
Honolulu HI 96809
Telephone: 808-587-0263
Fax: 808-587-0219
Web: <http://www.state.hi.us/dlnr/cwrm/>

Idaho
Idaho Geological Survey
University of Idaho

Morrill Hall Room 332
Moscow ID 83844-3014
Telephone: 208-885-7991
Fax: 208-885-5826
Web: <http://www.uidaho.edu/igs/igs.html>

Illinois

Illinois State Geological Survey
121 Natural Resources Building
615 East Peabody Drive
Champaign IL 61820-6964
Telephone: 217-333-5111
Fax: 217-244-7004
Web: <http://www.isgs.uiuc.edu/>

Indiana

Indiana Geological Survey
Indiana University
611 North Walnut Grove
Bloomington IN 47405-2208
Telephone: 812-855-5067
Fax: 812-855-2862
Web: <http://adamite.igs.indiana.edu/index.htm>

Iowa

Iowa Geological Survey Bureau/IDNR
109 Trowbridge Hall
Iowa City IA 52242-1319
Telephone: 319-335-1575
Fax: 319-335-2754
Web: <http://www.igsb.uiowa.edu/>

Kansas

Kansas Geological Survey
1930 Constant Avenue Campus West
The University of Kansas
Lawrence KS 66047-3726
Telephone: 785-864-2108
Fax: 785-864-5317
Web: <http://www.kgs.ukans.edu/>

Kentucky

Kentucky Geological Survey
228 Mining and Mineral Resources Building
University of Kentucky
Lexington KY 40506-0107
Telephone: 859-257-5500
Fax: 859-257-1147
Web: <http://www.uky.edu/KGS/home.htm>

Louisiana

Louisiana Geological Survey
University Station Box G
Baton Rouge LA 70893-4107
Telephone: 225-388-5320
Fax: 225-388-3662
Web: <http://www.lgs.lsu.edu/>

Maine

State Geologist
Maine Geological Survey
Natural Resources Information and Mapping Center
22 State House Station
Augusta ME 04333-0022
Telephone: 207-287-2801
Fax: 207-287-2353
Web: <http://www.state.me.us/doc/nrimc/mgs/mgs.htm>

Maryland

Maryland Geological Survey
2300 St Paul Street
Baltimore MD 21218-5210
E-mail: <ecleaves@mgs.md.gov>
Telephone: 410-554-5500
Fax: 410-554-5502
Web: <http://mgs.dnr.md.gov>

Massachusetts

Executive Office of Environmental Affairs
Massachusetts Geological Survey
100 Cambridge Street 20th Floor
Boston MA 02202
Telephone: 617-727-5830 extension 305
Fax: 617-727-2754
Web: <http://www.magnet.state.ma.us/envir/eoea.htm>

Michigan

Michigan Geological Survey Division
Department of Environmental Quality
Box 30256 735 East Hazel Street
Lansing MI 48909-7756
Telephone: 517-334-6923
Fax: 517-334-6038
Web: <http://www.deq.state.mi.us/gsd/>

Minnesota

Minnesota Geological Survey
University of Minnesota
2642 University Avenue W Room 104
St Paul MN 55114-1057

Telephone: 612-627-4780
Fax: 612-627-4778
Web: <http://www.geo.umn.edu/mgs/>

Mississippi
Mississippi Office of Geology
PO Box 20307
Jackson MS 39289-1307
Telephone: 601-961-5500
Fax: 601-961-5521
Web: <http://www.deq.state.ms.us/>

Missouri
Missouri Division of Geology and
Land Survey
PO Box 250
111 Fairgrounds Road
Rolla MO 65401-0250
Telephone: 573-368-2101
Fax: 573-368-2111
Web: <http:// www.dnr.state.mo.us/dgls/
homedgls.htm>

Montana
Montana Bureau of Mines and Geology
Montana Tech of the University
of Montana
1300 West Park Street
Butte MT 59701-8997
Telephone: 406-496-4180
Fax: 406-496-4451
Web: <http://mbmgsun.mtech.edu/>

Nebraska
Nebraska Conservation and
Survey Division
113 Nebraska Hall
University of Nebraska
Lincoln NE 68588-0517
Telephone: 402-472-3471
Fax: 402-472-4608
Web: <http://csd.unl.edu/csd.html>

Nevada
Director and State Geologist
Nevada Bureau of Mines and Geology
University of Nevada Reno MS 178
Reno NV 89557-0088
Telephone: 775-784-6691 extension 126
Fax: 775-784-1709
Web: <http://www.nbmg.unr.edu>

New Hampshire
Department of Environmental Services
PO Box 95
6 Hazen Drive
Concord NH 03302-0095
Telephone: 603-271-6482
Fax: 603-271-2867
Web: <http://www.des.state.nh.us/
descover.htm>

New Jersey
New Jersey Geological Survey
Division of Science Research
and Technology
Department of Environmental Protection
PO Box 427
29 Arctic Parkway
Trenton NJ 08625-0427
Telephone: 609-292-1185
Fax: 609-633-1004
Web: <http://www.state.nj.us/dep/njgs/>

New Mexico
New Mexico Bureau of Mines
and Mineral Resources
New Mexico Tech
801 Leroy Place
Socorro NM 87801-4796
Telephone: 505-835-5420
Fax: 505-835-6333
Web: <http://www.geoinfo.nmt.edu>

New York
New York State Geological Survey
New York State Museum
3140 Cultural Education Center
Albany NY 12230
Telephone: 518-474-5816
Fax: 518-473-8496
Web: <http://www.nysm.nysed.gov/
geology.html>

North Carolina
North Carolina Geological Survey
Department of Environment, Health
and Natural Resources
Division of Land Resources
1612 Mail Service Center
Raleigh NC 27699-1612
Telephone: 919-733-2423
Fax: 919-733-0900
Web: <http://www.geology.enr.state.nc.us>

North Dakota
North Dakota Geological Survey
600 East Boulevard Avenue
Bismarck ND 58505-0840
Telephone: 701-328-8000
Fax: 701-328-8010
Web: <http://www.state.nd.us/ndgs/>

Ohio
Ohio Division of Geological Survey
Department of Natural Resources
4383 Fountain Square Drive Building B
Columbus OH 43224-1362
Telephone: 614-265-6988
Fax: 614-447-1918
Web: <http://www.dnr.state.oh.us/odnr/geo_survey>

Oklahoma
Oklahoma Geological Survey
100 East Boyd Room N-131
Norman OK 73019-0628
Telephone: 405-325-3031
Fax: 405-325-7069
Web: <http://www.ou.edu/special/ogs-pttc>

Oregon
Oregon Department of Geology and Mineral Industries
800 NE Oregon Street #28
Suite 965
Portland OR 97232-2162
Telephone: 503-731-4100
Fax: 503-731-4066
Web: <http://sarvis.dogami.state.or.us>

Pennsylvania
Pennsylvania Bureau of Topographic and Geologic Survey
Department of Conservation and Natural Resources
PO Box 8453
Harrisburg PA 17105-8453
Telephone: 717-787-2169
Fax: 717-783-7267
Web: <http://www.dcnr.state.pa.us/topogeo>

Puerto Rico
Area Recursos de Agua y Minerales
Departamento de Recursos Naturales
Apartado 9066600
Puerta de Tierra Station
San Juan Puerto Rico 00906-6600
Telephone: 787-725-3852
Fax: 787-722-5586

Rhode Island
Office of Rhode Island State Geologist
Rhode Island Geological Survey
The University of Rhode Island
315 Green Hall
Kingston RI 02881
Telephone: 401-874-2265
Fax: 401-874-2190
Web: <http://nick.uri.edu/cels/gel/rigs.html>

South Carolina
South Carolina Geological Survey
5 Geology Road
Columbia SC 29210-4089
Telephone: 803-896-7708
Fax: 803-896-7695
Web: <http://water.dnr.state.sc.us/geology>

South Dakota
South Dakota Geological Survey
Department of Environment and Natural Resources
Akeley Science Center
University of South Dakota
414 East Clark Street
Vermillion SD 57069-2390
Telephone: 605-677-5227
Fax: 605-677-5895
Web: <http://www.sdgs.usd.edu/>

Tennessee
Tennessee Division of Geology
Department of Environment and Conservation
401 Church Street
L and C Tower 13th Floor
Nashville TN 37243-0445
Telephone: 615-532-1500
Fax: 615-532-1517
Web: <http://www.state.tn.us/environment/tdg/>

Texas
Texas Bureau of Economic Geology
University of Texas at Austin
University Station Box X

Austin TX 78713-8924
Telephone: 512-471-1534
Fax: 512-471-0140
Web: <http://www.beg.utexas.edu/>

Utah
Utah Geological Survey
1594 West North Temple
Suite 3110
PO Box 146100
Salt Lake City UT 84114-6100
Telephone: 801-537-3300
Fax: 801-537-3400
Web: <http://www.ugs.state.ut.us/>

Vermont
Vermont Geological Survey
Agency of Natural Resources
103 South Main Street Laundry Building
Waterbury VT 05671-0301
Telephone: 802-241-3608
Fax: 802-241-3273
Web: <http://www.anr.state.vt.us/geology/vgshmpg.htm>

Virginia
Virginia Division of Mineral Resources
PO Box 3667
Charlottesville VA 22903
Telephone: 434-951-6350
Fax: 434-951-6366
Web: <http://www.geology.state.va.us>

Washington
Division of Geology and Earth Resources
Department of Natural Resources
1111 Washington Street SE Room 148
PO Box 47007
Olympia WA 98504-7007
Telephone: 360-902-1450
Fax: 360-902-1785
Web: <http://www.wa.gov/dnr/htdocs/ger/ger.html>

West Virginia
West Virginia Geological and Economic Survey
Mont Chateau Research Center
PO Box 879
Morgantown WV 26507-0879
Telephone: 304-594-2331
Fax: 304-594-2575
Web: <http://www.wvgs.wvnet.edu/>

Wisconsin
Wisconsin Geological and Natural History Survey
3817 Mineral Point Road
Madison WI 53705-5100
Telephone: 608-262-1705
Fax: 608-262-8086
Web: <http://www.uwex.edu/wgnhs/>

Wyoming
Wyoming State Geological Survey
PO Box 3008
University Station
Laramie WY 82071-3008
Telephone: 307-766-2286
Fax: 307-766-2605
Web: <http://www.wsgsweb.uwyo.edu/>

Archaeological Information Sources
State Archaeologists

Alabama
Thomas O Maher PhD
State Archaeologist
Alabama Historical Commission
468 South Perry Street
Montgomery AL 36130-0900
Telephone: 334-230-2645
Fax: 334-240-3477
E-mail: <tmaher@mail.preserveala.org>
Web: Archaeological Services of Alabama Historical Commission <http://www.preserveala.org/arch.html>

Alaska
Robert D Shaw
State Archaeologist/Deputy SHPO
Alaska Department of Natural Resources
Office of History and Archaeology
3601 C Street Suite 1278
Anchorage AK 99503-5921
Telephone: 907-269-8727
Fax: 907-269-8908
E-mail: <bobsh@dnr.state.ak.us>

American Samoa
David J Herdrich
Territorial Archaeologist/Deputy HPO
Executive Offices of the Governor
American Samoa Historic Preservation
American Samoa Government
Pago Pago AS 96799

Telephone: 011-684-633-2384
Fax: 011-684-633-2367
E-mail: <herdrich@samoatelco.com>
Web: American Samoa Historic Preservation Office <http://www.ASHPO.org/>

Arizona

Paul R Fish
Curator of Archaeology
Arizona State Museum
University of Arizona
PO Box 210026
Tucson AZ 85721-0026
Telephone: 520-621-2556
Fax: 520-621-2976
E-mail: <pfish@email.arizona.edu>
Web: University of Arizona's Arizona State Museum
<http://www.statemuseum.arizona.edu>
Check their Archaeological Research
<http://www.statemuseum.arizona.edu/arch/>

Arkansas

Ann M Early
State Archeologist
Arkansas Archeological Survey
2475 North Hatch Avenue
Fayetteville AR 72704
Telephone: 501-575-3556
Fax: 501-575-5453
E-mail: <amearly@uark.edu>
Web: Arkansas Archeological Survey
<http://www.uark.edu/campus-resources/archinfo/>

California

J Charles Whatford and Steve Grantham
Associate State Archaeologists
Office of Historic Preservation
PO Box 942896
Sacramento CA 94296-0001
Telephone: 916-653-2716
and 916-653-8920
Fax: 916-653-9824
E-mail: <cwhat@ohp.parks.ca.gov>
and <sgran@ohp.parks.ca.gov>
Web: California State Parks' Office of Historic Preservation
<http://www.ohp.cal-parks.ca.gov/>

Colorado

Susan M Collins
State Archaeologist/Deputy
SHPO Archaeology
Colorado Historical Society
1300 Broadway
Denver CO 80203
Telephone: 303-866-2736
Fax: 303-866-2711
E-mail: <susan.collins@chs.state.co.us>
Web: Colorado Historical Society's Office of Archaeology and Historic Preservation
<http://www.coloradohistory-oahp.org/>

Connecticut

Nicholas Bellantoni
State Archaeologist
Office of State Archaeology
Box U-4214
University of Connecticut
Storrs CT 06269-4214
Telephone: 860-486-5248
Fax: 860-486-6364
E-mail: <nbell@uconnvm.uconn.edu>
Web: Connecticut Office of State Archaeology <http://www.archnet.uconn.edu/archnet/topical/crm/conn/ctosa.html>

Delaware

Faye Stocum
Delaware State Historic Preservation Office
#15 The Green
Dover DE 19901-3611
Telephone: 302-739-5685
Fax: 302-739-5660
E-mail: <fstocum@state.de.us>
Web: Delaware State Historic Preservation Office <http://www.state.de.us/shpo/index.htm>

District of Columbia

Nancy J Kassner
Archaeologist
Historic Preservation Division
614 H Street NW Room 305
Washington DC 20001
Telephone: 202-727-7360
Fax: 202-727-7211

Florida

James J Miller
State Archaeologist
Bureau of Archaeological Research
Division of Historical Resources
500 South Bronough Street

Tallahassee FL 32399-0250
Telephone: 850-487-2299
Fax: 850-414-2207
E-mail: <jmiller@mail.dos.state.fl.us>
Web: Florida Division of Historical Resources' Bureau of Archaeological Research <http://www.dhr.dos.state.fl.us/bar/arch.html>

Georgia
David Crass PhD
State Archaeologist
Historic Preservation Division
156 Trinity Avenue SW
Suite 101
Atlanta GA 30303-3600
Telephone: 404-656-9344
Fax: 404-657-1040
E-mail: <david_crass@mail.dnr.state.ga.us>
Web: Georgia Department of Natural Resources Historic Preservation Division <http://www.gashpo.org/>

Guam
Mr Victoriano April
Territorial Archaeologist
Historic Resources Division
Department of Parks and Recreation
PO Box 2950
Agana Guam 96910
Telephone: 671-475-6290 or 671-475-6291
Fax: 671-477-2822
E-mail: <vicapril@ns.gov.gu>
Web: Guam Department of Parks and Recreation's Historic Resources Division <http://www.ns.gov.gu/dpr/hrdhome.html>

Hawaii
Dr Ross Cordy
Head Archaeologist
State Historic Preservation Division
601 Kamokila Boulevard Suite 555
Kapolei HI 96707
Telephone: 808-692-8025
Fax: 808-692-8020

Idaho
Kenneth C Reid
State Archaeologist/Deputy SHPO
State Historic Preservation Office
Idaho State Historical Society
210 Main Street
Boise ID 83702
Telephone: 208-334-3847
Fax: 208-334-2775
E-mail: <kreid@ishs.state.id.us>
Web: Idaho State Historical Society's State Historic Preservation Office <http://www2.state.id.us/ishs/SHPO.html>

Illinois
Mark E Esarey PhD
State Archaeologist
Preservation Services Division
Illinois Historic Preservation Agency
500 East Madison Street
Springfield IL 62701
Telephone: 217-785-4999
Fax: 217-782-8161

Indiana
James Richard Jones III
State Archaeologist/Deputy SHPO
Division of Historic Preservation and Archaeology
Department of Natural Resources
402 West Washington Room W274
Indianapolis IN 46204
Telephone: 317-232-1646
Fax: 317-232-0693
E-mail: <rjones@dnr.state.in.us>

Iowa
William Green
State Archaeologist
Office of the State Archaeologist
University of Iowa
700 Clinton Street Building
Iowa City IA 52242-1030
Telephone: 319-384-0732
Fax: 319-384-0768
E-mail: <bill-green@uiowa.edu>
Web: University of Iowa's Office of the State Archaeologist <http://www.uiowa.edu/~osa/>

Kansas
Bob Hoard
State Archeologist
Cultural Resources Division
Kansas State Historical Society
6425 SW 6th Avenue
Topeka KS 66615-1099
Telephone: 785-272-8681 extension 268
Fax: 785-272-8682

E-mail: <rhoard@kshs.org>
Web: Kansas State Historical Society
<http://www.kshs.org/>

Kentucky
Dr Jim P Fenton
State Archaeologist
WS Webb Museum of Anthropology
and Office of State Archaeology
211 Lafferty Hall
University of Kentucky
Lexington KY 40506
Telephone: 859-257-8208
Fax: 859-323-3686
E-mail: <jpfent2@pop.uky.edu>
E-mail: <museum@pop.uky.edu>
Web: William S. Webb Museum of
Anthropology <http://www.uky.edu/AS/
Anthropology/Museum/museum.htm>

Louisiana
Thomas H Eubanks
State Archaeologist
Division of Archaeology
1051 North 3rd Street Room 405
PO Box 44247
Baton Rouge LA 70804-4247
Telephone: 225-342-8170
Fax: 225-342-4480
E-mail: <teubanks@crt.state.la.us>
Web: Louisiana Division of Archaeology
<http://www.crt.state.la.us/crt/ocd/arch/
homepage/index.htm>

Maine
Arthur E Spiess
Archaeologist
Maine Historic Preservation Commission
State House Station 65
Augusta ME 04333-0065
Telephone: 207-287-2132
Fax: 207-287-2335
E-mail: <arthur.spiess@state.me.us>
Web: Maine Historic Preservation Commission
<http://www.state.me.us/mhpc/>

Maryland
Richard B Hughes, Chief
Office of Archeology
Maryland Historical Trust
Division of Historical and
Cultural Programs
100 Community Place
Crownsville MD 21032-2023
Telephone: 410-514-7660
Fax: 410-987-4071
Email: hughesr@dhcd.state.md.us
Web: Maryland Historical Trust
<http://www.marylandhistoricaltrust.net/>
Office of Archeology <http://www.
marylandhistoricaltrust.net/xarcheol.html>

Massachusetts
Brona Simon
State Archaeologist/Deputy SHPO
Massachusetts Historical Commission
220 Morrissey Boulevard
Boston MA 02125
Telephone: 617-727-8470
Fax: 617-727-5128
TDD: 617-878-3889
E-mail: <brona.simon@sec.state.ma.us>
Web: Massachusetts Historical Commission
<http://www.state.ma.us/sec/mhc/>

Michigan
John R Halsey
State Archaeologist
Michigan Historical Center
Department of State
Lansing MI 48918-1847
Telephone: 517-373-6358
Fax: 517-241-4738
E-mail: <johnh@sosmail.state.mi.us>
Web: Michigan's Office of the State
Archaeologist <http://www.sos.state.mi.us/
history/archeol/>

Minnesota
Mark J Dudzik
State Archaeologist
Office of the State Archaeologist
Fort Snelling History Center
St. Paul MN 55111-4061
Telephone: 612-725-2411
Fax: 612-725-2729
E-mail: <mark.dudzik@state.mn.us>
Web: Minnesota's Office of the State
Archaeologist
<http://www.admin.state.mn.us/osa/>

Mississippi
Samuel McGahey
Chief Archaeologist
Department of Archives and History
PO Box 571

Jackson MS 39205-0571
Telephone: 601-359-6940
Fax: 601-359-6955

Missouri
Claire F Blackwell
Director/Deputy SHPO Historic Preservation Program
PO Box 176
Jefferson City MO 65102
Telephone: 573-751-7857
Fax: 573-526-2852
E-mail: <nrblacc@mail.dnr.state.mo.us>
Web: Missouri Department of Natural Resources' Historic Preservation Program <http://www.mostateparks.com/hpp/>

Montana
Stan Wilmoth PhD
State Archaeologist
State Historic Preservation Office
1410 8th Avenue
PO Box 201202
Helena MT 59620-1202
Telephone: 406-444-7719
Fax: 406-444-6575
E-mail: <swilmoth@state.mt.us>
Web: Montana's State Historic Preservation Office <http://www.his.state.mt.us/departments/shpo/>

Nebraska
Rob Bozell
Associate Director for Archeology
Nebraska State Historical Society
Archeology Division
1500 R Street
PO Box 82554
Lincoln NE 68501
Telephone: 402-471-4789
Fax: 402-471-3314
E-mail: <rbozell@mail.state.ne.us>
Web: Nebraska State Historical Society's Archeology Division <http://www.nebraskahistory.org/archeo/index.htm/

Nevada
Gene Hattori
State Archaeologist
State Historic Preservation Office
100 North Stewart Street
Capitol Complex
Carson City NV 89710
Telephone: 775-684-3448
via Barbara Prudic Program Manager
Fax: 775-684-3442
E-mail: <blprudic@clan.lib.nv.us>
Map: <http://www.dmla.clan.lib.nv.us/docs/shpo/map.htm>
Web: Nevada State Historic Preservation Office <http://www.dmla.clan.lib.nv.us/docs/shpo/>

New Hampshire
Gary W Hume
State Archaeologist
Division of Historical Resources
State of New Hampshire
Department of Cultural Resources
PO Box 2043
Concord NH 03301-2043
Telephone: 603-271-3483
or 603-271-3558
Fax: 603-271-3433
Voice/TDD: 1-800-735-2964
E-mail: <preservation@nhdhr.state.nh.us>
Web: New Hampshire Division of Historical Resources
<http://www.state.nh.us/nhdhr>

New Jersey
Lorraine E Williams
Acting Director
New Jersey State Museum
205 West State Street
PO Box 530
Trenton NJ 08625-0530
Telephone: 609-292-8594
Fax: 609-599-4098
E-mail:<lorraine.williams@sos.state.nj.us>

New Mexico
Glenna Dean
State Archaeologist
Historic Preservation Division
La Villa Rivera
228 East Palace Avenue
Santa Fe NM 87501
Telephone: 505-827-3989
Fax: 505-827-6338
E-mail: <gdean@oca.state.nm.us>
Web: New Mexico Office of Cultural Affairs' Historic Preservation Division
<http://www.museums.state.nm.us/hpd/>

New York
Dr Robert Kuhn
Office of Parks, Recreation,
and Historic Preservation
Peebles Island
PO Box 189
Waterford NY 12188
Telephone: 518-237-8643 extension 3255
Fax: 518-237-9049
E-mail: <robert.kuhn@oprhp.state.ny.us>

North Carolina
Stephen R Claggett
State Archaeologist
Office of State Archaeology
4619 Mail Service Center
Raleigh NC 27699-4619
Telephone: 919-733-7342
Fax: 919-715-2671
E-mail: <steve.claggett@ncmail.net>
Web: North Carolina Office of State Archaeology <http://www.arch,dcr.state.nc.us/fosa.htm>

North Dakota
Fern E Swenson
Chief Archeologist/Acting Deputy SHPO
State Historical Society of North Dakota
Archeology and Historic
Preservation Division
North Dakota Heritage Center
612 East Boulevard Avenue
Bismarck ND 58505-0830
Telephone: 701-328-2672
Fax: 701-328-3710
E-mail: <fswenson@state.nd.us>
Web: State Historical Society North Dakota <http://www.state.nd.us/hist/index.html>

Ohio
David Snyder
Ohio Historic Preservation Office
567 East Hudson Street
Columbus OH 43211-1030
Telephone: 614-298-2000
Fax: 614-298-2037
E-mail: <dsnyder@ohiohistory.org>
Web: Ohio Historical Society's Ohio Historic Preservation Office <http:www.ohiohistory.org/resource/histpres/>

Oklahoma
Robert L Brooks
State Archeologist
Oklahoma Archeological Survey
University of Oklahoma
111 East Chesapeake Room 102
Norman OK 73019
Telephone: 405-325-7211
Fax: 405-325-7604
E-mail: <rbrooks@mail.uoknor.edu>
Web: Oklahoma Archaeological Survey <http://www.ou.edu/cas/archsur/>

Oregon
Dr Leland Gilsen
State Archaeologist
State Historic Preservation Office
State Parks and Recreation
1115 Commercial Street NE Suite 2
Salem OR 97301-1012
Telephone: 503-378-4168 extension 232
Fax: 503-378-6447
E-mail: <leland.gilsen@state.or.us>
Web: Oregon's State Historic Preservation Office <http://www.prd.state.or.us/>

Pennsylvania
Kurt Carr, Chief
Division of Archaeology and Protection
PO Box 1026
State Museum Building
Harrisburg PA 17120-1026
Telephone: 717-783-9926
Fax: 717-783-1073
E-mail: <kcarr@state.pa.us>

Puerto Rico
Miguel Bonini
State Historic Preservation Office
La Fortaleza
PO Box 82
San Juan PR 00901
Telephone: 809-721-3737
Fax: 809-723-0957

Republic of the Marshall Islands
Richard Williamson
State Archaeologist
Historic Preservation Office
PO Box 1454
Majuro MH 96960

Marshall Islands
Telephone: 011-692-625-4476
Fax: 011-692-625-4476
E-mail: <rmihpo@ntamar.com>

Rhode Island
Paul Robinson
Principal/State Archaeologist
Historic Preservation Commission
Old State House
150 Benefit Street
Providence RI 02903
Telephone: 401-277-2678
Fax: 401-277-2968

South Carolina
Jonathan Leader
Christopher Amer
State and State Underwater Archaeologists
South Carolina Institute of Archaeology and Anthropology
University of South Carolina
1321 Pendleton Street
Columbia SC 29208-0071
Telephone: 803-777-8170
Fax: 803-254-1338
E-mail: <leader@sc.edu> and <amerc@sc.edu>
Web: South Carolina Institute of Archaeology and Anthropology <http://www.cla.sc.edu/sciaa/sciaa.htm>

South Dakota
James K Haug
State Archaeologist
Archaeological Research Center
2425 East Saint Charles Street
PO Box 1257
Rapid City SD 57709-1257
Telephone: 605-394-1936
Fax: 605-394-1941
E-mail: <Jim.Haug@state.sd.us>
Web: South Dakota State Historical Society's Archaeological Research Center <http://www.sdsmt.edu/wwwsarc/>

Tennessee
Nick Fielder
State Archaeologist
Tennessee Department of Environment and Conservation
Division of Archaeology
5103 Edmonson Pike
Nashville TN 37211-5129
Telephone: 615-741-1588 extension 13
Fax: 615-741-7329
E-mail: <nfielder@mail.state.tn.us>
Web: Tennessee Division of Archaeology <http://www.state.tn.us/environment/arch/>

Texas
Patricia Mercado-Allinger
State Archaeologist
Texas Historical Commission
Box 12276 Capitol Station
Austin TX 78711
Telephone: 512-463-8882
Fax: 512-463-2530
E-mail: <pat.mercadoallinger@thc.state.tx.us>
Web: Texas Historical Commission <http://www.thc.state.tx.us/>

US Virgin Islands
David M Brewer
Senior Territorial Archaeologist
Division for Archaeology and Historic Preservation
#17 Kongens Gade
St. Thomas USVI 00802
Telephone: 340-776-8605
Fax: 340-776-7236

Utah
Kevin T Jones
State Archaeologist
Antiquities Section
300 Rio Grande
Salt Lake City UT 84101
Telephone: 801-533-3524
Fax: 801-533-3503
E-mail: <kjones@history.state.ut.us>
Web: Utah Antiquities Section <http://www.history.utah.org/Services/arhome.html>

Vermont
Giovanna Peebles
State Archeologist
Division for Historic Preservation
National Life Drawer 20
Montpelier VT 05620-0501
Telephone: 802-828-3050
Fax: 802-828-3206
E-mail: <gpeebles@dca.state.vt.us>

Web: Vermont Agency of Transportation's

Archaeological and Historic Resources <http://www.aot.state.vt.us/archaeology/design/default.htm>

Virginia
M Catherine Slusser
State Archaeologist
Department of Historic Resources
2801 Kensington Avenue
Richmond VA 23221
Telephone: 804-367-2323
Fax: 804-367-2391
E-mail: <cslusser@dhr.state.va.us>
Web: Virginia Department of Historic Resources <http://www.dhr.state.va.us>

Washington
Dr Robert G Whitlam
State Archaeologist
Department of Community, Trade, and Economic Development
Office of Archaeology and Historic Preservation
PO Box 48343
Olympia WA 98504-8343
Telephone: 360-407-0771
Fax: 360-407-6217
E-mail: <robw@cted.wa.gov>
Web: Washington State Office of Archaeology and Historic Preservation <http://www.ocd.wa.gov/info/lgd/oahp>

West Virginia
Patrick Trader
Senior Archaeologist
West Virginia Division of Culture and History
The Cultural Center
1900 Kanawha Boulevard East
Charleston WV 25305-0300
Telephone: 304-558-0220 extension 179
Fax: 304-558-2779

Wisconsin
Robert A Birmingham
State Archaeologist
State Historical Society of Wisconsin
Madison WI 53704
Telephone: 608-264-6495
Fax: 608-264-6404
E-mail: bob.birmingham@mail.admin.wisc.edu

Wyoming
Dr Mark E Miller
State Archaeologist
Department of Anthropology
Box 3431
University of Wyoming
Laramie WY 82071
Telephone: 307-766-5301
Fax: 307-766-4052
E-mail: <mmiller@uwyo.edu>
Web: Office Wyoming State Archaeologist <http://www.spacr.state.wy.us/cr/arch/>

Archaeological Societies

Society for American Archaeology
SAA Headquarters
900 Second Street NE #12
Washington DC 20002
Telephone: 202-789-8200
Web: <http://www.saa.org/>

Arizona
Arizona Archaeological Society
Box 9665
Phoenix AZ 85068

Arkansas
Arkansas Archeological Society
PO Box 1222
Fayetteville AR 72702

California
San Diego County Archaeological Society
PO Box A-81106
San Diego CA 92138

Fort Guijarros Museum Foundation
PO Box 23130
San Diego CA 92193

Colorado
Colorado Archaeological Society
PO Box 1004
Attention: Larry Riggs
Boulder CO 80306

Telephone: 303-449-2852

Illinois
Illinois Association for
Advancement of Archaeology
RR 1 Box 63 c/o Glen R. Hanning
Huntsville IL 62344

Georgia
Society of Primitive Tech
2550 Elberton Road
The Woods
Carlton GA 30627

Louisiana
Louisiana Archaeological Society
Attention: David Jeane, Treasurer
305 Hickory Street
Springhil LA 71075

Maine
New England Antiquity Research Association
94 Cross Point Road
Edgecomb ME 04556

Maryland
Archaeological Society of Maryland
4302 Crow Rock Road
Myersville MD 21773-8826

Massachusetts
Massachusetts Archaeological Society
PO Box 700
Middleborough MA 02346-0700
Telephone: 508-947-9005

Missouri
Missouri Archaeological Society
PO Box 958
Columbia MO 65205
Telephone: 314-882-3544
Fax: 314-882-9410

Nevada
Nevada Archaeological Association
PO Box 73145
Las Vegas NV 89170-3145

New Mexico
Archaeological Society of New Mexico
PO Box 3485
Albuquerque NM 87110

New York
Suffolk County Archaeological Association
c/o Gaynell Stone PhD
PO Box 1542
Stony Brook NY 11792
Telephone: 516-929-8725
Fax: 516-929-6967

Oregon
Archaeological Society of Central Oregon
PO Box 8146
Attention: Susan Gray
Bend OR 97708-0146

Oregon Archaeological Society
PO Box 13293
Portland OR 97213

Pennsylvania
Society for Pennsylvania Archaeology
c/o Judy Duritsa
301 North Drive Beech Hills
Jeannette PA 15644

Texas
Texas Archeological Society
CAR/UTSA
6900 N Loop 1604 West
San Antonio TX 78249-0658
Telephone: 210-458-4343
Fax: 210-458-4870
E-mail: <txarch@onr.com>

Virginia
Archaeological Society of Virginia
PO Box 340
Attention: Russell Darden
Courtland VA 23837

Wyoming
Wyoming Archaeological Society
1617 Westridge Terrace
Casper WY 82604
Telephone: 307-268-2212

Appendix 3

Cave and Karst Management Reference List

About This Reference List

This Appendix contains a comprehensive reference list to further information on cave and karst management and on the development of cave management plans. Included are international cave and karst references, complete-to-date lists of proceedings from the International Sinkhole Conferences and from National Cave and Karst Management Symposia (NCKMS) held in the United States. References to Australasian Cave and Karst Management Conference Proceedings and references to other selected management reports and technical conferences are also included. We regret omission of other important cave management documents and resources. Please contact the editors so we are sure to correct any omissions in future editions.

Information about conference proceedings is formatted using data from OCLC Online Computer Library Center Inc., the producer of the Worldcat database. Formatting for each reference mirrors the OCLC Worldcat cataloging in order to make it easier to find through interlibrary loan. The Worldcat database includes more than 52 million records for books, maps, proceedings, and other resources from more than 37,000 member libraries in the world. References to journal articles were checked in the most appropriate subject database, including SciSearch, ArtIndex, BIOSIS, and Georef.

The Werkers thank Elery Hamilton-Smith, Diana E. Northup, Graham S. Proudlove, and George Veni for their assistance in compiling these references.

Cave and Karst Management Plans

Cave and karst management plans are essential tools for long-term conservation, protection, and in some cases, preservation. This volume addresses many of the factors to consider when making management decisions for cave systems. Much of the information presented in this book should be useful in developing cave management plans.

Sample cave management plans are included in Appendices 6, 7, and 8. We thank James R. Goodbar, Jerry L. Trout, and Jim C. Werker for providing these examples.

Plans and prescriptions vary tremendously and are designed to protect the specific resource values of a particular cave, cave system, or karst system. Some cave management plans focus on protecting particular endangered species or recovering an endangered species habitat. Karst management plans often address groundwater quality. The management prescription for a cave site with archaeologic or paleontologic remains should include special guidelines for protecting and managing specific sensitive resources.

For example, the specialized Cave Management Prescription for La Cueva de las Barrancas in Appendix 7 contains guidelines for maintaining the cave as a pristine underground laboratory reserved for speleological and geomicrobiological research.

The McKittrick Hill Caves Management Plan in Appendix 6 provides examples of general cave management prescriptions and guidelines for specific cave resources and values.

Included in Appendix 8 is a sample generic management prescription. The Joe Doe Cave Management Plan describes a cave that has received little visitation and the resources within require immediate protection.

Selected References to Cave and Karst Management

Addis R. 1979. *A Study for the NSS: Knox Cave*. Bulletin No 3. New York Cave Survey. 139 p.

Aley T. 1997. Caves in crisis. In: *1997 Yearbook of Science and the Future, Encyclopaedia*. Chicago (IL): Britannica, Inc. p 116-133.

Baichtal JF, Swanston DN, Archie AF. 1995. An ecologically-based approach to karst and cave resource management. In: Rea GT, editor. *Proceedings of the 1995 National Cave Management Symposium: Spring Mill State Park, Mitchell, Indiana, October 25-28, 1995*. Indianapolis (IN): Indiana Karst Conservancy. p 10-27.

Bell P. 1993. Moondyne Cave: Development to guided adventure. *Cave Management in Australasia* 10:9-11.

Bodenhamer H. 1995. Monitoring human-caused changes with visitor impact mapping. In: Rea GT, editor. *Proceedings of the 1995 National Cave Management Symposium: Spring Mill State Park, Mitchell, Indiana, October 25-28, 1995*. Indianapolis (IN): Indiana Karst Conservancy. p 28-37.

Buecher RH. 1993. Pre-development studies at Kartchner Caverns. In: Foster DL, Foster DG, Snow MN, Snow RK, editors. *1991 National Cave Management Symposium Proceedings: Bowling Green, Kentucky, October 23-26, 1991*. Horse Cave (KY): American Cave Conservation Association, Inc. p 144-163.

Buecher RH. 1996. Footprints, routes, and trails. In: Rea GT, editor. *Proceedings of the 1995 National Cave Management Symposium: Spring Mill State Park, Mitchell, Indiana, October 25-28, 1995*. Indianapolis (IN): Indiana Karst Conservancy. p 47-50.

Buecher RH. 1996. Monitoring the cave environment. In: Rea GT, editor. *Proceedings of the 1995 National Cave Management Symposium: Spring Mill State Park, Mitchell, Indiana, October 25-28, 1995*. Indianapolis (IN): Indiana Karst Conservancy. p 41-46.

Buecher RH, Hill CA, Hose LD, Pisarowicz JA, editors. 1999. *Kartchner Caverns: Research Symposium*. Huntsville (AL): National Speleological Society. 87 p. Published in: *Journal of Cave and Karst Studies* 61(2):40-123.

Bunting B, Balks M. 2001. A quantitative method for assessing the impact of recreational cave use on the physical environment of wild caves. *Australasian Cave and Karst Management Association Journal* 44:10-18.

Burri E, Castiglioni B, Sauro U, editors. 1999. Karst and agriculture in the world. *International Journal of Speleology* 28B:5-185.

Cigna AA. 1993. Environmental management of tourist caves: The examples of Grotta di Castellana and Grotta Grande Del Vento, Italy. *Environmental Geology* 21:173-180.

Cigna AA. 2002. Modern trend in cave monitoring. *Acta Carsologica* 31(1):35-54.

Culver DC, White WB, editors. 2004. *The Encylopedia of Caves*. Elsevier Science.

Despain J. 1995. The creation and implementation of restrictive cave management policies: Three examples from Sequoia and Kings Canyon National Parks. In: Pate DL, editor. *Proceedings of the 1993*

National Cave Management Symposium: Carlsbad, New Mexico, October 27-30, 1993. [Huntsville (AL)]: National Cave Management Symposium Steering Committee. p 175-183.
Doerfliger N, Jeannin PY, Zwahlen F. 1999. Water vulnerability assessment in karst environments: A new method of defining protection areas using a multi-attribute approach and GIS tools (EPIK method). *Environmental Geology*. 39(2):165–76.
Drew D, Hötzl H, editors. 1999. Karst hydrogeology and human activities: Impacts, consequences and implications. *International Contributions to Hydrogeology 20*. Rotterdam: A.A. Balkema Publishers. 322 p.
Dunlap K. 1996. Inexpensive temperature monitoring in caves. In: Rea GT, editor. *Proceedings of the 1995 National Cave Management Symposium: Spring Mill State Park, Mitchell, Indiana, October 25-28, 1995*. Indianapolis (IN): Indiana Karst Conservancy. p 76-87.
Elliott T, editor. 1994. *Knock Fell Caverns Cave Conservation Plan*. Peterborough (UK): English Nature.
Elliott WR. 2000. Conservation of the North American cave and karst biota. In: Wilkens H, Culver DC, Humphreys WF, editors. *Subterranean Ecosystems*. Ecosystems of the World 30. Amsterdam: Elsevier. p 665-689. Internet <http://www.utexas.edu/tmm/sponsored_sites/biospeleology/>. Accessed 2006 May 6.
Elliott WR. 2004. Protecting caves and cave life. In: Culver DC, White WB, editors. *The Encylopedia of Caves*. Elsevier Science. p 458-467.
Eckenfelder Inc. 1997. *Guidelines for Wellhead and Springhead Protection Area Delineation in Carbonate Rocks*. EPA 904/B-97/003, U.S. Environmental Protection Agency. Unpaginated.
Federation Francaise de Speleologie. 1989. Protection des cavernes et du milieu karstique. *Spelunca* No. 35.
Field MS. 1999. *A Lexicon of Cave and Karst Terminology with Special Reference to Environmental Karst Hydrology*. EPA 600/R-99/006 (January 1999). The National Center for Environmental Assessment-Washington Division and the U.S. Environmental Protection Agency's Office of Research and Development. Washington D.C.: U.S. Government Printing Office: 1999-750-101/00037. 194 p.
Gillieson DS. 1996. *Caves: Processes, Development, Management*. Oxford: Blackwell Publishers. 324 p.
Gillieson D, Smith DI, editors. 1989. *Resource Management in Limestone Landscapes: International Perspectives*. Department of Geography and Oceanography, University College. Canberra: Australian Defense Force Academy.
Goodbar JR. 1995. Cave permits and permitting systems. In: Rea GT, editor. *Proceedings of the 1995 National Cave Management Symposium: Spring Mill State Park, Mitchell, Indiana, October 25-28, 1995*. Indianapolis (IN): Indiana Karst Conservancy. p 99-107.
Gookin J. 1991. Secrecy and discretion as cave management tools. *NSS News* 55(4):114-115.
Günay G, Johnson AI, editors. 1997. *Karst Waters and Environmental Impacts*. Rotterdam: AA Balkema. 525 p.
Gunn J. 2004. *Enclyclopedia of Caves and Karst Science*. New York: Fitzroy Dearborn. 902 p.
Gurnee J, editor. 1989. *A Study of Fountain National Park and Fountain Cavern*. [Huntsville (AL)]: National Speleological Foundation. [Editors' Note: Working plan for preservation, protection, and development as a show cave archaeological site in Anguilla, British West Indies.]
Hamilton-Smith E, editor. 1976. *Cave Management in Australia*. Broadway (AU): Australian Speleological Federation.
Hildreth-Werker V, Werker JC. 1999. Restoration, trail designation, and

microbial preservation in Lechuguilla Cave. In: Stitt RR, editor. *Proceedings National Cave Management Symposium 1997: October 7-10 in Bellingham, Washington, and Chilliwack and Vancouver Island, B.C., Canada.* National Cave Management Symposium Steering Committee. p 181-189.

Hildreth-Werker V, Werker JC, editors. 2006. *Cave Conservation and Restoration.* Huntsville (AL): National Speleological Society. 500+ p. [Editors' note: Detailed principles and methods for cave management, conservation, restoration, and speleothem repair.]

Hill CA, Forti P. 1997. *Cave Minerals of the World.* 2nd edition. Huntsville (AL): National Speleological Society. 463 p.

Jones WK, Hobbs HH III, Wicks CM, Currie RR, Hose LD, Kerbo RC, Goodbar JR, and Trout J. 2003. *Recommendations and Guidelines for Managing Caves on Protected Lands*. Karst Waters Institute Special Publication No. 8. Charles Town: Karst Waters Institute. 95 p.

Kiernan K. 1988. *The Management of Soluble Rock Landscapes*. Sydney (AU): The Speleological Research Council. 61 p.

Klimchouk [Klimchuk] AB, Ford DC, Palmer AN, Dreybrodt W, editors. 2000. *Speleogenesis: Evolution of Karst Aquifers*. Huntsville (AL): National Speleological Society. 527 p.

Kranjc A, editor. 2002. *Monitoring of Karst Caves. Acta Carsologica* 31.

Limestone Research Group 1990. *The Impact of Agricultural Operations on the Scientific Interest of Cave SSSI.* Peterborough (UK): Nature Conservancy Council. [Editors' note: Includes impact matrix and extensive bibliography.]

Moore GW, Sullivan N. 1997. *Speleology: Caves and the Cave Environment.* Revised 3rd edition. St. Louis (MO): Cave Books. 176 p.

Mylroie JE. 1984. Hydrologic classification of caves and karst. In: LaFleur, editor. *Groundwater as a Geomorphic Agent*. Boston: Allen and Unwin, Inc. p 157-172.

National Outdoor Leadership School. 1998. *Leave No Trace Outdoor Skills and Ethics: Caving*. Tuohy L, editor; Austin A, illustrator. Boulder (CO): Leave No Trace, Inc. 20 p. Internet <www.lnt.org/ LNTPublications/s&e.pdf.docs/ S&ECaving.pdf>. Accessed 2003 Apr 5.

National Speleological Society, Inc. 2003. *A Guide to Responsible Caving*. Jones CJ and Anonymous, editors; Dale M, design. Huntsville (AL): National Speleological Society. 24 p. Internet < http://caves.org/ brochure/grc.pdf>. Accessed 2003 Dec 5.

Newbould RL. 1976. Steamcleaning of the Orient Cave, Jenolan. In: Hamilton-Smith E, editor. *Cave Management in Australia.* Broadway (AU): Australian Speleological Federation.

Nieland J. 1992. Inventory, evaluation, and management of publicly owned caves in the western United States and the impact of the Federal Cave Resources Protection Act. In: Halliday WR, Rea GT, editors. *6th International Symposium on Vulcanospeleology: Hilo, Hawaii, August 1991*. Huntsville (AL): National Speleological Society. p 273-279.

Northup DE, Mobley ED, Ingham KL III, Mixon WW. 1998. *A Guide to the Speleological Literature of the English Language 1794-1996*. St. Louis (MO): Cave Books. 539 p.

O'Donnell L, Elliott WR, Stanford RA. 1994. *Recovery Plan for Endangered Karst Invertebrates in Travis and Williamson Counties, Texas*. Region 2, Albuquerque (NM): US Fish and Wildlife Service.

Poulson TL. 1976. Management of biological resources in caves. In: Aley T, Rhodes D, editors. *1976 National Cave Management Symposium Proceedings: Albuquerque, New Mexico, October 6-10, 1975*. Albuquerque (NM): Speleobooks. p 46-52.

Price G, Wright R. 1992. Cave conservation moves forward. *Earth Science*

Conservation 30:26. [Editors' note: Outline of proposals for production of Cave Conservation Plans in UK.]
Price G, editor. 1999. *Cave Conservation Handbook*. London (UK): National Caving Association. [National cave conservation policy in Great Britain developed in association with statutory conservation agencies. Softbound 1999. First published 1997 in ringbound format.]
Rohde K, Kerbo R. 1978. Cave restoration: An underview of correcting overviewing. In: Zuber R, Chester J, Gilbert S, Rhodes D, editors. *National Cave Management Symposium Proceedings: Big Sky, Montana, October 3-7, 1977*. Albuquerque (NM): Adobe Press. p 55-58.
Sasowsky ID, Fong DW, and White EL, editors. 1997. *Conservation and Protection of the Biota of Karst*. Karst Waters Institute Special Publication 3. Charles Town: Karst Waters Institute. 118 p.
Smith DW, editor. 1998. *Cave and Karst Management in Australasia XII*. Carlton South, Victoria (AU): Australasian Cave and Karst Management.
Spate A, Hamilton-Smith E, Little L, Holland E. 1998. Best practice and tourist cave engineering. In: Smith DW, editor. *Cave and Karst Management in Australasia XII*. Carlton South, Victoria (AU): Australasian Cave and Karst Management Association. p 97-109. [*Cave and Karst Management in Australasia* 12:97-109.]
Tercafs, R. 2001. *The Protection of the Subterranean Environment: Conservation Principles and Management Tools*. Luxembourg: PS Publishers. 400 p.
Thornber N. 1953. Appendix A: Speleological code and ethics. In: Cullingford CHD, editor. *British Caving: An Introduction to Speleology*. London: Routledge and Kegan Paul, Ltd.
Tranel MJ. 1992. Changing to prevent change: Building a science and resource management program. *NSS News* 52(1):22-25.
Travous KE, Ream JP. 2001. Developing and managing an environmentally responsible tourist cave. *Proceedings of the 13th International Congress of Speleology, Brasilia*. Paper 086/S6, Abstract 269.
Uhl PJ. 1983. Photomonitoring as a management tool. In Mylorie JE, editor. *First International Cave Management Symposium Proceedings, Murray State University, July 15-18, 1981*. Murray (KY): Murray State University, College of Environmental Sciences, Department of Geoscience. p 123-128.
Veni G. 1997. Speleothems: Preservation, display, and restoration. In: Hill C, Forti P, editors. *Cave Minerals of the World*. 2nd edition. Hunstville (AL): National Speleological Society. p 301-309.
Veni G. 1999. A geomorphological strategy for conducting environmental impact assessments in karst areas. *Geomorphology* 31:151-180. Reprinted in: Giardino JR, Marston RA, Morisawa M, editors. 1999. *Changing the Face of the Earth – Engineering Geomorphology: Proceedings of the 28th Binghamton Symposium in Geomorphology: Held 28 August to 3 September 1997 in Bologna*. Amsterdam: Elsevier Publishers. p 151-180.
Veni G, DuChene H, editors. 2001. *Living with Karst: A Fragile Foundation*. AGI Environmental Awareness Series, no 4. Alexandria (VA): American Geological Institute. 64 p. Internet <http://www.agiweb.org/environment/publications/karst.html>, <http://www.agiweb.org/environment/publications/karst.pdf >. Accessed 2005 June 21.
Vermeulen J, Whitten T. 1999. *Biodiversity and Cultural Property in the Management of Limestone Resources: Lessons from East Asia*. Washington DC: The World Bank. 120 p.
Vories KC, Throgmorton D, Harrington A, editors. 2004. *Proceedings of Bat Gate Design: A Technical Interactive Forum held in Austin, Texas, March 4-6, 2002*. Carbondale (IL): USDOI Office of Surface Mining, Alton, Illinois and the Coal Research Center, Southern Illinois Univer-

sity at Carbondale. 434 p. Proceedings available on the Internet <http://www.mcrcc.osmre.gov/bats>. Accessed 2004 March 30.

Watson J, Hamilton-Smith E, Gillieson D, Kiernan K, editors. 1997. *Guidelines for Cave and Karst Protection*. Gland (SZ) and Cambridge (UK): IUCN International Union for Conservation of Nature and Natural Resources. 63 p.

Werker JC, Hildreth V. 1994. Streamlined photomonitoring. *NSS News* 52(2):66.

Werker JC, Hildreth-Werker V. 1996. New improved system for photomonitoring. In: Rea GT, editor. *Proceedings of the 1995 National Cave Management Symposium: Spring Mill State Park, Mitchell, Indiana, October 25-28, 1995*. Indianapolis (IN): Indiana Karst Conservancy. p 311-312.

Werker JC. 1996. Formation repair. *NSS News* 54(2-3):65-67.

Werker JC. 1996. Proven techniques in repairing formations. In: Rea GT, editor. *Proceedings of the 1995 National Cave Management Symposium: Spring Mill State Park, Mitchell, Indiana, October 25-28, 1995*. Indianapolis (IN): Indiana Karst Conservancy. p 308-310.

Werker JC, Hildreth-Werker V. 2003. Cave management plan for an underground laboratory: La Cueva de las Barrancas, prototype site for Mars studies. In: Rea GT, editor. *Proceedings of the 15th National Cave Management Symposium: Tucson, Arizona, October 16-19, 2001*. Coronado National Forest: USDA Forest Service. p 77-87.

Werker JC, Hildreth-Werker V. 2004. Stalactite repair: Large speleothem hangs again in Colossal Cave, Arizona. *NSS News* 62(4):98.

Werker, JC. 2004. Characteristics of materials used in cave and mine gates. In: Vories KC, Throgmorton D, Harrington A, editors. *Proceedings of Bat Gate Design: A Technical Interactive Forum held in Austin, Texas, March 4-6, 2002*. Carbondale (IL): USDOI Office of Surface Mining, Alton, Illinois and the Coal Research Center, Southern Illinois University at Carbondale. 434 p. Proceedings available on the Internet <http://www.mcrcc.osmre.gov/bats>. Accessed 2004 March 30.

White WB. 1988. *Geomorphology and Hydrology of Karst Terrains*. New York: Oxford University Press. 464 p.

Wilkens H, Culver DC, Humphreys WF, editors. 2000. *Subterranean Ecosystems*. Ecosystems of the World 30. Amsterdam: Elsevier.

Williams PW. 1993. *Karst Terrains: Environmental Changes and Human Impacts*. Catena Supplement 25. Germany: Cremlingen-Destedt. 268 p.

Wilmut J, editor. 1972. *Caves and Conservation*. Birmingham (UK): National Caving Association.

Wilson JM. 1989. Protection vs use: The first 10 years of PerCCams. In: Huppert GN, editor. *1987 Cave Management Symposia Proceedings: Rapid City, South Dakota*. Huntsville (AL): National Speleological Society. 146-160.

Wilson JM. 1992. Using cave registers to further the understanding of human effects on caves. *NSS News* 50(2):38-40.

Wright JR, Price G. 1992. *Cave Conservation Plans: An Integrated Approach to Cave Conservation*. Peterborough (UK): English Nature.

Zokaites C. 1997. *Living on Karst: A Reference Guide for Landowners in Limestone Regions*. Richmond (VA): Cave Conservancy of the Virginias. 26 p. Available on the Internet <http://www.dcr.virginia.gov/dnh/livingonkarst.htm>. Accessed 2005 June 21.

Internet Based Resources on U.S. Cave/Karst Conservation and Management

National Speleological Society

<http://www.caves.org/>
NSS Cave Conservation Division and Committee
<http://www.caves.org/committee/conservation/>
NSS Cave Conservation and Management Section
<http://www.caves.org/section/ccms/>
National Cave and Karst Management Symposia
<http://www.nckms.org/>
National Cave and Karst Research Institute
<http://www2.nature.nps.gov/nckri/>
American Cave Conservation Association
<http://www.cavern.org/>
Bat Conservation International
<http://www.batcon.org/>
Karst Waters Institute
<http://www.karstwaters.org/>

References to National Cave and Karst Management Symposia Proceedings Through 2003

Information about conference proceedings is formatted using data from the OCLC Worldcat in order to make it easier to find these references through interlibrary loan. Apparent changes in the reference formatting mirror periodic changes in cataloging rules.

1975 NCMS Proceedings
Aley T, Rhodes D, editors. 1976. *1975 National Cave Management Symposium Proceedings: Albuquerque, New Mexico, October 6-10, 1975*. Albuquerque (NM): Speleobooks. 146 p. [OCLC listing includes Cave Research Foundation.]

1976 NCMS Proceedings
Aley T, Rhodes D, editors. 1977. *1976 National Cave Management Symposium Proceedings: Mountain View, Arkansas, October 26-29, 1976*. Albuquerque (NM): Speleobooks. 146 p.

1977 NCMS Proceedings
Zuber R, [Chester J, Gilbert S, Rhodes D,] editors. 1978. *1977 National Cave Management Symposium Proceedings: Big Sky, Montana, October 3-7, 1977*. Albuquerque (NM): Adobe Press. 140 p.

1978/1980 NCMS Proceedings
Wilson RC, Lewis JJ, editors. 1982. *National Cave Management Symposia: Proceedings [for Carlsbad, New Mexico 1978 and Mammoth Cave, Kentucky 1980]*. Oregon City (OR): Pygmy Dwarf Press. 234 p.

1982 NCMS Proceedings
Thornton H, Thornton J, editors. 1985. *1982 Cave Management Symposia Proceedings: Harrisburg, Virginia*. Richmond (VA): American Cave Conservation Association. 119 p.

1984 NCMS Proceedings
Vandyke JE, editor. 1985. *Proceedings of the 1984 National Cave Management Symposium: Rolla, Missouri*. Rolla (MO): Missouri Speleological Survey. Missouri Speleology 25(1-4). 236 p.

1987 NCMS Proceedings
Huppert GN, editor. 1989. *1987 Cave Management Symposia Proceedings: Rapid City, South Dakota*. Huntsville (AL): National Speleological Society. 160 p.

1989 NCMS Proceedings
Jorden JR, Obele RK, editors. 1993. *Proceedings of the 1989 National Cave Management Symposium: New Braunfels, Texas*. Austin (TX): Texas Cave Management Association; Texas Parks and Wildlife Department. 157 p.

1991 NCMS Proceedings

Foster DL, Foster DG, Snow MN, Snow RK, editors. 1993. *1991 National Cave Management Symposium Proceedings: Bowling Green, Kentucky, October 23-26, 1991*. Horse Cave (KY): American Cave Conservation Association, Inc. 405 p. [Also listed as 1991 in OCLC].

1993 NCMS Proceedings

Pate DL, editor. 1995. *Proceedings of the 1993 National Cave Management Symposium: Carlsbad, New Mexico, October 27-30, 1993*. [Huntsville (AL)]: National Cave Management Symposium Steering Committee. 357 p.

1995 NCMS Proceedings

Rea GT, editor. 1995. *Proceedings of the 1995 National Cave Management Symposium: Spring Mill State Park, Mitchell, Indiana, October 25-28, 1995*. Indianapolis (IN): Indiana Karst Conservancy. 318 p.

1997 NCKMS Proceedings

Stitt RR, editor. 1999. *Proceedings of the 1997 National Cave and Karst Management Symposium: October 7-10 in Bellingham, Washington, and Chilliwack and Vancouver Island, B.C., Canada*. Huntsville [AL]: National Cave Management Symposium Steering Committee. 250 p. [Also listed as 13th National Cave Management Symposium in OCLC.]

1999 NCKMS Proceedings

Rea GT, editor. 2001. *Proceedings of the [1999]14th National Cave and Karst Management Symposium: Chattanooga, Tennessee, October 19-22, 1999*. Chattanooga (TN): Southeastern Cave Conservancy. 221 p.

2001 NCKMS Proceedings

Rea GT, editor. 2003. *Proceedings of the 15th National Cave and Karst Management Symposium: Tucson, Arizona, October 16-19, 2001*. Coronado National Forest: USDA Forest Service. 194 p.

2003 NCKMS Proceedings

Rea GT, editor. 2004. *Proceedings of the 16th National Cave and Karst Management Symposium: Gainesville, Florida, October 13-17, 2003*. Huntsville (AL): Cave Diving Section of the National Speleological Society. 122 p.

2005 NCKMS Proceedings

Rea GT, editor. (in prep). P*roceedings of the 16th National Cave and Karst Management Symposium: Albany, New York. October 31-November 4, 2005*. Huntsville (AL): National Cave Management Symposium Steering Committee.

Other Symposia Proceedings

Sims M, Sims L, editors. 1980. *Far West Cave Management Symposium Proceedings [for Redding California, 1979]*. Pygmy Dwarf Press. 109 p.

Sims M, Sims L, editors. 1981. *Far West Cave Management Symposium Proceedings [for Portland Oregon, 1981]*. Pygmy Dwarf Press. 99 p.

Mylroie JE, editor. 1983. *First International Cave Management Symposium Proceedings, Murray State University, July 15-18, 1981*. Murray (KY): Murray State University, College of Environmental Sciences, Department of Geoscience. p 123-128.

References to Australasian Cave and Karst Management Conference Proceedings Through 2005

Hamilton-Smith E, editor. 1976. *Cave Management in Australia*. Broadway, NSW (AU): Australian Speleological Federation. (Proceedings of Conference held in Jenolan, NSW. July 1973.) iii + 109 p.

Middleton GJ, editor. 1977. *Cave Management in Australia II*. Published jointly by Sandy Bay (TAS): National Parks and Wildlife Service and Broadway, NSW (AU): Australian Speleological Federation. (Proceedings of 2nd Australian Conference on Cave Tourism and Management, Hobart. May 1977.) iv + 119 p.

Robinson AC, editor. 1980. *Cave Management in Australia III*. Published jointly by Adelaide (SA): National Parks and Wildlife Service and Broadway, NSW (AU): Australian Speleological Federation. (Proceedings of 3rd Australian Conference on Cave Tourism and Management, Mt Gambier. May 1979.) vii + 170 p.

Watson JR, editor. 1982. *Cave Management in Australia IV*. Published jointly by Busselton (WA): Busselton Tourist Bureau and Broadway, NSW (AU): Australian Speleological Federation. (Proceedings of 4th Australian Conference on Cave Tourism and Management, Yallingup. September 1981.) vii + 82 p.

Hamilton-Smith E, editor. 1990. *Cave Management in Australia V*. Published jointly by Carlton South (VIC): Australasian Cave and Karst Management Association. (Proceedings of 5th Australian Conference on Cave Tourism and Management, Lakes Entrance, VIC. April 1983.) iii + 67 p.

William DR, Wilde KA, editors. 1985. *Cave Management in Australasia VI*. Published by Waitomo (NZ): Tourist Hotel Corporation of New Zealand. (Proceedings of 6th Australasian Conference on Cave Tourism and Management, Waitomo, NZ. Sept 1985.) 231 p.

Spate AP, Bell P, Henderson K, editors. 1987. *Cave Management in Australasia VII*. Published jointly by Hurstville (NSW) National Parks and Wildlife Service of New South Wales, Sydney (NSW) Tourist Commission of New South Wales and Broadway, NSW (AU): Australian Speleological Federation. (Proceedings of 7th Australian Conference on Cave Tourism and Management, NSW. May 1987.) iii + 49 p.

Wilde KA, editor. 1989. *Cave Management in Australasia VIII*. Published by Hokitika (NZ): Department of Conservation. (Proceedings of 8th Australasian Conference on Cave Tourism and Management, Punakaiki NZ. Sept 1989.) v + 183 + 72 p.

Bell P, editor. 1991. *Proceedings of the Ninth ACKMA Conference*. Published by Carlton South (VIC): Australasian Cave and Karst Management Assn. (Conference held at Margaret River, WA. September 1991.) i + 92 p.

Henderson K, Bell P, editors. 1993. *Cave Management in Australasia X*. Published by Carlton South (VIC): Australasian Cave and Karst Management Association. (Proceedings of 10th ACKMA Conference, Rockhampton, QLD. May 1993.) ii + 76 p.

Henderson K, Houshold I, Middleton G, editors. 1997. *Cave Management in Australasia XI*. Published by Carlton South (VIC): Australasian Cave and Karst Management Association. (Proceedings of 11th ACKMA Conference, Tasmania. May 1995.) iv + 233 p.

Smith DW, editor. 1998. *Cave and Karst Management in Australasia XII*. Carlton South, Victoria (AU): Australasian Cave and Karst Management Association. (Proceedings of 12th ACKMA Conference, Waitomo, NZ. May 1997.) iii + 159 p.

Henderson K, editor. 1999. *Cave Management in Australasia XIII*. Published by Carlton South (VIC): Australasian Cave and Karst Management Association. (Proceedings of 13th ACKMA Conference, Mt. Gambier, SA. April 1999.) vi + 158 p.

Henderson K, editor. 2002. *Cave Management in Australasia XIV*. Published by Carlton South (VIC): Australasian Cave and Karst Management Association. (Proceedings of 14th ACKMA Conference, Wombeyan, NSW. April–May 2001.) iv + 141 p.

Henderson K, editor. 2005. *Cave Management in Australasia XV*. Published by Carlton South (VIC): Australasian Cave and Karst Management Association. (Proceedings of 15th ACKMA Conference, Chillagoe, QLD. May 2003.) 91 p.

Smith DW, editor (in prep). *Cave Management in Australasia XVI*. Published by Carlton South (VIC): Australasian Cave and Karst Management Association. (Proceedings of 16th ACKMA Conference, NZ. April 2005.)

References to Sinkhole Conferences Through 2003 Multidisciplinary Conference on Sinkholes and the Engineering and Environmental Impacts of Karst

Beck BF, editor. 1984. *Sinkholes: Their Geology, Engineering, and Environmental Impact.* Rotterdam: A.A. Balkema. 429 p.

Beck BF, Wilson WL, editors. 1987. *Karst Hydrogeology: Engineering and Environmental Applications.* Rotterdam: A.A. Balkema. 467 p.

Beck BF, editor. 1989. *Engineering and Environmental Impacts of Sinkholes and Karst.* Rotterdam: A.A. Balkema. 384 p.

Beck BF, editor. 1993. *Applied Karst Geology*. Rotterdam: A.A. Balkema. 295 p.

Beck BF, editor. 1995. *Karst Geohazards*. Rotterdam: A.A. Balkema. 581 p.

Beck BF, Stephenson JB, editors. 1997. *The Engineering Geology and Hydrogeology of Karst Terranes*. Rotterdam: A.A. Balkema. 516 p.

Beck BF, Pettit AJ, Herring JG, editors. 1999. *Hydrogeology and Engineering Geology of Sinkholes and Karst*. Rotterdam: A.A. Balkema. 478 p.

Beck BF, Herring JG, editors. 2001. *Geotechnical and Environmental Applications of Karst and Hydrology.* Rotterdam: A.A. Balkema. 437 p.

Beck BF, editor. 2003. S*inkholes and the Engineering and Environmental Impacts of Karst.* Geotechnical Special Publication No. 122. American Society of Civil Engineers. 737 p.

Beck BF, editor. 2005. *Sinkholes and the Engineering and Environmental Impacts of Karst–Proceedings of the 10th Multidisciplinary Conference.* Geotechnical Special Publication No. 144. American Society of Civil Engineers. 676 p.

References to Additional Karst Conference Proceedings

Dilamarter RR, Csallany SC, editors. 1977. *Hydrologic Problems in Karst Regions*. Bowling Green, Kentucky: Western Kentucky University. 481 p.

Dougherty PH, editor. 1983. *Environmental Karst*. Cincinnati, Ohio: GeoSpeleo Publications. 167 p.

National Water Well Association, editor. 1986. P*roceedings of the Environmental Problems in Karst Terranes and Their Solutions Conference, October 28-30, 1986, Bowling Green, Kentucky*. Dublin, Ohio: National Water Well Association. 525 p.

National Water Well Association, editor. 1988. *Proceedings of the Second Conference on Environmental Problems in Karst Terranes and Their Solutions, November 16-18, 1988, Nashville, Tennessee*. Dublin, Ohio: National Water Well Association. 441 p.

Yuan D, editor. 1988. *Karst Hydrogeology and Karst Environmental Protection: Proceedings of the IAH 21st Congress*. 2 vols. Beijing, China: International Association of Hydrogeologists.

National Water Well Association, editor. 1992. *Proceedings of the Third Conference on Hydrogeology, Ecology, Monitoring, and Management of Ground Water in Karst Terranes, December 4-6, 1991, Nashville, Tennessee.* Dublin, Ohio: National Water Well Association. 793 p.

Hungarian Speleological Society. 1998. *Subcity '98: Meeting on Cave of Natural Origin Under Cities and Urban Areas.* Budapest: Hungarian Speleological Society. 150 p.

Research Institute of Geology and Mineral Resources, editor. 2004.

Appendix 4

Cooperative Agreement Between the BLM-NM and NSS-SWR

COOPERATIVE AGREEMENT
between the
BUREAU OF LAND MANAGEMENT
STATE OF NEW MEXICO
and the
NATIONAL SPELEOLOGICAL SOCIETY
SOUTHWESTERN REGION

I. Statement of Joint Objectives

A. Purpose

This agreement is made and entered into by the Department of the Interior, Bureau of Land Management, New Mexico State Office (BLM), and the National Speleological Society (NSS), Southwestern Region (SWR), for the purpose of providing a framework to facilitate cooperative projects for the inventory, monitoring, research, and management of caves on lands managed by the BLM in the State of New Mexico. The NSS and BLM have a nationwide Memorandum of Understanding (MOU) to work cooperatively in addressing management of caves, cave resources, public education, and resource protection. The NSS is the principal conservation group in the country to deal with caves and their resources.

B. Objective

This Agreement will cover BLM Field Offices in New Mexico. Specific projects will be outlined in Task Orders issued under this Agreement. The intent of such work is to provide information required for the proper management of caves and their resources on public lands.

C. Authority

1. Federal Land Policy and Management Act of 1976 (P.L. 94-579) (43 U.S.C. 1701 *et seq.*).
2. Federal Cave Resources Protection Act of 1988 (P.L. 100-691).
3. National Environmental Policy Act of 1969 (42 U.S.C. 4321-47).

D. Benefits

The BLM has interests and responsibilities for responsibly

managing caves and their resources on the public lands. BLM manages a wide variety of caves and karst areas that provide unique habitat for troglobitic ecosystems, discrete sources for groundwater recharge, and unique opportunities for scientific research and recreation. BLM seeks sufficient information to continue to identify these resources which may require special management attention, and other assistance necessary for the conservation, monitoring, and management of sensitive cave resources.

II. Definitions

A. Agreement: means this cooperative agreement
B. Assistance Officer (AO): means the BLM's Assistance Officer
C. Assistance Representative (AR): means the BLM's Assistance Representative.
D. BLM: means the Bureau of Land Management. May also be referred to as Bureau.
E. CFR: means Code of Federal Regulations.
F. Fiscal Year (FY): means the Federal fiscal year which extends from October 1 of one year through September 30 of the following year.
G. SWR: means the Southwest Region of the National Speleological Society. May also be referred to as recipient.
H. Not-to-Exceed Amount (NTE): means the maximum Federal funding amount.
I. OMB: means Office of Management and Budget.
J. Project Inspector (PI): means the BLM's project inspector.
K. Task Order (TO): means the order which is issued against the agreement to obligate funds for specific services or work to be accomplished.

III. Project Management Plan

A. SWR agrees to:
   1. Assist BLM personnel in the development of proposals for projects that are possible Challenge Cost-Share projects.
   2. Cooperate in the development of Task Orders which specify individual projects involving inventory, monitoring, protection, management, and cost of work mutually agreed to be performed.
   3. Fulfill the terms of any Task Order in compliance with the Federal Cave Resources Protection Act, and obtain all necessary State and Federal permits, when necessary, or not otherwise obtained by BLM.
   4. Provide labor, supplies and equipment in accordance with the specifications of individual Task Orders drawn supplemental to this Agreement.
   5. Prepare and submit to BLM a written report (when required by BLM) after completion of each project conducted in accordance with the specifications of individual task orders drawn supplemental to this Agreement.
   6. Fund 50% of the project cost, through donated labor involved in carrying out the project.

B. The BLM agrees to:

1. Assist SWR personnel in the development of proposals determined to be suitable for Cost-Share.
2. Provide 50% funding as available for SWR to undertake tasks covered under this agreement upon approval of a budget for the current fiscal year.
3. Provide advance payments or reimbursements to the SWR in accordance with Section VI, Financial Support, and Section VII, Payments, of this agreement and applicable OMB and Treasury Regulations.

C. SWR and the BLM mutually agree to:
   1. Cooperate in the development of Task Orders supplemental to this Agreement (which specify who, what, when, where, why, how), and cost of work mutually agreed to be performed.
   2. Coordinate efforts to ensure completion of one project prior to beginning another.

IV. Term of Agreement

This agreement shall become effective on the date of signature of the BLM Assistance Officer and shall remain in effect until terminated in accordance with the provisions of 43 CFR, Subpart F, Section 12.961.

V. Task Orders

A. Issuance. TO's will be issued in writing by the AO and must be signed by both the authorized SWR official and the AO to be effective.

B. Contents. A TO will contain:
   1. The specifications or statement of work that will be performed under that specific TO.
   2. A list of any deliverable items that are required.
   3. Any necessary drawings and/or location maps.
   4. The delivery schedule or completion time which has been negotiated based on the level of difficulty, site location, etc.
   5. A NTE amount for the task.
   6. Any other detail or information necessary.

VI. Financial Support

A. This agreement shall be funded by issuance of TO's based on the availability of BLM funding. SWR hereby releases the BLM from all liability due to failure of Congress to appropriate funds for this agreement.

B. Funds obligated for a specific TO but not expended in that FY can be carried forward and expended in subsequent FY's for that particular TO.

C. TOs will specify the NTE amounts. The BLM shall not be obligated to pay for nor shall SWR be obligated to perform any effort that will require the expenditure of Federal funds above the NTE amount specified in that TO.

D. Cost sharing for this agreement shall be in accordance with 43 CFR, Subpart F, Section 12.923.

VII. Payments

A. Electronic Funds Transfer Payments
   1. Payment under this agreement will be made by the Govern-

ment by electronic funds transfer (through the Treasury Fedline Payment System (FEDLINE) or the Automated Clearing House (ACH)).

2. After award, but no later than 14 days before an invoice or agreement financing request is submitted, the Recipient shall designate a financial institution for receipt of electronic funds transfer payments (SF-3881), and shall submit this designation to the following address:

   Bureau of Land Management
   Service Center, SC-616
   Denver Federal Center, Bldg. 50
   PO Box 25047
   Denver, CO 80225-0047

3. If a designation has been submitted to the BLM under a previous agreement it is not necessary to complete another SF-3881 unless you are changing your designation of financial institution.

B. SWR shall be entitled to reimbursement or advance payment at least quarterly upon submission of an original Request for Advance or reimbursement, Standard Form (SF) 270 to the AR. Payments shall be governed by the provisions of 43 CFR Subpart F, Section 12.922 and 12.952.

C. If advance payments are made SWR must submit a Federal Cash Transaction Report, SF 272 to the AO 15 working days following the end of each quarter.

D. Advance payments shall be made only in amounts necessary to meet current disbursement needs and shall be scheduled so that the funds are available immediately prior to their disbursement.

VIII. Key Officials

A. Assistance Officer (AO)
Mary Ann Crafton Williams
Bureau of Land Management
Administrative Support Center
435 Montana, NE
Albuquerque, NM 87107
Telephone Number (505) 761–8710

The AO is the only individual authorized to obligate funds, award, modify or terminate the agreement or any TO thereto. The AO is responsible for issuing TO's, monitoring agreement and task(s) and TO compliance, enforcing the agreement provisions, issuing timely performance and payment approvals, terminating the agreement or any TO thereto and closing out the agreement.

B. Assistance Representative (AR)
Doug Melton / Noe Gonzales
Bureau of Land Management
620 E. Greene Street
P.O. Box 1778
Carlsbad, NM 88220

Phone: (505) 887- 6544

The AR will be designated for the purpose of administering the technical aspect of the agreement. The AR is authorized to clarify technical requirements, and to review and approve work which is clearly within the scope of the work specified in this agreement. The AR is not authorized to issue changes or in any other way modify this agreement.

C. Project Inspector (PI)
James R. Goodbar
Bureau of Land Management
620 E. Greene Street
P.O. Box 1778
Carlsbad, NM 88220
Phone: (505) 887–6544

At the time of award, a BLM employee(s) may be appointed as the PI. If appointed, the PI will be responsible for providing on-site inspection of the work and for giving SWR representative any special instructions, guidance, or training necessary to complete or perform the work. The PI will not be authorized to issue changes or in any way modify the agreement

D. SWR representative(s) for this Agreement:
Dave Belski
408 Southern Sky
Carlsbad, NM 88220
(505) 885–6168

IX. Special Terms and Conditions

A. Order of Precedence
Any inconsistency in this agreement shall be resolved by giving precedence in the following order: (a) Any national policy requirements and administrative management standards; (b) requirements of the applicable OMS Circulars and Treasury regulations; (c) 43 CFR Part 12; (d) special terms and conditions; (e) all Agreement sections, documents, exhibits, and attachments; and (f) all TO sections, documents, exhibits, and attachments.

B. Modifications
This agreement may be modified by written agreement signed by both a SWR official and the Assistance Officer. Administrative changes (i.e. AO name change) which do not change the project management plan, NTE amount, etc. or otherwise affect the recipient may be signed unilaterally by the AO.

C. Officials Not to Benefit
No member of or delegate to Congress, or resident commissioner, shall be admitted to any share or part of this agreement, or to any benefit arising from it. However, this clause does not apply to this agreement to the extent that this agreement is made with a

corporation for the corporation's general benefit.

D. Endorsement Provision
Recipient shall not publicize or otherwise circulate, promotional material (such as advertisements, sales brochures, press articles, manuscripts or other publications) which states or implies governmental, departmental, bureau, or government employee endorsement of a product, service, or position which the recipient represents. No release of information relating to this award may state or imply that the Government approves of the recipient's work products, or considers the recipient's work products to be superior to other products or services.
All information submitted for publication or other public releases of information regarding this project shall carry the following disclaimer:
"The views and conclusions contained in this document are those of the authors and should not be interpreted as representing the opinions or policies of the U.S. Government. Mention of trade names or commercial products does not constitute their endorsement by the U.S. Government."
Recipient must obtain prior Government approval for any public information releases concerning this award which refer to the Department of the Interior or any bureau or employee (by name or title). The specific text, layout photographs, etc., of the proposed release must be submitted with the request for approval.
A recipient further agrees to include this provision in a subaward to any subrecipient, except for a subaward to a State government, a local government, or to a federally recognized Indian tribal government.

E. Procurement Procedures:
It is a national policy to place a fair share of purchases with minority business firms. The Department of the Interior is strongly committed to the objectives of this policy and encourages all recipients of its grants and cooperative agreements to take affirmative steps to ensure such fairness. Positive efforts shall be made by recipients to utilize small businesses, minority-owned firms, and women's business enterprises, whenever possible. Recipients of Federal awards shall take all of the following steps to further this goal:
   1. Ensure that small businesses, minority-owned firms, and women's business enterprises are used to the fullest extent practicable.
   2. Make information on forthcoming opportunities available and arrange time frames for purchases and contracts to encourage and facilitate participation by small businesses, minority-owned firms, and women's business enterprises.
   3. Consider in the contract process whether firms competing for larger contracts intend to subcontract with small businesses, minority-owned firms, and women's business enterprises.
   4. Encourage contracting with consortiums of small businesses, minority-owned firms and women's business enterprises when a contract is too large for one of these firms to handle indi-

vidually.

5. Use the services and assistance, as appropriate, of such organizations as the Small Business Development Agency in the solicitation and utilization of small business, minority-owned firms and women's business enterprises.

X. General Provisions

A. National Policy Requirements and Administrative Management Standards. All applicable national policy requirements and administrative management standards as set forth in the Office of Management and Budget, Financial Management Division, Directory of Policy Requirements and Administrative Standards for Federal Aid Programs are incorporated by reference.

B. 43 CFR Part 12, Administrative and Audit Requirements and Cost Principles for Assistance Programs is incorporated by reference.

C. OMB Circular A-133, Audits of Institutions of Higher Learning and Other Non-Profit Institutions is incorporated by reference.

D. OMS Circular A-122, Cost Principles for Nonprofit Organizations is incorporated by reference.

E. 43 Code of Federal Regulations (CFR) Part 12, Appendix A to Subpart D, Certification Regarding Debarment, Suspension, and Other Responsibility Matters—Primary Covered Transaction and completed Form DI-1953 are incorporated by reference.

F. 43 CFR Part 12, Appendix C to Subpart D, Certification Regarding Drug-Free Workplace Requirements, Alternate I (Grantees other than individuals) and completed Form DI-1955 are incorporated by reference.

G. Pursuant to Sec. 307 of the Department of the Interior and Related Agencies Appropriations Act, 1995, Public Law 103-332, 108 Stat. 2499, please be advised of the following:
In the case of any equipment or product that may be authorized to be purchased with financial assistance provided using funds made available in this Act, it is the sense of the Congress that entities receiving the assistance should, in expending the assistance, purchase only American-made equipment and products. Recipient agrees to follow the procedures in 43 CFR Part 12, Subpart E, Section 12.700—Buy American Requirements for Assistance Programs (See Federal Register, Vol. 59, No. 2437, December 20, 1995, Pages 65499- 65500).

Task Order
for
Cooperative Agreement
between the
Bureau of Land Management
and
The National Speleological Society (NSS)
Southwestern Region (SWR)

Southwestern Region (SWR) proposes to conduct bat population monitoring at four locations.

Bureau of Land Management (BLM) bat caves. The purpose of this monitoring is to develop data for long range studies to determine bat population trends and fluctuations.

This monitoring will be for the summer of 2000. The caves to be monitored are; Yellowjacket, Endless, McKittrick and Lair Cave. Bat populations will be counted between the months of June and September.

Each cave will be monitored twice each month using the visual count method and hand held tally counters at a minimum. If computerized infrared counters are available they may also be used. Monitoring will begin thirty to forty-five minutes before dusk and continue until nightfall or the bat flight is complete. Monitoring will be conducted by a minimum of two people at each cave. A bat count observation sheet will be completed for each night's monitoring activities. These sheets will be provided by BLM.

A year end report will be due October 31, 2000. The report will include a summary of the year's activities at each site. Data from the visual counts at each cave will appear in an appendix in the report. The appendix will include representative plots of the emergence flights over the course of the summer.

McKittrick Cave has multiple entrances. Counts will be conducted at the Main entrance and the Ledge entrance. SWR will also document the location of the visual counts for consistency of data collection in future years.

The fee for this work will be $2,420 and will include labor for the counts, transportation to and from the cave sites, equipment to conduct the counts, and a final report. At the end of the summer the equipment will be returned to the investigators.

Appendix 5

Guidelines for Entering Lechuguilla Cave

Appendix E in the 1995 Cave and Karst Management Plan for Carlsbad Caverns National Park, New Mexico

Guidelines For Entering Lechuguilla Cave

The reason behind developing guidelines for entering Lechuguilla Cave is to allow limited access for scientific research, survey when in association with exploration, and NPS management related trips while impacting the cave as little as possible. Of primary importance are the (1) impacts to the cave and (2) the safety of all who enter.

BEFORE ENTERING THE CAVE

- Everyone *must* sign a permit.
- Expedition leaders are ultimately responsible for the personnel on their expedition. Expedition leaders should do their best to recruit cavers who are willing to follow the guidelines that have been established. Before assigning anyone to the Far East, an Expedition leader should be reasonably sure that individual is fully prepared for such a trip.
- Every team entering the cave will have one designated team leader. Team leaders have tremendous responsibilities. They are responsible for the safety of their team and for the actions of their team members. If a team member is acting in an unsafe manner or not being careful and actually causing more damage to the cave, then it is the team leader's responsibility to correct that person's actions. If problems persist, then the team leader must abort the trip and have the team leave the cave. A team leader should gear team activities to the least experienced member of the team. This pertains to speed of travel as well as climbing leads. A team should also stay together unless an emergency requires different actions.
- Everyone entering the cave is responsible for their actions while in the cave. They are also responsible for reporting to the team leader acts by other team members that are unsafe or damaging to the cave. The overall goal is to allow limited access to the cave while minimizing all impacts. It is everyone's responsibility to assure that Lechuguilla Cave remains as pristine as possible and that each team member is very safety aware while in the cave.
- Clothing, boots, and caving gear should be clean before entering the cave to minimize the introduction of foreign bacteria, molds, and fungus into this unique ecosystem.
- *Boots must have non-marking soles.* If you are in doubt, scrape boot over white floor or limestone rocks. Marking boots will definitely leave marks.
- Using a frame pack in the cave can be a problem and a major nuisance. It is recommended that these NOT be used.
- Electric lights are a must. No carbide lights or open-flame lights will be allowed in Lechuguilla. One of the reasons for this is safety, because of the flammability of the massive sulfur deposits found in the cave.
- The use of gloves is recommended; however, there are numerous areas

in the cave that are signed "Gloves Off". Gloves get very dirty, particularly in the corrosion residue areas. This corrosion residue must not be smeared over clean or delicate areas.

- Everyone entering the cave must have experience in vertical caves and be proficient in all aspects of single rope techniques. Rebelay anchors may be encountered in different sections of the cave.
- The entrance to Lechuguilla Cave is at an elevation of about 4,640 feet. Cavers not used to this elevation should try to spend a couple of days at this elevation before entering the cave. The cave is warm and humid at 68 degrees fahrenheit. Heat exhaustion and dehydration can be a serious problem. Drink lots of water even if you are not thirsty and eat something every few hours. Those entering the cave for the first time are urged to talk to others concerning clothing to be worn.
- Everyone should be aware that Histoplasmosis has been found in the entrance chamber.
- Any accident of a serious nature should be reported to park service personnel as soon as possible. An incident report *must be filled out* for any accident that may happen in the cave.
- Any team entering the cave is required to fill out a trip report form and return it to our Cave Resources Office within two weeks. The report should include the date, personnel, places visited, and work accomplished including specifics such as types of information or samples collected. If survey was the objective, list the station numbers that were set. If inventory was the objective, list the station numbers that were inventoried.
- All teams entering Lechuguilla Cave must make the Cave Resources Office aware of their "out" time. Any team 6 hours late or later will have search crews initiated. In the case of multi-trip expeditions, expedition leaders should post sign-in, sign-out sheets for all trips into the cave. DO NOT BE LATE!

WHEN TRAVELING THROUGH THE CAVE

- All teams must have a minimum of 3 people. Survey and exploration teams can have no more than 4 members on a team unless given specific permission by the Cave Specialist. Science and management related teams must be limited to no more than 6 members.
- *All Solid Wastes Must Be Carried Out of the Cave Without Exception.* Trips of 2-day duration or less must carry out all liquid waste as well. On longer trips, everyone is encouraged to carry out all liquid wastes. Any excess liquid waste may be dumped in designated sites near the camps. *Liquid waste may not be deposited anywhere else in the cave.* Be sure to take enough containers to hold all liquid wastes.
- *Burrito bags may not be left along the trail* to be picked up on the way out. Carry these items to your camp and then remove from the cave.
- No tobacco products are allowed in the cave and no smoking of tobacco or other products in the cave.
- No consumption of alcohol.
- Safety lines are located throughout the cave as an aid to crossing traverses. Everyone must clip into all traverse lines. There have been several accidents directly due to failure to clip into the safety lines provided.
- Everyone entering the cave is responsible for the care and protection of all ropes and the subsequent rigging that is utilized. Wear spots or other problems should be brought to the attention of the trip leader, who should fix the problem immediately, if necessary, or notify the Cave Resource Office of the concern. Ropes should not be re-rigged without permission from the Cave Resource Office unless an immediate threat is perceived. Please notify the Cave Resource Office of any changes or potential problems noticed. If possible, leave a note for

other expedition members explaining the change in rigging and why it was necessary.

- When traveling through the cave, stay on established trails. If trails are hard to see, either re-flag them immediately or notify the Cave Resource Office so that we will know of the problem. Flagging use in the cave in the past has been as follows: Bright Orange, for marking the trail; White with Red Stripes, denotes sensitive areas such as gypsum crust, aragonite, etc.; Blue, utilized for survey stations; and White with Blue Stripes, indicates a lead. When flagging the trail, use plenty of flagging tape and flag both sides. This helps keep the trail as narrow as possible.
- Of primary importance is the preservation of all the resources in the cave. Everyone is cautioned to be very careful. Carelessness and disregard for the resources has already done damage in the cave and will not be tolerated. Intentional damage to make it easier to move through certain areas will be treated as serious violations and will be prosecuted. Everyone entering must do their utmost to preserve cave features. The Pearlsian Gulf in the Southwest Branch has already had mud introduced into a number of the cave pearls, as has the pool near them in several places. This is a direct result of carelessness and disregard for that resource. Persons displaying this attitude will be denied access to the cave. Continued occurrences will render that area off limits to everyone.
- No cave material (minerals, speleothems, bones, etc.) may be removed from the cave without a valid, existing scientific collecting permit. Collecting for someone else who has a valid collecting permit requires written authorization from the permit holder and from the Cave Specialist.
- No digging, blasting, hammering, or breaking of formations, rocks, etc. may be performed without prior permission from the Cave Specialist.
- Aqua socks or other clean, non-marking shoes must be worn when crossing flowstone areas. Wearing boots or walking barefoot across these areas is not allowed.
- Do not establish a new trail in any area that has a trail. This creates two impacted areas. The Deep Seas Camp, in particular, continues to have problems in this regard. Stop making new trails for convenience. Use existing trails only.
- Please take a couple of minutes to read and record evaporimeter levels, thermometer readings (if it hasn't been done in a day or two), and other science-related instruments. If you do not know or understand the instruments, then please leave them alone. Leave all dataloggers (enclosed in water-proof cases) and their instrumentation alone. Please do not open cases. This allows more moisture into the instrumentation and is ultimately very costly.
- There will be no wading or swimming in pools to reach cave passages or leads without prior consent of the Cave Specialist. This includes the Lake Castrovalva area.
- Pools in newly discovered areas must remain pristine for on-going microbiological research. Any contamination may compromise undiscovered microbes.

CLIMBING

- There are numerous climbs in the cave. Many have been done and some have not. Those wishing to do climbs should clear suggested climbs with the Cave Specialist before attempting any climb.
- The use of bolts, while not strictly prohibited, should be used sparingly.
- Any bolt that will not be used after an initial climb must be removed

and the hole covered. Any bolt or hanger that is to be left in place must be made from austenitic stainless steel.
- All climbs must be surveyed.
- When a climb is completed, all ropes should be derigged and a note on flagging tape should be left stating that the climb has been completed and does not go.
- Ice crampons and other technological implements must be approved by the Cave Specialist before attempting their use in the cave.

RESOURCE PROTECTION ZONES

• The following areas, due to their sensitive nature or on-going scientific research, are off-limits to all entering Lechuguilla Cave unless you have a valid reason to visit this area and you have prior permission from the Cave Specialist.
• Ongoing microbe research areas have been flagged off. You are cautioned not to go beyond that point. These areas are found throughout the cave. Please do not visit or disturb these areas:

Entrance To EF Junction

Pool in Sugarlands - This pool is well-marked.

Western Branch

Pink Dot Pool - Bio-hazard pool. Do not drink the water from here. Part of the Blanca Navidad Room. The off-limits area is well-marked. Also, all pools in the room should be left alone. Drinking water may not be obtained from any of these pools. This pristine room may hold microbes suitable for cancer research.

Pellucidar/Barsoom Areas - Very fragile area, very unique subaqueous helectites found here. Anyone climbing into the Barsoom area above Pellucidar threatens the helectites.

Hudson Bay - Very fragile area, also one where unique microorganisms have been found. Recommended as a “Microbial Preserve.”

Red Lakes Passage - Very fragile with soda straws dipping down to lake level. No attempt should be made to obtain water from this pool.

Oasis Pool - This area is being studied for its microbe population and may be removed from this list once the study is complete.

Chocolate Lake in the Chocolate Factory - Another microbe study pool.

Little Lake of the Clouds - Microbe study in progress. Anyone entering should wear aqua-socks and clean clothes.

Southwestern Branch

Dilithium Crystal Pool - Unique, very fragile pool with selenite crystals growing underwater. This is a very rare occurrence. Cave of the Swords in Chihuahua, Mexico has thousands of selenite crystals that evidently grew in an underwater environment.

YO Acres - Located beyond the Pearlsian Gulf and contains large amounts of corrosion residues. Special care must be taken to avoid contamination of the Pearlsian Gulf area.

Temple of Dagon - Located above the Pearlsian Gulf and has the potential for knocking debris into the pearl areas.

Lake Castrovalva - Very fragile, unique area. No one may swim in or enter the inner pool, not even for photographs.

Void Sulfur Area - Unique, very fragile geological and microbiological resource.

Darktown - Very long, very fragile gypsum hairs hang in this room. They are hard to see and easy to destroy.

Ultra Primo - This is a very heavily decorated, sensitive area.

Lechy's Lair - This is a very heavily decorated, fragile area.

Vesuvius - This is also a very heavily decorated, sensitive area.

Sulfur Shores - Rare and fragile folia are found in this area.

Tower Place (Pool near Dark Star) - Microbial research continues in this pool. It is well marked. No one should venture near it.

Western Side of the Sewing Room - Extremely fragile gypsum needles found here.

Eastern Branch

Ghost Town - (1) Sulfur area in small passage off the NE portion of the room is fragile as well as scientifically interesting. (2) Balcony overlooking the room is very fragile. Damage will occur by anyone going there.

Ghostbusters Hall - Contains large sulfur deposits.

Bryce Canyon - Contains fragile and delicate rillenkarren. Also a spectacular aragonite bush is in danger each time someone enters the area.

Boundary Waters - This is the upstream source for the Lost Pecos River drinking water. Also very delicate and fragile area.

H Survey - Very delicate, very fragile. Damage occurs anytime anyone goes through this area.

Happy Hunting Grounds - Requires a wade through water and is fragile.

Lake of the White Roses - Rare and fragile folia in a one-hundred foot vertical area.

Burning Lakes - A pristine area with pools discovered in 1993 and awaiting microbial studies. Once this study is completed this area may be taken off this list.

Quasimoto's Lair - This is a very fragile, decorated area.

Stud Lake - Closed for microbe studies.

SPECIAL ATTENTION AREAS

- The following areas, though not strictly off limits, require special precautions when entering them. If these precautions are not followed and unacceptable damage occurs, then they will be moved into a Resource Protection Zone.

- Anyone entering the following areas should wear very clean aqua socks or equivalent shoes and clean clothes. Anything carried into these areas should be clean also. General clothes that have been used for caving SHOULD NOT be worn into these areas. Dust, corrosion residues, mud, and other items of debris collected on caving clothes tend to drop off and spoil these pristine areas. Anyone entering these areas should make every effort to keep them pristine. The Pearlsian Gulf has already seen the unacceptable consequences of mud introduced into the cave pearl areas. This must not happen again.

Entrance To EF Junction - Slow Down! Especially in the Rim City area.

Southwest Branch - Pearlsian Gulf, Tower Place (except for area mentioned in the Resource Protection Zone), Underground Atlanta

Eastern Branch - Firefall Hall

- All areas of the cave should be traveled through carefully. However, EXTREME CAUTION should be exercised when moving through the Land of the Lost. Move at a very slow, careful pace, being aware of the gypsum hairs, threads, and beards in the area. Also, when moving through the China Shop, camp packs should be removed and passed carefully through the aragonite bushes.

CAMPING AND RESTING

- Camping is allowed only in designated camps. No bivouacking is allowed without permission of the Cave Specialist. Teams entering the cave should inform the Cave Resource Office where they plan on camping. Designated camps and maximum numbers allowed at those camps are as follows:

Eastern Branch
Grand Guadalupe Junction - 12
Rusticles - 6
Ghosttown - 6

Southwestern Branch
Chandelier Camp - 8
High Hopes Camp - 6
Big Sky - 6

Western Branch
Deep Seas - 12

- Every team is required to leave a note in a prominent place in camp with the following information on it: what day it is, who is on the team, and where in the cave you expect to be.
- Water should be collected from designated pools utilizing the pitchers that have been placed there. Under no circumstances should anyone place their canteen, hands, or any other object into the pool itself.
- Water should not be withdrawn from any pool containing subaqueous helictites or from the Dilithium Crystal Pool in the Southwestern Branch of the cave. It should also not be withdrawn from the Red Lakes Room pool due to the sensitive nature of the soda straws that touch the water's surface. Water should also not be taken from the Pink Dot Pool in the Western Branch of the cave.

- Small stoves may be used in camp areas only. Propane is the preferred fuel. Stoves may be used in other areas with the permission of the Cave Specialist.
- No caches of food, water or other items are allowed in the cave.
- When eating (either traveling through the cave or at camp) or preparing food, great care should be taken to not drop any crumbs or food particles in the cave. Eating should take place over a plastic sheet or other catchment type material.
- Toothpaste and other materials should not be drained into the cave. Rinsing off and allowing contaminated water to drain into the cave is not acceptable. Moist towelettes that can be removed from the cave is the only option for bathing.

UNEXPLORED AREAS AND SURVEYING

- Exploration in Lechuguilla Cave may not occur without surveying what you see. Survey as you go. *Exploration Without Surveying Is Strictly Prohibited.* Violators will be denied future access to the Lechuguilla Cave. Standards listed in the NPS flyer titled "Cave Survey Standards for Carlsbad Caverns National Park" must be adhered to when surveying.
- Discovery teams may name new areas, but names deemed inappropriate or distasteful, or named after living people will not be accepted.
- When moving into unexplored areas, trails should be established that minimize the damage to the cave. Persons running lead tape position should carefully evaluate the passage and choose the path that will do the least damage to the cave. Trails should be flagged immediately, so that those who will follow will not have a choice as to where to walk and will have to stay on the established trail.

****IN EXTREMELY SENSITIVE AREAS,** such as where aragonite bushes block the path (like the China Shop), or other noteworthy speleothems deter progress, stop and do not proceed. Notify the Cave Resource Office. The National Park Service will make a decision of this magnitude.

Appendix 6

Cave Management Plan: Caves of McKittrick Hill

Sample Cave Management Plan: Caves of McKittrick Hill
Bureau of Land Management, Carlsbad Resource Area

This plan is consistent with the Carlsbad Resource Management Plan of 1988 and other Bureau of Land Management (BLM) national, state, and district plans and policies. It shall become effective when signed and may be updated or amended when necessary.

Prepared by: James R. Goodbar, Cave Specialist

Recommended by: Original Signed by Scott Powers 8/18/1994
Chief, Multi-Resources

Approved by: Original Signed by Richard Manus 8/18/1994
Area Manager

McKittrick Hill Caves Management Plan Contents

Management Summary

Eight caves are located within the McKittrick Hill Special Recreation Management Area (SRMA). The management goals for the SRMA are to provide opportunities for quality recreation, scientific research and education, continued exploration, and ensure visitor safety. These caves are managed under the guidelines of the Federal Cave Resources Protection Act of 1988 (FCRPA), the Bureau of Land Management (BLM) National, State, and District cave management policies, and the Carlsbad RMP. This plan presents the cave management practices and policies established for the caves on McKittrick Hill to accomplish those goals. Management practices and activities are designed to resolve or avoid resource use conflicts.

Damaged areas of the caves will be restored as much as possible while delicate formations or historical sites will be flagged to decrease the threat of accidental destruction. The cave resources will be monitored in several ways including patrolling, photo monitoring, bat counts, visitor feed back, and register information.

McKittrick Hill Caves Management Plan

I. Introduction

The McKittrick Hill Caves Complex was designated in the Carlsbad Resource Management Plan (RMP) of 1988 as a Special Management Area (SMA). See Map 1 for a location of the SRMA. It was designated to protect numerous caves and their resources in the McKittrick Hill area. There are eight known caves in the SRMA containing a variety of resource values. These caves have significant resources which include; geologic, mineralogic, biologic, archaeologic, paleontologic, scenic, scientific, educational, and recreational values. Some of the caves are quite extensive and complex with as much as two miles of surveyed passage, while others are relatively short. The purpose of this plan is to establish the overall management goals and objectives for the SRMA and the specific management practices for each cave.

Management practices will be consistent with the broad programmatic policies, goals, and objectives established in the FCRPA, the BLM Cave Resources Manual (8380), State Office Supplemental Hand Book (Cave Inventory and Classification), and the Pecos District Caves Program plan and Resource Management Plan. Through the use of available management options, conflicts between cave resources and cave use are either avoided, resolved, or mitigated.

A. **Management Goal and Objectives:** The management goal for The McKittrick Hill Caves Complex SMA is two-fold. The first goal is to protect and maintain sensitive and fragile cave resources, and their surface to subsurface relationship. The second management goal is to provide for quality recreational, educational, and scientific use of the cave resources in the McKittrick Hill Caves Complex. These two management goals can be accomplished through a variety of management actions which utilize both physical and regulatory prescriptions. Management prescriptions include the use of protective structures and restrictions, discretionary and nondiscretionary lands actions, and public education. Cave specific management plans may be developed in subsequent years based on need and management priorities. Decision factors considered for setting priorities will include; use conflicts with high-value cave resources, threats to biological, archeological, or paleontological resources, and public health and safety issues.

B. **Background:** McKittrick Hill is located 12 air miles west of the city of Carlsbad, New Mexico. Legal public access to McKittrick Hill was acquired along the primary access route to the SMA in 1982 with the purchase of approximately 1½ miles of access easement.

The caves of McKittrick Hill were evaluated for designation as a National Natural Landmark in 1981 by Mr. Thomas Aley. His report was presented to the South Central Region of the Heritage Conservation and Recreation Service. The caves were determined to be eligible for designation as a National Natural Landmark and were proposed as such in the Draft Carlsbad RMP. This proposal was protested by the grazing allottee on the basis that it would increase the visitation to an undesirable level. The protest was resolved by not formally pursuing Landmark designation.

The area has been the site of cave exploration and use since the earliest known human occupation as evidenced by cultural remains. The earliest recorded visits by white settlers begin in 1889 and have continued to increase steadily. Some of the first extensive exploration to occur on McKittrick Hill began in the mid 1920s. Bob Nymeyer, a Carlsbad photographer, began exploring and photographing many of the caves as early as 1926. An interesting history of his exploration of the caves can be found in his 1978 book *Carlsbad, Caves, and a Camera*. Since the time of the earliest exploration most of the caves have been heavily vandalized and some of their speleothems removed for souvenirs or for sale as curios at Carlsbad Caverns. Because of the early photos that Mr. Nymeyer had taken, there is at least some record of the original beauty of the caves. A more complete history of the exploration of the caves of McKittrick Hill can be found in the History section of Appendix A (The Caves of McKittrick Hill).

C. **Visitor Use:** Currently, Endless, Sand, and McKittrick Caves account for 75% of the cave entry permits issued annually in the Carlsbad Resource Area. An average of 150 permits are issued to these three caves each year and represent an average of 870 visitor use days. Figures 1, 2, & 3 show the number of permits issued to Endless, Sand, and McKittrick caves from 1981 to 1993.

As a result of this heavy use and in order to maintain acceptable levels of visitor-use-impacts, restoration trips are a regular occurrence in the caves. The restoration work generally involves washing speleothems, graffiti removal, establishing trails, and repairing broken formations. All of the restoration work is conducted by volunteer groups. These groups include high school classes, colleges, explorer posts, caving organizations, and environmental education groups. At least two restoration trips are undertaken each year. These trips are normally supervised by a BLM employee to ensure maximum protection of the cave.

Due to the density and the complexity of the caves on McKittrick Hill, proposals for scientific research are received on a regular basis. Research work has been conducted for individual papers, for masters thesis, and for PhD dissertations. Research topics include speleogenesis, geology, mineralogy, radio isotope studies, biological diversity and inventory, bat population stability, archeology, and paleontology. For a more complete description of the paleontological resources of the caves refer to the paleontological section of Appendix A (The Caves of McKittrick Hill).

D. Geology: Structurally the area has been uplifted to the northeast as a result of the Carlsbad fault. Azotea Mesa, where McKittrick Hill is located, has also been highly folded and fractured in a northwest to northeast curving arch called the Waterhole Anticlinorium and by a set of folds generally running in a northwest to southeast direction (Kelley 1971, page 51,52). A complete description of the geology can be found in the geology section of Appendix A (The Caves of McKittrick Hill).

Several mineralogical studies have been conducted in the caves of McKittrick Hill. Collectively the caves contain a high diversity of rare and common types of speleothems. Some of the rare types of speleothems and speleogens include: subaqueous helictites, biogenetic honeycombs, cave pearls, helictites, gypsum vents, and others. The minerals have been described in *Cave Minerals of the World; The Mineralogy, Petrography, and Diagenesis of Carbonate Speleothems from Caves in the Guadalupe Mountains, New Mexico; Caves of the Guadalupe Mountains;* and in several periodicals and journals.

E. Biology: The biological resources of the caves of McKittrick Hill are generally common to most of the other limestone caves in the Carlsbad area. As recently as 1993 a contract for an intensive biological inventory was issued. Over the years cave biologists have collected numerous specimens. A compilation of this data up to 1977 can be found in Appendix A (The Caves of McKittrick Hill). Further data following 1977 is being compiled by several researchers.

II. General Management Practices

Management practices of the caves on McKittrick Hill are guided by the following laws, policies, and management plans:

- The Federal Cave Resources Protection Act of 1988,
- Washington Office Policies (W.O. Instruction Memorandum 84 541/ Cave Management Policy),
- New Mexico State Office Manual Supplement (Cave Inventory and Classification),
- Pecos District Caves Program Plan and Resource Management Plan,
- The Carlsbad Resource Management Plan.

A. Carlsbad RMP: The RMP prescribed the following management actions for the SRMA as protective measures:

1. No Surface Occupancy stipulations to future Oil and Gas leases (4,920 acres)
2. Withdraw from mining claim location and entry (1,210 acres)
3. Closed to solid mineral leasing (755 acres)
4. Closed to mineral material disposal (4,920 acres)
5. Restricted surface disturbance to conform with VRM Class II guidelines (4,920 acres)
6. Limited ORV use (4,920 acres)
7. Right-of-Way avoidance area (3,440 acres)

The withdrawal from surface entry and mining of 1,210 acres within the SRMA was completed and published in the *Federal Register* on September 1, 1992. All of the other RMP management prescriptions are discretionary actions and are in place.

B. Research: The goal of scientific research is to increase our knowledge and understanding of the caves. Responsible research is encouraged in

BLM caves by any competent researcher with a legitimate project. BLM employees should be provided time and support to conduct studies or assist other researchers in their studies. Guidelines for cave research within the SMA can be found in Appendix B.

C. **Restoration:** The goal of restoration is to reduce the amount of impact on heavy traffic areas and lower the cumulative impacts on the cave. Restoration trips will be conducted as per the guidelines in the Roswell District Cave Plan. All restoration trips will be accompanied by at least one BLM employee unless other arrangements are made for a BLM approved leader. Trips will concentrate on areas that show the most need or are threatened by nearby traffic. All restoration participants will be signed up as volunteers.

If observations and monitoring indicate that remedial actions are necessary to counteract visitor use impacts, appropriate restoration trips will be scheduled. Types of restoration activities that can take place are; graffiti removal, formation washing, trail building and flagging, speleothem repair, and trash removal.

D. **Monitoring:** The goal of monitoring is to keep management informed of resource changes and needs that may affect the overall health of the cave ecosystem. A number of monitoring activities will take place at each of the caves. These monitoring activities will be ongoing and performed as a regular function. Continued monitoring work is dependent on available budget, BLM personnel, and volunteers.

1. Visitor Monitoring - Monitoring by visitors is the most frequent and immediate method of monitoring conducted on The Hill. Much of the feedback is received from the post trip comments written on the back of the returned cave permits. Information regarding the status of locks, bat populations, restoration needs, and acts of vandalism are reported. Conscientious cave visitors provide the backbone for cave monitoring on McKittrick Hill especially in times of low budget and Bureau staffing.
2. Photo Monitoring - Photomonitoring points have been established in Endless, Sand, and McKittrick Caves. Points are being installed in Dry Cave as of April 1994. Photo points will be established at each of the caves as budget and staffing allows. Points will be tied into the cave surveys for exact locations. Photos will be taken annually or semiannually depending on volunteer availability. The photos or slides will be stored in the CRA. Photos should be reviewed annually or biannually to monitor for cumulative impacts in the caves. Photo points should be located in areas receiving heavy traffic or containing delicate speleothems or other sensitive resources.
3. Law Enforcement Patrol - Patrolling by law enforcement rangers and BLM personnel is periodic. More frequent patrols occur during incidences of vandalism or break-ins.
4. Bat Monitoring - Bat monitoring occurs annually during the spring and summer months. Bat counts are conducted at Endless and McKittrick Caves beginning thirty minutes before sundown and continuing for the duration of the flight or until it gets too dark to see. The standard method of counting is to use tally counters. It is recommended that at least two persons count and preferably three. The numbers are averaged. If there is one number that is "wild" it is thrown out. It is not recommended to have more than four persons at the entrance counting as human presence definitely affects the bat flight. Counts should be conducted weekly or bimonthly between

April 1 and September 15. Volunteers generally conduct the bat counts.

5. Visitor Use Monitoring - Using data from returned permits and signatures on registers provides a good profile of the actual visitor use. *Registers should be placed at each of the caves and maintained on a regular basis.* Information from the registers is used to determine the amount of traffic in relation to the amount of visitor use impacts. Using this information, visitor use limitations can be adjusted to better meet management goals and objectives.

E. **Continued Exploration:** The goal of continued exploration is to learn and understand more about the extent of the resources we are managing. Therefore, the continued exploration and mapping of the caves of McKittrick Hill is strongly encouraged. Since 1986 several thousand feet of new passage has been discovered. This has greatly added to the understanding of the cave systems and to the discovery of many new cave resources. Much of the new exploration and discoveries have been aided by geologic studies and the perseverance of cavers in their quest for the unknown.

The most significant new discoveries have been made as a result of digging activities. See Appendix C for the Guidelines for Cave Digging Operations.

III. Ongoing Management Activities

There are several actions that occur on a regular basis as part of the overall cave management program. These actions are in conformance with the Roswell District Cave Management Plan. The following list of ongoing management activities at McKittrick Hill are conducted as time, staffing, and budget allows:

- Issue recreational use permits,
- Review and make recommendations on research and digging proposals,
- Change the lock combinations at regular intervals,
- Maintain cave gates, locks, and signs,
- Maintain and direct monitoring programs,
- Maintain positive caver contacts and relationships,
- Maintain a viable volunteer constituency and,
- Maintain a progressive public outreach and environmental education initiative using the caves at McKittrick Hill.

IV. Cave Specific Management Practices

The following are descriptions and current management practices for each of the individual caves in the McKittrick Hill Cave Complex. The class of each cave is determined by using the inventory and classification system found in the New Mexico State supplemental handbook for cave management. The inventoried contents (i.e., biological, geological) of the caves are noted with letters (A,B,C, etc.); hazards are noted with Roman numerals (I,II,III, etc.); and management classes are designated with numbers (1,2,3, etc.).

A. **Endless Cave:** Class C,III,3. Endless cave is probably the most popular of the caves on McKittrick Hill. General descriptions of the cave with maps are in Appendix A & D (Special Stipulations and Hazards and map).

1. West Maze Passage of Endless Cave: Class D,II,2. There is a nursery colony of bats that roost in the west maze near the entrance of the cave. In order to protect the bats from human disturbance this area of the cave is "off limits" for use during April 15 to September

30. The two passages that lead to the West Maze are signed and flagged to notify visitors and prevent access. Bat counts are conducted each season as time and staffing allows.

- The general and specific hazards and stipulations for Endless Cave are listed in Appendix D.
- Because of popular demand and the impacts of heavy use, entry permits to Endless Cave are limited to one permit per week. The permit week runs from Monday through Sunday.
- Registers have been placed in the Gypsum Room and in the War Club Room.
- Photo monitoring points have been established by the Lubbock Area Grotto. Photo points are monitored annually. The photo monitoring points are identified on a map of the cave. The photos are maintained in a binder with the map in the CRA cave management program files.
  * Management Needs: No specific management needs.

B. Sand Cave: Class C,III,3. General descriptions and maps of Sand Cave can be found in Appendix A & E.

- Specific hazards and stipulations are defined in Appendix E.
- Permits are limited to one group in the cave at a time which generally means one permit per day. Visitor use impacts have not increased to the point where permit limitations are necessary.
- Photo monitoring points have been established by the Lubbock Area Grotto. Photo points are monitored once each year.
- A register is placed in the Sandy Floored Room.
  * Management Needs: No specific management needs.

C. McKittrick Cave: Class C,III,3. General descriptions and maps of McKittrick Cave can be found in Appendix A & F.

- Specific hazards and stipulations are defined in Appendix F.
- Permits are limited to one group in the cave at a time which generally means one permit per day. Visitor use impacts have not increased to the point where additional permit limitations are necessary.
- Photo monitoring points for McKittrick Cave have been established by the Lubbock Area Grotto. Photo points are monitored once each year. The photo monitoring points are pinpointed on a map of the cave. The photos are maintained in a binder with the map and are stored in the CRA cave management program files.
- A register is placed in the Lower Maze near the Main (Sink) Entrance.
- There is usually a summer bat roost near the pit entrance. This area is flagged off and signed to prevent visitor traffic from disturbing the bats. Additionally, the combination to the pit entrance is not given out during the roosting months.
  * Management Needs: No specific management needs.

D. Little Sand: Class B,II,4. General descriptions and maps of Little Sand can be found in Appendix A & G.

- Specific hazards and stipulations are defined in Appendix G.
- Permits are no longer required for Little Sand due to the lower resource values there. The lock has been taken off the gate and visitors are allowed unrestricted access.
  * Management Needs: Photo monitoring points, place register.

E. Dry Cave: Class E,IV,5. General descriptions and maps of Dry Cave can be found in Appendix A.

- Dry Cave has been designated as a 420 acre Research Natural Area (RNA) in the RMP. The cave was designated as an RNA for its paleontological resources which are up to fifty thousand years old.
- The cave is generally closed to recreational use and is primarily used for research and environmental education.
- Occasional restoration trips are taken into the cave for graffiti removal from past years.
- Survey and exploration trips are still being conducted in the cave by the Cave Research Foundation (CRF) as a volunteer project under the national Memorandum of Understanding between the CRF and the BLM.

  * Management Needs: No specific management needs.

F. Phalangid Cave: Class II-B-4.

- A description of the cave with maps is in the CRA Cave Program case file.
- The cave is ungated at this time and does not require a permit.
- A register is placed at the entrance to track visitation. The cave receives approximately 8 visitors per year.

  * Management Needs: Photo monitoring points.

G. Spider and Little Spider Caves: Class I-B-4.

- Both of these caves are ungated and do not require permits.
- These caves are generally less than 100 feet in length but have not been surveyed or fully inventoried.
- Locational information can be found in the case file.

  * Management Needs: Photo monitoring points, place registers.

V. List of Appendices

A. The Caves of McKittrick Hill
B. Guidelines for Cave Research
C. Guidelines for Cave Digging Projects
D. Endless Cave Special Stipulations and Hazards, Map
E. Sand Cave Special Stipulations and Hazards, Map
F. McKittrick Cave Special Stipulations and Hazards, Map
G. Little Sand Cave Special Stipulations and Hazards

Appendix A for The Caves of McKittrick Hill (Excluded here)

Appendix B for The Caves of McKittrick Hill Guidelines for Cave Research

1. A written proposal for each research project must be submitted to the Carlsbad office.

2. Approval must be obtained from the Carlsbad office before research can proceed, if any of the following are anticipated:
   a. Collecting of any kind (see additional guidelines for collecting),
   b. Disruption of the natural state of the cave,
   c. Entry to restricted areas or during restricted times. In this case the proposal must state explicitly how the impact of the research on the site will be minimized.

3. A biannual summary of research projects must be submitted to the Carlsbad office. Any findings or publications from research results will be made available to BLM and provided in a document.

4. No research will be permitted that endangers visitors, biota, or environmental quality, or which does not relate to speleology.

5. Researchers with permission to enter restricted areas, or during restricted times, may only be accompanied by minimum assisting personnel, as determined by BLM.

6. Guidelines for collecting:
   a. Approval for collecting will be granted only to qualified persons with demonstrable credentials (including prior publication of related research), or by persons under the direct supervision of such authority (i.e., college students, college faculty).
   b. Sampling should not impair the natural aspect of the cave.
   c. No more than 5% of any population (e.g., biota) or material (e.g., sediment) should be removed. Exceptions include materials that are not indigenous to the site, and which are in danger of being damaged if left in place, or which must be entirely removed to be studied (e.g., bones).
   d. No endangered species may be collected, except under extraordinary circumstances. Any doubt as to the status of a species should be clarified in coordination with BLM.
   e. No in situ speleothems or macroscopic fossils in bedrock may be collected, except under extraordinary circumstances.
   f. Bedrock samples should consist of loose fragments whenever possible. No visible scarring of cave walls is permitted.
   g. No collecting will be permitted simply for display purposes.
   h. Samples must be located explicitly, both vertically and horizontally, with the aid of surveys or maps, to ensure that collecting is systematic and not haphazard.
   i. Any remaining samples after the project is complete must be returned to the BLM. Any samples retained by the researcher must be made available for use by other researchers.

Appendix C for The Caves of McKittrick Hill
Guidelines for Cave Digging Projects

1. All digging must be approved by the BLM prior to the activity beginning.

2. A written proposal for digging must be submitted to the BLM and approved by the Area Manager.

3. Proposals should include the following points:
   a. Location of dig on surface and subsurface maps,
   b. Reasons for conducting the dig (geologic and physical evidence that suggests the dig and location will be productive),
   c. An operation plan (how the dig will be conducted, anticipated personnel, anticipated scheduling, disposal of spoils, mitigations),
   d. Safety concerns and practices in accordance with appropriate requirements to be followed by digging participants.

4. All digging proposals will be screened for possible conflicts with other resources, i.e., paleontology, archeology, biology, resource degrada-

5. A report will be submitted to BLM within five days after each digging event describing the progress and safety concerns or efforts. Any new discoveries will be reported within 24 hours. All injuries will be reported to the BLM immediately.

6. **NO** blasting or explosive devices will be used in digging operations.

7. All new discoveries will be surveyed as they are explored. Copies of all survey notes and sketches will be made available to BLM within 14 days.

8. Management needs and recommendations will be discussed with BLM for inclusion in a management plan.

Appendix 7

Cave Management Prescription: La Cueva de las Barrancas

Cave Implementation Schedule
La Cueva de las Barrancas

USDA Forest Service
Southwest Region (Region 3)
Lincoln National Forest
Guadalupe Ranger District

[Editors' Note: This Individual Cave Implementation Schedule is a working document to set specific guidelines and aid in the management of La Cueva de las Barrancas. As a result of this Schedule, quality management practices have been put into place to protect the unique, valuable, and finite cave resources contained in La Cueva de las Barrancas. Scientific research, exploration, inventory, and survey are being conducted in a well-planned, purposeful manner as outlined in this Schedule. *This document was approved and signed in February, 1999.*]

Section 1
Introduction

La Cueva de las Barrancas presents a rare opportunity for scientists, cavers, and the USDA Forest Service to investigate and establish baseline data on a pristine cave environment. Discovered by Jim Werker and Mike Reid in November of 1991, Barrancas was first entered by enlarging a fist-sized opening. Because Barrancas may have had little or no exposure to surface biota and the passages and pools were isolated with no evidence of human entry prior to 1991, potential exists for finding unique microbial communities in the cave.

This Schedule for La Cueva de las Barrancas focuses on unique opportunities for scientific research in a pristine underground environment on the Lincoln National Forest.

As Barrancas is carefully studied and explored, the search for new knowledge should be balanced with precautions to prevent unnecessary changes in the cave's ecosystem. Data gathered from initial research in this cave is facilitating management decisions regarding exploration and scientific study. Photographic documentation will be conducted before entry into each new area of the cave. Photomonitoring points will be established to record changes in the cave over time. Microbiologists have initiated ongoing studies in Barrancas and microbial sampling will provide useful monitoring information.

Scientific investigation and analysis of speleologic resources will continue in La Cueva de las Barrancas. Microbe investigators will be the first to enter virgin pool areas and establish study sites. Geologic and mineralogic resources will be carefully inventoried and mapped without disturbing biota. Paleontologic studies and invertebrate inventories will be conducted. Water analyses and hydrologic studies will be initiated.

Research methods will be designed to address preservation of native biota. Scientists have discovered new microbial communities in the pristine pools of Barrancas. Initial investigation has established Barrancas as one of several caves in the Guadalupe Mountain region that contain isolated microbial communities. These microbes are being studied to determine whether they produce toxins that are useful to astromicrobiological and/or pharmacological research. La Cueva de las Barrancas must be thoughtfully protected and that exploration and research be carefully managed.

Description

The cave entrance is located on the southern end of the Guadalupe Ranger District. Access to Barrancas is limited by rugged desert canyon terrain. Access requires travel on four-wheel-drive roads, then a moderately strenuous hike. The cave is entered through a solid steel gate. After a tight 15-foot crawl-way, the passage drops down a 350-foot pit. The descent is divided into three rappels and permanent bolts have been set for anchors and rebelays. Water flowing into the cave has deposited mud and bone fragments. Some of the known cave passages are coated with mud up to one foot thick. Many of the cave formations are naturally mud coated. Some formations show unique patterns of solutioning and redeposition. Unusual mud formations are scattered through the known passage; at least one of these may be a unique or undescribed speleothem. Airflow at the entrance, often in excess of 40 mph, indicates potential for Barrancas to be a large cave. To date, approximately one-quarter mile of passage has been entered. Because of the scientific potential offered by the pristine passages of La Cueva de las Barrancas, extensive exploration and survey has not been initiated.

Classification

Based on the management classification system listed and described in the Lincoln National Forest Cave Ecosystem Management Direction, La Cueva de las Barrancas, C-095, has been listed as Class 5-E-IV. Barrancas is classified as a hazardous vertical cave with biologic, paleontologic, geologic, mineralogic, and scientific significance.

Current Objectives

A. Identify, protect, and preserve the natural cave system. Identify, protect, and preserve the ecosystems and native microbial communities within the cave.

B. Establish a system of photomonitoring stations. Use photoinventory, photodocumentation, and photomonitoring in managing, protecting, and preserving the cave's resources.

C. Facilitate scientific study of the cave's resources.

D. Develop new conservation protocol for cave exploration, inventory, and research.

E. Assure that anyone who enters the cave is fully aware of and agrees to follow the "Policies and Guidelines for Entering La Cueva de las Barrancas."

Acknowledgements

As cavers and cave researchers learn more about spelean environments, we are in a continuing process of evaluating and redefining techniques for protecting cave resources. This Schedule, including the Policies and Guidelines, and the Minimum Impact Caving Code, comes from the experiences and thoughtful contributions of many cavers and speleologists. Policy statements are developed with input from many sources: The Cave Management Plan for Lincoln National Forest, 1972; Lincoln National Forest Cave Ecosystem Management Direction, 1995; and the Cave and Karst Management Plan for Carlsbad Caverns National Park, Appendix E: Guidelines for Entering Lechuguilla Cave, 1995. Jim Werker drafted initial ideas for this Schedule in 1992. Ransom Turner and Kevin Glover compiled a beginning draft in the spring of 1996. Val Hildreth-Werker collected additional information and ideas from collaborators Brent Botts, Jerry Trout, Richard Carlson, Mike Baca, Johnny Wilson, Penny Boston, Larry Mallory, Diana Northup, Dave Jagnow, Dale Pate, and Jim Werker. Using this input, Hildreth-Werker revised the Schedule into working drafts during 1996-1998.

Section 2
Policies and Guidelines for Entering La Cueva de las Barrancas
Research Directives

A. Implement minimum impact techniques for all activities in the cave. Encourage standards of excellence in speleological research and in minimum impact protocol.
B. Continue microbial investigations and pool studies as the highest priority for developing research and protocol techniques in Barrancas.
C. Continue photographic documentation and photomonitoring. Photographs will be used as management tools for tracking and evaluating changes in the cave. Speleothems, cave passages, pools, paleontological resources, areas of impact, etc., will be inventoried and monitored. Virgin areas, geologic resources, etc., will be photodocumented upon discovery, and periodically thereafter.
D. Allow appropriate research projects that will not interfere with microbial studies. Research will be conducted by experienced, careful cave investigators with scientific and conservation expertise.
E. Proposals will be submitted to the Forest Supervisor for approval. Proposals will include projected time frames for successful completion. Projects will require advance planning to focus on minimizing the number of cave entries.
F. Forest Supervisor decisions concerning La Cueva de las Barrancas will be based on review and comment provided by the National and Region 3 Cave Coordinator, Forest Service Cave Management Specialists, cave researchers, and interested parties.
G. Collection permits will be approved through the Forest Supervisor before taking samples from Barrancas. Written authorization is required from the collection permit holder if collecting is to be conducted by another researcher.
H. Inventories will be done from the trail rather than by walking across pristine surfaces. Inventory photographs will be made from the trail. Permission to extend trails to specific study sites may be requested and approved through the Forest Supervisor.

Cave Entries

A. Permits are required for cave entry. Permits will be issued only for conducting approved research projects or if necessary for emergency rescue.
B. Prior to entry, each person who intends to enter the cave must read and agree to follow the "Policies and Guidelines for Entering La Cueva de las Barrancas." Prior to entry, each person must sign the cave permit.
C. The Hazard Rating of Barrancas (IV) requires that teams entering the cave have a minimum of four (4) people. Permit requirements state that no more than six (6) cavers will occupy a single permit. However, if special needs for research can be proven, the Forest Supervisor may approve more than one permit per day. All trips shall have four (4) people as a minimum, with an exception being made for two teams of three (3) cavers simultaneously entering as part of one expedition.
D. A Job Hazard Analysis will be reviewed during safety meetings prior to cave entries.
E. Expedition Leaders are responsible for the actions of each person on the expedition.

F. Team Leaders are responsible for the actions of people in their group. The cave entry and activities must be geared to the least experienced member of the team. Each person is ultimately responsible for his or her own individual safety.
G. Trip reports from Team Leaders will be submitted to the Forest Supervisor immediately following each entry. Trip reports will include date, time in cave, names of personnel, sites visited, work accomplished, brief explanation or information about samples collected, and survey numbers referenced. Trip reports will be submitted before leaving the District or within twenty days of entry.
H. Detailed reports from Expedition Leaders will be provided to the Forest Supervisor and the National and Region 3 Coordinator/Cave Resources, during the course of every project. Detailed reports will include cave entries, photodocumentation, research results, and future objectives.
I. Currently, there is no need for overnight camping in Barrancas. Each expedition will plan to exit the cave on the day of entry. If distance becomes a factor, see the section below titled Policy Changes.

Exploration

A. Exploration will be conducted with prudence and deliberateness for the purpose of discovering new microbial research sites. Microbe investigators will be given access priority to enter virgin areas for testing. Through approval by the Forest Supervisor, research areas may become off-limits until scientific investigation in those areas is completed.
B. Virgin passages are valued resources for the undisturbed microbial communities they contain. Science teams will precede exploration teams in unexplored areas of Barrancas. Virgin passage will be reserved for science teams and will be protected from human impact or human entry prior to biologic investigation.
C. When entering unexplored areas, trails will be established immediately to minimize impact to the cave. A path will be marked that will cause the least impact. Cavers will not be allowed off the trail unless approved by the Forest Supervisor in order to achieve specific management or research objectives.
D. When sensitive areas are discovered, cavers will stop and should not proceed. If aragonite bushes block the path or if other noteworthy speleothems deter progress, cavers will stop and report to the Forest Supervisor for decisions on how and whether to proceed.
E. The primary objective of this Schedule for La Cueva de las Barrancas is to provide unique opportunities for scientific research in a pristine underground environment. Survey will support scientific research and exploration. Exploration will proceed in a slow, prudent and deliberate manner. The first priority of exploration will be to identify potential sites for microbial research.
   1. The first team entering an area will take photos of the pristine passage, carefully choose a path, and lay double flagging tape to define the trail. Trail width generally should be 18 inches or less, depending on the purpose, destination, and speleothems present in the path. Consideration for wider trail definition shall be given to include handholds on climbs and crawls.
   2. Exploration will stop upon finding areas with potential for microbial studies.
   3. The microbiologist will be the first to enter the area and will set up testing sites and do photographic documentation.
   4. Photomonitoring points will then be installed as determined appropriate.

5. Inventories and surveys will eventually be initiated using technology that allows all participants to stay on the flagged trail. All survey stations will be accessible without getting off the flagged trail. Inventory notes and maps will refer to survey stations. All survey stations will be set permanently. Some permanent stations will require an offset and reference to protect resource values. All cave surfaces off the flagged trail will be preserved in the original pristine state.

F. Advanced technologies in survey, mapping, and cartography will be used in order to achieve the highest standards for minimum human impact in the passages of Barrancas. Survey, a discipline within speleological research, will also require the proposal, review, and approval process.
   1. Survey in Barrancas will not be conducted until it can be done exclusively from the flagged trails. Precision mapping of Barrancas will begin when survey instruments and techniques become available to surveyors and cartographers so they can remain on the flagged trails. In order to preserve microbial resources within the cave, off-trail survey and mapping will be conducted using improved technologies such as range finders, laser devices, 3-D imaging equipment, etc.
   2. Carlsbad Caverns National Park (CCNP) has established acceptable survey standards for traditional tape and compass survey technique. These standards are listed in the most current revision of Appendix F: Cave Management Plan for CCNP. In the event that policy changes require survey to be initiated using tape and compass, the CCNP survey standards will be used in Barrancas until advanced technologies become available.

Policy Changes

A. Management policies, guidelines, and codes will be evaluated and adjusted as necessary to protect the resource. Changes will be approved through the Forest Supervisor. Approval will be based on review and comments made by the National and Region 3 Cave Coordinator, Forest Service Cave Management Specialists, cave researchers, and interested parties.

B. All parts of the Schedule for La Cueva de las Barrancas shall be reviewed and updated annually, and/or as necessary to protect the resources.

Section 3

Minimum Impact Caving Code for La Cueva de las Barrancas

The overall goal of the USDA Forest Service for La Cueva de las Barrancas is to allow limited scientific access and to identify and minimize impacts to the cave. Every person entering the cave is responsible for his or her own actions and safety and for the actions of team members. Expedition Leaders and Team Leaders have tremendous responsibility for the caving ethics of their personnel and for impacts to the cave. If problems persist, the Leader must abort the trip and the team will leave the cave.

As more is learned about cave environments, there is a continuing process of evaluating and redefining caver ethics. The following statement of conduct for Barrancas comes from the experiences and thoughtful contributions of many cavers. Think safety; take care of yourself and your team. Move with stewardship; avoid microbial, biological, and environmental impacts; and give utmost importance to the preservation of all cave resources.

- All clothing and equipment must be freshly washed to avoid transfer of microbes from other environments. Additionally, research is being conducted to determine whether boot soles and gloves should be treated with a disinfecting solution just prior to cave entry.
- Use boots and flowstone shoes with non-marking soles. If in doubt, scrape the boot over a white floor, concrete, or limestone rock. Marking soles will definitely leave a mark.
- The cave entrance is at an elevation of 6000 feet. Cavers not accustomed to the area should plan on spending a couple of days at this elevation to acclimate before entering the cave.
- Only cave packs or internal frame packs will be used. No external frame packs or ammunition boxes.
- Electric lights are required. No carbide is allowed.
- Always travel through the cave with your team. Do not get separated. Only an emergency might require different actions.
- Each team must have a minimum of four (4) cavers. Regulations on cave permits state that no more than six (6) people may enter the cave per permit.
- Be willing to discuss and report unsafe or damaging behavior so it can be corrected. It is every caver's responsibility to ensure that Barrancas remains as pristine as possible and that every team member is safe.
- Drink plenty of water. Watch for signs of dehydration.
- Use layered clothing and insulating pads to protect yourself from cold. Watch for signs of hypothermia and fatigue. Take corrective measures before symptoms escalate.
- Do not enter the cave if you know you are sick or injured.
- Do not enter the cave if you are not well rested.
- Report any accidents to the Forest Supervisor as soon as possible. Fill out an incident report for any injury or accident.
- Wear gloves. Check your gloves for mud, dirt, and holes to avoid extra impact. Rather than grabbing handholds along the trail, use a gloved knuckle for balance where possible.
- Pack in powder-free, non-latex surgical gloves for use in gloves-off areas and in pristine sections.
- Carry freshly washed flowstone shoes and protective covers for boots. Some trails in Barrancas are very muddy. Do not wear muddy boots across clean or pristine areas. Do not use bare feet or socks. Always use clean flowstone shoes. Check and clean mud from flowstone shoes frequently.
- Move carefully through the entire cave. Move slowly and gently through delicate areas. Always move slowly enough to avoid kicking up dust. Avoid new impacts to floors, walls, and muddy areas.
- Stay on established flagged trails. Do not impact the cave beyond designated trails. Sit within the trail. Be careful not to set your pack outside the trail. Always look for and use the most impacted areas of the trail when stopping.
- Trails with double flagged boundaries will be marked immediately upon entering any new area of the cave.
- Approval must be obtained from the Forest Supervisor before entering virgin territory, making new trails, or flagging new areas.
- No smoking and no use of tobacco in the cave.
- No consumption of alcohol.
- No illegal drugs.
- Obtain experience in vertical caving practices and become proficient in Single Rope Technique before entering the cave. Rebelay anchors will be encountered in the cave. Clip into all safety and traverse lines.
- Austinetic stainless steel bolts and hangers will be used exclusively when bolting. Any bolt that will not be used again will be removed.

- Check ropes and rigging before clipping in. Everyone entering the cave is responsible for the care and safety of ropes, bolts, carabiners, etc. Notify the Trip Leader of any problems. The Trip Leader will fix the problem immediately and/or notify the Forest Supervisor of the concern or change. If necessary, leave a note with the rigging to explain the problem or change.
- Leave all scientific instruments alone. Avoid touching instruments or cases. Avoid going near flagged-off microbe research areas. Remember, thousands of flakes of skin and debris fall from each of our bodies every hour!
- Assume all pools are off-limits. Avoid touching pools. Avoid standing over pools. Water may be collected only if the Forest Supervisor has validated a collecting permit. Pools must remain pristine for microbial research. Contamination may destroy valuable microbial resources.
- Do not enter off-limits areas unless you have specific authorization from the Forest Supervisor. Be certain you know which areas are off-limits resource protection zones.
- Special-attention areas require clean clothes, shoes, and gear. Do not enter special-attention areas wearing general caving attire. Perform extra efforts to keep these areas pristine. Clean tyvek suits may be required when entering virgin areas or research sites.
- No cave materials, minerals, speleothems, bones, etc. may be removed without a valid collecting permit approved by the Forest Supervisor. No digging may be performed without a permit from the Forest Supervisor approving such actions.
- Remove all solid and liquid wastes from the cave. Contain and carry feces, urine, vomit, spit, etc., out of the cave and dispose of properly. Plan for adequate container space. Never leave burrito bags along the trail while traveling. Adequate wrapping will make travel more pleasant for everyone. Always ask for updates on proper procedures for disposal of burrito bags outside of the cave.
- Care must be taken to avoid dropping crumbs or food particles in the cave. Always eat over a large disposable plastic bag. Carry out all crumbs and debris. Do not eat on the move.
- If stoves are needed for scientific application, use only alcohol or propane fueled stoves.
- Do not comb or brush hair in the cave. Use nylon swim cap, hair net, or bandanna to contain long hair and catch sweat.
- Avoid spreading pencil eraser particles in the cave.
- Develop caving practices that will reduce the input of organic carbons.
- An "out time" must be left with a responsible person and the Forest Service. All teams must inform the District Cave Specialist or other designated FS representative of an "out time"; the specific time they intend to be out of the cave and back at their vehicles or at the Administration Site. Search will be initiated for any team that is six hours late. Don't be late.

I have read and agree to act by the Policies and Guidelines for Entering Barrancas set forth in the Implementation Schedule for La Cueva de las Barrancas. I have read and agree to act by the Minimum Impact Caving Code for La Cueva de las Barrancas. By signing this document, I agree to act by these standards while in Barrancas.

Signature ____ Date __________

Appendix 8

Sample Cave Management Plan: Joe Doe Cave

Generic Cave Management Plan

Wunderling National Forest
Upper Wunderling Ranger District

2003

Prepared by: ______________________________ __________
Cave Specialist Date

Recommended by: __________________________ __________
Recreation Staff Date

Approved by: ______________________________ __________
District Ranger Date

1. Introduction

1.1 Purpose and Need:

Joe Doe Cave has been designated as a "significant cave" through the nomination and evaluation process as defined and described in the Federal Cave Resources Protection Act of 1988 (FCRPA–1988), (PL100-691). The Act mandates that caves on USDA Forest Service lands are, in part, regulated by the Act, which states that:

(1) significant caves on Federal lands are an invaluable and irreplaceable part of the Nation's natural heritage; and
(2) in some instances, these significant caves are threatened due to improper use, increased recreation demand, urban spread, and lack of specific statutory protection.

The stated purposes of the act are:

(1) to secure, protect and preserve significant caves on Federal lands for the perpetual use, enjoyment, and benefit of all people; and
(2) to foster increased cooperation and exchange of information between governmental authorities and those who utilize caves located on Federal lands for scientific, educational, or recreational purposes.

The Washington Office Manual amendment 2300-91-3, (2356–CAVE MANAGEMENT) states that: caves are dynamic natural systems affected by surface and subterranean environmental changes. While similar in many respects to surface resources, cave resources present some unusual management challenges because of the nonrenewable nature of cave contents and the sensitivity of cave ecosystems to man-caused changes.

The Forest Service is mandated to manage Joe Doe Cave in a manner that protects and maintains, to the extent practical, significant caves under the following Authorities (2356.01):

(1) The Organic Administration Act of June 4, 1897. 16 U.S.C. 551 (36 CFR 261.9a, 9b, 9g, and 9h), (36 CFR 294.1), and (36 CFR 261.53) mandates regulation of occupancy, use, protection, and preservation of Joe Doe Cave and gives authority to regulate visitation and installation of a control gate, and gives authority for special closures to protect threatened cave resources.

(2) The Antiquities Act of 1906 (34 Stat.225; 16 U.S.C. 431 et seq.) provides for protection of historic or prehistoric remains or any object of antiquity within Joe Doe Cave. Uniform rules and regulations pursuant to this Act are in FSM 1530.12.

(3) Archaeological Resources Protection Act (ARPA) October 31, 1979 (16 U.S.C. 470aa) prohibits removal of archaeological resources and authorizes confidentiality of site location, authorizes permit procedures for study and investigation of archaeological resources by qualified individuals. The Act supplements, but does not replace the Antiquities Act of 1906.

(4) Endangered Species Act of 1973 (87 Stat. 884m as amended; 16 U.S.C. 1531) describes the process for determining endangered and threatened species, and the protection of those species using the entrance and other portions of Joe Doe Cave.

(5) Federal Cave Resources Protection Act of 1988 (290.4) mandates confidentiality regarding the location of Joe Doe Cave.

1.2 Description/Background/History:

Description:

(Insert description of Joe Doe Cave.)

Background:

Primary issues are protection/preservation and safety. These issues have been discussed with Forest Service personnel, cave specialists, cavers and caving groups over the last two (2) years through personal contacts, group meetings, cave advisory "friends" meetings, on-site inspections, etc. This management prescription is intended to balance use in keeping with protecting a known, well-decorated pristine cave, and other delicate resources within the cave, while allowing certain amounts of recreational trips and scientific studies.

History:

(Insert history of Joe Doe Cave.)

1.3 Definitions:

The Federal Cave Resources Protection Act of 1988 defines a cave as: Any natural occurring void, cavity, recess, or system of inter-connected passages which occurs beneath the surface of the earth or within a cliff or ledge (including any cave resource therein, but not including any vug, mine, tunnel, aqueduct, or other man-made excavation), and which is large enough to permit an individual to enter, whether or not the entrance is naturally formed or man-made. Such term shall include any natural pit, sinkhole, or other feature which is an extension of the entrance.

The following definitions are to be used by the USDA Forest Service in order to:

Delineate the difference between a cave and a karst feature.

Cave – A cave must be traversable by a human, must be a minimum of 50 feet long and its entrance cannot be as wide as the cave is long. A vertical cave must be at least 20 feet deep. Features that do not meet this criteria will be considered to be karst features.

Karst Feature – Cavities, sinkholes, or other solution features that seem to be a cave, but do not quite fit the definition of cave, i.e., a hole that blows air, but is too small to enter.

2. Cave Management Objectives and Policies

2.1 Objectives for management of Joe Doe Cave are:

(1) Protect and perpetuate the natural cave and karst systems from internal and external threats.
(2) Classify Joe Doe Cave in a management category based on resource and hazard characteristics.
(3) Provide limited educational, recreational, and scientific study opportunities of the Cave's resources and systems.
(4) Establish regulations, guidelines, and permit stipulations that ensure maximum safety for the cave visitor and protection of wildlife habitat within the cave.
(5) Develop monitoring techniques that will determine if management activities are protecting and preserving all aesthetic, recreational, and scientific resources within the cave.
(6) Encourage conservation of cave resources during visitor contacts.
(7) Develop a funding source for maintenance of cave features and conservation materials. List operation/maintenance tasks and expand these tasks to include volunteers. This may include volunteer activities such as "adopt the cave" and/or challenge cost share programs.
(8) Analyze projects and funds needed to resolve, within the wilderness area, identified problems in managing the cave. Include estimated needs for manpower, materials and equipment, volunteer help, interpretation, and specialists.

2.2 Policy: (WO Amendment 2300-91-3, 2356.03)

(1) Manage Joe Doe Cave as a nonrenewable resource to maintain its geological, scenic, educational, cultural, biological, hydrological, paleontological, and recreational values.
(2) Emphasize wild cave management with few or no facilities to aid or facilitate use.
(3) Coordinate surface and cave/karst resource management activities.
(4) Protect threatened, endangered, proposed and sensitive, species in accordance with the Endangered Species Act (16 U.S.C. 1531) and FSM 2670.
(5) Protect cultural and Nation's natural heritage sites in accordance with FSM 2361.03.
(6) Develop and foster communications, cooperation, and volunteerism with interested publics, Federal agencies, States, and local governments.

3. Resource Evaluation

The District cave management plan presented a rating system for several resource values. These ratings go from none to outstanding. For Joe Doe Cave, resources were evaluated as follows:

3.1 Biological Values: Good

Components are present and moderately sensitive to disturbance.

Fauna present is uncommon in other sites and may include rare, threatened, and endangered species. Habitat in the cave may also be important to these species occurring on a local level. Research opportunities at the site are uncommon in the area and may be regionally significant.

3.2 Cultural and Paleontological Values: Good
(Insert paleontology statement.)

3.3 Geologic/Mineral Content Values: Outstanding
Preliminary geologic/mineral inventory indicates components are numerous and/or sensitive to disturbance. Speleothems include: stalagmites, stalactites, curtains, flowstone, travertine dams, cave coral, cave popcorn, helictites, and other calcite formations. These features are of regional significance. They could be seriously disturbed or destroyed by the poorly informed visitor. There is relatively little damage. Almost all damaged features can be restored. There are unique opportunities for scientific study.

3.4 Hydrological Values: Poor
(Insert hydrology statement.)

3.5 Recreational Values: Outstanding
Scenic appeal is very high. There are unique opportunities for interpretation. The entrance drop is 42 feet, the Big Room requires technical climbing and there are crawlways less than one-foot high. Special caving expertise and equipment are required. While most values described above fall in the "good" rating, the scenic appeal, interpretation opportunities and the anomaly of a limestone cave in an area in which solutional caves are rare, warrant an "outstanding" rating.

3.6 Educational and Scientific Values: Outstanding
The Cave is virtually in a pristine state with little evidence of human disturbance or impact. This pristine condition qualifies Joe Doe Cave as a possibly unique situation for scientific study.

4. Classification

Based upon the management classification identified in the Forest's approved Cave Management Plan and the Forest Service Manual, Joe Doe Cave is classified as a "Class-4-D-IV" Cave, which is described in the Forest Service "Cave Inventory System", in part, as:

4.1 Management Class 4 caves are closed to general use pending further evaluation for designation in another management category. Caves are designated Class 4 because:

(1) they are newly discovered and require further exploration, research, and/or inventory to evaluate how they should be managed;
(2) they have been explored and known for years but have not been sufficiently studied or inventoried; or
(3) they are potential Class 2 or 3 caves that have been well-explored and inventoried, but are being withheld from reclassification pending the results of resource impact studies.

Class 4 caves that have been explored and inventoried may be open to small, guided groups. Such trips will be authorized only for groups with bona fide instructional need, or for USDA Forest Service personnel involved in cave interpretation, or approved, qualified, and designated volunteers. Class 4 cave entry is approved only for minimum administrative purposes and research.

4.2 Resource Class D caves contain formations that are of unusual quality or are extremely delicate and susceptible to breakage, or resources of scientific value that could be seriously disturbed or destroyed by cavers. Examples of Class D cave formations would be: selenite needles, gypsum flowers, epsonite formations or crystals, cave helictites, etc.

4.3 Hazard Class IV caves are the most difficult to traverse. Visitation should be conducted by no less than three cavers, all of whom have considerable caving experience (including rope descent and ascent). All team members must observe caving safety and vertical safety rules, and use the following basic equipment: hard hats, three light sources per person with no loose or protruding attachments that might become entangled while doing vertical work. Each caver should have a complete set of climbing equipment for descent and ascent of vertical pitches.

The following are characteristics of Class IV caves:

a. "Maze" passageways
b. Vertical drops over 15 meters
c. Loose ceiling rocks on crawlways under 2 meters
d. Tight crawlways of less than one foot
e. Technical climbing areas

5. Visitor Use

Management of visitor use will be:

5.1 A caving permit will be required for entry into the gated portions of the cave.
5.2 Group size will be not less than 3 and not more than 4 unless special permission is obtained from the North Wunderling Ranger District Cave Specialist or designated USDA Forest Service employee for special projects, etc.
5.3 The number of trips, through the year 2008, will be limited to 4 trips per year, excluding trips for administrative purposes.

6. Research

A permit will be required for research activities. Guidelines for issuing a permit are found in the Region 3 Cave Management Plan. For Joe Doe Cave the following will also be considered:

6.1 Research is of a regional or national nature.
6.2 Research will emphasize nondestructive activities.
6.3 A Forest Service employee or a qualified and approved volunteer will be present during data collection in the cave.
6.4 The specific location of the cave will not be revealed.

Most research activities will not affect the number of trips allowed per year. The approval of an intensive study, however, may require a Special Use Permit. A policy has been established of science first and exploration/recreation later.

7. Administrative Use

Administrative visits by Forest Service employees and volunteers will normally be limited to trips necessary for monitoring, maintenance of

materials such as trail markers and gates, checking the register, training of guides, and inventory such as mapping and bone identification. Group size is normally 3 or 4. Trip frequency is normally one per month unless emergencies arise. If possible, and to reduce the number of administrative trips, some maintenance and monitoring may be done during approved research trips.

Occasionally, (and probably not more than once a year), visits will be for review of the management prescription and for solicitation of funds from other levels of the Forest Service and from other sources.

8. Restoration

There will be one restoration project per year. The project will usually be a weekend work session with 2 days in the cave. Only 4 people will be allowed in each section of the cave at one time. A Forest Service employee or a qualified and approved volunteer will be present at all times.

9. Search and Rescue

Search and rescue is primarily the responsibility of the county and state Search and Rescue Associations. Due to the sensitive nature of Joe Doe Cave, a District representative will be present during an operation to ensure resource values are protected. When a District employee is not available, a previously approved volunteer will represent the District.

Procedures for emergency use of key volunteers will be established.

10. Regulations

One regulation is currently in effect. A permit is required for entrance to Joe Doe Cave and other caves with gates. Additionally, the permit will describe the responsibilities of visitors.

11. Monitoring

Monitoring will determine if activities permitted by the standards and regulations are maintaining or improving the conditions of the Pristine and Primitive recreation classes and the resources evaluation ratings. They will also be the basis for making changes to the management objectives and the standards and regulations. Monitoring of some conditions will probably require long-term efforts before trends can be established.

Donations – Most uses of Joe Doe Cave will not require a fee, however, a donation will be accepted for the conservation of caves on the North Wunderling Ranger District. A donation of $25.00 per group is suggested on the permit. Monies donated will be used for District Caves and not for salaries. Examples of uses are:

Replacement materials for gates

- Gate locks
- Trail markers
- Registers
- Restoration tools
- Cave conservation literature
- Monitoring equipment

Appendixes

A. Cave Classification System
  1. Management Classes
  2. Resource Classes
  3. Hazard Classes

B. Cave Use Statistics

C. Mandatory and Recommended Equipment

D. General Observations/Recommendations

E. Action Plan

F. Sign-off /Accomplishment List

Biographical Notes

Biographical sketches of contributing authors are included below. The codes and awards that follow the National Speleological Society (NSS) member numbers are explained here:

HM Honorary Member of the NSS
OS William J. Stephenson Outstanding Service Award
RL Regular Life Member
CM Certificate of Merit
CO Conservation Award
LB Lew Bicking Award
PH Peter M. Hauer Spelean History Award
FE Fellow of the Society

James F. Baichtal, NSS #33277 FE
Jim Baichtal has 23 years experience with the U.S. Federal Government as an engineering and resource geologist. He graduated from Washington State University with a B.S. in Geology (1980) and an M.S. in Geology (1982). Since 1990, Baichtal has worked as the Forest geologist on the Tongass National Forest with responsibilities for developing karst and cave resource management strategies addressing the effects of timber harvest and road construction on the karst systems of Southeast Alaska. He has worked both nationally and internationally on karst resource issues. Baichtal is involved with an active public education program where he teaches about the geologic and glacial history of Southeast Alaska and the karst ecosystem and cave resources on the Forest.

Hazel A. Barton, Ph.D., NSS #38864
Hazel Barton received her Ph.D. in microbiology from the University of Colorado Health Sciences Center. Following post-doctoral work studying microbial ecology and physiology, Dr. Barton is presently the Ashland Endowed Professor of Integrative Science and an Assistant Professor of Biological Sciences at Northern Kentucky University. Her work includes a National Institutes of Health funded study examining *Salmonellae* adaptation to starvation, and the interactions of microbial communities in starved cave environments, funded in part by the National Science Foundation KY EPSCoR program. Barton is also a director of the National Speleological Society and the Quintana Roo Speleological Survey. She is an accomplished cave cartographer. Barton's work has been featured in *Sports Illustrated, Forbes, National Geographic Explorer, Outside Magazine,* on National Public Radio, and in an IMAX documentary.

Jonathan B. Beard, NSS #21408 RL FE
Jon Beard has been employed as an industrial training designer since 1989. He began caving in 1970 in Kansas and moved to the Missouri Ozarks in 1978. In 1983, vandalized Breakdown Cave was gated by local cavers. Beard directs the management of the cave as a restoration laboratory and classroom. Using many types of adhesives produced at the factory where he works, he has experimented with ways to reattach broken speleothems. He uses a variety of methods to remove calcified muddy handprints and graffiti. Beard currently helps manage several gated caves located on private properties.

Thomas B. Bemis, NSS #16184
Tom Bemis has been caving since 1970 and has worked for the National Park Service at Carlsbad Caverns National Park since 1979. He currently works for the cave resources office at Carlsbad Caverns, overseeing cave rescue, recreational caving, cave restoration, and cave impact monitoring programs within the park. Bemis also teaches

vertical caving and cave rescue for the National Park Service and New Mexico State University, and teaches cave rescue for the High Guads Restoration Project, National Cave Rescue Commission, New Mexico State Police, and New Mexico Emergency Service Council. He is past chairman of the Southwest Region of the NSS, past co-editor of the Southwestern Caver, and past editorial assistant for the NSS News.

Barbara Bilbo, NSS #39800
Barbara Bilbo attended San Diego State and University of New Mexico, receiving a dual B.S. degree in Geology and Geography. She then received an M.S. in Library Sciences from the University of Illinois, Urbana. While employed at Texas Tech University as a Science Reference, Government Documents, and Map Librarian, she completed a course of study with the International Center for Arid and Semi-Arid Land Studies (ICASALS) and received her second M.S. Bilbo's principal interests include Holocene and recent environmental development and change. Her research papers have been published in the *Geological Society of America Bulletin*, El Paso Archaeological Society Bulletin (*The Artifact*), Southwest Federation of Archaeological Societies proceedings, annual papers of the American Rock Art Research Association (*American Indian Rock Art*), and the National Cave and Karst Management Symposia.

Michael Bilbo, NSS #14994
Mike Bilbo has been a member of the NSS since 1974 and the El Paso Archaeological Society since 1968. Bilbo served in the U.S. Army Infantry overseas and the U.S. Army Reserve totaling 30 years. He received a B.S. from Texas Tech University in Park Administration. In the early 1990s he began working in New Mexico and Nevada as an outdoor recreation planner/cave specialist with the Bureau of Land Management (BLM). For seven years, Bilbo has managed Fort Stanton and Torgacs Caves National Landmarks and other New Mexico BLM caves. He coordinated initial science and survey trips into the Fort Stanton Cave Snowy River discovery and wrote comprehensive environmental assessments regarding digging and documentation of resources in Fort Stanton Cave. Bilbo helped found four NSS Grottos, and he has published in many archaeological journals on identifying and restoring dark zone rock art in New Mexico, west Texas, and northern Mexico.

Hans G. Bodenhamer, NSS #16668 FE
Hans Bodenhamer worked as a cave specialist for the National Park Service and Forest Service for many years. He developed Visitor Impact Mapping in 1985 while working at Grand Canyon National Park. He has a B.S. in civil engineering, but teaches biology and geology at Bigfork High School in northwestern Montana. As a teacher Bodenhamer continues his work in natural resource management by involving his students in numerous projects with local agencies. His students have won national awards for their efforts to conserve and enhance populations of threatened leopard frogs. Currently his students are mapping and developing monitoring strategies for caves in Glacier National Park. Bodenhamer enjoys teaching and spends summers mapping and studying the geology of Montana's wilderness caves.

Penelope J. Boston, Ph.D., NSS #44478
Penny Boston is Director of Cave and Karst Studies and Associate Professor of Earth and Environmental Sciences at New Mexico Institute of Mining and Technology, in Socorro, New Mexico. Boston is also Associate Director of the National Cave and Karst Research Institute in Carlsbad, New Mexico. She received her Ph.D. from University of Colorado, Boulder. Her research areas include geomicrobiology and astrobiology in extreme environments; human life support issues in space and planetary environments; and use of robotics to assist exploration and science in extreme Earth and extraterrestrial environments. Boston is author of over 100 technical and popular publications, editor of 4 volumes, and author of two upcoming popular books. Her work has been featured in over 100 print and broadcast media outlets over the past dozen years.

Debbie C. Buecher, NSS #13590 FE
Debbie Buecher is a self-employed consulting bat biologist in Tucson, Arizona. She has a B.S. from the University of Arizona in Ecology and Evolutionary Biology and is currently completing a Master's degree in Wildlife Ecology in the School of Natural Resources at the same institution. Her research focuses on acoustic monitoring and inventories for bats, particularly emphasizing community ecology of bat populations. She has conducted bat inventories for Bureau of Land Management, National Forest Service, U.S. Bureau of Reclamation, National Park Service, and Arizona Game and Fish. Because bats are highly disturbed by human visitation in their roosts, her work emphasizes low-disturbance techniques. As a long-time member of the NSS and Cave Research Foundation, Buecher supports both cave exploration and science and has been studying caves since 1970.

Paul Burger, NSS #26452 FE

Paul Burger has been caving since 1984. He graduated from the Colorado School of Mines in 1991 with a B.S. in Geological Engineering and in 1999 with a Masters of Engineering, Geological Engineering. Since 1999, he has been the Park Geologist/Hydrologist for Carlsbad Caverns National Park. His main areas of study are the speleogenesis and sedimentology of alpine caves in the Lime Creek Karst in Colorado and the controls on cave development in Lechuguilla Cave and other caves of the Guadalupe Mountains, New Mexico. Burger is the co-author of *Deep Secrets: The Discovery and Exploration of Lechuguilla Cave* and *Cave Exploring*, an introductory manual on caving.

Alvin D. Collier, NSS #14546 FE

Al Collier is retired from the Denver and Rio Grande Western Railroad after working in the mechanical department for 39 years. Al started caving in 1970 in Colorado. He has served as conservation and restoration chair for the five Colorado Grottos, Cave of the Winds, and Glenwood Caverns. Collier has worked on cave restoration and speleothem repair in Colorado, Utah, Wyoming, and New Mexico.

George M. Crothers, Ph.D., NSS #24150

George M. Crothers is Director of the Museum of Anthropology and Assistant Professor at the University of Kentucky. He has a Ph.D. in Anthropology from Washington University in St. Louis. He has been active in cave archaeology since 1980, and a member of the NSS since the early 1980s. His long-term research interest is in the prehistoric use of Mammoth Cave and the Central Kentucky Karst. He is currently on the Cave Research Foundation Board of Directors.

Robert R. Currie, NSS #20615 FE

Bob Currie is a biologist for the U.S. Fish and Wildlife Service in Asheville, North Carolina. He has been a member of the NSS and North Carolina's Flittermouse Grotto since the late 1970s. His main interests are the protection and conservation of cave and mine dependent bats. He is the Service representative on the National Cave and Karst Management Symposium Steering Committee, a director on the board of the Southeastern Bat Diversity Network, and has been actively involved in mine, cave, karst, and bat management training programs for private, state and federal land managers since the mid-1980s.

Joseph C. Douglas, Ph.D, NSS # 20059 FE PH

Joe Douglas is Associate Professor of History at Volunteer State College in Gallatin, Tennessee. He received his Ph.D. in American History from the University of Houston, specializing in the cultural history of the cave environment. An experienced public historian, he is an NSS Fellow and received the prestigious Peter Hauer Spelean History Award. He is a member of the University of Tennessee Cave Archaeology Research Team. His research has appeared in numerous historical, archaeological, and speleological publications. His current research focuses on identification, distribution, and analysis of prehistoric and historic material culture in caves. Douglas has participated in numerous cave preservation projects involving cultural and biological resources. He is currently Social Science Editor for the Journal of Caves and Karst Studies.

Harvey R. DuChene, NSS #6318 FE

Harvey DuChene is a consulting petroleum geologist in Denver, Colorado specializing in the Rocky Mountain and Permian basin regions. He earned B.A.(1968) and M.S.(1973) degrees in geology from the University of New Mexico. His research interests include sulfuric acid speleogenesis and the relation-ship between cave development and tectonism along the Rio Grande Rift of Colorado and New Mexico. He has studied the geology and mineralogy of caves in the Guadalupe Mountains, and from 1991—2000, was principal investigator for the Lechuguilla Cave Geology and Mineral Inventory Project for the National Park Service at Carlsbad Caverns. He is the representative for the NSS to the American Geological Institute and is a former Board Chairman of the National Cave and Karst Research Institute (NCKRI), headquartered in Carlsbad, New Mexico.

William R. Elliott, Ph.D., NSS #10847 FE

William R. Elliott ("Bill") is the cave biologist for the Missouri Department of Conservation. He was trained at UT/Austin and received his Ph.D. from Texas Tech University. His work includes cave conservation, millipede taxonomy, macrophotography, management, education, recreation, and work with private and public cave owners. He has studied cave life in Belize, Mexico, and in many states from Texas to Alaska. He has published numerous scientific articles, and he trains people in cave ecology and management. Bill is an NSS Fellow, a board member of the Missouri Caves and Karst Conservancy, and is active in Bat Conservation International, Missouri Speleological Survey, and the NSS Biology Section. He is a past director of the American Cave Conservation Association and former editor of the Texas Speleological Survey.

Joseph H. Fagan, NSS#10666 FE
Joey Fagan started caving in 1966. He works as a Karst Protection Specialist for the Virginia Department of Conservation and Recreation Division of the Natural Heritage Karst Program. He serves on the Board of the Cave Conservancy of the Virginias and as Secretary of the Virginia Cave Board. Fagan is an NSS Fellow, a founding member of the Blue Ridge Grotto, Conservation Chair of the VPI Cave Club, and is active in the NSS Virginia Region. He is the Caving Coordinator for the Girl Scouts of the Virginia Skyline Council Adventurers Program and is a member of the NSS Youth Group Liaison Committee.

John A. French, NSS #7846 FE
John French dates his earliest cave memory to about 1930. Active caving continues to the present and includes discovery, exploration, and (or) reporting about 60 new findings in Alabama. French led the Alabama Cave Law Drive. He has participated in at least a dozen cave rescues. International cave activities include China, Hungary, and Germany.

James R. Goodbar, NSS #9715 RL FE HM
Jim Goodbar began caving with his parents and two sisters in central Texas at 9 years of age in 1960. He is still actively caving with his wife and son in the Guadalupe Mountains of New Mexico. He earned a B.S. in Park and Recreation Management, Texas A&M 1973, and attended Western Kentucky University graduate program in Cave/Karst Geology and Geomorphology 1979–1981. Much of his career has been with the Bureau of Land Management (BLM) Cave Manage-ment Program. Goodbar currently serves as the BLM's Washington Office Senior Cave and Karst Specialist. He assists the BLM in developing national cave management policy, agreements, manuals, cave management training courses, and assisted the development and implementation of the Federal Cave Resources Protection Act, Regulations, and Implementation Procedures. Caving and cave management assignments have taken him to 15 foreign countries.

Val Hildreth-Werker, NSS #28963 FE
Val Hildreth-Werker has been photographing caves since the 1970s and brings professional experience in medical and commercial photography, animation, multi-image, video production, and broadcast media. She has B.S. and B.A. degrees from Texas Tech University with graduate studies in photography and ecology. For 20-plus years she was the Advisor of a coed High Adventure Explorer Post, leading teenagers on monthly expeditions to mountains, rivers, and caves. Hildreth-Werker started doing cave restoration in 1980. She and Jim Werker study exploration, conservation, and management in caves around the world and conduct seminars and workshops on cave conservation, restoration, and speleothem repair. They work as a team in developing environmental photomonitoring systems for cave and surface environments. Together, they co-chair the NSS Conservation Division.

Rodney D. Horrocks, NSS #25107 CM FE
Rod Horrocks has been the Cave Management Specialist at Wind Cave National Park since 1999. Starting in 1992, he was employed as the first Cave Management Specialist at Timpanogos Cave National Monument and Great Basin National Park. Horrocks is the current editor of the on-line publication, Inside Earth, the cave management newsletter for the National Park Service. He has been an NSS caver for 23 years, is an NSS Fellow, and was awarded the Certificate of Merit. He currently chairs the Cave Conservation and Management Section.

George N. Huppert, D.A., NSS #7717 CO FE
George Huppert, Doctor of Arts in Geography, was Professor of Geography and Earth Science at the University of Wisconsin—La Crosse until his death in 2001, having taught 23+ years. He started caving in 1959 with the Boy Scouts. He joined the NSS in 1964. Huppert worked for the renowned Jim Quinlan at Mammoth Cave National Park in 1975. His Master's thesis was a study of the physical systems of Papoose Cave in Idaho. His Doctoral dissertation was a history of cave conservation in the U.S. Huppert shared his cave management and conservation expertise with Boy Scouts, students, and adults on hundreds of cave trips; and through his prolific publications and presentations at cave/karst conventions throughout North America and on five other continents. He was an invited member of the famous Explorer's Club, and received the 1996 NSS

Conservation Award. He was an early proponent of the concept of Underground Wilderness. Huppert served on the Board of Directors of the American Cave Conservation Association from 1982 to 2001. He volunteered thousands of hours to the ACCA and NSS.

Jay R. Jorden, NSS #14356 RL FE
Jay Jordan has worked in the news business and public relations for nearly 30 years. A graduate of The University of Texas at Austin and University of Houston with journalism and law degrees, respectively, he has covered politics, courts, education, environmental issues and general assignments as a reporter and editor for Texas newspapers and The Associated Press. His freelance writings and photographs have also appeared in books and other publications worldwide. He is currently working in corporate marketing and communications for a Fortune 100 company. Jay has won a number of writing and editing awards during his career. As a volunteer, he chairs the NSS Cave Vandalism Deterrence Reward Commission and the NSS Public Relations Committee. Jay and his wife, Sheila, live in Celina, Texas with their son, Liam.

Jim Kennedy, NSS #26791 RL FE
Jim Kennedy ("Crash") is Cave Resources Specialist at Bat Conservation International (BCI) in Austin, Texas. He has been caving since 1973. At BCI he teaches Bat Conservation and Management workshops and Cave Gating work-shops. He advises federal and state land managers on cave and karst inventory and protection methods, promotes wise cave management, and coordinates a multi-state study of critical hibernacula for the endangered Indiana Bat. Kennedy is an NSS Fellow, former editor of the Texas Speleological Survey, past chairman of the Texas Speleological Association and Mid-Appalachian Region, and serves on the National Cave and Karst Management Symposium steering committee.

Ronal C. Kerbo, NSS #11539 HM FE
Ron Kerbo is the National Cave and Karst Management Coordinator for the U.S. National Park Service, working out of Denver, Colorado. He started diving in 1959 in gypsum karst sinkholes in New Mexico and started his NPS career at Carlsbad Caverns National Park in 1976 as the Park's cave specialist. Kerbo has been diving and caving for over 40 years, and is a 35+ year member of the NSS. He is currently an adjunct professor at Red Rocks Community College in Lakewood, Colorado teaching a course on the geology and evolution of caves.

Kathleen H. Lavoie, Ph.D., NSS #17033
Kathy Lavoie is a Professor of Biology and Dean of the Faculty of Arts and Sciences at the State University of New York College at Plattsburgh. She began caving in Indiana and quickly became interested in the life of the cave. She did her doctoral work with Tom Poulson, using the cave as a natural lab comparing the microbial and invertebrate communities in the deep cave vs. entrance areas. She studies cave cricket biology, microbiology, and invertebrate communities. Over the last ten years she have been active as part of the team of scientists studying Cueva de Villa Luz in southern Mexico. Lavoie has been an active member of NSS and CRF since the 1970s, has been an editor of the NABN, and represents the NSS biology section to the AAAS.

Thomas Lera, NSS #14821 RL FE
Thomas Lera has been caving for over 35 years and resides in Falls Church, Virginia. He authored *Bats in Philately* (1995), is a contributor to various journals and magazines including the *American Philatelist, Journal of Cave and Karst Studies, Journal of Spelean History, NSS News,* and *American Caves*. He is past Editor of *The Underground Post, f*ormer Conservation Editor of the *NSS Bulletin* and is a contributing Editor to *Speleophilately International*. He served as the Administrative Vice-President of the NSS for several years and is currently a Board Member of the Virginia Cave Board. Lera's articles and papers are available on the Web, including *The Legal Protection of Caves and Bats* which lists the cave and bat legislation in the United States.

Douglas M. Medville, NSS #70730 RL OS LB CM FE
Doug Medville is a retired research manager and is a member of several committees of the National Research Council, for which he is studying innovative technologies for the destruction of chemical weapons. He has an M.S. in Operations Research from New York University. Medville is a past Administrative Vice President of the National Speleological Society. In that capacity, he was responsible for the Society's activities in cave management, conservation, ownership, and education. His research has included the study of stratigraphic and structural controls on cave development in the Appalachians, alpine karst landforms in the northern Rockies, and the development of

lava tubes in the Hawaiian Islands. He is a Director of the National Speleological Society, the West Virginia Speleological Survey, and the Hawai'i Speleological Survey.

Paul J. Meyer, NSS #34328 RL FE
Paul Meyer has been caving for approximately 12 years and is the conservation committee chair for the Huntsville Grotto and preserve manager for the NSS Shelta Cave property. He co-developed a cave formation repair technique with fellow caver John French and utilized these techniques in several caves of north Alabama. As preserve manager of Shelta, he was responsible for the successful effort to construct a perimeter fence around the cave in favor of the restrictive cave gates that had discouraged occupation of the cave by bats. Since the gates were removed in 2002, bats have been steadily returning to the cave. Professionally, he holds masters degrees in meteorology and computer science and works for NASA.

Stephen R. Mosberg, M.D., NSS #20444 FE
Stephen Mosberg started caving with the Boy Scouts in 1970 in Baltimore. After Medical School he relocated to West Virginia for a Residency in Family Practice. As a founding member of the Parkersburg Area Grotto (PAG), he was active in the exploration of the Nepal section of Scott Hollow Cave and, later, Lechuguilla. He remains active in caving and cave rescue after changing professional venues to Emergency Medicine. A Fellow of the NSS and certified cavern diver, Mosberg serves as Chairman of the PAG, Chairman of the NSS Medical Section, Instructor and Medical Coordinator of NCRC, and Instructor, Eastern Region, NCRC. He has contributed to other NSS publications, including *On Call* and the NCRC text book.

Diana E. Northup, Ph.D., NSS #11561 FE
Diana Northup has been studying organisms that live in caves since 1984. She has a Ph.D. in Biology from the University of New Mexico (UNM). She and her colleagues on the SLIME (Subsurface Life In Mineral Environments) Team investi-gate microbial interaction with rock surfaces in caves and in desert varnish. Her research of cave microbial ecology using molecular and microscopy techniques has been featured on NOVA. Northup is a Professor Emerita at the University of New Mexico, lectures as a Visiting Associate Professor of Biology; teaches Communi-cating Science, the Freshman Learning Community, World of Microbes, and information search skills in the UNM Centennial Library. She and other UNM librarians are creating a Karst Information Portal with the University of South Florida and the National Cave and Karst Research Institute.

Marc Ohms, NSS #31288 FE
Marc Ohms is a Physical Science Technician at Wind Cave National Park, in the Black Hills of South Dakota where he assists in cave management duties and is responsible for air and water quality monitoring in the park. In 1989, he began caving in his home state of Iowa. He graduated from the University of Wisconsin—Platteville. Ohms is an NSS Fellow, chairman of the Paha Sapa Grotto since 1998, and is the Rocky Mountain Regional Coordinator of the National Cave Rescue Commission.

Rick Olson, NSS #13432 FE
Rick Olson is an ecologist at Mammoth Cave National Park. Over the past twelve years he has focused on ecological restoration in park caves, forests, prairies, and rivers. He is a Fellow of the Cave Research Foundation and the National Speleo-logical Society, has authored many papers, and contributed to four books.

William D. Orndorff, NSS #28413
Wil Orndorff runs the Virginia Karst Program in the DCR Division of Natural Heritage, where he has worked since Y2K. A native of the Shenandoah Valley, Orndorf holds a B.S. from Johns Hopkins and a Masters Degree in Geology from Virginia Tech. Wil's prior work experience includes a stint at the Smithsonian Environmental Research Center studying riparian denitrification, service as interim professor of Geology and department chair of physical sciences at West Virginia's Concord College, employment at the Princeton Review as a test preparation specialist, and performance of hydrological, biological, and environmental investigations in karst settings as a self-employed consultant. In addition to his work, he enjoys family life with his wife and 2 daughters, cave exploration, cycling, and music.

Dale L. Pate, NSS #12704 FE
Dale Pate began his caving career in the summer of 1970 in central Texas. In the fall 1970 he joined the Southwest Texas Grotto in San Marcos and became an NSS member in December 1970. Pate spent the next 21 years working in caves in Texas and Mexico before becoming the Cave Specialist at Carlsbad Caverns National Park in July 1991. For the past 15 years Pate has been involved with the cave and karst management of

this significant cave park with world-class cave resources including Carlsbad Cavern and Lechuguilla Cave.

Victor J. Polyak, Ph.D., NSS #26681 FE
Victor Polyak received his Ph.D. from Texas Tech University. He is a Research Scientist at the University of New Mexico. His studies focus on cave geology. He manages the Radiogenic Isotope Laboratory for Yemane Asmerom in the depart-ment of Earth and Planetary Sciences at the University of New Mexico. Currently funded by the National Science Foundation, they use geochronology on speleo-thems to study paleoclimates of the southwestern U.S., as well as other regions globally. Polyak is co-founder of the Lubbock Area Grotto, and is an NSS Fellow. He and his wife, Paula Provencio, manage a Sandia Grotto volunteer project surveying lava tube caves for El Malpais National Monument, New Mexico.

Paula P. Provencio, NSS #38769 FE
Paula Provencio is a member of the technical staff at Sandia National Laboratories. Her work focuses on nano-structural, crystallographic, and compositional characteri-zation of nano-scale materials using high resolution electron microscopy, and she applies her expertise to cave research projects. Provencio is an NSS Fellow and a member of the Sandia Grotto. She and her husband, Victor Polyak, manage a Sandia Grotto El Malpais National Monument lava tube cave survey and inventory project.

John E. Roth, NSS #45774 RL
John Roth has been caving since 1966. His 1977 thesis was on the effect of rock type on cave enlargement at Jewel Cave. He has taught cave groups ranging from junior high to the university level at Jewel, Carlsbad, Ape, and Oregon Caves. During the past 22 years, he has also been a resource manager at Carlsbad and Oregon Caves. Roth maintains cave bibliographies, directories, glossaries, and species lists for caves north of Mexico. Writings include ethnographic and ethnohistorical studies of caves and protocols for cave restoration, inventory, and monitoring.

Patricia E. Seiser, Ph.D., NSS #28650 FE
Pat Seiser has been caving since 1985. She has a Ph.D. in Cave and Karst Steward-ship from West Virginia University. Her research focus was on cave wilderness. Pat currently works with the National Cave and Karst Research Institute, and serves as an adjunct professor at New Mexico Tech. She volunteers for the National Park Service, the US Forest Service, and the Bureau of Land Management. In 1999, during National Women's History Month, she was named as one of twenty women recognized for significant contributions via exploration, survey, conservation, and scientific work throughout the history of Carlsbad Caverns National Park. Seiser is an NSS Fellow. She has been an instructor for the National Cave Rescue Commis-sion (NCRC) since 1998. She also serves as an Advisor for the Speleo Venture Crew, a caving program for teens, in Carlsbad, New Mexico.

Bernard W. Szukalski, NSS #17780 RL FE
Bernard Szukalski is a senior staff member of Environmental SystemsResearch Institute (ESRI), and lives in Redlands, California. Since joining ESRI in 1986 he has been involved in many aspects of Geographic positions that have covered a broad spectrum of GIS projects and activities. In addition to other responsibilities, he is the ESRI Cave and Karst Program Manager. Szukalski started caving in 1976, and serves on the Board of Directors of the National Speleological Society, Cave Research Foundation, American Cave Conservation Association, and the Hawai'i Speleological Survey.

Michael Ray Taylor, NSS #21969
Michael Ray Taylor is a professor of mass media at Henderson State University in Arkadelphia, Arkansas, where he lives with his wife and three sons. He is the author of the books *Caves*, *Dark Life* and *Cave Passages*. He has written about caves and cave conservation for *Audubon*, *Outside*, *Reader's Digest*, *National Geographic Traveler, Sports Illustrated*, *Wildlife Conservation* and other magazines.

Rickard S. Toomey III, Ph.D., NSS #39607
Rick Toomey is the director of the Mammoth Cave International Center for Science and Learning at Western Kentucky University and Mammoth Cave National Park. Previously he was the Science and Research Manager and Cave Resources Manager for Arizona State Parks. He has a B.S. in Geological Sciences from Brown Univer-sity (1985) and a Ph.D. in Geological Sciences from The University of Texas at Austin (1993). His research includes cave paleontology, bats, and cave resource management. His work at Mammoth Cave has included work on reconstructing and restoring past cave environments. Toomey is a past president of the Cave Research Foundation and

the Illinois Speleological Survey. He worked on the Illinois Department of Natural Resources' Karst Working Group in developing management policies and protection strategies.

Jerry L. Trout, Ph.D, NSS #4279 RL FE
Jerry Trout is the National Coordinator/Cave and Karst Resources for the USDA Forest Service, working out of Tucson, Arizona. He began caving in 1946. His career in cave management started in 1972 as the Cave Specialist for the Guadalupe Ranger District, Carlsbad, New Mexico. He teaches classes in Cave and Bat Management and Cave Conservation and Awareness at the University of Arizona, Cochise College, and Elderhostel, as well as government agencies and various civic and conservation organizations. Trout is a founding member of the Guadalupe Grotto, Eastern New Mexico University Student Grotto, and Cochise County Cavers, and was the former editor of the *Guadalupe Caver*. He is on the Board of Directors for National Cave and Karst Research Institute.

George Veni, Ph.D., NSS #17322 RL CM FE
George Veni is a caver and hydrogeologist specializing in caves and karst terrains. He received his Master's degree from Western Kentucky University in 1985 and his Ph.D. from Pennsylvania State University in 1994. Since 1987 he has owned and served as principal investigator of George Veni and Associates. He has explored caves and conducted extensive karst research throughout the United States and in other countries. He is President of the Texas Speleological Survey, past Executive Secretary of the NSS Cave Geology and Geography Section, and serves on the governing board of the International Union of Speleology. Veni is a doctoral committee advisor at The University of Texas and teaches karst geoscience courses as an adjunct professor for Western Kentucky University. He has published over 120 papers and 4 books on hydrogeology, biology, and environmental management in karst terrains.

Jim C. Werker, NSS #31653 FE
Jim Werker has been exploring caves for 40-plus years. During his 34-year career as a mechanical engineer at Sandia National Laboratories he worked in underground nuclear research and testing at the Nevada Test Site and applies that expertise to speleothem repair, environmental monitoring installations, and tool design for scientific experiments in cave environments. Since childhood, Werker wandered the mountains of New Mexico, exploring caves and mines. In the early 1980s, he started repairing speleothems in the Guadalupe Mountains. He and Val Hildreth-Werker have installed environmental photomonitoring systems in caves on federal, state, and private lands. They study exploration, conservation, and management in caves around the world and conduct workshops for land managers, cavers, and public outreach. Together, they co-chair the NSS Conservation Division.

John M. Wilson, NSS #13112 RL OS FE
John M. Wilson has worked professionally and as a volunteer for environmental and social welfare organizations since 1969. He as a B.A. from Coe College in Cedar Rapids, Iowa and an M.Ed. from Virginia Commonwealth University. He has served as president of the Appalachian Cave Conservancy since 1977. He is the founder and the honorary president of the American Cave Conservation Associa-tion. Wilson has actively participated in many leadership roles in the National Speleological Society including sections, conservancies, committees, grottos, a commission, a region, a survey, an NSS executive officer, and member of the NSS Board of Governors. He was recognized as an NSS fellow in 1977 and with the William J. Stephenson Outstanding Service Award in 1997. He has published many studies and articles, and presented numerous papers at conventions and symposia.

Index

A

B

C

D

G

H

I

J

K

L

M

N

O

P

Q

R

S

U

V

W

Y

Z